EFFICIENT ELECTRICITY USE

A Reference Book on Energy Management for Engineers, Architects, Planners, and Managers

Second Edition

EFFICIENT ELECTRICITY USE

A Reference Book on Energy Management for Engineers, Architects, Planners, and Managers

Second Edition

CRAIG B. SMITH
EDITOR

Pergamon Press, Inc.
New York/Toronto/Oxford/Sydney/Frankfurt/Paris

Pergamon Press Offices:

U.S.A. Pergamon Press Inc., Maxwell House, Fairview Park, Elmsford, New York 10523, U.S.A.

U.K. Pergamon Press Ltd., Headington Hill Hall, Oxford OX3 0BW, England

CANADA Pergamon of Canada, Ltd., 75 The East Mall, Toronto, Ontario M8Z 5W3, Canada

AUSTRALIA Pergamon Press (Aust.) Pty. Ltd., 19a Boundary Street, Rushcutters Bay, N.S.W., 2011, Australia

FRANCE Pergamon Press SARI, 24 rue des Ecoles, 75240 Paris, Cedex 05, France

FEDERAL REPUBLIC OF GERMANY Pergamon Press GmbH, 6242 Kronberg/TS West Germany

Library of Congress Cataloging in Publication Data
Main entry under title:

Efficient electricity use.

"Applied Nucleonics Company, Inc., Santa Monica, California, was selected by the Electric Power Research Institute to conduct a research program to compile efficient methods of using energy, particularly electricity."

Includes bibliographies and index.

1. Energy conservation–Handbooks, manuals, etc.
2. Electric power–Handbooks, manuals, etc. 3. Energy consumption–Handbooks, manuals, etc. I. Smith, Craig B. II. Applied Nucleonics Company. III. Electric Power Research Institute.
TI163.3.T4 1976 621.3 75-44373
ISBN 0-08-023227-2

Prepared Under Contract No. RP 211-1 to the
Electric Power Research Institute, Palo Alto, California

Legal Notice: This report was prepared by Applied Nucleonics Company, Inc. (ANCO) as an account of work sponsored by the Electric Power Research Institute, Inc. (EPRI). Neither EPRI, members of EPRI, nor ANCO nor any person acting on behalf of either:

a. Makes any warranty or representation, express or implied, with respect to the accuracy, completeness, or usefulness of the information contained in this report or that the use of any information, apparatus, method, or process disclosed in this report may not infringe privately owned rights; or

b. Assumes any liabilities with respect to the use of or for damages resulting from the use of, any information, apparatus, method or process disclosed in this report.

Printed in the United States of America

CONTENTS

CONTENTS

CONTENTS

PRINCIPAL AUTHORS AND CO-AUTHORS

M.K.J. Anderson . . . . Chapter 8
R.A. Fazzolare . . . . Chapter 2
G.E. Howard . . . . Chapter 10
P. Ibáñez . . . . Chapter 11
K. Iyengar . . . . Chapter 7
L. Martin . . . . Appendix D
R.L. Rudman . . . . Chapter 1
R. Schoen . . . . Chapter 13
C.B. Smith . . . . Chapters 9 & 14
R.B. Spencer . . . . Chapter 5
R.T. Taussig . . . . Chapter 3
G.B. Taylor . . . . Chapters 6 & 12
S.L. Westfall . . . . Appendix E
T.T. Woodson . . . . Chapter 4

LIST OF CONTRIBUTORS

The following individuals and organizations either acted as reviewers, contributed information and ideas, or information has been extracted from their works:

Individuals

R.A. Allen
P.E. Andersen
H.E. Barr
D.C. Bauer
C.A. Berg
E. Binsacca
A.H. Bonnett
C. Booth
W. Brenan
J. Brock
W.J. Cavagnaro
Charles Chern
M.H. Chiogioji
M.J. Christiansen
A.T. Churchman (United Kingdom)
D. Conn
Fred S. Dubin
D.E. Elliott (United Kingdom)
C.V. Engle
L. Erickson
T. Farfagilia
N.S. Fink
M. Finniston (United Kingdom)
H.E. Gearhart
J.E. Gibbons
H.M. Glass (United Kingdom)
B.G. Gossett
M. Grupa
W.E. Gundy
Elias P. Gyftopoulos
H.B. Hansell
Frank Harvey
Virgil Haynes
T.J. Healy
R.D. Heap (United Kingdom)
T.H. Highley
A.S. Hirshberg
O. Jesperson
R.R. Jones
Steven I. Kaplan
T. Kaporch
W.F. Kelber
G.J. Kenagy
R.S. Keowen
R. Klein (Federal Republic of Germany)
F. Komita
E.G. Kovach
J. Lawton (United Kingdom)
E.A. Lisle (France)
Quentin Looney
B. Loring
Juan Lovett
O.R. Lunt
R. Marble
A.J. Marino, Jr.
H.D. McFerson
B. Milan
J.P. Millington (United Kingdom)
R.E. Montgomery
H.D. Nash
C. Nelson
Frank O. Osman, Jr.
B.V. Palk
N. Patterson
Henry Perkins
R.P. Perkins
William G. Pollard
B. Prassas
H.W. Prengle
W.W. Pritsky
Courtland Randall
K. Riegel
Willard L. Rogers
D. Rosoff
Marc H. Ross
R. Rowberg
M.D. Rubin
William Rudoy
M.J. Sehnert
Allen C. Sheldon
B.J. Sheridan
T. Silson
Jorge Vieira da Silva (France)
M.F. Simon (France)
E.F. Slattery
R.J. Smith
Irving G. Snyder, Jr.
William T. Snyder
Robert H. Socolow
B.R. Steele (United Kingdom)
J.G. Sunderland (United Kingdom)
S.W. Swan
G.P. Thompson
M. Wachs
W.B. Walton
A.F. Waterland
J.C. Wilburn
R.H. Williams
R.G. Winegerter
W.G. Wonka
A. Ziegler (Federal Republic of Germany)

LIST OF CONTRIBUTORS

Organizations

Aluminum Company of America
Pittsburgh, Pennsylvania

American Physical Society
New York, New York

Bunderministerium für
Forschung und Technologie
West Germany

Carnation Company
Los Angeles, California

Dow Chemical Company
Midland, Michigan

DuPont Energy Management Service
Wilmington, Delaware

E.I. DuPont de Nemours Company
Wilmington, Delaware

Electricité de France
Paris, France

Facility Systems Engineering
Company
Los Angeles, California

Fluor Corporation
Anaheim, California

General Electric Company
Cleveland, Ohio

IBM General Systems Division
Atlanta, Georgia

ITT Gilfillan Company
Van Nuys, California

Johnstown Sanitary Dairy
Johnstown, Pennsylvania

Libby-Owens-Ford Company
Toledo, Ohio

Lightolier Company
Jersey City, New Jersey

Lockheed Missiles and Space
Company
Sunnyvale, California

Los Angeles Department
of Water and Power
Los Angeles, California

Martin-Marietta Aluminum
Torrance, California

Massachusetts Institute of
Technology
Cambridge, Massachusetts

Menasco Manufacturing, Inc.
Burbank, California

Harvey Mudd College
Claremont, California

Norman Engineering Company
Los Angeles, California

North Atlantic Treaty Organization
Belgium

Owens/Corning Fiberglas Corp.
Bloomington, Minnesota

Oak Ridge Associated Universities
Oak Ridge, Tennessee

PPG Industries
Pittsburgh, Pennsylvania

Raytheon Company
Wayland, Massachusetts

San Diego Gas and Electric Company
San Diego, California

Smith Electric Company, Inc.
Glendale, California

Southern California Edison Company
Rosemead, California

The Aluminum Association, Inc.
New York, New York

The Edison Electric Institute
New York, New York

The Electricity Research Council
United Kingdom

The Farr Company
El Segundo, California

The Trane Company
La Crosse, Wisconsin

TRW Systems
Redondo Beach, California

U.S. Electric Motors Corporation
Prescott, Arizona

US Government Agencies or
Institutions:

Department of Commerce
Department of Interior
Federal Energy Agency
Federal Power Commission
General Services Administration
National Bureau of Standards
Oak Ridge National Laboratory

University of Arizona
Tucson, Arizona

University of California
Los Angeles, California

University of California
Agricultural Extension Service

University of Houston
Houston, Texas

University of Pittsburgh
Pittsburgh, Pennsylvania

University of Tennessee
Nashville, Tennessee

VSI Corporation
Culver City, California

PREFACE TO THE SECOND EDITION

The availability of adequate supplies of energy is a matter of great concern throughout the world. Energy is linked to food production, industrial output, wealth, health, and lifestyle. Since an increasing portion of energy requirements is met in the form of electricity, the efficient use of electricity has become an urgent international goal. At a time when fuel shortages are occurring, prices are escalating, and environmental constraints on energy conversion are increasing, improved efficiency of utilization is the best short-term alternative for insuring that adequate supplies of electric energy remain available.

Applied Nucleonics Company, Inc., Santa Monica, California, was selected by the Electric Power Research Institute to conduct a research program to compile efficient methods of using energy, particularly electricity. The primary objective of the program was to develop a technical handbook, suitable for use by engineers, architects, planners, managers, and the public, which summarized current technology for improving electrical energy use efficiency.

A secondary objective was to evaluate the impact of energy efficiency improvements which may be economically feasible during the next several decades.

The methods discussed in this book are based on currently available technology or technology which is believed to be implementable in the next five to ten years.

In addition to our own research and resources, we consulted with energy experts in diverse fields throughout the world. Information was obtained from industry, consultants, academicians, government agencies, and electric utilities. A subcontract was given to the American Physical Society which conducted an institute during the summer of 1974 at Princeton University on the topic "Technical Aspects of Efficient Energy Utilization."

Technical areas considered in the program included: 1) improving end use efficiency (by industrial, residential, commercial, and agricultural users); 2) improving use efficiency by the equipment and process designer (heating, lighting, mechanical processes, electrolytic and electronic processes); 3) improving use efficiency by the architect and urban planner (municipal and urban systems, urban and suburban dwelling units, commercial establishments, and public assembly buildings).

The program did not include consideration of alternative generation methods, transmission and distribution techniques, or storage methods. The possibility of substituting other energy sources for electricity or electricity for other sources was secondary to the main thrust of this study.

In the United States and other parts of the world, electricity prices today (1977) range from 0.01 to 0.05 $/kWh for industrial and commercial users. This is the *effective energy cost* which includes *both* energy and demand charges. For residential users, the range is 0.02 to 0.09 $/kWh. Actual rates vary widely depending on the type of service, geographical area, and fuel costs.

Since further increases in electricity rates may be anticipated in the future, in some sections of the book we have used 0.05 $/kWh as the anticipated (future) cost of electricity. This is roughly double the rate currently paid by the small-to-moderate sized business and industry.

Metric (SI) units have been used consistently throughout this book, except in those few instances where it was considered clearer to use American/British units. In the SI, the joule is the unit of energy and the watt is the unit of power. The gigajoule (GJ) is 1,000,000,000 J or 10^9 J and is a convenient unit for many engineering purposes.* The American/British equivalent values are generally indicated also. (See Appendix A for appropriate conversion factors and a discussion of SI units.)

In the book an effort has been made to avoid the use of the terms "energy consumption" and "energy conservation" since these technically inaccurate expressions confuse the issue. Energy is never consumed and is always conserved. What is at stake in efficient energy use is to use the least amount of energy to perform a given task and, in doing so, to minimize degradation of the *quality* of the energy form. This concept is developed in detail in the book (see particularly Chapter 2 and Appendix C). While these distinctions may seem trivial, the editor and authors felt that they should be pointed out in the interest of sharpening the focus on important aspects of inefficiency.

Each chapter has been reviewed by several experts in the field who were not directly associated with the program. We hope in this way to have avoided errors and to have represented diverse viewpoints. Final technical responsibility rests with the individual authors. Also, the opinions expressed are those of the authors and do not necessarily represent the views of the Electric Power Research Institute.

A principal difficulty in this program, in addition to the breadth of the subject, was the injunction to restrict the investigation to technological factors--omitting the influence which factors such as economics, institutional barriers and political considerations have in determining efficient use of energy. The fact that these factors are not developed in more detail herein does not imply that they were considered less important, but is because a detailed treatment was outside the scope of this investigation.

*Multiply GJ by 0.947 to get MBtu (10^6 Btu)

We recognize the dynamic nature of energy technology today and anticipate new developments in the near future. For this reason we plan to update the information contained in the handbook. We would appreciate receiving suggestions, comments, criticism, additional case studies, information on new technologies for efficient energy use, or comments as to how this book has been helpful or could be made more useful. Any remarks or contributions should be addressed to the editor. Contributions accepted for future editions of this handbook will, of course, be acknowledged.

The preparation of this book would have been impossible without the assistance of many people. Especially significant were the efforts of Marcia Untracht, *presentation*; Mildred Hoffman, Karen Smith, and Harriet Palk, *production*; Tamara Bender and Jill Hardwick, *bibliographers*; M.S. Kuppaswamy Iyengar, John Newcomb, and John Johnson, *graphics*.

The efforts of R.T. Crow, René Malès, and L.J. Williams at EPRI, as well as R. Maxwell and R.N. Miranda of Pergamon Press, are appreciated. Special thanks must be given to the American Physical Society for its assistance.

The Project Manager for the program was C.B. Smith; Associate Project Managers were G.B. Taylor and R.B. Spencer. We gratefully acknowledge the efforts of our staff members and the consultants, subcontractors, utility personnel and industrial experts who have contributed to this project.

Craig B. Smith, Editor
P.O. Box 24313, Village Station
Los Angeles, California USA 90024

PROLOGUE

PRINCIPLES OF EFFICIENT ELECTRICITY USE

by

Chauncey Starr, President
Electric Power Research Institute

This book deals with methods for using electrical energy more efficiently. By purpose it places emphasis on electrical energy and on ways in which currently available technology can be applied to improve the efficiency of electricity use.

Because of the relatively long time constant (roughly 50 years) for establishing new sources of energy, they are of no use for resolving short-term shortages. The best short-term solution is simply to make more effective use of the energy we already have.

The basic approaches to improving energy use efficiency fall into three general categories: (1) the development of manufacturing processes to reduce their energy requirements and to reduce wastage of energy intensive materials; (2) the selection and development of materials in relation to their end use; and (3) the design and management of systems, subsystems and components so as to reduce the total integrated energy used to accomplish specific end purposes. The criterion of energy resource conservation, as contrasted with the historical criterion of minimizing monetary costs, leads to substantially different technological development and performance goals.

Several broad principles of energy management are apparent in this report. They are: (1) the integration and aggregation of processes, industries and energy using activities; (2) the sequential use of energy forms and energy-intensive material, starting at their highest performance level and proceeding gradually to the lowest; (3) the substitution of more efficient technology or devices where indicated on a life cycle basis; and (4) the removal of built-in obsolescence.

The design and management of systems, subsystems, and components, so as to reduce the total integrated energy used to accomplish specific end purposes, is a most complex issue. The basic difficulties arise from the separation, both in space and time, of the various components and processes that sequentially interact with each other before any end purpose is achieved.

In the manufacturing process industries the physical separation of individual steps and the storage in inventory between steps tend to disaggregate the energy investment and induce substantial energy waste. In the delivery of products and services to end users, the physical separation of activities, suppliers, and consumers results in a similar energy waste. And finally, our socially common life style separates our industrial complexes from our residential and these from our recreational and leisure activities.

The basic point is that energy efficiency is best achieved by aggregating sequential and interacting activities both in space and time so that energy systems may supply requirements in a continuous manner from the highest temperature level in small steps to the lowest temperature heat sink. Disaggregation (i.e., separation in space or time) tends to force large temperature changes and consequent energy wastage as well as energy use for transportation and handling both of which are not fundamentally essential to the system. Appendix C on available work describes the theoretical background for this concept.

To achieve the benefits of aggregation, industrial complexes which concentrate processes and permit continuous temperature sequences for process purposes are suggested.

Some typical examples of disaggregation are: (1) the upgrading of impure ores with rejection of impurities with high energy content; (2) wasteful process heat management during separate stages, each requiring high temperatures; and (3) the scrapping of material, with its invested energy, during cold working manufacturing processes.

For residential and commercial environmental conditioning, there appear to be energy advantages in building complexes which would permit the use of integrated utility systems (electricity, heat, light, water and sewage), heat pumps, and a variety of common heat sources and sinks. The creation of such complexes for residential, commercial and industrial purposes would appear to raise new problems of transportation and associated energy demands which certainly would require further study.

The sequential use of energy forms and energy intensive materials starting at their highest performance level is perhaps a more subtle concept. It involves minimizing the degradation of use after each level of performance is completed. For example, in an ideal energy economy, wood should be used first as a structural material to take full advantage of the free solar energy and nature's cementing of its individual fibers. Next, it could be used for manufactured fiberboards or synthetic cloth, and finally as a basic material for paper pulp.

The idealized sequence does not end here. Waste paper may be recycled as feed stock for the chemical industry or, in a final step, as a source of heat energy.

This idealized sequence is not fully applicable with our present forestry practices in which much of the growth is not suitable for construction lumber and is therefore converted directly to paper pulp. Nevertheless, an energy efficiency goal would suggest reexamination of forest managerial options, as well as an evaluation of long-term planning implications and the energy requirements of substitute construction materials and paper fibers.

Coal is similar in this respect to wood. It contains many complicated hydrocarbons in addition to pure carbon, and therefore a portion of its content is energetically significant as a starting point for valuable petrochemical manufacturing processes. The direct combustion of coal prior to removal of its useful hydrocarbons represents an unnecessary resource degradation if efficient energy use is the primary criterion.

The energy investment in alternative materials for specific application is a field of study that is just beginning. As examples of such alternatives, we have synthetic versus natural fibers for cloth, steel beams versus laminated wood, plastic versus metal sheet, aluminum versus steel cans and plastic versus glass containers. The list is long, but our quantitative knowledge of the energy investment in these alternatives is too limited now for significant conclusions.

The recycling of materials may or may not lead to energy savings. Recycling will have to be examined on a case-by-case basis. The energy investment in collecting waste material, separating it into components, and then upgrading it into a reconstituted form is not obviously less than the energy required to start from initial resources. Mineral resource conservation may not always be compatible with minimum energy use, although clearly both are socially and environmentally desirable.

There are many areas where more efficient energy use technology exists or could be developed easily. A number of examples are described in this book such as:

1. The use of presently available high-pressure sodium lamps. They provide the same light output as fluorescent or mercury vapor lamps but use half the electricity.
2. The use of microwave energy for both industrial purposes (paint drying, baking) and cooking.

3. The use of heat pumps and heat pipes as energy recovery devices.

Regarding built-in obsolescence, for example, the energy input required to rustproof motor cars by appropriate coatings is small compared to the total energy input to the vehicle. Obviously, if a long life for energy-intensive products can be achieved at low additional energy cost (or perhaps by better design only), there appears to be no question that it should be done.

Further, anticipated technological obsolescence may place an upper limit on the lifetime objective of a complicated device. For example, a house may properly be designed for a forty-year lifetime, but it is doubtful that a telephone should be. The criterion here should be the minimum average annual energy use projected for the performance required, taking into account the initial energy resource investment, maintenance, change in performance expectations, and end-of-life recycle or disposal options.

These concepts are leading to an examination of the rationale for our present life style which encourages individual dwellings, separation of work and industrial activities, and deurbanization of our high density population centers. The outcome could be an "energy-constrained" society as discussed in Chapter 14.

If one assumes that visual amenities, privacy, and avoidance of cross-pollution are the basic motivations for such separation, then an extreme solution, such as the underground cities described in Chapter 13 might provide an environmentally and energetically acceptable solution. The surface would be devoted to parks, food production, leisure facilities and natural scenery. The underground would provide a stable heat reservoir with a long time constant for a heat pump system and privacy would now be compatible with a concentration of energy-using functions. Transport requirements would be drastically reduced and made compatible with increased accessibility to leisure and recreational activities. The question remains as to whether such underground cities require less total integrated energy for their construction and operation than that required by our present, apparently inefficient, surface cities.

The suggestion of a compact underground city, with completely integrated and aggregated activities, represents a wide departure from presently accepted life style concepts. Nevertheless, this extreme scenario illustrates most of the concepts which are a consequence of optimizing energy and resource conservation. It deserves serious thought as the end point limit of the spectrum of options between present energy-abundant life styles and the most energy-constrained system.

A more homely example of these principles of aggregation is the concept of a single intense light source, of the highest efficiency, located in a house energy center. The light would be distributed by "light pipes" to the many points of use. Such a concept could be engineered today and could use less energy than now needed to produce the same light. The unavoidable waste heat, being concentrated, could easily be converted to an end use such as hot water supply.

These concepts illustrate the relationship between the scientific and engineering aspects of energy use efficiency and the problems of urban planning and residential design. The interaction between relevant fields such as architecture and sociology and the technology of energy resource conservation becomes apparent upon reading this book. It is evident that present practices in urban planning and building design are not compatible with an energy-constrained society.

For society to assess the many real options for systems which thoroughly meet energy and resource constraints, the present data base and analytical models are inadequate. The time elements of social costs and social benefits, long-term and short-term, are not adequately understood. Economic models and technological system models need to be integrated more fully and with greater insight into the key parameters. To achieve a balance between man, machines, money, environment, and social objectives, we require a more complete and quantitative analysis of the complex interactions of future societies in our ecosystem.

LIST OF ABBREVIATIONS

A Ampere
Å Angstrom
ac Alternating Current
Atm Atmosphere

bbl Barrel
Btu British Thermal Unit

°C Degree Centigrade
cal Calorie
cd Candela
CFM Cubic feet per minute
cm Centimeter
COP Coefficient of Performance
cu Cubic

DB Dry Bulb
dc Direct Current

E Energy (joules)
ev Electron Volt

°F Degree Fahrenheit
fc Footcandle
fL Foot-Lambert
fpm Feet per minute

G Giga (10^9)
g Gram
gal Gallon
Gcal Gigacalorie
GJ Gigajoule
GL Gigaliter
gpm Gallons per minute

ha Hectare
hp Horsepower
hr Hour
Hz Hertz (cycle per second)

I Current (electrical)
in Inch

J Joule
Je Joule electric

°K Degree Kelvin
k Kilo (10^3)
kcal Kilocalories
kft Kilofeet
kg Kilogram
km Kilometer
kVA Kilovolt-ampere
kW Kilowatt
kWh Kilowatt hour

ℓ Liter
L Lambert
lb Pound
lbf Pound force
lbm Pound mass
lm Lumen
lx Lux

M Mega (10^6)
m Meter
m Milli (10^{-3})
ma Milliampere
Mbpd Mega barrels of oil per day
MBtu Mega Btu
Mcf Mega cubic feet
Mha Megahectare
mi Mile
min Minute
MJ Megajoule
mm Millimeter
mph Miles per hour
mrt Mean radiant temperature
mt Metric ton (1000 kg)
MW Megawatt
MWe Megawatt electrical
MWh Megawatt hour
nm Nanometer (10^{-9} meters)

P Power (watts)
pers Person
psi Pounds per square inch
psia Pounds per square inch absolute
psig Pounds per square inch gauge

r Resistance
RH Relative Humidity
rpm Revolutions per minute

s Second
sq Square
$ Dollar

ton American/British ton-2000 lbm

V Volt

W Watt
WB Wet Bulb
We Watt electrical
Wh Watt hour

yd Yard
yr Year

LIST OF ABBREVIATIONS

A-E	Architect-Engineer
ARI	Air Conditioning Refrigeration Institute
AHAM	Association of Home Appliance Manufacturers
AIA	American Institute of Architects
ANCO	Applied Nucleonics Company
ANSI	American National Standards Institute
ASHRAE	American Society of Heating, Refrigerating, and Air Conditioning Engineers
AWG	American Wire Gauge
BER	Building Energy Ratio
CCH	Commerce Clearing House
CIE	Commission Internationale de l'Eclairage
DOC	US Department of Commerce
DOE	US Department of Energy
EEI	Edison Electric Institute
EER	Energy Efficiency Ratio
eff	Efficiency
EPRI	Electric Power Research Institute
ERC	US Energy Resources Council
ERDA	US Energy Research and Development Administration
FEA	US Federal Energy Administration
FHA	US Federal Housing Administration
FPC	US Federal Power Commission
GNP	Gross National Product
GSA	US General Services Administration
HUD	US Department of Housing and Urban Development
HVAC	Heating, Ventilating, Air Conditioning System
IEEE	Institute of Electrical and Electronics Engineers
kWe	Kilowatt electrical
kWeh	Kilowatt electrical hour
LED	Light Emitting Diode
LP	Liquid Petroleum
MF	Maintenance Factor
MIUS	Modular Integrated Utility System
NASA	US National Aeronautic and Space Administration
NBS	US National Bureau of Standards
NEC	National Electric Code (US)
NEMA	National Electrical Manufacturers Association
NRC	US Nuclear Regulatory Commission
NSF	US National Science Foundation
PP&L	Pennsylvania Power and Light Company
PUC	Public Utilities Commission
R&D	Research and Development
SBA	US Small Business Administration
SCR	Silicon Controlled Rectifier
SI	International System (i.e., the International System of Units)
SIC	Standard Industrial Classification
UCLA	University of California at Los Angeles
UK	United Kingdom
US	United States
USSR	Union of Soviet Socialist Republics
VCP	Visual Comfort Probability

ACKNOWLEDGMENTS

We gratefully acknowledge the following for permission to use material from the following publications:

American Association for the Advancement of Science, *Science*, Copyright © 1974
American Institute of Architects, *AIA Journal*, Copyright © 1974
American Nuclear Society, *Nuclear Technology*
American Society of Heating, Refrigerating, and Air Conditioning Engineers, *ASHRAE Journal*
American Society of Mechanical Engineers, *Mechanical Engineering*
Association of Iron and Steel Engineers, *Iron and Steel Engineer*
Babcock and Wilcox, *Steam/Its Generation and Use*
Ballinger Publishing Company, *Potential Fuel Effectiveness in Industry*, E.P. Gyftopoulos, et al., Copyright © 1974, The Ford Foundation
George Braziller, Inc., *Village Planning in the Primitive World*, Douglas Fraser
Cahners Publishing Company, *Specifying Engineer* and *Professional Builder*
Canadian Institute of Mining and Metallurgy, *The Canadian Mining and Metallurgical Bulletin*
Citizens' Advisory Committee on Environmental Quality, *Citizen Action Guide to Energy Conservation*
Consumers Union of United States, *Consumer Reports*, Copyright © 1975
E.P. Dutton and Company, *The Titanic Effect*, Kenneth E.F. Watt, Copyright © 1974
Duxbury Press, *Energy: Sources, Use, and Role in Human Affairs*, Carol E. Steinhart and John S. Steinhart, Copyright © 1974, Wadsworth Publishing Co., Inc.
Family Circle, Inc., *Family Circle*, and American Plywood Association
Gulf Processing Company, *Hydrocarbon Processing*, Copyright © 1973, All rights reserved
Harper and Row, *Collected Essays*, Aldous Huxley
Honeywell, Inc., for an article in *ASHRAE Journal*, February 1974 by L.W. Nelson and J.R. Tobias
Illuminating Engineering Society, *IES Lighting Handbook*, 5th ed., Edited by J.E. Kaufman
Institute of Electrical and Electronics Engineers, Inc., *IEEE Recommended Practice for Electric Power Systems in Commercial Buildings*, and *IEEE Spectrum*
A.D. Little, Inc., *Prospects for Electric Vehicles--A Study of Low-Pollution Vehicles-Electric*, J.H.B. George, L.J. Stratton, and R.G. Acton
McGraw-Hill, Inc., *Chemical Engineering*, Copyright © 1974, *Metals Week*, and *Standard Handbook for Electrical Engineers*, Edited by D.G. Fink, Editor-in-Chief and J.C. Carrol, Associate Editor
MIT Press, *Energy and Form, An Ecological Approach to Urban Growth*, Ralph E. Knowles
National Academy of Sciences, *Legal, Economic, and Energy Considerations in the Use of Underground Space*
National Society of Professional Engineers, *Professional Engineer*
Penton Publishing Division, Penton, Inc., *Machine Design*
Praeger Publishers, *The Video Telephone: Impact of a New Era in Telecommunications*, Edward M. Dickson, Copyright © 1973, Cornell University
Prentice-Hall, Inc., *Fundamentals of Tool Design*, American Society of Tool and Manufacturing Engineers, Copyright © 1962
W.B. Saunders Company, *Fundamentals of Ecology*, E.P. Odum
Scientists' Institute for Public Information, *Environment*, Copyright © 1974
Society of Automotive Engineers, *SAE Journal of Automotive Engineering*, Copyright © 1972, All rights reserved
Stanford Research Institute, *Patterns of Energy Consumption in the United States*
Technical Publishing Company, *Consulting Engineer Magazine*, March 1973, Volume XL, Number III, page 120, Copyright © 1973
Times-Mirror, Inc., *Popular Science*, June 1974
University of Houston, *Potential for Energy Conservation in Industrial Operations in Texas*
Viking Press and McIntosh and Otis, Inc., *The Challenge of Man's Future*, Harrison Brown
John Wiley and Sons, *American Civil Engineers Handbook*, Edited by Thaddeus Merriman and *Manufacturing Processes*, M.L. Begeman, Copyright © 1957

PART I

BENEFITS AND LIMITATIONS OF ENERGY MANAGEMENT AND THE ENERGY CONVERSION PROCESS

THE RAPIDITY WITH WHICH ENERGY RESOURCES ARE CONSUMED IN THE FUTURE WILL DEPEND ON THE RAPIDITY WITH WHICH REGIONS OF THE WORLD INDUSTRIALIZE, THE RATE OF POPULATION GROWTH, THE ULTIMATE LEVEL OF HUMAN DESIRES TO POSSESS MATERIAL GOODS--AND THE EFFORT THAT IS MADE TO ACCELERATE PRODUCTION TO FULFILL THOSE DESIRES.

--Harrison Brown, 1954

THE WORLD HAS ALWAYS BEEN RESOURCE LIMITED. THIS LIMIT HAS BEEN OVERCOME TRADITIONALLY BY GEOGRAPHIC EXPANSION OR TECHNICAL INNOVATION.

--C. Starr, 1977

CHAPTER 1

PRACTICAL SAVINGS ACHIEVABLE WITH EFFICIENT ENERGY USE

R.L. Rudman*

CHAPTER CONTENTS

KEY WORDS

Efficient Energy Use
Energy Management
Government Policy
Institutional Barriers

SUMMARY

Improved energy use efficiency has become a national objective of highest priority. To evaluate the success of programs aimed at improvements in efficiency, a target is needed against which actual performance can be compared. That target is an estimate of the savings in energy use that are sound economically and technically.

The distinction between conservation measures that are economic and those which are possible is made in Figure 1.1. Many energy saving methods for which the necessary technology is at hand are simply too expensive to implement at the current price of energy. Many other methods of improving energy efficiency save both energy and money; these are the methods considered in this study.

Energy management technologies which are currently available and appear to be economic could reduce total US fuel consumption by 28 to 46 percent by the year 2000, compared with an extrapolation of present usage patterns. About one-third of this savings, or 9 to 18 percent of total energy use, could be achieved by using electricity more efficiently, thereby reducing fuel needed by generating

---

*Director, Planning Staff, Electric Power Research Institute

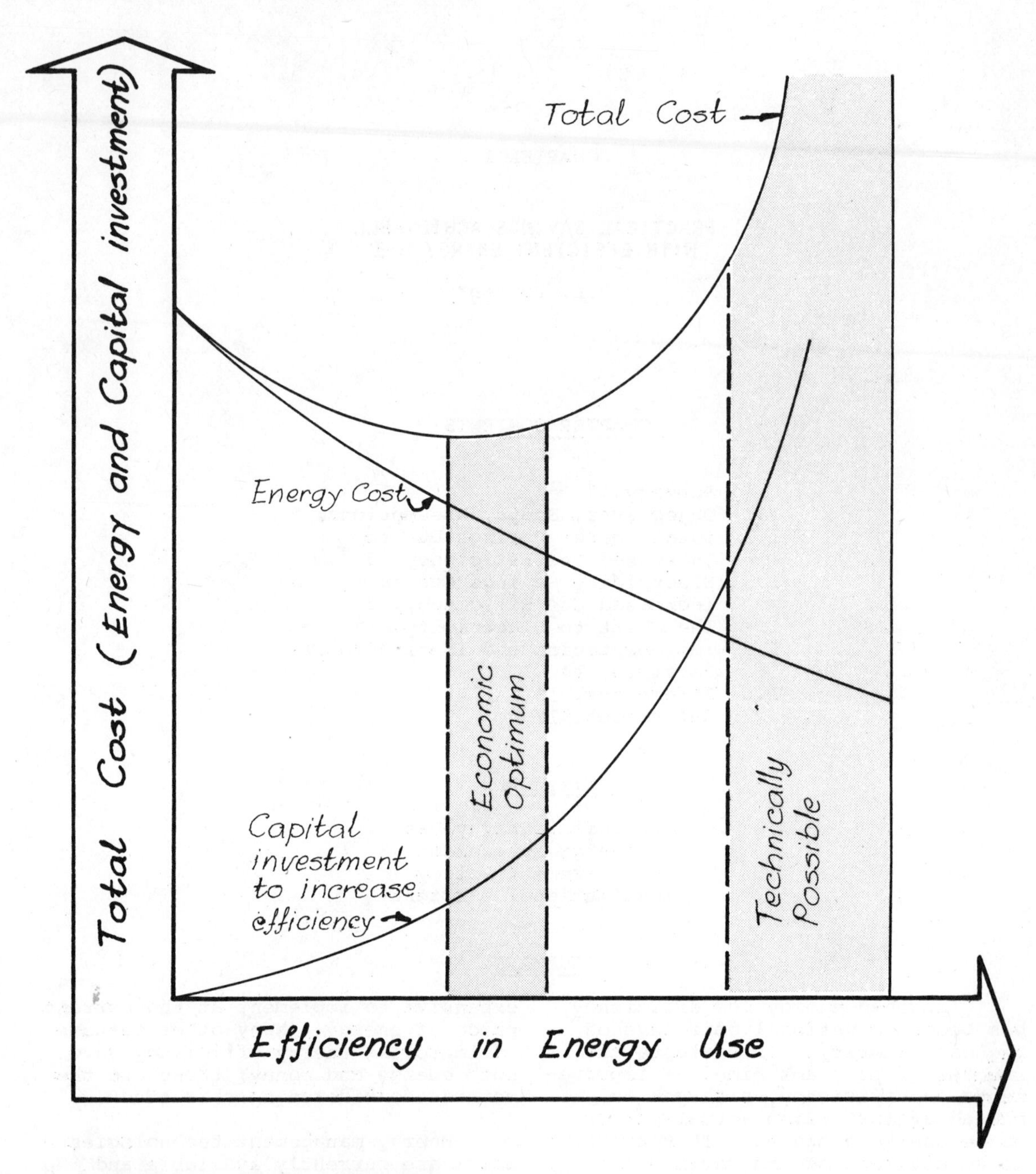

A BENEFIT—COST CURVE FOR ENERGY MANAGEMENT

FIGURE 1.1

plants. Accelerated shifts of other energy uses to electricity are likely to occur during the balance of this century due to dwindling supplies of petroleum and natural gas. These new demands will tend to offset some of the savings resulting from any measures which improve end-use efficiency, thus underscoring the need to use energy efficiently. These estimates of electricity and total energy savings have been based in part on engineering judgments of the economic viability of the technical options described in Chapters 2,3,4,5, and 6, and in part on the Case Studies included in this book.

Efficient technology alone is not enough to accomplish energy savings. Technological solutions must be economic, understandable, and acceptable. Certain national and local codes and standards create difficulties since they restrict efficient energy use. These institutional barriers will require resolution if efficient energy use options are to be fully implemented. Furthermore, special skills, new materials and equipment, capital investment, and time are also necessary for improving energy use efficiency.

1.1 OBJECTIVES, SCOPE, ASSUMPTIONS

The objective of this chapter is to evaluate the impact of energy efficiency improvements which are economically feasible during the next several decades, in order to assess both fuel savings and the practicality of reducing growth in electrical generating capacity.

The scope of this work is primarily electrical energy, but since many situations arise in which electrical energy use cannot be considered apart from the fuels used to generate it, energy use in the broader sense is also considered to some extent. This is made necessary by the fact that substitutions between different types of fuels and different energy forms are occurring; these substitutions must be evaluated in the broader context to assess their impact on natural resource availability and life style.

For this reason, two conventions have been used in the book. First, when considering overall fuel efficiencies, basic fuel quantities are used, rather than end-use energy. Thus, electrical quantities (GJe or kWh) are converted to equivalent amounts of fuel supplied to generating stations. The assumption is made that the average heat rate for electricity generation will be approximately 10.8 MJ (10,200 Btu) of fuel used to generate 3.6 MJe for a conversion efficiency of $3.6 \div 10.8 = 0.33$, or 33 percent, or 3.0 J/Je. This number has been used throughout the book in calculations where electricity was combined with other energy forms. Second, when consumer end-use decisions are being analyzed, end-use energy flows are employed. This more nearly parallels the kinds of analyses typical users would make. (For a more detailed discussion of this question and the problems involved in making such comparisons, see Appendix A.)

The principal motivation for this convention is to provide a consistent basis for evaluating shifts to electricity and to identify fuel substitution opportunities. For example, nuclear fuels theoretically could be used to generate process steam rather than electricity, coal theoretically could be liquefied or gasified, etc. Although a heat rate is undefined for hydroelectricity and nuclear electricity, the same conversion efficiency is assumed on the rationale that together they account for a small fraction of total electricity generation and that as displacers of fossil fuels this is a useful measure of their value. Also, this convention is consistent with national practice, e.g., as used by the Federal Energy Administration and Department of Commerce in their publications, and in the energy statistics reported by the US Department of Interior. It provides an indication of the impact of electrification on resource use. In the final analysis, however, price, availability, and convenience ultimately determine the substitutability of one fuel for another.

1.2 TOTAL ENERGY SAVINGS DUE TO INCREASED USE EFFICIENCY

A base case energy demand estimate can be calculated for a continuation of the trends in energy use and the economy over the past 80 years. If past trends were to continue, energy use in the year 2000 would be 180 x 10^9 GJ (170 x 10^{15} Btu). This is equivalent to a growth rate (in an exponential manner) of 3 percent per year from 1973, or 3.5 percent from the recession year 1975.

The fraction of this fuel used to generate electricity has been growing exponentially at 2.6 percent per year. A continuation of this trend projects a year 2000 electric fraction of 53 percent (up from the 28 percent of 1975). This base case exponential growth in electricity is at 6 percent per year from 1975.

In addition to the modifications to this base case that will result from energy management, there are projected losses of efficiency in electrical generation that stem from environmental requirements.

The extrapolation of the historical trends in energy use that provides the base case implicitly includes the assumption that the efficiency with which primary fuels are converted into electricity is increasing. Were this a reasonable assumption, the heat rate (which measures the ratio of fuel input to electrical output) would drop from its present value of about 3.1 J/Je (10,400 Btu/kWh) to 2.9 J/Je (10,000 Btu/kWh). This trend is not expected to continue for several reasons. First are the energy losses that accompany the use of environmental control equipment such as scrubbers, precipitators, and dry cooling towers. A second factor will be the increasing use of light water reactors (which have slightly higher heat rates than modern fossil units due to lower operating temperatures). A final factor which would reduce conversion efficiency is the use of the developing coal conversion technologies. Current methods for obtaining clean liquid or gaseous fuels from coal require 20 to 50 percent of the input energy. The overall impact of these factors could lead to a year 2000 heat rate of 3.2 J/Je (11,000 Btu/kWh). This represents a 10 percent increase in fuel requirements for electric generating stations, but does not affect historical projections for installed capacity or for electricity use.

This modification to the historical projection of the demand for fuel for generating electricity is an increase from 95.4 x 10^9 GJ/yr to 105 x 10^9 GJ/yr. The projected usage of electricity is unchanged at 32.8 x 10^9 GJe/yr (9.1 x 10^{12} kWh/yr) in the year 2000.

The expected range of savings based on available technology and current practices has been summarized at the end of the appropriate chapter. (See, for example, Chapter 2, Section 2.4; Chapter 3, Section 3.5, etc.) Estimates have been made for both total energy and electricity. Estimated values have been rounded, yielding the values listed in Table 1.1.

The organization of the data in Table 1.1 is important because it reflects both the significance of energy use in each sector (and therefore the relative priority for improving the technology of energy efficiency) and the time span in which improvements can be expected. The time span is important because while some improvements can be made immediately, most require a certain amount of time. The time span has been divided into the following categories:

- *Immediate savings* are those which can be accomplished in one year or less and involve modified methods of operation or housekeeping improvements and little or no capital expenditure. The same equipment and processes are employed but methods are modified to improve efficiency.

- Over the *near-term*, a slightly longer period of two to five years, additional improvements can be gained but generally only with capital expenditure. An example might be improved insulation of residences or installation of more efficient lighting. Not only is time required to make these changes but a certain period must pass before the changes are made in a sufficient number of facilities or installations to have a

(A) US TOTAL ENERGY SAVINGS (1975-2000), IN PERCENTAGE OF SECTOR ENERGY USE COMPARED TO BASE CASE

Sector	*Immediate* (0-1 yr) [Operational and house-keeping changes]	*Near-Term* (2-5 yr) [Some invest-ments and pro-cess and equip-ment changes]	*Long-Term* (5-25 yr) [Major invest-ments and process and equipment changes]	*Annual Savings* in the year 2000
Industry (Chapter 2)	5-10	10-15	10-15	25-40
Commerce (Chapter 3)	10-15	10-15	10-20	30-50
Residential (Chapter 4)	5-10	10-15	10-20	25-45
Transporta-tion (Chapter 5)	10-15	10-15	10-20	30-50

(B) US ELECTRICAL ENERGY SAVINGS (1975-2000) IN PERCENTAGE OF SECTOR ELECTRICITY USE COMPARED TO BASE CASE

Sector	Immediate	Near-Term	Long-Term	Annual Savings
Industry (Chapter 2)	5-10	5-10	5-10	15-30
Commerce (Chapter 3)	5-10	5-10	10-20	20-40
Residential (Chapter 4)	5-10	5-10	15-25	25-45
Other*	0-5	5-10	5-10	10-25

*Includes estimated savings in agriculture (Chapter 6) and transportation and communication (Chapter 5).

TECHNOLOGY LIMITED POTENTIAL SAVINGS WITH IMPROVED USE EFFICIENCY

TABLE 1.1

significant effect.

- Over the *long-term*, which has been arbitrarily defined as five to twenty-five years, additional improvements are possible but nearly all require capital expenditures for new equipment, new facilities, or changed technology.

The time associated with the *long-term* is that required for the design, development, and acquisition of new equipment and systems. An example would be the conversion of residences to solar heating technology. Even though the technology is essentially available today, it would take a period of time--say 30 years--before the nation's stock of millions of homes could be converted. For example, simply to establish the manufacturing capability and tooling to produce the needed heat exchangers and components would take a period of five to ten years.

Table 1.2 shows the impact of these savings on base case energy use in the United States. In Table 1.2, the annual savings in the year 2000 (from Table 1.1) are combined with the sector importance to get the weighted annual savings (again referenced to the year 2000). These savings are then multiplied times the base case energy use to determine the potential savings. As the table indicates, 28 to 46 percent of base case energy use could conceivably be saved by increased use efficiency. In physical units, this is 50 to 82 x 10^9 GJ/yr (47 to 78 x 10^{15} Btu), or an energy equivalent of 22 to 36 Mbpd of crude oil.

No attempt has been made to estimate the total cost of these improvements. Those which can be achieved immediately are for the most part obtainable with minimal capital expenditure. But those which extend over the *near-* and *long-term* will require capital expenditures, as indicated in the Case Studies at the end of this book. In many instances the *immediate* and *near-term* savings in energy cost due to improved efficiency will pay for the investment in a period of one to a few years.

Table 1.2 includes fuels and energy used as direct fuels, process heat, and feedstock, as well as the fuels needed to generate electricity. Thus, the potential electricity savings to be discussed later are included in the total energy savings shown in Table 1.2.

An estimate of the worldwide potential for improved efficiency is of interest. If areas outside the US take credit for potential efficiency improvements only 50 percent as great as in the US, or 14 to 23 percent, a total saving roughly equal to that which can be achieved by the US is obtained. In terms of total world energy use in the year 2000 (approximately 600 x 10^9 GJ/yr), the potential saving is 84 to 138 x 10^9 GJ/yr, (80 to 130 x 10^{15} Btu), which is equivalent to 38 to 62 Mbpd of oil.

1.3 ELECTRICITY SAVINGS DUE TO INCREASED USE EFFICIENCY

Table 1.3 shows the potential savings in electrical energy. The annual savings (Table 1.1) have been combined with the sector importance in Table 1.3 to obtain estimates of possible electricity savings. These estimates show that the potential savings could reach 17 to 34 percent of electricity use in the year 2000.

As a convenient means of visualizing the significance of these savings, an equivalent number of power plants can be computed. For this illustrative example, a plant capacity of 60 percent has been assumed. The selection of 1000 MWe is arbitrary and does not signify that all future plants will have this capacity. Likewise, 60 percent is an arbitrary choice for the average capacity factor, although it is a reasonable target for the year 2000. The results are shown in Table 1.4.

Clearly, improved efficiency has an important payoff, since the potential lifetime economic savings from not having to fuel several hundred power plants would exceed several hundred billion (100 x 10^9) dollars. When the capital costs of new construction which could be avoided are added to this, the savings are even greater.

Sector	Annual Savings (in percent of sector energy use for the year 2000)(1) (%)	Sector Importance(2) (Year 2000) (%)	Weighted Annual Savings (in percent of total US energy use in year 2000)(3) (%)	Potential Savings(4) (Year 2000) 10^9 GJ/yr	Mbpd(5)
Industry	25-40	40	10-16	18-29	8-13
Commerce	30-50	15	5-8	9-14	4-6
Residential	25-45	20	5-9	9-16	4-7
Transportation	30-50	25	8-13	14-23	6-10
TOTALS	---	100	28-46	50-82	22-36

Notes:
(1) See text and Table 1.1.
(2) Estimated for year 2000 base case.
(3) Calculated by multiplying sector importance times annual savings.
(4) Based on year 2000 total energy use of 180 x 10^9 annual GJ/yr (170 x 10^{15} Btu/yr, 80 Mbpd). This corresponds to an annual growth rate of 3.25 percent (22-year doubling time). Numbers are rounded.
(5) Energy equivalence expressed in 10^6 barrels of oil per day. Table uses the conversion that 1 Mbpd = 2.24 x 10^9 GJ/yr to calculate energy use in equivalent quantities of crude oil.

TECHNOLOGY LIMITED POTENTIAL TOTAL ENERGY SAVINGS FOR U S

TABLE 1.2

Sector	Potential Savings 1975-2000(1) (in percent of sector electricity use) %	Sector Importance(2) (Year 2000) (as a percent of total electrical energy) %	Weighted Annual Savings (as a percent of total electricity use in year 2000)(3) %
Industry	15-30	45	7-14
Commerce	20-40	15	3-6
Residential	25-45	20	5-9
Other	10-25	20	2-5
Totals	---	100%	17-34

Notes:
(1) See text and Table 1.1B.
(2) Projected for year 2000 base case.
(3) Calculated by multiplying sector importance times annual savings. Totals are rounded. These electricity savings (17 to 34 percent) are equivalent to 10 to 20 percent of total energy for the base case where electricity accounts for 60 percent of total energy use.

TECHNOLOGY LIMITED POTENTIAL ELECTRICITY SAVINGS FOR U S

TABLE 1.3

Description	Base Case Electricity Savings, Year 2000 (%)	Equivalent No. of 1000 MWe Power Plants (3)
Base Case, no conservation, year 2000 (1)	--	1700
Improved efficiency case with sector savings of: (2)		
industry	7-14	120-240
commerce	3-6	50-100
residential	5-9	85-155
other	2-5	35-85
Total Savings	17-34	290-580

Notes: (1) The equivalent number of power plants for the base case is estimated using a 60 percent capacity factor, e.g., 9×10^{12} kWh ÷ [(0.6)(10^6 kW)(8760 hr/yr) = 1700 plants. Note that if a 50 percent capacity factor is used, the equivalent number of plants for the base case is ∿2000 plants.

(2) Numbers shown are from Table 1.3.

(3) Except for base case (see Note 1), numbers are obtained by multiplying base case by percent savings.

TECHNOLOGY LIMITED IMPROVED USE EFFICIENCY AND INSTALLED CAPACITY

TABLE 1.4

1.4 THE SHIFT TO ELECTRICITY

Electricity use has been growing more rapidly than total energy use virtually since its introduction. This shift to electricity has occurred during periods in which no thought was given to diminishing oil and gas supplies. The securing of supply associated with electricity (because it can be generated from a number of different fuels) will quite likely offer an added incentive to use electricity. If the substitution of coal for oil and gas occurs, it is quite likely that much of this coal will be used to generate electricity.

Conversion efficiency is sometimes cited as a disadvantage to increased electricity use on the grounds that heat rejected from generating plants is "wasted." Ultimately each possible approach to meeting a specific energy need must be evaluated on its own merits, and economics will normally govern.

There are, however, valid technical reasons why the electricity growth rate is expected to continue to exceed that of non-electrical energy forms. One of these--the fact that electricity can be generated from many different fuels--was mentioned above. Several other points should also be noted:

- Practical efficiency of electricity generation is high.

  The central generating station is actually a highly efficient device for recovering the work potential of fuels. As shown in Chapter 2 and Appendix C, electricity generation produces a high quality energy form and recovers most of the available work of the fuel. "Waste" heat, while amounting to many joules (Btu), has a small potential for doing useful work. This is analogous to a kilogram of gasoline and a swimming pool of lukewarm water. While they could each have the same energy content in joules (Btu), the *quality* of the energy form (a measure of its ability to do useful work) is obviously very different (see Chapter 2).

- Better control permits efficient use.

  Electricity is easily and precisely controlled with inexpensive equipment as compared to all other energy forms. This means that waste is reduced; energy is used only when and where it is needed. A switch costing less than a dollar can control an electric heater; compare this to the valves, piping, insulation, steam traps, and noise and leaks associated with a steam heating system. Fuel combustion devices also have transmission and distribution losses, and must be vented, resulting in a loss of heat up the stack. When all losses are taken into account in these alternate systems, practical "efficiencies" comparable to electricity are found.

- Electricity is environmentally clean.

  Fuel is burned at a single location (the power plant) when electricity is generated. This permits more effective pollution control than attempting to do it at a thousand or million points of use. Also, it is generally easier for the utility to make the needed capital investment in pollution control equipment than for the individual users.

Rapid price increases for oil and problems with availability of gas are of recent origin and the impacts of these factors may not be properly represented in the historical data. A concerted push to conserve oil and gas could result in a growth in the electricity fraction that exceeds the historical rate. Examples of these additional (beyond the base case) shifts from non-electric to electric energy use which could occur in the next 25 years are shown in Table 1.5. The table indicates that these shifts could be equivalent to 204-519 1000 MWe power plants, and would result in an electric fraction of 60 percent or greater.

Sector	Energy Shift Description	Fossil Fuel Displaced 10^9 GJ/yr (4)	Substitution Factor (5)	No. of 1000 MWe Power Plants (6)
Industry	Shift of 10-20% of industrial heat to electricity (1)	4.3 - 8.6	2	140 - 280
Commerce	Shift of 10-20% to commercial heat to electricity (2)	1.3 - 2.6	1.5	32 - 64
Residential	Same as commercial	1.3 - 2.6	1.5	32 - 64
Transportation	Shift of 5-15% of transportation energy to electricity (1% autos per year, for example) (3)	2.3 - 6.8	1.0	0 - 111 (7)
	TOTALS	9.2 - 20.6		204 - 519

Notes:

(1) Industrial energy, year 2000 = (40%)(180 x 10^9 GJ) = 72 x 10^9 GJ (Table 1.2). Since heat is 60% of total and shift is assumed to be 10 to 20% of heat, 4.3 - 8.6 x 10^9 GJ shifted.

(2) Commercial energy, year 2000 = (15%)(180 x 10^9 GJ) = 27 x 10^9 GJ. Since heat is 48% and the total shift is assumed to be 10 to 20% of heat, 1.3 - 2.6 x 10^9 GJ shifted.

(3) Transportation energy, year 2000 = (25%)(180 x 10^9 GJ) = 45 x 10^9 GJ; 5 to 15% of this is 2.3 - 6.8 x 10^9 GJ/yr.

(4) See notes 1-3.

(5) This factor adjusts for the relative efficiency (on a primary fuel basis) of fossil fuels and electricity. The conversion losses of electricity generation are partially compensated by higher end-use efficiency, giving a factor of 2 J fossil for 1 J fuel for electricity. Some use of heat pumps for commercial and residential reduces this factor to 1.5 for these sectors. Electric autos are assumed to be as efficient (on an overall fuel basis) as conventional autos, due to lighter vehicles, more efficient motors, etc.

(6) The power plant requirements are calculated by recognizing that one 1000 MWe power plant with a heat rate of 3.2 J/Je uses 61 x 10^6 GJ/yr of fuel to produce 19 x 10^9 GJe/yr (5.3 x 10^9 kWh/yr).

(7) The plant capacity needed depends on the type of battery system developed and the fraction of users who will take advantage of off-peak hours for recharging their vehicles.

SHIFTS OF OTHER ENERGY FORMS TO ELECTRICITY

TABLE 1.5

Figure 1.2 summarizes the options for more efficient use of total energy and electricity in the US over the period 1975-2000. The total energy graph (upper curve) shows the upper limit anticipated (a "historical extrapolation" based on 3.25 percent growth per year and a 22-year doubling time). The shaded area indicates the possibilities for reducing energy use with improved efficiency, as is discussed in Section 1.2 of this chapter. The electricity curve is based on an extrapolation of historical growth (7.2%/yr and 10-year doubling time). The shaded area indicates the potential effect of improved efficiency on electricity use for this base case.

1.5 IMPLEMENTATION AND INSTITUTIONAL BARRIERS

Programs to improve energy use efficiency could be implemented on several levels--in industry and commerce, in homes, on farms, and by governments. Programs could be initiated by managers, by corporations, by local and municipal authorities, by state and federal agencies, and even by private citizens. In each case, it is important to recognize the three critical elements involved:

- awareness of the need,
- access to solutions, and,
- visibility of economic benefits.

The last point is most important. It is vital to recognize that the economic payoff must be there before a significant savings can result.

A number of institutional hindrances make particular efficiency investments unattractive to the individual consumer though attractive to society as a whole. Consider an apartment building in which the tenants pay their own fuel bills. The owner of the apartment has no incentive to invest capital that would result in energy savings because the savings would accrue to the tenants. In addition, there are a number of cases in which cost-effective energy management measures could be implemented, but the cost of the bother exceeds the benefit--as is the case for the homeowner who desires attic insulation, must find a contractor, and then apply for a loan to cover the cost of the work. If all that this effort might produce is a monthly savings of a few dollars, with two bills to pay instead of one, chances are the homeowner will not make the improvements. Such extra "transaction costs" must be considered.

Availability of capital also affects the rate of adoption of more efficient technology. Any particular company is limited in its ability to raise capital; sometimes a choice must be made between expansion and energy efficiency. As a result, in some instances cost-effective efficiency investments will not be made, particularly if alternative investments with a higher rate of return are available.

Uncertainty is another factor because in times of uncertainty people prefer to have their money in the bank rather than tied up in capital equipment. Uncertainty about future prices of fuel and other materials affects both the homeowner considering additional insulation and the industrial manager considering refitting his plant or replacing machinery. In short, "frictional" forces in the economy--institutional patterns, capital stock, and transaction costs--will reduce the amount of fuel conservation and energy management investments which ultimately are made.

Some measures, however, such as tax incentives tied to more efficient energy use, may lead to beneficial savings for the manager who is quick to seize opportunities when he sees them. Longer-term benefits may come about from increased industrial and government research aimed at developing alternatives or more efficient technologies.

As mentioned above, the complexity of new and existing laws and regulations dealing with energy supply and use does not always facilitate the most efficient energy use. Certain types of standards, codes, rules, or regulations may actually impose inefficient methods of energy use due to traditional or new social needs.

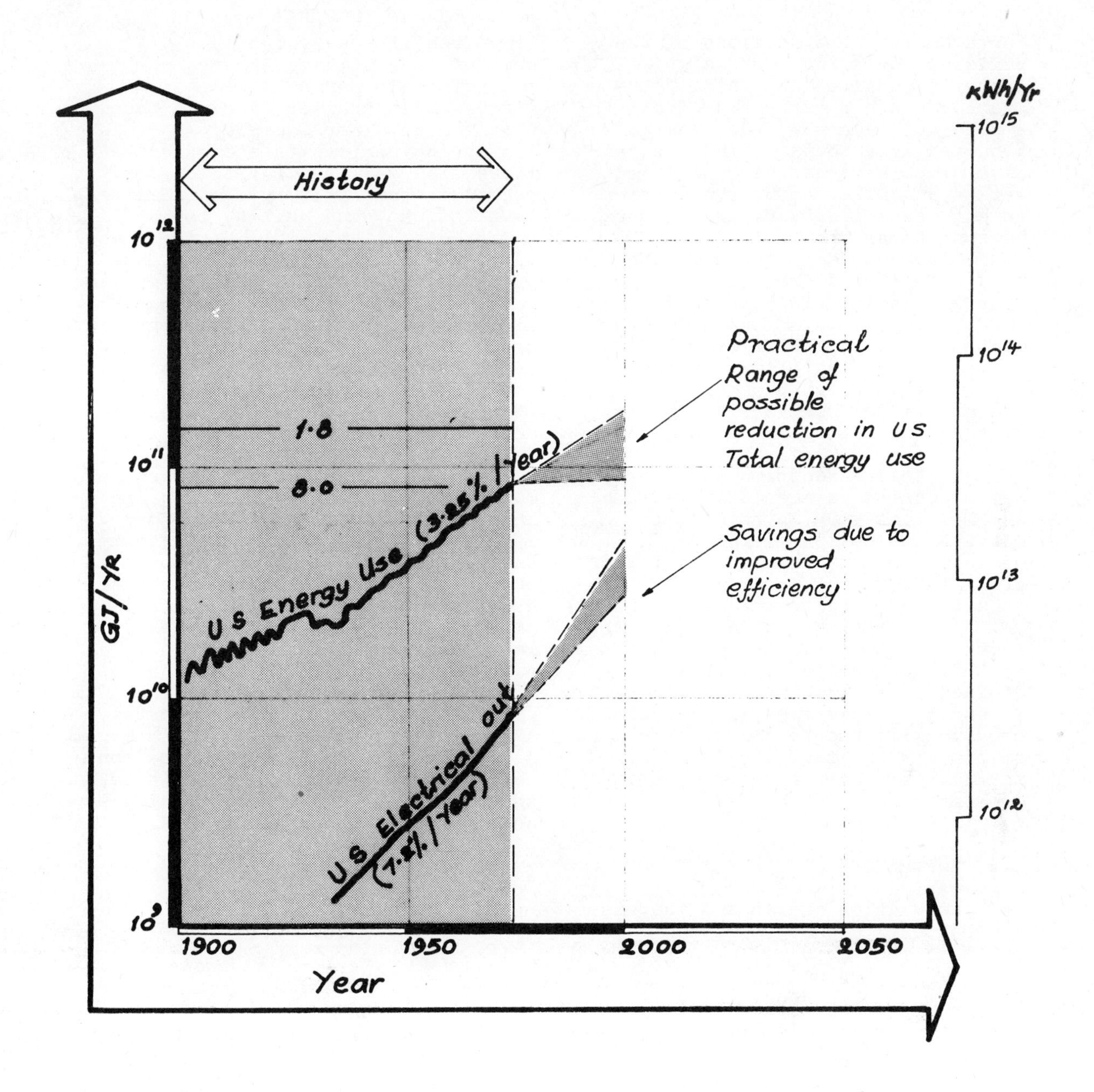

POTENTIAL IMPACT OF IMPROVED EFFICIENCY

FIGURE 1.2

For example, some sections of the building codes require the use of energy-intensive materials. These practices are a carry-over from a previous time when the materials required were abundant and inexpensive or no alternative materials were available. Among the barriers to efficient energy use which must be considered are individual preferences and choices molded by long-established traditions or customs; engineering standards; and codes, laws, regulations and rules. The problem is further complicated by the multi-disciplinary nature of energy use and the historical preference for considering initial cost rather than life cycle cost.

Increased adoption of life cycle costing in the future will be an important step in helping to clarify the options for more efficient energy use. New voluntary programs for appliance and air conditioner labels are a first step in this direction (see Appendix D). Another example is the increased use of this approach by industrial engineers and architect-engineers. As energy prices rise, life cycle costing will become a necessary part of process and facility designs, particularly as related to selection of energy forms to be used.

1.6 INFERENCES

Population growth and increasing per capita energy use in many parts of the world will contribute to an increased need for energy for the next several decades. This, in combination with rising costs and eventual scarcity of non-renewable energy resources, is the motivation for seeking ways of using energy more efficiently.

The next fifty years may be considered a time of transition between reliance on non-renewable energy resources such as oil and future reliance on sources such as breeder reactors, solar energy, or fusion. Improved use efficiency will smooth this transition and will maximize the benefit to be obtained from non-renewable resources,

Even if energy *availability* is removed as a critical issue, one or more of the other constraints discussed in Chapter 14 could become significant. For example, an increasing per capita energy use and an increasing quality of life do not correlate indefinitely. At some high level of use, the ability to use energy effectively will become constrained as the social costs become excessive or as physical limits such as cooling water availability or waste heat disposal are reached. Efficient energy use will reduce the impact of these types of problems, hopefully contributing to the highest quality of life possible for nations in an energy-constrained world.

As a general approach energy management programs will bring immediate payoffs. In many cases just an *awareness* of energy use can lead to savings of a few percent. When housekeeping improvements and other steps to improve operational efficiency (such as those described in later chapters) are added to this awareness, savings in the range of 5 to 10 percent with little financial expenditures are commonly reported. After this, further savings are more difficult and generally require concerted effort, capital expenditure, and the use of efficient technologies such as those reported in this book.

* * *

REFERENCES

1. Gyftopoulos, Elias P.; Lazaridis, Lazaros J.; and Widmer, Thomas F., *Potential Fuel Effectiveness in Industry,* (Cambridge, Massachusetts: Ballinger Publishing Company, 1974).

2. American Physical Society Study Group, "Efficient Use of Energy," *Physics Today,* (August 1975): 23-33.

3. Starr, Chauncey, "A Strategy for National Electricity Production," Presented to the Joint Committee on Atomic Energy, Washington, D.C., 10 July 1975.

4. Beller, M.; Cherniavsky, E.A.; Hoffman, K.C.; and Williamson, R.H., *Interfuel Substitution Study--The Role of Electrification,* BNL 19522, ESAG-17, (Upton, New York: Brookhaven National Laboratory, November 1974).

PART II

ELECTRIC ENERGY MANAGEMENT FROM THE PERSPECTIVE OF THE END-USER

HARDLY LESS IMPORTANT THAN REGIONAL SELF-SUFFICIENCY IN FOOD IS SELF-SUFFICIENCY IN POWER FOR INDUSTRY, AGRICULTURE, AND TRANSPORTATION.

--Aldous Huxley, 1946

Chapter 2

INDUSTRY

Rocco Fazzolare*

CHAPTER CONTENTS

KEY WORDS

Chemicals Industry
Combustion
Computers
Cost/Benefit Analysis
Efficient Energy Use
Energy Accounting
Energy Conversion
Energy Intensiveness
Heat Exchangers
Heat Recovery
Industrial Use
Manufacturing
Paper Industry
Petroleum Industry
Power Factor
Primary Metals Industries
Process Modification
Steel Industry

SUMMARY

Industry as a whole utilizes about 40 percent of US energy resources, of which roughly 38 percent is natural gas, 25 percent petroleum, 29 percent coal, and 8 percent is from hydro and nuclear sources. Approximately 65 percent of the industrial sector's energy resources is used for process heat application (40 percent for process steam and 25 percent for direct heat); an additional 25 percent is used to produce electric energy primarily for motive operations. The remaining 10 percent is used as feedstock materials.

This chapter shows that overall energy savings in the range of 25 to 40 percent are possible in industrial operations over the time period 1975-2000. Immediate reductions (5 to 10 percent) in energy use can be anticipated through improved housekeeping and minor modifications in operation with a small financial outlay. An additional 10 to 15 percent savings in the next 2 to 5 years will require capital expenditures for process redesign and new equipment. Over the long-term (5 to 25 years) additional savings in the range of 10 to 15 percent are possible, but only with process changes involving large investments.

Improved efficiency of energy use depends on the application of energy management principles. Energy audits

---

*Associate Professor of Engineering, University of Arizona

and economic analysis of energy management options identify economically feasible alternatives. Initial technical implementation is based on improved maintenance procedures and the elimination of waste points in operation. Systems and process retrofit or redesign can lead to further opportunities in fuel and electricity savings for the longer term. Improvements in combustion processes, efficient motive operations, heat recovery and recycle, combined power cycles and steam generation, power recovery, and energy effective product design are some of the technical means to achieve efficient energy use in industry.

2.1 INTRODUCTION

In 1975 the US Department of the Interior estimated that industry used 39 percent of the energy used in the United States. The actual numbers were: US total energy inputs, 75 x 10^9 GJ (71 x 10^{15} Btu); industry, 29 x 10^9 GJ (27 x 10^{15} Btu).[1] These numbers include the fuel consumed to generate electricity in both categories. Industrial energy sources have changed since 1973, with coal and electricity increasing, and gas and oil decreasing. The energy used was less in 1975 than it was in 1973 by about 7 percent. The downward trend reflects in part the recession and in part industry's response to recent fuel price increases, and also indicates that energy management is a fact of life in industry today.

In 1972, the industrial sector spent a total of 12 x 10^9 $ for energy. Given the escalation of energy prices since 1972, we can estimate that today these costs are over 30 x 10^9 $ per year. The price tag in the future will certainly surpass these figures. Clearly, businesses and industry have a large stake in the efficient use of energy; the incentive for private enterprise is simply a question of economics and profit.

This chapter describes methods to improve efficiency and achieve savings in industry. The presentation is technically oriented without undue reliance on mathematical or engineering jargon and equations in order to appeal and respond to the needs of a wide spectrum of activity and interest which characterizes the industrial sector. Energy management practices consistent with presently available technology are emphasized. It is understood, however, that significant electricity and fuel savings in industry over the long-term will require the introduction of innovative concepts, major changes in process design, and capital ivestment.

It is not within the scope of this chapter to address the many varied and diverse activities which characterize industrial operations. Each operation, industry, or facility is unique and must necessarily be examined on its own merit. Instead, the chapter presents some generalized concepts, strategies and energy management principles of broad applicability. In addition, the case studies for this chapter illustrate the methods and provide specific applications.

It is appropriate at this point to make some preliminary remarks on the concepts of energy conservation and efficiency as they are generally understood and misunderstood. In any process energy is always conserved, contrary to the popular notion, it cannot be wasted or lost. The initial energy content of fuel going into a process can always be accounted for in the products and by-products of that process. In any actual process, the quality of the energy form changes. Energy quality can be measured by the ability of the system to perform work, and it can be characterized quantitatively by a thermodynamic concept called *availability*. This concept is introduced here and is further developed in Appendix C.

The concept of efficiency as it is generally accepted refers to the ratio of work or heat output to energy input. For the most part, this notion of thermal efficiency will be referred to often in this chapter in the interests of facilitating communication. However, when addressing the question of efficient fuel utilization, this is not the most appropriate measure of performance. It would be more meaningful to examine performance on the basis of utilizing the maximum work potential or the thermodynamic availability of the fuel.

Availability analysis is based on the concept of *effectiveness*, which is the ratio of the theoretical minimum work needed for a given process and the actual useful work required. The difference between the popular notions of efficiency and effectiveness is illustrated by steam boiler operation.

An acceptable boiler is one which achieves efficiencies of about 90 percent. That is, only 10 percent of the input energy is dissipated in the flue gas or by heat transfer losses. We are satisfied then with 90 percent efficiency and consider we are doing the best by present technological standards. Yet, this overlooks consideration of whether we have utilized the fuel to its maximum potential. On the basis of thermodynamic availability, this "efficient" operating boiler has an *effectiveness* of only 40 to 45 percent, indicating some work was lost unnecessarily in making the steam.

Finally, it is appropriate to consider some practicalities as they relate to implementing energy management programs in industry. The chapter presents diverse techniques for using electricity more efficiently and for saving fuels. Furthermore, the theoretical potential for savings, based on thermodynamic principles, is discussed. However, the motivations for action in a specific situation are influenced by a number of counteracting forces which relate to technical, economic, and financial factors. In addition, institutional regulations and constraints can be dominant factors.

Thus, while it may be more efficient to co-generate steam and electric power, legal, economic, or reliability considerations may make it impractical for many situations. The reader must make the final evaluation; the fact that a more efficient technology has been found of value in certain applications (and reported herein) does not mean it will be appropriate in all applications.

Obviously, conservation practices in industry will depend heavily on the cost and availability of fuel. In many cases increased fuel costs will simply be passed on to the consumer in the price of the product. On the other hand, when fuel costs become a significant part of production or when fuel or electricity savings will facilitate competition, conditions exist for decisive action on the part of management. Moreover, any commitment of funds for conservation measures must compete with other priorities and opportunities available to management; financial resources are necessarily limited and are applied to options which maximize economic return.

2.2 ENERGY USE PATTERNS

In the US basic data on industrial energy use are compiled by the US Department of Commerce during its census of wholesale trade, retail trade, service industries, manufacturers, and mineral industries, and by the US Department of Interior, Bureau of Mines. Examples of these periodic reports are listed in the references. [1,2,3,4] Tables 2.1 through 2.3 portray current utilization patterns.

The distribution of energy sources (Table 2.1) shows the large dependence (38 percent) on natural gas in spite of the fact that gas is the scarcest of US fossil fuel reserves. Interruptions in supply of natural gas to industrial customers are now a common occurrence in most areas of the country. The shortage of natural gas in the future will present a serious problem for industry. Many operations must be capable of shifting to an alternate energy source, generally oil or electricity. In any case, dramatic increases in fuel costs, possibly greater than three times present figures, can be anticipated.

An important use of fuel in industry (Table 2.2) is for production of process steam ($\sim$40 percent). Note also the higher growth rate trend (6.3 percent annually) for electrical energy as compared with the overall trend (4.7 percent annually). Even as

	10^{15} Btu	10^9 GJe	% of Total Industry
Coal	4.3	4.5	15.6
Oil	5.6	5.9	10.4
Gas	9.0	9.5	32.9
Hydro	0.1	0.1	0.3
Electrical†	8.4	8.9	30.8
TOTAL	27.4	28.9	100.0

†Electrical Fuels Equivalent:

Fuel Input	
Coal	44%
Nuclear	8%
Oil	16%
Gas	16%
Hydro	16%

Reference 1

INDUSTRY ENERGY SOURCES (1975)

TABLE 2.1

	Annual Growth Rate (%)	% of Sector Energy Requirements
Process Steam	3.6	39
Direct Heat	2.8	26
Feedstock	6.1	9
Electricity†	6.3	26
Average / Total	4.7	100
†Electricity		
Electric Drive	5.3	20
Electrolytic	4.7	3
Direct Heat	---	2
Other	---	1

†Purchased electricity except for 1.5% generated on site.

Reference 4

END USE OF FUEL IN INDUSTRY

TABLE 2.2

SIC No.	Industry Group	Total Energy (%)	Electricity (%)
33	Primary Metals	21	23
28	Chemical and Allied Products	20	29
29	Petroleum Refining and Related Industries	11	4
26	Paper and Allied Products	5	5
20	Food and Kindred Products	5	6
32	Stone, Clay, Glass and Concrete Products	5	5
	Total	67	72

Energy Intensive Individual Industries

Industry	Specific Fuel Consumption (GJ/kg)	Percentage of Industrial Sector Fuel
Iron and Steel	30.7	15.2
Petroleum Refining	5.1	11.4
Paper and Paperboard[†]	28.5	5.4
Aluminum	180	2.8
Copper	30	0.4
Cement	9.2	2.5
Total		37.7

[†]Does not include heating value of waste products (bark and spent pulp liquor).

Reference 6

MAJOR ENERGY USERS IN INDUSTRY

TABLE 2.3

conservation efforts move ahead (these efforts will tend to decrease the historical overall growth trends in energy utilization), there is much reason to believe that electrical growth trends will be maintained and could, in fact, increase as a result of substituting electricity for scarce fuels. Other reasons are:

- conservation measures themselves (heat pumps, solar heating, microwave process heat, etc.);
- environmental factors (less combustion, need for electricity to reduce air and water pollution); and,
- a technology with increasing electric and electronic dependence (materials handling equipment, computers, electric motors).

Approximately two-thirds of energy utilization in industry is in six SIC code groups, as shown by Table 2.3. Fuel requirements for six individual, energy intensive industries are also shown in Table 2.3. These industries use over 0.2 GJ/$ of value added, i.e., per $ of capital, labor, and raw materials. Average use in manufacturing is about 0.06 GJ/$.[5]

A recent Federal Reserve Board study of energy use indicated that for industry as a whole the expenditure of energy per unit of production declined approximately 15 percent between 1958 and 1971.[7] Additionally, Kaplan has determined that for the top 11 energy intensive industries during the same period, reductions of about 28 percent were obtained.[8] These trends indicate that industry has increased its energy use efficiency.

More recently, the rapidly escalating cost and reduced availability of energy resources have provided additional incentives for improving industrial energy use efficiency.

2.3 GENERAL STRATEGIES FOR ENERGY MANAGEMENT

A. Organizing An Energy Management Program [12,13,14,15,16]

The effectiveness of energy utilization varies with specific industrial operations because of diversity of the products and in the processes required to manufacture them. The organization of personnel and operations involved also varies. Consequently, an effective conservation program should be custom designed for each company and its plant operations. There are some generalized guidelines, however, for initiating and implementing an energy management program. Many of the large companies have already instituted energy management programs and have realized substantial savings in fuel utilization and costs. Smaller industries and plants, however, often lack the technical personnel and equipment to institute and carry out effective programs. In these situations reliance on external consultants may be appropriate to initiate the program. Internal participation, however, is essential for success. A well planned, organized, and executed energy management program requires a strong commitment by top management.

Assistance can also be obtained from local utilities. Utility participation would include help in getting the customer started on an energy management program, technical guidance, or making information available. Some utilities today have active programs which include training of customer personnel or provision of technical assistance.

Table 2.4 summarizes the elements of an effective energy management program. These will now be discussed in more detail.

- Phase 1: Management Commitment.

A commitment by the directors of a company to initiate and support a program is essential. An Energy

Phase 1: Management Commitment

1.1 Commitment by management to an Energy Management Program.
1.2 Assignment of an Energy Management Coordinator.
1.3 Creation of an Energy Management Committee of major plant and department representatives.

Phase 2: Audit and Analysis

2.1 Review of historical patterns of fuel and energy use.
2.2 Facility walk-through survey.
2.3 Preliminary analyses, review of drawings, data sheets, equipment specifications.
2.4 Development of energy audit plans.
2.5 Conduct facility energy audit, covering:
a) processes, and
b) facilities and equipment.
2.6 Calculation of annual energy use based on audit results.
2.7 Comparison with historical records.
2.8 Analysis and simulation step (engineering calculations, heat and mass balances, theoretical efficiency calculations, computer analysis and simulation) to evaluate energy management options.
2.9 Economic analysis of selected energy management options (life cycle costs, rate of return, benefit-cost ratio).

Phase 3: Implementation

3.1 Establish energy effectiveness goals for the organization and individual plants.
3.2 Determine capital investment requirements and priorities.
3.3 Establish measurement and reporting procedures. Install monitoring and recoring instruments as required.
3.4 Institute routine reporting procedures ("energy tracking" charts) for managers and publicize results.
3.5 Promote continuing awareness and involvement of personnel.
3.6 Provide for periodic review and evaluation of overall energy management program.

ELEMENTS OF AN ENERGY MANAGEMENT PROGRAM — TABLE 2.4

Coordinator is designated and an energy management committee is formed. The committee should include personnel representing major company activities utilizing energy. A plan is formulated to set up the program with a commitment of funds and personnel. Realistic overall goals and guidelines in energy savings should be established based on overall information in the company records, projected activities, and future fuel costs and supply.

- Phase 2: Audit and Analysis

Energy Audit of Equipment and Facilities

Historical data for the facility should be collected, reviewed, and analyzed. The review should identify gross energy uses by fuel types, cyclic trends, fiscal year effects, dependence on sales or work load, and minimum and maximum energy use ratios. Figure 2.1 shows historical monthly energy use data in an agricultural chemical producing facility.

Historical data assist in planning a detailed energy audit and alert the auditors as to the type of fuel and general equipment to expect. A brief facility "walk-through" is recommended to establish the plant layout, major energy uses, and primary processes or functions of the facility.

The energy audit is best performed by an experienced or at least trained team, since visual observation is the principal means of information gathering and operational assessment. A team would have from three to five members, each with a specific assignment for the audit. For example, one auditor would check the lighting, another the HVAC system, another the equipment and processes, another the building structure (floor space, volume, insulation, age, etc.), and another the occupancy use schedule, administration procedures, and employees' general awareness of energy management.

The audit's objectives are to determine how, where, when and how much energy is used in the facility. In addition, the audit helps to identify opportunities to improve the energy use efficiency of the facility and its operations.

Some of the problems encountered during energy audits are determining the rated power of equipment, determining the effective hours of use per year, and determining the effect of seasonal, climatic, or other variable conditions on energy use. Equipment ratings are often obscured by dust or grease (unreadable nameplates). Complex machinery may not have a single nameplate listing the total capacity, but several giving ratings for component equipment. The effect of load is also important because energy use in a machine operating at less than full load may be reduced along with a possible loss in operating efficiency.

The quantitative assessment of fuel and energy use is best determined by actual measurements under typical operational conditions using portable or installed meters and sensing devices. Such devices include light meters, ammeters, thermometers, air flow meters, recorders, etc. In some situations sophisticated techniques such as infrared scanning or thermography are useful. The degree of measurement and recording sophistication naturally depends on available funds and the potential saving anticipated. For most situations, however, nameplate and catalog information are sufficient to estimate power demand. Useful information can be obtained from operating personnel and their supervisors--particularly as it relates to usage patterns throughout the day. Figure 2.2 is a sample form which can be used for recording audit data.

The first three columns of the form are self-explanatory. The fourth column is used for the rated capacity of the device, e.g., 5 kW. The seventh column is used if the device is operated at partial load.

Usage hours (column 8) are based on all work shifts, and are corrected to account for the actual operating time of the equipment. The last three columns are used to convert energy units to a common basis (e.g., MJ or Btu).

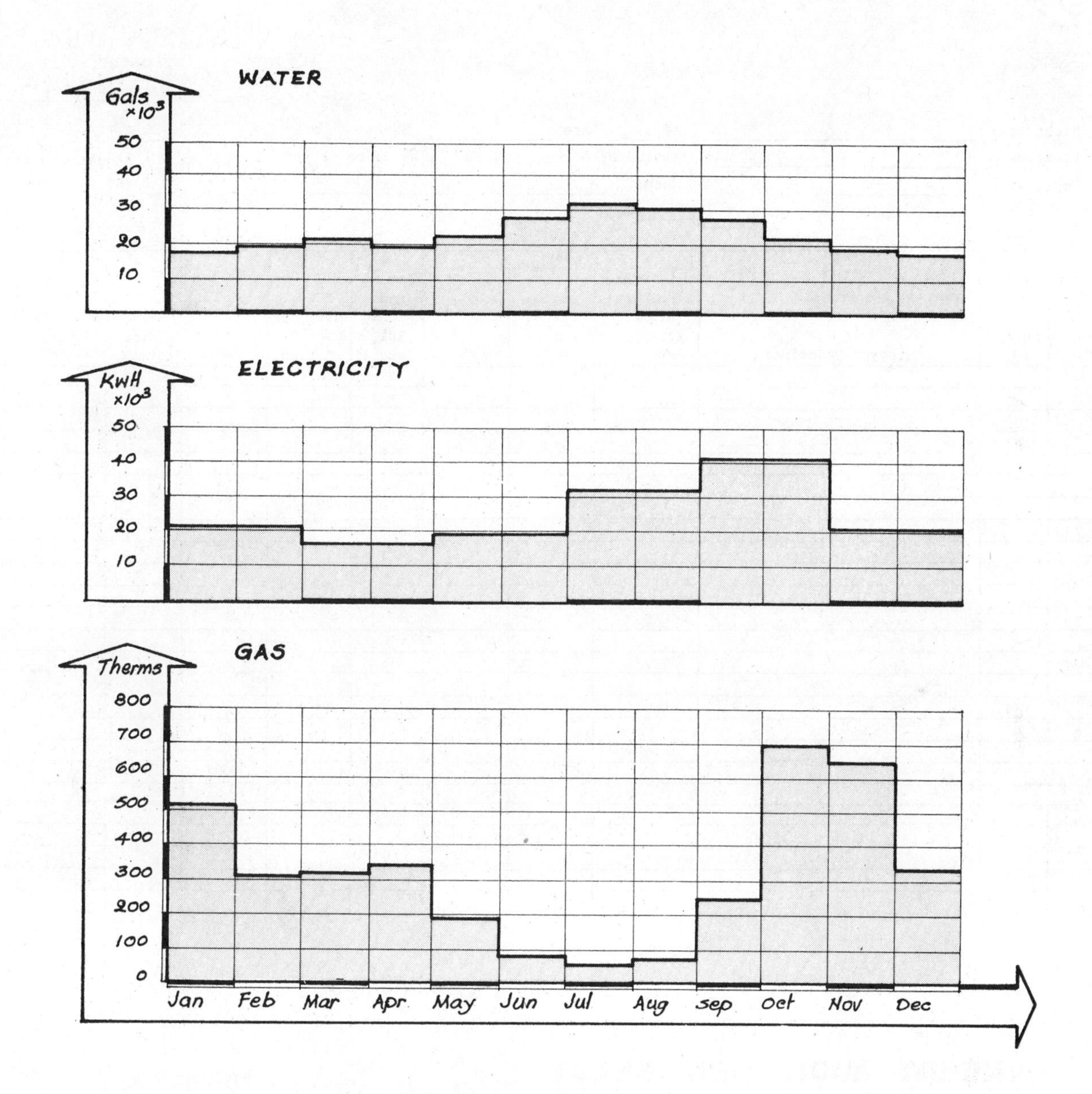

HISTORICAL ENERGY USE IN A SMALL MANUFACTURING FACILITY

FIGURE 2.1

Symbols: k = 10^3
M = 10^6

Conversion Factors

Multiply	by	to get
kWh	3.6	MJ
Btu/hr	0.000293	kW
hp	0.746	kW

Plant Name ____________________ By ____ Date ____________ Sheet ___ of ___

Location ____________________ Period of Survey: 1 day 1 wk 1 mo 1 yr

Department ____________________ Notes ____________________

			Power							
Type Fuel	Equip. ID #	Equipment Description	Name Plate Rating (Btu/hr, kW, hp, etc.)	Conv. Factor to kW	kW	Est. % Load (100%, 50%, etc.)	Est. Hrs Use Per Period	kWh	Conv. Factor	Total Energy Use Per Period (MJ)

ENERGY AUDIT DATA SHEET

FIGURE 2.2

Data recorded in the field are reduced easily either by hand or computer. The advantage of using a computer is that uniform results and summaries can be obtained easily in a form suitable for review or for further analysis. Computer analysis also provides easy modification of the results to reflect specific reporting requirements of management or to present desired comparisons for different energy use, types of equipment, etc.

A Special Case: Energy Audit of a Process

In some manufacturing and process industries it is of interest to determine the energy content of a product. This can be done by a variation of the energy audit techniques described above. Since this approach resembles classical financial accounting, it is sometimes called *energy accounting*.

In this procedure the energy content of the raw materials is determined in a consistent set of energy units. Then, the energy required for conversion to a product is accounted for in the same units. The same is done for energy in the waste streams and the by-products. Finally, the net energy content per unit produced is used as a basis for establishing efficiency goals. A guide and procedures for calculating the energy content of a product are shown in Table 2.5.

Table 2.5 is divided into six sections:

- description of product,
- raw materials energy,
- conversion energy,
- waste disposal energy,
- by-products energy, and,
- energy content goal.

In the raw materials section, all materials used in the product or to produce it are determined. Input raw materials used in any specific period are normally available from plant records. Approximations of specific energy content for some materials can be found in Appendix B or can be obtained from the US Department of Commerce or other sources. The energy content of a material includes that due to extraction and refinement as well as any inherent heating value it would have as a fuel prior to processing. Consequently, non-fuel type ores in the ground are assigned zero energy, and petroleum products are assigned their alternate value as a fuel prior to processing in addition to the refinement energy. The energy of an input metal stock would include the energy due to extraction, ore refinement to metal, and any milling operations.

Conversion energy is an important aspect of the energy audit since it is under direct control of plant management. All utilities and fuels coming into the plant are accounted for. They are converted to consistent energy units (joules or Btu) using the actual data available on the fuels or the suggested conversions given in Appendix A.

As noted in Chapter 1, electrical energy is assigned the actual fuel energy required to produce the electricity. This accounts for power conversion efficiencies. A suggested approach is to assume (unless actual values are available from your utility) that 10.8 MJ (10,200 Btu) is used to produce 3.6 MJe (1 kWh), giving a fuel conversion efficiency of $3.6 \div 10.8 = 0.33$ or 33 percent. See Appendices A and B for additional discussion of fuel conversion factors.

The energy content of process steam includes the total fuel and electrical energy required to operate the boiler as well as line and other losses. Some complexities are introduced when a plant produces both power and steam since it is necessary to allocate the fuel used to the steam and power produced. One suggested way to make this allocation is to assume that there is a large efficient boiler feeding steam to a totally condensing vacuum turbine. Then, one must determine the amount of extra boiler fuel that would be required to permit the extraction of steam at whatever pressure while maintaining the constant load on the generator. The extra fuel is considered the energy content of the steam being extracted.

"DO IT YOURSELF KIT" For Calculating The Energy Content of A Product

GUIDELINES OF THE NATIONAL INDUSTRIAL ENERGY CONSERVATION COUNCIL

SUGGESTED PROCEDURE FOR CALCULATING ENERGY CONTENT (BTU'S) OF A PRODUCT

FOR THE PERIOD BEGINNING January 1, 1974, PERIOD ENDING February 1, 1974.

COMPANY	RESPONSIBLE MANAGER
PRODUCT 1	PRODUCT I.D. NO. 2

TOTAL UNITS PRODUCED 3

UNITS OF PRODUCTION (LB, GAL, PIECE, ETC.)

RAW MATERIAL ENERGY (LIST MAJOR RAW MATERIALS)

4 RAW MATERIAL	5 TOTAL UNITS	x 6 BTU'S UNIT	= 7 TOTAL BTU'S
A:			
B:			
C:			
D:			
E:			

TOTAL BTU'S 8

CONVERSION ENERGY (LIST ALL MAJOR UTILITIES)

9 UTILITY	10 TOTAL UNITS	x 11 BTU'S UNIT	= 12 TOTAL BTU'S
A:			
B:			
C:			
D:			
E:			

TOTAL BTU'S 13

WASTE DISPOSAL ENERGY

14 WASTE	15 TOTAL DISPOSAL BTU'S	17 TOTAL WASTED UNITS
A:		
B:		
C:		
D:		
E:		

TOTAL BTU'S 16

GROSS ENERGY CONTENT OF PRODUCT (SUM OF ITEMS 8, 13 AND 16) BTU'S 18

BY-PRODUCT ENERGY CREDIT (LIST ALL MAJOR BY-PRODUCTS)

19 BY-PRODUCT	20 TOTAL UNITS	x 21 BTU'S UNIT	= 22 TOTAL BTU'S
A:			
B:			
C:			
D:			
E:			

TOTAL BTU'S 23

NET ENERGY CONTENT OF PRODUCT (ITEM 18 LESS ITEM 23) 24 BTU'S

ENERGY CONTENT PER UNIT OF PRODUCTION (ITEM 24 DIVIDED BY ITEM 3) 25 BTU'S UNIT

GOAL (TARGETED ENERGY CONTENT FOR THIS PERIOD) BTU'S UNIT — 26

IF ITEM 26 IS EQUAL TO ITEM 25, GOAL WAS MADE (CHECK ITEM 27) — 27 MADE GOAL

IF ITEM 26 IS NOT EQUAL TO ITEM 25, COMPUTE DEVIATION FROM GOAL:

ITEM 26 LESS ITEM 25 — 28

ITEM 28 DIVIDED BY ITEM 26 — 29

MULTIPLY ITEM 29 BY 100 — 30

IF ITEM 26 IS GREATER THAN ITEM 25, COPY ITEM 30 HERE — 31 + % BEAT GOAL

IF ITEM 26 IS LESS THAN ITEM 25, COPY ITEM 30 HERE — 32 − % MISSED GOAL

ENERGY CONTENT CHART TABLE 2.5

1. Finished product ready for shipment.
2. Product I.D. No. is the numerical identification of the product.
3. Units of the product (item 1) made during this time period.
4. The material that goes into producing and packaging the product (includes fuels used as raw material).
5. Units of the raw material (item 4) that were used during this time period.
6. Every material has a specific energy content. Energy content is measured in terms of Btu's. Raw material supplier may provide this number or an approximation is available for most materials from the U.S. Department of Commerce. If unavailable from these sources, it can be estimated as the heat of combustion of the material. This estimate is always low.
7. (Item 5) multiplied by (item 6).
9. Utilities include primarily electricity, fuel oil and natural gas.
10. Units of utility (item 9) used during this time period.
11. For fuel, this is the heat of combustion of the fuel. This number is available from supplier. For other utilities, this is the energy necessary to generate one unit of the utility (e.g., 1 kWh). See Appendices A and B for typical values or refer to your utility.
12. (Item 10) multiplied by (item 11).
14. Waste is that material which requires additional Btu's to dispose of.
15. Estimted energy to dispose of the waste (item 14). This may be the energy to truck away and bury a solid, the energy to burn some scrap or the energy to run a waste disposal plant.
17. Units of waste produced during this time period. Unit of waste is not needed for the calculation, but may be recorded for later reference.
19. By-products are those saleable materials which are made incidental to the production of the desired product or products.
20. Units of by-product (item 19) made during this time period.
21. The usable energy in the by-product. As an approximation, use of the ratio of the value of the by-product to the value of the product multiplied by the gross energy content of the product (item 18).
22. (Item 20) multiplied by (item 21).

GUIDE FOR USING CHART IN TABLE 2.5

TABLE 2.5 CONT'D

In another method, steam at an intermediate pressure is assumed to be the base; its theoretical enthalpy is divided by a reasonable boiler efficiency to determine its energy content. The energy content at other pressures of steam is then related to this base case on the basis of the steam's ability to perform work. Steam at the condensing temperature is considered to have no ability to perform work. As an example, if we assume that the base is 100 psia, the enthalpy is 2.77 MJ/kg (1194 Btu/lbm), and the generator efficiency is 85 percent. The base case energy allocation then is:

$$2.77 \div 0.85 = 3.26 \text{ MJ/kg} \quad (1900 \text{ Btu/lb})$$

The energy content of steam at 2.07×10^5 N/m^2 (30 psia) would then be computed in the following way:

$$3.26 \times \frac{(2.70 - 2.57)}{(2.77 - 2.57)} = 2.12 \text{ MJ/kg} \quad (914 \text{ Btu/lb})$$

where the enthalpy at 2.07×10^5 N/m^2 is equal to 2.70 MJ/kg (1164 Btu/lb), and the enthalpy at 6985 N/m^2 (1 psia) equals 2.57 MJ/kg (1106 Btu/lb). Finally, another method would be to allocate energy content according to thermodynamic availability.

Waste disposal energy is that energy required to dispose of or treat the waste products. This includes all the energy required to bring the waste to a satisfactory disposal state. In a case where waste is handled by a contractor or some other utility service, it would include the cost of transportation and treatment energy.

If the plant has by-products or co-products, then an energy credit is allocated to them. A number of criteria can be used. If the by-product must be treated to be utilized or recycled (such as scrap), then the credit would be based on the raw material less the energy expended to treat the by-product for recycle. If the by-product is to be sold, the relative value ratio of the by-product to the primary product can be used to allocate the energy. In the final section of the audit chart, goals are set and compared with actual performance.

Analysis of Audit Results, Identification of Energy Management Opportunities

Often the energy audit will identify immediate energy management opportunities, such as unoccupied areas which have been inadvertently illuminated 24 hours per day, equipment operating needlessly, etc. Corrective housekeeping and maintenance action can be instituted to achieve short-term savings with little or no capital investment.

An analysis of the audit data is required for a more critical investigation of fuel waste and identification of the potentials for conservation. This includes a detailed energy balance of each process, activity, or facility. Process modifications and alternatives in equipment design should be formulated, based on technical feasibility and economic and environmental impact. Economic studies (see Section 2.3 C and Appendix E) to determine payback, return on investment, and net savings are essential before making capital investments.

- Phase 3: Implementation

At this point goals for saving energy can be established more firmly and priorities set on the modifications and alterations to equipment and the process. Effective measurement and monitoring procedures are essential in evaluating progress in the energy management program. Routine reporting procedures between management and operations should be established to accumulate information on plant performance and to inform plant supervisors of the effectiveness of their operations. Time-tracking charts of energy use and costs can be helpful. Involvement of employees and recognizing their contributions facilitate the achievement of objectives. Finally, the program must be continually reviewed and analyzed with regard to established goals and procedures. (See Case Study 2-2 for an illustration of an energy audit.)

- Energy Management Methods Are Applicable to All Sectors

The methodology described above is not limited to the industrial energy user. The procedures are quite general and can be used effectively in the commercial, residential, and agricultural sectors as well. Refer to Chapters 3, 4, and 6 for additional details and examples.

B. The Concept of Available Work

Until recently, the economic value of input fuels and combustible by-products has been relatively unimportant for all but the most energy-intensive industries (chemical products, primary metals, paper mills, cement). In addition, efficient energy utilization has often knowingly been penalized at the expense of increased operating costs in the interest of lower initial investments. As a result, process equipment and facility design innovations have been realized with the partial exclusion of improved energy efficiency. Some specific fuel consumption values (energy per unit mass) in the manufacture of products and mining are listed in Appendix B.

To appreciate more the potential utility of fuel energy, one must come to grips with the notion that all energy forms are not equivalent and that their ability to do work varies. A useful measure of energy utility is embodied in the concept of *available work, thermodynamic availability*, or *exergy* as it is called in Soviet and European literature.(6,9] The available work of a system can be shown to be given approximately by:*

$$B = E + \rho_o V - T_o S - \sum_{i}^{n} \mu_{io} n_i$$

where:

E = initial energy,

V = volume,

S = entropy,

μ_{io} = potential energy,

n = number of moles,

ρ_o = pressure of the atmosphere,

T_o = temperature of the atmosphere, and

i = i^{th} molecular component.

For a given energy, volume and system composition, it can be seen that B decreases as the system entropy increases. If applied to a hydrocarbon fuel, B is the minimum useful work required to form the fuel in a given state from the water and carbon dioxide in the atmosphere. Since the minimum is also the useful work of a reversible process, B also represents the maximum useful work which could be obtained by oxidation of the fuel and return of the products to the atmosphere.[6]

To put it another way, B provides a measure of not only the *quantity* of energy, but the *quality* as well. Thus one kg of gasoline has the same energy content (48 MJ/kg) as 2,296 kg of water warmed to 5 °C above ambient, but the ability of gasoline to perform useful work is much greater.

If gasoline in an idealized system were burned with no losses and all of its energy were used to warm water, no energy would be "consumed" (fuel, however, would be!). Clearly, the benefit of tepid water is less than gasoline; the potential for work has been lost and the quality of the energy has been degraded.

A measure of thermodynamic *effectiveness* or *efficiency of energy use* for a process can be determined by the ratio of the increase in available work attained by the products in the process, divided by the available useful work of the fuel consumed. If the products leave the process at high temperature, as in a blast furnace, the process should not be charged with available work in the products by virtue of the elevated temperature. If this work is lost, the deficit is not in the process but in the means by which the products are cooled.[6] A recent study of certain energy intensive industries based on a theoretical thermodynamic analysis indicates the saving which can be achieved in the

*See Appendix C for a more detailed discussion of available work as well as the terms neglected in the above equation.

production of paper, aluminum, steel and cement; a summary of the results is presented in Table 2.6.

The concept of available work provides a useful measure of efficiency. In addition, it suggests process steps or areas where improvements in efficiency are possible. (See Case Study 2-1 for an illustration of the thermodynamic analysis of a process.) *A word of caution* is appropriate with regard to interpreting the concept of the theoretical minimum energy required. Theoretically, a car driving from Denver to Los Angeles should use no fuel and, as a matter of fact, should yield useful energy (due to the difference in elevation). Thus, while the theoretical minimum is a useful concept with which to gauge the potential for fuel conservation, there is no implication that the attainment of such a goal is practical or even possible.

Recent evidence indicates that energy savings can be realized without decreasing industrial production, yields, or product quality. Data provided by DuPont on case histories involving 75 energy intensive industrial plants suggest that on the average, an immediate savings in excess of 15 percent of energy costs can be realized through improved energy management techniques and operational modifications (see Table 2.7). The data in Table 2.7 are based on energy management techniques and operational modifications which required non-capital expenditures and capital expenditures respectively. These gains in energy use efficiency were obtained with existing facilities and did not require fundamental changes in process concepts, product quality, yields, production levels, or other basic design modifications.

Another study to establish the conservation potential for some energy-intensive industries in Texas (based on a survey of more than 200 companies) is summarized in Table 2.8. The projected savings were based on engineering case studies in each of the industries examined. In general, savings on the order of 10 to 30 percent were projected that could be implemented over 5 to 7 years. Combustion savings amounted to one-third to two-thirds of the total amount. Capital requirements were estimated to average from 7 to 15 x 10^6 dollars for the average medium to large plant surveyed (production greater than 3.6 x 10^5 mt/yr [4 x 10^5 ton/yr]).

The case studies for this chapter provide further evidence of the magnitude of potential savings which can be achieved with an effective energy management program.

C. Economic Analysis of Energy Management Opportunities.

Economics is the prime incentive for energy management. Fuel availability is another important consideration, but it is reflected to a greater or lesser extent in the cost of fuel or electricity. The extent to which capital expenditures will be made is determined largely by the economic return. Each firm has its own approaches to this problem; a common element is the estimation of costs and prediction of savings or economic benefits.

Simple calculations of rate of return and payback period are often used to assess economic opportunities, but they do not adequately reflect all factors bearing on investment decisions. The simplified methods are generally acceptable for equipment replacement or projects with economic lives of two to three years. For longer project lifetimes, the current dynamic market conditions involving high interest rates and escalating costs must be considered. For private firms, taxes and investment tax credits are important factors.

Figure 2.3 illustrates several of these points. The maximum justified capital expenditure that can be invested per annual dollar saved is shown in the figure. Zero escalation and no tax consideration cases are shown for comparison. The figure demonstrates the dramatic impact escalation can have, particularly for project lifetimes greater than five years. See Appendix E for additional details regarding economic analysis.

	1968 Specific Fuel Consumption (MJ/kg)	Potential Specific Fuel Consumption-Using Technology Existing in 1973 (MJ/kg)	Theoretical Minimum Specific Fuel Consumption Based Upon Thermodynamic Availability Analysis (MJ/kg)
Iron and Steel	30.8	19.9	7.0
Petroleum Refining	5.1	3.8	0.5
Paper	45.3*	27.6*	¶Greater than -0.2 Smaller than +0.1
Primary Aluminum Production†	221	177	29.3
Cement	9.2	5.5	0.4

*Includes 16.8 MJ/kg of paper produced from waste products consumed as fuel by paper industry.

†Does not include effect of scrap recycling.

¶Negative value means that no fuel is required.

Reference 6

COMPARISON OF SPECIFIC FUEL CONSUMPTION OF KNOWN PROCESSES WITH THEORETICAL MINIMUM FOR SELECTED U S INDUSTRIES

TABLE 2.6

	Annual Energy Value	Savings/yr Noncapital Expenditures	With Capital Expenditures	Total Savings	%
Chemical & Material Process Industries	9.81	0.51	0.94	1.45	14.8
Primary Metals	2.20	0.297	0.231	0.526	24.0
Manufacturing	3.47	0.295	0.373	0.674	19.4

Reference 10

INDUSTRIAL ENERGY SAVINGS IN MILLIONS OF DOLLARS PER YEAR PER FACILITY

TABLE 2.7

	1973 Production (mt/yr) Energy Use (GJ/yr)	Specific Energy Content (MJ/kg)	Estimated Savings* (%)	Time (yr)
1. Refining	183×10^6 0.85×10^9	4.6	30.1	6.6
2. Chemicals	73×10^6 1.4×10^9	19	31.1	6.5
3. Pulp and Paper	2.0×10^6 0.074×10^9	37	19.7	10
4. Metals	5.8×10^6 0.30×10^9	51	11.0	5

*Calculated as % reduction in specific energy content

Reference 11

ENERGY USE BREAKDOWN & SAVINGS FOR INDUSTRIAL OPERATIONS IN TEXAS

TABLE 2.8

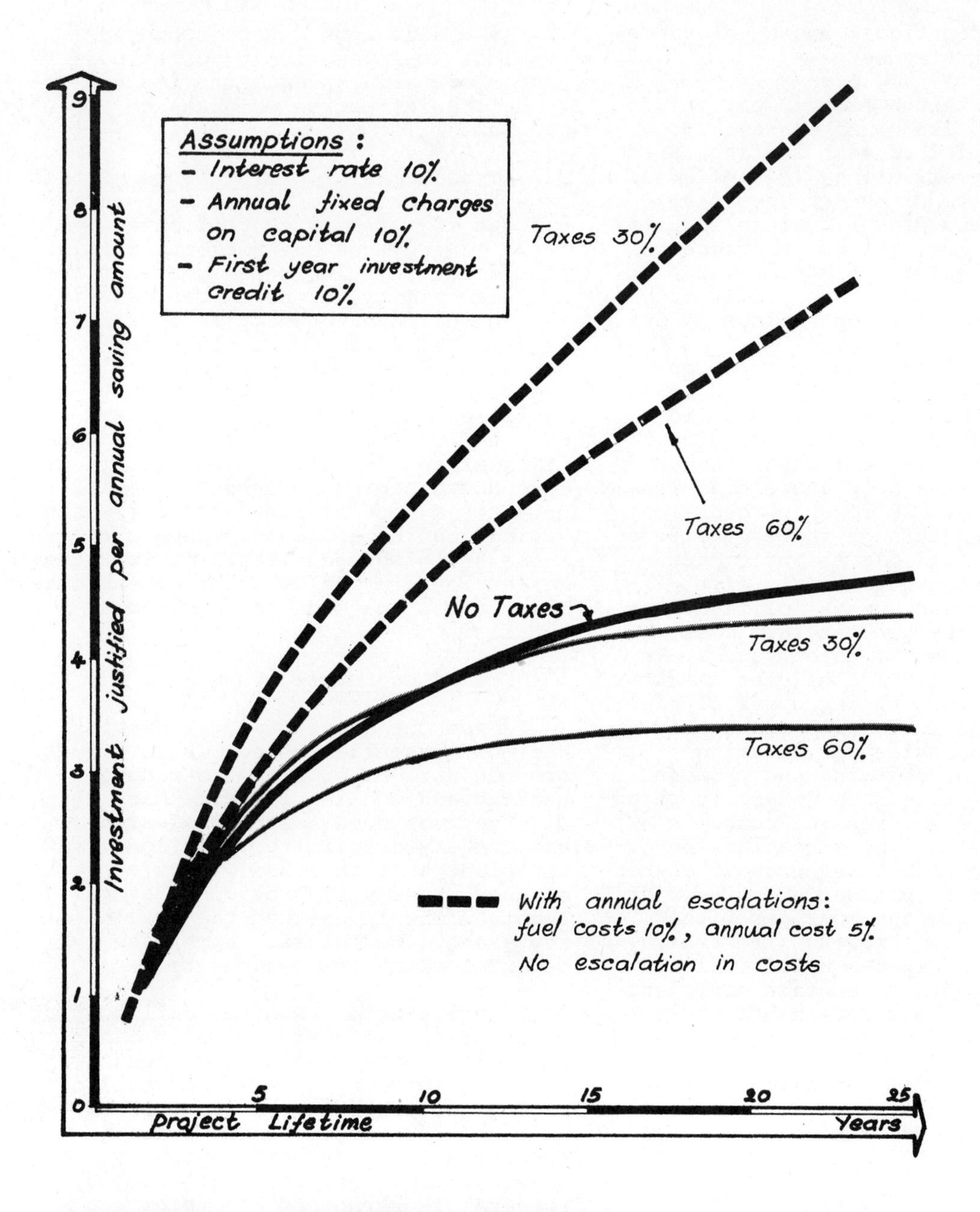

MAXIMUM JUSTIFIED CAPITAL EXPENDITURE FOR ENERGY SAVINGS

FIGURE 2.3

D. Operational and Maintenance Strategies

Although each industrial facility and its operation are unique, the potential of each for increased energy use efficiency is similar. Irrespective of the product character, the end use of fuel in industry generally involves either process heat in the form of steam (∿40%) or direct heat (∿25%) and electricity usually in motive operations (∿25%).

The initial measures to conserve fuel and electricity in existing plants can result in 5 to 10 percent savings and will require little or no capital investment. Higher savings in plant energy use--an additional 20 to 30 percent--can be achieved through the expenditure of capital and can be implemented in the next two to twenty-five years. In many cases modifications are repaid by current fuel cost savings in one to three years. In other cases it will only be a matter of time before rising costs of electricity and fuels will adequately justify modifying the existing operation or changing the process. In this section some generally recognized opportunities and concepts applicable to many situations in industry are reviewed without elaborating on the applicable engineering details. For the most part, the technology and know-how are available and many system modifications and components to enhance efficient energy use are standard off-the-shelf items.

In this and the next two subsections, energy management opportunities are grouped into three categories:

- operational and maintenance strategies,
- retrofit or modification strategies, and,
- new design strategies.

The above order corresponds to increasing capital investment and increasing implementation times.

In most industrial operations embarking on an energy management program, the immediate savings in energy utilization can be achieved by minor improvements in operation and, above all, through the implementation of an effective maintenance program.

Traditionally, maintenance efforts have been motivated primarily by the concern for preserving capital equipment. Energy use has not been a primary consideration. Increased costs of fuel and electricity, however, can result in a significant penalty in operating costs. The situation is further complicated by poor equipment design and lack of performance indicators. Experience has shown that a minimum of 5 to 10 percent improvement in energy utilization can be achieved through effective housekeeping measures. Table 2.9 provides a partial list of areas where savings can be achieved. Some of these areas are discussed briefly below; the reader is referred to other appropriate chapters of the book for additional details.

Building Envelope and Site

From the housekeeping viewpoint the most essential aspect is to provide a tight envelope to reduce heating and cooling losses. Many alternatives need to be considered and there are infinite variations. In one plant skylights and window areas can save energy by providing natural light, while in another they would represent a heat loss. Each case must be examined on its own merits.

Infiltration is often an important cause of excess energy use; it can be controlled by upgrading weatherstripping, using door closers, and sealing piping penetrations. Insulation can be installed in some situations; in other cases it may be very costly. Site considerations are sometimes helpful. For example shrubbery and trees can shade areas exposed to sun and thereby reduce cooling needs. Refer to Chapter 7 for details.

Lighting

Industrial lighting needs range from low level requirements for assembly and welding of large structures (such as shipyards) to the high levels needed for manufacture of precision mechanical and electronic components (e.g., integrated circuits).

System	Typical Opportunities	For additional details, refer to Chapter:
Building Envelope	Reduce infiltration, improve insulation; modify paint colors, etc.	7
Lighting	Task lighting, more efficient lamps, improved controls	9
HVAC Systems	Heat recovery, better controls use of outside air, modified ventilation	8
Combustion Processes	Optimize excess air, heat recovery	2
Steam Systems (hot water also)	Cascade energy use; heat and power recovery, reduce losses	10
Process Heat	Heat recovery, more efficient equipment, reduce losses	10
Compressed Air Systems	Optimize equipment type and size, reduce losses, heat recovery	2
Electric Power Systems	Reduce i^2r losses; modify power factor, reduce peak demand, decrease losses	11
Cooling Towers	Optimize flow rates and fan speeds, improve controls and heat transfer	2
Material Transport	Reduce handling steps and distances; provide more efficient equipment	5
Industrial Processes	Reduce scrap; use different materials; use water-base paints, etc.	2

PLANT SYSTEMS TO BE REVIEWED FOR HOUSEKEEPING ENERGY MANAGEMENT OPPORTUNITIES

TABLE 2.9

There are four basic housekeeping checks which should be made:

- Is a more efficient lighting application possible? (Remove excessive or unnecessary lamps.)
- Is relamping possible? (Install lower wattage lamps during routine maintenance.)
- Will cleaning improve light output? (Fixtures, lamps, and lenses should be cleaned periodically.)
- Can better controls be devised? (Eliminate turning on more lamps than necessary.)

Chapter 9 should be consulted for additional approaches.

Heating, Ventilating, Air-Conditioning Operation

Much energy can be saved in the environmental conditioning of buildings and structures associated with manufacturing operations. Opportunities and the procedures to improve these energy use patterns are described and discussed in Chapters 3 and 8.

The environmental needs in an industrial operation can be quite different than in a residential or commercial structure. In some cases strict environmental standards must be met for a specific function or process. More often the environmental requirements for the process itself are not severe; however, conditioning of the space is necessary for the comfort of operating personnel, and thus large volumes of air must be processed. Quite often opportunities exist in the industrial operation where surplus energy can be utilized in environmental conditioning. A few suggestions follow:

- Building heating and cooling controls should be examined and preset.
- Ventilation, air, and building exhaust requirements should be examined. (A reduction of air flow will result in a saving of electrical energy to motor drives and additionally reduce the energy requirements for space heating and cooling.) Due to the fact that pumping power varies as the cube of the air flow rate, substantial savings can be achieved by reducing air flows where possible. Figure 2.4 shows that a 20 percent reduction in air flow cuts power in half.
- Review air-conditioning and heating operations, seal off sections of plant operations which do not require environmental conditioning, and use air-conditioning equipment only when needed. During nonworking hours the environmental control equipment should be shut down or reduced. Automatic timers can be effective.
- Insure that all equipment is operating efficiently. (Filters, fan belts, and bearings should be in good condition.)
- When multiple units are available, examine the operating efficiency of each unit and put their operation in sequence in order to maximize overall efficiency.

See Chapter 8 for additional suggestions.

Combustion Processes

Fired heaters, boilers and other units which rely on combustion to liberate heat can contribute significantly to fuel savings through proper operations involving a minimum of capital expenditure. Electricity is used to operate motors, fans and controls for combustion processes. Changes in the operation of combustion systems will also affect use of electricity.

- *Provide correct amounts of excess air*--Too much excess air can be as costly as not enough. Excess air should be maintained at about 10 percent, with results in about 2 percent oxygen in the flue gas. The potential fuel savings for systems operating with higher flue gas oxygen content can be obtained from Figure 2.5. The amount of oxygen in the flue gas can be determined by analyses which should be

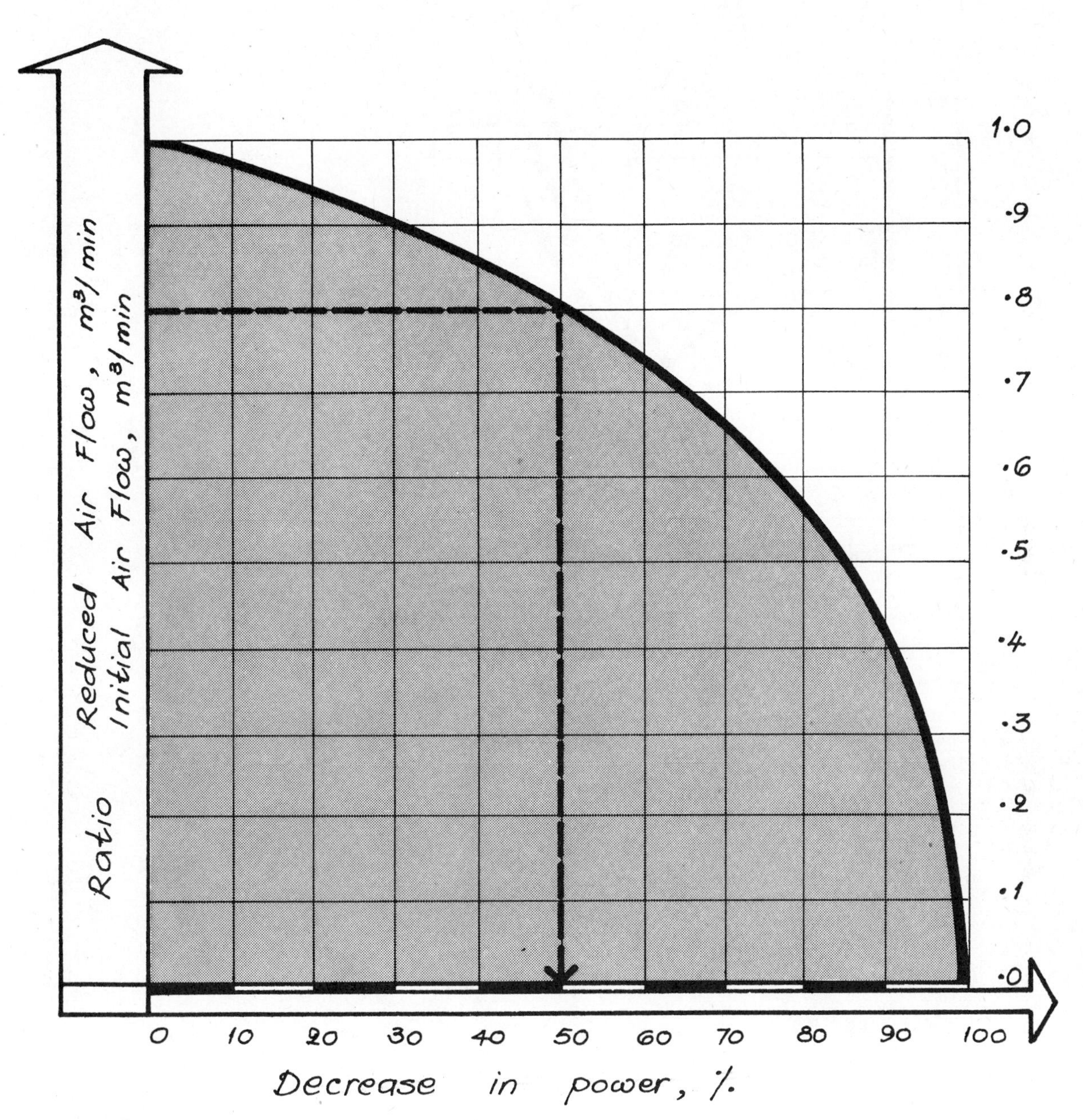

Reference 12

TYPICAL DECREASE IN POWER BY REDUCING FAN SPEED

FIGURE 2.4

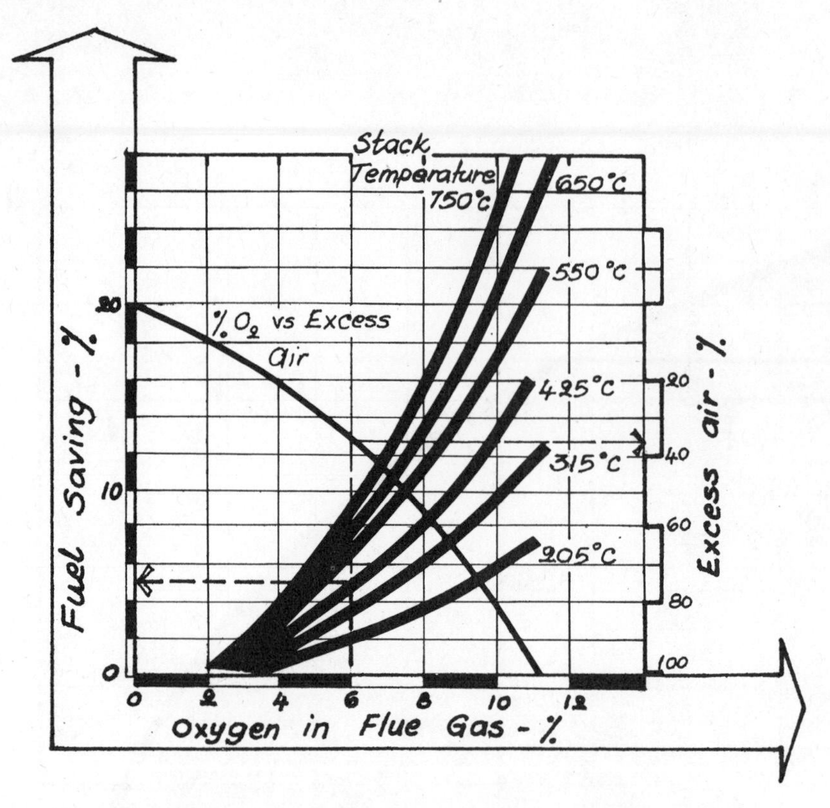

The improvement in efficiency of radiant and combination radiant & convection heaters or boilers without air pre-heaters that can be realized by reducing Excess air is 1·5 times the apparent efficiency improvement from air reduction alone due to the accompanying decrease in flue gas temperature.

Stack temperature = 425°C and O_2 in Flue gas = 6%. Saving = 5·0%. Excess air = 36%.

Reference 12

HYDROCARBON GASEOUS FUEL SAVING AVAILABLE BY REDUCING EXCESS AIR TO 10%

FIGURE 2.5

conducted routinely. Forty percent excess air would correspond to 6.2 percent oxygen in the flue gas. Assuming that the exhaust gases were at 360 °C (600 °F), the potential fuel savings in reducing the excess air from 40 to 10 percent would be 4 percent. Assuming that a plant were using 2.83 x 10^3 m^3/hr (10^5 ft^3/hr) of natural gas and were operating for 8,000 hrs/yr, 4 percent savings translates to a saving of 33.8 x 10^3 GJ/yr (32 GBtu/yr). Assuming gas priced at 1.2 $/GJ, this translates to saving 40,000 $/yr. Table 2.10 shows excess air required.

- *Establish procedures for routine maintenance and cleaning of burners*--The temperature at which fuel oil is delivered to burners contributes to proper atomization and efficient combustion.

- *Consider including automatic control systems and monitoring equipment*--In the combustion process, for example, boiler fuel costs are directly related to combustion air as discussed previously. The simplest system for proper operational maintainenance requires a determination of oxygen and combustibles in the flue gas in order to evaluate combustion efficiency. The proper ratio of air to fuel can then be adjusted, but such a method depends on vigilant and continuous operator attention. A fuel and air metering system would achieve better results and could maintain the excess air at 10 percent. Such a system would cost in the neighborhood of $30,000 to $35,000. The savings would be on the order of $40,000/year for an ∿50 GJ/yr (50 MBtu/hr) boiler based on 2.00 $/GJ fuel costs. Similarly, many units in the industrial operation could profit from more advanced instrumentation and measuring devices along with control systems.

  Metering devices for gas, fuels, electricity, and water should be installed at strategic locations to facilitate obtaining energy analysis information for specific facilities and operations in the plant. Automatic timers are the simplest control devices to help reduce waste during times of plant and equipment inactivity. They can be preset to turn on or shut off equipment. Alarms also help to make personnel aware of an energy waste point requiring corrective action.

- *Use waste heat to preheat combustion air*--The approximate improvement in efficiency when heating combustion air is shown in Figure 2.6. Air preheaters can be of the tubular type, as shown in Figure 2.7, or the rotary type shown in Figure 2.8. In the latter system a rotor revolves slowly through both the flue gas and the air streams with heat transfer surfaces continuously absorbing heat from the flue gas and releasing it to incoming air. Preheating the incoming air is convenient for forced draft furnaces.

Many other savings are possible in boilers and direct-fired industrial equipment. These subjects, however, are outside the scope of this handbook and are not discussed further here. The reader should refer to an appropriate handbook for additional information.

Steam Systems

Inasmuch as approximately 40 percent of the energy utilized in industry goes toward the production of process steam, it presents a large potential for energy misuse and fuel waste from improper maintenance and operation. For example:

- *Steam leaks from lines and faulty valves result in considerable losses*--These losses depend on the size of the opening and the pressure of the steam, as illustrated by Figure 2.9.

Fuel	Type of Furnace or Burners	Excess Air % by Weight
Pulverized coal	Completely water-cooled furnace for slag-tap or dry-ash-removal	15-20
	Partially water-cooled furnace for dry-ash removal	15-40
Crushed coal	Cyclone furnace--pressure or suction	10-15
Coal	Spreader stoker	30-60
	Water-cooled vibrating-grate stoker	30-60
	Chain-grate and traveling-grate stokers	15-50
	Underfeed stoker	20-50
Fuel oil	Oil burners, register-type	5-10
	Multifuel burners and flat-flame	10-20
Acid sludge	Cone and flat-flame type burners, steam-atomized	10-15
Natural, coke-oven, and refinery gas	Register-type burners	5-10
	Multifuel burners	7-12
Blast-furnace gas	Intertube nozzle-type burners	15-18
Wood	Dutch oven (10-23% through grates) and Hofft-type	20-25
Bagasse	All furnaces	25-35
Black liquor	Recovery furnaces for Kraft and soda-pulping processes	5-7

Reference 17

USUAL AMOUNT OF EXCESS AIR SUPPLIED TO FUEL-BURNING EQUIPMENT

TABLE 2.10

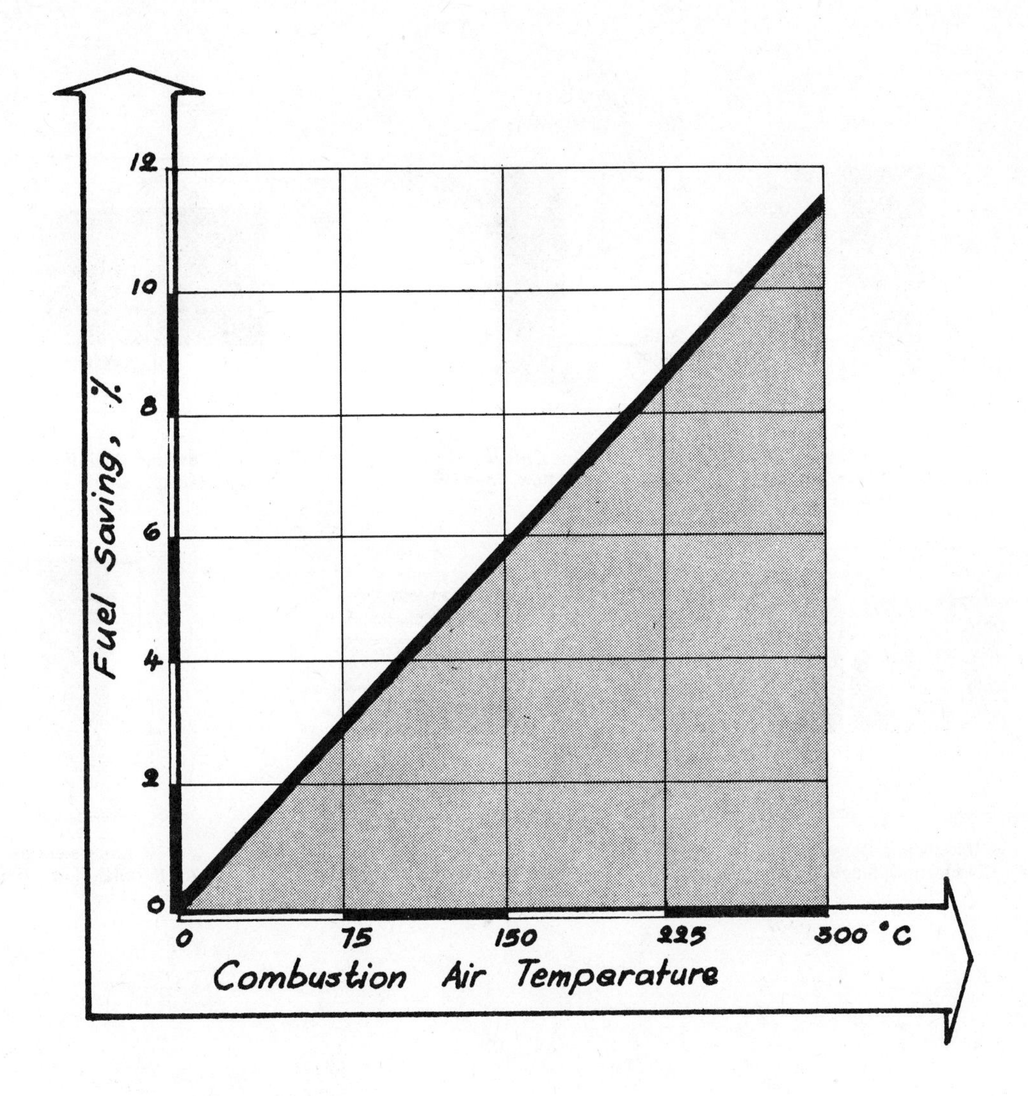

APPROXIMATE FUEL SAVINGS WHEN HEATED COMBUSTION AIR IS USED IN BOILER UNITS

FIGURE 2.6

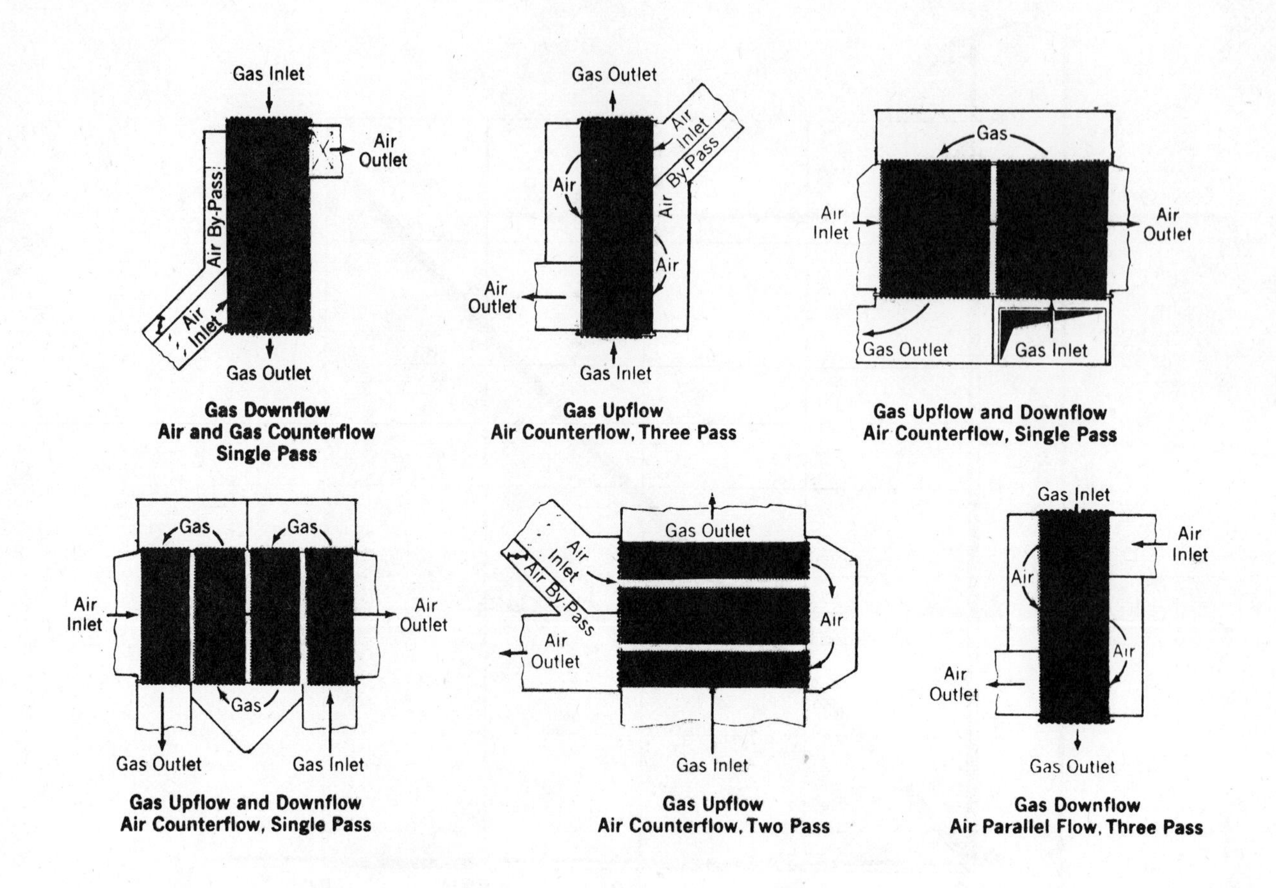

Reference 17

SOME ARRANGEMENTS OF TUBULAR AIR HEATERS TO SUIT VARIOUS DIRECTIONS OF GAS AND AIR FLOW

FIGURE 2.7

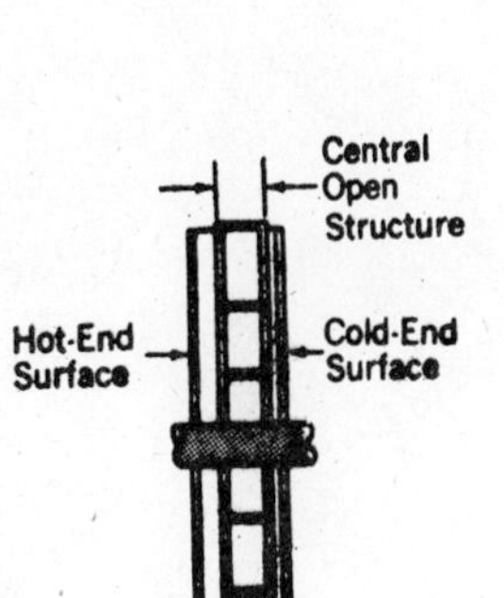

section
through rotor

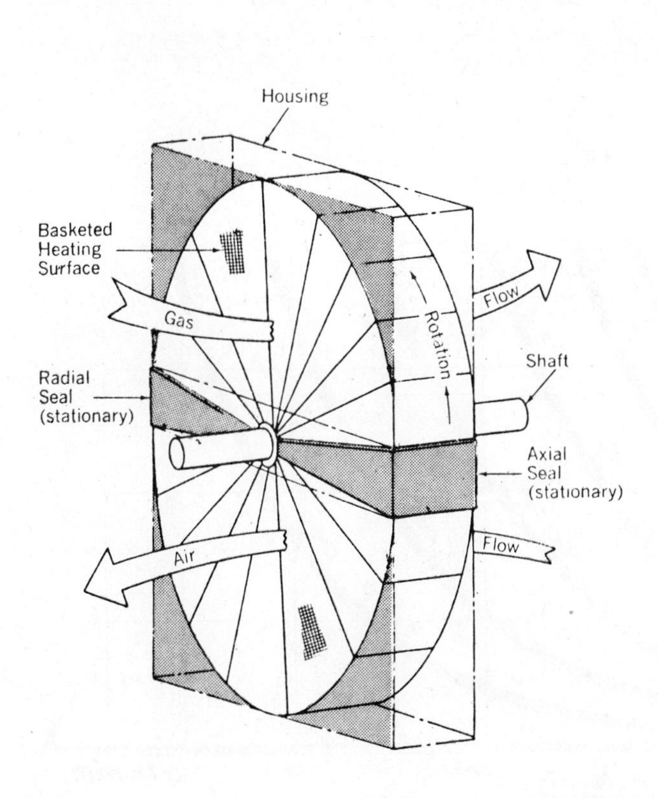

Reference 17

FIGURE 2.8

ELEMENTS OF A ROTARY REGENERATIVE AIR HEATER

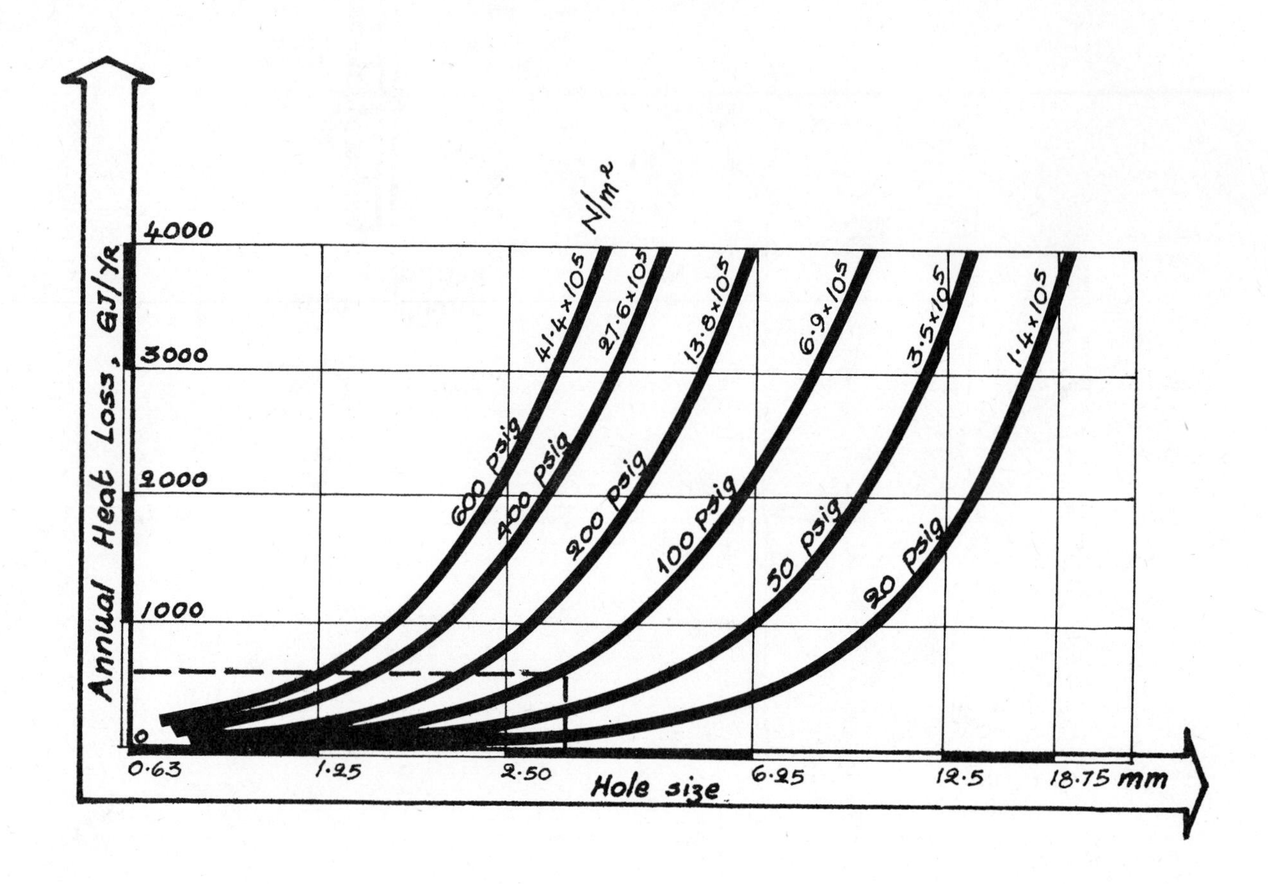

Reference 12

HEAT LOSS FROM STEAM LEAKS

FIGURE 2.9

- *Steam traps are major contributors to energy losses when not functioning properly*--A large process industry might have thousands of steam traps which could result in large costs if they are not operating correctly. Steam traps are intended to remove condensate and non-condensable gases while trapping or preventing the loss of steam. If they stick open, orifices as large as 6 mm (0.25 in.) can allow steam to escape. Such a trap would allow 5.7 GJ/yr (6 MBtu/yr) of heat to be rejected to the atmosphere on a 6.89×10^5 N/M^2 (100 psi) pressure steam line. Many steam traps are improperly sized, contributing to an inefficient operation. Routine inspection, testing, and a correction program for steam valves and traps are essential in any energy program and can contribute to cost savings.
- *Poor practice and design of steam distribution systems can be the source of heat waste up to 10 percent or more*--It is not uncommon to find an efficient boiler or process plant joined to an inadequate steam distribution system. Modernization of plants results from modified steam requirements. The old distribution systems are still intact, however, and can be the source of major heat losses. Large steam lines intended to supply units no longer present in the plant are sometimes used for minor needs such as space heating and cleaning operations which would be better accomplished with other heat sources.
- *Steam distribution systems operating on an intermittent basis require a start-up warming time to bring the distribution system into proper operation*--This can extend up to two or three hours which puts a demand on fuel needs. Not allowing for proper ventilating of air can also extend the start-up time. In addition, condensate return can be facilitated if it is allowed to drain by gravity into a tank or receiver and is then pumped into the boiler feed tank.
- *Proper management of condensate return*--Proper management, can lead to great savings. Lost feed water must be made up and heated. For example, every 0.45 kg (1 lb) of steam which must be generated from 15 °C feedwater instead of 70 °C feedwater requires an additional 1.056×10^5 J (100 Btu) more than the 1.12 MJ (1,063 Btu) required or a 10 percent savings of fuel. A rule of thumb is that a 1 percent fuel saving results for every 5 °C increase in feedwater temperature. Maximizing condensate recovery is an important fuel saving procedure.
- *Poorly insulated lines and valves due either to poor initial design or a deteriorated condition*--Heat losses from a poorly insulated pipe can be estimated using Figure 2.10. This curve shows that a poorly insulated line can lose ∿1000 GJ/yr (10^9 Btu/yr) or more per 30 m (100 ft) of pipe. At steam costs of 2.00 $/GJ, this translates to $2,000 savings per year.
- *Improper operation and maintenance of tracing systems*--Steam tracing is used to protect piping and equipment from cold weather freezing. The proper operation and maintenance of tracing systems will not only insure the protection of traced piping but also saves fuel. Occasionally these systems are operating when not required. Steam is often used in tracing systems and many of the deficiencies mentioned above apply (e.g., poorly operating valves, insulation, leaks).

Process Heat

Industrial process heat applications can be divided into four categories: direct-fired, indirect-fired, fuel, or electric. Here we shall

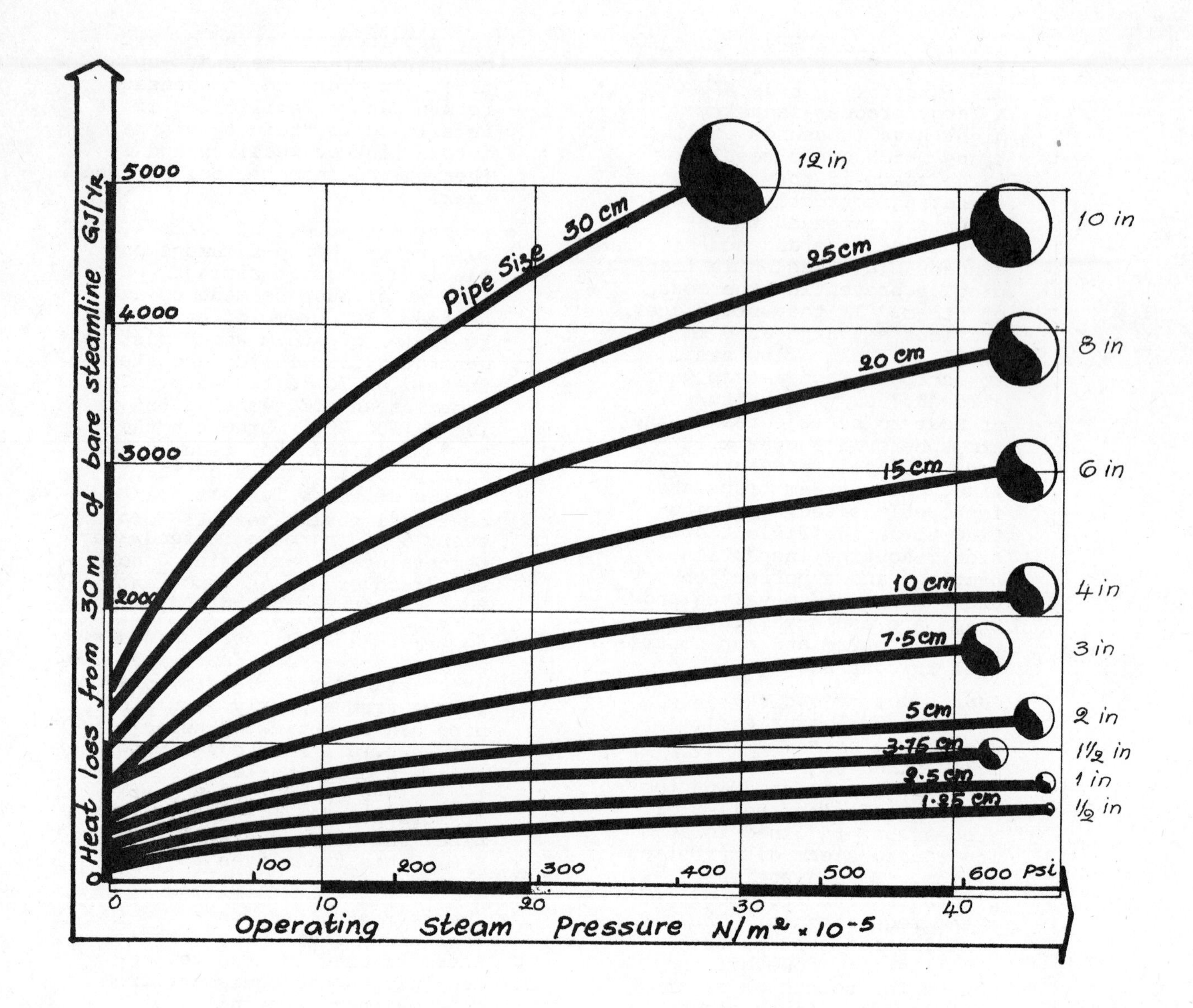

Reference 12

FIGURE 2.10

HEAT LOSS FROM BARE LINES

consider electric direct-fired installations (ovens, furnaces) and indirect-fired (electric water heaters and boilers) applications.

From the housekeeping and maintenance point of view, typical opportunities would include:

- *Repair or improve insulation*--Operational and standby losses can be considerable, especially in larger units. Remember that insulation may degrade with time or may have been optimized to different economic criteria.
- *Provide finer controls*--Excessive temperatures in process equipment waste energy. Run tests to determine the minimum temperatures which are acceptable, then test instrumentation to verify that it can provide accurate process control and regulation.

See Chapter 10 for additional details on electric process heat.

Compressed Air Systems

Compressed air is a major energy use in many manufacturing operations. Electricity used to compress air is converted into heat and potential energy in the compressed air stream. Efficient operation of compressed air systems, therefore, requires the recovery of excess heat where possible, as well as the maximum recovery of the stored potential energy.

Efficient operation is achieved by:

- *Selecting the appropriate type and size of equipment for the duty cycle required*--Process requirements vary, depending on flow rates, pressure, and demand of the system. Energy savings can be achieved by selecting the most appropriate equipment for the job. (See Table 2.11.)

  The rotary compressor is more popular for industrial operations in the range of 20-200 kW, even though it is somewhat less efficient than the reciprocal compressor. This has been due to lower initial cost and reduced maintenance.[18] When operated at partial load, reciprocating units can be as much as 25 percent more efficient than rotary units. However, newer rotary units incorporate a valve that alters displacement under partial load conditions and improves efficiency. Selection of an air-cooled versus a water-cooled unit would be influenced by whether water or air were the preferred media for heat recovery.
- *Proper operation of compressed air systems can also lead to improved energy utilization*--Obviously, air leaks in lines and valves should be eliminated. The pressure of the compressed air should be reduced to a minimum. The percentage saving in power required to drive the compressor at a reduced pressure can be obtained from Figure 2.11. For example, for a compressor operating at 6.89×10^5 N/m^2 (100 psi) and a reduction of the discharge pressure to 6.20×10^5 N/m^2 (90 psi), a 5 percent decrease in brake horsepower would result. For a 373 kW (500 hp) motor operating for one year, the 150,000 kWh savings per year would result in about $5,000/yr at current electric power costs.
- *The intake line for the air compressor should be at the lowest temperature available*--This normally means outside air. The reduced temperature of air intake results in a smaller volume of air to be compressed. The percentage horsepower saving relative to a 21 °C (70 °F) intake is shown in Table 2.12.
- *Leakage is the greatest efficiency offender in compressed air systems*--The amount of leakage should be determined and measures taken to reduce it. If air leakage in a plant is

Type	Power (kW)	Power (hp)	Speed (rpm)	Volume m^3/min	Volume scfm	Pressure kN/m^2	Pressure psig	1977 Equipment Cost $
Rotary, air-cooled	19	25	3550	2.77	98	689	100	5,100
	56	75	3550	9.76	345	793	115	14,000
	112	150	1775	19.8	700	793	115	21,500
Rotary, water-cooled	19	25	3550	2.77	98	689	100	5,100
	56	75	1775	9.91	350	689	100	13,000
	93	125	2500	17.8	630	689	100	19,500
	298	400	1390	58.6	2070	689	100	51,000
Reciprocating	56	75	450	9.84	348	689	100	13,000
	298	400	590	63.5	2245	689	100	60,000

Reference 18

CHARACTERISTICS OF TYPICAL AIR COMPRESSORS

TABLE 2.11

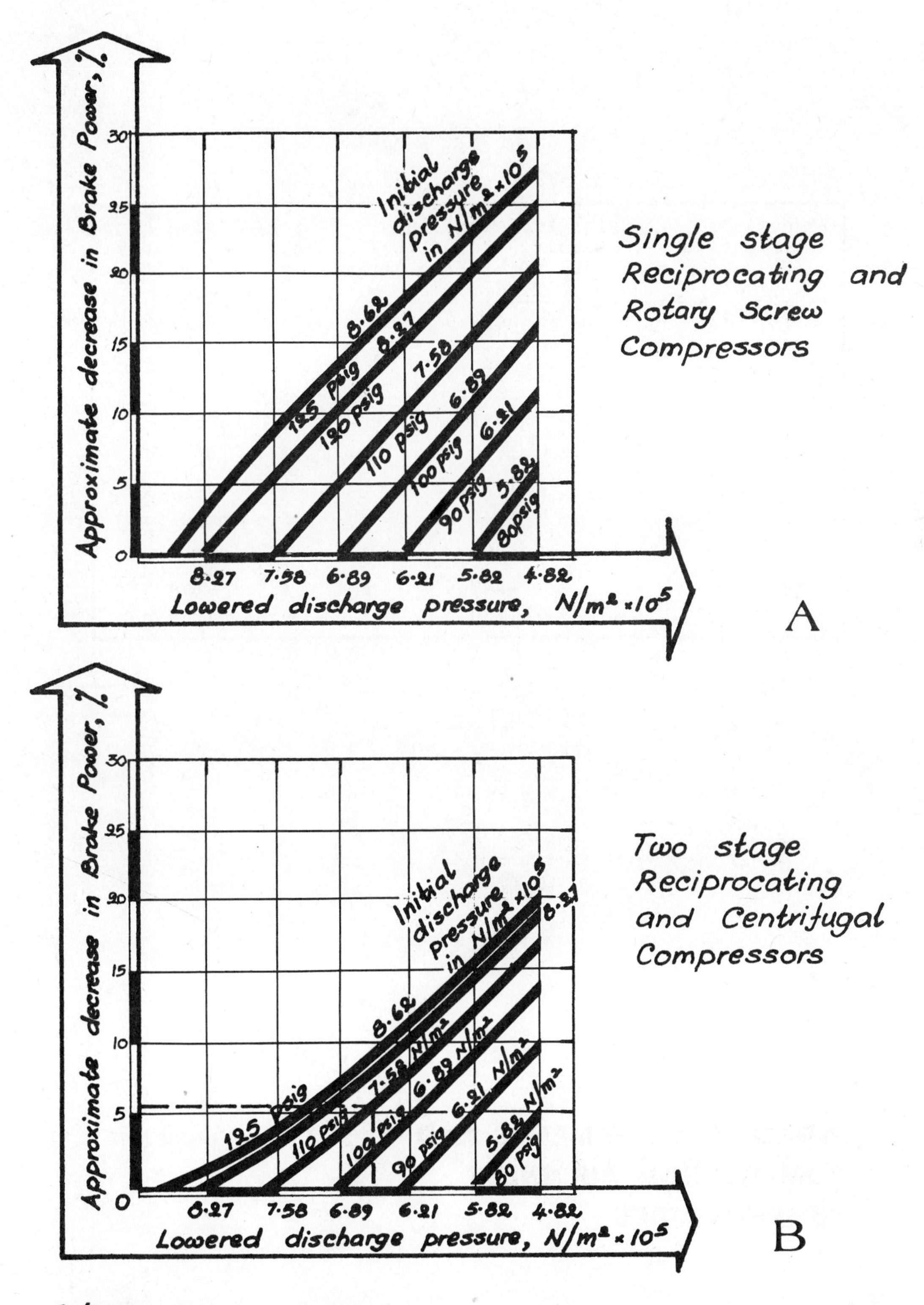

Reference 12

EFFECTS OF LOWERING PRESSURE

FIGURE 2.11

Temperature of Air Intake °C	°F	Intake Volume in m^3 Required to Deliver 1000 m^3 of Free Air at 21°C (70°F)	% kW Saving or Increase Relative to 21°C (70°F) Intake
-1	30	925	7.5 Saving
5	40	943	5.7 Saving
10	50	962	3.8 Saving
16	60	981	1.9 Saving
21	70	1000	0
27	80	1020	1.9 Increase
32	90	1040	3.8 Increase
37	100	1060	5.7 Increase
43	110	1080	7.6 Increase
49	120	1100	9.5 Increase

POWER REQUIREMENTS AND COMPRESSOR AIR INLET TEMPERATURES

TABLE 2.12

more than 10 percent of the plant demand, a poor condition exists. The amount of leakage can be determined by a simple test during off production hours by noting the time that the compressor operates under load compared with the total cycle. This indicates the percentage of the compressor's capacity which is used to supply the plant air leakage. Thus, if the load cycle compared with the total cycle were 60 seconds compared with 180 seconds, the efficiency would be 33 percent; or 33 percent of the compressor capacity is the amount of air leaking in m^3/min (ft^3/min).

- *Recover heat where feasible*--This topic is discussed in more detail in Section 2.3 E of this chapter, and in Chapter 10 (also see Case Study 2-9). It is mentioned here, however, because one sometimes encounters situations where water-cooled or air-cooled compressors are a convenient source of heat for hot water, space heating, or process applications. As a rough rule of thumb, about 300 J per m^3/min of air compressed ($\sim$10 Btu per standard cubic foot per minute) can be recovered from an air-cooled rotary compressor.[18]

Electric Power Systems

Electric power systems include substations, transformers, switchgear, distribution systems, feeders, power and lighting panels, and related equipment. Motors are not included since they are discussed separately in Chapter 11. Possibilities for energy management include:

- *Use highest voltages which are practical*--For a given application, doubling the voltage cuts the required current in half and reduces the i^2r losses by a factor of four.

- *Eliminate unnecessary transformers*--They waste energy. Proper selection of equipment and facility voltages can reduce the number of transformers required and cut transformer losses. Remember, the customer pays for losses when the transformers are on his side of the meter. Examples: it is generally better to order equipment with motors of the correct voltage, even if this costs more, than to install a special transformer.

- *Energy losses are an inherent part of electric power distribution systems*--primarily due to i^2r losses and transformers. The end use conversion systems for electrical energy used in the process also contribute to energy waste. Proper design and operation of an electrical system can not only minimize energy losses but contribute to the reduction of electricity bills. Oversized motors, pumps, and equipment increase energy usage. Most equipment, including electric motors, operate more efficiently and have a better power factor near rated capacity. Motors and transformers, even though they are not loaded, are also responsible for energy losses. Equipment not in use should be turned off. Where long feeder runs are operated at near-maximum capacities, check to see if larger wire sizes would permit savings and be economically justifiable.

- *The overall power factor of electrical systems should be checked for low power factor*--which could increase energy losses and the cost of electrical service, in addition to excessive voltage drops and increased penalty charges by the utility. Electrical system studies should be made and consideration should be given to power factor correction capacitors. In certain applications as much as 10 to 15 percent savings can be achieved in a poorly operating plant.

- *Check load factors*--This is another parameter which measures the plant's ability to use electrical power efficiently. It is defined as the ratio of the actual kWh used to the maximum demand in kW times the total hours in the time period. A reduction in demand to bring this ratio closer to unity without decreasing plant output means more economical operation. For example, if the maximum demand for a given month (200 hours) is 30,000 kWe and the actual kWh is 3.6 x 10^6 kWh, the load factor is 60 percent. Proper management of operations during high demand periods which may extend only 15 to 20 minutes can reduce the demand during that time without curtailing production. For example, if the 30,000 kWe could be reduced to 20,000 kWe, it would increase the load factor to about 90 percent. Such a reduction could amount to a $15,000 to $25,000 reduction in the electricity bill.

- *Reduce peak loads wherever possible*--Many nonessential loads can be shed during high demand peak without interrupting production. These loads would include such items as air compressors, heaters, coolers, and air conditioners. Manual monitoring and control is possible but impractical because of the short periods of time that are normally involved and the lack of centralized control systems. Automatic power demand control systems are available.

- *Provide improved monitoring or metering capability, submeters, or demand recorders*--While it is true that meters alone will not save energy, plant managers need feedback to determine if their energy management programs are taking effect. Often the installation of meters on individual processes or buildings leads to immediate savings of 5 to 10 percent by virtue of the ability to see how much energy is being used and to test the effectiveness of corrective measures.

Cooling Towers

Many process and air conditioning systems reject heat to the atmosphere by means of wet cooling towers. Poor operation can contribute to increased power requirements.

- *Water flow and air flow should be examined to see that they are not excessive*--The cooling tower outlet temperature is fixed by atmospheric conditions if operating at design capacity. Increasing the water flow rate or the air flow will not lower the outlet temperature.

- *The possibility of utilizing heat which is rejected to the cooling tower for other purposes should be investigated*--such as preheating feedwater, heating hot water systems, space heating, or other low temperature applications. If there is a source of building exhaust air with a lower wet bulb temperature, it may be efficient to supply this to a cooling tower.

Material Transport

Movement of materials through the plant creates opportunities for saving energy.

- *Combine processes or relocate machinery to reduce transport energy*--Sometimes merely relocating equipment can reduce the need to haul materials.
- *Turn off conveyors and other transport equipment when not needed*--Look for opportunities where controls can be modified to permit shutting down of equipment not in use.
- *Use gravity feeds wherever possible*--Avoid unnecessary lifting and lowering of products.
- *Apply maintenance procedures and lubrication to minimize friction losses*--Refer to Chapter 5 for additional information.

Industrial Processes

The variety of industrial processes is so great that detailed specific recommendations are outside the scope of this book. Useful sources of information are found in trade journals, vendor technical bulletins, and manufacturers' association journals. These suggestions are intended to be representative, but by no means do they cover all possibilities.

- *In machining operations, eliminate unnecessary operations and reduce scrap*--This is so fundamental from a purely economic point of view that it will not be possible to find significant improvements in many situations. The point is that each additional operation and each increment of scrap also represent a needless use of energy.
- *Review air usage in paint spray booths*--In paint spray booths and exhaust hoods, air is circulated through the hoods to control dangerous vapors. Makeup air is constantly required for dilution purposes. This represents a point of energy rejection through the exhaust air.

  Examination should be made of the volumes of air required in an attempt to reduce flow and unnecessary operation. Possible mechanisms for heat recovery from the exhaust gases should be explored using recovery systems.
- *Examine electrolysis and plating operations for savings*--Review rectifier performance, heat loss from tanks, and the condition of conductors and connections. Also see Chapter 12.
- *Reduce losses in process hot water systems*--Electrically heated hot water systems are used in many industrial processes for cleaning, pickling, coating, or etching components. Hot or cold water systems can dissipate energy. Leaks and poor insulation should be eliminated. Water temperature requirements should be examined and lowered or raised in order to minimize energy use in heating and cooling operations. Evaluate points in the plant where heated and cooled waters are being rejected and could be recovered and reused.

E. Retrofit Strategies

In the preceding section, emphasis was placed on operational modifications which could be achieved in a short time at a low or moderate capital cost. In this section the energy management concepts presented will generally require a few months to a few years to implement, and the expenditure of capital funds. They are applicable to a wide variety of industrial applications.

Building Envelope, HVAC, and Lighting Modifications

A wide variety of modifications can be made in industrial facilities. These range from major revamping of buildings, including new HVAC systems which permit heat recovery, to the installation of more efficient, high intensity lighting systems. For details, refer to Chapter 7--for building envelope changes (also see Chapter 3); Chapter 8--for modifications to heating, ventilating, and air-conditioning systems; and Chapter 9--for lighting systems.

Heat Recovery

This is a major possibility for energy savings in all types of industrial facilities. It should receive careful consideration because of the potential for large economic benefits which frequently exists.

Flue gases from boilers and furnaces and other units which depend on combustion provide excellent opportunities for heat recovery. Depending on the flue gas temperatures, exhaust heat can be used to raise steam or to preheat the air to the combuster. An example of such a system is shown in Figure 2.12 where an ammonia reformer heater is designed to conserve fuel by using a steam generator and an air preheater

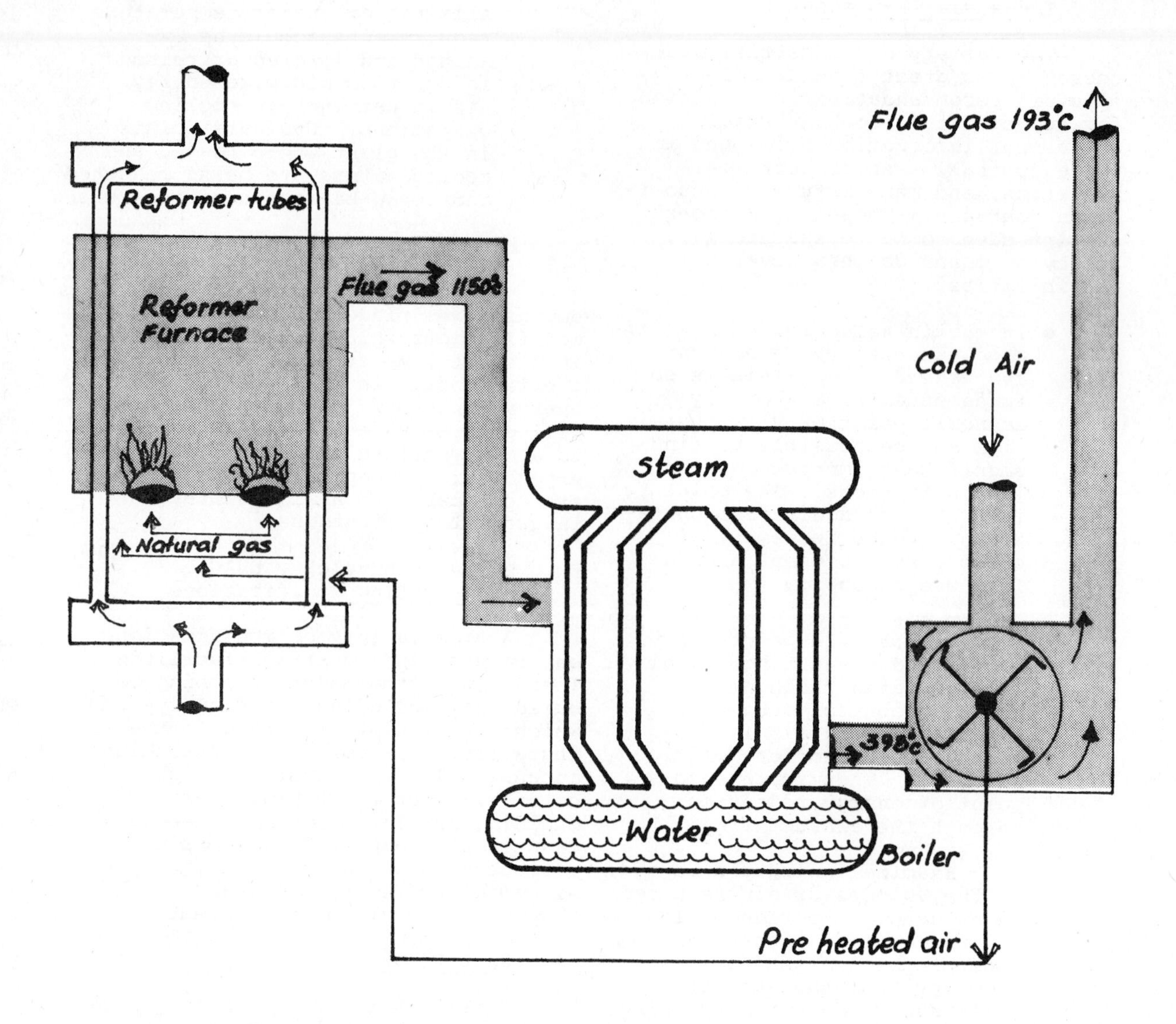

HEAT RECOVERY USING AN AIR PREHEATER

FIGURE 2.12

to recover heat from the stack gas. The technical and economic advantages of air preheating in boiler operations were discussed above in Section 2.3 C.

Another potential source of waste heat recovery is the exhaust air which must be rejected from industrial operations in order to maintain health and ventilation safety standards. If the reject air has been subjected to heating and cooling processes, it represents an energy loss inasmuch as the makeup air must be modified to meet the interior conditions. One way to reduce this waste is through the use of recovery wheels which operate in a fashion similar to the rotary air recuperator described above. Both heat and moisture can be transferred with these systems. Figure 2.13 shows the operation of the recovery wheel.

Energy in the form of heat is available at a variety of sources in industrial operations, many of which are not normally derived from primary heat sources. Such sources include electric motors, crushing and grinding operations, air compressors, and air thickening and drying processes. These units require cooling in order to maintain proper operation. The heat from these systems can be collected and transferred to some appropriate use such as space heating. An example of this type of heat recovery is shown in Figure 2.14. All of the energy supplied to the motor in electrical form is ultimately transformed into heat, and nearly all of it is available to heat buildings or for domestic water or mine air heating.

The heat pipe is gaining wider acceptance for specialized and demanding heat transfer applications. The transfer of energy between incoming and outgoing air can be accomplished by banks of these devices. A simplified view of a heat pipe and a counterflow heat exchanger made of these devices is shown in Figure 2.15. A refrigerant and a capillary wick are permanently sealed inside a metal tube setting up a liquid-to-vapor circulation path. Thermal energy applied to either end of the pipe causes the refrigerant to vaporize. The refrigerant vapor then travels to the other end of the pipe where thermal energy is removed. This causes the vapor to condense into liquid again and the condensed liquid then flows back to the opposite end through the capillary wick. These units promise to be highly efficient and require minimal maintenance.

Industrial operations involving fluid flow systems which transport heat such as in chemical and refinery operations offer many opportunities for heat recovery. With proper design and sequencing of heat exchangers, the incoming product can be heated with various process steams. For example, proper heat exchanger sequence in preheating the feedstock to a distillation column can reduce the energy utilized in the process.[11]

Heat pumps should also be considered for use as heat recovery equipment. Refer to Chapters 8 and 10. For additional information on heat recovery, refer to Chapter 10.

Power Recovery

Power recovery concepts are essentially an extension of the concepts presented in the previous section. Many industrial processes today have pressurized liquid and gaseous streams at 150-375 °C (300-700 °F) that present excellent opportunities for power recovery. In many cases high pressure process stream energy is lost by throttling across a control valve.

The extraction of work from high-pressure liquid streams can be accomplished by means of hydraulic turbines (essentially diffuser-type or volute-type pumps running backwards). These pumps can be either single or multistage. A number of such installations have been placed into service in refinery operations. Power recovery ranges from 170-1340 kW (230-1800 hp). The lower limit of power recovery approaches the minimum economically justified for capital expenditures at present power costs. In most hydraulic turbine applications the units are used in conjunction with pump-drive trains (see Case Study 2-8).

Substitute Electric Motors for Air Motor (Pneumatic) Drives

Electric motors are far more efficient. Typical vaned air motors range in size from 0.15 kW to 6.0 kW (0.2 to 8 hp), cost 200 to 1200$, and

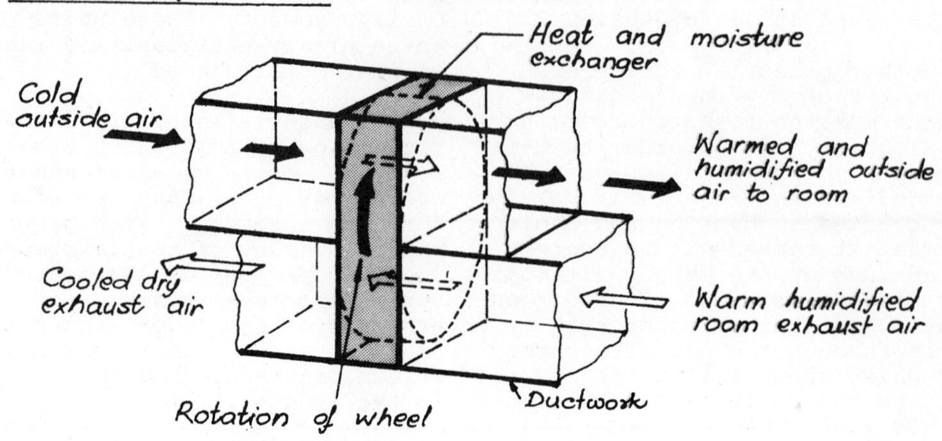

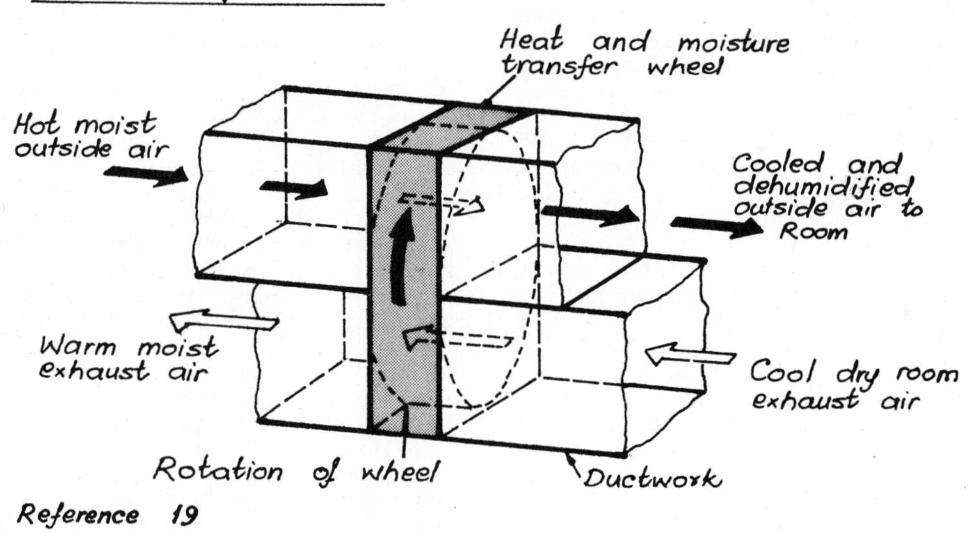

Reference 19

OPERATION OF HEAT WHEELS

FIGURE 2.13

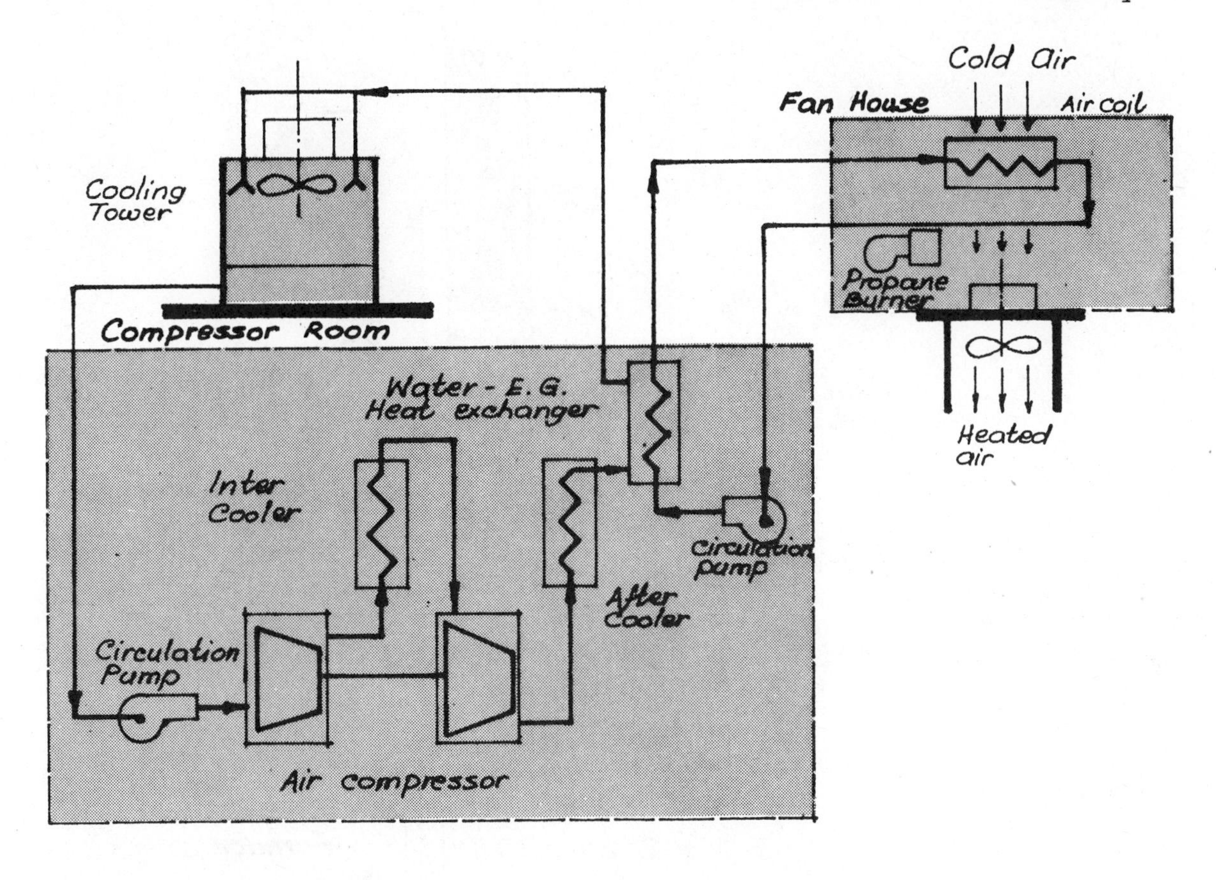

Heat Balance

Power Source	Power (MW)
Electrical power to Motor	1·60
Motor losses	0·16
Radiation from Compressor	0·08
Cylinders	0·41
Intercooler	0·48
Aftercooler	0·41
Compressed air	0·06
TOTAL	1·60

Reference 20

FLOW DIAGRAM OF MINE AIR HEATING & COMPRESSOR COOLING CYCLE

FIGURE 2.14

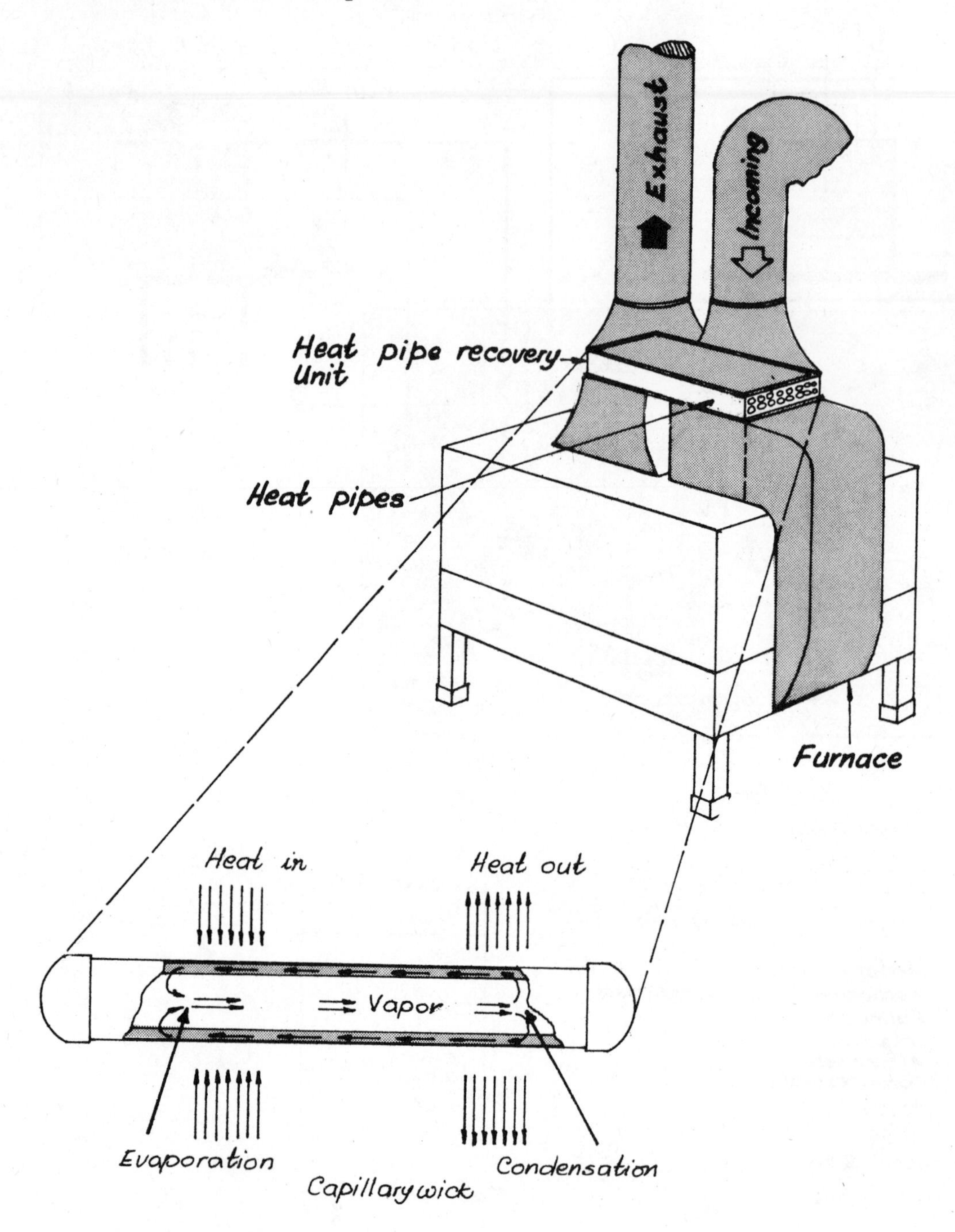

TYPICAL HEAT PIPE INSTALLATION

FIGURE 2.15

produce 1.4-27 N-m (1 to 20 ft-lbs) of torque at 620 kN/m^2 (90 psi) air pressure. These are used in manufacturing operations where electric motors would be hazardous, or where light weight and high power are essential. Inefficiency results from air system leaks and the need (compared to electric motors) to generate compressed air as an intermediate step in converting electrical to mechanical energy.

Instrumentation, Computer Control, and Simulation

The capability to monitor and measure performance and operation of units and processes is essential to any management program. Most simply, this would entail periodic recording of appropriately installed metering and measuring devices. A higher level of energy management is achieved by automatic control systems which not only monitor performance indicators but regulate operations. Most equipment today is adequately instrumented and has control devices to maintain minimum safety and environmental standards. However, adequate instrumentation and control equipment for measuring and minimizing electricity and fuel use are not generally available in plants or on equipment. Not only is it difficult to ascertain how well an individual unit is operating, but it is also difficult to measure the general flow of energy through the plant. For energy-intensive processes, it may be justified to install additional instrumentation.

Computers have played a major role in the financial management and operations of industrial and commercial facilities. This experience can be applied to the management of energy in the same facilities. In some industrial operations energy utilization can be complicated, involving many interacting forces such as energy supply, prices, availability, and a complex demand pattern in the manufacturing operations themselves. The situation is further complicated by governmental policies, contractual arrangements, prior commitments, and supply interruptions. Traditional methods of analysis and control are inadequate to develop proper energy strategies and analyses.

Computer techniques are being applied today in two areas of energy management:

- to actively interact and control industrial operations, and,
- to simulate plant operations and energy flow.

The interactive computer system can monitor power and fuel demand within a plant and either display the information for manual control of operations or actively control plant operation and energy use according to predetermined limits. An interactive system can be applied to a single process or device, such as a blast furnace.

Computer power management can also be applied to an industrial operation in order to control power demand within specified limits. Computer control can assist in the dispatching of power supply to the fluctuating demands of plant facilities. Large, electrically based facilities are capable of forcing large power demands during peak times which exceed the limits contracted with the utility or cause penalties in increased costs. Computer control can "even out" the load and minimize power costs. In times of emergency or fuel curtailment, operation of the plant can be programmed to provide optimum production and operating performance under prevailing conditions. Furthermore, computer monitoring and control provide accurate and continuous records of plant performance.

In any given operation, boilers, coolers, air compressors, air conditioners, and other large, energy using equipment can be monitored. Cumulative power requirements can be computed and projections made of power needs in any future interval. If the projected power exceeds predetermined limits, the system can alert the plant controller to shut down appropriate machinery and equipment. A computer system can also implement these procedures automatically.

In summary, automatic computer power control and monitoring systems can provide the following benefits:

- improved efficiency of load dispatch;
- minimization of power and energy use while permitting maximum production;
- control of power demand within predetermined limits during normal operation and in emergencies;
- curtailment of wasteful operations such as equipment and units operating when not needed;
- reduction of peak power demands and providing for a more even load utilization;
- a real-time display of plant performance and furnishing complete records of energy uses; and,
- provision of data that can be used to analyze energy flow and potential conservation opportunities.

Active computer monitoring and control can also be applied to a specific unit such as a furnace. Again, by proper management, the unit's performance can be measured, analyzed, and controlled.[21] If multiple units in a specific process are involved, "load-shedding" can be applied as required. The performance of an individual unit can be actively recorded and operations can be optimized to minimize energy and fuel demands. If fuel switching is necessitated because of availability for environmental limitations, the switch can be planned and controlled by the computer in an orderly way.

It should be stressed here that many of these same functions can be carried out by manual controls, time clocks, microprocessors, or other inexpensive devices. Selection of a computer system must be justified economically on the basis of the number of parameters to be controlled and the level of sophistication required. Many of the benefits described above can be obtained in some types of operations without the expense of a computer.

Plant simulation models can be useful in the management of energy utilization in a facility. The plant energy system is modeled using engineering principles along with regression analysis of actual data from plant performance. Individual components, processes, and plants can be described by mathematical expressions related to energy parameters. Energy parameters can be quantities of electric power, natural gas, and fuels. Demand parameters can be building-floor space, degree-days, and units of production. An example of a steel plant energy model is provided in Case Study 2-3. Operations research programming techniques can be used to optimize and analyze performance.

Depending on the needs of the analysis, the computer energy model can be based primarily on empirically developed equations which describe overall performance of operations and units. This would involve equations developed by utilizing historical data and curve-fitting techniques in order to establish fuel and electrical requirements in accordance with the production output. Note that a more detailed engineering analysis simulating unit performance based on physical and scientific principles may also be appropriate and can have the advantage that efficiency alternatives and design modifications can be explored.

Regardless of the complexity or simplicity of the manufacturing operation, a generalized energy analysis based on engineering principles can be developed in most cases. An engineering model is also suitable for thermodynamic analysis of availability and effectiveness. Case Study 2-4 illustrates the development and use of an engineering simulation model.

For additional information on the use of computers for energy management, also refer to Chapter 5.

F. New Design Strategies

The following methods generally are not suited for retrofit applications because of the high capital cost involved. They would be applicable in new plant designs when major new facilities are required.

Cascaded Energy Use

The thermodynamic availability function (see section 2.3 B and

Appendix C) ascribes the potential for performing work to any system: the higher the availability, the more potential there is for work. Theoretically, one can transfer the availability from one component in the system to another, i.e., availability of fuel can be transferred from the combustion products to steam. Practically, however, the transfer is never perfect and availability is lost or wasted. In a "reversible process" available useful work is conserved, but this is not the case in an actual process in which irreversibilities occur and in which there is a decrease of useful available work. Fuel conservation, therefore, should be implemented by conserving availability insofar as possible. For example, it makes little sense to use high quality fuels capable of combustion temperatures above 1000 °C to heat water and air to temperatures of 50 to 100 °C.

In an ideal industrial process design, each step would be planned so that the loss of availability would be minimized. In casting, rolling, or forging operations, for example, each step is planned so that the product heat is most effectively utilized without cooling and reheating. This is the concept employed in "continuous" steel rolling mills. Another example is to combust fuel in a turbine or engine, extract useful work, and *then* use the exhaust heat (rather than the fuel directly) to produce low temperature steam or hot water.

Additional examples of this concept are so-called "combined cycles" and co-generation (combined steam and electricity production) which are discussed later. As will be shown, cascaded energy use can increase efficiency but does so at increased capital cost. Other problems are sometimes introduced: more complicated equipment, decreased reliability, and mismatching of loads and energy sources.

Still, because of its potential advantages, cascaded energy use should be evaluated in industrial processing operations. The same concept can be extended to the processing of energy intensive materials. The degradation of material quality should be planned so as to permit reuse of scrap at the highest possible quality. The final step should then be melting down or reconstitution of the raw materials into the original form. In an analogous manner, by-product hydrocarbons would preferentially be used as feedstock or products to the greatest extent possible, and only in the last resort combusted, since this represents the greatest loss of availability for these organic molecules.

Co-generation and Other Combined Cycles

Figure 2.16 schematically illustrates the availability concept as applied to combustion of fossil fuels. Ideally, nearly all of the fuel's energy content or heating value is initially available to do work. In subjecting the fuel to combustion, however, there is a loss of thermodynamic availability. The loss is due to diffusion of combustion products (H_2O and CO_2) from their chemical potential in the atmosphere. This loss is more than 25 percent of the initial value of available work. Consequently, whenever combustion is involved in a conventional sense, only 70 percent of the fuel work potential can be used. The only hope of recovering this "loss" of available work is through the use of fuel cells, but their technology is insufficiently developed at this time for large-scale use. Conceptually, the hot combustion products near 2000 °C (3600 °F) could be used through a gas turbine to extract work. Practically, however, such a high temperature cannot be maintained and in fact could not be used anyway. Turbine technology today will permit temperatures of about 1100 °C (2000 °F). Gas turbines today, operating between an inlet temperature of 1000 °C (1800 °F) and an outlet temperature of 620 °C (1130 °F), give an efficiency of ∿30 percent. It is hoped, however, that with improved turbine designs and the use of high temperature materials, inlet temperatures of 1425 °C (2600 °F) will be obtainable. Efficiencies could then increase to ∿55 percent. Turbine exhaust gases, by virtue of their high temperature, still have the potential to perform additional work.

If, however, we skip the extraction of work from the combustion

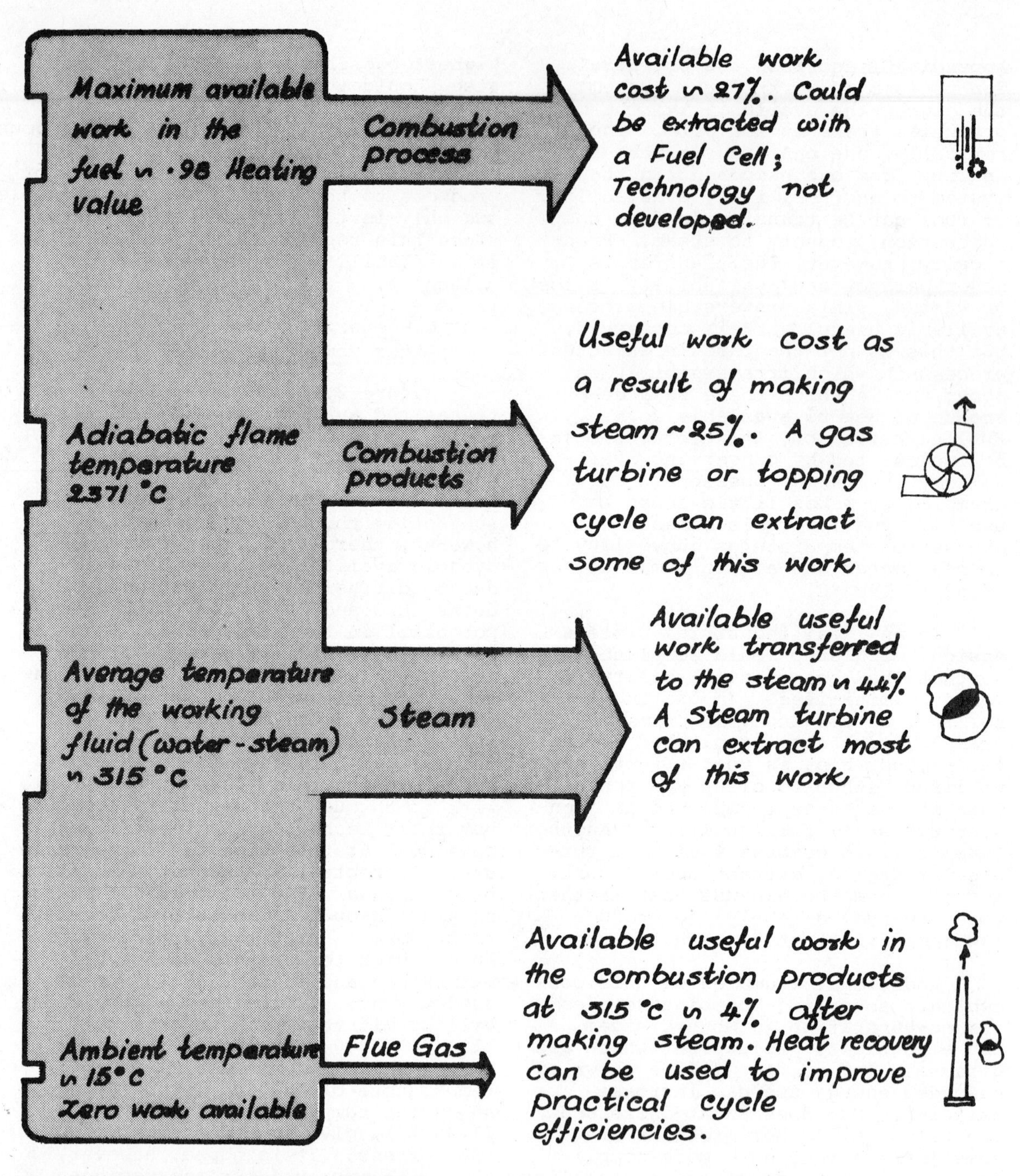

Maximum available work in the fuel ∼ ·98 Heating value
Combustion process
Available work cost ∼ 27%. Could be extracted with a Fuel Cell; Technology not developed.
Adiabatic flame temperature 2371 °C
Combustion products
Useful work cost as a result of making steam ~25%. A gas turbine or topping cycle can extract some of this work
Average temperature of the working fluid (water-steam) ∼ 315 °C
Steam
Available useful work transferred to the steam ∼ 44%. A steam turbine can extract most of this work
Ambient temperature ∼ 15°C Zero work available
Flue Gas
Available useful work in the combustion products at 315 °C ∼ 4% after making steam. Heat recovery can be used to improve practical cycle efficiencies.

AVAILABLE USEFUL WORK IN STEAM GENERATION BY COMBUSTION

FIGURE 2.16

products directly and generate steam (as is conventionally done in a steam power plant), we effectively transfer work availability from the combustion products to the steam. Steam temperatures today, due to materials and corrosion problems, are limited to approximately 550 °C (1000 °F). The large temperature drop between the combustion products and the steam results in a loss of work potential. Moreover, in a real boiler system only a small portion of the steam is at 550 °C (1000 °F). The temperature differential between combustion gases and steam is much greater throughout the boiler; in fact, we should assign an average temperature of ∿315 °C (600 °F) to the working fluid for a modern steam plant. Accordingly, another 25 percent of available work is lost through this operation. Approximately 45 percent of the fuel's initial available work is transmitted to the steam. After electrical generation only 35 to 40 percent of the useful available work is converted to electrical energy for distribution. Thus, while the availability of electrical energy is high and practically all of the energy can be converted to work, an amount of available work twice as great is dissipated at the power plant.

Co-generation is an approach which permits some of this availability to be recovered, although a higher capital cost is required and several additional complexities are introduced.

Consider the hypothetical cycle shown in Figure 2.17. The requirement is to generate electricity and 14×10^5 N/m^2 (200 psi) steam at 195 °C (382 °F). Since 300 kW (∿1 GJ/hr or 1 MBtu/hr) will produce around 100 kWe of electricity, an efficient boiler (90%) will need about 1.1 GJ/hr of fuel to produce 1 GJ/hr of steam or approximately 360 kg/hr (800 lb/hr) at the specified pressure. Now consider combining power generation with steam production as shown by the three alternate approaches in the figure. In case (b), process steam is generated from the exhaust of a gas turbine. In (c), steam is extracted after passing through a steam turbine. In (d), the gas turbine exhaust is used to produce steam which is passed through a steam turbine and then used for process heat. Assume approximate efficiencies approriate to current technology and the specified conditions: gas turbine efficiency = 22%; steam turbine efficiency operating between 550 °C (1000 °F) and 175 °C (350 °F) = 15%; when combined to the gas turbine exhaust = 12%; boiler efficiency = 90%. (Actual efficiencies would depend on the cycle design.) Comparing fuel requirements for the combined steam raising and power production with generation only, we note that fuel savings from 30 to 40 percent are possible compared to separate cycles.

These savings are not obtained for free. As Figure 2.17 shows, the combined cycle generally requires more equipment and a greater capital investment. Reliability and load matching are other concerns in a practical application. If the plant is operated by industry, a backup electrical supply must be available for those times when the plant must undergo maintenance. If the plant is operated by a utility, a constant use for co-generated steam is required. If the cooperating industry can not use the steam for part of the day or year (due to seasonal or other production variations), the steam must be dumped and much of the gain in efficiency is lost.

There have been a few successful co-generation applications in the United States, but many more in Europe. Part of the difficulty in the US has been the large distances between the towns and power plant siting problems. It is not economical to transport steam for long distances, so the power plant must be located close to the industrial steam user. Environmental restrictions on power plant siting impose severe limits on the number of locations where this is feasible. For the future, the most likely application of this concept will be the construction of a power plant at a site where new industries can also be located.

As an example of an industrial application, Case Study 2-7 shows several generations of power plants used by the Aluminum Corporation of America.[23]

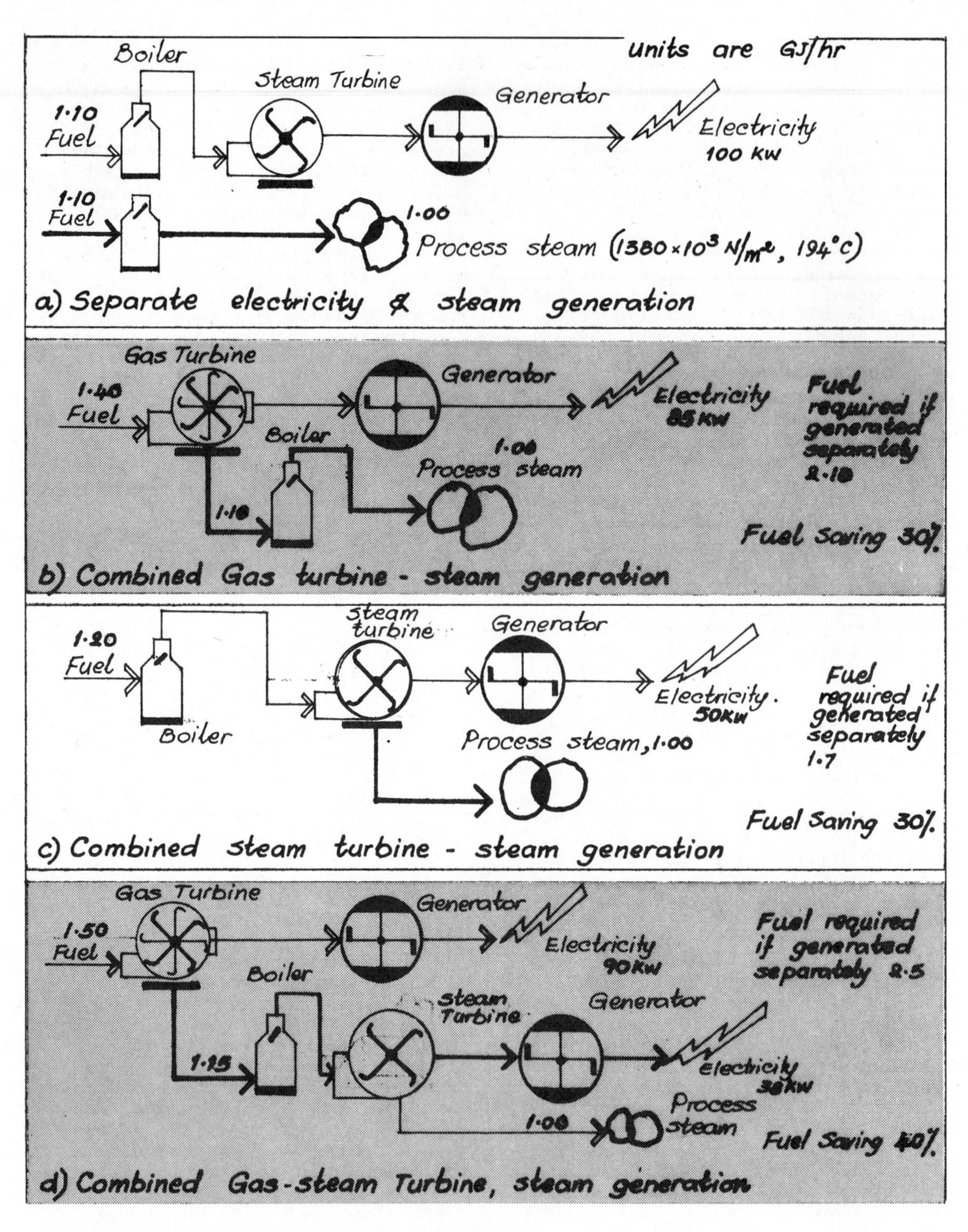

COMPARISON OF POWER CYCLES FIGURE 2.17

In the above, emphasis was on power generation but the same concepts can be applied to any industrial process (see Case Study 2-6). Again, availability can be used as a measure of effectiveness or efficiency as discussed in section 2.3.

Power generation (in the form of either electrical energy or mechanical energy) and thermal heating processes should be viewed as concomitant operations for an effective fuel conservation program. A process heating operation and low temperature space heating should be accompanied by some type of work extraction system if economically feasible.

Equipment and Process Modification[22]

Much of the energy required to process and finish metal products depends upon decisions made by the product designer. It is at the design stage that processes can be selected to use energy efficiently. Many processes require preheating and postheating and the merits of alternate cold processes should be considered. Heat treating processes such as carburizing are energy intensive. In this process, carbon must diffuse into the surface of the steel, with diffusion requiring between 4 and 12 hours at temperature. In addition, the carburizing atmosphere in the furnace is made up of unburned petroleum products which are wasted after each cycle. Several alternatives to carburizing can be considered.

A hard surface can be produced by induction heating, which is a more efficient energy process. Plating, metalizing, flame spraying, or cladding can substitute for carburizing although they do not duplicate the fatigue strengthening compressive skin of carburization or induction hardening.

Pre-hardened and pre-annealed steel should be specified whenever possible since steel mill operations are generally more efficient than individual plant operations. Sometimes hardened steel can be replaced by a high strength steel that meets specifications and requirements without heat treatment. Cold work products can also replace the need for hardening. Elimination of machining after heat treating reduces the case hardening depths required.

Paint finishing in industrial operations often leads to inefficient utilization of energy. The paints are normally solvent based and must be cured at elevated temperatures in ovens which are only 30 to 50 percent efficient. In addition, if afterburners are used to remove harmful solvents from exhaust air, typical fuel consumption is increased by 75 to 100 percent. Alternative painting processes should be considered, not only to reduce fuel consumption but also to minimize environmental problems.

Powder coating is a substitute process in which no solvents are used. Powder particles are electrostatically charged and attracted to the part being painted so that only a small amount of paint leaving the spray gun misses the part and the overspray is recoverable. The parts can be cured rapidly in infrared ovens which require 42 percent less energy than standard hot air systems.

Water based paints and high solids coatings are also being used and are less costly than solvent based paints. They use essentially the same equipment as the conventional solvent paint spray systems so that the conversion can be made at minimum costs. New water based emulsion paints contain only 5 percent organic solvent and require no afterburning. High solids coatings are already in use commercially for shelves, household fixtures, furniture, and beverage cans and require no afterburning. They can be as durable as conventional finishes and are cured by either conventional baking or ultraviolet exposure.

Hot forging may require a part to go through several heat treatments. Cold forging with easily wrought alloys may offer a replacement. Lowering the pre-heat temperatures may also be an opportunity for savings. Squeeze forging is a relatively new process in which molten metal is poured into the forging dye. The process is nearly scrap free, requires less press power, and promises to contribute to more efficient energy utilization.

Forming operations can be made more efficient by the use of stretch forming. In this process sheet metal or extrusions are stretched 2 to 3 percent prior to forming, which makes the material more ductile so that less energy is required to form the product. Finally, the finished part is stronger and therefore thinner sheets and lighter extrusions can be used.

Welding operations can also be made more efficient by the use of automated systems which require 50 percent less energy than manual welding. Manual welders deposit a bead only 15 to 30 percent of the time the machine is running. Automated processes, however, reduce the no-load time to 40 percent or less. Different welding processes should be compared in order to determine the most efficient process. Electro-slag welding is suited only for metals over 1 cm (0.5 in.) thick but is more efficient than other processes.

Machining itself is not particularly demanding on energy use. Machining processes, however, generally produce the most scrap. As much as 50 percent of the raw material can be wasted as scrap during typical machining operations. The designer should specify raw material sizes as close as possible to the finished sizes, or require a casting or forging which requires a minimum amount of machining.

Scrap production from several metal working processes is shown in Figure 2.18. Scrap recycling programs should be instituted whereever possible, not only to lower costs in the operation but also to improve overall energy efficiency.

Powder metallurgy generates the least amount of scrap, but parts must be sintered which is usually accomplished in inefficient convection-type heating furnaces. Induction sintering will reduce energy requirements by two or three times.

Over-design is another area for improving material and energy utilization. Design practice has generally tended to overstate safety factors and uncertainties and often is used as a hedge against poor and improper design and materials selection. The case against over-design and waste of fuels is shown in Figure 2.19.

Finally, a designer must consider the materials selected for the part or finished product. Energy is required to process the raw material and the materials used dictate how much energy must be used in manufacturing the product (see Appendix B for typical values). Energy costs in the selection of materials have not been a serious consideration in the past. Too often these costs are hidden and are not obvious to the designer. The rising price of fuels, however, will have an impact on the future prices of materials.

2.4 POTENTIAL ENERGY SAVINGS

An overview of the overall strategies for an effective program to improve energy efficiency in industry suggests the need for two simultaneous approaches: (1) effective energy management, and (2) a technical implementation program. The elements of an effective energy management effort parallel conventional industry procedures and include:

- commitment and leadership of a company director;
- an effective plan and organization with goals and procedures;
- establishment of accounting and reporting procedures on energy use patterns; and,
- priorities for implementation and review of programs and initiatives.

The technical implementation of industrial conservation efforts can be simplified to the following basic approaches:

- analyze and evaluate energy flow and utilization in the process and operation;
- establish effective maintenance

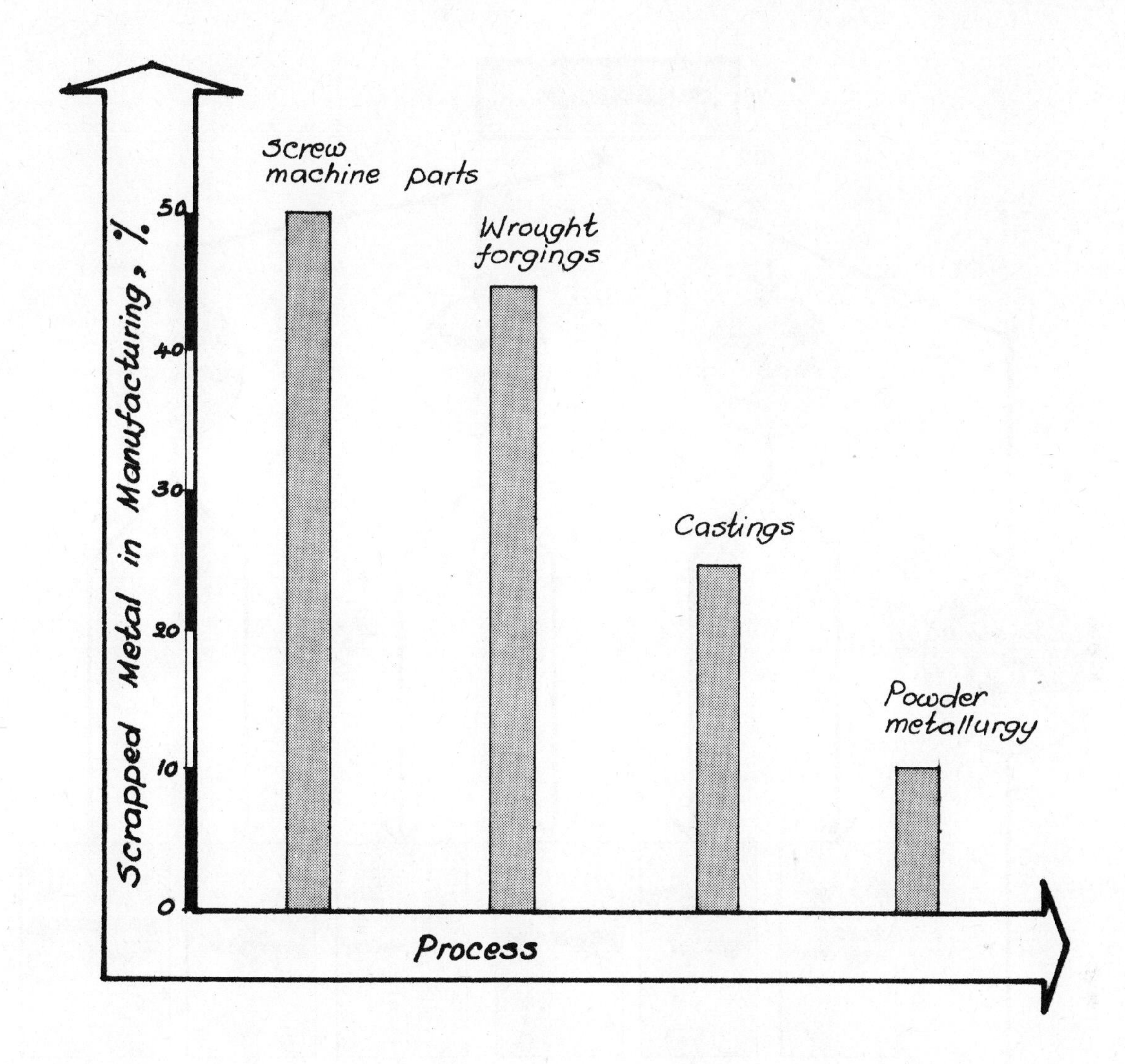

Reference 22

SCRAP PRODUCED BY METAL WORKING PROCESSES FIGURE 2.18

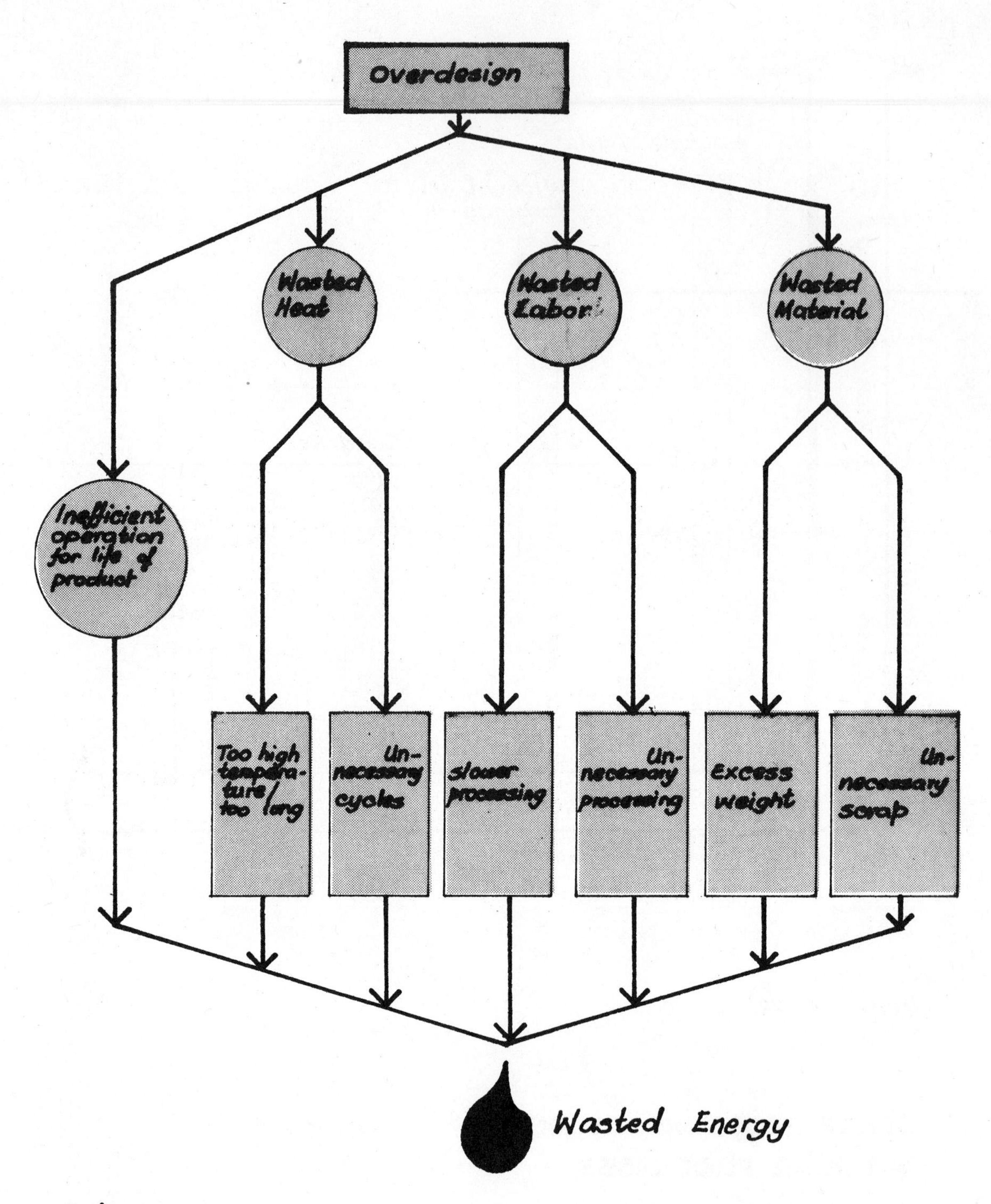

Reference 22

OVERDESIGN WASTES ENERGY IN MANY WAYS

FIGURE 2.19

9. US Federal Power Commission, Office of the Chief Engineer, *A Technical Basis for Energy Conservation*, Staff Report, (Washington, D.C.: April 1974).

10. E.I. DuPont de Nemours & Co. (Inc.), *DuPont Energy Management Services*, packet containing variety of materials, (Wilmington, Delaware: 1974).

11. Prengle, H. William Jr.; Crump, Joseph R.; Fang, C.S.; Grupa, M.; Henley, D.; and Wooley, T., *Potential for Energy Conservation in Industrial Operations in Texas*, Report No. S/D-10, (Houston, Texas: University of Houston, Department of Chemical Engineering, Cullen College of Engineering, November 1974).

12. Gatts, Robert R.; Massey, Robert G.; and Robertson, John C., *Energy Conservation Program Guide for Industry and Commerce, (EPIC)*, NBS Handbook 115, (Washington, D.C.: US Department of Commerce, National Bureau of Standards, September 1974).

13. "Industrial Energy Management Workshop," (Lecture Notes), University of Arizona, Tucson, Arizona, 12-16 May 1975.

14. Snyder, W.T.; Miller, W.A.; and Schmidt-Bleek, F., *How to Cut the Energy Budget in Business and Industry, Guidelines and Suggestions*, (Knoxville, Tennessee: University of Tennessee Environment Center and Center for Industrial Services, February 1974).

15. Sheldon, Allen C., "Organizing an Energy Conservation Effort," Paper presented at the "Industrial Energy Management Workshop," University of Arizona, Tucson, Arizona, 12-16 May 1975.

16. Snyder, Irving G., Jr., "Energy Accounting," Paper presented at the "Industrial Energy Management Workshop."

17. Babcock & Wilcox Company, *Steam/Its Generation and Use*, 38th edition, (New York: 1972).

18. Edwards, Paul L., "Energy Costs Weigh on Compressor Field," *Energy User News*, 11 April 1977.

19. Zugman, Howard, "Reduce Energy Waste with Recovery Wheels," reprint from *Specifying Engineer*, (n.d.).

20. Lambert, E.H., "Energy Conservation for the Mining Industry," *Canadian Mining and Metallurgical Bulletin* (October 1973): 69-74.

21. Crowell, W.H., "Cut Your Electric Bill Without Cutting Production," *Industry Week* (15 April 1974): 47-49.

22. Bittence, John C., "Processes that Conserve Power," *Machine Design* 46 (4 April 1974): 94-102.

23. Robertson, J.C., "Energy Conservation in Existing Plants," *Chemical Engineering* (21 January 1974): 104-111.

CHAPTER 3

COMMERCE

By R.T. Taussig*

CHAPTER CONTENTS

KEY WORDS

Air Conditioning
Banks
Building Codes
Commercial Buildings
Commercial Equipment
Commercial Use
Environmental Conditioning
Government Buildings
Heat Recovery
Hotel
Illumination Practices
Illumination Standards
Insulation
Refrigeration
Window Losses

SUMMARY

Total energy supplied to the household and commercial sectors was 35 percent of the nation's energy budget in 1974. This includes the fuel required to generate electricity used in these sectors (see Table 3.1). The household and commercial sectors are major electricity users accounting for about 58 percent of the nation's electricity use.[1] Industry used the balance (42%) in 1974.

---

*Principal Research Scientist, Mathematical Sciences Northwest, Inc.

Sector	Direct Fuel Use	Electricity Conversion Loss	Electricity	Total Use	Percentage of Total Use
Household and Commercial	14.6	8.28	3.99 (58%)	26.8	34.8
Industrial	21.5	6.01	2.86 (42%)	30.4	39.5
Transportation	19.4	0.04	0.02 (0.3%)	19.5	25.3
Miscellaneous	0.3	-	-	0.3	0.4
Totals	55.8	14.3	6.87	77.0	100.0

Units are 10^9 GJ

Reference 1

ENERGY USE IN COMMERCE, 1974 **TABLE 3.1**

	Energy (10^{15} Btu)	Energy 10^9GJ	Percent of Sector Energy	Percent of U.S. Energy
Space Heating	2.3	2.4	41	3.4
Air Conditioning	0.5	0.5	8	0.7
Lighting	1.3	1.4	23	1.9
Refrigeration	0.4	0.4	7	0.6
Other	1.2	1.3	21	1.8
Total	5.7	6.0	100	8.4

Reference 2

U S COMMERCIAL ENERGY USE, 1970 **TABLE 3.2**

Commerce uses 30 to 40 percent of energy in the residential and commercial sectors, depending on how the data are interpreted. The uncertainty results from different statistical reporting techniques. Major energy uses within the commercial sector in 1970 were space heating (41% of sector energy); lighting (23%); air conditioning (8%); refrigeration (7%); and other (21%).[3] Electricity use for various purposes is shown in Table 3.2.

Many modern buildings are designed for total environmental conditioning, regardless of outside weather. For example, air conditioning often must be used in office buildings to remove heat generated by people and lights even when the outside temperature is lower than the building temperature. In addition to this characteristic, many modern buildings make extensive use of glass surfaces which can result in a large solar heat gain in the summer and large heat losses in the winter.

Modern commercial establishments generally use fluorescent lighting systems. In the past, with low-cost electricity widely available, lighting system designs and controls were not optimized for efficiency. More efficient lighting systems can be designed, however, and more efficient use of light is possible.

Other options for improving the efficiency of commercial electricity use involve new heating, ventilating and air-conditioning system designs, heat recovery and retrofit installation in existing buildings, heat pumps, shading, insulating and reflecting glass, and other architectural considerations.

Because of projected building trends, near-term energy savings will occur principally in old, not new buildings, through changes in operation and by corrective modifications, repair and maintenance of existing equipment. The case studies for this chapter show that savings of up to 30 percent can be achieved with payback periods in the range of one year.

Several options for energy savings are also examined in the chapter text. The results indicate an overall potential for saving 30 to 50 percent of commercial energy use, which corresponds to approximately 5 to 8 percent of US energy use (see section 3.5). The needed improvements in efficiency are assumed to take place over the period 1975-2000.

Potential electricity savings could be offset, however, by new construction and increased substitution of electricity for scarce or interruptable fuels. This might occur in regions where unavailability of oil or gas requires an increased use of electricity generated from coal, nuclear, geothermal, or hydroelectric sources.

3.1 INTRODUCTION AND OVERVIEW

Electricity use in the commercial sector grew at the rate of 9.6 percent per year from 1962 to 1972, increasing faster than either the residential or industrial sectors during the same period.[4] This rapid growth was due to an increased use of air conditioning and refrigeration, increased interior lighting, more elevators, and more elaborate air handling equipment (fans, pumps, etc.).[5] A combination of economics and architectural factors have contributed to the more widespread use of these systems. For example, as urban land values have increased, so have high-rise buildings.

The tall, glass-faced, sealed commercial building has become an identifiable building type with standard equipment, catalog components, and design methodology. The emphasis has been on speeding the design process rather than exploiting the benefits of individualized building components integrated with the local site and environment. Modern buildings of this type have been designed to separate interior and exterior environments, relying on mechanical-electrical systems to do most of the work. The most important determinants of such a design are traditional building techniques, compliance with specific building codes, ease of construction, reduction of first costs, and marketability.

Technical options exist which can greatly reduce electrical energy use in commercial structures.[6] Besides saving energy, many of these alternatives are economically desirable when included in the initial design and operation of buildings. More importantly, energy-saving schemes are available which apply to existing buildings. Energy-saving methods may be grouped as:

- operation and maintenance strategies which the tenants and building owner may employ;
- modifying and retrofitting existing structures and equipment--to be carried out by building owners, constructors, and equipment manufacturers; and,
- new design of equipment and building components by equipment manufacturers, constructors, engineers, and architects.

This chapter presents a basic description of energy-using devices, energy-saving remedies, and a methodology for reducing energy use in commercial structures by employing more efficient methods or equipment and eliminating waste. A detailed evaluation of HVAC systems for commercial buildings is deferred to Chapter 8. Also see Chapter 5 for a discussion of computers and applications in commercial buildings and Chapter 11 for a discussion of motors.

3.2 OBJECTIVE, SCOPE, ASSUMPTIONS

A. Objective

The main objective of this chapter is to provide background data on commercial electricity use in order to indicate sensible, specific, and effective energy-saving schemes. Energy savings will be presented from the user's viewpoint (in this case the tenant or building owner) since the user must live or work in the resulting environment, will often be the final arbiter in accepting or rejecting a particular scheme, and will probably pay the costs.

B. Scope

The commercial sector consists of a diverse group of energy users:

Office Buildings
Department Stores
Schools
Restaurants
Hotels
Warehouses
Churches
Banks
Hospitals
Theaters
Libraries
Government and Other Public Buildings

These are some of the largest users. Public works such as street lighting, water and sewage treatment plants, garbage disposal, and harbor and airport facilities are excluded from this chapter because they involve individual energy use characteristics.

The emphasis of this chapter is on electrical energy savings and its effect on overall energy savings. Specifically, direct electrical energy use by equipment, modes of operating such equipment, the efficiency of equipment, and equipment costs are considered. Examples are taken from:

Heating Systems
Air Conditioning
Water Heating
Large Cooking and Refrigeration Facilities
Lighting
Elevators and Escalators

Individual items which should be considered are discussed in this chapter (or other chapters of the book) and include:

Motors
Pumps
Fans
Compressors
Stokers
Dampers

Fluorescent Lighting
Resistance Heaters
Transformers
Switches
Voltage Control Devices
Motor-Generator Sets
Heat Pumps
Synergistic Systems (e.g., solar-assisted heat pumps)

C. Assumptions

Energy savings are measured against an extrapolation of long-term historical trends in US energy use and in commercial sector electricity demand. These trends clearly do not include near-term economic effects on energy, such as an economic recession. They should, however, serve as a common yardstick with which one can measure the relative merits of alternative energy-saving techniques.

In presenting possible energy savings, *immediate* savings are defined as those applying within zero to two years; *near-term* savings are those possible in two to five years; and *long-term* savings are those achievable by the end of the century (5 to 25 years). Energy usage and energy savings have been computed on the assumption that recommended guidelines for design and equipment operation (as set forth by ARI, ASHRAE, IES, ANSI, EEI, etc.) are being followed. We also assume that the reader is familiar with energy-using equipment, but possibly not with the potential for improving their utilization of energy.

Indirect energy use, for example, in the manufacture of a particular piece of equipment, is discussed elsewhere, and will appear in this chapter only when it affects the level of energy use in a commercial structure measured over the lifetime of the structure.

3.3 USEFUL FACTS AND BACKGROUND

A. Introduction

The purpose of this section is to provide background information which will suggest a rationale for choosing between energy saving alternatives. For example, data on the fastest growing uses of electricity indicate where efforts to modify design and equipment must be made in new buildings. Large regional differences in winter temperature impose additional constraints on the utility (or practicality) of strict insulation requirements. Most importantly, the facts presented below show where the greatest use of electricity occurs in existing buildings. Disregarding renovation of older buildings, the construction rate of new commercial structures has been approximately 5 percent per year. Hence, in the immediate- and near-term the largest market for energy savings exists in buildings already standing. Most of the economies suggested in Section 3.4 apply, therefore, to older buildings.

B. Energy Use Details

Energy use in a six-story commercial building is shown in Table 3.3. In this example the end use electricity was 39 percent of total energy end use and natural gas was 61 percent. The major use of electricity was for lighting (50% of all electricity), followed by equipment (33%) and cooling (17%). Annual use of energy amounts to 1,436 $MJ/m^2 \cdot yr$ (126 $kBtu/ft^2 \cdot yr$).

In a recent report the National Petroleum Council identified building types which accounted for most of the energy used in the commercial sector.[7]

Colleges	Supermarkets
Hotels and Motels	Schools
	Stores
Offices	Hospitals

Even though electric space heating still represents a small fraction of energy used in the commercial sector, its rate of growth has been rapid over the past decade. Over a three-year period ending in 1969, electrically heated areas in office buildings increased 3000 percent; in libraries, 1100 percent; in colleges and universities, 1230 percent; and in non-food stores, 1260 percent.[8] A continuation of current growth rates could bring electric space heating to a significant fraction of the total energy use (e.g., 10 percent) in the commercial sector by 1985. It is worthwhile noting that the FEA Project

	User End		
Natural Gas	MWh/yr	MBtu/yr	GJ/yr [(2)]
Heating	-	9,217	9,733
Hot Water	-	422	445
Electricity			
Cooling	312.8	1,068	1,128
Lights	916.5	3,129	3,304
Office Equipment, Fans, Pumps, etc.	607.5	2,074	2,190
TOTALS	1,836.8	15,910	16,800

EUPF = 126,000 Btu/ft^2yr
= 1,436 MJ/m^2yr

Notes: (1) Area = 11,700 m^2 (126,000 ft^2)
(2) Multiply MBtu by 1.056 to get GJ

TABLE 3.3

ENERGY USE IN A SIX STORY BUILDING

Independence Report considered two scenarios involving an increase in electric space heating (by the replacement of all oil and gas heat in new buildings with electric heat) as part of the plan for achieving energy self-sufficiency.[11] As an alternative, heat produced by electric heat pumps was considered in the same fashion.

Lighting represents one of the largest single end uses of electricity in the commercial sector. The fixtures producing the light and the light itself release heat into the building, either adding in a useful way (i.e., during the winter in a cold room) to the building's heating requirements or in a bothersome way (i.e., to an already warm room) when the building must be cooled. By designing efficient lighting one may also reduce energy use and save on equipment costs (e.g., by using smaller units) required for space conditioning. With proper design, heat from lights can also be used constructively throughout a building (see Chapter 9). This particular example is treated in more detail in Case Study 3-2.

To account properly for the relative impact of different fuel conservation schemes, one must consider regional and seasonal variations of the environmental heating and cooling loads due to temperature and solar heat gain. These effects are discussed in Chapter 8. The annual mean solar insolation is also given in Chapter 8.

Several building indices have been proposed to measure the relative effectiveness of energy use in a building. This concept is similar to the Energy Utilization Index used to relate a unit of industrial production to energy used in processing or manufacturing. [12] Guidelines for determining the energy content of products have been published (see Chapter 2).[13]

For buildings several approaches can be considered. The simplest approach sums the energy inputs (kWh, m^3 or ft^3 of gas, liters or gallons of fuel oil) and converts them to a common basis, e.g., joules or Btu. This value is then normalized per unit area of floor space or per unit volume. A similar method is frequently used for power estimates. For example, lighting loads may be expressed in terms of watts (electrical) per square meter. Typical ranges for commercial buildings might be 10-50 We/m^2 (1-5 We/ft^2). This approach is described in the literature.[14]

Each of these approaches has the disadvantage of representing an oversimplistic view of energy use. On a per unit area basis, for example, the measure does not adequately account for important differences such as height to floor ratio (involving differences in building volume); percentage of window space (which affects the heat transfer characteristics of the building skin); building aspect ratio (surface to volume ratio, affecting the interior heat distribution as well as heat exchange across the building skin), and percentage of lighting available from natural light. Refer to Chapter 7 for a discussion of the impact of the building envelope and site on energy usage.

Buildings are constructed primarily for one purpose--to serve their human occupants. This suggests that the most appropriate building energy index might be the total energy used, normalized to the number of persons occupying or using the building. Thus, total energy use would fall in the range of 10 to 100 GJ/person-year. This index could be useful as a basis for comparison and evaluation of alternative designs, or for monitoring progress of energy management program efforts (see Case Study 3-1).

Such an index would not be appropriate for certain classes of structures such as warehouses, but would be appropriate for offices, stores, banks, apartments, hospitals, hotels, etc. For some types of commercial operations, a measure based on units of production is suggested. For example, in a hotel this would be guest-nights and meals served; in a laundry, kg of cleaned product; in a hospital, patient-days, etc.

The energy use cycle over 24 hours and for different seasons of the year, otherwise known as the *building duty cycle,* is an extremely important factor in determining how to save energy. Different classes of commercial structures obviously have widely divergent duty cycles and some generalizations must be made to reduce the discussion of this topic to reasonable dimensions. Buildings can be roughly divided according to day-time versus full-time occupancy (e.g., hospitals) and by stationary occupancy (e.g., an office) versus circulating occupancy (restaurants, banks, supermarkets). Day-time buildings have distinctly reduced loads over part of their cycle, and the sizing, costs and operating procedures for building equipment must reflect the load factor or relative proportion of time that full equipment performance is utilized. For full-time buildings more expensive equipment is justified if the running costs are reduced appropriately; in energy terms it is cost-effective because running expenses are proportional to energy use. Circulating occupants may endure somewhat more uncomfortable conditions (e.g., lower winter indoor temperatures) than stationary occupants because they only spend a limited time in the building environment. Hence the standards affecting energy use for space heating and cooling may be relaxed under such circumstances. As an example, air-conditioning requirements at different temperature and load levels for selected types of commercial buildings are shown in Table 3.4. These data present almost a factor of four difference between the maximum and minimum equivalent full-load operating hours among the various types of commercial buildings surveyed (last column, Table 3.4).

C. Starting An Energy Management Program [16]

The building operator or management must first establish the need for an energy management program. Once this *management commitment* has been obtained, the program can be formulated.

The next step is to prepare an energy management plan (Figure 3.1). The figure outlines the essential elements of the energy management program, including:

- Review of historical data;
- Building surveys or energy audits;
- Analysis of audit results;
- Identification of energy management opportunities;
- Economic evaluation of opportunities; and,
- Implementation and follow-through.

A review of historical data is an obvious first step since it provides information on the past record of energy use. (This will not be discussed further since an example is provided in the case studies for this chapter.)

A building survey provides, at a minor cost, a preliminary review of building energy use. Table 3.5 shows a one-page form which evolved after experience was gained in conducting surveys. The organization of the form permits a quick evaluation of building energy use through several "Energy Use Performance Factors" which are defined in the table. If energy use seems to be higher than it need be, the other information can be used to help isolate the cause.

Table 3.6 provides sample results obtained in building surveys. While the spread in building energy use is only about 24:1, the spread in *occupant* energy use is 35:1, reflecting the range of needs of different occupancies. The building survey can also be used to evaluate a particular energy form, as has been done in Table 3.7 for electricity. Compilation of data in this format is also useful in providing guidelines for new projects or designs.

The *energy audit* is the most effective tool for identifying energy management opportunities (see Chapter 2, section 2.4; also Case Study 3-3 for additional details).

Type of Building	Operating Schedule: Days per week	Operating Schedule: Hour per day	Outside Temperature at which Air Conditioning Is Required (°F min)	DUTY CYCLE — Annual Operating Hours at Load: One Quarter	One Half	Three Quarters	Full Load	Total	Equivalent of Full Load Operating Hours
Hospital	7	24	60	610	1,472	1,212	155	3,449	2,053
Hotel	7	24	65	499	1,205	992	128	2,824	1,595
			70	365	881	724	93	2,063	1,168
Office	5-1/2	12	55	418	684	404	71	1,577	820
			60	359	588	347	61	1,355	701
Department Store	6	12	55	26	901	655	138	1,720	1,087
			60	22	775	563	118	1,478	934
College	7	13	60	495	811	478	84	1,868	972
Library	7	13	65	405	664	392	69	1,530	996
			70	296	485	286	50	1,117	581

Reference 17

EFFECT OF DUTY CYCLE ON AIR CONDITIONING REQUIREMENTS

TABLE 3.4

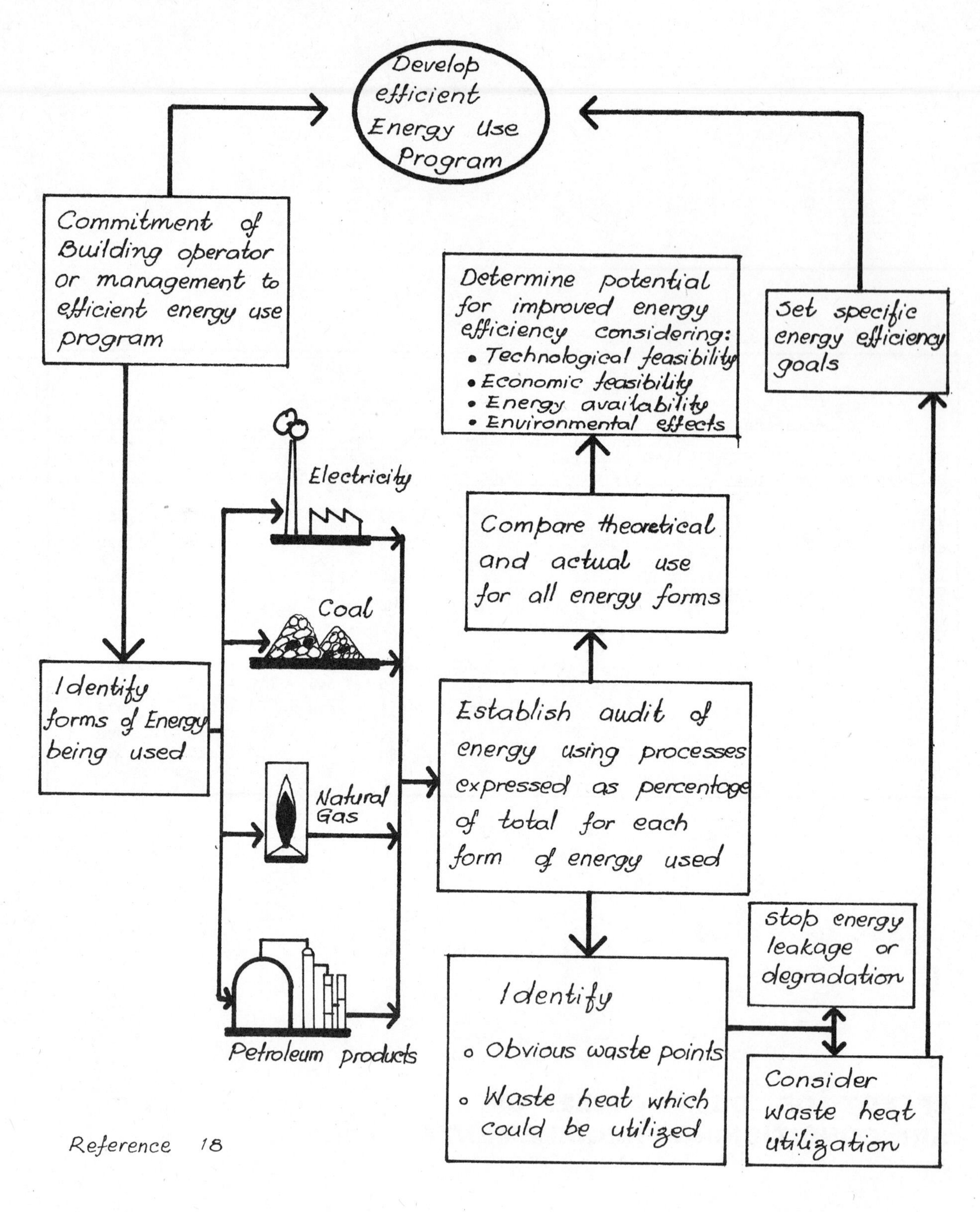

FIGURE 3.1

ENERGY MANAGEMENT FLOW CHART

Building Description

heating degree days ____

- Name: Age: __ years
- Location:
- No. of floors __ Gross floor area __m^2 (ft^2) Net floor area __m^2 (ft^2)
- Percentage of surface area which is glazed __% cooling degree days ____
- Type of air conditioning system; heating only __ evaporative __ dual duct __ other (describe) ____
- Percentage breakdown of lighting equipment: Incandescent __% fluorescent __% high intensity discharge (type) __% other ___%

Building Mission

- What is facility used for:
- Full time occupancy (employees) __ persons
- Transient occupancy (visitors or public) __ persons
- Hours of operation per year ___
- Units of production per year ___ Unit is ________

Installed Capacity

- Total installed capacity for lighting __kW
- Total installed capacity of electric drives greater than 7.5 kW (10 hp) (motors, pumps, fans, elevators, chillers, etc.) __hp x 0.746 = ___kW
- Total steam requirements __lbs/day or __kg/day
- Total gas requirements __ft^3/day or Btu/hr or ___m^3/day
- Total other fuel requirements ________

Annual Energy End Use

Energy Form	x Conversion		kBtu/yr	Metric Units		Conversion		MJ/yr
• Electricity	__kWh/yr x 3.41	=	____	__kWh/yr	x	3.6	=	____
• Steam	__lb/yr x 1.00	=	____	__kg/yr	x	2.32	=	____
• Natural gas	__cf/yr x 1.03	=	____	__m^3/yr	x	38.4	=	____
• Oil	__gals/yr x{#2 139, #6 150}	=	____	__ℓ/yr	x	{#2 38.8, #6 41.8}	=	____
• Coal	__tons/yr x 24,000	=	____	__kg/yr	x	28.0	=	____
• Other	__ x	=	____	__	x	____	=	____
	TOTALS		____					____

Energy Use Performance Factors (EUPF's) for Building

- EUPF #1 = MJ/yr (kBtu/yr) ÷ Net floor area = ____MJ/m^2yr (kBtu/ft^2yr)
- EUPF #2 = MJ/yr (kBtu/yr) ÷ Average annual occupancy = ___MJ/person·year (kBtu/person·year)
- EUPF #3 = MJ/yr (kBtu/yr) ÷ Annual units of production = MJ/unit·yr (kBtu/unit·yr)

BUILDING ENERGY SURVEY FORM TABLE 3.5

Facility Description[2]	Net Area (m^2)	EUPF #1[3] kBtu/ $ft^2 \cdot yr$	EUPF #1[3] MJ/ $m^2 \cdot yr$	EUPF #2 MBtu person·yr	EUPF #2 GJ person·yr
Small Engineering Office*	279	43.6	496	11.6	12.2
Engineering Office and Laboratory*	929	28.7	326	14.4	15.2
Elementary School	4,000	72.7	826	4.86	5.13
Hospital Laundry	4,200	687	7,810	404	427
Elementary School	7,000	61.4	698	3.84	4.06
City Administration Building*	16,250	119	1,350	16.1	17.0
Office Building with Private Clubs	16,700	175	1,990	52.7	55.6
Police Administration Building*	41,500	202	2,300	-	-
Office Building	46,500	106	1,210	-	-
Office Building	48,000	139	1,580	-	-
Los Angeles City Hall Complex *	138,600	162	1,840	51.5	54.4

Notes: (1) Buildings are located in California, Wisconsin, Louisiana, Virginia, and Pennsylvania. Data are for period 1973-1976.

(2) Buildings marked with an asterisk have had energy management programs which have reduced energy use by 20 to 30 percent compared to pre-1973 values.

(3) Multiply kBtu/ft^2 by 11.37 to get MJ/m^2

TABLE 3.6

SAMPLE ENERGY USE PERFORMANCE FACTORS (EUPF's) [1]

Facility Description	Net Area (m²)	Net Area (ft²)	Total Electricity Use kWh/(m²·yr)	Total Electricity Use kWh/(ft²·yr)	Total Electricity Use kWh/(Occupant·yr)	Lighting Installed Capacity W/m²	Lighting Installed Capacity W/ft²
Small Engineering Office	279	3,000	67	6.3	1,670		
Engineering Office and Laboratory	929	10,000	22	2.0	750	16	1.5
Elementary School	4,000	43,500	229	21.3	1,430	31	2.9
Hospital Laundry	4,200	45,000	242	22.6	11,950	26	2.4
Elementary School	7,000	75,000	194	18.0	1,130	30	2.8
City Administration Building	16,250	175,000	234	21.7	2,920		2.0-4.0
Office Building with Private Clubs	16,700	180,000	345	32.1	9,620	33	3.1
Police Administration Building	41,500	447,000	194	18.0	---	-	-
Office Building	46,500	500,000	127	11.8	---	-	-
Office Building	48,000	517,000	243	22.6	---	-	-
Los Angeles City Hall Complex	138,600	1,492,000	202	18.7	4,940	-	-

REPRESENTATIVE ELECTRICITY USE IN BUILDINGS

TABLE 3.7

The objectives of the energy audit are to establish the details of energy use under actual operating conditions in the "as-built" (and perhaps modified!) building. Audit results provide insights which cannot be obtained from historical data; in some cases, immediate identification of waste occurs, leading to near-term savings.

Another objective of the audit is to compare predicted energy use with actual values in order to verify analytical or design results and mathematical models.

Once the audit has been accomplished, additional steps can be taken. More detailed energy audits which measure when and where energy is used in the building and which attempt to correlate the data with building use, occupancy patterns, weather conditions, and so forth, can help to identify specific areas where energy savings may be achieved. Lighting intensity and room temperature can be measured at various points throughout the building and over several daily cycles. Static pressure profiles in ducts and across heat exchange equipment can be measured. Large electrical equipment can also be monitored to determine variations in the power factor. The composite of these measurements gives a rough picture of the building's energy duty cycle and will reveal equipment or applications which are inefficient.

Knowledge of the energy duty cycle is especially useful in those larger commercial structures in which the internal climate is automatically controlled, for example by computer. By reprogramming the control system appropriately, a whole spectrum of energy-saving changes in building operation can be incorporated as a well-integrated scheme. A variety of building systems simulation routines are now available which can be used to experiment with alternative energy duty cycles on a computer and to make an optimum choice for the building before subjecting the building itself to specific changes.[19-24]

In conducting a detailed building energy audit, it is useful to consult with an engineering specialist familiar with electrical and mechanical building equipment so that a progressive, cost-effective audit can be conducted, beginning with the most likely areas for improved efficiency and proceeding step-by-step to the more difficult areas on the basis of successive re-evaluation of the data and costs. The same technique can be used to construct a realistic goal for overall, long-term energy savings. Periodic audits can be used to measure subsequent improvements in specific energy utilization; long-term improvements should become evident in the life-cycle costs of the building operation. Typical fees for an audit by a consulting engineer would be in the range of 1 to 10 percent of the annual energy bill.

The next step after an energy audit is to identify energy management opportunities, carry out engineering and economic analyses as required, implement the program, and provide continuous monitoring to verify that the expected savings are actually being realized.

Numerous guidelines in addition to energy efficiency influence suitable alternatives for saving energy. Cost, ease of installation, operation and maintenance, comfort, and a conducive working environment are foremost among these criteria. The American Society of Heating, Refrigerating, and Air Conditioning Engineers has long published a series of guidelines concerning the design of appropriate equipment for buildings.[25] Similarly, the Illumination Engineering Society recommends values of illumination for visual tasks.[26]

A variety of government agencies (e.g., NBS, GSA, FHA, etc.)have also published changes in long-standing practices in order to encourage efficient energy use.[27,28,29,30].

3.4 ENERGY ECONOMIES IN THE COMMERCIAL SECTOR

A. Introduction

Immediate energy savings can be accomplished through changes in building operation and maintenance and by selective modification of existing equipment. The near- and long-term savings continue these initial efforts and add to them the effects of more extensive equipment retrofitting and the capabilities of new design in building components and equipment. This section treats specific energy saving schemes with emphasis on those approaches which directly affect electricity usage.

In the sub-sections that follow, methods for economizing on energy use in commercial buildings are discussed. These have been categorized as operational changes, modifications, and new designs. In addition, options are further subdivided into those suited to existing buildings compared with those achievable for new buildings, as well as those which can be achieved without capital investments compared to those which require some capital investment (For a summary see Table 3.8 at the end of this section.)

B. Operations

1. Operational Changes

Operational changes can often be implemented immediately with little or no additional cost. Typically these are changes which building managers can accomplish by "fine-tuning" systems, by instituting an on-going preventive maintenance program, by implementing a closer watch on equipment, or by informing operating personnel or building occupants of the need to use energy more efficiently. Reference 14 is typical of the types of "checklists" which provide detailed suggestions. Reference 31 is a pamphlet with a number of useful suggestions. Both documents are readily available and cost less than one dollar.

Typical operational changes would include: (Note that the savings percentages discussed below are not additive, since some of these operations are mutually exclusive.)

- reduction of thermostat settings in the winter, raising them in the summer, and using a night set-back if the system permits. *(Caution: some systems will use more energy unless modified [see Section 3.4C and Chapter 8]);*
- limiting lighting intensity in halls, storage areas, etc. where close work is not required; control of night lighting for cleaning purposes;
- use of natural ventilation or shielding windows to prevent excessive heat losses or gains on buildings where this is feasible;
- rescheduling deferrable electrical equipment to non-peak hours; turning off equipment when not in use;
- covering frozen food cases at night, limiting entry into walk-in freezers;
- cleaning all heat transfer surfaces in boilers, refrigerators, ducts, etc.; and,
- fixing leaky faucets, water and steam valves, and pipes.

Several of these measures will now be discussed in detail.

2. Environmental Conditioning (HVAC)

Environmental conditioning includes space heating, cooling, and ventilation. A number of options exist for reducing energy use in space heating and cooling. Since the technical details are described in detail in Chapter 8, only a brief discussion will be presented here.

- Cooling

Turning off non-essential equipment is the first energy management option in cooling systems. This requires a systematic review to insure that adequate comfort is still provided for working personnel, without cooling unoccupied and other non-essential areas. The amount of savings depends, of course, on the design of the specific facility and on weather conditions. Typically, savings can be in the range of 5 to 10 percent.[32]

Raising thermostat settings to 26 °C (78 °F) and humidistat settings to 60 percent RH is the second operational measure. Refer to Chapter 8 for several cautions and further details, particularly if the system uses reheat. Potential savings can reach 10 to 15 percent compared to conventional use.

Maintaining equipment in good working order is the third operational measure. Dirty filters, poorly maintained heat exchange surfaces, etc. can lead to as much as a 10 percent loss in system efficiency.

Reduction of unnecessary heat loads wherever possible (unneeded lighting, solar loads which can be shaded, warm air infiltration, etc.) is the fourth measure.

The fifth measure is to make as effective use as possible of outside air for cooling whenever conditions permit. (Technical aspects of this approach are discussed in Chapter 8.) This measure alone can result in a 20 percent reduction in energy required for air conditioning. [33]

- Heating

Heating energy can be saved using methods similar to those described above for cooling. For example, heating thermostats can be set down to 20 °C (68 °F) during the day and to 13 to 15 °C (55 to 60 °F) at night. *Caution: when setting heating thermostats back, be sure that controls are interlocked so air conditioning is not used to reach the lower settings.*

Unoccupied spaces should not be heated, cold air infiltration should be minimized, equipment should be properly maintained and adjusted, inside air should be recycled to the extent permissible, and heat recovery techniques should be used where feasible. Chapter 8 discusses these concepts in detail.

- Ventilation

Air flows should be reviewed and compared with code requirements. If excessive they should be reduced.

3. Demand Control

In most large commercial buildings demand charges are an important part (20 to 50 percent) of the electricity bill. It pays to give some thought to the operation of large motorized equipment to prevent situations in which high demands result for short periods of time. Usually such operations can be deferred to reduce the peak demand. This can be done manually or with computers. (See Chapter 5 for a discussion of computer control systems.)

By careful scheduling of electrical equipment, additional energy and cost savings can be obtained. For example, shutting off air-conditioning systems at night (except for the case when the following day is anticipated to be exceptionally hot) and on weekends saves energy. On hot days it pays to pre-cool the building in the early morning hours and to let the building temperature rise in the afternoon; air-conditioning equipment operates most efficiently when the outside temperatures are lowest. This strategy also avoids large electrical loads during peak demand hours and therefore will lower electrical demand costs for the building. Reducing the number of required air changes in those parts of the building where feasible also cuts down on fan power and on heating and cooling loads for the building. Fans and pumps for air-handling equipment in an air conditioning system may account for as much as 50 percent of the total energy used in the system (the other 50 percent is for the refrigeration cycle when cooling is required).[35]

Rescheduling the operation of all large motors to avoid peak demand periods may not necessarily save in building operating energy, but it can ease the requirement for peak utility generating capacity. It also represents a cost saving for building operators. For instance, domestic water pumps can be scheduled for night time, freezer motors can be shut down during peak periods provided the doors are kept closed, dehumidifiers can be shut off at the same time, air conditioners can be turned on a little earlier and run at partial load all day to avoid morning peak power demands (see Figure 3.2) and if different chillers are used during different portions of the day, their operation should not overlap.[36]

4. Lighting

A whole series of operational strategies have been used successfully to reduce lighting energy use while maintaining lighting quality. Included are task-oriented lighting, better controls, more efficient lamps and luminaires, use of lighter paints and wall surfaces, and use of natural light where practical. In some cases conservative design procedures lead to higher-than-required illumination levels. This can be established by measurements and corrected by delamping, relamping, use of low-wattage, or "phantom" tubes, etc. Ballasts are significant (10 to 20 percent of lamp load) and should be disconnected if lamps are removed. Lamp maintenance is another important energy economy measure. Merely cleaning lamp fixtures can increase light output by 10 to 20 percent, reducing the number of lamps needed. Reducing lighting energy use may ease air-conditioning requirements, providing further savings. Reduction of heat input, however, may have adverse effects during cold weather if supplementary heating is then required.

See Chapter 9 for detailed recommendations concerning lighting systems.

5. Other Equipment (Water Heaters, Refrigerators, Elevators, etc.)

Other major energy uses in commercial buildings include water heating, appliances, and equipment such as typewriters, duplicating machines, refrigerators, and elevators. While less important than environmental conditioning, these energy uses also reflect opportunities for significant savings.

Water heaters and other appliances or equipment which are continually operating should be reviewed to establish the feasibility of turning them off during evening or weekend hours. Operational procedures for using such types of equipment should also be reviewed to establish whether energy savings can be accomplished. Examples include reducing the number of equipment units in service and increasing the use factor for each one, reviewing heating-cooling requirements or equipment for possible heat recovery or integration into building energy requirements, and providing coverings or seals on devices such as refrigerators and freezers to reduce heat losses and waste.

In commercial buildings with large expanses of open freezers or refrigerated display cases, an excessive use of energy sometimes results. Energy is used to operate the refrigeration cycle; then additional energy is required to heat the building which has been cooled by leakage from the refrigeration system. Covers, air curtains, or other thermal barriers can help mitigate this problem. Also, commercially available heat recovery systems recover heat from the refrigeration cycle for use in space heating.

Elevator energy usage can be reduced by insuring that elevators operate fully loaded and are not used for short hauls. Practical methods include personnel training, signs, and controls which permit the number of operational elevators to be varied in accordance with the expected

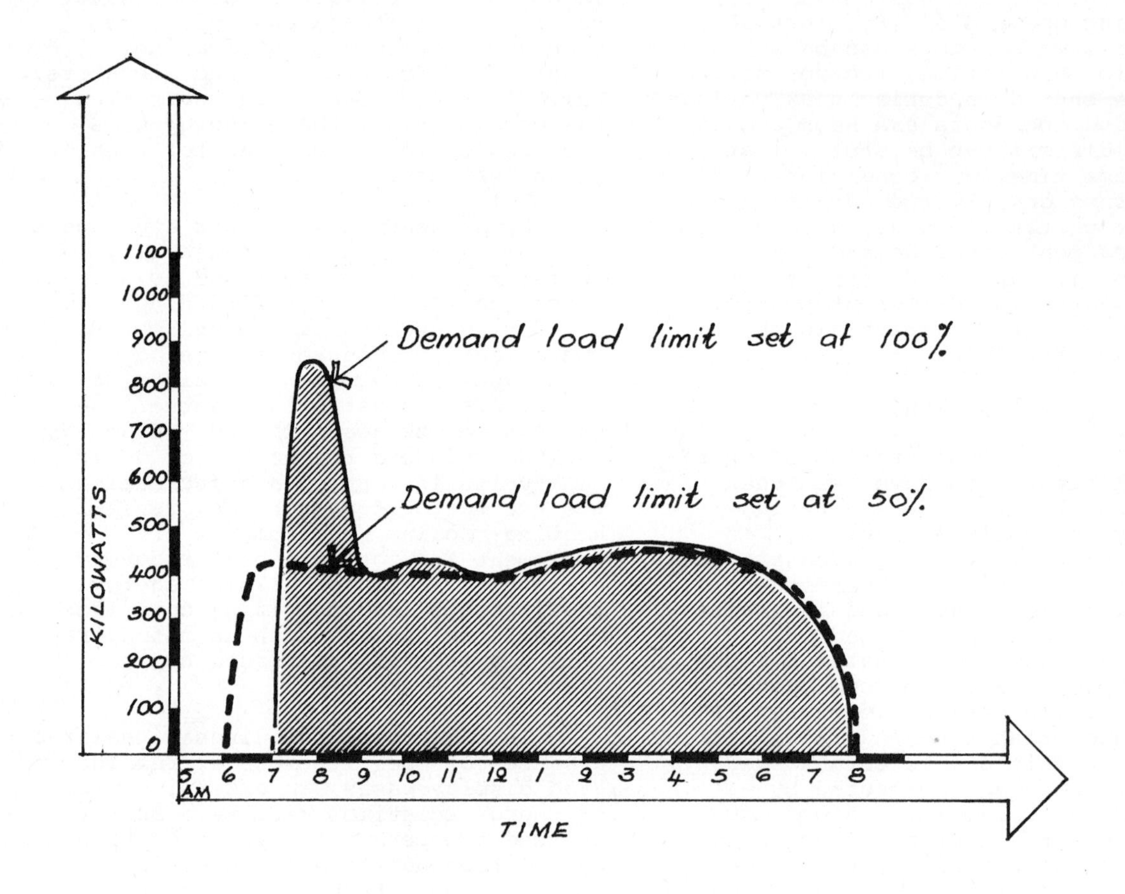

Reference 34

USE LOAD LIMITING CONTROLS TO REDUCE PEAK DEMAND

FIGURE 3.2

passenger load. Note that building codes and fire regulations must be observed.

C. Modifications and Retrofitting

Suggestions in this subsection deal with changes which can be made to existing buildings but which involve some additional capital cost. Some of these changes result in immediate payoffs, while others are long-term investments which require evaluation in terms of life cycle costing in order to select the optimum alternative. Not all of these suggestions are suitable for all buildings and the payoff for each suggestion will vary from building to building depending on the type of use and the building design. Obviously some of these measures are in conflict with each other, which points out the need for an effective energy management plan before beginning substantial changes to the building. The discussion below centers on those schemes which have the largest potential savings in electrical energy and for which the cost may be retrieved in these savings. These schemes principally affect heating, ventilating, and air-conditioning systems (HVAC) and lighting.

- HVAC System Modifications

  Modification of HVAC systems can range from revision of existing systems all the way to gutting the building interior and installing a new system. The latter approach is being employed in older buildings which are no longer rentable because their interior climate control is not competitive in today's market.

  Modifications to existing HVAC systems include, for example, the improvement of existing controls, dampers, by-pass ducts, and peripheral zone heat transfer characteristics; the selective reduction of exhaust air volumes; and the use of air filters to remove odors and dust so that more interior air can be recirculated.

  A major possibility involves installation of heat recovery equipment. (Refer to Chapters 8 and 10). Waste heat in exhaust air can be recovered by use of heat wheels, run-around systems, heat pipes, static heat exchangers, and other types of equipment.

- Building Envelope Modifications

  The building envelope (windows, walls, doors, roof, etc.) is an important source of heat loss and gain. These can be reduced by added weather stripping, insulation, shading, double glazing, and other techniques described in Chapter 7.

- Lighting System Modifications

  The most important changes involve task lighting, more efficient lamps and luminaires, and improved controls.

  If lighting systems are to be revised, relocatable luminaires are available for some ceiling types. These permit installation of luminaires in those locations where illumination is required. As needs change (desks or equipment are moved, etc.) these devices can be moved to a new location. Thus, uniform illumination throughout the building is not needed.

  More efficient lamps and luminaires permit achievement of equivalent illumination levels with less electricity. There can be other advantages as well. Incandescent systems should be replaced with fluorescent or high intensity discharge lamps wherever possible. Besides the energy savings, there can be a substantial reduction in maintenance costs with the much longer-lived lamps (10,000 hours vs 1,000 hours).

  Better controls permit savings by allowing only those lamps which are required to be used at any specified time.

For example, zoned switching (rather than turning on entire floors at once) will lead to substantial savings when the building is only partially occupied.

Refer to Chapter 9 for a more detailed description of lighting system improvements.

D. New Designs

Many options exist for designing more efficient use of energy in new buildings and a number of authors have addressed this question.[37,38,39] New buildings offer the possibility not only of saving on energy and operational costs but of saving on the energy expended in the construction of the building itself. The overriding theme which appears in the literature dealing with this subject is the need for integration of the site, the building, and its intended use (on this, see Chapter 13). Selection of the site and orientation of the building on the site affects not only the cost and energy required to construct the building but the energy expended in operating the building both directly and indirectly. The orientation of the building with respect to the site will affect the heating and cooling requirements directly. Site layout and location of the site relative to occupant residential areas will affect the transportation energy expended in commuting to and from work indirectly.

Construction of the building itself has a variety of options ranging from brick and mortar to glass and steel. In some cases there are trade-offs or alternatives among the materials considered which are decided on bases other than economics, although this is the usual criterion. Another concern is the esthetics of the building. This suggests the possibility that esthetic considerations be met with the most energy-efficient material, since many materials in common use in the construction industry today are energy intensive. Careful planning and layout of the building itself relative to functional work spaces, location of equipment, services within the building, orientation of corridors, elevators, stairs, windows, and finally environmental conditioning systems can affect total energy use. Optimization of these design parameters requires a synthesis of building functions and the application of sophisticated methods of design and analysis. Preferably these methods should be dynamic; that is, they should be capable of taking into account variations in time, both daily and seasonally, to provide a more accurate representation of the needs of the building and its occupants.

Besides conventional energy sources in use today, future buildings will see an increased use of alternate energy sources such as solar or wind. Prototype systems are in development now and in some cases have been installed in experimental buildings.[40]

Another development which may be anticipated is that in the future, building codes and design standards will tend to be more performance-based rather than stipulating design practices. Thus, the building designer will be required to achieve a certain degree of performance for the energy available; the means for achieving this will be left up to the designer.

For example, in the United States the American Society of Heating, Refrigerating, and Air Conditioning Engineers has recently published ASHRAE Standard 90-75, "Energy Conservation in New Building Design." This standard spells out minimum thermal performance standards for buildings and describes analysis procedures. It has been adopted by several states and others are in the process of modifying it for their use.

Widespread use of ASHRAE 90-75 will increase commercial building energy use efficiency. In a study of five building types, it was found that application of the standard (compared to conventional designs) led to lower building costs in addition to saving energy.[41,42] Although building envelope costs were greater because of the standard, lighting and HVAC system costs were reduced, resulting in savings. In the study, energy savings on an office building, school building, apartment, and retail store ranged from 40 to 60 percent compared to conventional structures. Reductions in space heating resulted in 60 to 75 percent of the savings.

E. Government Buildings

Government and public buildings constitute a significant fraction of commercial buildings. Executive departments and agencies occupy and control approximately 2.3×10^8 m^2 (2.5×10^9 ft^2) of building space. Eight percent of this space, or about 19×10^6 m^2 (200×10^6 ft), is controlled by the Public Building Service of the GSA.[43

Public buildings provide a means whereby the government may set an example in improved energy use efficiency in buildings and where new ideas can be tested and new technical options developed and proven. In the US the GSA is responsible for many publicly owned or leased buildings ranging from one story in height to more than 25 and having usable floor areas ranging from 2,000 m^2 to 200,000 m^2. The Federal Government has instituted a number of programs aimed at encouraging efficient energy use in government facilities. A government energy conservation group has been created involving participation by a number of federal agencies with leadership provided by the GSA. Energy management programs which have been developed include measures to reduce electricity use, heating, air conditioning, lighting, and automobile transportation. Plans have been developed for energy reduction programs in the event of power shortages or outages. The government has also been active in developing contingency plans for state and local governments. The Director of the Office of Emergency Preparedness has distributed these plans to governors and mayors and the assistance of GSA field representatives has been offered to help implement the plans.[44,45]

The new GSA policies provide for maximum illumination levels of 540 lux (50 fc) at work stations. Uniform lighting policies are established which provide for lower light levels in non-working areas. As a general guide a maximum of 10 watts per square meter (1 watt per square foot) may be sufficient.[46] However, additional lighting may be needed for more difficult visual tasks. It should be noted that the above is the position of the GSA; however, the actual minimum requirements are in debate among experts (see Chapter 9).

The GSA directives require that energy use for cooling government-owned and leased space be reduced. During the seasonably hot months air cooling systems are to be held at not lower than 26 to 27 °C (78 to 80 °F) during working hours. Necessary adjustments are to be made to cooling system controls so that the temperature in the space is maintained at these values with no reheat. Requirements for humidity control and special types of space or locations are to be handled on a case-by-case basis by officials responsible for operation and maintenance of the facility with the concurrence of the GSA Energy Conservation Coordinator.

During seasonably cold months heating temperature control devices are to be set to maintain temperatures of 18 to 20 °C (65 to 68 °F) during working hours and are to be set to maintain temperatures of not more than 13 °C (55 °F) during non-working hours. Temperatures in warehouses and similar space during working hours are to be adjusted lower than the 18 to 20 °C (65 to 68 °F) range depending upon the type of occupancy and activity in the space. Cooling energy is specifically prohibited from use to achieve temperatures for heating. Further measures have been proposed to cut down heat losses and improve efficiency; these are similar to measures previously discussed in this chapter, however, and will not be described here.

The GSA has also imposed a requirement for energy usage analyses for each government-owned building. These are to be completed and submitted for each building type for the entire fiscal year showing energy usage on a monthly basis by type of energy category.

As a target for its new office buildings, the GSA has established an Energy Use Performance Factor (EUPF) of 625 $MJ/m^2 \cdot yr$ (55 kBtu/ $ft^2 \cdot yr$) and one of 850 $MJ/m^2 \cdot yr$ (75 $kBtu/ft^2 \cdot yr$) for existing buildings. This is an ambitious (but not impossible) goal, as can be seen by comparison with Table 3.6.

To implement these measures, the GSA, in cooperation with other government agencies such as the National Bureau of Standards (NBS), has sponsored studies and research programs on methods and technology which can be used to secure energy savings in the design, construction, and operation of commercial buildings. One report describes opportunities for efficient energy use in building siting, construction, and planning. Also discussed are design of ventilation and infiltration, planning of heating and air-conditioning systems, lighting and power, transportation in buildings, hot and cold water systems, solid waste management, and operation and maintenance. Over 150 options for using energy more efficiently are described.[47] The FEA has prepared a manual containing an extensive discussion of energy conservation options in existing buildings.[48]

F. Summary

Table 3.8 presents an overall summary outlining some of the ideas developed in this chapter. The table indicates major options available to building users and designers for achieving more efficient use of energy.

3.5 POTENTIAL ENERGY SAVINGS

In round numbers 1973-74 energy use in the commercial sector was 10^{10} GJ/yr, of which approximately 4 x 10^{11} kWhe/yr was electricity. The electricity figure corresponds to 1.4 x 10^9 GJ/yr. Based on the techniques described in this chapter, it is estimated that savings in total energy for space heating, water heating, cooking, and other uses could reach:

- *immediate*--operational and housekeeping changes, 10 to 15 percent;
- *near-term*--some investment and process changes, 10 to 15 percent; and,
- *long-term*--major investments and process and equipment changes, 10 to 20 percent.

Total potential savings by the year 2000 would therefore be in the range of 30 to 50 percent.

Others predict long-term total savings in the US commercial sector of 15 to 20 percent.[39,49] In the short-term, savings in the range of 1 to 3 percent of US total energy use could be achieved.[33]

For a 1973-74 base year, commerce used approximately 15 percent of US energy. Thus the potential savings in the commercial sector alone (30 to 50 percent) amount to 4 to 8 percent of national energy use in the year 2000, if the assumption is made that commerce's share of total energy use remains constant at 15 percent.

The potential savings of commercial electricity have also been estimated:

- *immediate*--operational and housekeeping changes, 5 to 10 percent;
- *near-term*--some investments and process changes, 5 to 10 percent; and,
- *long-term*--major investments and process and equipment changes, 10 to 20 percent.

These estimates are based on information presented in the chapter and case studies and include possible improvements in air conditioning, refrigeration, lighting, and other electricity uses. Chapter 1 has a more complete discussion of the potential savings due to increased use efficiency.

Many factors could alter these potential savings. For example, if 10 to 20 percent of the commercial heating load by the year 2000 is assumed to shift from fuel to electricity, the effect of improved electricity use efficiency would be nearly cancelled (see Chapter 1). Shifts of energy demand from other

Option	Operations	Modifications, Renovation, and Retrofitting
1. Energy management	Management concern; energy audits, fuel use, wastes, computer simulation.	Return on investment, payout period, economic studies, life cycle costing.
2. Environmental conditioning • General	Turn off systems when not in use; disconnect unoccupied areas; reduce usage in infrequently occupied areas; reduce nighttime use.	Improve existing systems or install new ones; improve existing controls, dampers, ducts; add insulation; reduce air volumes; add filters to permit recirculation.
• Cooling	Raise thermostat settings to 26 °C (78 °F) and humidstat settings to 60% RH; improve zoning, maintain equipment, move or adjust thermostats, balance system; reduce lighting; reduce infiltration; use outside air.	Disconnect terminal reheat when cooling is required; reduce losses through building "skin" by sealing doors and windows, adding low conductivity window frames, reducing losses through windows.
• Heating	Setback thermostats to 20 °C (day) and 13-15 °C (night); reduce ventilation and window losses; limit night heating.	Recover waste heat where possible using heat wheels, run-around systems, heat pipes, and static heat, exchangers; use electric igniters rather than pilot lights; add insulation or heat pumps.
3. Lighting	De-lamp or re-lamp; substitute more efficient lamps; reevaluate light levels; clean fixtures, cool fixtures, avoid unnecessary nighttime use.	Employ light heat recovery by using total return, bleed-off systems, or water-cooled luminaires; use more efficient lamps or luminaires.
4. Other	Schedule equipment use for off-peak hours where possible; turn off non-essential equipment at night and on weekends.	Properly size motors and pumps; change out oversized equipment; improve power factor.

TABLE 3.8

SUMMARY – ENERGY MANAGEMENT OPTIONS FOR COMMERCIAL BUILDINGS

Option	New Design	Government Buildings
1. Energy management	Consider synthesis of building use, planning, site, layout, materials, and design.	Develop plans for energy reduction programs, contingency plans for energy shortages or outages, and building energy use reports.
2. Environmental conditioning • General	Evaluate energy and cost of construction materials; where options exist, use less energy-intensive materials.	Encourage employees to wear warmer clothing, control window and door opening, reduce energy use.
• Cooling	Evaluate additional insulation; better windows, use of shades, awnings, site orientation to optimize heating and cooling; consider waste heat utilization.	Reduce energy for cooling, raise air conditioning temperature minimum to 26-29 °C; eliminate humidity control.
• Heating	Use computer simulation studies to evaluate alternatives; weigh impact of future fuel and electricity prices; use heat recovery techniques.	Reduce energy for heating; setback thermostats to 18-20 °C (maximum) in working spaces; eliminate humidity control; prohibit portable heaters.
3. Lighting	Design appropriate task-oriented lighting systems; optimize for the intended use; use high efficiency lamps and luminaires.	Reduce illumination to acceptable task levels.
4. Other	Investigate alternative energy systems; solar, wind, etc.; recycle wastes for fuel.	Spell out guidelines on transportation (parking and car pools) and on use of electrical equipment in buildings.

Caution: Any action must be considered from a systems point of view, e.g., reduction of lighting levels may impact building heating and air conditioning.

TABLE 3.8
CONTINUED

fuels to electricity can be anticipated if scarcities develop, and this should be considered in planning for commercial energy use.

* * *

REFERENCES

1. US Department of the Interior, Bureau of Mines, "Annual US Energy Use Drops Again," News Release (5 April 1976), see Table 2.

2. US Federal Energy Administration, *Residential and Commercial Energy Use Patterns, 1970-1990*, (Washington, D.C.: November 1974).

3. Rothberg, Joseph E., "Energy Consumption and Conservation in the US," Paper presented at the U.K. Science Research Council Summer School on "Aspects of Energy Conversion." Lincoln College, Oxford, England, 14 July 1975.

4. "The 24th Annual Electrical Industry Forecast," *Electric World*,(15 September 1973): 45-54.

5. Stanford Research Institute, *Patterns of Energy Consumption in the United States*, Report prepared for the Office of Science and Technology, (Menlo Park, California: January 1972).

6. US Office of Emergency Preparedness, *The Potential for Energy Conservation*, A Staff Study, (Washington, D.C.: US Government Printing Office, 1972), pp. J-6, J-18.

7. National Petroleum Council, *Potential for Energy Conservation in the United States: 1974-1978*, (Washington, D.C.: 10 September 1974).

8. McKenzie, J.J. and Osherenko, G., "Regional Patterns and Trends in the Generation and Consumption of Electricity in the U.S.," in *Towards an Energy Policy*, ed: Keith Roberts, (San Francisco, California: The Sierra Club, 1973), pp. 20-34.

9. Dubin-Mindell-Bloome Associates, P.C., (in cooperation with AIA/ Research Corporation and Heery and Heery, Architects), *Energy Conservation Design Guidelines for Office Buildings*, (Washington, D.C.: General Services Administration and Public Buildings Service, January 1974), pp. 4-4ff, 10-1.

10. Miller, A.J., et al., *Use of Steam-Electric Power Plants to Provide Thermal Energy to Urban Areas*, Report No. ORNL-HUD-14, (Oak Ridge, Tennessee: Oak Ridge National Laboratory, January 1971), p. 20.

11. US Federal Energy Administration, *Project Independence Report*, (Washington, D.C.: US Government Printing Office, November 1974), pp. 186-191.

12. Kaplan, S.I., *Energy Demand Patterns of Eleven Major Industries*, Report No. ORNL-TM-4610, (Oak Ridge, Tennessee: Oak Ridge National Laboratory, September 1974), pp. 9-10.

13. Gatts, Robert R.; Massey, Robert G.; and Robertson, John C., *Energy Conservation Program Guide for Industry and Commerce (EPIC)*, NBS Handbook 115, (Washington, D.C.: US Department of Commerce, National Bureau of Standards, September 1974), pp. 2-25, 2-38.

14. National Electrical Contractors Association: *Total Energy Management*, (Washington, D.C.: 1976).

15. Grumman, David L., "Buildings that Conserve Energy-I," *Building Systems Design* , 71(February/March 1974): 10.

16. Smith, C.B. "Energy Management," presented at American Public Power Association 1976 Services and Communication Section Workshop, San Francisco, October 20, 1976.

17. Stanford Research Institute, *Patterns of Energy Consumption*, p. 74.

18. Snyder, W.T.; Miller, W.A.; and Schmidt-Bleek, F., *How to Cut the Energy Budget in Business and Industry, Guidelines and Suggestions*, (Knoxville, Tennessee: University of Tennessee Environment Center and Center for Industrial Services, February 1974), p. 40.

19. US Department of Commerce, National Bureau of Standards, *Technical Options for Energy Conservation in Buildings*, NBS Technical Note 789, (Washington, D.C.: July 1973).

20. Kasuda, T., ed., *Use of Computers for Environmental Engineering Related to Buildings*, Proceedings of a Symposium held at the National Bureau of Standards, Gaithersburg, Maryland, November 30-December 2, 1970, (Washington, D.C.: National Bureau of Standards, Buildings Science Series No. 39, October 1971).

21. Edison Electric Institute, *AXCESS Energy Analysis Reference Manual*, (New York: Edison Electric Institute, n.d.).

22. American Gas Association, *E-Cube (Energy Conservation Utilizing Better Engineering)*, (Arlington, Virginia: American Gas Association, 1975).

23. Trane Company, *TRACE (Trane Air Conditioning Economics), A Computer Program of the Trane Company*, (La Crosse, Wisconsin: 1973).

24. Kusada, Tamami, *NBSLD, The Computer Program for Heating and Cooling Loads in Buildings*, (Washington, D.C.:US Government Printing Office, 1977).

25. American Society of Heating, Refrigerating and Air Conditioning Engineers, Inc., *ASHRAE Handbook of Fundamentals, ASHRAE Guide and Data Book - Applications, ASHRAE Guide and Data Book - Systems, ASHRAE Guide and Data Book - Equipment*, (New York: 1971 +).

26. Kaufman, J.E., ed., *IES Lighting Handbook*, 5th edition, (New York: Illuminating Engineering Society, 1972).

27. Dubin-Mindell-Bloome Associates, P.C., *Energy Conservation Design Guidelines for Office Buildings*.

28. General Services Administration, Public Buildings Service, *Energy Conservation Guidelines for Existing Office Buildings*, (Washington, D.C.: February 1975).

29. General Services Administration, Public Buildings Service, *Conservation of Utilities*, (Washington, D.C.: 1974).

30. US Department of Commerce, National Bureau of Standards, and General Services Administration, *$E=MC^2$: Roundtable on Energy Conservation in Public Buildings*, (Washington, D.C.: July 1972).

31. US Department of Commerce, *Energy Conservation Handbook for Light Industries and Commercial Buildings*, (Washington, D.C.: May 1974).

32. US Department of Commerce, National Bureau of Standards, *Technical Options*.

33. US Department of Commerce, National Bureau of Standards, and General Services Administration, *$E = MC^2$: Roundtable on Energy Conservation*, p.56.

34. General Services Administration, Public Buildings Service, *Energy Conservation Guidelines for Existing Office Buildings*, pp. 3-5, 3-11.

35. Dubin-Mindell-Bloome Associates, P.C., *Energy Conservation Design Guidelines for Office Buildings*, p. 9-31.

36. US Department of Commerce, National Bureau of Standards, and General Services Administration, *$E = MC^2$ Roundtable on Energy Conservation*, p. 57.

37. Dubin-Mindell-Bloome Associates, P.C., *Energy Conservation Design Guidelines for Office Buildings.*

38. Dubin, F.S.; Mindell, H.S.; and Bloome, S., *How to Save Energy and Cut Costs in Existing Industrial and Commercial Buildings: An Energy Conservation Manual,* (Park Ridge, New Jersey: Noyes Data Corporation, 1977).

39. "Energy Management 1977," *Professional Engineer* (Special Insert) 47 (February 1977): 31-42.

40. Kreider, Jan F. and Kreith, Frank, *Solar Heating and Cooling: Engineering, Practical Design and Economics,* (Washington, D.C.: Hemisphere Publishing Corporation, 1975).

41. "ASHRAE Standard 90-75 Seen to Affect Design Engineer, Building Product Industry," *Professional Engineer,* 46 (February 1976): 35-37.

42. Patterson, N.R. and Alwin, James B. "The Impact of Standard 90-75 on High Rise Office Building Energy and Economics," *Heating/Piping/Air Conditioning,* (January 1976): 38-44.

43. General Services Administration, *Public Buildings and Space: Action Plan for Conservation of Electricity,* Bulletin FPMR D-91, (Washington, D.C.: 16 May 1972); and General Services Administration, *Public Buildings and Space: Conservation of Energy, Particularly Heating Fuel, in Public Buildings,* Bulletin FPMR D-101, (Washington, D.C.: 27 November 1973).

44. General Services Administration, Office of Federal Management Policy, *Federal Management Circular: Federal Energy Conservation,* FMC 74-1, (Washington, D.C.: 21 January 1974).

45. US Office of Emergency Preparedness, *The Potential for Energy Conservation,* Appendix H.

46. General Services Administration, *Conservation of Energy, Particularly Fuel Oil, in GSA-Operated Buildings,* GSA Order PBS P 5800.34, (Washington, D.C.: 28 November 1973).

47. Dubin-Mindell-Bloome Associates, P.C., *Energy Conservation Design Guidelines for Office Buildings.*

48. Dubin, Mindell, and Bloome, *How to Save Energy and Cut Costs.*

49. US Federal Energy Administration, Office of Conservation and Environment, *Lighting and Thermal Operations: Energy Management Action Program for Commercial, Public, Industrial Buildings; Building Energy Reports, Case Studies,* (Washington, D.C.: 1974).

CHAPTER 4

RESIDENCES

T.T. Woodson*

CHAPTER CONTENTS

KEY WORDS

Air Conditioning
Cooking
Cost/Benefit Analysis
Energy Management
Entertainment Use
Environmental Conditioning
Household Appliances
Illumination Practices
Laundry
Product Modification
Refrigeration
Residence Modification
Residential Use
Water Heating

SUMMARY

Energy use in the residential sector accounts for about 20 percent of total US energy use. Of this, natural gas represents 40 percent; fuel oil represents 27 percent; and electricity accounts for the remaining 33 percent.[1] These statistics are based on the primary fuels required. Four-fifths of the energy input to the residential sector is used as either fossil fuel or electric heating. The remaining fifth generates electricity which is used for mechanical drives and lighting. The average per capita direct residential energy use throughout the country is 60-70 GJ/person-year.

A breakdown of this direct residential energy use for United States residences in 1968 is found in Reference 2:

Space heating	57.5%
Water heating	14.9%
Food refrigeration	6.0%
Cooking	5.5%
Air conditioning	3.7%
Lighting	3.5%
Television	3.0%
Food freezing	1.9%
Clothes drying	1.7%
Other	2.3%
	100.0%

Energy use in this sector could be reduced significantly by increasing thermal insulation, reducing infiltration, installing more efficient lamps, modifying appliance operation methods, and improving understanding of energy use facts.

This chapter discusses changes which apply to the following:

*Director, Engineering Clinic, Harvey Mudd College

- *operating strategies* which the *user* can readily apply to reduce energy demand significantly;
- *on-site modifications* which the *owner* or *lessor* may undertake practically, resulting in energy and money savings; and,
- *original equipment redesign* which the *manufacturer* or *builder* may develop, leading to appreciable savings in operating cost, in total life cycle costs, and possibly in first cost.

The major actions suggested within the above headings are:

- control settings, schedules of use, ventilation and shading, maintenance, cost comparisons, appropriate lamp selection, careful purchasing, clothing adaptation, color choice, cold water laundry;
- added insulation, weather stripping, duct and piping changes, fluorescent lamps, sector light switching, attic ventilation, other thermal insulation, awnings, pilot light changes, evaluation of central versus room air conditioning, solar pool and water heating, tree planting, roof changes, other site considerations; and,
- substitution of less energy-intensive materials, designs promoting material salvage, mechanism efficiency data, tabulation of available principles, scanning lists of suggestions, process alternatives, and revision of present methods of materials utilization.

Reductions in residential energy use through the changes described in this chapter are estimated to be: 5 to 10 percent immediately; 10 to 15 percent over the near term (2 to 5 years); and an additional 10 to 20 percent over the long term (5 to 25 years). For the year 2000, annual savings in the range of 25 to 45 percent of the residential energy base case appear economically and technically feasible.

4.1 INTRODUCTION AND OVERVIEW

This chapter presents the basic data and methodology necessary to improve energy use efficiency in residences. The suggestions included are derived from many sources--government, utilities, institutions, and the private sector, and range from simple ideas to sophisticated analyses. They have been segregated by potential audience.

The primary agent for conservation varies from topic to topic. Common goals that unite them are (1) prudently designed equipment with salvageable material utilization; (2) economical manufacture and compatible distribution; (3) efficient design and installation; (4) informed operation; (5) equitable energy costs and supply; and (6) overall legislation balancing the interests of consumer, producer, nation, and the environment.

In addition to the applications , design, sales or service engineer, the following people should find interest in this chapter:

apartment manager
appliance service man
building inspector
householder
landlord
legislators and staff
plumber, electrician, etc.
utility representative.

4.2 OBJECTIVE, SCOPE, ASSUMPTIONS

A. Objective, Scope

The objective of this chapter is to present background information and methodologies for more efficient residential energy use. The material included applies primarily to residential electricity and secondly to other energy forms (as gas, oil, etc.). All direct-energy using household equipment and their installations are considered.

The median home of interest typically has three bedrooms, uses 7,000 to 10,000 kWh/yr and has an installed investment in electrical and gas equipment of $3000 to $4000. Domiciles range, of course, from one-room apartments and mobile homes to conventional houses, condominiums, and mansions.

Only direct energy uses in and around the home are included in this chapter. Excluded is the energy embodied in food, clothing, utensils, rugs, furniture, interior furnishings, maintenance supplies, tools, packaging and the like. Those are defined as indirect energy users and are treated separately in Chapters 2, 3, and 6.

For the owner or builder, information is presented to aid in planning economizing changes and modifications, especially in the important area of space heating/cooling.

B. Assumptions

Consumer skills assumed are those of the average urban or suburban family members with little technical background (for operating strategies and modifications) while "designers" (for manufacturing redesign) are assumed to be product engineers or the equivalent. Public officials and lawyers working with engineers will be those chiefly concerned with legal policy alternatives.

The constraints governing an on-site modification are primarily economic (will it pay off in one or two years?) and practical (can changes be made by an electrician or carpenter or by the owner?). The constraints limiting original equipment design changes are primarily:

- continuation of the rule of economics in design, manufacture, sale, use, and salvage;
- use of non-exotic materials or processes;
- retention of generally standard manufacturing methods, although possibly with some changed concepts of fabrication and waste recovery;
- conformity to current US safety standards and legal codes; and,
- the heretofore-used economics of manufacture (material, labor, overhead, etc.) modified by factors accounting for salvage, scarcity, previous energy-to-produce, and environmental impact.

Data for the several residential fuels (electricity, gas, petroleum, etc.) are cited separately. Infrequently, they have been combined and the data appropriately footnoted.

The American National Standards Institute (ANSI), Underwriters Laboratories (National Electric Code), and American Gas Association standards or practices have been assumed, and the usages and recommendations of the various trade and manufacturing associations (NEMA, AHAM, EEI, etc.) are noted.

4.3 USEFUL FACTS AND BACKGROUND

A. General

The following data are included for several reasons: they give a general picture of residential energy growth rates; they provide a factual base for projections; and they serve to suggest the variety of information sources available.

Historical patterns of residential electricity use are shown in Table 4.1. Note that per capita use in 1970 is almost five times what it was in 1950. The economics of energy use (sub-section B) plus the general population growth explain the significant increase in residential electricity sales in recent years. Further detail is provided by Table 4.2. This table illustrates the increasing saturation of energy intensive appliances in US households. Table 4.3 applies this information to summarize residential energy use for all major appliances. Between 1950 and 1970, energy use by each appliance increased an

	1950	1960	1970
Total electricity sales (10^9 kWh)	281	683	1391
Residential electricity sales (10^9 kWh)	70.1	196	448
Residential sales/total sales	24.9%	28.7%	32.2%
Resident population (10^6)	152	180	204
Wired households (10^6)	38.9	51.4	64.0
Residential sales per capita (kWh)	461	1089	2196
Residential sales per household (kWh)	1800	3820	7000
Average size of household	3.37	3.33	3.17

Reference 3

ELECTRICITY SALES, POPULATION, AND WIRED HOUSEHOLDS

TABLE 4.1

	Saturation (%)			Average annual use in households having the appliance (kWh/household)		
	1950	1960	1970	1950	1960	1970
Refrigerators	82.8	98.1	99.8	345	780	1,300
Air Conditioning						
Room	0.8	10.9	26.5	1,402	1,663	1,946
Central	0.1	1.9	11.3	3,560	3,560	3,560
Lighting	100.0	100.0	100.0	500	600	750
Space Heating	0.7	1.8	7.6	10,000	12,945	14,588
Water Heating	10.9	21.0	25.2	3,675	4,010	4,500
Clothes Drying	1.05	12.3	29.1	520	935	993
Ranges	16.2	31.7	40.3	1,250	1,225	1,175
Television	12.9	90.1	94.7	290	335	417
Food Freezers	6.2	19.0	28.0	620	888	1,384
Clothes Washers	73.6	75.9	70.5	45	60	363
Dishwashers	1.7	6.7	18.9	120	347	363
Irons	79.8	88.5	99.6	110	132	144

Reference 3

APPLIANCE SATURATION AND AVERAGE ANNUAL ELECTRICITY USE

TABLE 4.2

	Total Usage[a] (billions of kWh)			Total Usage (%)		
	1950	1960	1970	1950	1960	1970
Total Residential Sales	70.1	196.4	447.8	100.0	100.0	100.0
Refrigerators	11.1	39.3	82.9	15.8	20.1	18.5
Air Conditioning						
Room	0.4	9.3	32.9	0.6	4.7	7.3
Central	0.1	3.6	25.8	0.2	1.8	5.7
Lighting	19.5	30.8	48.0	27.8	15.7	10.7
Space Heating	2.8	12.1	71.2	3.9	6.2	15.9
Water Heating	15.6	43.3	72.5	22.3	22.1	16.2
Clothes Drying	0.2	5.9	18.5	0.3	3.0	4.1
Cooking	7.9	20.0	30.3	11.2	10.2	6.8
Television	1.5	15.5	25.3	2.1	7.9	5.6
Food Freezers	1.7	10.7	27.7	2.5	5.5	6.2
Other (Clothes Washers, Dishwashers, Irons, Radios, etc.)	9.3	5.9	12.7	13.2	3.0	2.8

[a]Calculated using the relationship:

(electricity consumption per appliance) = (saturation) x (No. of wired households) x (average electricity use per appliance)

with the number of wired households as 38.9×10^6 in 1950 51.4×10^6 in 1960 and 64.0×10^6 in 1970.

Reference 3

SUMMARY: RESIDENTIAL USAGE OF ELECTRICITY

TABLE 4.3

average of 50 percent. Total electricity use due to these appliances, however, increased by a factor of six.

B. Economic Base

Consumer economics fueling this increase in energy use between 1950 and 1970 can be seen in Figure 4.1. Disposable income, electricity cost and appliance price drops combined to yield an increasingly favorable atmosphere for owning appliances and using electricity.

Energy used in US households varies considerably with income. This is illustrated in a study by the Washington Center for Metropolitan Studies in 1973. Figure 4.2 shows household electricity, natural gas and gasoline use by income group. Differences are smallest in the use of natural gas which is used for necessities such as cooking and space and water heating. Electricity use shows a greater difference as non-essential appliances such as air conditioners and automatic dishwashers become more common. Gasoline usage shows the largest difference among income groups.

Table 4.4 indicates the cost of this direct energy use relative to family incomes. Although the "poor" use less, they pay a higher percentage of their total income for energy. The "well off" spend roughly 4 percent of their income on energy, while the "poor" pay in the range of 15 percent. This implies that the benefits of energy efficiency will be comparatively greater for those families which earn less money. These are the people, however, who can least afford improvements in their homes.

C. Basic Facts Related to Operations, On-Site Modifications and Re-Design

This sub-section presents a range of information expected to be of possible conservation value to the user, the owner, and the design engineer. The user is either seeking energy savings per se or striving simply for an optimum cost/benefit balance with existing appliances, furnishings, and structures. Such a user's options are really comprised of:

- selecting appropriate means to an end (e.g., shades versus storm windows, pots versus ovens);
- scheduling usage (e.g., duration, cycling, time of day);
- controlling an operating variable (e.g., temperature, water quantity, detergent);
- choosing place of use (e.g., laundry versus kitchen versus porch);
- removing from service (e.g., disconnect or turn off equipment.)

Options for the owner (modifier) are:

- replacing installed equipment with new equipment (lower life-time cost);
- improving installation efficiency (e.g., add insulation, relocate, recircuit); and,
- providing emergency or preventive maintenance (e.g., clean heat exchanger, paint).

The re-designers' options are widespread, but they demand unfettered examination of the alternatives and their total life cycle cost. These options are:

- extended design life of the product, beyond that now in practice;
- new or refined operating principles, new materials, new processes;
- inclusion of material salvage criteria;
- simplified manufacturing processes when production machine maintenance will benefit;
- conversion of "disposables" to "re-usables" where economics indicate;
- minimization of environmental impact of manufacture and use; and,

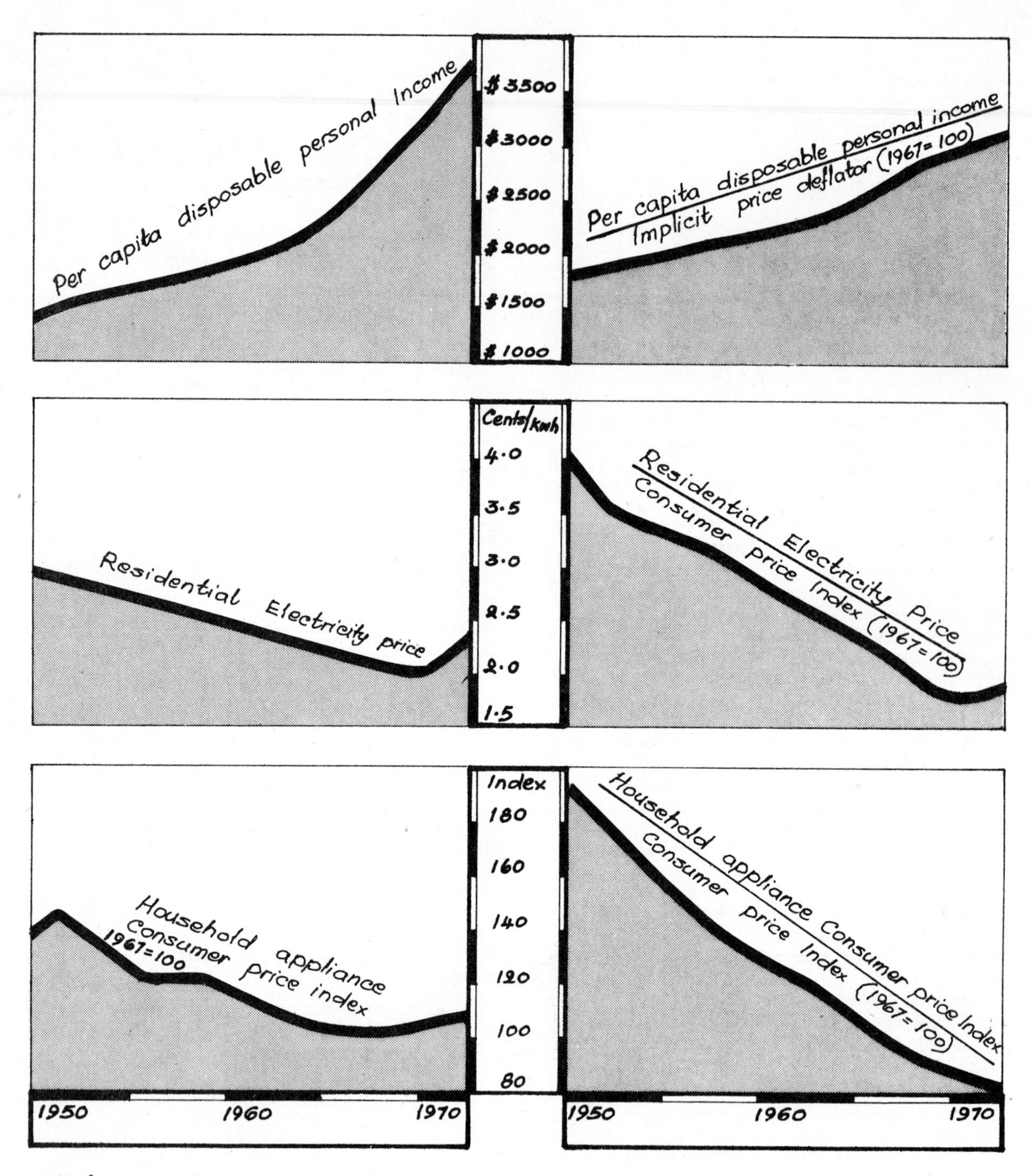

Reference 3

ECONOMIC FACTORS DETERMINING RESIDENTIAL ELECTRICITY USE

FIGURE 4.1

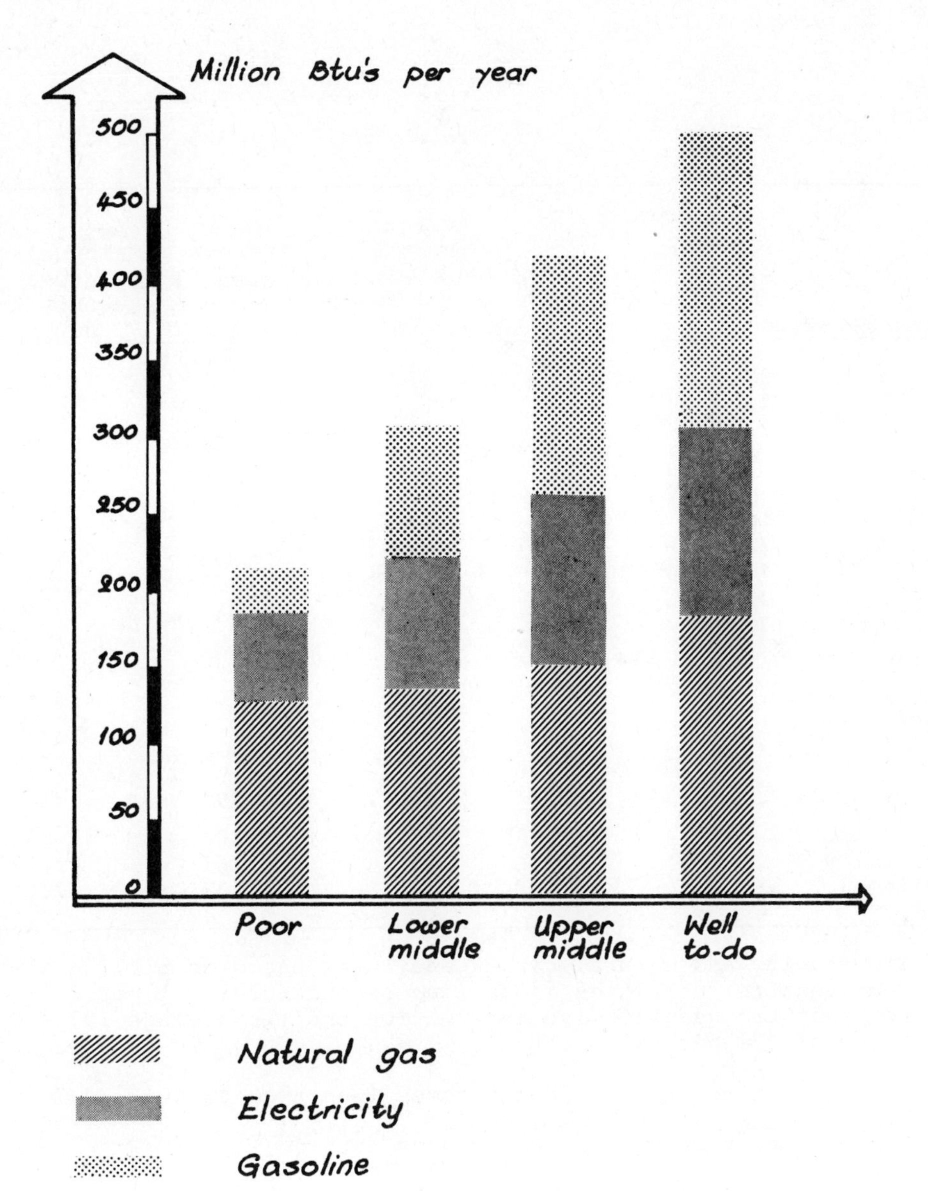

HOUSEHOLD ENERGY USE BY INCOME GROUP

FIGURE 4.2

Income Status	Average Income	Average Annual Btus (Millions per Household)	Average Annual $ Cost per Household	Percent of Total Annual Income Spent on Energy
Poor: Total	$2,500[a]	207	379	15.2
Natural gas		118	147	5.9
Electricity		55	131	5.2
Gasoline		34	101	4.0
Lower middle: Total	$8,000	294	572	7.2
Natural gas		129	153	1.9
Electricity		80	167	2.1
Gasoline		85	252	3.2
Upper middle: Total	$14,000[b]	403	832	5.9
Natural gas		142	166	1.2
Electricity		108	213	1.5
Gasoline		153	453	3.2
Well off: Total	$24,500[b]	478	994	4.1
Natural gas		174	200	.8
Electricity		124	261	1.1
Gasoline		180	533	2.2

Note: Electricity and natural gas expenditures based on billing data received from utilities. Gasoline expenditures estimated from respondents' quantitative information and the average 1972-73 price of 37¢ per gallon.

[a]77 percent of the poor had incomes less than $3,000.

[b]Calculated from unpublished census data

Source: Washington Center for Metropolitan Studies Lifestyle and Energy Survey, 1972-73 (Reference 4)

THE PERCENTAGE OF FAMILY INCOME SPENT ON ENERGY DECLINES AS INCOME INCREASES

TABLE 4.4

• coordination with emerging governmental and industrial performance standards.

Over all of the above options are economic constraints; what is the pay-out period? Have tooling and inventories been taken into account? What are the probable fuel or energy cost changes over the life of the equipment?

The information and data presented below will assist in decisions for some of the many options above. Some references are listed for further research.

D. Appliances

The following pages discuss the energy use of major appliances and changes that can be made to improve their efficiencies. Table 4.5 shows typical energy use by a variety of residential appliances. For a list of modifications which can be made in a residence, refer to Section 4.4, Suggested Economies.

Space Heating

The use of fossil fuels for residential space heating is common throughout most of the US. A diagram of a typical natural gas system efficiency is shown in Figure 4.3. It reveals the major heat loss to be out of the heater's exhaust stack. Uninsulated ducts could cause even greater loss if they are affected by outside weather.

There are some actions which can be taken to improve the efficiency of an already existing gas or oil space heater--such as reducing stack losses by the dampers and igniters; cleaning and maintenance; reducing capacity to match load by orificing, and direct ducting of combustion air. Most effort should go toward improving insulation and decreasing infiltration of the house itself, as discussed in Chapter 7.

Where electric resistance heating is used, it can provide the ability for room-by-room control, possibly reducing total heating requirements. However, modifications may be economically desirable in some cases. A heat pump can supply roughly twice as much heat for the same input as electric stripheaters. (See the case studies in Chapter 15.)

Several facts should be remembered when installing heat pumps. First, because they generate lower temperature air than strip heating, the heat pump and duct work must be properly sized for maximum efficiency. Second, heat pumps can also be used as air conditioners; thus the initial investment is justified not only by heat savings, but by increased comfort during the summer months.

In most areas of the US, a supplemental heater will be needed because heat pump efficiencies drop off at low temperatures. The heat pump can still supply the majority of heat needed, however. For example, in the colder climates of the US, a heat pump using outside air as a heat source can supply roughly two thirds of the seasonal heating needs.

Air Conditioners

Residential air conditioners are gaining popularity in US homes. Their efficiency is measured by the EER (energy efficiency ratio) which is the number of Btu's per hour of cooling resulting from an electrical input of one watt. Typical EER's for modern air conditioners are shown in Table 4.6. The higher the EER, the more efficient the unit.

Although the amount of energy used by air conditioners varies as the "heat load" changes, the EER is only measured at one point--full load. Some new units use multiple speed motors to draw less energy at lower loads.[5] The EER of these multi-speed units at partial load would be improved compared to the value at full load. Since they are designed to operate at low speed for all but a few hours during the year, these air conditioners are more efficient than EER's indicate.

Another variable in the EER of residential air conditioners is the

ELECTRIC APPLIANCES*

	Power (Watts)	Typical Use (kWh/yr)	Typical Use (GJe/yr)
FOOD PREPARATION			
Blender	300	1	0.0036
Broiler	1,140	85	0.31
Carving Knife	92	8	0.03
Coffee Maker	1,200	140	0.50
Deep Fryer	1,448	83	0.30
Dishwasher	1,201	363	1.31
Egg Cooker	516	14	0.05
Frying Pan	1,196	100	0.36
Hot Plate	1,200	90	0.32
Mixer	127	2	0.0072
Oven, microwave (only)	1,450	190	0.68
Range			
with oven	12,200	700	2.52
with self-cleaning oven	12,200	730	2.63
Roaster	1,333	60	0.22
Sandwich Grill	1,161	33	0.12
Toaster	1,146	39	0.14
Trash Compactor	400	50	0.18
Waffle Iron	1,200	20	0.07
Waste Dispenser	445	7	0.03
FOOD PRESERVATION			
Freezer			
Manual Defrost - 16 cu. ft.	--	1,190	4.28
Automatic Defrost - 16.5 cu. ft.	--	1,820	6.55
Refrigerators/Freezers			
manual defrost, 12.5 cu.ft.	--	1,500	5.40
automatic defrost, 17.5 cu. ft.	--	2,250	8.10
LAUNDRY			
Clothes Dryer	4,856	993	3.57
Iron (hand)	1,100	60	0.22
Washing Machine (automatic)	512	103	0.37
Washing Machine (non-automatic)	286	76	0.27
Water Heater	2,475	4,219	15.19
(quick recovery)	4,474	4,811	17.32
HOUSEWARES			
Clock	2	17	0.06
Floor Polisher	305	15	0.05
Sewing Machine	75	11	0.04
Vacuum Cleaner	630	46	0.17
COMFORT CONDITIONING			
Air Cleaner	50	216	0.78
Air Conditioner (room)	860	860*	3.10
Bed Covering	177	147	0.53
Dehumidifier	257	377	1.36
Fan (attic)	370	291	1.05
Fan (circulating)	88	43	0.15
Fan (rollaway)	171	138	0.50
Fan (window)	200	170	0.61
Heater (portable)	1,322	176	0.63
Heating Pad	65	10	0.04
Humidifier	177	163	0.59
HEALTH AND BEAUTY			
Germicidal Lamp	20	141	0.51
Hair Dryer	600	25	0.09
Heat Lamp (infrared)	250	13	0.05
Shaver	15	0.5	0.0018
Sun Lamp	279	16	0.06
Tooth Brush	1.1	1.0	0.0036
Vibrator	40	2	0.0072
HOME ENTERTAINMENT			
Radio	71	86	0.31
Radio/Record Player	109	109	0.39
Television			
black & white			
tube type	100	220	0.79
solid state	45	100	0.36
color			
tube type	240	528	1.90
solid state	145	320	1.15

GAS APPLIANCES

	Typical Use (kft^3/yr)	Typical Use (GJ/yr)
Clothes dryer	5.0	5.3
Furnace	65	69
Gas light	18	19
Pool heater	50-150	53-158
Range	10	11
Water heater	30	32

*Source: Edison Electric Institute

RESIDENTIAL ENERGY USAGE - TYPICAL APPLIANCES

[ELECTRICITY AND GAS]

TABLE 4.5

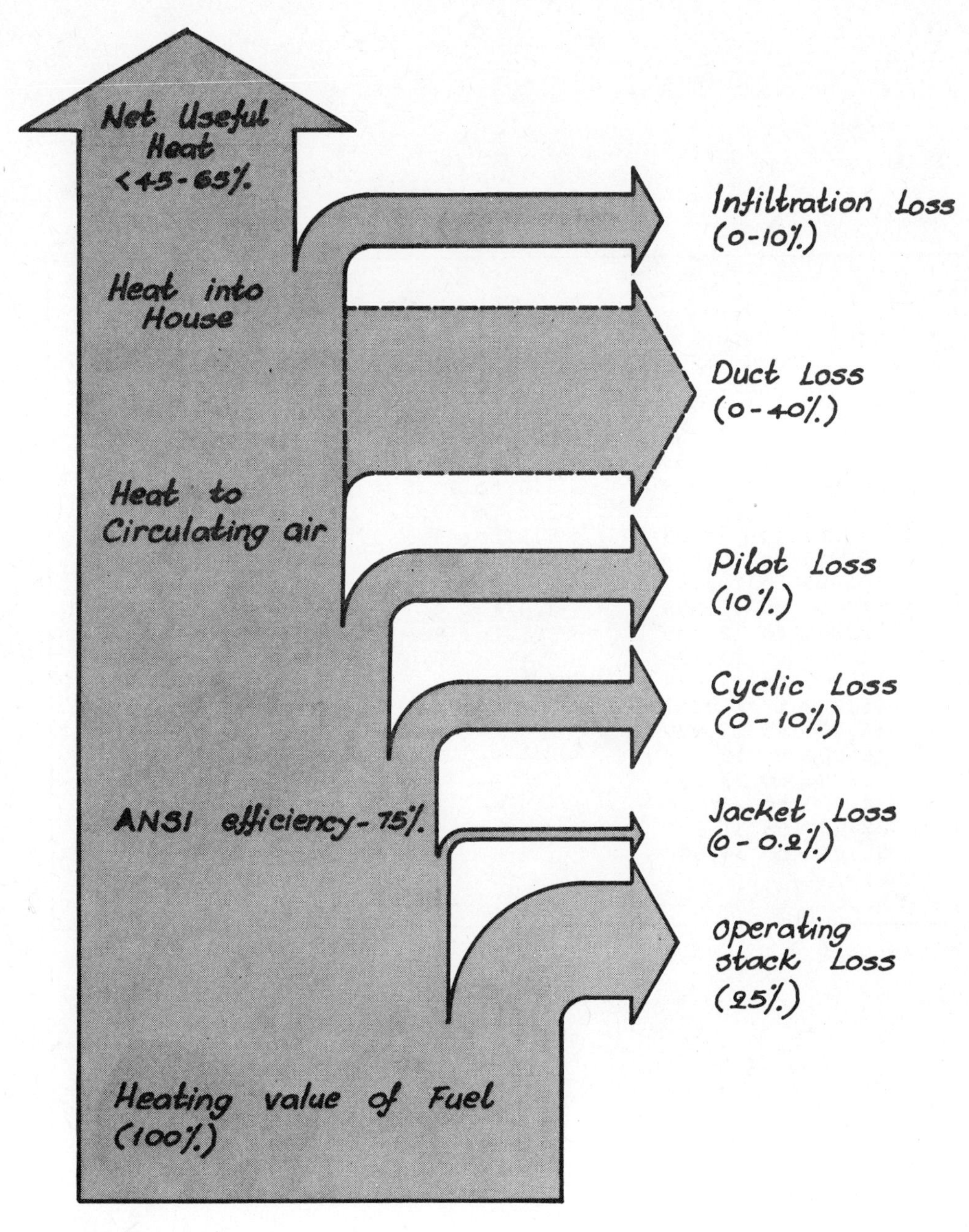

Reference 5

ENERGY FLOW FOR A GAS FURNACE SYSTEM

FIGURE 4.3

Cooling Capacity Range (Btu/hr)	EER Range 115-V Units	EER Range Higher Voltage Units
Up to 4,999	5.2 to 5.4	--
5,000 to 5,999	5.1 to 8.8	--
6,000 to 6,999	5.6 to 10.5	5.4 to 6.1
7,000 to 7,999	5.2 to 9.2	6.3 to 6.3
8,000 to 8,999	5.8 to 9.9	4.9 to 6.7
9,000 to 9,999	6.4 to 11.5	4.8 to 8.0
10,000 to 10,999	6.3 to 12.0	5.3 to 8.0
11,000 to 11,999	8.0 to 8.5	4.7 to 7.4
12,000 to 12,999	8.7 to 9.6	4.7 to 8.3
13,000 to 13,999	9.4 to 10.0	4.4 to 8.5
14,000 to 14,999	10.1 to 10.3	5.1 to 8.0
15,000 to 15,999	--	4.8 to 8.0
16,000 to 17,999	--	5.8 to 8.5
18,000 to 19,999	--	5.8 to 9.3
20,000 to 23,999	--	5.7 to 8.1
24,000 to 27,999	--	6.1 to 7.6
28,000 to 31,999	--	6.0 to 7.8
32,000 to 36,000	--	6.2 to 7.1

Reference 6

EER RANGE CHART FOR AVAILABLE 60-Hz UNITS **TABLE 4.6**

use of automatic distribution fans. Normal fans remain on while the compressor cycles on and off at low loads. This creates a low EER, as shown in the theoretical curves of Figure 4.4, because the fan is using just as much energy while much less cooling is being done. Automatic fans which cycle off with the compressor give a much higher EER at low loads though they may cause stratification and less precise temperature control within the residence.

An energy saving technique adapted from commercial air conditioning is the use of outside air when its enthalpy (or wet bulb temperature) is less than the air recycled from the residence. This can be advantageous in areas of high humidity and relatively cool nighttime temperatures though it is fairly expensive to install. A simple damper can automatically open and close to minimize energy required to cool the air. Most residences have an advantage over commercial buildings in that their windows can be opened to utilize this free cooling directly.

Since building codes require only a small amount of attic ventilation, it may be advantageous to supplement this if the attic is not insulated. A small fan controlled by a thermostat can exhaust the hot air during the summer and retain it during the winter. Heat gain through the ceiling, a significant part of the heat load, is usually reduced 15 to 40 percent through this modification. It should, of course, be combined with adequate ceiling insulation.

Evaporative cooling is a viable alternative to refrigeration in some climates. It can be made more attractive by designs such as the dry evaporative cooler which uses moist, cool air to cool dry air. The results can be a comfortable atmosphere with only a fraction of the energy necessary for refrigerated air conditioning.

Water Heaters

Water heaters are generally fired by natural gas or electricity. Typical losses during operation are electric, 17 percent, gas-fired, 51 percent.[5] Since electric water heaters use less energy (measured from the outlet) than gas water heaters and are generally better insulated, their standby insulation loss is less than that of gas heaters.

Hot water pipes are an additional source of losses. By reducing the distance between the water heater and the point of use, and by insulating the pipes, such losses can be reduced.

These numbers indicate that the largest savings in the operation of water heaters will result from improvement of burner efficiency, reduction of standby losses, and elimination of the pilot. These and other modifications are detailed in Table 4.7.

Industry standards for heat loss in electric water heaters prescribe a maximum heat loss of 4-6 watts/ft^2. (Natural gas heat loss limits are somewhat higher.) A typical conductivity is 0.17 Btu/hr/ft^2 °F for a 50 gallon electric water heater wall. This can be used to estimate the effects of improvement in insulation.[5]

Refrigerator/Freezers

Between 1925 and 1975 the design of refrigerator/freezers changed considerably. Typical sizes increased from 0.14-0.28 m^3 (5-10 ft^3) to 0.34-0.68 m^3 (12-24 ft^3). The required energy input has also increased from roughly 210 to 490 W/m^3 (6 to 14 W/ft^3).

The theoretical heat loss of a modern refrigerator/freezer is tabulated in Table 4.8. A standard

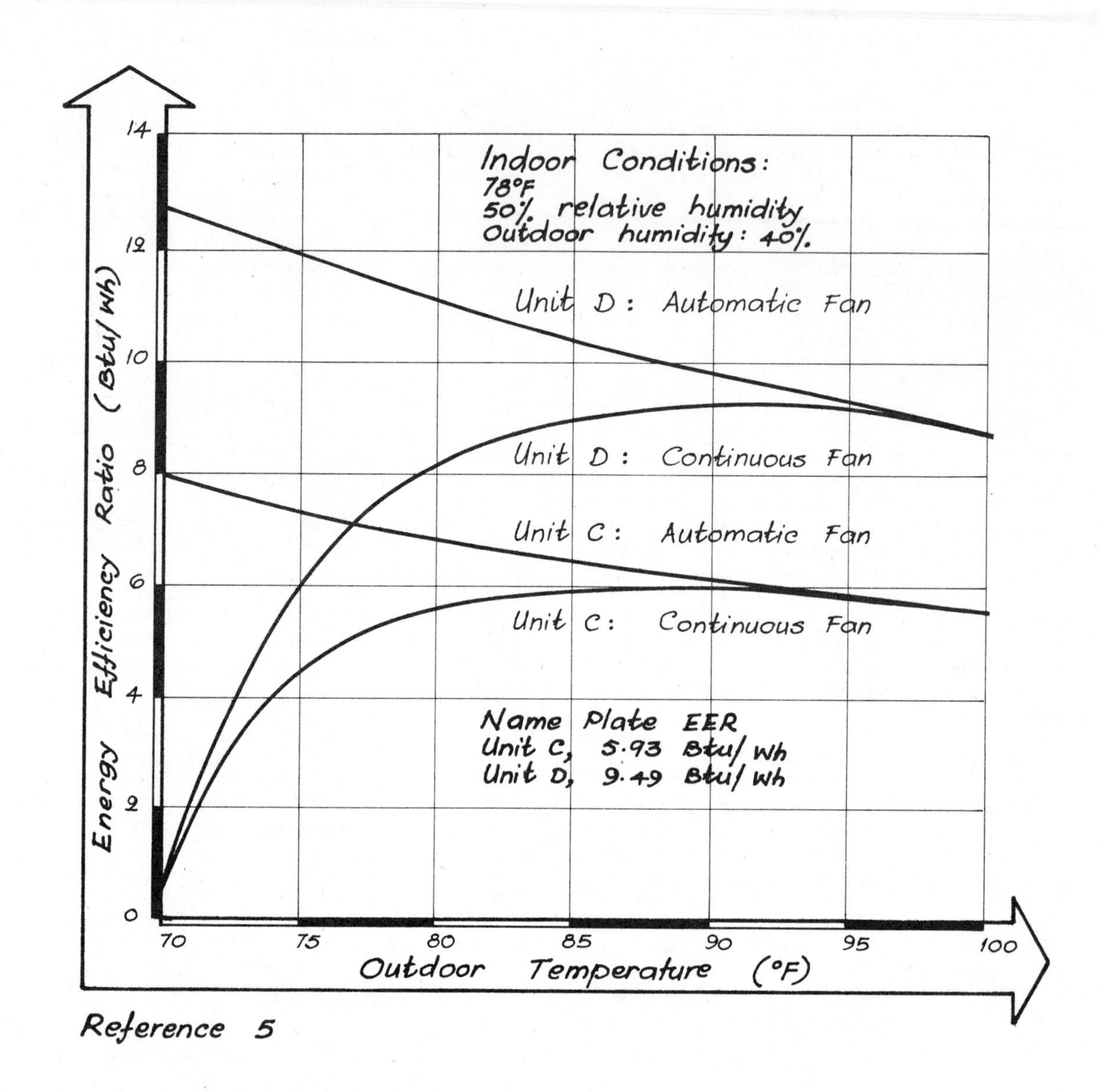

Reference 5

THEORETICAL EER VARIATION WITH OUTDOOR TEMPERATURE FOR CONTINUOUS AND AUTOMATIC FAN OPERATION FIGURE 4.4

	Percent Reduction in Annual Fuel Use	
	Gas	Electric
Operating Strategies		
Lower hot water temperature (per 10 °F reduction)	5.8%	4.5%
Modifications		
Increased tank insulation		
10 cm (4 in.), justified at 1¢/kWh		8.2%
18 cm (7 in.), justified at 4¢/kWh		11.0%
8 cm (3 in.), justified at 10¢/therm	21.6%	
13 cm (5 in.), justified at 40¢/therm	25.0%	
Hot water plumbing insulation (7.5 m (25 ft) of exposed pipe)	1.6%	1.6%
Solar preheat tank	20-60% gas/electric	
Design		
Heat recovery from air conditioning	20-80% gas/electric	
Low excess air	8.0%	
Electric ignition with damper	13.0%	
Reduced flue temperatures	10.0%	
Reduced burner rate	2.0%	
Instantaneous heater with small tank	8.0%	
Heat pump water heater	50.0%	
Programmed off periods	10.0%	
Indirect heater with electric pilot	17.0%	
Ambient preheat tank	8.0%	
Condensation of flue gases	9.0%	

Reference 5

ENERGY SAVING MODIFICATIONS FOR WATER HEATERS

TABLE 4.7

Compartment	Heat Transfer through Walls	Door Openings	Total
Freezer	42 W	4 W	46 W, 33%
Refrigerated Compartment (GRC)	58 W	14 W	72 W, 51%
	100 W, 71%	18 W, 13%	118 W, 84%

Other Items	Total
Anti-Sweat Heaters (40% of 20 W)	8 W, 5.5%
Air Circulation Fan	7 W, 5.0%
Defrost Heaters (1.6% of 500 W)	8 W, 5.5%
	141 W, 100%

Notes:

1. GRC = general refrigerated compartment.
2. 27 °C (80 °F) Ambient, 3 °C (37 °F) GRC, -18 °C (0 °F) freezer.
3. 50 door openings/day for GRC.
4. 25 door openings/day for freezer.
5. GRC volume = 0.31 m^3 (10.8 ft^3), freezer volume 0.12 m^3 (4.2 ft^3).
6. Door openings assume 60% relative humidity and complete air change for each opening.

Reference 5

HEAT LOSS OF A TYPICAL REFRIGERATOR / FREEZER

TABLE 4.8

sized unit is used (with a freezer volume of 0.12 m^3 [4.2 ft^3] and a refrigerator volume of 0.31 m^3 [10.8 ft^3]). It is apparent from the results shown in the table that the largest heat loss results from thin walls and insufficient insulation.

Electrical input to the refrigerator is used to remove this acquired heat. Table 4.9 gives a breakdown of the energy which is necessary to operate the same typical refrigerator. The operation of the compressor is responsible for about three-quarters of the total energy use.

Anti-sweat heaters are used in many locales to prevent condensation on the exterior of refrigerator/freezers. Heat is either supplied by a resistance heater or part of the condenser loop waste heat. When the source is electrical resistance heating, it can generally be turned off by a "Power Saver" switch during dry weather. If the resistance heater is left off for two thirds of the time or more, it will use less energy than the condenser loop which cannot be turned off.

Table 4.10 shows the sensitivity of electricity use to variations in the operation of a refrigerator. For example, by increasing the temperature setting 1 °C, energy use drops by about 8 percent. In addition, opening the door three times will increase the energy use by roughly 1 percent.

In the purchase of a new refrigerator, efficiency should be an important consideration. Units are available with added insulation and efficient compressors which decrease the yearly energy use by 25 to 45 percent. Information on energy efficiency of specific models is available from the Association of Home Appliance Manufacturers (AHAM).[7]

Proper temperatures for storing a variety of foods are shown in Tables 4.11 and 4.12. With this information the consumer can ensure that the maximum safe storage temperatures are used for the greatest efficiency.

Split phase motors with a capacitor start are commonly used in domestic refrigerator compressors. They have an aluminum cast rotor with carbon steel laminations and a copper-wound stator. Modifications to improve efficiency could include the following:

- add more copper to the stator,
- add more iron to the magnetic circuit,
- add another capacitor for the run mode.

The efficiency would be increased roughly 12 percent by these improvements.

<u>Ovens/Ranges</u>

This sub-section contains a brief description of cooking equipment design. For a discussion of cooking techniques, please refer to the case studies in Chapter 15.

Heat loss from a typical oven occurs in four major areas:

Oven door edge	200 W
Oven window	70 W
Air circulation	80 W
Insulation	520 W

Improvements in the oven door may be possible by using titanium bolts to join the inner and outer walls. These offer a lower thermal conductivity than common metals.

Standard oven insulation is 3.8 cm (1.5 in) of fiberglas. Improvements can be made by using other materials or thicker walls. Self-cleaning ovens normally are built with more insulation than common ovens. Due to improved efficiency during normal usage, their average energy use is not significantly greater, despite periods of high demand.

Much of the heat loss in ranges results from inefficiencies in burner design. Effective surface burner efficiencies are 30 to 50 percent for gas and 50 to 70 percent for electricity. Capacity for improvement exists, possibly in the design of pots and pans to couple more effectively with the burners.

	Average Watts
Anti-sweat heater	20
Fan motor	7
Defrost heater	8
Compressor motor, electrical and mechanical losses (based on 66.7% efficiency)	58
Compressor friction loss (based on mechanical efficiency of 80%)	23
Vapor compressor power: Lower limit [based on ideal coefficient of performance of 4.72 for R-12 refrigerant between -18 °C (0 °F) and 27 °C (80 °F)]	30
Extra power required for:	
Lower evaporator temperature -28 °c (-18 °F)	12
Higher condenser temperature 6 °C (+42 °F)	26
Superheat at compressor limit 71 °C (+160 °F)	11
Miscellaneous inefficiencies	15
Total Input Power	210

Reference 5

INPUT POWER DISTRIBUTION OF A REFRIGERATOR/FREEZER

TABLE 4.9

Operation Change	Resulting Change in Electricity Use	
Temperature setting	-7 to 9%/°C	(-4 to 5%/°F)
Room temperature	+3.6%/°C	(+2%/°F)
Door opening (at 60% RH)	+0.3%/Opening	
Food load at ambient	+3.5%/kg	(1.6%/lb)

Reference 5

REFRIGERATOR / FREEZER ENERGY USE VARIABLES

TABLE 4.10

Article	Temperature °F	Article	Temperature °F
Game, to freeze	0	Flowers, cut	36
Poultry, to freeze	0	Ginger Ale	36
Fish, to freeze	5	Grapes	36
Game, after frozen	10	Cucumbers	38
Poultry, after frozen	10	Lemons	38
Butter	14	Sauerkraut	38
Fish, salt water, not frozen	15	Berries, fresh	40
Ice cream	15	Cantaloupes, short carry	40
Scallops, after frozen	16	Fish, dried	40
Fish, fresh water, frozen	18	Fruits, canned	40
Cabbage	20	Fruits, dried	40
Hams, not brined	20	Meats, canned	40
Livers	20	Nuts, in shell	40
Oleomargarine	20	Sardines, canned	40
Fish, not frozen, short carry	28	Watermelons, short carry	40
Furs	28	Buckwheat flour	42
Game, short carry	28	Cornmeal	42
Apples	30	Oatmeal	42
Beef, fresh	30	Tomatoes, ripe	42
Eggs	30	Wheat flour	42
Poultry, dressed, iced	30	Meats, salt, after curing	43
Beans, fresh	32	Oysters, in shell	43
Celery	32	Beans, dried	45
Cider	32	Beer, in bottles	45
Lambs	32	Corn, dried	45
Onions	32	Honey	45
Plums	32	Maple syrup and sugar	45
Cantaloupes, long carry	33	Oils	45
Carrots	33	Peas, dried	45
Cranberries	33	Sugar	45
Cream	33	Syrup	45
Pears	33	Peaches	50
Tenderloins	33	Wines	50
Oranges	34	Bananas	55
Potatoes	34	Dates	55
Cheese	35	Figs	55
Milk, short carry	35	Raisins	55

Reference 8

APPROXIMATE COLD STORAGE TEMPERATURES

TABLE 4.11

Product	Months
Meat, fresh	
Beef, steaks or roasts	9-12
Beef, ground	4-6
Beef or lamb liver	3-4
Lamb	9-12
Pork	4-6
Pork, ground	1-3
Pork, sausage	1-3
Meat, smoked	
Bacon, slab (Do not freeze bacon)	1-3
Frankfurters	1-3
Ham, whole	1-3
Sausage	1-2
Poultry, fresh	
Chicken, ready-to-cook	12
Ducks, geese, ready-to-cook	6
Fish and Shellfish	
Lean fish	4-6
Fatty fish, clams, oysters	3-4
Cooked crabmeat and lobster	2-3
Meat	2-3
Cooked shrimps	1-2
Other Foods	
Butter and cheese, except cottage cheese (do not freeze cream cheese)	6-8
Eggs, yolks and whites separated	12
Fruits and vegetables	10-12
Fruit juices	up to 16
Ice cream	1-2
Cook or Prepared Foods	
Baked pies, biscuits, muffins	2
Baked yeast bread and rolls	6-8
Cakes (unfrosted), fruit cakes, unbaked fruit pies	6-8
Leftovers, fried foods, newburgs, thermidors, pasta dishes	1
Roast beef, lamb, veal, and chicken	4-6
Roast pork and turkey, stews	2-4
Soups	6

Reference 8

MAXIMUM FROZEN FOOD STORAGE TIMES AT −17.7°C

TABLE 4.12

Finally, microwave ovens offer not only more efficient heating but also shorter cooking times. (See the case studies in Chapter 15.)

Clothes Washers

Most of the energy used by clothes washers is employed in heating water rather than in the electrical drives. This is illustrated in Figure 4.5 which gives the energy efficiency of several competitive washing machines.

The energy used by washers, therefore, is best limited by controlling water quantity and temperature in the wash and rinse cycles. A hot water wash is not generally required, but may be advantageous for special purposes such as cleaning heavily soiled white clothes or removing oil and grease stains. Generally, it is necessary to use water at 50 °C (122°F) and detergent to emulsify animal oils and fats during the wash cycle. Clothes are just as clean (in terms of bacteria count) after a 20 °C (70 °F) wash as after a 50 °C (122 °F) wash. If there is concern for sanitation (e.g., a sick person in the house), authorities recommend use of a chlorine bleach.

Many manufacturers contend that there is little or no need for anything other than a cold rinse. Typically, a warm rinse capability is provided solely because of consumer demand. Some machines do not even allow a hot wash/cold rinse or warm wash/cold rinse control option.

In the United States the use of colder temperatures in clothes washers is beginning to be accepted by the public. The following data from Reference 5 show cycle selection for a washer with five combinations of wash/rinse temperatures.

Wash/Rinse Temperature Setting	1971 Survey Use Factor	1975 Survey Use Factor
Hot/warm	.25	.18
Hot/cold	.15	.12
Warm/warm	.30	.30
Warm/cold	.20	.25
Cold/cold	.10	.15

Water level is also important in the efficiency of a washer. A Proctor and Gamble study revealed that the average load size is 2.5 kg (5.4 lb) in a 6.14 kg (14 lb) washer and 2.7 kg (5.9 lb) in 7-9 kg (16-20 lb) washers.[5] At the same time 79 percent of all loads in normal size machines and 60 percent of washes in larger machines are done with the maximum water fill. This indicates the consumers' lack of awareness concerning the operating capabilities and constraints of their clothes washers.

Any measure of the actual efficiency of a washing machine should also include its efficiency in removing soils. The Association of Home Appliance Manufacturers (AHAM) specifies tests through which this ability can be compared for different machines. It should be taken into account along with energy considerations when purchasing new equipment.

Clothes Dryers

Clothes dryers represent a large portion of a typical household's appliance energy use. Similar to washers, dryers operate most efficiently when fully loaded. This is illustrated in Figure 4.6, which graphs a comparison of six competitive models of electric dryers. Operating with one-third to one-half load costs roughly 15 percent in energy efficiency.

Locating dryers in heated spaces can save 30-40 percent during winter operation. This must be traded off against heat gain to the space from the walls of the dryer and heat loss from the space through the air exhausted to the outside.

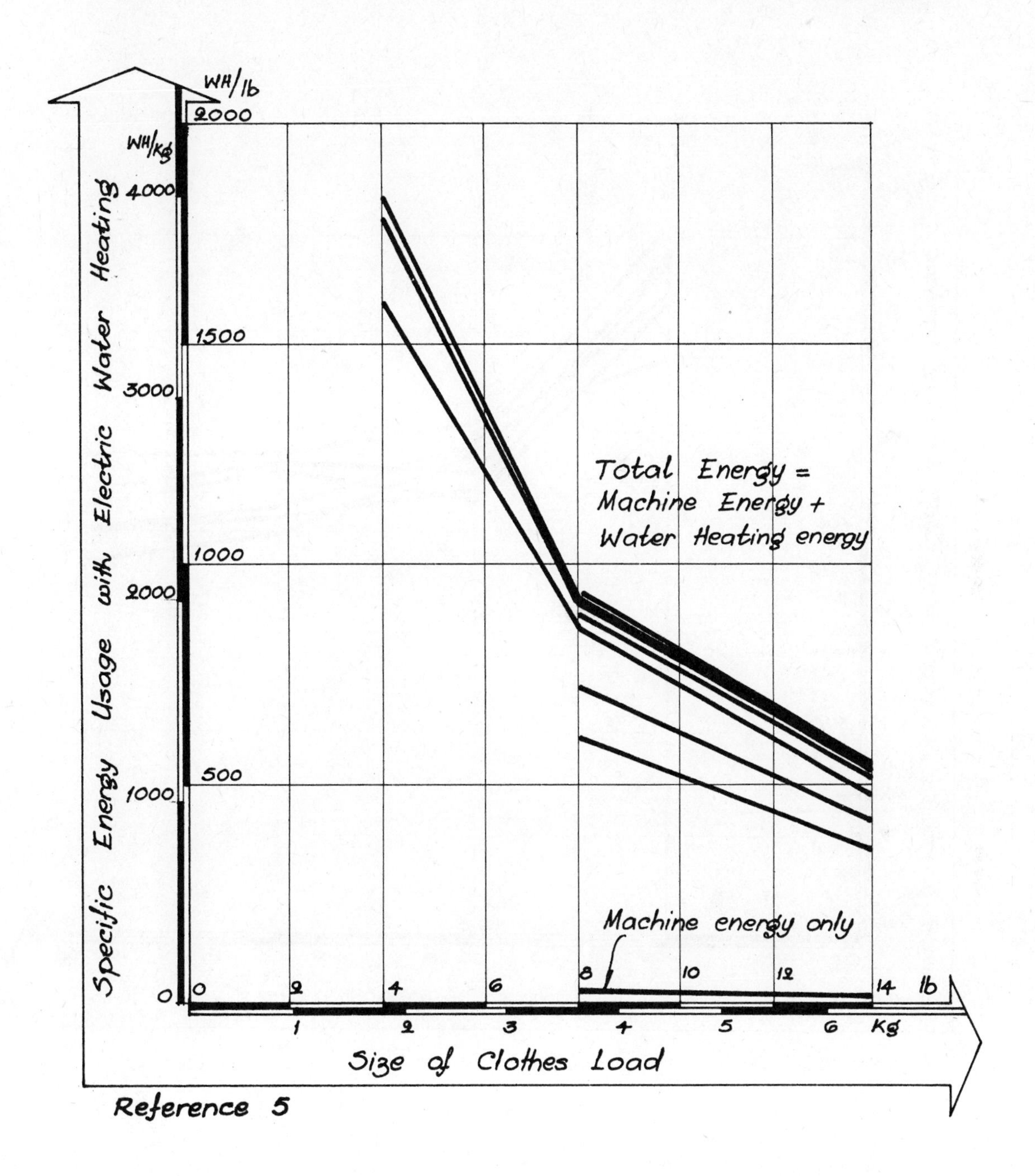

TYPICAL CLOTHES WASHERS—SPECIFIC ENERGY USE

FIGURE 4.5

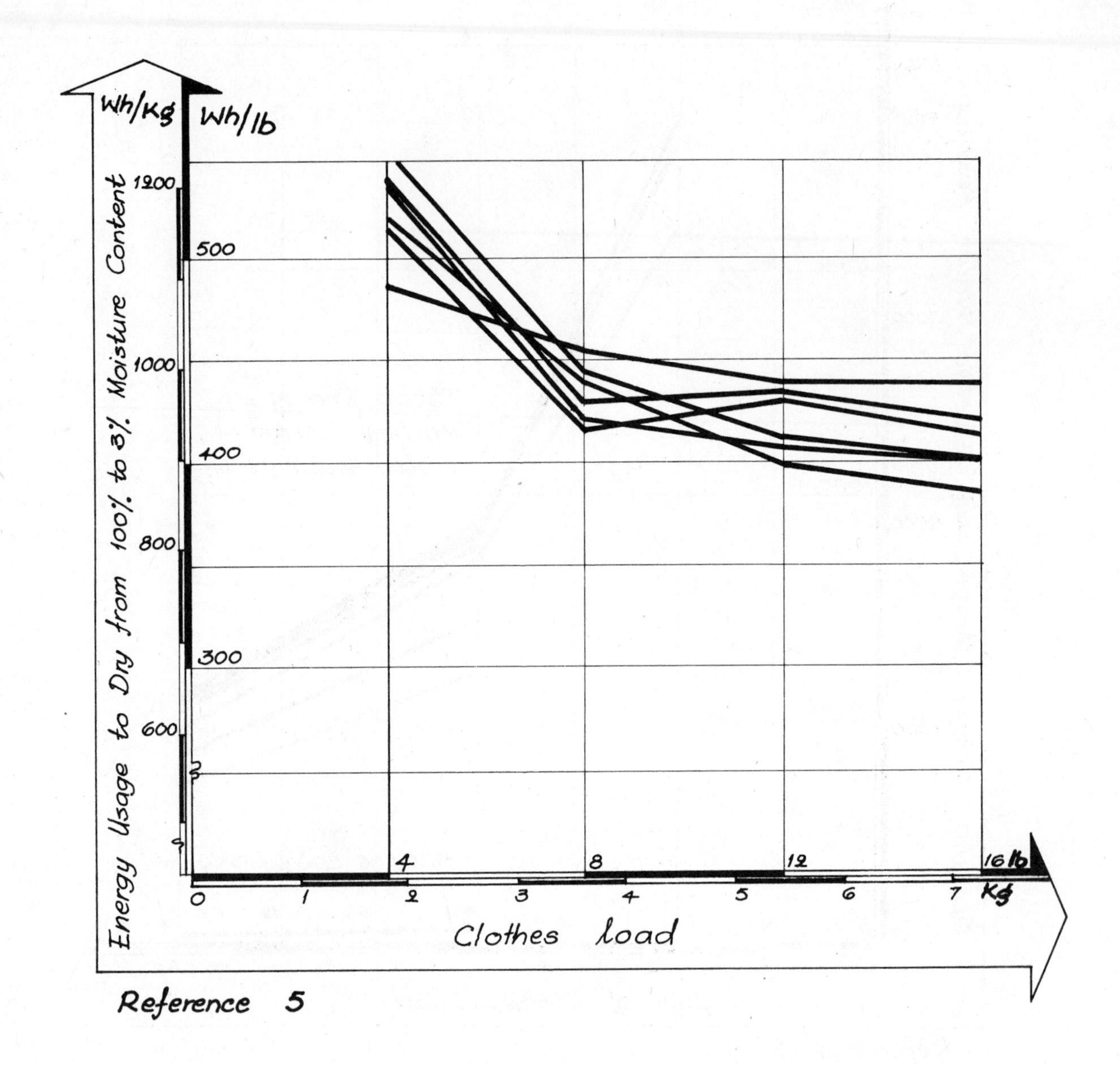

TYPICAL ELECTRIC CLOTHES DRYERS—SPECIFIC ENERGY USAGE

FIGURE 4.6

From the manufacturer's vantage point, the generally acknowledged method to improve efficiency is to decrease the heat rate.[5] This results in a longer dry time, however, which may not be acceptable to some consumers.

Manufacturers can also supply automatic sensors to shut down the dryer either when a specified moisture level or exhaust temperature is achieved. (Dryers typically drop moister content from 80 percent to 3 - 5 percent.) Theoretically, automatic controls should prevent overdrying and thereby increase efficiency. It is contended, however, that some automatic controls are set too low, indicating that simple timed dryers are preferable in some cases. This control capability should be examined when purchasing a new clothes dryer.

Dishwashers

The two major uses of energy in a dishwasher are the heating of water and the dry cycle. Energy used in pumps and fans is not a significant proportion.

The volume of hot water used ranges from 45 to 61 liters (12 to 16 gallons). This can be controlled by the operator on some machines through selection of the proper wash and rinse cycle. The minimum number of cycles that will clean the dishes should always be chosen.

Temperature requirements of dishwashers often dictate the hot water temperature for the entire residence. Residential dishwashers require water at 55-60 °C (130-140 °F) while most functions need water at 38-43 °C (100-110 °F). (As mentioned above, 50 °C [122 °F] is necessary to emulsify animal fats and oils when used with detergents.) In some areas it would be economical to add a booster heater and lower the temperature of the main water heater.

Lighting

Lighting is discussed in detail in Chapter 9. However, two basic ideas concerning efficient lighting in residences will be mentioned here.

First, turn off all unneeded lights. Turning on and off fluorescent lamps, even for very short periods of time, will save energy. However, it will also shorten the life of the lamp. As an economic tradeoff between the cost of energy and new lamps, it is generally recommended that fluorescent lamps be turned off if they can be left off for 5 minutes or more.

Second, use higher efficiency light sources. Incandescent lamps, the most commonly used in residences, are the least efficient lamps available. Wherever possible they should be replaced with fluorescent lamps. Fluorescent lamps generally provide 60 to 80 lumens of light per watt while incandescent lights put out 15 to 25 lumens per watt. Fluorescent fixtures are now available in a variety of designer styles suitable for use throughout the home.

4.4 SUGGESTED ECONOMIES

A. Digest of Householder Operator Options

The suggestions in Table 4.13 are addressed primarily to the direct residential user of the various equipment. Some suggestions may apply to the owner, however. Apartment managers, home economists, architects, product design and service engineers, manufacturers' agents, governmental agencies, and educational institutions may all find these suggestions relevant.

B. Digest of Home-Owner Installation and Change Economies

The suggestions in Table 4.14 are addressed to the homeowner--or any individual interested in saving energy and expense either personally or for the occupants. Some of the ideas apply to present equipment, some to replacements, and some to structural accessories and installation. The major savings relate to space conditioning, where generally the objective of these changes is to reduce heat transfer outward in

Reduce Use of Equipment	Reduce Losses	Substitutions for Same Function	Maintenance and Prevention	Reduce Connected Load and Other
		Space Heating or Cooling		
Turn heater thermostat to 20°C (68°F) in day, to 15°C (60°F) at night, (4% day, 6% night) Set AC thermostat to 26°C(78°F) (15% fuel) Operate A/C only when needed Modify clothing to accommodate to wider comfort limits (2%) Use openable windows to cool (3%) Reduce attic ventilation in winter (maintain acceptable humidity limit) (1%) Maintain a constant thermostat setting (<1%)	Install and use shades, drapes, awnings, louvers, on south, west (2%) Keep leaking down around windows, doors, fireplace (3%) Use storm windows or tack on clear plastic sheet (3%) Where glass area is large add a second pane (3%) Use light color drapes facing outward (<1%) Use shrubs, vines outside as shades for sunlight (2%) Insulate floor of attic (1%) Shut off pilot lights in summer when away (1%)	Use ventilating fans where appropriate, particularly for kitchens or laundry	Keep filters cleaned regularly (<1%)	Shut off unused rooms and spaces (3%)
		Refrigeration Freezing		
Pre-cool heated foods prior to loading in refrigerator/freezer (2%) Remove no more ice cubes than needed (<1%)	Reduce door openings to minimum (5%) Be sure refrigerator is not against hot walls, or insert a sheet of fiberboard and an extra air space (2%)		Keep rear coils (heat exchanger) clean (<1%) Make sure door gaskets close tightly all around, replace if torn (1%)	Check actual need for the use of your second refrigerator (8%)

DIGEST OF HOUSEHOLDER OPERATIONAL OPTIONS

TABLE 4.13

Reduce Use of Equipment	Reduce Losses	Substitutions for Same Function	Maintenance and Prevention	Reduce Connected Load and Other
Empty refrigerator, leave off and open for vaction (2%) Set thermostat to keep food at 4-5°C (39-41°F) (ref.) or -17°C to 12°C (0°-10°F) (3%) (freezer)			Make sure interior lamp extinguishes when door closes (push door switch with finger) (2%)	
		Water Heater		
Set thermostat to 49°C (120°F) (10%) Don't wash dishes under running hot water; run dishwasher only when fully loaded, or wash by hand	If water heater is located in a cold cubicle, add fiber sheet, or other insulation (10-20%) Check leaky hot water faucets (& pipes) (1%)	Wash clothes in cold water with cold-water soap (4% fuel) Use lower flow shower heads (5%)	Check flame adjustment annually, if gas (1%) Add pipe insulation, if convenient (3%) Check and repair heater insulation (1%)	
		Lighting		
Turn off lights when not used (signs: "last out-lights out") (15%) Concentrate light in reading and work areas; cut general room lighting (5%) Remember heat from lights adds to the A/C load (1%)	Where possible, use fluorescent lamps in place of incandescent (40 W vs 100 W) (15%) Use one large bulb in place of several small ones (e.g., 100 W vs 2-60 W) (1%) Turn off outside night lamps during day (<1%)	Use sunlight at periphery of house (2%) Light colored walls and ceilings, reflective screens reduce the need for artificial lighting (2%)	Replace fluorescent tubes as soon as they begin "blinking" (<1%) Keep lamps and fixtures clean (<1%) Reconnect large areas so as to be served by more than one switch (2%)	In clusters, remove one bulb permanently (3%) Remove lamps used for decorative purposes only (3%)

TABLE 4.13

Reduce Use of Equipment	Reduce Losses	Substitutions for Same Function	Maintenance and Prevention	Reduce Connected Load and Other
		Cooking and Related Appliances		
Use all-oven-cooked meals (3%) Use oven self-cleaning option sparingly (2%) Plug in counter-top roasting appliances only as long as cooking (<1%) Turn down heat when pot bubbles (<1%) Turn off heater several minutes before food is done (1%) Use a single small appliance (e.g., casserole cooker) in place of a larger device (2%)	Cover the pans; use pans that cover the heating element (<1%) Permit the cooking appliance to heat the kitchen, as a consequence of its use; vent the range in the summer to reduce A/C load (1%) Do not pre-heat oven more than 2-3 minutes (<1%) Do not use range to heat the house (<1%) Pre-scrape dishes to permit use of short cycle (<1%)	Use hand mixing, etc., where appropriate (<1%)	Keep pans flat bottomed for electric heating units (<1%) Keep range exhaust filter clean (<1%) Check whether a defective appliance is at end of useful life (<1%) If gas, check setting for blue-flame; adjust pilots (<1%)	Plan use of heavy load appliances before 8 a.m. or after 6 p.m. when possible (e.g. vacuum clean or iron up a batch on week-ends)
		Laundry and Dishwasher		
Use washers with full loads of mixed sizes of articles ($3/bill) Where available use "suds-saver" (1%) Use dryer only with load that has gone through a full spinout cycle of washer (2%)	Match detergent to the water hardness, save rinses (<1%) Open dishwasher door for air drying after last rinse (1%) Use only the number of cycle portions necessary (e.g., skip soak cycles) (<1%) Avoid over drying (<1%)	Dry clothes on rack or outdoors, when suitable (2%) Hand wash single items (<1%)	Keep water and air filters cleaned out (<1%)	Run machines at off-peak times, such as before 8 a.m. and after 6 p.m.

TABLE 4.13

Reduce Use of Equipment	Reduce Losses	Substitutions for Same Function	Maintenance and Prevention	Reduce Connected Load and Other
Set dryer times for proper time and temperature (2%)				
Partially dry clothes, fold and place on dryer during next load (2%)				
TV, Radio and Other Appliances				
Keep sets off except when actually attended (1%)	Unplug "instant-on" TV sets when not in use (3%)	Use Small Screen TV sets where possible (1%)	Keep cutting edges sharp for motorized tools (<1%)	
Watch swimming pool pump and heater for unnecessary use (1%)	Check percolator, electric blankets, and heating pads when not in use (1%)	Develop other forms of family home entertainment (2%)		
	Turn off shop tools, etc., when not in use (<1%)			

Notes: ( ) indicates % savings of energy used for sector or function indicated.
<1% means less than 1%

TABLE 4.13

	Reduce Losses	Reduce Connected Load
Space Heating/Cooling		
For moderate winters	Insulate ceilings 6", walls 3-1/2" Weatherstrip, caulk windows, add storm windows or double panes Add wind screens for west and north doorways	Replace inefficient heaters, boilers, air-conditioners with new properly sized units (get quotations from experts) Select system with lowest life cycle cost
For severe winters/summers	Insulate ceilings 9", walls 3-1/2" Add entrance vestibules for doors Plant deciduous trees or shrubs on south and west side for shade Use light colored roofing for reflection Use reflectively coated double panes interchangeable between summer/winter	Buy air conditioners with maximum Energy Efficiency Ratio rating (from the nameplate): $E.E.R. = \frac{BTU/hr\ cooling}{KW\ required}$ Use experts for installation, test, and operating instructions Install new or replacement equipment considering ambient temperatures, duct and pipe runs, thermostat locations, possible room isolation, etc.
On new homes	Limit windowed area to 10% of floor area In cold climates, minimize windows to the north, vice versa for hot climates Maximize use of natural light Permit windows to use natural ventilation Insulate heating/cooling pipes and ducts	
Water Heating	Add exterior shell insulation Add a tempering tank if exceptionally cold intake water	Select size for lowest life cycle cost (electricity vs. gas?)
Lighting	Review lamps in each fixture so as to meet actual need Use high reflectance (80-90%) ceiling finishes, medium reflectance finishes for walls (40-60%)	Disconnect decorative lighting or remove lamps Replace suitable incandescent lamps with fluorescent, as appropriate. For higher powers, consider mercury vapor or metal halide types.

SUMMARY OF INSTALLATION AND CHANGE ECONOMIES **TABLE 4.14**

winter and inward in summer.

C. Digest of Manufacturer's Potential Design Changes

This group of suggestions is for product development and the design engineers and associates. The viewpoint argued here urges the engineer to go beyond the "first-stated" product function, its immediate mechanical implementation, and the concentration on lowest "first cost." Clearly, manufacturing energy depends upon design choices of materials and processes. Also, life cycle energy usage (and expected cost) can complicate both the designer's and the consumer's choices. Certain companies are voluntarily moving toward life cycle costing. Industry associations and regulatory agencies are cooperating in taking the first steps in this direction (see Appendix D).

Major impacts in energy reduction (see also Chapter 2) lie along the avenues of:

- extending the life of the product 25 to 50 percent and improving efficiency;
- providing maximum access for wear-out parts (belts, lamps, gaskets);
- replacing disposable items with reusable ones, whenever economical (filters);
- specifying materials which can be recycled or are biodegradable;
- considering the maintenance of manufacturing machinery during product design (i.e., extend tool life);
- designing for optimum scrap salvage, when possible (ferrous versus non-ferrous); and,
- increasing use of standardized components and accessories.

The Federal Energy Administration appliance efficiency program has replaced the NBS voluntary program which established targets for efficiency improvements in major appliances. These may be seen in Table 4.15. Industry in general considers that these goals can be achieved though there is some fear that sales may suffer.

4.5 POTENTIAL ENERGY SAVINGS

As in Chapters 2 and 3, potential savings have been categorized as *immediate*, *near-term* (2-5 yr), or *long-term* (5-25 yr). The main areas for saving residential energy are space conditioning, water heating, cooking, refrigeration, and lighting.

Space conditioning (heating and cooling) accounted for 60 percent of residential energy use in the 1973-74 base year. Table 4.16 illustrates the potential improvement possible in existing houses; over a 2 to 16 year period the total change is approximately 30 percent. For new construction, the impact of proposed new federal, state and private performance standards is as significant as the technological options and must not be ignored (see Appendix D). Other possible technological changes are solar heating and heat pumps. Heat pumps could lead to a net increase of electricity for heating purposes.

In the short term, water heating losses could be reduced in several ways; for example by design of more efficient equipment and by insulating tanks and pipes. In the long-term the possibility exists for using recovered heat or solar energy for water heating. Lighting could be improved immediately by development of consumer-acceptable fluorescent lamps for homes; with proper design this could have application both in retrofitting existing residences and in new construction, with an ultimate potential for reducing lighting electricity use by 50 percent.

New purchases of major equipment must be evaluated on a life cycle cost basis to obtain the maximum of the above benefits.

In round numbers 1973-74 energy use in the residential sector was 1.5×10^{10} GJ/year (1.4×10^{10} MBtu/year). Based on the information

APPLIANCE CATEGORY	FEDERAL REGISTER PUBLICATION DATE		ENERGY EFFICIENCY MEASURE			INDUSTRY GOAL IN PERCENT	
	Proposed	Final	Title	Abbrev.	Units	Reduction in Consumption	Increase in Measure
Room Air Conditioners	6/3/75	--	Energy Efficiency Ratio	EER	Btu/Wh	22	28.2
Electric Clothes Driers	7/9/75	--	Energy Factor	EF	STL/kWh	6	6.4
Gas Clothes Dryers	7/9/75	--	Energy Factor	EF	STL/T	12	13.6
Clothes Washers	7/10/75	--	Energy Factor	EF	STL/kWh	10	11.1
Dishwashers	7/10/75	--	Energy Factor	EF	STL/kWh	18	22.0
Electric Ranges	7/9/75	--	Range Thermal Efficiency	E_t	%	10	11.1
Gas Ranges	7/10/75	--	Range Thermal Efficiency	E_t	%	30	42.9
Freezers	6/23/75	--	Energy Factor	EF	$ft^3/(kWh/d)$	25	33.3
Refrigerators & Combination Refrigerator Freezers	6/23/75	--	Energy Factor	EF	$ft^3/(kWh/d)$	30	42.9
Color Television Receivers	7/8/75	--	Receiver Energy Efficiency	REE	%	42	72.4
Monochrome Television Receivers	7/8/75	--	Receiver Energy Efficiency	REE	%	48	92.3
Electric Water Heaters	6/23/75	--	Tank Efficiency	E_t	%	9	9.9
Gas Water Heaters	6/23/75[a]	--	Tank Efficiency	E_t	%	25	33.3

[a]Correction published 7/10/75

NBS APPLIANCE EFFICIENCY PROGRAM STATUS REPORT

TABLE 4.15

Source of Savings	Action Required	Probable Saving/Unit	% Units Applicable	Assumed % Compliance	Percent Savings Per Year	Percent Savings Total
1. Reduce conduction convection a. Insulation	Insulate ceiling, crawl spaces & improve insulation where marginal	15%	50	80	1.5	6.0
b. Convection	Storm windows, weather strip doors, seal cracks.	10%	50	80	1.0	4.0
2. Heating plant a. Heating System	Install more efficient heating system, p.e. switch operated electric starter, insulate ducts, etc.	10%	70	80	0.3	3.0 (10 yrs.)
b. Maintenance & Repair	Maintain & repair heating systems, clean filters, etc.	15%	70	80	4.2	8.4
3. Personal & family habits a. Temperature reduction	Lower temperature 1+° during day; 10° during night.	4-5% for day; 6% for night	Day 100% Night 60%	80 80	2.0 1.4	4.0 2.8
4. Other Habits	a. Handling blinds & drapes in unoccupied rooms b. Closing dampers when fireplace is not used. c. Closed door discipline. d. Other conservation measures.	6%	8	80	1.8	3.6

Reference 9

PRE-1974 HOMES: POSSIBLE REDUCTIONS IN ENERGY REQUIRED FOR SPACE HEATING

TABLE 4.16

presented in the chapter and in the case studies, the possible range of savings due to increased energy use efficiency can be estimated. For the year 2000, annual savings in the range of 25 to 45 percent (total energy) and 25 to 45 percent (electricity) appear economically and technically feasible. (See Table 4.17 for a breakdown of these estimates.)

These values are similar to other published data. For example, a US government study indicates short-, mid-, and long-term savings in the residential/commercial sector of 10 percent, 14 percent, and 30 percent respectively.[8]

As has been noted in Chapters 2 and 3, there are several events which could affect these potential savings. Perhaps most significant is the prospect of a shift of use based on some other energy form (fuel oil or natural gas, for example) to electricity. Chapter 1 should be reviewed for a more complete discussion of potential savings and a possible impact on these projections of energy shifts.

* * *

REFERENCES

1. Stanford Research Institute, *Patterns of Energy Consumption in the United States*, Report Prepared for the Office of Science and Technology, (Menlo Park, California: January 1972).

2. Steinhart, Carol and Steinhart, John, *Energy: Sources, Use and Role in Human Affairs*, (North Scituate, Massachusetts: Duxbury Press, 1974), p. 228.

3. Tansil, John, *Residential Consumption of Electricity, 1950-1970*, Report No. ORNL-NSF-EP-51, (Oak Ridge, Tennessee: Oak Ridge National Laboratory, July 1973).

4. Freeman, David S., et al., *A Time to Choose: America's Energy Future*, Final Report of the Energy Policy Project of the Ford Foundation, (Cambridge, Massachusetts: Ballinger Publishing Company, 1974), pp. 117, 118.

5. California Energy Resource Conservation and Development Commission, *Appliance Efficiency Program*, (Sacramento, California: 15 March 1976).

6. Jacobson, Richard A., "Efficiency by Decree," *Machine Design*, 46(2 May 1974): 98-104.

7. Association of Home Appliance Manufacturers, *1977 Directory of Certified Refrigerators and Freezers*, (Chicago, Illinois: 1977).

8. Tighe, Eileen, ed., *Woman's Day Encyclopedia of Cookery*, Vol. 5 (New York: Fawcett Publications, Inc., 1966).

9. Havron, M. Dean, *Projecting the Impact of Energy Conservation Measures in the Home*, (McLean, Virginia: Human Sciences Research, Inc., December 1973), p. 32.

Period	Potential Savings (%) Total Energy	Electricity
• *Immediate*--Operational housekeeping changes	5-10	5-10
• *Near-Term*--Some investments and process equipment changes	10-15	5-10
• *Long-Term*--Major investments and process and equipment changes	10-20	15-25
• *Annual savings*--in the year 2000	25-45	25-45

POTENTIAL SAVINGS IN RESIDENTIAL ENERGY USE

TABLE 4.17

CHAPTER 5

TRANSPORTATION, COMMUNICATION, AND COMPUTERS

R.B. Spencer*

CHAPTER CONTENTS

KEY WORDS

Automobiles
Communication Use
Computers
Electric Vehicles
Energy Conversion
Energy Storage
Motors
Transportation Use

SUMMARY

About one-fourth of the energy used in the United States today is expended for communication, computers, and transportation, the latter accounting for the majority of this fraction. Direct combustion of fossil fuels presently provides over 90 percent of the energy used for transportation, communication, and computers. Electricity accounts for less than 1 percent of transportation energy; essentially all communication systems (except messenger services) are electric, however, as are all computer installations.

It is expected that the electric share of transportation energy use will expand. High fuel prices, fuel scarcity, and urban pollution are stimulating renewed interest in electric vehicles and mass transit systems. The recent development of battery systems having more attractive power and energy densities is improving the prospects for electric vehicles. Continuing technological advances in telecommunications and computers (e.g., cost reductions, reduced power, and solid-state miniaturization) are causing rapid growth rates in applications of these technologies. The near future will see electronic mail services and electronic communication systems begin to replace personal transportation, with an associated savings in total energy use.

5.1 INTRODUCTION

A. Perspective

Transportation annually contributes about $200 billion to the US

*Principal, Applied Nucleonics Company, Inc.

economy, roughly 20 percent of the Gross National Product.[1,2] More than 10 million people, or nearly 15 percent of the civilian labor force, are employed in transportation and related industries.[2] Intercity freight traffic is sufficient each year to move 10 mt (11 tons) of freight 1600 km (1000 miles) for every person in the US. Total passenger traffic amounts to about 16,000 km (10,000 miles) per person annually.[1] Table 5.1 shows typical fuel use values for passenger transport.

On the other hand, transportation-caused accidents claim 60,000 lives and cause more than 5 million injuries a year.[2] Transportation is also responsible for more than half the total weight of air pollution emissions in the country.[3] Finally, transportation contributes to other environmental problems such as urban congestion and noise pollution.

Between 1950 and 1970 annual fuel use for transportation (almost entirely petroleum) grew from 9.2×10^9 GJ to 17.4×10^9 GJ (8.7×10^{15} to 16.5×10^{15} Btu), with an average annual growth rate of 3.2 percent.[4] In 1970, transportation accounted for more than half of national petroleum use, 23 percent of which was imported.[5] The National Petroleum Council projected that oil imports would account for 57 percent of domestic petroleum consumption in 1985 based on pre-1971 information.[6]

US electrical energy use in transportation has decreased in this century due to an increasing use of gasoline and diesel engines. In Europe, however, where historically fuel costs have been higher, there is a proportionately greater use of electricity for transportation purposes.

Electricity is used extensively today for communication purposes. Although the energy cost is low using electronic media--typically 10^{-2} kWh/min for telephone conversations--the energy investment in equipment is significant.

Electricity related to computers is required for actual computing power requirements as well as the associated cooling and humidity control for the computer environment. Approximately equal amounts of electrical energy are used for these two purposes.

B. Objectives and Scope

The material in this chapter is intended to serve as a guide to further the efficient use of electrical energy. The technical depth of the information is sufficient only to perform preliminary comparative analyses between competing concepts. The equations are intended primarily to convey the relationship between important parameters involved in calculating energy and power requirements. The information is not meant for the design engineer or analyst requiring accurate design calculations. More detailed texts and specific references should be consulted when performing detailed analysis and design of actual systems.

C. Fundamentals of Transportation Energy Use

A convenient starting point is to consider the power requirement for a vehicle, since virtually all transportation systems can be idealized as vehicles. The propulsion power required at the wheels of a vehicle can be expressed as the sum of four terms:

- power to overcome rolling resistance or friction;
- power to lift vehicle against force of gravity (up-grade);
- power to accelerate vehicle; and,
- power required to overcome aerodynamic drag.

An equation relating these quantities can be written.[7]

$$P = (k_o + k_1 v)w \cos[\tan^{-1} b] + vw \sin[\tan^{-1} b] + k_2 avw + k_2 Av^3 \cong (k_o + k_1 v)w + bvw + k_2 avw + k_3 Av^3 \quad (5.1)$$

where:

	MJ/passenger-km	Assumed Load Factors
Bicycle	0.13	1.0
Walking	0.20	1.0
Buses	0.80	0.5
Railroads	1.10	0.5
Automobiles	3.00	0.5
Airplanes	6.50	0.5

Reference 9

TYPICAL FUEL ENERGY USE FOR PASSENGER TRANSPORT **TABLE 5.1**

k_o, k_1 = constants which depend on the type of wheel or tire, road surface, or nature of levitation system;

v = velocity;

w = weight of vehicle;

b = slope of grade (expressed as a percentage);

k_2 = a dimensional constant including rotational inertia of wheels;

a = vehicle acceleration;

k_3 = a constant which depends on vehicle shape and air density; and,

A = effective frontal area of vehicle.

As equation 5.1 indicates, vehicle weight is the most critical factor affecting energy use. Next in importance, particularly at higher speeds, is velocity. Under certain conditions (such as regenerative braking) energy can be recovered economically. Additional data on various types of transportation systems can be found in the literature.[7,8]

Since the primary object of this book is to discuss the technology of improving electrical use efficiency, the area of non-electrical (primarily fossil fuel) transportation has not been considered except to put the total energy use for transportation in perspective. In the balance of the chapter, emphasis will be placed on a discussion of electric vehicles and transportation systems, and on electronic communication and computer systems.

5.2 ELECTRIC MODES OF TRANSPORTATION

A. Electric Vehicles--Technology Considerations[10]

Mounting pressures to conserve petroleum and reduce pollution are changing traditional automobile design. Alternate power plants are being evaluated and, of all the alternatives, electric propulsion presently appears to have many advantages. While the energy conversion efficiency of the motor may not be superior to other engines, the important advantages include: flexibility of primary fuel sources in electricity generation (e.g., nuclear, geothermal, etc.), possibility of link-up with centralized power distribution systems under or alongside roads, greater ease of designing and implementing automatic control of individual vehicles on highways and freeways, and higher system efficiency since in a central power station the waste heat could be beneficially utilized.

For reliable and practical electric vehicles for urban and suburban use to become a reality, several technological improvements in the present designs must be made. A realistic goal for a commuting electric vehicle is to have a range of 160 km (100 miles) and speeds of 80 to 100 km/hr (50 to 60 mph). To achieve this goal, the following technological improvements are needed:

- efficiency of the energy storage devices (batteries);
- engineering to reduce road-load and accessory losses;
- efficiency of electric motors (both ac and dc);
- efficiency of motor controllers;
- efficiency of electric vehicle transmissions; and,
- efficiency of electrical connectors and cables.

The following is a summary of some of the more important technological areas and future needs for practical electric vehicle development.

1. Electric Vehicle Design

Because of the serious power limitation of present batteries for use in electric vehicles, the engineering design of the vehicle to reduce friction, drag, and weight becomes all-important. Major engineering efforts must be made to reduce energy wasted on road-load energy losses and accessory losses. In addition to minimizing energy dissipation, continuing engineering improvements must be made in the areas of electric vehicle braking, and packing and protection of batteries.

Road-load energy losses include aerodynamic drag, chassis losses, and inertia losses. An equation for the power lost to drag is given below (equation 5.2):

$$P_d = \frac{C_d A v^3}{C_f} \qquad (5.2)$$

where: P_d = power lost to drag, watts;

C_d = drag coefficient;

A = frontal area, m^2;

v = velocity, m/sec; and,

C_f = a conversion factor (constant)

The drag coefficient reflects the effect of the vehicle shape on air resistance. At the high end of the drag range, a flat disc, perpendicular to an air stream, has a drag coefficient of 1.17. At the low end, a highly streamlined profile with minimum drag has a coefficient of about 0.05. Most modern cars have drag coefficients in the range of 0.45 to 0.55. Desirable drag coefficients for electric vehicles should be in the range of 0.2 to 0.3, a difficult but not impossible target.

Chassis losses, the second component of road-load, are made up of tire loss, bearing loss, and power train loss. By far the dominant factor is tire loss which accounts for about 85 percent of chassis loss. The use of belted radial tires can drop tire losses by about 20 percent. Experimental tire designs can provide still lower losses through such techniques as special rubber compounding, reduced width, stiffer sidewalls, and lower aspect ratios (the aspect ratio being defined as the tire minor diameter to major diameter). Tires with a low aspect ratio can reduce rolling resistance by almost 60 percent.

Inertia losses, the third component of road-load, are losses associated with accelerating the car. These losses, which to a large extent represent the useful work of a car, can be broken down into components for linear acceleration of the whole vehicle and rotational acceleration of components such as wheels, motor armatures, and driveline components. The momentum of the accelerating masses actually represents stored kinetic energy which can be used by the vehicle during acceleration unless it is wasted in the brakes. Equation 5.3 describes the power potentially lost to inertia:

$$P_i = mav + \sum_{i=0}^{n} I_i \alpha_i \omega_i \qquad (5.3)$$

where: P_i = power lost to inertia, watts;

m = total vehicle mass, kg;

a = linear acceleration, m/sec^2;

v = linear velocity, m/sec;

I = rotational moment of inertia, $kg\text{-}m^2$;

α = angular acceleration, rad/sec^2;

ω = angular velocity, rad/sec;

i = i^{th} rotating mass; and,

n = number of rotating masses.

The left-hand term in the equation representing linear acceleration is typically about ten times larger than the term for rotational acceleration.

The brake systems of electric vehicles are presently the subject of controversy. Regenerative brakes, disc brakes, and plain, old-fashioned drum brakes have their proponents. Dynamic brakes (which differ from regenerative brakes in that they dissipate energy in resistor banks instead of pumping it back to the batteries) are not favored since they combine the disadvantages of complicated controllers, costly resistors, and energy dissipation.

Probably the most interesting aspect of electric vehicle braking is regenerative braking. The basic concept is simple--instead of dissipating energy mechanically, brake control circuits change the drive motor to a generator that pumps energy back into the batteries. The major benefit of regenerative braking is the extended range provided by battery recharge.

Theoretical estimates of the amount of recoverable energy range from 10 to 40 percent, but the maximum obtained in service for lead-acid battery systems seems to be in the 10 to 15 percent range. A secondary benefit of regenerative braking for lead-acid batteries is that it keeps the electrolyte from forming layers of different acidity within the battery. As a result, more of the acid is effectively used in both charging and discharging, and battery life and efficiency are increased.

Although many different batteries are under development for electric vehicles, the only realistic type at present is the lead-acid battery. This presents some unusual automotive design problems in terms of structure and arrangement. Structural problems are related to the sheer mass of the batteries involved. The curb weight of most electric vehicles (those that have been designed as electrics) is divided about equally between batteries and the balance of the vehicle. Compounding the storage problem are safety, heating, and maintenance problems.

Three key factors of importance in housing batteries are low center of gravity, easy accessibility, and proper temperature. Placing the batteries low in the chassis keeps the vehicle center of gravity low. Good accessibility is needed for electrolyte replacement, corrosion removal, and battery replacement. Finally, batteries must be kept warm to deliver rated power and current.

2. Electric Vehicle Motors

Various types of electric motors can be considered for use in electric vehicles. In the past, interest in electric motors for electric vehicles focused on series-wound dc motors which inherently provide the desirable characteristics of high torque at low speeds and high speeds at low torque demand. Furthermore, the series-wound motor is simple and reliable, does not need sophisticated controls, has a high starting torque, operates directly from battery power, and the series field reduces current ripple when the motor operates at reduced power on time ratio pulse controls. A dc motor produces high tractive effort at low speed for acceleration and grades by drawing more current from the battery. This condition increases system losses due to electrical resistance heating. Components become hot and less efficient and the battery drains rapidly. To reduce these effects during heavy tractive loads, changeable gear ratios can be used with electric drives. In a system with a small battery capacity and a high battery internal resistance--conditions that limit available peak power and stalled (high) motor torque--gearing also helps. Here, increasing the gear ratio increases both low speed tractive effort and available power.

Since speed, torque, and thermal characteristics of electric motors are different from internal combustion engines, an electric vehicle motor cannot be selected by the same criteria as an internal combustion engine. The starting point for approximating an electric motor size can be obtained from the following relation which gives the power requirement for level road vehicle operation.

$$P = C_1 W v + C_2 W v^2 + C_d A v^3 \qquad (5.4)$$

where: P = power requirement, watts;

v = vehicle full speed, m/sec;

W = vehicle weight, newton;

A = vehicle frontal area, m^2 (typically 0.5 to 1.4 m^2);

C_1 = tire rolling resistance and chassis losses constant (typically 0.03 to 0.9 N of drag per N of vehicle weight);

C_2 = constant for tire heat generation from flexing (typically 0.06 to 0.12 N of drag per N of vehicle weight per m/sec); and,

C_d = aerodynamic drag coefficient (typically 0.2 to 0.5).

Maximum motor speed is limited by allowable mechanical stress of the rotor at its periphery. In general, ac rotor diameters are smaller than dc rotors and thus can handle higher peripheral speeds as shown in Figure 5.1. Traction motor

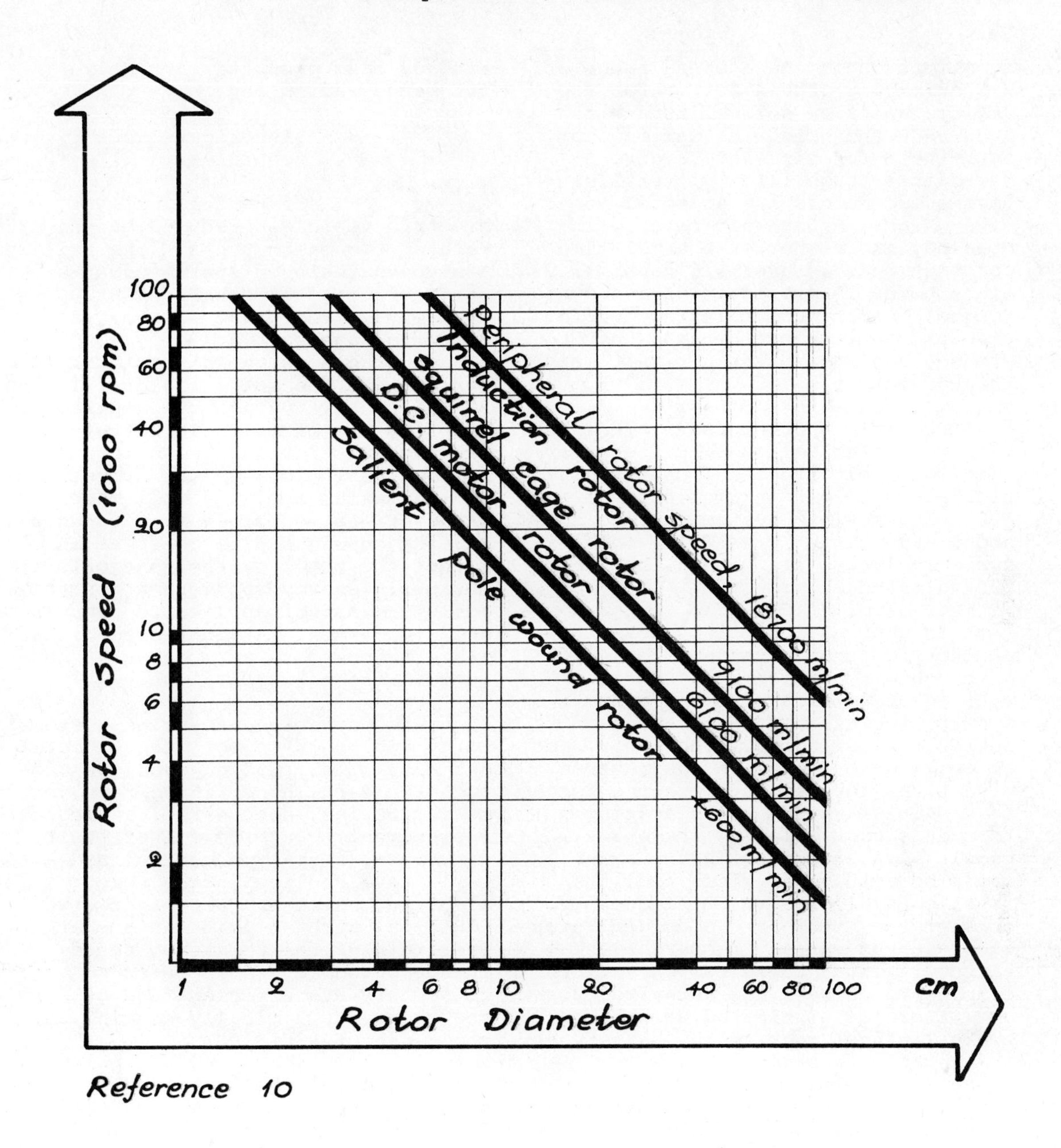

Reference 10

AC-DC MOTOR SPEED CHARACTERISTICS

FIGURE 5.1

suppliers recommend a speed range of approximately 4000-4500 rpm for top vehicle speed on a level road with 5000-6000 rpm being the maximum to optimize motor size and to give a reasonable brush life. If smaller high-speed motors are selected to save weight, brush-wear rate is increased, and a shorter thermal time constant reduces overload capability. Since power is a product of speed and torque, ac motors are attractive for electric vehicle use because they produce a higher power per unit weight than dc motors.

Electric current requirements depend on system voltage. Battery paralleling limitations and safety considerations limit maximum current to 500 A and maximum system voltage to about 400 volts. Presently, most system voltages are within the 72-200 volt range.

In general, dc motors have another inherent advantage for electric vehicle duty. The electric vehicle brake mechanism can actuate a switching circuit that reverses the series field winding. The motor, now attempting to rotate in the reverse direction, applies decelerating torque to the drive shaft. Decelerating torque is controlled by the brake pedal. Depressing the brake pedal can also actuate a switch that transposes controller circuit elements to convert the motor into a dc generator whose output current recharges the battery. Thus, these two operations combine to deliver regenerative energy (otherwise dissipated as heat in brake linings) back to the battery as the vehicle is stopped.

Traction motor cost is relatively high and is another inhibiting factor in the development of a practical electric vehicle. The cost of a 15 kW (20 hp) motor ranges from $700-$900. This is the result of limited production. In fact, all traction motors presently manufactured in the US are hand assembled. One motor manufacturer has stated that if a demand existed for 100,000 units annually, thereby justifying the cost of automated production equipment, the price per motor would be cut in half. This would bring the cost well within the range of comparable internal combustion engines already mass produced in highly-automated facilities.

3. Electric Vehicle Controllers

Control systems for electric vehicles produced before 1920 were of two basic types: one utilized a stepped resistor system and the other utilized a mechanical stepping system. In the first system, the motor voltage was varied by coupling an adjustable (stepped) resistor in series with the motor fields. In the second, motor voltage was applied in steps by use of multiple battery taps and mechanical relays. The resistor system was seriously inefficient. With the batteries applying full power whenever the vehicle was switched on, all the energy had to be spent in either the motor or the controller. Obviously energy spent in a control resistor was not applied to propulsion and was wasted. At low speed, this waste amounted to as much as 90 percent of the energy available.

The mechanical stepping system is noisy and rough on both the vehicle and batteries. Sudden overloads imposed by each change in voltage degrade the batteries, necessitating frequent replacement. Despite all of their drawbacks, step-type controllers still have a place in the spectrum of electric vehicles. For lightweight vehicles, such as golf carts that operate near their maximum speed, these controllers are preferred because they are efficient and by far the cheapest, typically costing in the range of $30-$100.

The development that makes multi-stop, low-speed, road-worthy vehicles feasible again is the power semiconductor. Appearing in the early 1960s, it is used in controllers basically as a switch. It pulses full power to a motor, then cuts it off repeatedly, far faster than a mechanical switch could operate. In this way, batteries are not lost through resistor dissipation. The inductance of the motor field as well as the inertia of the motor has the effect of a low-pass filter on the chopped power pulses, and the motor responds as though it were getting a smooth signal (Figure 5.2). Motor speed depends on the rate at which the motor is switched onto the

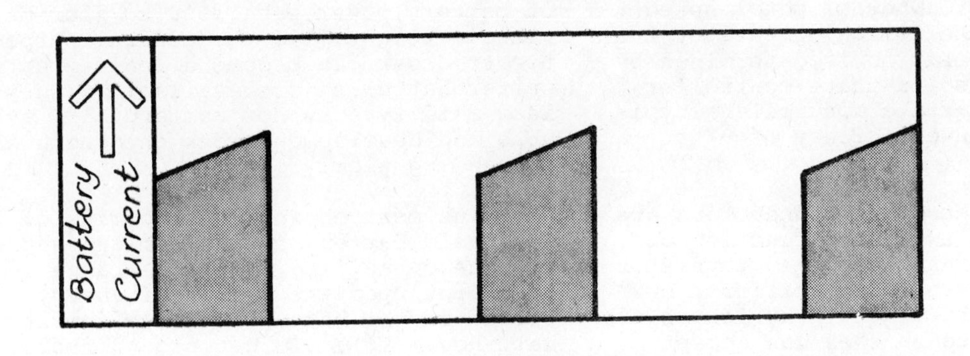

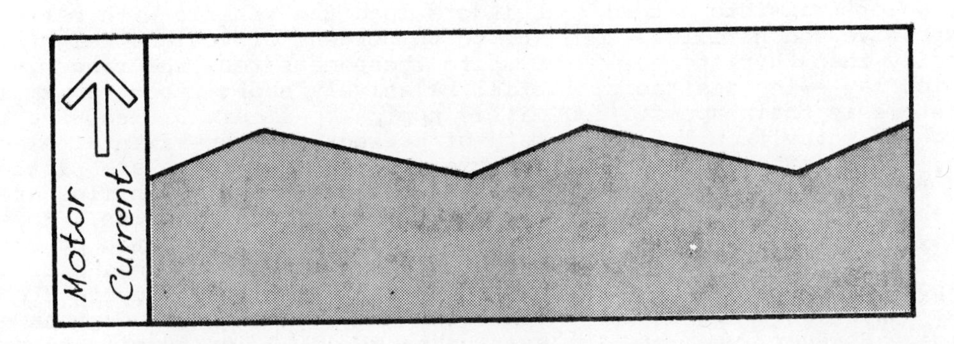

Battery current is chopped by SCR controller (top). But energy stored in the field winding maintains a current through the flyback diode and armature until the battery is switched on again. Thus, the motor receives a more uniform current (bottom).

Reference 10

EFFECT OF MOTOR FIELD ON CHOPPED POWER PULSES **FIGURE 5.2**

battery or on the duration of the power pulses. Whether motor speed depends on the switching rate or the pulse rate duration is determined by the type of solid-state controller used. Solid-state controllers typically have an efficiency of 97 percent.

Two types of semiconductors are used for the switching function of controller-transistors and thyristors (silicon controlled rectifiers, or SCRs). At the moment, thyristors are preferred because they can accept higher loads (typically, 48 volts and 600A). Also, power transistors tend to be more expensive and have less overload capacity than thyristors. One factor which may swing designers toward transistors is their speed. Most can switch at rates from 1 to 20 kHz. SCRs, on the other hand, rarely switch faster than 1 kHz which means a mild shudder in electric vehicles during startup.

For low-power, short-range vehicles, step controllers are still the best choice. However, when precise control of speed, good acceleration throughout the speed range, multiple stops, and greatest range are required, a semiconductor-based controller is the best solution. Most designers agree that a vehicle power limit of 3.75 kW (5 hp) is the cutoff point in the stepper versus semiconductor controversy. Vehicles over this limit should be fitted with a semiconductor controller. See Chapter 11 for further information related to solid-state electric motor controllers.

4. Electric Vehicle Power Sources

The main problem area in the practical design of electric vehicles is that of high-performance, long-lived batteries. Even if new electro-chemistry systems are developed, electric vehicle batteries are not likely to be able to store enough energy to make possible vehicle performance and occupant comfort matching those of the big, accessory-equipped family cars of today. In order to operate safely in traffic, a certain minimal acceleration is required (0 to 50 km/hr in 10 sec or less is probably acceptable).[10] This means a power level of about 25 kW of battery power per 1,000 kg (per mt) of vehicle weight.[11] No general purpose electric car can become a reality until better batteries are available. Only the lead-acid type is now commercially available and development work on others is proceeding slowly.

The most obvious shortcoming of the lead-acid battery is its limited energy storage capability. Vehicle range is dependent upon the amount of energy stored in the battery (energy density in watt-hours/kg). To get around problems of low-energy density, present electric builders load the vehicle with batteries (up to 50 percent of vehicle weight). Despite these measures, the range is still relatively short; about 80 km (50 mi) is typical. Figure 5.3 shows a summary of present-day experimental electric vehicles. Speed and range are plotted on coordinates. All of the vehicles are powered by lead-acid batteries. Some have an excessively high ratio of battery weight-to-curb weight. By means of an averaging technique for grouping the vehicles, the data indicate that an average range of 80 km (50 mi) and a speed of 40 km/hr (25 mph) comprise the expected performance of the electric vehicles on the market today.

The energy required for average urban driving needs is typically 0.10 to 0.12 kWh/mt-km (at the axle), and the average power (at the axle) is 3 to 3.5 kW/mt.[11] Therefore, the battery power source should deliver 0.14 to 0.17 kWh/mt-km x 100 km = 14 to 17 kWh of energy for a 100 km (60 mi) range urban vehicle.[11] This takes into account efficiency losses between the battery and the axle and assumes an average power level of 4 to 5 kW. Peaks of 25 kW might be needed for a 1,000 kg vehicle to accomplish a full day's urban travel.[11]

A test bed electric vehicle that has seen considerable service uses 341 kg (750 lb) of lead-acid batteries, almost 46 percent of its curb weight. Twelve batteries are mounted in a tray that can be removed quickly for reloading. Suppose such a car were marketed and the batteries cost $50 each. If they were recharged every night and if they had a life of 1,000 cycles, $600 worth of batteries (minus trade-in price) would have to be replaced every three years. Therefore,

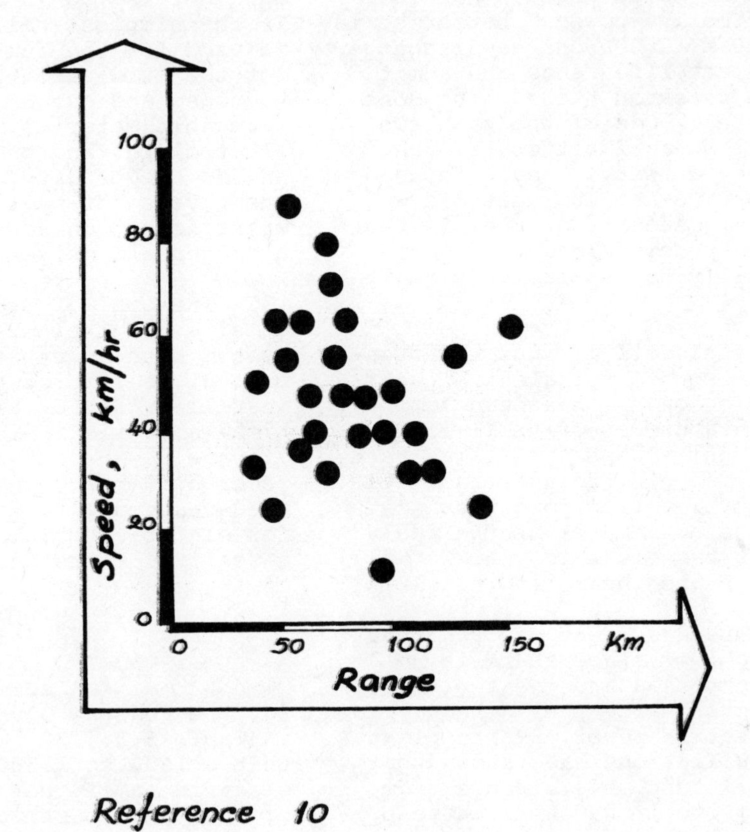

Reference 10

SUMMARY OF EXPERIMENTAL ELECTRIC VEHICLES TO DATE FIGURE 5.3

both first cost of the batteries and cycle life are critical to the future of the electric car.

Battery life depends on a number of factors including how deeply the batteries are discharged. Manufacturers of motive-powered batteries indicate 300 to 2,000 charge-discharge cycles as the life range and admit that much research needs to be done in this area. One of the problems with the lead-acid battery is that high energy density, long life, and cyclic service are incompatible parameters. Another problem is that the less expensive lead-acid batteries generally have relatively short lives.

The fuel cell is another potential power source for electric vehicles. This concept has been demonstrated in a number of small-scale applications. Chapter 12 illustrates an industrial electrolyzer/fuel cell concept for leveling power demands. Some of the advantages of the fuel cell for urban electric vehicle application are the following:[12]

- There are fewer environmental problems since there is no combustion and the fuel cell converts its fuel directly to electric power. Power plant applications have shown that pollutant levels range from 1/10 to 1/50,000 of those produced by a fossil-fuel power plant.
- No cooling water is needed for fuel cell operation.
- The fuel cell makes no objectionable noise, providing a quiet ride for passengers.
- Fuel cells can be built in a wide variety of sizes to suit the desired application.
- Present fuel cells have a relatively high operating efficiency of greater than 35 percent. (If a fuel cell system can be integrated with a heat pump to utilize the waste heat, a total system efficiency of 94 percent has been proven possible.) See Figure 5.4.
- A wide variety of fuels may ultimately be used with the fuel cell.

The operational principle of the fuel cell is similar to a refuelable battery. As shown in Figure 5.5, fuel is supplied at the anode, and the oxidizer is fed in at the cathode.[12] In the simplest cell, pure hydrogen is used for the fuel and pure oxygen for the oxidizer. In the fuel cell hydrogen and oxygen pass through permeable material, interact in an electrolyte, and produce water, heat and dc electricity. The reaction is the reverse of hydrolysis, in which water is broken down electrically into hydrogen and oxygen.[12]

The goal of present research for on-site power plants is a fuel cell system with an efficiency of 37 to 40 percent, a life of 40,000 hr, and an installed cost of $250/kW.[12] Practical electric vehicle design would dictate lower cost per kW and high power and energy densities with associated reduced life times. The fuel cell for electric vehicle use could not be a reality before the mid-1980's.

5. Future Electric Vehicle Requirements

A summary of one study for the Federal Government is presented in Table 5.2.[13] The table indicates that a 1600 kg (3500 lb) six-passenger electric family car with special lightweight construction would have an allowable battery weight of 637 kg (1402 lb) and would require a battery energy density of 191 Wh/kg (87 Wh/lb) and a power density of 132 W/kg (60 W/lb). At the present time, the best that the lead-acid battery can do is to produce energies of 35-40 Wh/kg (16-20 Wh/lb) at low drains and power densities of about 77 W/kg (35 W/lb) for brief periods. Clearly the current state-of-the-art lead-acid battery will not be a candidate power source for the full-size family car of the future.

Other batteries now being developed also fall short of the power and energy density requirements for the 1600 kg family car. Figure 5.6 presents a comparison of the power and energy capabilities of various types of batteries in addition to a comparison with the internal combustion engine. Data for the familiar battery

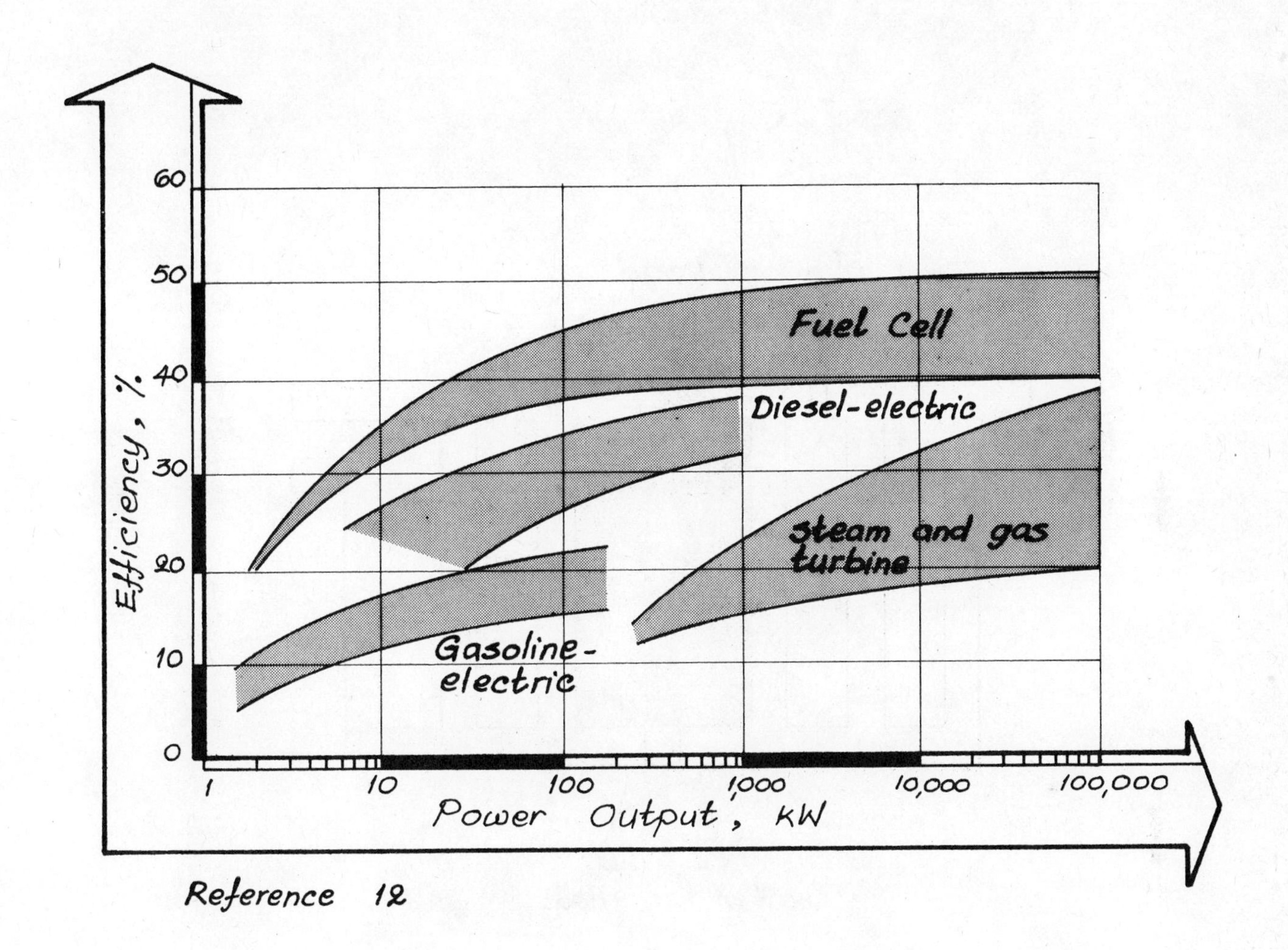

Reference 12

FUEL CELL EFFICIENCY RELATIVE TO VEHICLE POWER SOURCE ALTERNATIVES

FIGURE 5.4

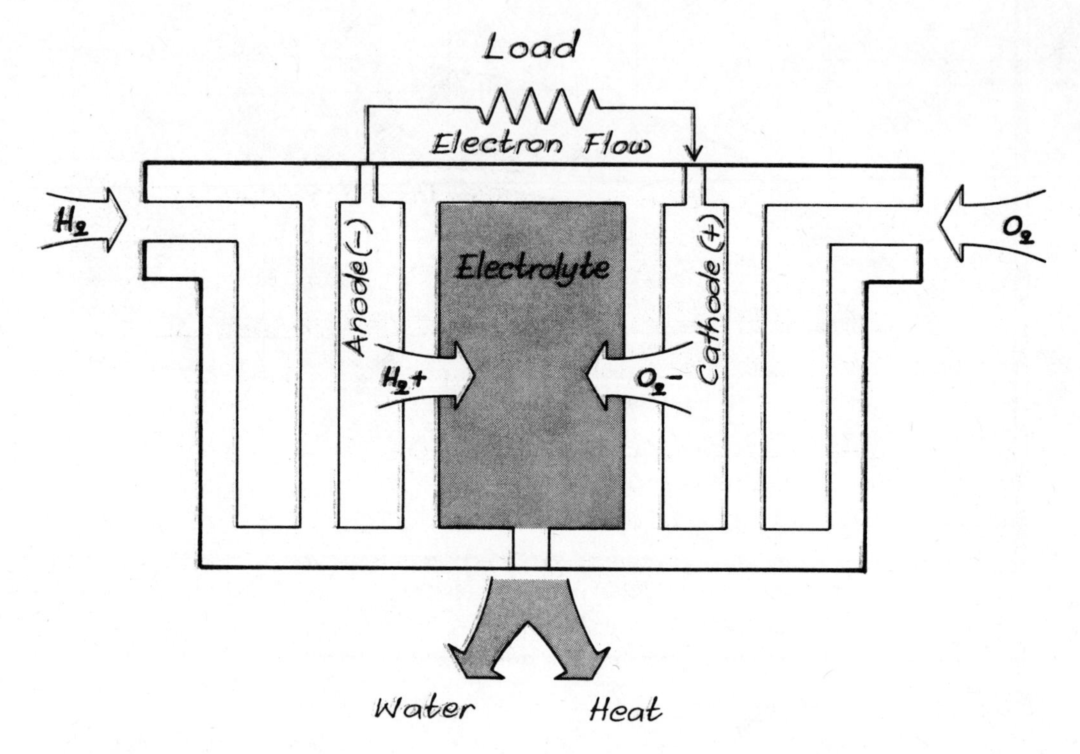

Reference 12

OPERATION OF A FUEL CELL **FIGURE 5.5**

	Family Car	Com-muter Car	Utility Car	Delivery Van	City Taxi	City Bus
Maximum Velocity (kmph)	161	129	105	89	121	89
Acceleration to (kmph)	97	97	48	64	64	48
In (sec)	15	30	10	20	15	15
Range (km)	322	161	80	97	241	193
Seats or Payload (kg)	6	4	2	1,134	6	4,536
Curb Weight (kg)	1,588	953	635	2,041	1,588	9,072
Stored Energy (kWh)	122	26	11	57	99	353
Conventional Construction						
Weight Assignable to Power Source (kg)	409	287	190	518	472	1,989
Power Source Requirements:						
Energy Density (Wh/kg)	298	90	57	110	212	179
Power Density (W/kg)	207	101	88	121	99	79
Lightweight Construction						
Weight Assignable to Power Source (kg)	636	423	280	790	699	2,896
Power Source Requirements:						
Energy Density (Wh/kg)	192	62	40	73	141	121
Power Density (W/kg)	132	68	62	79	66	55

Reference 13

RECOMMENDED ELECTRIC VEHICLE SPECIFICATIONS

TABLE 5.2

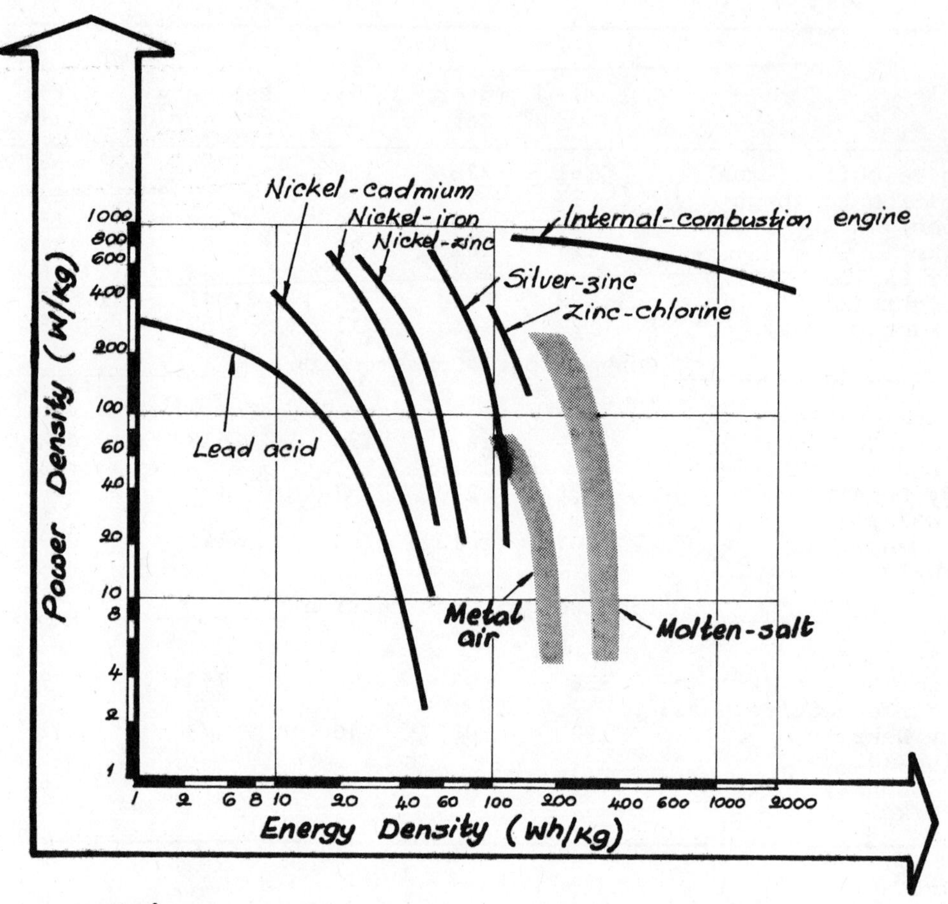

Reference 10

POWER AND ENERGY CAPABILITIES OF VARIOUS BATTERIES

FIGURE 5.6

types are well established, although information is changing as improvements are made. Data for advanced batteries are less trustworthy, representing the best educated guesses the battery industry can make at this time. The figure indicates that the molten salt battery is the only one which would fulfill the requirements for a full-size family electric vehicle of the future. Other types of personal road cars fare better in the analysis, even though most cannot be powered by the lead-acid battery. The lightweight version of the tiny two-seat utility car appears to be a reasonable near-term prospect if the lead-acid battery can be improved.

6. Batteries Under Development

Since the lead-acid battery is inadequate for the requirements of future vehicles, better performance is required in a new battery. The report described above provides calculations which compare the range of a commuter car powered by an 88 Wh/kg (40 Wh/lb) energy density battery system with the same vehicle powered by lead-acid batteries with an assumed 26 Wh/kg (12 Wh/lb) energy density. The calculations revealed that a modest increase in energy density allows a substantial performance improvement in the vehicle, as illustrated in Figure 5.7.

Some of the more salient characteristics of the various types of candidate batteries are summarized in Table 5.3. A review of this table along with Figure 5.6 indicates that the molten salt batteries have the greatest potential for meeting the requirements for electric vehicle power sources.

One of the main contenders in the molten salt category is the sodium-sulfur battery because of the abundant world supply of the constituents. Mass production of batteries would not deplete reserves of sodium and sulfur. A distinctive feature of the sodium-sulfur cell is that a solid electrolyte separates a liquid sodium electrode from a liquid sulfur electrode (see Figure 5.8).[14] The electrolyte acts as an ion filter so that sodium passes through it to react with sulfur to form sodium sulfide, but only when an electrical current passes around an electrical circuit such as in the case of an electric vehicle motor. The sodium used in the reaction is contained in a stainless-steel sodium reservoir connected to the solid electrolyte.

Sulfur is contained in the space between the outside of the solid electrolyte tube and the outer case of the cell. Since sulfur is an insulator, this space is filled with porous carbon to provide electrical conduction for efficient operation of the electrode. The electrode is partially filled with sulfur when fully charged but is completely filled with sodium sulfide when fully discharged. When the cell is charged, sodium and sulfur are regenerated from the sodium sulfide.

Road tests conducted by the Electricity Council in England have shown that a sodium-sulfur-powered van has a range between 96 and 160 km (60 and 100 mi), depending on road and driving conditions. Recent technical developments have led to substantial reductions in overall battery size and weight, and it now appears feasible to increase the capacity of single cells to about 550 Wh. Ninety such cells will be used to construct a 50 kWh van battery. Table 5.4 presents a summary comparison of present and projected sodium-sulfur and lead-acid batteries.

7. Summary for Electric Vehicles

With sufficient research into the critical areas of electric vehicle design, electric vehicle production could become a reality in five to ten years. Table 5.5 summarizes critical areas for achieving efficient electricity use with electric vehicles. Case Study 5-1 discusses one recent development of a prototype electric vehicle.

B. Electric Trains

This section is concerned with the use of electricity to provide energy to propel large commuter and freight vehicles such as trains and

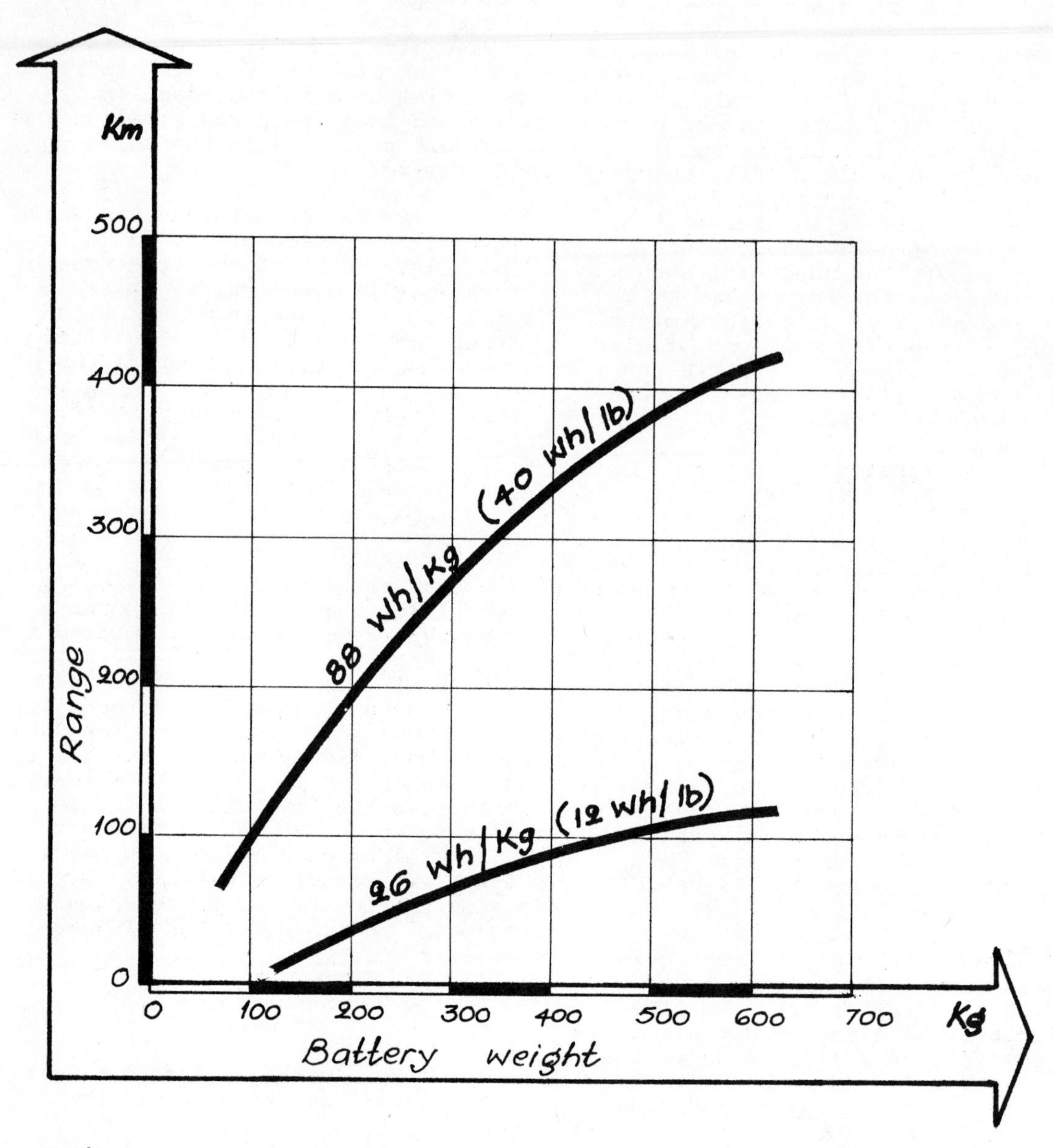

Reference 10

VEHICLE PERFORMANCE COMPARISON

FIGURE 5.7

Battery Type	Energy Density Range Wh/kg)	Power Density Range (W/kg)	Operating Temperature (°C)	Cell Life	Projected Development Time - Yrs	Relative Initial Cost
Lead-acid	35-44	77	Ambient	300-2000 cycles	0	Low
Nickel-cadmium	11-44	18-33	Ambient	?	?	High
Nickel-iron	22-44	44-440	Ambient	?	2 yrs	Moderate
Nickel-zinc	22-66	44-440	Ambient	?	5 yrs	Moderate
Silver-zinc	66-110	44-440	Ambient	Low	?	Very high
Zinc-chlorine	110-150	110-330	Ambient	500 cycles	4 yrs	Very low
Metal-air: Zinc-air	99	?	Ambient	Short	?	Moderate
Molten salt: Lithium-sulfur	>220	>220	400	Short	5 yrs	Moderate
Lithium-chlorine	229-330	220-620	650	Very short	>7 yrs	Moderate
Lithium-chloride/potassium chloride	150	220	300-500	?	?	Moderate
Sodium-sulfur	>220	>220	300	5 yrs (1000 cycles)	?	Moderate

SUMMARY OF CANDIDATE BATTERY CHARACTERISTICS

TABLE 5.3

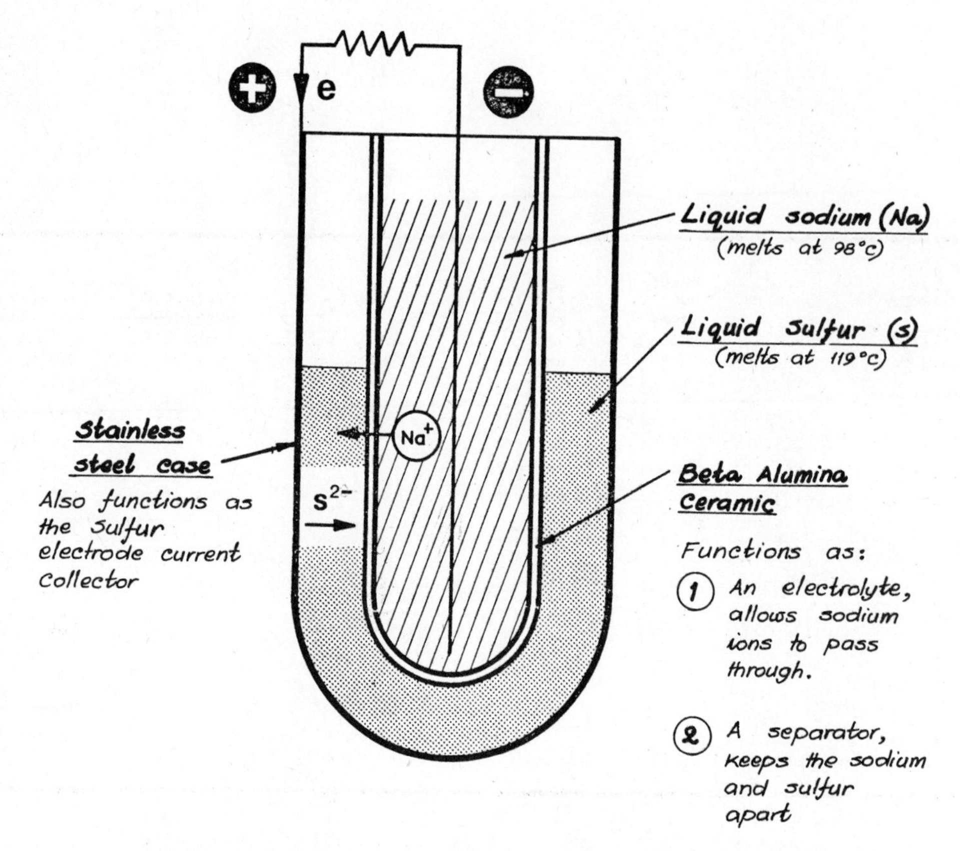

Liquid Reactants, Solid Electrolyte

Current is carried by Sodium ions (Na^+) which give up electrons (e) to the external circuit, pass through the solid electrolyte and react with sulfur (S)

Reference 14

SODIUM SULFUR CELL **FIGURE 5.8**

POWER SOURCE	Weight of 50 kWh	Volume of 50 kWh	Cycle Life
Commercial lead acid traction battery	2000 kg	1.06 m^3	2000
Electricity Council sodium-sulfur cell	313 kg	0.156 m^3	2500 to date
Electricity Council experimental battery No. 1 (NAS 1)	800 kg	1.5 m^3	---
Electricity Council experimental battery No. 2	600 kg	1.2 m^3	---
PROJECTIONS			
POWER SOURCE	Weight of 50 kWh	Volume of 50 kWh	Cycle Life
Developed lead acid traction battery	1250 kg	?	1000
Developed NaS cell	170 kg	0.11 m^3	>2000
Developed NaS battery	250 kg	0.25 m^3	>1500

Reference 14

SUMMARY COMPARISON OF SODIUM–SULFUR POWER SOURCES

TABLE 5.4

Electrical Vehicle Design Improvements

Reduce aerodynamic drag:

- Minimize vehicle frontal area by possibly seating passengers single file and using semi-reclining position.
- Enclose underbody with a smooth pan.
- Make glass areas flush with vehicle.
- Use aircraft style air inlets.
- Minimize all protuberances such as door handles, windshield wipers, and bumpers.
- Employ body designs that minimize lift and flow separation.

Reduce chassis losses:

- Minimize tire losses by using belted radial tires, special rubber compounding, reduced width, stiffer sidewalls, and lower aspect ratios.
- Minimize bearing losses by using optimum viscosity lubricants.
- Minimize power train losses by using efficient gearing and mechanical energy transfer mechanisms.

Reduce inertia losses:

- Minimize total vehicle weight.
- Optimize vehicle acceleration requirements.
- Minimize number and weights of rotating masses on vehicle such as wheels, motor armatures, and driveline components.

Reduce accessory energy losses:

- Optimize energy requirements for headlights and windshield wipers by using most efficient lamps and considering possible semi-manual wipers (e.g., spring windup or hand-pumped compressed air system).
- Minimize vehicle power source drain for comfort heating by considering the use of:
  1) waste heat transfer from the motor and controller for vehicle heating requirements,
  2) electrically powered radiant heat panels and in-seat heating coils, or
  3) fossil-fueled heaters.
- Eliminate unnecessary accessories such as air-conditioning, power windows, and other power systems within the vehicle.

Reduce braking power losses:

- Avoid use of dynamic brakes which dissipate energy in resistor banks.
- Investigate carefully the weight, drag, and maintenance considerations for selecting drum vs disc brakes.
- Incorporate regenerative braking if possible (note that synchronous reluctance motors with disc-shape rotors and stators require the least controller modification for regenerative braking). Consider attaching an alternator to the drive motors to provide regenerative braking.

Optimize packaging and protection of batteries:

- Place batteries low in the chassis to maintain a low center of gravity and to the front of the vehicle for safety of handling.
- Maintain batteries at proper temperature to maximize efficiency and capacity (proper temperature is dependent on the type of battery used).
- Provide easy accessibility of batteries for electrolyte replacement, corrosion control and removal, in-vehicle charging, and battery replacement.
- Consider auxiliary battery racks on coasters for easy removal and external vehicle charging of batteries.

Electric Vehicle Motor Considerations

- To specify the proper motor, the motor manufacturer must know the following vehicle performance requirements: desired level running top speed, required acceleration, frequency of starts, desired gradability, maximum emergency tractive effort, gross vehicle weight, available gear ratios, rolling radius of drive wheels, vehicle drag characteristics, battery and internal resistance, cable resistances, and type of system control.

Electric Vehicle Controller Considerations

- The stepped resistor control system should be avoided since it is seriously inefficient by dissipating energy as heat in resistors, particularly at low speeds.
- Mechanical stepping control systems have the disadvantages of being noisy and rough on both the vehicle and batteries by imposing sudden overloads at each change in voltage. Their advantages are that for lightweight vehicles, they operate efficiently near their maximum speed and are comparatively inexpensive.
- Solid-state controllers have efficiencies of about 97 percent, batteries are not subjected to sudden overloads, power is not lost through resistor dissipation, precise control of speed is achievable, and good acceleration throughout the speed range is possible.
- Thyristors (SCR's) are presently preferred over transistors as the semiconductor used in solid-state switching function controllers because they can accept higher loads (typically 48 V and 600 A), are less expensive and have greater overload capacity, although they have slower switching speeds (1 kHz compared with up to 20 kHz for transistors).

Electric Vehicle Power Source Considerations

- The power source requirements for a lightweight construction six-passenger family car are for batteries with an energy density of 191 Wh/kg (87 Wh/lb) and a power density of 132 W/kg (60 W/lb).
- Table 5.3 and Figure 5.6 give a summary of candidate battery characteristics.

IMPORTANT EFFICIENCY CONSIDERATIONS FOR ELECTRIC VEHICLES

TABLE 5.5

articulated buses. The method of locomotion is different from that in the previous section since electrical energy is supplied from outside the system, as opposed to self-contained energy supplied from batteries or fuel cells. This type of transportation is referred to here as electric traction. The distinguishing feature of electric traction is the separation of the source of motive power from the rolling stock.

Certain classes of traffic are suitable for electric traction. Street and inter-urban systems are particularly suitable because they permit the generation of motive power to be concentrated in a single economical plant instead of being distributed over a number of comparatively uneconomical ones, as in the case of steam or diesel traction for trains.

The fractions of both traffic carried and energy used by railroads in the US declined markedly between 1950 and 1970.[1] In 1950, railroads accounted for 47 percent of intercity freight traffic, 7 percent of intercity passenger traffic, and 25 percent of transportation energy use.[1] By 1970, these values had declined 35, 1, and 3 percent, respectively.[1] (Table 5.6 summarizes traffic data and energy use for railroads between 1950 and 1970.) This raises questions concerning the efficiency of the transportation industry. In fact, it has been said that the health of a country cannot be better than that of the transportation industry, of which railroads are a major element.[15]

A potential use for electrified railways is in mass rapid transit systems. Urban mass transit includes gasoline-powered buses and electric subways, elevated trains, trolleys, and surface railways. Mass transit in 1970 accounted for 0.5 percent of transportation energy use and 3 percent of urban passenger traffic. In 1950 the comparable figures were 1.8 percent and 15 percent, respectively.[1] About 60 percent of mass transit traffic is carried by bus. Table 5.7 summarizes energy use and traffic data for mass transit between 1950 and 1970.

One solution to the transportation fuel crisis is the electrification of railways, including in urban transportation systems. Although electric rail has existed in the US since 1880, the nation has never adapted the system as extensivley as have most other industrialized nations: Switzerland has almost 100 percent, Sweden has 60 percent, Japan has 40 percent, and the Soviet Union has 25 percent and is reported adding up to 1100 km each year of electrified rail.[16] The US has less than 1 percent of its 332,000 km (206,000 mi) network electrified.[16] In fact, the Los Angeles area had the largest electric interurban railway system in the world, over 1930 km (1200 mi) of tracks with more than 7000 trains operated daily in 1925.[17] The entire system was eliminated from 1951 to 1961; its demise is credited with being the direct cause of the region's present transit problems.[17]

There is a renewed interest in railway electrification with the role of the diesel-electric locomotive being reevaluated.[16,18] For example, the Canadian Pacific Railway has conducted extensive studies to determine the feasibility of converting its main line between Calgary, Alberta, and Vancouver, B.C., from diesel-electric to all electric operation.[18] The rising cost of diesel fuel, coupled with the hydroelectric development of the Columbia River and its tributaries, has provided a new perspective for considering the economic potential for electrification.[18]

In 1976 the US government enacted the Four-R Act (Railroad Revitalization and Regulatory Reform) which authorized $1.75 billion to improve freight and passenger service in the Northeast Corridor.[16] In this five-year program, the already existing electrified line between Washington and Boston will be improved, and the tracks from Boston to New Haven, Connecticut will be electrified.[16] In the US there are added advantages to all electric rail transportation, including reduced environmental pollution and the ability to satisfy future needs for high speed transportation.

On electric train lines, the generating and distributing systems

	IC Freight		IC Passenger		Total
	Traffic 10^9 mt·km	EI kJ/mt·km	Traffic 10^9 Pkm	EI kJ/Pkm	Energy 10^9 GJ
1950	920	2,240	53	12,560	2,320
1955	960	870	47	6,280	940
1960	880	570	35	4,920	570
1965	1,050	520	29	4,580	600
1970	1,120	480	18	4,920	580

Note: IC = intercity
mt·km = metric ton-kilometer ≈ 0.68 ton-mile
kJ = kilojoules ≈ 0.95 Btu
Pkm = passenger-kilometer ≈ 0.62 passenger-mile
EI = energy intensiveness

Reference 1

RAILROAD TRAFFIC AND ENERGY USAGE IN THE US — TABLE 5.6

	Electric		Bus		Total		Average
	Traffic 10^9 Pkm	EI kJ/Pkm	Traffic 10^9 Pkm	EI kJ/Pkm	Traffic 10^9 Pkm	Energy 10^9 GJ	EI kJ/Pkm
1950	35	6,620	39	5,260	74	170	5,940
1955	21	6,450	29	5,770	50	120	5,940
1960	14	6,620	26	5,770	40	94	6,110
1965	12	6,620	24	5,940	35	85	6,280
1970	12	6,960	21	6,280	32	80	6,450

Note: mt·km = metric ton-kilometer ≈ 0.68 mile
kJ = kilojoules ≈ 0.95 Btu
Pkm = passenger-kilometer ≈ 0.62 passenger-mile
EI = energy intensiveness

Reference 1

URBAN MASS TRANSIT TRAFFIC AND ENERGY USAGE IN THE US — TABLE 5.7

must have large capacities in order to handle the heavy trains, but are not kept fully employed due to the comparative infrequency of traffic. This results in heavy capital investments which are idle a large part of the time and have high fixed charges. The following conditions are exceptions to this generalization:[19]

- when the cost of fuel is high, the operating costs may be sufficient to counterbalance the fixed charges;
- when nearby hydroelectric power is available, a similar condition may occur;
- when it is desired to increase the freight weight of trains hauled on steep grades, eliminating the power generating system from the locomotive by electric traction permits an increased concentration of motors in the locomotive, thereby obtaining the desired increase in hauling power, at a lower cost than by the elimination of grades;
- when it is necessary to eliminate smoke and gases in tunnels or city streets, the economic element being necessarily subordinated to safety or convenience;
- when traffic density is unusually great, as in certain suburban zones; and,
- when by the elimination of smoke and gases, urban trains can be run underground and the ground surface utilized for other purposes.

Electrification of railroads cannot occur on a small scale since economics dictate that large sections of rail be electrified at one time.[16] Present estimates are that electrification would cost between $56,000 and $81,000 per track kilometer ($90,000 and $130,000 per track mile).[16] Signal and communication systems would account for about 20 percent of this amount with the remainder allocated for overhead wire and power stations. An electric locomotive costs about 30 percent more than a diesel-electric; on a cost per unit power basis, however, the electric locomotive has a slight advantage.

Electric railway stationary equipment includes generation, transmission, and distribution systems. These systems must have sufficient capacity to provide for peak loads and therefore are likely to be relatively idle during a major part of the operation time. The relative activity with time of these systems is dependent upon the number of cars in operation during various periods of operation.

The moving equipment portion of an electric train system consists of motors, controllers, and other parts of the train necessary for electric operation. The relative activity with time of these components does not depend on the number of cars in service, but only upon the activity of each car of the train.

There are several systems for electric trains and these differ in the relative distribution of the investment between the stationary and moving equipment.[19] Stationary equipment is less expensive than moving equipment of the same capacity because it is concentrated in larger units and its basic design is not limited by the construction of the car on which the equipment is mounted. Thus, electric train systems requiring a large number of cars for service will have a relatively large investment in stationary equipment. Conversely, railway systems using few cars will economize in stationary equipment and concentrate the investment in a limited quantity of moving equipment.

The newer electrified rail systems in the US operate at 50 kV and 60 Hz.[16] Power was fixed at 60 Hz so that utility power could be used without costly frequency coversion. The 50 kV system has the advantage of lower current than previously used 25 kV systems. Higher voltage and lower current means that 1) a lighter weight, less costly catenary is required; 2) current loss is less; and 3) greater spacing between substations is allowed so that fewer stations need be constructed.

New advances in rapid transit equipment will improve the performance of electric rail systems. One example is the use of magnetic levitation systems to improve the ride and reduce friction losses and noise. The results of one demonstrated small scale system showed that actual levitation power amounts to only 0.35 to .70 watts per kg (0.5 to 1.5 watts per pound) of gross weight.[20]

Case Study 5-2 gives an historical review of electric rail in the US and explains the basic operation of the electric locomotive.

C. Other Modes of Electric Transportation

In addition to the two major forms of electric powered transportation (i.e., electric vehicles and electric trains), other modes of short distance transportation are based on electric power. These are "people movers," usually located within structures to replace walking from floor to floor or over horizontal distances within large facilities.

Typical examples of people movers are elevators, escalators, moving sidewalks, small electric carts, intra-mall trams, etc. People movers almost exclusively use electric power for locomotion, either directly through ac electric motors or indirectly under charged battery operation.

Elevators are often the target of energy use reduction economies in commercial buildings. Normal working time of an elevator in a commercial building is estimated at 10 hours per day and 22 days per month.[21] A rule of thumb for the average elevator system capacity is: full capacity 6 percent of time, 33 percent capacity 75 percent of time, and empty 19 percent of time.[21] This results in an average of approximately 31 percent capacity continuously over the normal working time period.

The most efficient elevator systems are counter-weighted. With a 40 percent counter-weighted elevator system, the electrical section--motor generator set and hoist motor--experiences a working equivalency of 33 percent of the continuous full-load duty during loading.[21] With the motor generator concept for elevators, the power used to transport people up is returned when the people are brought down. Moving people to the twentieth floor may be considered nothing more than storing them as potential energy; this potential energy is converted to kinetic energy when they are returned to the ground floor and the electrical elevator system regenerates power through the motor generator.[21]

Typical energy losses in an elevator system are in three major areas: 1) motor-generator set: the efficiency of a motor-generator set is between 83 and 87 percent at full rated load; 2) hoisting machine and motor: the motor alone will have an efficiency usually between 88 and 92 percent, the gear unit (if used) will average 85 percent efficiency; and 3) the hoistway: this is usually associated with a loss of 10 percent up to 300 m/sec (1000 fpm) and a 12 percent loss above 300 m/sec.[21]

Case Study 5-6 indicates that not much energy can be saved by reducing the number of elevators in buildings except in cases where the building has excessive capacity. Reducing the number of elevators will result in less passenger service, and possibly higher maintenance on the remaining cars because of their extra burden of operation and misuse, e.g., double button pushing, holding doors, etc.[21]

One effective energy management option is to interlock elevators with the lighting system in order to be able to turn both off when not in use. The security and fire protection services should also be interlocked with the elevators but independent of the lighting/elevator interlock. Another effective measure is to request people to "walk up one, walk down two floors, please." This can save approximately 10 to 15 percent of the energy used in an elevator system.[21] In a 15-floor building, the average run is three floors; so if 50 percent of the one-floor runs are saved, a significant fraction of the energy used by the elevator system will be reduced.

5.3 ENERGY USE DISPLACEMENT BY COMMUNICATION

In the past 20 years telecommunications and electronic data processing have increased their relative utilization of electrical energy at a faster rate than most of the other current electricity users. At the same time, the efficiency of their electricity use per unit function has also increased. This efficiency increase is primarily a result of the development and growing use of low-power solid-state electronics for control and logic elements.

Telecommunications have the potential for saving energy by substituting electronic communication for personal transportation. In situations requiring primarily information transmittal, telecommunications offer a cost effective method of accomplishing the transmittal.

The potential for urban telecommunications in the US has been under intensive study during the past several years.[22] It has been spurred on by the rapid development of cable television systems and the future possibility that these systems in combination with the present telephone communication system, could be used to create a system of urban telecommunications.

Telecommunications are defined as any transactions, emissions, or reception of signs, signals, written images and sounds, or intelligence of any nature by wire, radio, visual, or other electromagnetic systems, including any intervening processing and storage.[23] The common interpretation is any of a wide variety of applications of two-way audio/visual communication systems. Some examples of the wide range of applications are remote educational and medical services, automated transportation vehicle control, remote monitoring of human activities, automated data gathering, and instantaneous polling of the public on current issues.[22]

A recent concept that is receiving increased attention is "electronic mail service" for transmission of written information.[24,25,26] Up to 80 percent of first-class mail could soon be carried electronically. A rudimentary national electronic mail network that already exists with more than 100,000 teletypewriters, 15,000 communicating word processors, 1 million computer terminals and a multitude of telecopiers and telefax machines, could be coupled with existing telephone systems.[25] The price of the service would be competitive since even today one can send a page of text via telephone lines a distance of 1600 km (1000 mi) for a cost of 5 cents, with delivery accomplished within a few minutes.[25] Figure 5.9 shows the trend of costs for various information transfer systems. Continued reduction in the cost of using electronic mail systems will have the impact of increasing electricity use and decreasing motorized mail transportation.

A prime prospect for the substitution of telecommunications for transportation exists with members of the information industries (insurance, finance, education, government agencies, and administrators in general) for work and business related trips.[27] Using a local insurance company as an example, it is estimated that for each 1 percent of the US urban commuter work force which substitutes telecommunications for transportation, a net energy savings of 5.4×10^7 GJ/yr (5.1×10^{13} Btu/yr) might be possible. Preliminary calculations indicate that telecommunications substitution needs would use approximately 3 percent of the energy consumed by the private automobile it would replace. The same substitution would use about 6 percent of the energy required by urban mass transit with normal load factors, and about 35 to 40 percent of that required by urban mass transit with a 100 percent load factor.[27]

The "white collar" work roles which have a high degree of conceptual and transactional activity show the best telecommunications/transportation tradeoff.[28] Estimated dates when the work trip and other trip substitutions might be introduced in the US are:[29]

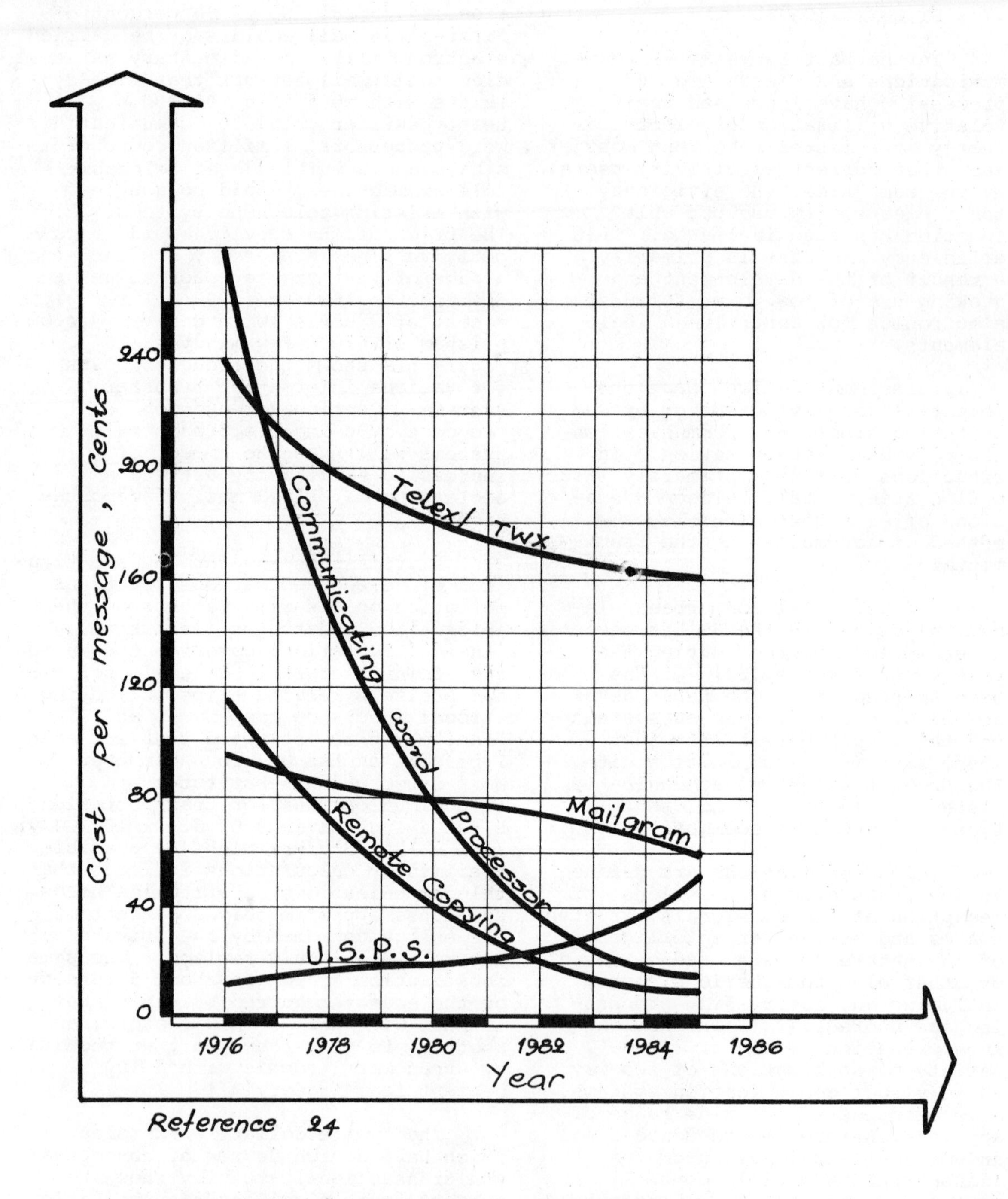

TIME AND COST FOR ELECTRONIC MAIL AND MESSAGE SYSTEMS FIGURE 5.9

Telecommunications Substitution	Estimated Dates
Work Trip	1985-1990
Education - Adult	1980-1985
Shopping - General	1985-1990
Shopping - Grocery	1980-1990
Education - Library	1985-1990
Work - Company File Access	1985-1990
Education - Tutor	1980-1990
Education - Computer Aided	1982-1987

Thus, possibilities exist for telecommunications/transportation substitutions within the next five to fifteen years. The costs and benefits of such substitutions, however, are not yet defined.

Crucial differences exist between the nature of transportation and telecommunications technology.[30] Because transportation involves the physical movement of objects, it has greater limitations in speed and concentration. Progress in transportation presently seems to be nearing a point of saturation; speed and convenience are no longer steadily increasing.[31,32] In some circumstances, transportation has even become less convenient because of increased congestion.[33] In many respects the state of telecommunications technology today resembles transportation technology in the 1920s when the potentials of the automobile and airplane were just being perceived and exploited. The decades ahead could offer several new telecommunications options, such as video telephones, fast telefacsimiles, and two-way cable television.[30]

Moves to substitute video telecommunications for travel and to substitute telefacsimile and office information systems for mail and archives are likely to arise from constraints growing out of concern for environmental quality, dwindling energy resources, and land use planning.[32,34,35] Long distance video telephone utilization will require deployment of significant amounts of new transmission capacity and can lead to lower transmission costs for long distance audio telephone.[30] Transmission today accounts for about 17 percent of the cost of long distance service, and new transmission facilities will have significantly lower cost per unit bandwidth.[36]

Energy Considerations

Communication by video telephone could require less energy than personal travel. A transcontinental journey by a Boeing 747 "jumbo jet" for the purpose of a face-to-face meeting requires roughly eight times as much energy as a conversation of the same length over the anticipated intercity video telephone network.[30] A breakdown of this comparison for a transcontinental interaction between New York City, New York, and Los Angeles, California, is displayed in Case Study 5-3. For shorter distances, an energy comparison shows that the energy content in 4 liters of gasoline (1 gal) is sufficient either to propel an automobile about 25 km (15 mi) or to provide about 66 hours of conversation on a video telephone.[30]

Some basic data that can be used to calculate the energy use of telecommunications systems are given in Reference [22]. An approximate estimate of the amount of power dedicated to long distance telephone calls, for equipment only, is 0.826×10^{-4} watts/call-km. Assuming that local telephone calls use five times more energy and that present telecommunications systems require approximately 25 more channels than telephone systems, roughly 10 milliwatts/call-km is the power required for a telecommunication interaction. Converting this to other units yields the following telecommunications power requirements:

10.3 milliwatts/call-km =

1.46×10^{-6} gal/hr per call-mile

Another number has been obtained for the electrical energy required for an average phone call in the US: 1.0×10^{-3} kWh/min.[37] In order to account for the energy used for all associated equipment, this number should be multiplied by 10 giving 1.0×10^{-2} kWh/min.[37]

Besides the energy used by a product, the energy used in its production should also be considered (see Chapter 2 for discussion of product energy content). Compilation of the energy inputs required in each step leading to a finished industrial product is now in its infancy, and figures are not yet available for comparing a video telephone network with a transportation system. It is believed, however, that the energy required to produce, deploy, and maintain a video telephone system (including terminal devices, switching stations, transmission networks, and repeater stations) is substantially less than the energy required to construct air or ground transportation systems (including vehicles, earthmoving, paving, navigation, and terminal facilities).[30]

Environmental and social costs borne by society at large rather than by the producer or consumer are usually termed "external costs."[30] One of the tasks presently being tackled by government in an effort to maintain a quality environment is the development of regulations and institutions designed to convert external costs that have led to overuse and degradation of common property resources (such as water and air) into internal costs. The rising costs of energy, increased costs of pollution abatement, and scant prospect for any significant increase (by more than a factor of 2) in the energy efficiency of transportation will tend to make the total cost of travel rise. Simultaneously, complex electronic devices using integrated circuits appear destined to continue their rapid decline in costs. The expense of long distance transmission bandwidth is also anticipated to decline significantly. Furthermore, the increased use of integrated circuits is substantially reducing the energy used by electronic apparatus.[30]

5.4 COMPUTERS AND ENERGY

Since their inception digital computers have continuously become more efficient, expanding into new applications with each major technological improvement. Today, digital computers can be grouped into two major categories (primarily by function and size, although there are exceptions). The first group consists of generally smaller computers used for on-line, real time functions such as process control; the second group includes the generally larger computers used primarily for off-line computation and analysis such as payroll, engineering analysis, etc. The current trend is to use smaller computers to do the work of larger computers so that for some applications large, expensive computers are being replaced with less expensive machines.

A. Digital Control

Computer controlled systems are used widely to automate production lines and to speed data acquisition. Energy savings can be achieved by using computers to monitor and control various energy using activities. The efficiency of a large building's environmental system, for example, can be more than doubled with a network of sensors and a minicomputer.[38] The sensors monitor environmental conditions inside and outside the building (time of day, sun position, etc.), and feed a minicomputer which controls heat and cooling sources separately. By contrast, many heating and cooling sources now run continuously year-round, with a flap valve mixing heat and cold for desired temperatures--*wasting* an average of 30 to 40 percent and, in some cases, up to 80 percent of the fuel energy.

A second variation on this approach is to utilize the external air surrounding a building to supplement the required heating and cooling of the building by means of computer controlled fans. For example, if the outside air is cooler than the inside air and the building temperature is too high (such as in the late afternoon or early evening), outside air can be used to reduce the temperature. The outside air can be conditioned by using filters and dehumidifiers, if desired (see Chapter 8 for more discussion on control of HVAC systems).

A third method is to directly monitor a facility's power consumption and demand. This approach is advocated by some large computer manufacturers and is used in various power management systems. These are built around small computers which monitor power demand and use a control program that operates on a demand leveling principle. As demand limits are approached, units are shut down or throttled back for variable times. The restraint times are varied by time of day and are defined by the user for his environment. Since control for the system is a stored program, the demand limits and the shutdown or throttle-back times for each unit can be changed by the user with relative ease. In addition, multiple meters may be monitored concurrently.

The power management system operates by having the computer continuously measure demand level via demand records which measure peak electrical power usage (demand) during a standard time interval. These actual demand level measurements are then compared with the user target demand level once every minute or more often as desired. The system initiates corrective action (throttling back air conditioners, shutting off air delivery fans, etc.) to maintain actual demand below the present target level. As a result of this process, energy use savings also occur which may be larger than the value of the demand savings. In addition, the system may be used to provide orderly start-up and shutdown of equipment at the beginning and end of the day to avoid short period, high-peak demands.

Appropriate implementation of a computer-controlled environmental conditioning system can result in substantial savings in both demand and total energy usage. By monitoring energy demands and interrupting selected groups of electrical devices for short periods of time, utility bills can be reduced. Computer-controlled systems presently are being used by thousands of companies to reduce their electricity bills. They are already at work in a wide range of industries, shopping centers, universities, and office buildings, providing savings in electricity of 10 to 20 percent and more.

Utilities are also turning to electronic load control to conserve energy and minimize future plant expansion requirements.[39]

Computer power management can also be applied to an industrial operation in order to control power demand within specific limits. The following benefits can be provided by automatic computer power control and monitoring systems for industrial operations:

- improved efficiency of load dispatch;
- minimization of power and energy use while permitting maximum production;
- control of power demand within predetermined limits during normal operation and in emergencies;
- curtailment of wasteful operations such as equipment and units operating when not needed;
- reduction of power peak demands and providing for a more even load utilization;
- real-time display of plant performance and the furnishing of complete records of energy uses; and,
- provision of data that can be used to analyze energy flow and potential energy management opportunities.

Active computer monitoring and control can also be applied to a specific unit such as a furnace. Again, by proper management, the unit's performance can be measured, analyzed, and controlled.[40] If multiple units in a specific process are involved, "load-shedding" can be applied as required. Because the performance of an individual unit can be actively recorded, operations can be optimized to minimize the energy and fuel demands. If fuel switching is necessitated because of availability for environmental limitations, the switch can be planned and controlled by the computer in an orderly way. One example of the application of computers to reduce energy use in a large firm is given in Case Study 5-4.

Incorporating smaller components into the computers themselves can also save energy. Silicon-on-sapphire (SOS) microprocessor chips have been developed which are only 16 mm^2 in size.[38] This represents a substantial size reduction and an order-of-magnitude reduction in direct power requirements over previous components.[38] Replacement of other computer components can also reduce energy requirements. For example, semiconductors use less energy than core memories and complementary metal-oxide-silicon (CMOS) and SOS technologies use even less power than discrete semiconductors.

Table 5.8 shows a comparison of the size and capabilities of the first electronic digital computer, ENIAC, with a recent commercial microcomputer.[41] Researchers are working with new techniques that promise far smaller devices; for example, grand scale integration with over 2500 logic gates on a single 200-mil chip; bubble memory technology which can provide storage for approximately 70,000 bits of data on a 0.64 cm^2 (0.25 $in.^2$) surface; charged-coupled device (CCD) logic circuits which occupy 50 percent less space and dissipate 1/20 the power dissipated by the best semiconductor integrated circuit (N-channel metal-oxide semiconductor field effect transistor, N-MOSFET); Josephson junction superconducting magnetic devices employing cryogenic techniques; simple molecular device circuits that may lead to devices so tiny as to be virtually invisible; and mass storage optical memories combining developments in laser, holography, and electro-optical technologies.[42,43,44]

Semiconductors can replace vacuum tubes in most electronic products, such as television receivers, to reduce power requirements by about 80 percent. Secondary power reductions come from decreased requirements on power supplies and cooling equipment. One other example of solid-state technology impact on energy utilization efficiency is that liquid-crystal displays, used in hand held calculators, allow up to 100 hours of operation with a single throw-away battery where previous operating time ranged from 5 to 50 hours.

B. Digital Simulation

Digital computers can also play a significant role in providing a fast and accurate means of analytical simulation of energy intensive systems. In recent years interest has increased in computer applications for energy system analysis by means of developed computer programs.

Computer programs are available which will perform the following analyses:

- calculation of hour-by-hour thermal and electrical loads for buildings and simulation of the operation of the air distribution system in meeting these loads;
- simulation of the operation of various pieces of equipment as they respond to loads imposed by a building's air-side system (as opposed to coolant-side system) to determine monthly and annual energy usage for the various systems being evaluated;
- calculation of monthly energy costs and other annual operating costs with initial investment and associated owning cost factors to find the total owning and operating costs of each candidate system on a year-by-year basis for any period up to 30 years; and,
- calculation of design point heating/cooling loads on a zone-by-zone basis using 24-hours of weather and solar data per month.

Table 5.9 indicates some of the computer programs available for simulation of energy systems for use in improving the design of equipment, systems and facilities. By means of these and other simulation programs, energy system designers can perform complex comparative engineering and economic analyses for a facility involving several thermal sections, each with various types of air distribution systems, using alternative mechanical plant configurations including the actual local utility rates. The results of such comparative analyses

Item	Parameter	ENIAC	F8*	Comments
1	Size	3,000 cubic feet	0.011 cubic feet	300,000 times smaller
2	Power consumption	140 kilowatts	2.5 watts	56,000 times less power
3	ROM	16K bits (relays and switches)	16K bits	Equal amount
4	RAM	1K bits (flip-flop accumulators)	8K bits	Eight times more RAM in F8
5	Clock rate	100 kilohertz	2 megahertz	Twenty times faster clock rate with F8
6	Transistors or tubes	18,000 tubes	20,000 transistors	About the same
7	Resistors	70,000	None	F8 uses active devices as resistors
8	Capacitors	10,000	2	5,000 times less
9	Relays and switches	7,500	None	
10	Add time	200 μsec (12 digits)	150 μsec (8 digits)	About the same
11	Mean time to failure	Hours	Years	More than 10,000 times as reliable
12	Weight	30 tons	<1 pound	

*For comparison in this table, the F8 microcomputer is assumped to consist of one 3850-CPU, one 3856-PSU (2K bytes of ROM), eight IK RAM packages, and the static memory interface chip 3853, plus teletype terminal interface circuitry. (ROM = read only memory; RAM = random access memory.)

Reference 41

COMPARISON OF PARAMETERS OF ENIAC WITH THE FAIRCHILD 8 (F8) MICROPROCESSOR

TABLE 5.8

Code Name	Originator	Applications
AXCESS	Edison Electric Institute	Comparison of a building's energy requirements using alternate mechanical/ electrical systems. Allows designer to compare total building energy usage and demand for each subsystem using any combination of available energy sources (i.e., electricity, gas, oil, coal).
ECUBE 75	The American Gas Association	Analysis of new buildings and various retrofit or replacement projects for existing buildings by predicting impact of thermostat setback and setup, ventilation air reduction schedules, and equipment shutoff in economic terms.
TRACE and Version 200	The Trane Company	Energy analysis of new and retrofit building designs for alternative HVAC equipment to achieve energy use reduction and life cycle cost optimization.
ESOP	NASA-Johnson Space Center	Analysis of energy requirements, water consumption, solid waste consumption, and waste water effluent from Integrated Utility Systems by calculating facility load requirements and evaluating yearly operational characteristics.
MACE	McDonnell Douglas Corporation	Determines monthly and annual energy requirements for a proposed building. Performs cost analysis of various heating and cooling systems, building construction, and prime energy sources using methods recommended by ASHRAE whenever possible.
NECAP	NASA-Langley Research Center	Determines building energy consumption and performs cost analysis. The set of six programs calculates a variety of energy use aspects for buildings following standard ASHRAE procedures.
NBSLD	National Bureau of Standards	Heat load analysis of new or existing buildings which can simulate a variety of indoor temperature modes: fixed temperature, night setback, night setback with equipment limitations or floating temperature.
CAL-ERDA	Energy Research and Development Administration and University of California, Berkeley	A public domain program under development with streamlined input procedures and shorter running time than NBSLD. It is a calibration for other programs to be used for building heat load analysis in California.

Note: A number of other computer programs are available for energy system analysis which are used and supplied by industrial companies and architect-engineering firms.

TYPICAL COMPUTER PROGRAMS FOR ENERGY SYSTEM SIMULATION **TABLE 5.9**

will indicate the most energy efficient and economic facility design as well as the optimum operating procedures. Thus, without having to construct a facility, engineers can test their energy use efficiency concepts by using actual statistical weather data, equipment performance data, and prevailing utility rates.

C. Computer Energy Use

The energy used by computers and their associated peripheral equipment is also an important consideration. Heat generated by the electronic equipment must be removed from a computer facility. Computer equipment manufacturers recommend closely controlled humidity and temperature environments for reliable operation of their equipment. The heating, ventilating, and air-conditioning (HVAC) requirements can be a substantial fraction of the energy use (∿50 percent) for facilities supporting large computers. Case Study 5-5 provides a typical example of a city owned computer facility, showing its historical and audited energy use profile and indicating the potential reduction of energy use and economic savings that can be achieved by applying energy management concepts.

5.5 POTENTIAL ENERGY SAVINGS

As pointed out in the introduction, nearly all of the energy currently used for transportation is derived from fossil fuels rather than electricity. Herein we consider two possibilities:

- responses to fuel scarcity or shortage, and,
- effects of improved efficiency or fuel shifts.

In the event that fuel shortages occur, both direct transportation and generation of electricity will be affected. One alternative that has been considered is a fuel priority allocation system.[7] In this approach, priority would be given to vehicles or transit systems which preserve the vitality of the economic structure with minimum energy use.[7] Both short-term and long-term strategies would be employed, depending on the nature and duration of the shortage:[7]

Short-term:

- car-pooling,
- switch from automobiles to mass transit,
- ration or restrict motor vehicle fuels, and,
- maintain public transportation fuel and energy supplies.

Long-term:

- stimulate development and use of small vehicles;
- develop energy efficient mass transit;
- allocate fuels and energy, with high priority given to energy efficient systems; and,
- provide incentives for development of electric or other vehicles capable of using alternative energy sources.

The potential for improved efficiency has been examined for three time periods: *immediate, near-term* (2-5 years), or *long-term* (5-25 years). Savings are expected to result from:

- *Immediate:*

  Reduced speed limits (now 88 km/hr or 55 mph in the US), greater use of smaller cars, car pools, fewer airline flights, trucking regulation changes, and reduced driving due to threat of increased gasoline taxes. Continued and increased emphasis on these measures will produce additional savings. Total potential savings = 10-15 percent.

- *Near-term:*

  Begin mandatory changeover to small cars, increase public transportation opportunities, some substitution of electric vehicles, introduction of video-telephone conference calls, initiation of electronic mail service, new trucking and railroad legislation and regulations. Total potential savings = 10-15 percent.

- *Long-term:*

  Widespread use of public transportation, adoption of electric vehicles on a commercial scale, use of communication alternatives to transportation, widespread use of electronic mail and message services, increased use of high speed railroads and other less energy-intensive transport modes. Total potential savings = 10-20 percent.

In summary, the cumulative potential for savings over the period 1977-2000 is estimated to be 30 to 50 percent. Note that automobiles account for over 50 percent of transportation energy use in the US. Thus a complete conversion to 1 mt, 4 cylinder autos (vs 2 mt, 8 cylinder types currently used) would in itself account for a 40 to 50 percent savings in automobile energy use, or a 20 to 25 percent savings in US transportation energy.* Other savings would be required such as car pools, mass transit systems, and in trucking, aircraft operations, trains, waterways, etc. to achieve the additional 5 to 25 percent. Certainly the lower limit--30 percent--is achievable by the end of this century; the higher limit would depend on more speculative developments.

The effect of these potential savings was discussed in Chapter 1 (see Table 1.2), where the range of savings is weighted by the sector importance to obtain projected savings in the year 2000. For the transportation sector, these savings amount to 8 to 13 percent of total US energy use in the year 2000. With current practice, only a small part of these savings would be reflected in reduced electricity demand.

More likely is an increase in electricity use, as part of the transportation demand is shifted from fossil fuels to electricity. This shift to electricity could come about from national efforts to expand electrified mass transit systems such as the Bay Area Rapid Transit System (BART), San Francisco, California, from efforts to substitute electronic mail service and electronic communication means for personal transportation, or from the gradual development and conversion to electric vehicles. A recent study sponsored by EPRI investigates the impact of projected electric vehicle use on the generation requirements of electric utilities in the 1985-2000 time frame.[45] Since none of the changes will occur quickly, we have used the following estimates:

Immediate	*Near-Term*	*Long-Term*
0-1 year	2-5 years	5-25 years
0%	0-5%	5-10%

These give a cumulative shift in transportation energy of 5 to 15 percent (see Chapter 1 for additional details and an estimate of the impact of this shift on future electricity needs).

It appears that any savings in electricity will be more than offset by a trend toward substitution of electrical energy for fuels which will become increasingly scarce in the future. Nevertheless, energy use efficiency can and should be improved in the transportation, communication, and data processing sectors, for both fuels and electricity. Given the fact that energy prices can only continue to rise, the benefit of energy management programs both in the US and other nations, is obvious.

* * *

*Much research has been conducted in the field of gasoline and diesel powered transportation. One recent article concerning the fuel consumption of automobiles and the potential for increasing their efficiency by 40 percent by 1980 is given in Reference 46.

REFERENCES

1. Hirst, Eric, *Energy Intensiveness of Passenger and Freight Transport Modes, 1950-1970*, Report No. ORNL-NSF-EP-44, (Oak Ridge, Tennessee: Oak Ridge National Laboratory, April 1973).

2. Transportation Association of America, *Transportation Facts & Trends*, 9th edition, (Washington, D.C.: 1972).

3. US Department of Commerce, Bureau of Census, *Statistical Abstract of the United States: 1971*, 92nd edition, (Washington, D.C.: 1971).

4. Morrison, W.E. and Reading, C.L., *An Energy Model of the United States Featuring Energy Balances for the Years 1947 to 1965 and Projections and Forecasts to the Years 1980 and 2000*, Bureau of Mines Information Circular 8384, (Washington, D.C.: US Department of the Interior, Bureau of Mines, 1968).

5. US Department of the Interior, Bureau of Mines, "U.S. Energy Use at New High in 1971," News Release (31 March 1972).

6. National Petroleum Council, *U.S. Energy Outlook: An Initial Appraisal 1971-1985*, (Washington, D.C.: 1971).

7. Healy, Timothy J., *Energy Use of Public Transit Systems*, A Report Prepared for the State of California Business and Transportation Agency, Department of Transportation, (Santa Clara, California: University of Santa Clara, August 1974).

8. Sanders, D.B. and Reynen, T.A., *Characteristics of Urban Transportation Systems, A Handbook for Transportation Planners*, Prepared for the US Department of Transportation, (Washington, D.C.: De Leuw, Cather and Company, May 1974).

9. Hirst, Eric, "How Much Overall Energy Does the Automobile Require?" *SAE Journal of Automotive Engineering* 80 (July 1972): 36-38.

10. "Special Report: EV Revival, A Full-Issue Review of Electric-Vehicle Technology," *Machine Design* 46 (17 October 1974). Because of the wealth of informative material in this excellent state-of-the-art review, a major portion of the text in section 5.2.A., "Electric Vehicles - Technology Considerations," has been taken from this reference.

11. Cairns, Elton J. and McBreen, James, "Batteries Power Urban Autos," *Industrial Research* 17 (June 1975): 56-60.

12. Aronson, Robert B., "Fuel Cells--A Sleeper in the Energy Race," *Machine Design* 49 (24 February 1977): 20-24.

13. George, J.H.B.; Stratton, L.J.; and Acton, R.G., *Prospects for Electric Vehicles--A Study of Low-Pollution Potential Vehicles-Electric*, (Cambridge, Massachusetts: A.D. Little, Inc., 15 May 1968).

14. The Electricity Council Research Centre, *Sodium Batteries for Electric Vehicles*, (Capenhurst, Chester, Great Britain: 1974).

15. Dellacanonica, O.G., "Electric Locomotives to Meet Today's Needs," *Mechanical Engineering* (October 1972): 17-23.

16. Aronson, Robert B., "Will Electricity Power Tomorrow's Trains," *Machine Design* 49 (7 April 1977): 20-25.

17. Myers, William A., "The Big Red Cars: When Rapid Transit Meant the Electric Railway," in *Transpo LA: Economic Leverage for Tomorrow*, Proceedings of the 4th Annual Symposium under the Auspices of the Los Angeles Council of

Engineers and Scientists, American Institute of Aeronautics and Astronautics, L.A. Section Monograph Series, Volume 18, ed: A.D. Emerson, (North Hollywood, California: Western Periodicals Company, 1975), pp. 326-333.

18. Friedlander, Gordon D., "Railroad Revival: On the Right Track," *IEEE Spectrum* (August 1972): 63-66.

19. Parker, Harry (editor-in-chief), *Architects' and Builders' Handbook*, 18th edition (New York: John Wiley and Sons, 1941).

20. Lilienstein, Fred and Rowe, Irwin, "A Backyard Magnetic Levitation System," in *Transpo LA: Economic Leverage for Tomorrow*, pp. 5-20.

21. Cleminson, Cedric A. and Rogers, R.L., "Can You Save on Elevator Power Consumption (And Is It Worth It?)," *Elevator World* (March 1974).

22. Goss, William P. and Carlson, Tage G.C., "Energy Savings by Substituting Telecommunications for Various Sectors of Metropolitan Area Travel," *Proceedings of the Ninth Intersociety Energy Conversion Conference*, San Francisco, California, 26-30 August 1974.

23. National Academy of Engineering, Committee on Telecommunications, Panel on Telecommunications Research, *Telecommunications Research in the United States and Selected Foreign Countries: A Preliminary Survey*, Vol. 1: *Summary*, (Washington, D.C.: June 1973).

24. Caswell, Stephan A., "Electronic Mail Delivers," *Computer Decisions* 9(April 1977): 32,36,39.'

25. "Postal Services Going Out of Business?" *Machine Design* 49(24 March 1977): 18.

26. Potter, R.J., "Electronic Mail," *Science* 195(18 March 1977): 1160-1164.

27. Nilles, Jack M.; Carlson, Frederick R.; Gray, Paul; and Hanneman, Gerhard, *Telecommunications--Transportation Tradeoffs*, Report No. NSF-RA-5.74-020, (Los Angeles, California: University of Southern California, December 1974).

28. Jones, D.W. *Must We Travel? The Potential of Communication as a Substitute for Travel*, (Stanford, California: Stanford University, March 1973).

29. Baran, P., "30 Services that Two-Way Cable Television Can Provide," *The Futurist* (October 1973): 202-210.

30. Dickson, Edward M., (in association with Raymond Bowers), *The Video Telephone: Impact of a New Era in Telecommunications*, (New York: Praeger Publishers, 1974).

31. Aviation Advisory Commission, *The Long Range Needs of Aviation*, (Washington, D.C.: January 1973).

32. "The Airport Crisis: No Place to Land," *Business Week* (30 September 1972): 43

33. "How to Make it From City to the Airport," *Business Week* (19 August 1972): 67-68.

34. "Moving to Regulate Auto Commuters," *Business Week* (16 December 1972): 34.

35. "Limited Auto Use: A Cure for Pollution?" *Chemical and Engineering News* (22 January 1972): 2.

36. Martin, James, *Future Developments in Telecommunications*, (Englewood Cliffs, New Jersey: Prentice-Hall, Inc., 1971).

37. Personal communication from Mr. E.W. Greninger, University of California, Los Angeles, Spring 1975.

38. Herzog, Raymond E. and Dann, Richard T., "Designing the Energy Miser," *Machine Design* 46(21 February 1974): 97-106.

39. Jack, Charlie F., "Conservation of Resources Through Utility Load Control and Peak Shaving; Practical Implementation and Implications for R & D," in *Proceedings of an EPRI Workshop on Technologies for Conservation and Efficient Utilization of Electric Energy*, prepared by Applied Nucleonics Company, Inc. (Palo Alto, California: Electric Power Research Institute, July 1976), pp. 4-33 to 4-42.

40. Crowell, William H., "Cut Your Electric Bill Without Cutting Production," *Industry Week* (15 April 1974): 47-49.

41. Linville, John G. and Hogan, C. Lester, "Intellectual and Economic Fuel for the Electronics Revolution," *Science* 195(18 March 1977): 1107-1113.

42. Comella, Thomas M., "The Incredible Shrinking Circuit," *Machine Design* 49(21 April 1977): 212-217.

43. Giordmaine, J.A., "Solid-State Electronics: Scientific Basis for Future Advances," *Science* 195(18 March 1977): 1235-1240.

44. Rajchman, J.A., "New Memory Technologies," *Science* 195 (18 March 1977): 1223-1229.

45. Mathtech (The Technical Research and Consulting Division of Mathematica, Inc.), "The Impact of Electric Passenger Automobiles on Utility System Loads, 1985-2000," Draft Final Report, (no date).

46. Pierce, John R., "The Fuel Consumption of Automobiles," *Scientific American* 232 (January 1975): 34-44.

Chapter 6

AGRICULTURE

By G.B. Taylor*

CHAPTER CONTENTS

KEY WORDS

Agricultural Use
Agricultural Machinery
Agricultural Water Use
Cattle
Corn
Crop Production
Efficient Energy Use
Energy Input/Output
Environmental Impact
Food Processing
Irrigation
Livestock Production
Paper Industry
Wheat

SUMMARY

In the United States there has been a marked shift away from the small family farm or ranch to industrialized farms and ranches. Accompanying this trend toward "industrial agriculture" are new solutions to age old problems of crop production, insect pests and soil yields. The use of specialized machinery such as automatic pickers and sorters, in-field vegetable packing, sprinkling systems, refrigeration, chemical fertilizers and pesticides has greatly increased agricultural production efficiency. However, all the new innovations in agricultural equipment, food processing, and storage require increasing amounts of energy.

Proper fertilization, soil conditioning, and correct use of insecticides have resulted in impressive increases in crop yields. Cattlemen now innoculate their cattle and feed them special growth foods in feed lots. After slaughter, the meat is shipped to wholesalers, retailers, and finally to the consumer. Chain meat wholesalers are building automated meat cutting and processing systems costing millions of dollars; retailers use attractive refrigerated display counters to sell prepackaged, plastic-wrapped cuts of meat.

As has happened in many other industries, the past abundance and low cost of energy have led to energy intensive practices in the agricultural industry. The examples already mentioned--fattening beef cattle in feed lots, packaging and displaying of food for possible consumer appeal, over-production without proper long-

*Principal, Applied Nucleonics Company, Inc.

range planning--must today be carefully reviewed and adjusted to insure a continuing supply of agricultural products consistent with energy prices and availability.

In summary, the demand for electrical energy is increasing as irrigation, food processing, food preservation and other agricultural uses increase. Solutions to the problems facing this industry will place new demands on electricity production. A significant potential exists for more efficient agricultural electricity use; realization of this potential will require new attitudes and approaches in both small farms and agribusinesses.

> *We have thought of other countries as having far more energy to sell us than they actually do, and they have thought of us as having far more food to sell than we actually do. But the truth will all too soon be clear to everyone.*
>
> --K.E.F. Watt, 1974

6.1 INTRODUCTION

The "Energy Revolution" in Agriculture

After World War II industrialized countries underwent a change in agricultural practices that has been called the "energy revolution."[1] Cheap and seemingly abundant forms of energy (electricity and fossil fuels) allowed industrialized countries to build up an agricultural technology based on mechanization, intensive use of chemicals (fertilizers, pesticides, and herbicides), and in many areas large, sophisticated, and expensive irrigation schemes. These measures resulted in spectacular increases in agricultural yields. An idea of the intensification of energy inputs can be obtained from available statistics. Figure 6.1, for example, shows farm output as a function of energy input to the US food system between 1920 and 1970. Table 6.1 presents the evolution of energy use in US farming after World War II and Table 6.2 shows the growing energy inputs in the US food system.

This evolution of cropping systems has had the effect of minimizing basic differences between agriculture (an essentially biological activity) and industry. Agriculture today may be regarded as one form or type of industrial activity and farms are assumed to be just "factories in the field."[4]

The following points summarize the evolution of agriculture in the US and other industrialized countries between the 1940s and 1970s:

- fewer and larger farms (trend to agribusiness and conglomerate agriculture);[3,5]
- increased growth of energy demands (direct and indirect);[6]
- more diversity of choice, greater freshness, less spoilage, and higher quality foods for consumers; and,
- diminishing returns in agricultural yields (slowdown in the rate of increase in output vs energy input).[6]

Suggested future trends for agriculture are:

- an awareness of the nonrenewability of fossil fuels and the need to plan for energy-efficient agriculture throughout the world;
- a serious concern for the future prospects and ultimate benefits of energy-intensive agricultural technologies while facing limited energy resources.

Recent developments in ecology and environmental science already provide a much better understanding of ecosystem manipulation and the functioning of agricultural systems. These developments also promise to build up holistic theories which until now have been missing in the development of our technologies, particularly as related to agriculture.

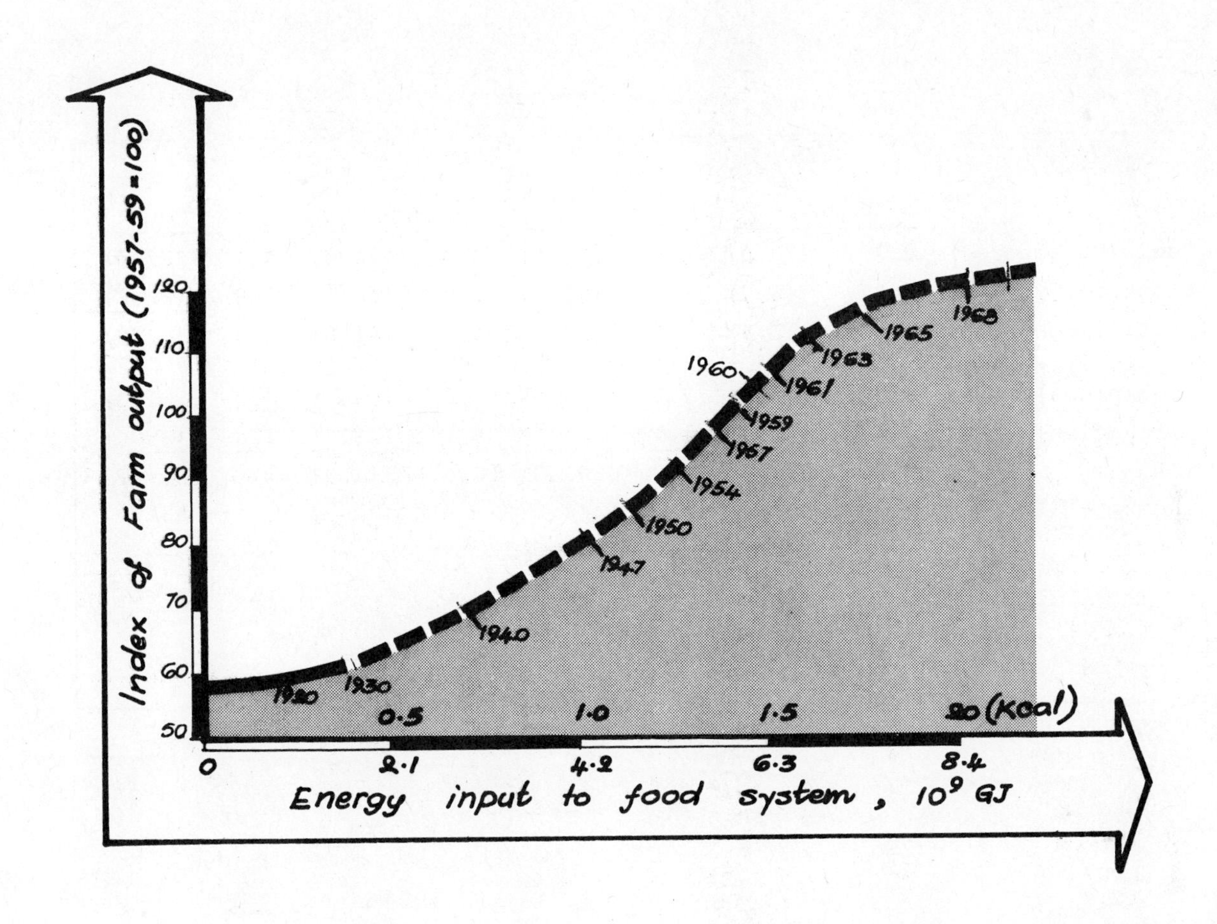

Reference 2

ENERGY INPUT AND FARM OUTPUT OF THE US

FIGURE 6.1

	1940		1950		1960		1970	
Fuel (Direct Use)	2.9	(70.0)	6.6	(158.0)	7.9	(188.0)	9.7	(232.0)
Electricity *	0.029	(0.7)	1.4	(32.9)	1.9	(46.1)	3.9	(92.2)
Fertilizers	0.52	(12.4)	1.0	(24.0)	1.7	(41.0)	3.9	(94.0)
(Mechanization Equipment)	0.98	(23.4)	2.7	(63.5)	2.7	(65.5)	4.2	(101.3)
Irrigation	0.75	(18.0)	1.1	(25.0)	1.4	(33.3)	1.5	(35.0)
Totals	5.2	(124.0)	13.0	(303.0)	16.0	(374.0)	23.0	(554.5)
Index		100		244		301		424

*Based on primary fuels.

Reference 3

POST WORLD WAR II EVOLUTION OF ENERGY IN US FARMING

GIGAJOULES x 10^8 (KCAL x 10^{12})

TABLE 6.1

	Gigajoules x 10^9	(kcal x 10^{12})	%
On Farm	2.2	(526.1)	24.0
Food Processing and Packaging Industry	2.2	(520.0)	24.0
Food Distribution	1.3	(320.9)	15.0
Refrigeration and Cooking (Commercial and Home)	3.4	(804.0)	37.0
Total *	9.1	(2,171.0)	100.0

Note: *Includes the following electricity use (see Reference 6, pp. 16,49,78,81,84,88, and 90.):

	10^9 kWh
On Farm	36.8
Food Processing	38.1
Feed	3.0
Fertilizer	2.7
Farm Machinery	1.3
Pesticides	0.3
Petroleum Products	2.9
Commercial Refrigeration	71.5
	156.6 = 1.69 x 10^9 GJ

Reference 3

US FOOD SYSTEM ENERGY INPUTS — 1970

TABLE 6.2

6.2 OBJECTIVE, SCOPE, ASSUMPTIONS

A. Objective

The primary objective of this chapter is to point out means by which energy, electrical energy in particular, can be saved in the agricultural sector. The time interval over which energy savings can be implemented ranges from immediate to the next 25 years. The agricultural sector includes all agriculturally related operations--farming, ranching, transportation, processing, packaging, distribution, and marketing (see Figure 6.2). Non-food agricultural production, such as wood, pulp, and textile products, are also included in this study.

B. Scope

Of primary importance is the presentation of energy saving procedures and recommendations which can be applied immediately or implemented at the "grass roots" level. The text, therefore, includes some background information to establish the operating context. Emphasis is placed on a set of energy saving measures directed to the farmer, rancher, food processor, distributor, marketing personnel, consumer, and all other individuals who play an active role in the agricultural production chain.

C. Assumptions

This is not an agricultural manual, nor a substitute for detailed agricultural information available from farm cooperatives, agricultural extension services, or government agencies. The text presumes some knowledge of basic science and engineering on the part of the reader. Although the background emphasizes current US practice, the concepts discussed herein are considered applicable to any industrialized or energy-intensive agricultural system.

6.3 USEFUL FACTS AND BACKGROUND

A. Crop Production

During the past 25 years, farm output has nearly doubled, caloric content of food has increased approximately 100 percent, and output per man hour has risen six-fold (Table 6.3).[7] This change in productivity is the result of new techniques and a 200 percent energy input increase (Figure 6.3).

Figure 6.1 indicated that in recent years increasing energy input to the food system is having less effect on food output. If one couples the relatively energy-intensive methods used in high productivity farming, food processing, and food distribution with energy costs, it is obvious that improved energy efficiency is needed. A comparison of the curves shown in Figure 6.4 with rising energy prices indicates that the cost of food to consumers will continue to increase since the food demand is relatively inelastic.[4]

Conservation of all fuel inputs must be implemented in the near future in order to avoid overtaxing scarce energy resources. Farmers will become aware of methods for more efficient use of equipment and processed materials such as fertilizer, pesticides, herbicides, farm equipment, lubricating oil, and fuels as energy costs continue to increase.[11]

The present trend of substituting diesel fuel for gasoline will continue in farming. By 1980 diesel fuel use in agricultural production is expected to increase from 8.3 Gigaliters (2.2 billion gallons) to between 10.6 to 11.0 Gigaliters (2.8 to 2.9 billion gallons), an increase of 27 to 32 percent.[12]

B. Livestock Production[12,13]

Consumers have become accustomed to meat, milk, and eggs in their daily diet, and wool and

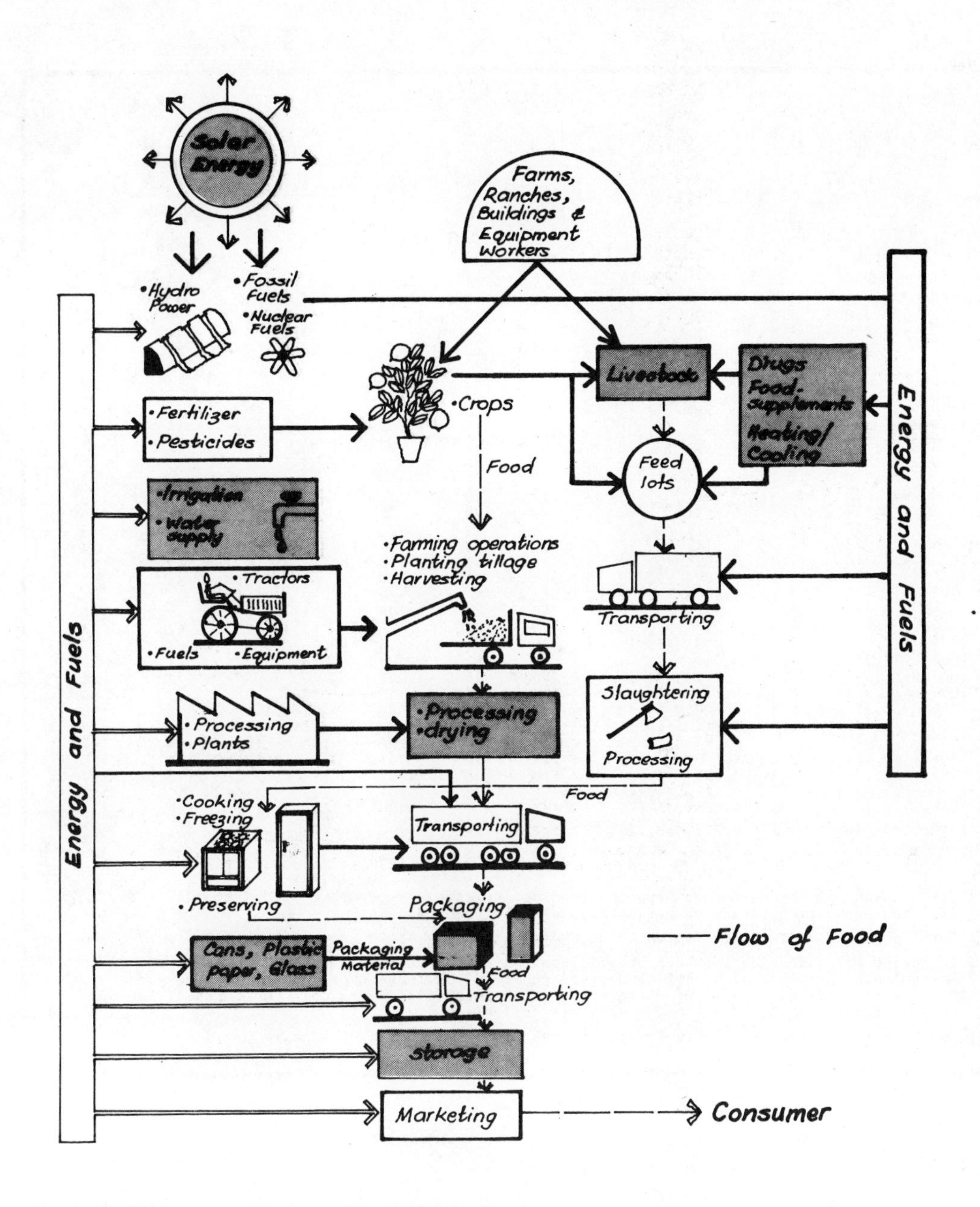

ENERGY FLOW IN AN INDUSTRIALIZED FOOD SYSTEM

FIGURE 6.2

(Index, 1967=100)

Year	Farm population[1] (April 1) Number (thousands)	Farm population As per-cent of total population[2]	Farm employment[3] (thousands) Total	Family workers	Hired workers	Farm Output[4] Total	Per unit of total input	Per man-hour Total	Crops	Live-stock and products	Crop production per acre
1940	30,547	23.1	10,979	8,300	2,679	60	62	21	22	27	62
1941	30,118	22.6	10,669	8,017	2,652	62	64	22	24	28	64
1942	28,914	21.4	10,504	7,949	2,555	69	69	24	26	30	70
1943	26,186	19.2	10,446	8,010	2,436	68	68	24	26	32	64
1944	24,815	17.9	10,219	7,988	2,231	70	69	25	27	31	68
1945	24.420	17.5	10,000	7,881	2,119	69	70	27	29	31	68
1946	25,403	18.0	10,295	8,106	2,189	71	72	29	31	32	70
1947	25,829	17.9	10,382	8,115	2,267	69	70	29	31	33	67
1948	24,383	16.6	10,363	8,026	2,337	75	76	32	35	34	75
1949	24,194	16.2	9,964	7,712	2,252	74	73	33	36	36	70
1950	23,048	15.2	9,926	7,597	2,329	73	73	35	39	37	69
1951	21.890	14.2	9,546	7,310	2,236	75	73	36	38	39	69
1952	21.748	13.9	9,149	7,005	2,144	78	76	39	42	40	73
1953	19,874	12.5	8,864	6,775	2,089	79	77	41	43	41	73
1954	19,019	11.7	8,651	6,570	2,081	79	78	43	45	43	71
1955	19,078	11.5	8,381	6,345	2,036	82	81	47	48	46	74
1956	18,712	11.1	7,852	5,900	1,952	82	82	50	52	48	77
1957	17,656	10.3	7,600	5,660	1,940	80	83	53	56	51	77
1958	17,128	9.8	7,503	5,521	1,982	86	89	59	65	55	86
1959	16,592	9.4	7,342	5,390	1,952	88	90	62	66	59	86
1960	15,635	8.7	7,057	5,172	1.885	90	93	67	71	62	88
1961	14,803	8.1	6,919	5,029	1,890	90	94	70	73	67	92
1962	14,313	7.7	6,700	4,873	1,827	91	95	73	77	71	95
1963	13,367	7.1	6,518	4,738	1,780	95	98	80	82	77	97
1964	12,954	6.8	6,110	4,506	1,604	94	97	83	85	83	95
1965	12,363	6.4	5,610	4,128	1,482	97	100	91	92	87	100
1966	11,595	5.9	5,214	3,854	1,360	96	97	94	95	93	99
1967	10,875	5.5	4,903	3,650	1,253	100	100	100	100	100	100
1968	10,454	5.2	4,749	3,536	1,213	102	101	106	106	105	104
1969	10,307	5.1	4,596	3,420	1,176	103	101	112	112	112	107
1970	9,712	4.7	4,523	3,348	1,175	102	101	113	110	119	102
1971	9,425	4.6	4,436	3,275	1,161	110	108	125	122	130	111
1972	9,610	4.6	4,373	3,227	1,146	112	110	131	126	138	115
1973	9,500	4.5	4,395	3,232	1,163	116	112	129	127	127	114

1 Farm population as defined by Department of Agriculture and Department of Commerce, i.e., civilian population living on farms, regardless of occupation.

2 Total population of United States as of July 1 including Armed Forces overseas.

3 Including persons doing farmwork on all farms. These data published by the Department of Agriculture, Statistical Reporting Service, differ from those on agricultural employment by the Department of Labor because of differences in the method of approach, concepts of employment and in time of month for which the data are collected.

4 Computed from variable weights for individual crops produced each year.

Reference 7

FARM POPULATION, EMPLOYMENT, AND PRODUCTIVITY, 1940-1973 **TABLE 6.3**

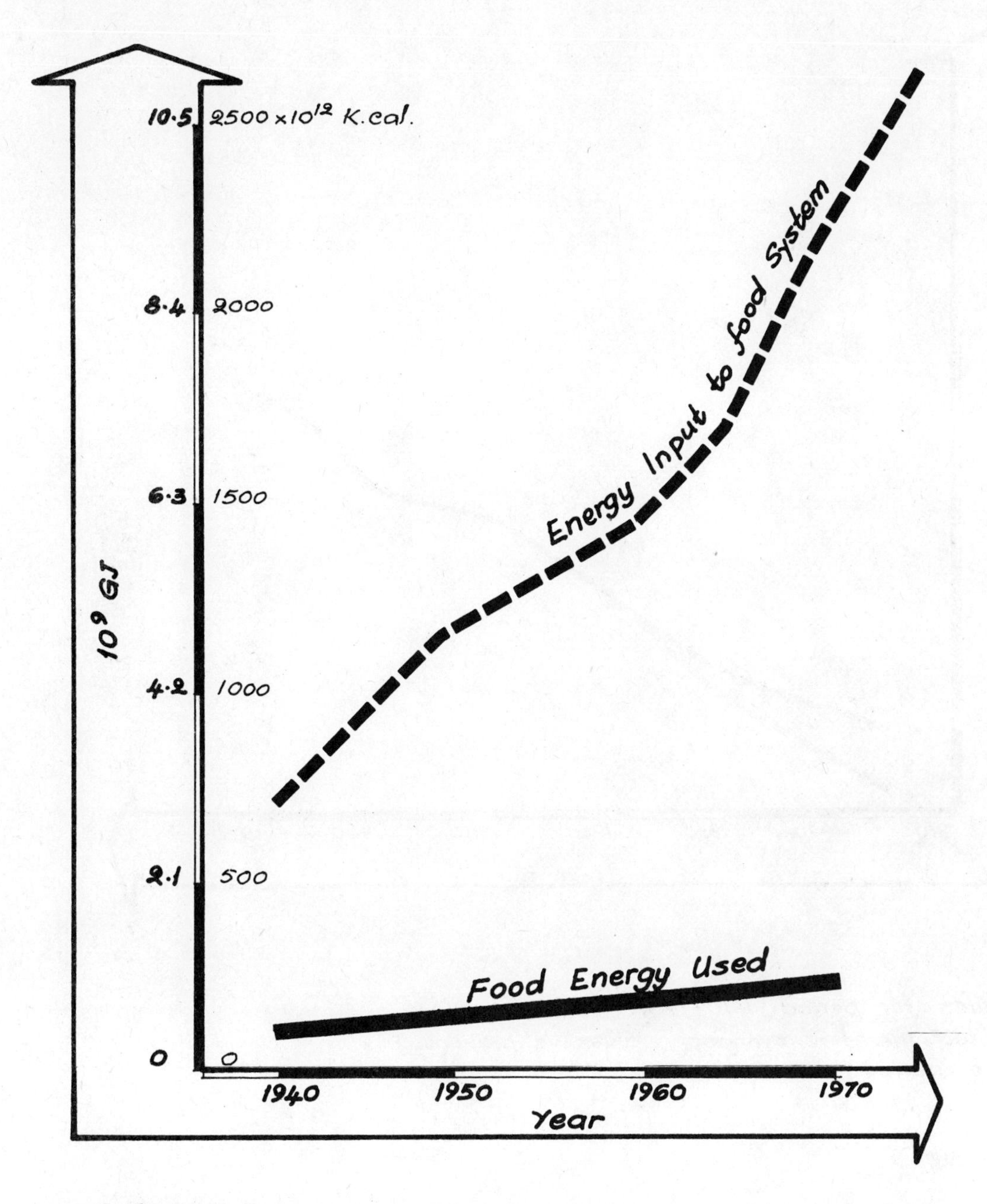

Reference 8

ENERGY USE IN THE US FOOD SYSTEM

FIGURE 6.3

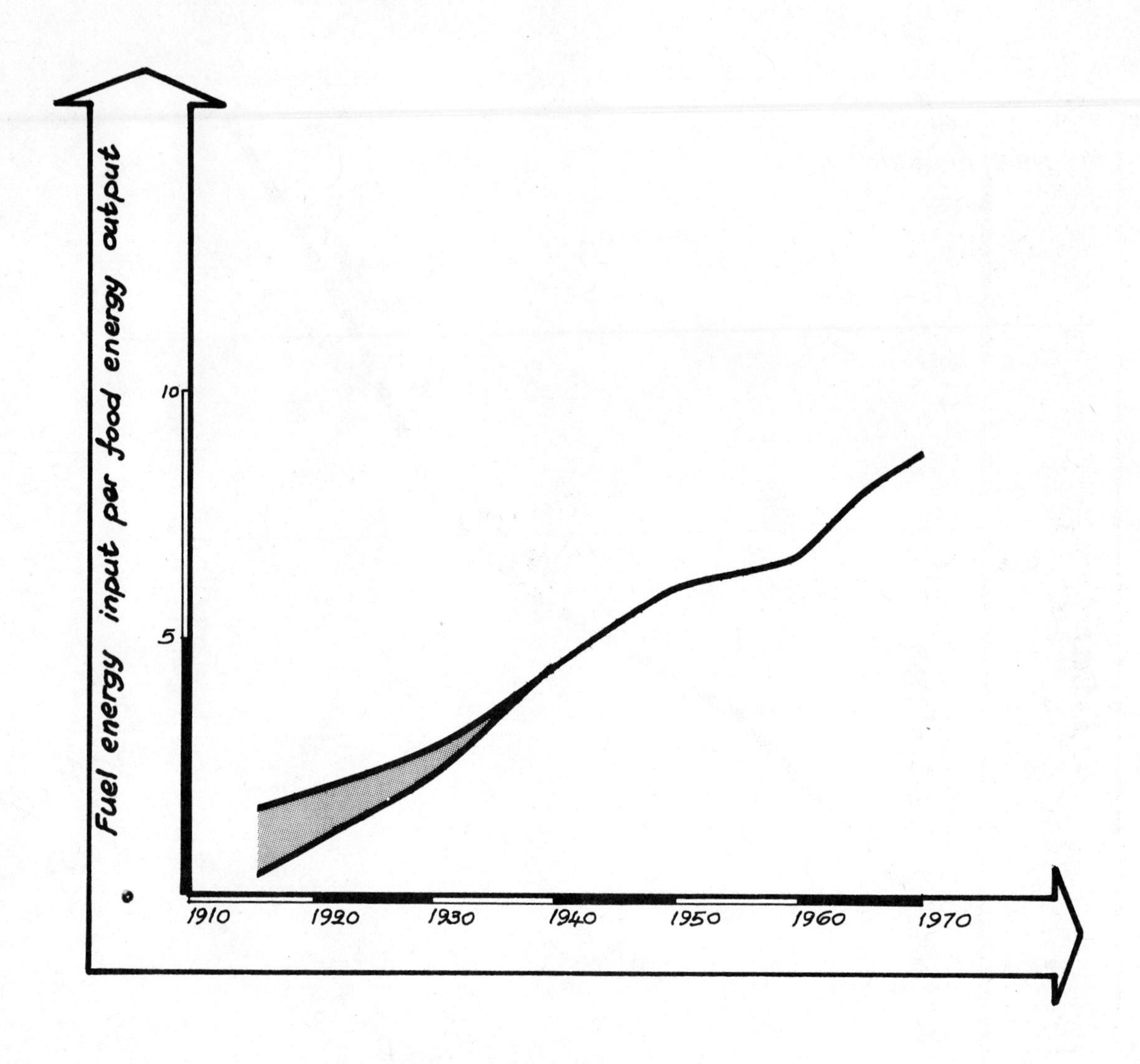

The values for period 1910-1937 cannot be fully documented, therefore a range of values is presented for that period.

Reference 9

ENERGY SUBSIDY TO FOOD SYSTEM

FIGURE 6.4

leather in their wardrobes. The huge demand for these animal products has resulted in a high output, energy-intensive animal production system. For example, virtually every milk-producing dairy cow in the US is milked twice daily using electricity, the milk is processed electrically, and animals are kept healthy electrically. Livestock production today requires a high degree of environmental and nutritional control with mixed and processed rations.[14] This controlled feeding of animals requires much more energy than traditional self-feeding operations.

Use of electricity and various fuels has been widely adopted for control of animal environments. Forced ventilation and the addition of heat are commonly used. Temperature control is necessary to maintain high productivity of young pigs. The current high efficiency in feed conversion (feed/kg gained) in the production of poultry has been achieved by manipulation of the growing environment (temperature, humidity, lighting, and air composition) to keep stresses associated with these parameters as low as possible.

In the event that energy, particularly electricity, were to become limited or unavailable in agriculture, severe and disastrous reductions in agricultural output would result. The short-term effects of energy constraints in animal agriculture can be seen in the following:

- Reduction of the energy available to dairies for milking, cooling, and related clean-up could result in a decrease in milk production.
- Reduction of energy for environmental control for young animals could reduce the number of pigs or chicks either through planned reduction or through mortality of those animals which could not survive lower temperatures.[15]
- Intermittent loss or reduction of available energy through planned electric power outage could result in mortality of chicks and pigs, delayed milking with subsequent reduction in milk production, and spoilage of animal products.
- Reduction of energy available for feed processing and animal production.

Table 6.4 indicates the amount of electricity utilized in California to produce livestock in various categories.

Turkeys

Turkey-raising requires electricity in several areas of production, depending upon the amount of environmental control desired. Turkey hens require a minimum of 21-22 lx (∿2 fc) to brood.[16] Although conditions vary widely across the country, the practical optimum is between 21.5 and 54 lx (2-5 fc) at bird height. Sixty watt bulbs without reflectors placed 10 feet apart and 5 feet from the sides of the building and 6 feet above the backs of roosting birds provide 27 to 32 lx (2.5-3 fc). If reflectors are used, power can be reduced by 20 percent.[17]

Low-intensity lighting is about as effective as high-intensity lighting provided the day length is 14 to 15 hours for brooding (egg-laying) hens.[18] Providing no more light than is needed (about 20 lx or 2 fc) will save a significant amount of energy in many cases.

Chickens

In open-type housing, electricity is required for:

- lights, and
- miscellaneous motors, feeders, collection systems, refrigeration, grading machines, and water pumps.

In environmental housing, electrical requirements are:

- lights;
- ventilation - a 373 W (0.5 hp) motor will move 283 m^3 (10,000

Category	% of State Use of Electricity	% of Agriculture Use of Electricity
Dairy (milk)	0.5249	6.71
Cow-Calf	0.0918	1.17
Feedlots	0.2095	2.68
Hogs	0.0108	0.14
Broilers, Fryers	0.0713	0.91
Eggs	0.1631	2.08
Sheep, Lambs	0.0086	0.11
TOTAL	1.08%	13.8%

Reference 12

ELECTRICITY REQUIRED TO PRODUCE LIVESTOCK IN CALIFORNIA (1972)

TABLE 6.4

ft^3) of air per minute.[19] If an average chicken requires 0.17 m^3/min (6 ft^3/min), 1,667 chickens are serviced by one 373 W motor; and,

- automation--mechanical feeders, egg collectors, fans.

Egg-processing Equipment, Refrigeration Equipment

In California, chicken production requires 4 x 10^{-3} GJe/yr (1.3 kWh/yr) per chicken. There are 38 x 10^6 chickens in California (November, 1974) of which 26 x 10^6 are in open housing. For these birds, therefore, 104 x 10^3 GJe (34 x 10^6 kWh) per year are used. In environmental housing, 309 x 10^3 GJe (85.8 x 10^6 kWh) per year must be provided.[20]

Cattle in Feedlots

From a case study conducted in the summer of 1974 on a feedlot in the 2-5 x 10^3 head-size category, it was determined that typical electrical usage is only 0.4 percent of total energy inputs, both direct and indirect, to the feedlot. Petroleum energy used was 20 times greater than electricity. Adding the energy contained in manufactured parts and supplies purchased for the feedlot doubled the directly accountable energy used by the feedlot, i.e., electricity, petroleum products, feed, etc. The direct use of electricity included primarily operation of the cattle feed mill which utilizes electric motors to aid in crushing, winnowing, conveying, mixing, and loading the various feeds used by the animals. In addition, a small portion was allocated to pumping water and general lighting on the premises and in the office.[21]

The greatest energy source in feedlots is the feed and potential energy stored in the cattle.[22,23] Energy involved in transporting cattle, commodities, and supplies can be significant--4.5 percent of the total energy flowing into the system.

Cow-Calf Operations

When they do not have a ready market for their cattle, cattle ranchers tend to leave them on the range where natural feed is available. As the animals are maintained on the range for longer periods of time, correspondingly fewer will be grazed to avoid depleting the grass. As a result, additional protein (70 kg/yr) must be fed to range cattle. Cotton seed meal, a cheap high-protein supplement, and molasses by-products from sugar refining are fed when available.

Significant electrical energy is used by pumps which water irrigated range land, increasing costs to the cattlemen. A cattle ranch that operates on irrigated pasture land and uses electric power for pumping shows an increased cost of $4-$6 per cow per year, resulting in a cost increase of 0.028-0.042 $/kg of beef sold.[24] Use of unirrigated range land can, therefore, provide a significant savings for the farmer and the cattle buyer.

Sheep and Hog Production

Sheep and hog production is similar to beef and poultry production and similar energy-efficient practices can be used. For example, as in raising chickens and turkeys, temperature control for young pigs is essential. Environmental housing, water and food supply systems are thus required.[25]

Sheep are transported to grasslands on a seasonal basis, requiring trucks and associated equipment. The indirect energy associated with the transport equipment, medicines, and vaccines is important and adds significantly to production costs.

Fish

Catfish and other fish raised commercially require a continuous supply of 27 °C (80 °F) water which has been adequately oxygenated. A processing plant is also required for raising catfish, adding a significant cost to the operation. The use of adequately heated water in catfish production is important. Neglect of proper water aeration results in severe losses. Water heating equipment and oxygenators require large energy inputs. Evaporation and conduction of heat from the ponds represent the major energy losses. Use of cooling water from electrical power generating stations represents a significant potential savings in water preheating costs.

C. Food Processing, Distribution and Consumption

Among the most energy-intensive aspects (38.5 percent) of the agricultural system are the packaging of food, its processing, transportation, and retail marketing.[26] Ultimately, even the consumer uses energy through preparation of food for the table, and freezing and refrigerating food. Many farm products are not ready for direct consumption but must be processed before they are edible. Other products, such as fruits and vegetables, are highly perishable in the raw state and must be preserved by canning or freezing. In 1971, for instance, energy used for heat and power in the food processing industry totalled 1.10 x 10^9 GJe (302 x 10^9 kWh), or 7.8 percent of the heat and power energy used in all manufacturing industries.[27] Table 6.5 shows 1971 purchased fuels and electric energy used in the food and kindred products industry by group.

Food Processing

More than 18,000 plants across the country are engaged in some phase of food processing, making it one of the largest and most diverse industrial sectors of the US economy.[29] Included are slaughterhouses, meat packers, grain mills, fruit and vegetable canning and freezing operations, sugar refineries, and cheese and dairy product operations. The total amount of food processed in the US has more than doubled in the last 25 years as a result both of population growth and an increase in the proportion of many crops converted into processed products.

Extensive food processing generates large quantities of waste material. Some of the residual waste materials may be more than 50 percent of the weight of the raw produce. Much of this waste contains a high nutrient and organic content and results in solid waste disposal problems. If released into water supplies, it also results in serious water pollution problems. Recently the food reprocessing industry has made significant advances toward reducing such problems by finding profitable commercial uses for its wastes.

For example, a survey taken in 1970 by the National Canners Association, encompassing the canned, frozen foods, and dehydrated products sectors of the industry, found that of a total of 30.4 x 10^6 mt (33.5 x 10^6 tons) of raw produce processed annually, 8.45 x 10^6 mt (9.3 x 10^6 tons) were disposed of by either land spreading or sanitary land fills. The remainder was burned: 7 percent was discharged into sewage systems and about 8 percent was discharged into the waterways without treatment. Of course, these figures represent only a small portion of the industry. Other sectors of the industry, such as slaughterhouses, also have high waste recovery whereby wastes are converted into useful by-products. Their conversion into economically attractive products not only increases profits but saves energy directly and indirectly and also decreases environmental pollution.

Food Marketing and Distribution

Fresh and processed foods must be moved quickly and efficiently, through the various marketing channels which requires rapid, dependable transportation and adequate cold and freezer storage.[30] Regional and local wholesale warehouses, supermarkets and other grocery stores, and all eating places other than the home (such as restaurants and clubs) have similar requirements.

Intercity transportation costs about 8 percent of the total food marketing bill. Part of the transportation costs are for fuel, amounting to approximately 5.45 x 10^8 GJ (157 x 10^{12} Btu) or 3.1 percent of total transportation energy usage. This is equivalent to 12.8 gigaliters

Industry Group	Value (millions)	As a percentage of Value of Shipments	As a percentage of Cost of all Materials
Meat Products	$130.8	0.5	0.6
Dairy Products	123.2	0.8	1.1
Canned and Preserved Fruits and Vegetables	124.4	1.0	1.7
Grain Mill Products	118.1	1.1	1.5
Bakery Products	66.5	0.9	2.1
Sugar	61.1	2.3	3.0
Confectionary Products	27.3	0.8	1.5
Beverages	109.2	0.8	1.6
Miscellaneous Food Preparations and Kindred Products	136.0	1.1	1.1
TOTAL	896.6	0.9	1.3

Reference 28

PURCHASED FUELS AND ELECTRIC ENERGY USED IN THE FOOD AND KINDRED PRODUCTS GROUP, BY INDUSTRY GROUP, 1971

TABLE 6.5

(376 million gallons) of gasoline. When energy needs for transporting farm products are included, transportation needs will probably approach 5 percent of total transportation energy.

In summary, agricultural markets and their transportation requirements are expected to continue expanding through the year 1980. Marketing volumes in 1969 to 1971 were estimated at 3.75 x 10^6 mt (4.13 x 10^6 tons) of products. Expressed as final demand by 1980, 4.65 x 10^6 mt (5.11 x 10^6 tons) are projected to be marketed, a 24 percent increase over the 1969 to 1971 levels.

Food Retailing

Many important changes have occurred in grocery retailing in the past few decades. Although the total number of retail grocery stores has been declining (from 355,000 in 1940 to 200,000 in 1973), grocery sales have become increasingly concentrated in large stores. For example, supermarkets comprising 20 percent of all grocery stores in 1973 accounted for 79 percent of all grocery store sales in that year.[31]

Some indication of the extent of energy use in supermarkets for the individual departments as well as the entire store may be found in Table 6.6. For the entire supermarket, energy costs an average of 0.36 percent of sales; significant differences exist among major departments, however. Produce displays are usually quite large, requiring refrigeration and ice and consequently a significant energy input. The energy impact of frozen foods is more dramatic since display cases containing these products need compressors for freezing. Meat and dairy products are usually displayed in open chilling cases which increase the amount of energy used. Dry, less perishable groceries require less energy input than items in other departments.

Food Consumption

The residential sector uses energy in storing food (refrigerators and freezers), in preparing food (stoves), and in transporting food from storage to homes (auto). In addition, energy is used by producers in delivering fuels to homes and in manufacturing and selling household kitchen equipment. Altogether these activities cost $6.1 billion and used 1.97 x 10^9 GJ (1.87 x 10^{15} Btu) in 1963. Of this, 3.74 x 10^8 GJe (1.04 x 10^9 kWh) of electricity was used. Approximately 85 percent of this energy was associated with operation of freezers, stoves, and refrigerators. The remainder was split evenly between appliance production and food shopping by car. This energy figure is 50 percent greater than the direct energy used in growing food on farms.[32]

D. Water Supply and Irrigation

Introduction

In 1970 irrigated land in the US was approximately 16 x 10^6 ha (40 x 10^6 acres), of which California (the leading state) had about 18 percent.[33] A significant amount of energy is required to pump water for crop irrigation.[34] The electrical energy required for pumping compares in magnitude with the energy used in manufacturing either fertilizer, tractors, or farm machinery for farming operations (refer to Figure 6.5). Whereas a large fraction of fossil fuel is used in the production and operation of tractors and farm machines, electrical energy is used to operate pumps for irrigation. Where electricity is not available, some fossil fuel is used by stationary engines for pumping operations.[33]

Crop irrigation can be divided into three categories. First, natural irrigation or rainfall is the most important means of crop irrigation and represents the primary means for the vast majority of land in the US.

Next in importance is irrigation with runoff water. Many rivers in the

(Percentage of sales)[1]

Item	Meat	Produce	Dry grocery	Dairy[2]	Frozen foods	Total store
Merchandise	78.80	69.00	86.73	80.79	75.60	82.53
Labor	11.51	15.58	6.92	8.78	10.91	8.94
Packaging	1.72	.31	.31	.19	.80	.63
Repairs	.34	.96	.38	.30	.92	.43
Utilities	.74	2.70	.27	.75	3.10	.73
Energy[3]	(.37)	(1.35)	(.14)	(.38)	(1.55)	(.36)
Depreciation	.49	1.41	.56		1.35	.63
Business taxes	.76	1.46	.50		1.22	.64
Rent	1.01	2.88	1.15		2.77	1.29
Interest	.13	.17	.08		.05	.12
Advertising[4]	1.80	1.80	1.80		1.80	1.80
Other	1.68	2.38	.68		1.12	1.32
Profit before taxes	1.02	1.35	.62		.36	.94
Total	100.00	100.00	100.00		100.00	100.00

Notes: (1) In-store margins exclude warehousing and delivery costs and headquarters expense.

(2) Includes ice cream and other refrigerated items such as bakery products, fruit juices, and dips.

(3) Assumes energy in one-half of utilities.

(4) Includes 0.05 cent for labor.

Reference 6

ESTIMATED IN-STORE GROSS COSTS, MARGINS, AND PROFITS OF SUPERMARKETS BY MAJOR DEPARTMENTS, 1972

TABLE 6.6

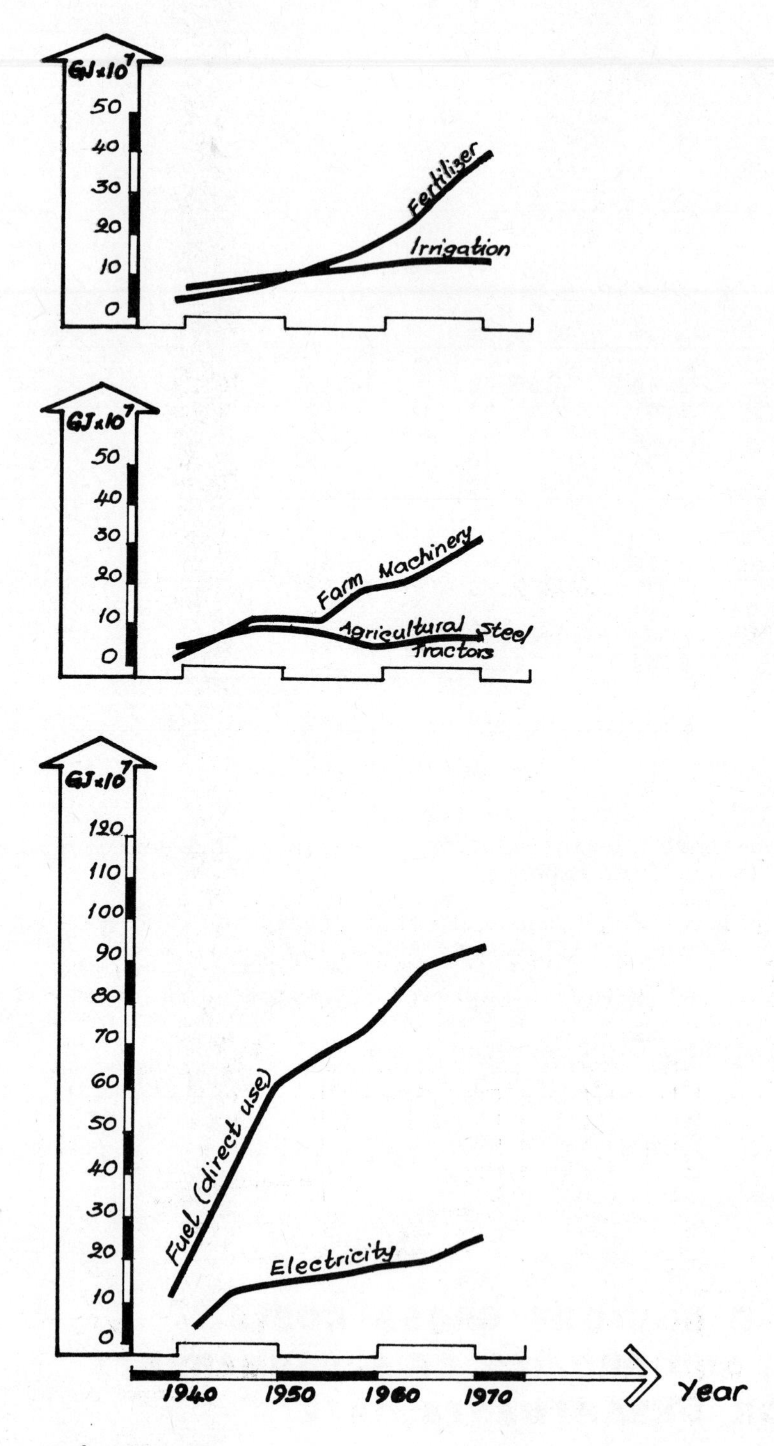

Reference 35

ENERGY USE ON FARMS FIGURE 6.5

US have been utilized to irrigate agricultural land. Dams have been placed at strategic points along river locations and the runoff water has been channeled through ditches and pipe lines to agricultural areas. A portion of the runoff can be used simultaneously to produce hydroelectric power.

The third and most energy-intensive form of irrigation involves the use of pumps to lift water from underground aquifers.

Irrigation using underground water tables is especially prevalent in arid regions such as the western and southwestern United States. During the growing seasons electrical or fossil fuel energy must be supplied to the farmer without interruption. Loss of power and subsequent loss of irrigation water would spell disaster.

In many parts of the US, particularly the southwestern region, the growing season lasts a substantial part of the year. Double and even triple cropping is possible in these regions, placing a greater demand on energy use.

Natural Irrigation (rainfall)

In the early history of the US, before the advent of electrically operated pumps and large developments for use of runoff water, farming was totally dependent on natural irrigation, or rainfall. The amount of crop production per acre was dependent, therefore, on the amount of rainfall during the year; and the extent of time over which the crop could be grown depended on the seasonal period of rainfall.

Rainfall still provides irrigation for a vast majority of acreage in the US today for several reasons. For example, with the exception of using sprinklers, it is extremely difficult if not impossible to irrigate contoured land. Secondly, in many parts of the country the rainfall is sufficient for adequate irrigation of a variety of crops. Only in the western and southwestern states do man-made methods of irrigation become necessary.

Irrigation with Runoff Water

Technological developments have made it possible to build large dams along major US rivers capable of containing large quantities of water. As the water level increased to heights sufficient for irrigation, it also became possible to control the water flow downstream. This greatly reduced the chance of flooding and allowed a controlled supply of water to percolate from the river to underground aquifers throughout the year, rather than only during spring floods. Besides protection of property, dams provide navigation and recreational benefits.

Building giant dams along the rivers, along with many benefits, brought about some undesirable changes. For example, by decreasing the river flow above the dams, sediment normally transported from the mountains down through the river is allowed to settle. As a result a sedimentary delta forms above the dam. This rich top soil sediment no longer flows downstream to farm lands to replenish the top soil which is continually being degraded and lost. It is also anticipated that the large dams will eventually fill with silt and become less useful.

When rivers are allowed to run free, adjacent farmlands are flushed and cleansed with flood waters which wash accumulated salt into the ocean. Dams prevent this flushing action. The problem is further aggravated when farmers distribute more and more fertilizers on the farm land to produce greater yields. For example, the Tulare basin in California, one of the largest agricultural producing areas in the US, is experiencing salt buildup in farming regions close to the delta at the mouths of the San Joaquin and Sacramento Rivers.[36] As a result, the abnormally high salt concentrations are causing a decrease in crop yields. This phenomenon is being experienced in other regions of the US and if corrective action is not taken, large areas of farm land may be rendered useless.

Pumped Irrigation

During the many years before pumped irrigation became a significant agricultural technique, windmills were used to pump water from a shallow aquifer. With limited pumping, the water table in many areas was encountered a few meters below the surface of the ground. Trees edged the rivers, shading a network of sloughs and waterways.

In recent years large quantities of water have been pumped from the aquifers and used for irrigation and city water supplies. Water tables have begun to recede. Wells have been drilled to lower level aquifers to obtain sufficient water for irrigation. As the water table began to recede, natural plant life began to die and water available for irrigation began to diminish. To alleviate the problem, dams were built to allow control of water flow and provide surface water for irrigation. After a number of dams were completed throughout the country, the water tables began to increase or at least kept pace with increased pumping levels. In many areas, however, the abnormally low water tables allowed encroachment of salt water into the underground aquifers and, in some cases, rendered the water from such aquifers unsuitable for irrigation and consumption.

In the future, silting of dams, salt buildup on farm land, and increased pumping action may again produce major difficulties in water districts throughout the world. Even greater changes, such as altering courses of major rivers to allow redistribution of water, may be necessary. Such a project has already been completed in California where the Feather River Water Project transfers excess water from Feather River in Northern California to arid Southern California. On route many storage basins allow regulation of the water supply and provide electrical power generation.

Irrigation Efficiency

Irrigation of farm land uses many different methods which are dependent upon land contour, water availability, the type of crop to be irrigated, and seasonal requirements of the crop. If water is not available at critical times during the growing season, damage can result and crop yield decreases. The soil also plays an important role in the type of irrigation methods employed. Structure, texture, and compaction control water percolation rates. In areas where sufficient runoff is available for irrigation, the farmer normally uses a less efficient method, such as flooding the land. The land must be relatively level for an even distribution of water. Use of checks or furrows will allow even distribution when flooding land that is slightly contoured. If the land is not sufficiently level, water may be stored in a reservoir and pumped through a sprinkler system. This method requires more pumping energy and greater capital investment and indirect energy use.

In areas where runoff water is insufficient, water must be pumped from underground aquifers, increasing costs and resulting in more energy use. In such cases sprinkler systems are often used or the land is face leveled to allow efficient surface irrigation. In areas where water is extremely valuable, drip irrigation may be used. This is the most efficient method since it waters only the area around the plant root to an easily controlled depth, thus minimizing surface evaporation and water loss through transpiration. Drip irrigation, however, requires an even larger initial investment (compared to a sprinkler system) since the water must be piped to each plant. This method only becomes feasible, therefore, in areas where water is in low supply. Drip irrigation also suffers from some major defects, the most common being plugging of the small openings in the pipe by silt or chemical deposits.

Hydroponic irrigation, which involves the growth of plants in nutrient-rich water without soil, is a specialized farming method. The water is used as a medium for containing the nutrients required for plant growth. Hydroponic farming normally requires a much larger investment than other methods of farming and is generally used under special conditions in which more conventional methods are not feasible. For example, in the far north the weather is much too cold to support the growth of many vegetables and fruits. As a result plants are grown in closed areas without soil, thus reducing the need to recondition soil each time a crop is harvested. In other areas specialized plants are grown hydroponically by utilizing treated waste water. The nutrient-rich water from sewage treatment plants is used to irrigate a large variety of plants without additional input of fertilizer. Utilizing the water in this way reduces the nutrient content and renders the water less hazardous as a waste material and less liable to pollute the environment.

In conclusion, the development of modern, efficient irrigation methods over the past half century is allowing the farmer to increase crop yields many fold. In recent years, however, adverse effects from some of these methods are becoming better understood, such as silting of rivers, salt buildup on high production land, and recession of water tables due to increased pumping pressure.

6.4 ENERGY SAVING PRACTICES

A. Introduction

Energy management on the farm is an important goal in the industrialized world since it involves saving large quantities of energy in all forms. One can appreciate the effect on our nation's economy of saving energy in the agricultural sector by noting that in the US, 17 million people (one worker out of five) are involved in that sector.[37] For example, the sub-sector which provides items such as fertilizers, pesticides, machinery, fuel, fencing, and other inputs used in farm production employs about 2 million people. The farm production subsector employs about 2.8 million farmers and somewhat less than 3.3 million operators and family workers. Some 10 million persons are employed in food processing, storage, distribution, marketing, and food preparation centers away from home.

This section presents a compilation of energy saving practices listed in order of their importance. Where possible, a rough estimate of percentage savings has been included. The divisions within this section include crop production, animal production, irrigation, food processing, transportation, consumption, and finally, conclusions. Energy saving tips common to most or all of the subsectors are discussed and listed first. Tips which apply to particular subsectors are listed under the appropriate subsector and discussed separately.

In the conclusions, Table 6.18 summarizes the energy saving tips described in each subsection. This table can be used to provide a reference to each method, allowing the reader to see quickly the relative merits of various energy saving practices.

B. General Energy Saving Practices

Many energy saving practices can be applied equally to all of the agricultural subsectors. For example, the maintenance of vehicle engines and mechanical systems is important in all subsectors. Badly fouled spark plugs can increase gas consumption by as much as 6 to 10 percent. A study indicated that 46 percent of all tractors studied were improperly adjusted (lowering performance by ∿9.5 percent).[38] A 59 kW (79 hp) gasoline engine wastes approximately 2 ℓ (0.5 gal) of gasoline each hour it is idling; this loss is the same as that for a 108 kW (145 hp) diesel engine.[38]

Proper storage of fuel, such as gasoline, is an extremely important energy saving practice and safety measure. For example, evaporation losses as high as 36 ℓ (9.69 gal) per month have been reported for dark-colored 1130 ℓ (300 gal) gasoline tanks stored above ground. Painting the tank white and locating it in the shade could reduce losses to 9 ℓ (2.4 gal) per month. By installing a pressure-release valve on a white, shaded tank, one can reduce this loss to 4.9 ℓ (1.3 gal) per month, with a total savings of 31.1 ℓ (8.3 gal) per month of gasoline. Storage in an underground tank would reduce the losses even further.[39]

Proper building insulation and temperature control offer the greatest savings in all subsectors. For example, in raising young chickens and turkeys energy can be saved by filling in all cracks and holes in the brooder building, applying insulation to walls and ceilings, installing proper thermostat controls on the heating systems, and locating the heating systems where they will blow warm air down onto the poultry rather than up to the roof.[40] Table 6.7 summarizes general energy saving practices.

C. Crop Production

Modern farming methods in agriculture today depend on properly maintained equipment. When equipment is not utilized efficiently, a large energy loss results. For example, tractors with faulty governors can lose as much as 80 percent of normal power; neglected maintenance can reduce tractor rated power by 25 percent. Table 6.8 lists some reminders for saving energy with farm equipment. (Refer to Chapters 7 and 8 for additional information.)

As an aid to production planning, all operations involved in the production of each crop should be listed and analyzed (see Table 6.9). This procedure will also save energy.[41]

Buildings which house livestock, poultry, or horticultural operations have the potential for energy savings. Table 6.10 lists some specific ideas for particular kinds of buildings; however, as mentioned earlier, some practices mentioned here apply to other types of buildings as well. (Refer to Chapters 7, 8, 9 and 10 for additional information.)

Corn, one of the largest crops in midwestern United States, requires special concern. It is usually harvested and handled dry and shelled. However, it can be harvested and handled in other ways to match the amount of fuel available. Table 6.11 lists points which can save considerable amounts of energy in this regard.

For example, corn harvested at 30 percent moisture will require 233 ℓ/ha (25 gal/acre) of liquid petroleum gas for drying, which is two to three times the fuel required in field operations. If corn is harvested at 30 percent moisture, 216 kg (476 lbm) of water must be evaporated for each ton of corn being discharged from the dryer at 13.5 percent. This requires 9.5 m^3 (335 ft^3) of liquid petroleum gas. If, however, it is harvested at 25 percent moisture, only 7.3 m^3 (260 ft^3) of liquid petroleum gas is required with a saving of 2.2 m^3 (78 ft^3) gas--a considerable cost savings.[42] Drying in the field, however, runs the risk of bad weather, that is, possibly losing part of the crop to wind and rain damage.

As natural gas becomes more limited in supply, the use of electric drying will increase. Maintaining moderate temperatures in well-insulated storage and processing equipment and using higher air velocities will lead to an efficient use of energy and a higher quality corn product.

The amount of fertilizer supplied for a particular crop and soil condition should be regulated. In many cases, it is advisable to call in a soil consultant for this purpose. The price paid for insufficient or excessive amounts of fertilizer is quite high and a large amount of energy can be wasted. In addition to following the points listed in Table 6.12, it is advisable to spend a few hours obtaining the proper information which can save many weeks of time and uncertainty and a considerable amount of money and energy.[43]

D. Livestock and Poultry Production

The energy required for livestock production varies widely depending on the animal and the conditions under which it is raised. For instance, beef cattle raised on open grasslands are nearly self-sufficient. They must be supplied with adequate water, supplemental food supplies upon occasion, and a salt-lick. If proper grazing methods are observed--in other words without

Procedure	% Energy Savings*
• Insulate buildings	10 - 15%
• Use proper thermostats and place them in proper locations	8 - 10%
• Place dampers on building roof ventilators	8 - 10%
• Place a ventilation fan in attic rather than install an air conditioner	6 - 8%
• Substitute supplementary energy systems such as solar water heaters where applicable	5 - 6%
• Proper engine maintenance	2 - 4%
• Store gasoline in properly designed tanks	1 - 3%
• Do not idle engines for long periods	1 - 2%

*Estimated percentage savings based on total expenditure of energy use. Actual savings will vary depending on specific applications, but should fall in these ranges. These savings are not necessarily additive.

GENERAL ENERGY SAVING PROCEDURES **TABLE 6.7**

Procedure	% Energy Savings*
• Use the right equipment for the job	5 - 20%
• Operate tractors near capacity	10 - 15%
• Use smaller tractors for lighter loads	10 - 12%
• Move large equipment by truck rather than driving it several miles	8 - 10%
• Gear up and throttle back on partial loads	5 - 10%
• Reduce tire slippage by adding wheel weights	6 - 8%
• Keep electrical equipment cleaned and lubricated properly	5 - 8%
• Use preventive maintenance	4 - 6%
• Shut the tractor off rather than let it idle unnecessarily	4 - 6%
• Keep all engines tuned up	2 - 4%
• Set controls properly	2 - 4%
• Couple equipment to reduce trips over the field	2 - 4%
• Check equipment manuals for tips on maintenance	2 - 3%
• Keep blades sharp	1 - 3%
• Double check machinery adjustments	1 - 2%
• Clean or replace air filters	1 - 2%

* Estimated percentage savings based on total expenditure of energy use. Actual savings will vary depending on specific applications, but should fall in these ranges. These savings are not necessarily additive.

FARM EQUIPMENT ENERGY SAVING HINTS **TABLE 6.8**

Procedure	% Energy Savings*
• Do not plow quite as deep unless there is a plow pan	10 - 20%
• Cultivate to control weeds	5 - 8%
• Plow when soil moisture is favorable, if possible	4 - 6%
• Harrow fields diagonally when two passes are needed	2 - 5%
• Plow around fields instead of across	2 - 4%
• Keep plow shares sharp	2 - 4%
• Work the long way of the field	1 - 2%

*Estimated percentage savings based on total expenditure of energy use. Actual savings will vary depending on specific applications, but should fall in these ranges. These savings are not necessarily additive.

ENERGY SAVING HINTS FOR PRODUCTION PRACTICES **TABLE 6.9**

Procedure	% Energy Savings*
Broiler Houses	
• Close up holes and cracks, insulate, maintain adequate air circulation to insure healthy animals	10 - 20%
• Brood as near the center of the house as you can	5 - 8%
• Brood as many chicks as the brooder can handle	4 - 6%
• Adjust brooding temperature to conserve heat	3 - 6%
• Keep litter dry	2 - 4%
Livestock Buildings	
• Cover windows with plastic if they are not essential for ventilation	10 - 12%
• Coordinate fan and heat thermostats	5 - 10%
• Clean and adjust thermostats	5 - 8%
• Use correct ventilation	4 - 6%
• Shutter fan outlets	2 - 5%
• Clean heaters	2 - 4%
• Use correct humidity	2 - 3%
Greenhouses	
• Adjust burners and stokers	10 - 15%
• Repair greenhouse openings	8 - 10%
• Paint greenhouses white	5 - 8%
• Be sure temperature instruments are working correctly	4 - 8%
• Put plastic on side walls and in walls of greenhouse	4 - 6%
• Do not steam sterilize more than necessary	2 - 5%

*Estimated percentage savings based on total expenditure of energy use. Actual savings will vary depending on specific applications, but should fall in these ranges. These savings are not necessarily additive.

ENERGY SAVING HINTS FOR FARM BUILDINGS — **TABLE 6.10**

Procedure	% Energy Savings*
• Harvest and handle as high-moisture shelled corn if an outlet is available	20 - 25%
• Harvest as corn silate if an outlet is available	10 - 20%
• Use natural air drying and low temperature drying as much as possible	10 - 15%
• Delay harvest as long as it is practical to dry on the stalk	5 - 10%
• Consider storing this year's harvest on the farm	5 - 10%

*Estimated percentage savings based on total expenditure of energy use. Actual savings will vary depending on specific applications, but should fall in these ranges. These savings are not necessarily additive.

ENERGY SAVING IN CORN PRODUCTION

TABLE 6.11

Procedure	% Energy Savings*
• Have a soil analysis run each year	10 - 20%
• Conserve the bulk of fertilizer supplies for application to lower-testing soils, or to prorate over a greater acreage and on crops which will make the maximum use of fertilizer	10 - 15%
• Starter fertilizer applied through the planter or drill is more efficient than broadcast fertilizer †	10 - 12%
• Adjust nitrogen application to the date of planting	8 - 10%
• Include credit for manure in rates adjustment .	5 - 10%
• Question whether fall application of nitrogen is right for the soil and the likely losses from leaching	5 - 8%

† Be sure to check with seed supplier or farm advisor regarding the application of starter fertilizer.

*Estimated percentage savings based on total expenditure of energy use. Actual savings will vary depending on specific applications, but should fall in these ranges. These savings are not necessarily additive.

ENERGY SAVINGS IN FERTILIZER USE

TABLE 6.12

overcrowding grasslands--the energy required per kilogram of beef produced is low. Sheep can also be pastured with low energy cost provided that they are not allowed to overgraze the area and kill the grass.

Poultry require more confined areas possessing feed and litter, plus adequate supplies of heat and ventilation in colder climates. During brooding a large source of heat is required to maintain an adequate temperature for chicks. Grouping chickens can reduce the heat requirement since each chicken, through its own body heat, can help warm its neighbor.

Carefully regulated temperature for pigs is also important as they are susceptible to disease. A large operation requires sterile conditions, with great care paid to the prevention of disease within the area.

In recent years, especially in the beef production system, feedlots have been used extensively to produce choice beef. The energy input to animals in feedlots is quite high per unit weight of animal produced. They must be totally supplied with feed, adequate water, vitamins, medicines, and minerals. Manure must be removed from the pen in order to avoid excessive buildup which can cause disease and deaths. Thus, feedlot operations are not only energy intensive but require careful maintenance and control.

Associated with many feedlots are sophisticated feeding systems, conveyor storage systems, and other energy intensive machines and processes. A large staff is often required to manage the cattle and the feeding. In addition, machinery is necessary to remove cattle waste and adequately store it for use as fertilizer, or for other means of disposal.

Table 6.13 lists points to follow in daily feedlot operations for significant energy savings.

The modern dairy industry is also energy intensive, requiring large amounts of water, electricity, and heat. Large refrigeration systems for milk storage, machines, various pumps, feed conveyors, and lighting are all used by the dairy industry. In addition, adequate feed storage must be available during the winter when cheap cattle feed supplies are not available. Silos and storage areas for the feeds and grains should be carefully designed and kept dry to avoid spoilage. Methods of compacting hay, such as baling, grinding, and perhaps cubing, are recommended since compacted hay requires much less space per ton and is easier to use. Supplementary feed such as sorghum and concentrated vitamins and minerals are recommended to provide adequate milk butterfat and quantity. Some specific points by which dairymen can save energy are listed in Table 6.14.[43,44]

E. Food Processing

Processing all agricultural products requires 8 to 10 percent of the total energy used in manufacturing in the US, and large quantities of energy can be saved if efficient procedures are employed. Refer to Chapters 2, 3, and 4 for information concerning energy saving practices in food processing, since many practices in the industrial, commercial, and residential sectors can be applied to the food sector.

Table 6.15 lists energy saving procedures, some of which are elaborated on here. In process operations, try to recapture waste heat for useful purposes such as heating office buildings or preheating ingredients in mixing processes. Chemical wastes which are not being utilized should be reviewed as possibly profitable by-products. In all manufacturing and process systems, the energy priorities of each subsystem should be reviewed to be sure that the subsystem is operating at its peak efficiency.

In wholesale and retail markets dealing in perishable items, improve methods for predicting the market to avoid excessive spoilage and waste of food. In cases where food has spoiled, try to find a means of utilizing it most efficiently rather than disposing of it. For example,

Procedure	% Energy Savings*
• When hauling feed or cattle locally, match carrier to size of load and try to operate the vehicle with a full load .	15 - 20%
• Check personnel to be sure that they are making efficient use of energy intensive and costly inputs such as feed, machinery and, of course, the cattle	10 - 20%
• Avoid placing cattle in excessively soft and muddy lots and provide shade in hot months, if possible. Construct concrete aprons in front of watering troughs	12 - 15%
• Find a market for manure to facilitate removal from feedlots. Meanwhile, stack in center of lot to serve as an elevated dry area. Also try to locate lot in crop production area to minimize hauling distances	10 - 12%
• Follow tips listed for farm equipment while operating heavy machinery	5 - 10%
• Take care in handling feed that none is lost	5 - 10%
• Keep dust, and hence incidence of infection, down by sprinkling with water during summer	5 - 10%
• Keep feed and silage dry and free from insects and rodents . .	6 - 8%
• Be aware of local crop surpluses which can be used as supplementary feed	5 - 8%
• Move cattle as few times as possible. The change in surroundings increases energy expenditure and usually takes weight off the cattle. Brand, inject, implant and dehorn all in one operation if possible	4 - 8%
• Select the right type and size of motor for the job. A motor operating at half rated load may lose only 1 percent in efficiency but may lose 10 percent or more in power factor. This increases current and power losses over that of properly matched motors and loads	4 - 6%
• When replacing machinery requiring fossil fuel for its operation, try to substitute diesel for gasoline engines . . .	2 - 5%
• Keep electric motors clean; clogged air ducts retard ventilation and may cause overheating and excessive energy use as well as possible need for replacement of the motor.	2 - 4%
• Shut off electric motors when equipment is not in use	1 - 2%

*Estimated percentage savings based on total expenditure of energy use. Actual savings will vary depending on specific application, but should fall in these ranges. These savings are not necessarily additive.

ENERGY SAVING HINTS FOR FEEDLOTS

TABLE 6.13

Procedure	% Energy Savings*
• Insulate buildings that are cooled by refrigeration or heated	15 - 20%
• Consider using mercury vapor or high-pressure sodium vapor lamps instead of incandescent lights for outside lighting	8 - 10%
• Consider the feasibility of using recovery heat to supply or supplement hot water requirements	5 - 10%
• Use infrared heat rather than convected heat in barns where semi-open or open structures are used	5 - 10%
• Check animal water supplies and be certain that they are free of contamination and disease	4 - 8%
• Use hot water wisely	4 - 6%
• Make sure the milk refrigerator is operating properly	4 - 6%
• Place lights so that they provide maximum illumination in areas where needed	4 - 5%
• Provide adequate drainage to avoid mudholes in which the cattle can pick up disease or decrease milk production	3 - 5%
• Keep pipes carrying hot water well insulated	3 - 5%
• Have available alternate energy sources such as stationary engine to supply the required power and fuel if shortages occur	2 - 5%
• Avoid excess washing of cows	3 - 4%
• Plan to conserve on water use, making maximum use of the water for cleaning the barns, etc.	2 - 3%
• Consider the possibility of re-using wash water	2 - 3%
• Set the hot water thermostat no higher than necessary	2 - 3%
• Put the timer switch on the pump control	1 - 2%
• Repair leaks in water lines and faucets	1 - 2%
• Reexamine water and power supply systems. Be sure that the pressure tank has an adequate amount of air. Turning the pump off and on requires additional electrical energy	1 - 2%

*Estimated percentage savings based on total expenditure of energy use. Actual savings will vary depending on specific applications, but should fall in these ranges. These savings are not necessarily additive.

ENERGY SAVING HINTS FOR DAIRY OPERATIONS

TABLE 6.14

Procedure	% Energy Savings*
• Place thermostats away from refrigerated displays to avoid excessive heating of buildings	10 - 15%
• Direct heat ducts away from ventilated refrigerated areas to avoid producing drafts which will extract cold air from refrigerated displays, causing excessive operation of the cooling system and simultaneously causing excessive operation of the heating system	10 - 15%
• Insulate buildings to avoid excessive heat losses through the walls and roof	10 - 15%
• Coordinate air-conditioning and heating requirements with the refrigerated systems	10 - 15%
• Check electrical power distribution panels, conduits, wires, and electric motors for proper sizing	10 - 12%
• Whenever possible close doors to the outside environment. . . .	6 - 8%
• In warm areas such as the southwestern US, paint the roof of buildings white to avoid generation of excessive heat. Conversely, in cold areas such as the northeastern and northern states, paint the roof black to absorb heat for heat generation purposes	6 - 8%
• Avoid the use of incandescent lamps in all areas	4 - 6%
• Use fluorescent lamps for inside lighting; use high-pressure mercury vapor or sodium lamps for outside lighting or even interior lighting (See Chapter 8)	4 - 6%
• Re-examine the distribution of lighting fixtures inside and outside the building, light levels needed and hours of operation, to reduce electricity wastes	3 - 5%
• In summer, turn up the thermostat temperature to decrease the energy used by air-conditioning systems	3 - 5%
• To contain warm or cold air, close all windows and where possible insulate the windows by placing curtains over them	2 - 5%
• Turn down thermostats in industrial and commercial buildings during the winter to decrease the load on heating systems	2 - 4%

*Estimated percentage savings based on total expenditure of energy use. Actual savings will vary depending on specific applications, but should fall in these ranges. These savings are not necessarily additive.

ENERGY SAVING IN FOOD PROCESSING

TABLE 6.15

wastes can be used as concentrated fertilizer for gardening or crop production.

Table 6.16 lists possible areas for saving energy in transportation. Refer also to Chapter 5.

Many of the energy saving practices in food retailing are similar to those used in processing and distribution. Again, the most energy intensive systems in retail stores are the heating, refrigeration, and air-conditioning systems. Great care should be exercised to be sure that these systems are not interacting excessively with each other and using more energy than is required.

Public eating establishments share many of the problems common to both the residence and the marketing systems. All heating and cooling requirements apply, as well as energy saving methods for the preparation of food in the proper quantities. (Energy saving practices in food preparation are covered in Chapter 4.)

Pumped irrigation is a particularly high user of energy due to pump operation. A considerable amount of electricity and fossil fuel is used in pumping water from underground aquifers to crop elevations. Additional energy is expended to pressurize sprinkler systems. Anything that can be done to improve crop production per unit of irrigation water applied will contribute significantly to the energy efficiency of the crop (see Table 6.17).

To summarize, increasing fuel and energy prices dictate improved energy use efficiency in all sectors of the agricultural system. Many energy reductions can be realized by relatively minor changes in present practices (as summarized in Table 6.18). In addition, surveys of new equipment and new processes should be made on a continual basis to take advantage of new innovations in energy saving systems. Redesign of equipment, development of new processes and production systems, and utilization of more efficient modes of transportation can all contribute to overall energy savings. In any business operation, and certainly in agriculture, energy savings mean cost savings and, therefore, greater profits.

6.5 POTENTIAL ENERGY SAVINGS

The agricultural sector (as described in this chapter) used approximately 13 percent of total US energy (9×10^9 GJ/yr) in 1970 (see Table 6.2). Of this, about 1.69×10^9 GJ/yr was used to produce 0.56×10^9 GJe/yr of electricity. Part of this energy use is included in the industrial, commercial, and residential energy use which is discussed in Chapters 2, 3, and 4.

A small part--the "on-farm" energy use--is not included in the savings discussed in Chapters 2, 3, and 4. In 1970 on-farm use of energy and electricity amounted to 2.2×10^9 GJ/yr and 0.13×10^9 GJe/yr (36.8×10^9 kWh).

The potential energy savings in the agricultural sector are estimated to be as follows: *immediate* (little or no capital costs), 15-20 percent; *near-term* (2-5 years, modest capital investment), 10-15 percent; *long-term* (5-25 years, heavy capital investment, process, and equipment changes), 5-10 percent (refer to Tables 6.7 to 6.18). Combined, these savings could reach 30-45 percent of total energy use in the agricultural sector over the 25-year period 1975-2000.

A potential for similar savings in electricity exists. There is a strong possibility, however, that any savings in agricultural electricity will be offset by shifts from scarce fuels to electricity. At the present time, natural gas supplies about twice as much (30 percent) of agricultural energy use as electricity.[45] For example, natural gas is extensively used for heating and drying of crops. In the event of gas scarcities, some of this demand will be shifted to electricity, emphasizing the need for widespread application of the energy management technologies described in this chapter.

Procedure	% Energy Savings*
• Try to carry a full load on the return trip rather than to come back empty	25 - 50%
• Consolidate processing of agricultural products to minimize transportation requirements from one point to another	20 - 40%
• Utilize the least energy intensive mode of transportation whenever possible, such as the railroad	20 - 30%
• Try to plan distribution of products so that minimum distances are required for transportation	10 - 20%
• Maintain transportation systems in top mechanical condition	10 - 15%

*Estimated percentage savings based on total expenditure of energy use. Actual savings will vary depending on specific applications, but should fall in these ranges. These savings are not necessarily additive.

ENERGY SAVINGS IN TRANSPORTATION

TABLE 6.16

PROCEDURE	% ENERGY SAVINGS*
• Consider whether flooding in rows or checks, sprinkler, or drip method best fits your conditions	15 - 20%
• Trickle irrigation can conserve considerable energy	4 - 8%
• Operate electric motor-driven irrigation wells during off-peak hours when possible	3 - 5%
• Install a re-use (tail-water) system on all surface irrigation systems	3 - 5%
• Adjust or re-engineer irrigation pumping plants to meet recommended performance standards	2 - 5%
• In hot dry areas, consider irrigating at night to reduce evaporation	3 - 4%
• When using grated pipe and siphon tubes with irrigation, consider installing automated grated pipe systems with a re-use system	3 - 4%
• Know how plants respond best to different soil moisture conditions	3 - 4%
• Feel and appearance of soil moisture and rooting conditions can be judged by using a soil spade and auger. Tensiometers and blocks work well in certain soils and crops	3 - 4%
• The water budget approach is also used as an additional aid in scheduling irrigation application	3 - 4%
• Hold pipe line and fitting losses to a minimum	2 - 3%
• Irrigation practices for different crops vary. Check with farm advisors for pre-season and main season, and cut off irrigation practices	2 - 3%
• Use aids to schedule correct months for irrigation water	2 - 3%
• With manually operated surface irrigation systems, use a water meter or some other method of measuring water	1 - 3%
• Plan to conserve water use. Less water pumped means that less energy will be used	4 - 5%
• Wells need to be designed carefully and then tested so that pumps and motors best fit the pumping situation	3 - 5%
• Efficiency of any pumping unit should be tested occasionally to determine excessive wear and energy usage. Pumping tests will assist in determining if changes are needed. (Most power suppliers offer a test pumping service)	1 - 3%
• Consider land-leveling needs. Well-prepared land surfaces are particularly important in increasing water use efficiencies for surface irrigation	4 - 5%
• Plan land use in cropping systems to fit soil and water conditions. Certain crops utilize water more efficiently than others in various soils	3 - 4%
• Check on deep percolation losses in conveyance ditches, in storage areas and in field application	3 - 5%

*Estimated percentage savings based on total expenditure of energy use. Actual savings will vary depending on specific applications, but should fall in these ranges. These savings are not necessarily additive.

ENERGY SAVINGS IN IRRIGATION

TABLE 6.17

Energy Savings Hints For	See Table	Range of Savings in Total Energy Use
General Energy Saving Procedures	6.7	1 - 15%
Farm Equipment	6.8	1 - 20%
Production Practices	6.9	1 - 20%
Farm Buildings	6.10	2 - 20%
Corn Production	6.11	5 - 25%
Fertilizer Use	6.12	5 - 20%
Feedlots	6.13	1 - 20%
Dairy Operations	6.14	1 - 20%
Food Processing	6.15	2 - 15%
Food Transportation	6.16	10 - 50%
Irrigation	6.17	1 - 20%
Total Possible Savings in Agriculture		5 - 25%

SUMMARY OF ENERGY SAVING TIPS

TABLE 6.18

REFERENCES

1. Blaxter, Kenneth, "Power and Agricultural Revolution," *New Scientist* (Great Britain), (14 February 1974): 400-403.

2. Steinhart, Carol E. and Steinhart, John S., *Energy: Sources, Use, and Role in Human Affairs,* (North Scituate, Massachusetts: Duxbury Press, 1974), p. 77.

3. Steinhart, John S. and Steinhart, Carol E., "Energy Use in the U.S. Food System," *Science* 184(19 April 1974): 307-316.

4. University of California Food Task Force, *A Hungry World: The Challenge to Agriculture,* (University of California, Division of Agricultural Sciences, July 1974). Also see McWilliams, Carly, *Factories in the Field,* (Santa Barbara, California: Peregrine Press, 1971).

5. Personal communication between Perry Stout and Eduardo Cruz de Carvalho at the University of California, Davis, 1975.

6. US Department of Agriculture, The Economic Research Service, *The U.S. Food and Fiber Sector: Energy Use and Outlook,* (Washington, D.C.: US Government Printing Office, 29 September 1974).

7. Ibid., p. 6.

8. Steinhart and Steinhart, *Energy: Sources, Use, and Role in Human Affairs,* p. 76.

9. Ibid., p. 80.

10. Ibid., p. 78.

11. Spomer, R.G., Piest, R.F., Heinemann, H.G., "Soil and Water Conservation with Western Iowa Tillage System," *Transactions of the ASAE* 19(January-February 1976): 108-112.

12. US Department of Agriculture, The Economic Research Service, *The U.S. Food and Fiber Sector: Energy Use and Outlook,* p. 23.

13. Cervinka, V.; Chancellor, W.J.; Coffelt, R.J.; Curley, R.G.; and Dobie, J.B., *Energy Requirements for Agriculture in California,* (Sacramento, California: California Department of Food and Agriculture and the University of California, Davis, January, 1974).

14. Coote, D.R.; Haith, D.A.; Zwerman, P.J., "Modeling the Environmental and Economic Effects of Dairy Waste Management," *Transactions of the ASAE* 19(March-April 1976): 326-331.

15. Albright, L.D., "Air Flow Through Hinged Baffle Slotted Inlets," *Transactions of the ASAE* 19(July-August 1976): 728-732, 735.

16. Williams Turkey Breeding Farms, Breeding and Research Department, *Modern Turkey Breeder Management,* (Oakdale, California: April 1967), p. 23.

17. Teter, N.C.; DeShazer, J.A.; Thompson, T.L., "Operational Characteristics of Meat Animals. Part IV: Turkeys, Large White and Bronze," *Transactions of the ASAE* 19(July-August 1976): 724-727.

18. Marsden, Stanley J., *Turkey Production,* Agriculture Handbook No. 393, (Washington, D.C.: US Department of Agriculture, Agricultural Research Service, March 1971), p. 24.

19. Reece, F.N.; Deaton, J.W.; Harwood, F.W., "Roof Insulation and Its Effect on Broiler Chicken Mortality in Hot Weather," *Transactions of the ASAE* 19(July-August 1976): 733-735.

20. Personal communication from Donald Bell, Riverside County Farm Advisor, to E.W. Greninger, Fall of 1974.

21. Larson, C.L.; James, L.G.; Goodrich, P.R.; Bosch, J., "Performance of Feedlot Runoff Systems in Northern Climates," *Transactions of the ASAE* 19 (November-December 1976).

22. Teter, N.C.; DeShazer, J.A., "Calculation of Protein and Water and Energy Input Requirements of Beef Animals," *Transactions of the ASAE* 18 (November-December 1975): 1155-1157.

23. Riskowski, G.L.; DeShazer, J.A., "Work Requirement for Beef Cattle to Walk Through Mud," *Transactions of the ASAE* 19 (January-February 1976): 141-144.

24. Personal communication from Chester A. Perry, Livestock Farm Advisor, Los Angeles County, to E.W. Greninger, 12 November 1974.

25. University of Arizona, Council for Agricultural Science and Technology (CAST) Task Force, "Energy in Agriculture," Preliminary Draft, Kenneth K. Barnes, Chairman, 26 November 1973.

26. US Department of Agriculture, The Economic Research Service, *The U.S. Food and Fiber Sector,* p. 2.

27. US Department of Commerce, Bureau of the Census, *1972 Census of Manufactures Special Report Series: Fuels and Electric Energy Consumed,* Report No. MC72(SR)-6, (Washington, D.C.: US Government Printing Office, 1973).

28. US Department of Agriculture, The Economic Research Service, *The U.S. Food and Fiber Sector,* pp. 47-48.

29. Hoove, Sam R. and Jasewiez, Lenore B., "Agricultural Processing Wastes: Magnitude of the Problem," in *Agriculture and the Quality of Our Environment,* ed. Nyle C. Brody, (Washington, D.C.: American Association for the Advancement of Science, 1967), pp. 187-204.

30. Singh, R.P.; Heldman, D.R.; Cargill, B.F.; Bedford, C.L., "Weight Loss and Chip Quality of Potatoes During Storage," *Transactions of the ASAE* 18 (November-December 1975): 1197-1200.

31. US Department of Agriculture, The Economic Research Service, *The U.S. Food and Fiber Sector,* p. 75.

32. Hirst, Eric, "Food Related Energy Requirements," *Science,* 184(12 April 1974): 134-138.

33. US Department of Agriculture, *Agricultural Statistics 1974,* (Washington, D.C.: US Government Printing Office, 1974), p. 425.

34. Stetson, L.E.; Watts, D.G.; Corey, F.C.; Nelson, I.D., "Irrigation System Management for Reducing Peak Electrical Demands," *Transactions of the ASAE* 18(March-April 1975): 303-306,311.

35. Steinhart and Steinhart, *Energy: Sources, Use, and Role in Human Affairs,* p. 73.

36. Pillsbury, A.F., and Blaney, H.F., "Salinity Problems and Management in River Systems," Proceedings Paper No. 4733, *Journal of the Irrigation and Drainage Division, Proceedings of the American Society of Civil Engineers* 92(1966): 77-90.

37. US Department of Agriculture, The Economic Research Service, *The U.S. Food and Fiber Sector,* p. 1.

38. US Department of Agriculture, Office of Communication, *Energy Management on the Farm,* (Washington, D.C.: March 1974), p. 2.

39. Ibid., p. 4.

40. Curley, R.G., *Energy Saving Tips,* (Berkeley, California: University of California, Agricultural Extension, February 1974).

41. Peterson, C.L.; Milligan, J.H., "Economic Life Analysis for Machinery Replacement Decisions," *Transactions of the ASAE* 19 (September-October 1976): 819-822, 826.

42. US Department of Agriculture, Office of Communication, *Energy Management on the Farm*, p. 5.

43. Curley, p. 6.

44. US Department of Agriculture, Office of Communication, *Energy Management on the Farm*, p. 5.

45. US Department of Agriculture, The Economic Research Service, *The U.S. Food and Fiber Sector*, p. x.

PART III

ENERGY MANAGEMENT TECHNOLOGIES

ENERGY REQUIREMENTS ARE RISING RAPIDLY AND WILL DOUBTLESS CONTINUE TO RISE FOR DECADES BECAUSE OF POPULATION INCREASES, THE CONTINUOUS SPREAD OF INDUSTRIALIZATION, THE RISE IN PER CAPITA DEMANDS FOR MATERIAL GOODS; AND THE INCREASE IN ENERGY REQUIREMENTS NECESSARY TO PROCESS LOW-GRADE ORES AND TO OBTAIN THE DIMINISHING FOSSIL FUELS THEMSELVES.

--Harrison Brown, 1954

CHAPTER 7

BUILDING ENVELOPES AND SITES

K.M.S. Iyengar*

CHAPTER CONTENTS

KEY WORDS

Architect-Engineer Services
Architect Services
Building Materials
Energy Conscious Design
Environmental Conditioning
Heat Losses
Infiltration
Insulation
Residential Buildings
Solar Energy
Sun Shading
Ventilation

SUMMARY

This chapter describes efficient energy use in building envelopes and sites. The building envelope (or shell) has an influence on energy use which is distinct from the equipment contained within and from the patterns of use established by the occupants.

Moreover, the building envelope and site have a pronounced impact on the energy incident on the structure and its occupants, and on heat exchange with the building's surroundings. By closely controlling the design of the building envelope and the orientation of the building on the site, the designer can influence the energy needs of the building in a profound way. Efficient designs seek to maximize the heating potential (due to solar energy) of the site in winter, and minimize it during summer, all in a cost-effective way.

This chapter discusses the theory and practice of determining heat losses and gains, and concludes with a section concerning retrofitting of buildings for efficient energy use.

*Member of the Technical Staff and Chief Architect, Applied Nucleonics Company, Inc.

7.1 INTRODUCTION

Those who enclose any human activity attempt to efficiently use materials, space, and energy through architecture, engineering, and construction. Traditionally, the goals in designing a new building have been to cause minimal disturbance to the land mass and, to whatever extent possible, to enrich the natural environment of the building's location.[1,2]

There are many simple ways in which buildings can be designed to use energy efficiently. Perhaps the most fundamental is to avoid overdesigning in terms of heating, cooling, lighting, and strength. With a greater sensitivity initially to local climatic conditions and topography, it is possible to achieve long-term economic benefits by saving non-renewable fuels. At present, the more serious problem of energy efficient design is in retrofitting existing buildings which were built prior to the present concern with energy-efficient building design. In addition to efficient building design and energy management, some designers, architects, and engineers are looking to new sources of energy for buildings which will reduce dependence on non-renewable fuels and diminish the environmental degradation related to their use.

Of all energy used in US buildings, 14 percent is used in 24 billion square feet of commercial space, and 19 percent in 73 million dwelling units (see Figure 7.1).[3] With respect to end-use, as shown in Figure 7.2, 64 percent of building energy is used for space heating and air conditioning, 24 percent for equipment (including hot water heating, home appliances, and office equipment), and 12 percent for lighting. The annual national increase in number of buildings is estimated at less than 3 percent. Available technology offers significant opportunities for energy management in new buildings and equipment, but at the nominal rate of building replacement, this will not affect the majority of structures before the end of the century. Therefore, major energy savings, if they are to occur by the year 2000, must come from retrofitting existing structures in addition to modifying equipment and improving operating procedures. Extensive efforts are necessary to encourage building owners to implement energy efficient features.

It is probably fair to say that building forms for many modern structures do not respond to the physical environment. Given the present necessity to design with economy of resources--space, money, structural materials and esthetics--it has become increasingly important to emphasize power and energy economy.

Space heating is the most substantial component of energy use in buildings, and it is the most likely candidate for savings in the residential and commercial sectors. Figure 7.3 illustrates how energy dissipates through the skin of a building. Minimizing these leaks requires making the skin less permeable. While it is possible to make a building very tight, it is not advisable from an economic point of view. It is important, therefore, to decide on an acceptable level of heat loss from the building skin. There are a number of additional methods to effect energy economy; for example, improved insulation, orientation, glazing and increased efficiency of furnaces and boilers, etc.

The following sections will discuss the mechanics of designing and evaluating the efficiency of a building envelope, in terms of building energy use.

7.2 FACTORS AFFECTING ENERGY USE IN BUILDINGS

Energy need in a building is a function of many individual, interrelated factors. Their interrelationships must be clearly understood in order to achieve energy efficiency while satisfying occupants' needs and esthetic and economic goals.

A. Building Users

Larger volume or space than actually required for the functional and psychological needs of the occupants increases costs as more materials

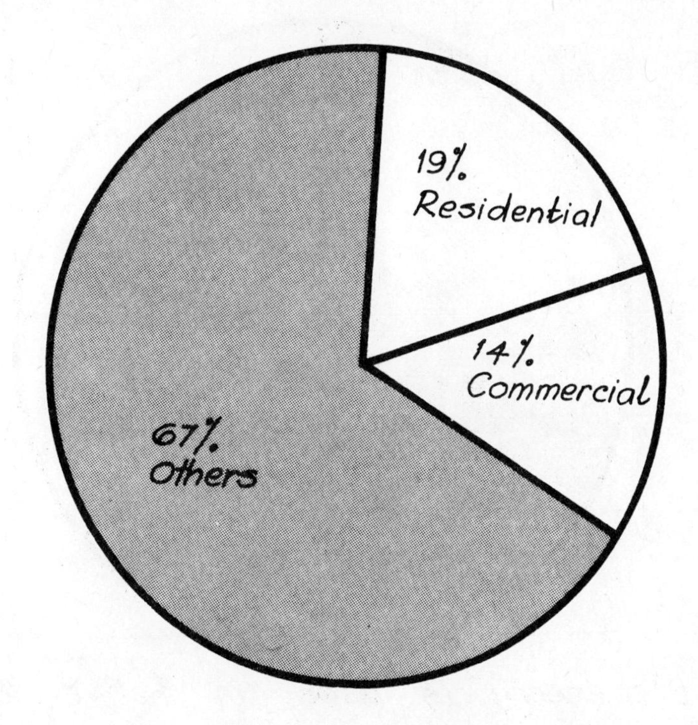

Reference 3

US ENERGY USE IN BUILDINGS FIGURE 7.1

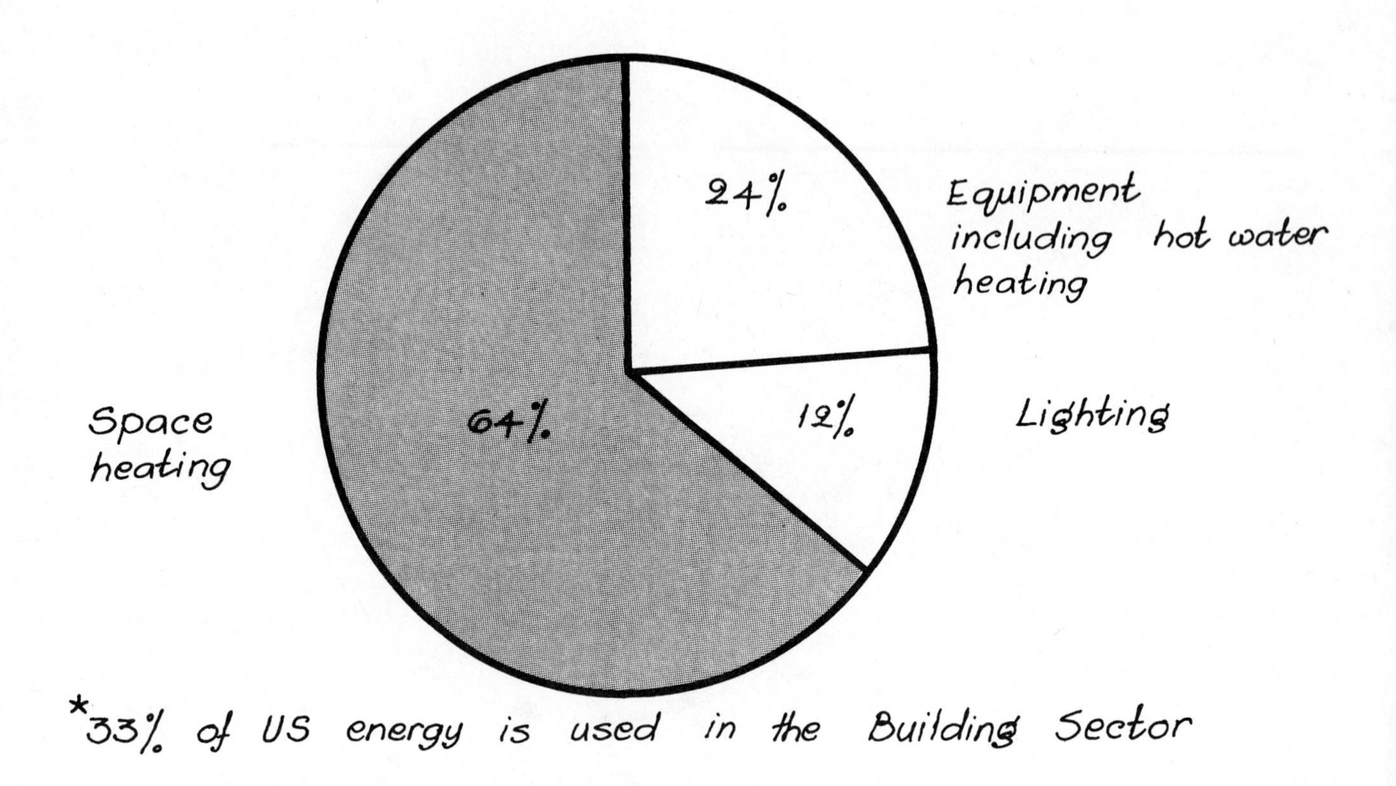

Reference 3

ENERGY USE IN US BUILDING SECTOR*

FIGURE 7.2

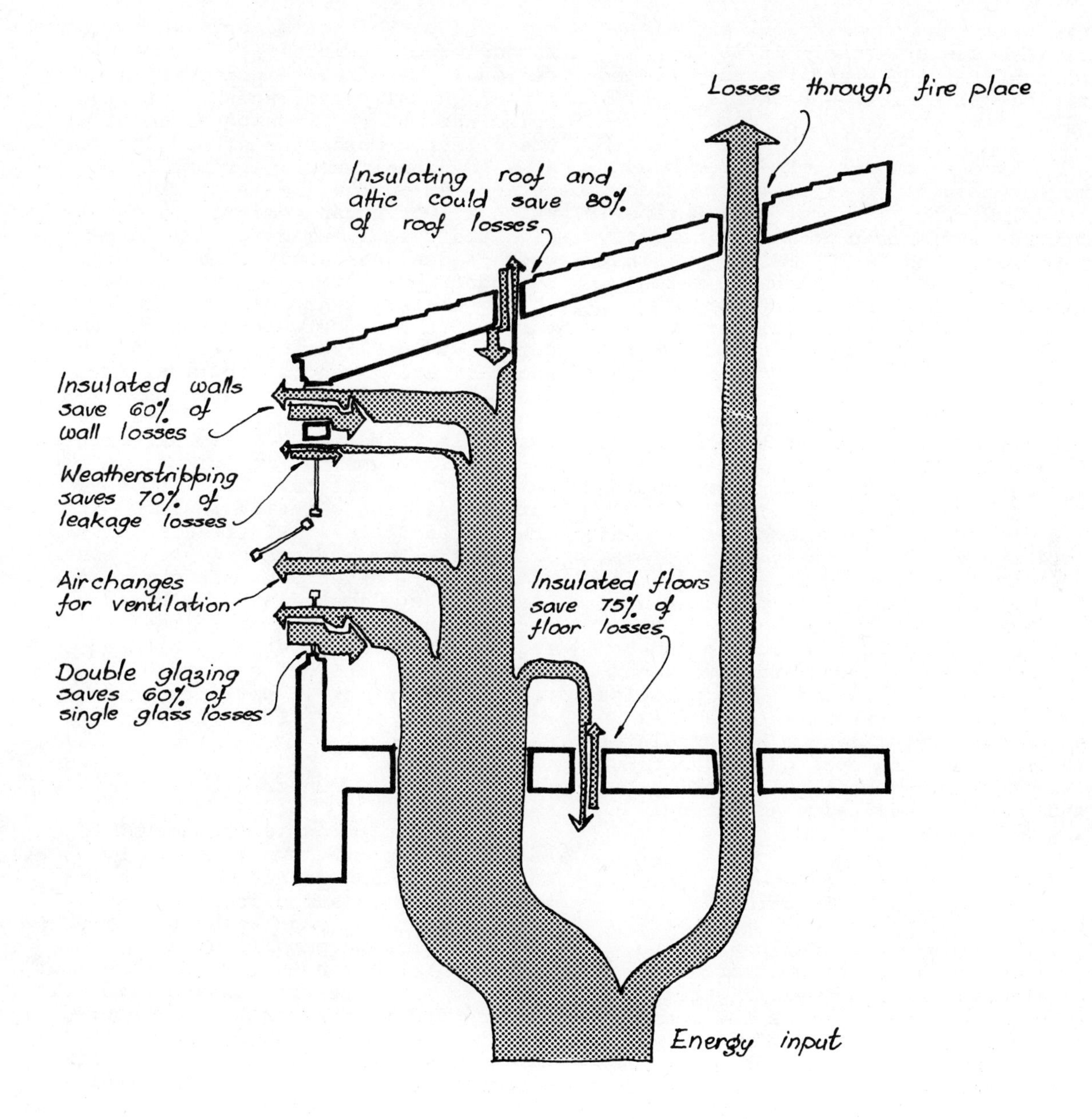

DIAGRAM OF BUILDING HEAT LOSS

FIGURE 7.3

are used for construction, and also increases energy use. It is important to understand the users' needs, biological comfort, and physiology.[4, 5]

Food taken into the human body may be thought of as fuel subjected to a low-grade "combustion" (metabolic) process sufficient to maintain a body temperature of 37 °C (98.6 °F). There is a wide range of metabolic rates dependent on physical activity. Figure 7.4 illustrates different rates of metabolism for various types of human activity.[6] Note that heat given off by the body at the rate of about 5 x 10^6 MJ/hr (4.75 kBtu/hr) while running is almost ten times that of the hourly heat loss of a person who is doing light work, e.g., typing.

Energy use in environmental control systems is directly proportional to demand and demand is almost entirely under the control of the user of a given building system. Thus, the challenge lies in limiting demand on an economically sound and psychologically acceptable basis. The principal factor used to establish demand is an environmental system which relates to heat loss and heat gain. Ventilation, air movement, air quality, and humidity will also affect building environmental needs.

B. Climate and Geographical Location

Climate is defined as the average condition of weather at a particular location over a period of years. Since weather is the momentary state of the atmospheric environment (temperature, wind velocity and precipitation) at a particular location, climate could be defined as the sum total of all the weather that occurs at any location.[6] Climates are comparatively constant and, despite short-term but significant and rapid changes, they have weather patterns that repeat themselves over and over again at given time intervals. Cold days occasionally occur in hot climates, dry climates also have rainy periods and wet climates sometimes have extended periods of drought. Nevertheless, every place on earth exhibits its particular combination of heat and cold, rain and sun.

The earth's climate is shaped by thermal and gravitational forces, regional pressure, temperature, and topographical differences. Existing buildings and their shapes also affect the climatic condition of any locality. A building designer is primarily interested in the limits of climate which affect human comfort and the design and use of buildings. Information which a designer would like to have includes temperature averages, changes in temperature extremes, the temperature difference between day and night, humidity, infiltration, exfiltration, snowfall and its distribution, sky conditions, hail, hurricanes, and thunderstorms. Climate data are available from the US National Weather Service for numerous locations. A number of systems have been proposed for classifying climatic regions within the United States. Figure 7.5 illustrates the temperature and solar conditions of a given day. In the United States, the climate has been broadly divided into four regions: cool regions, temperate regions, hot and arid regions, and hot and humid regions. Figure 7.6 portrays these regions with brief descriptions of each.[6]

C. Insolation

Insolation is the amount of radiation from the sun received by a surface. At any particular point on the earth, insolation is affected by several factors. Radiation received per unit area perpendicular to the incoming radiation is greater than that received per unit area on a horizontal surface due to the earth's curvature and the tilt of its axis. It is for this reason that solar collectors are generally placed on an angle. Radiation reaching the earth's surface is also affected by the atmospheric condition, its purity, vapor, dust, and smoke content. For example, radiation is absorbed and scattered by ozone vapor and dust particles in the atmosphere.[5,6]

One obvious factor affecting the amount of radiation striking a particular location is the length of the daylight period, which is dependent upon the day of the year. Figure 7.7 illustrates the annual mean daily insolation on a horizontal surface in the US.

Maximum Energy Capacity of a normal healthy 20 year old male.

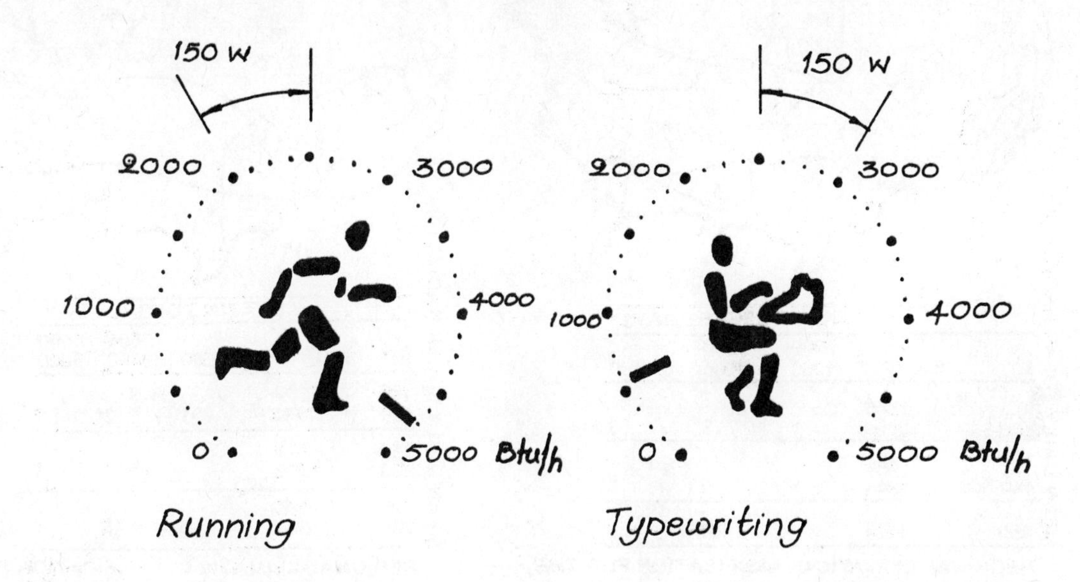

Reference 6

BODY HEAT PRODUCTION IN BTU's PER HOUR AND WATTS FOR AN AVERAGE PERSON

FIGURE 7.4

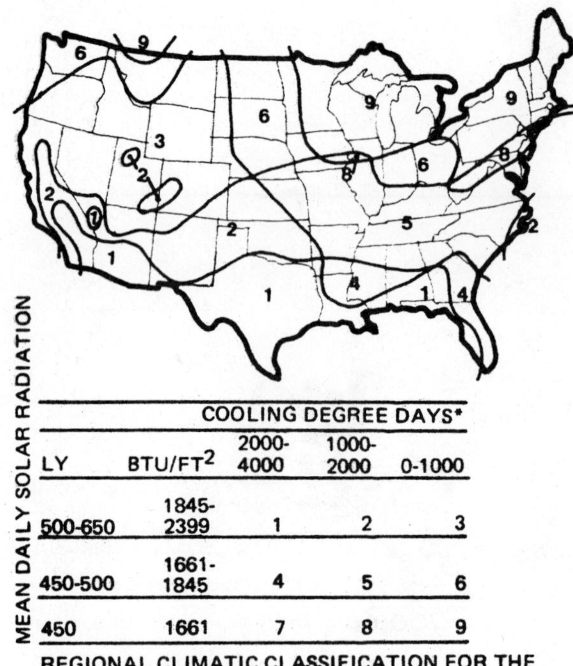

MEAN DAILY SOLAR RADIATION		COOLING DEGREE DAYS*		
LY	BTU/FT^2	2000-4000	1000-2000	0-1000
500-650	1845-2399	1	2	3
450-500	1661-1845	4	5	6
450	1661	7	8	9

REGIONAL CLIMATIC CLASSIFICATION FOR THE COOLING SEASON (MAY-OCTOBER)

MEAN DAILY SOLAR RADIATION		HEATING DEGREE DAYS		
LY	BTU/FT^2	0-2500	2500-5000	5000-9000
350-450	1292-1661	1	2	3
250-350	923-1292	4	5	6
175-250	646-923	7	8	9

REGIONAL CLIMATIC CLASSIFICATION FOR THE HEATING SEASON (NOVEMBER-APRIL)

LY = Langleys

Reference 6

REGIONAL CLIMATIC CLASSIFICATION FOR HEATING AND COOLING SEASONS IN US

FIGURE 7.5

Hot-arid regions: Hot-arid regions are characterized by clear sky, dry atmosphere, extended periods of overheating, and large diurnal temperature range. Wind direction is generally along an E-W axis with variations between day and evening.

Hot-humid regions: High temperature and consistent vapor pressure are characteristic of hot-humid regions. Wind velocities and direction vary throughout the year and throughout the day. Wind velocities of up to 120 mph may accompany hurricanes which can be expected from E-SE directions.

Cool regions: A wide range of temperature is characteristic of cool regions. Temperatures of minus 30° F(-34.4°C) to plus 100° F (37.8°C) have been recorded. Hot summers and cold winters, with persistent winds year round, generally out of the NW and SE, are the primary identifiable traits of cool regions. Also, the northern location most often associated with cool climates receives less solar radiation then southern locations.

Temperate regions: An equal distribution of overheated and underheated periods is characteristic of temperate regions. Seasonal winds from the NW and S along with periods of high humidity and large amounts of precipitation are common traits of temperate regions. Intermittant periods of clear sunny days are followed by extended periods of cloudy overcast days

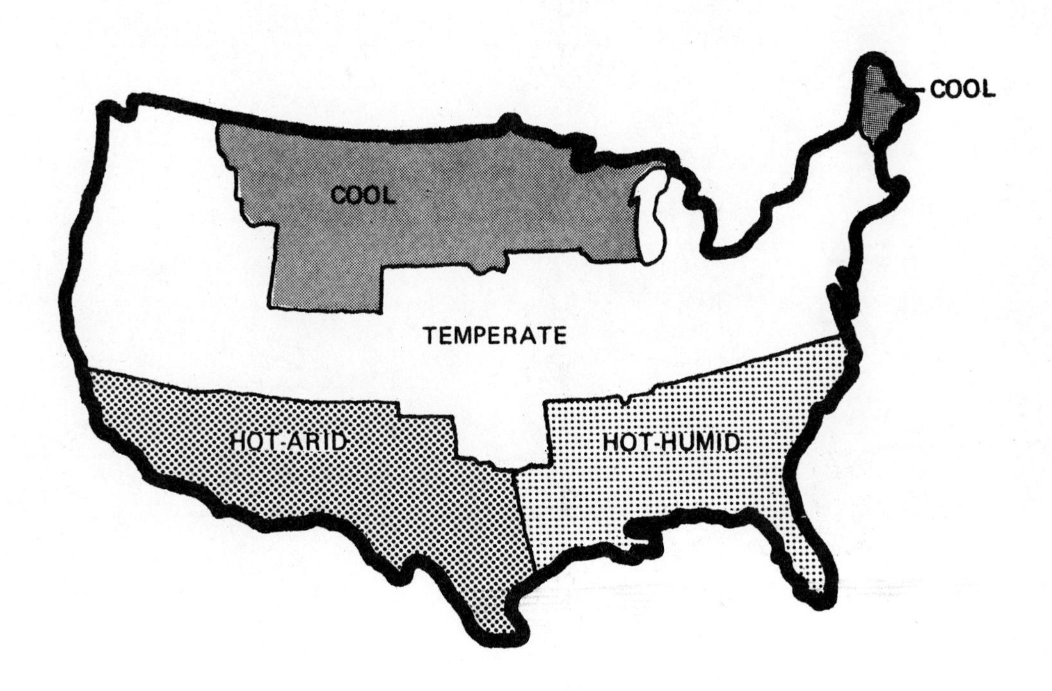

Reference 6

CLIMATIC REGIONS OF THE US **FIGURE 7.6**

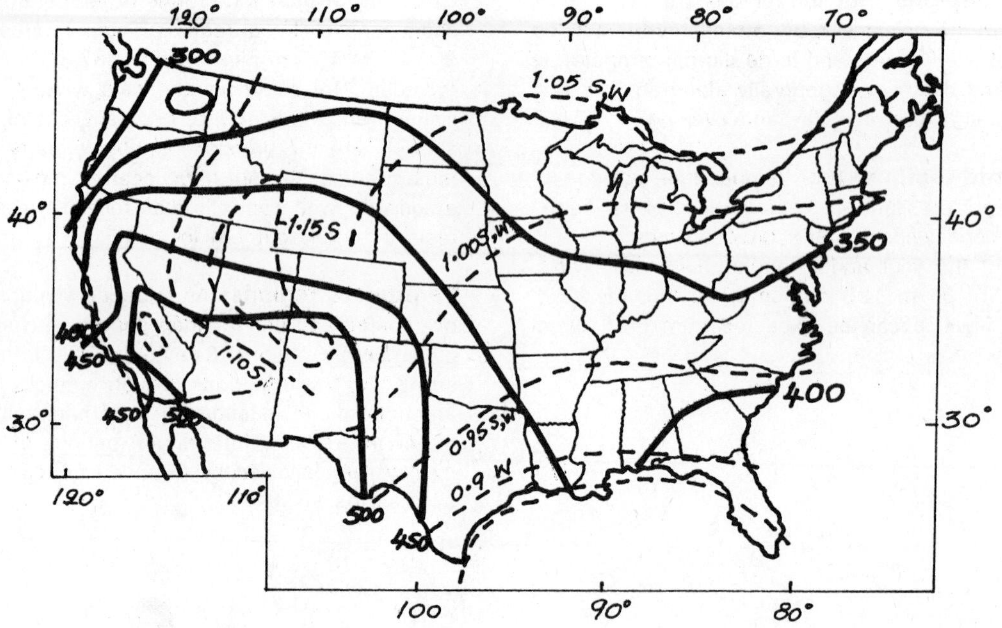

S Summer

W Winter

Summer & Winter Clearness Numbers

Annual mean daily Insolation in Langleys / Day

1 Langley $= 4.184 \times 10^4$ Joules/m^2
$= 1$ Cal/cm^2

Reference 7

ANNUAL MEAN DAILY INSOLATION AND CLEARNESS NUMBERS **FIGURE 7.7**

Heat gain or heat loss in any building is affected by the position of the sun in addition to several other factors. The earth rotates around its own axis, one complete rotation in 24 hours. The axis of this rotation is tilted at an angle, 23.5° to the plane of the earth's orbit, and the direction of this axis is constant. Figure 7.8 illustrates the earth's tilt and path around the sun. Maximum intensity of solar radiation is received on a plane normal to the direction of radiation. Due to the tilt of the earth's axis, the area receiving maximum solar radiation moves north and south, between the Tropic of Cancer and Tropic of Capricorn. This is the primary cause of seasonal changes.[6]

D. Siting, Orientation and Shape

Olgyay's calculations of the optimum shape for residential buildings, taking into account both heat loss in winter and heat gain in summer, show that for simple rectangular plans the most effective shapes will be those elongated in the east-west direction.[2,8] The east and west faces receive the greatest amount of summer radiation and therefore should be reduced in area. The southwest face receives radiation in the winter, but not so much in the summer and, thus, should be increased in size. Figure 7.9 shows a house in New York whose design, by Henry Niccols Wright, was based on these principles.[2]

Buildings can also be shaded to prevent solar radiation from reaching each surface, particularly the windows. Masonry or hard top surfaces surrounding a building may act to reradiate the heat received onto the building itself. This effect can be avoided by the use of shrubs and grass. The absorption or reflection of heat is also affected by the colors of the buildings it faces; for example, light colors, particularly for roofs, are effective in reducing heat gain. Figure 7.10 illustrates "heliothermic" site planning by Henry Niccols Wright. This concept extends his and Olgyay's thinking about building orientation and solar radiation to the large-scale problems of summer cooling and winter heating in residences.[2]

For energy efficiency, the location of the building on the site in relation to adjacent structures and natural features should be optimized. Sites for rectangular buildings may have a length-to-depth relationship which would permit or hinder the ability of the building configuration to take maximum advantage of the sun and wind. The site also may influence the direction of the building axis so that the most adverse effects of sun and wind will be felt.

In order to maximize the energy benefits from a particular site, it may be desirable to allow changes in the configuration of the building by increasing or decreasing the height. The orientation of each of the four or more facades and of the roof has a unique effect upon energy requirements, depending upon the climate, site, and geographic location. For instance, in any climate in the United States, a sloping roof facing south would be subjected to more solar radiation. In cold climates wind velocities are greatest on the north and west facades of the building.

In summary, each surface of a building is subjected to different environmental influences, depending upon the geographical location, climate, insolation, siting, orientation, and building shape.

E. Configuration

Besides its orientation and envelope, a building's configuration also determines the amount of energy used. If energy can be saved by use of natural illumination, the building perimeter should be increased and its interior space proportionately decreased. This will result in different building forms such as multiple courtyards, atriums, lightwells, or tall buildings or low buildings with skylights.

Configurations that resist unwanted heat gains and losses result in less energy use. With low buildings, a square configuration has less surface than a rectangular one of equal area and so experiences less thermal effect due to the environment (all other factors being the same). The number of stories, however, alters this relationship for the building

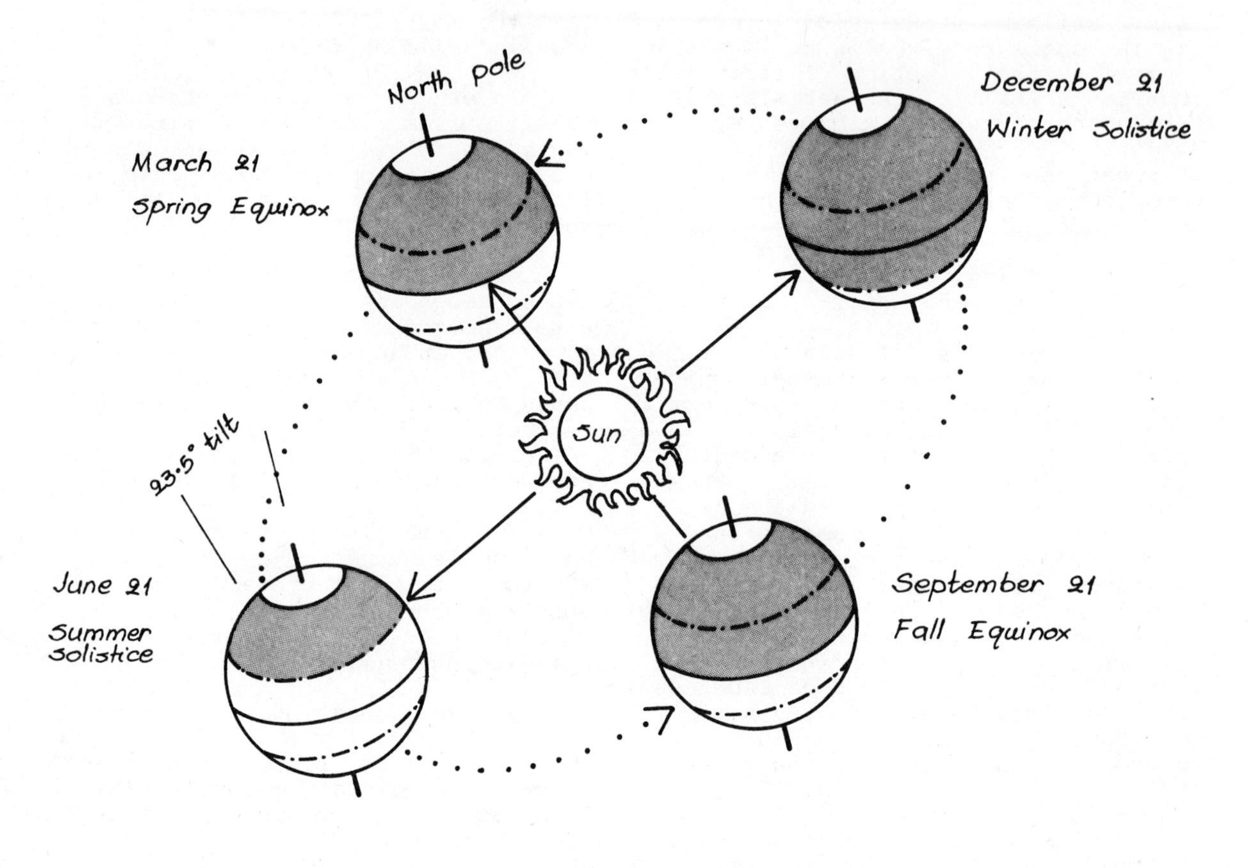

Reference 6

DIAGRAM OF THE EARTH'S PATH FIGURE 7.8

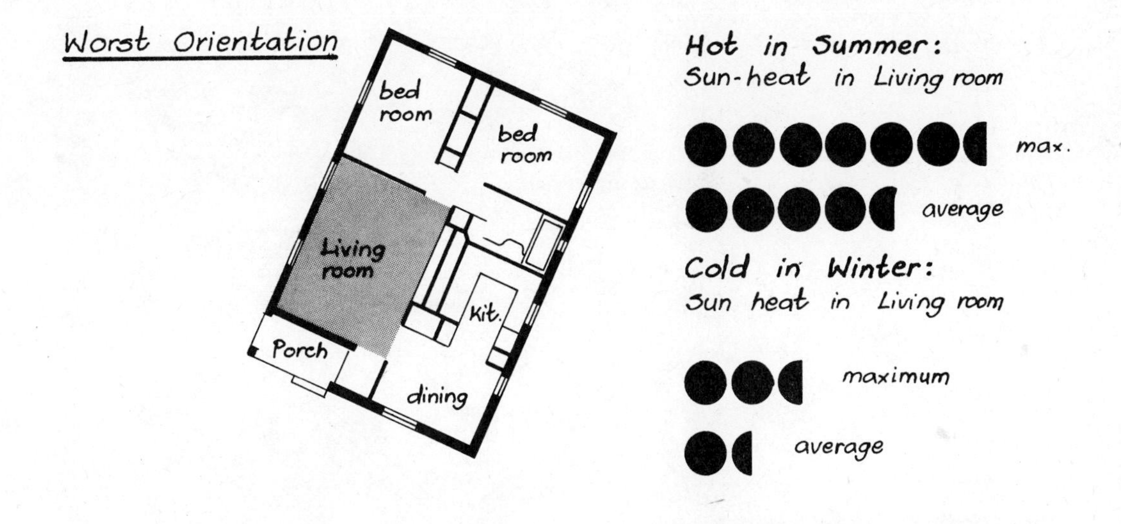

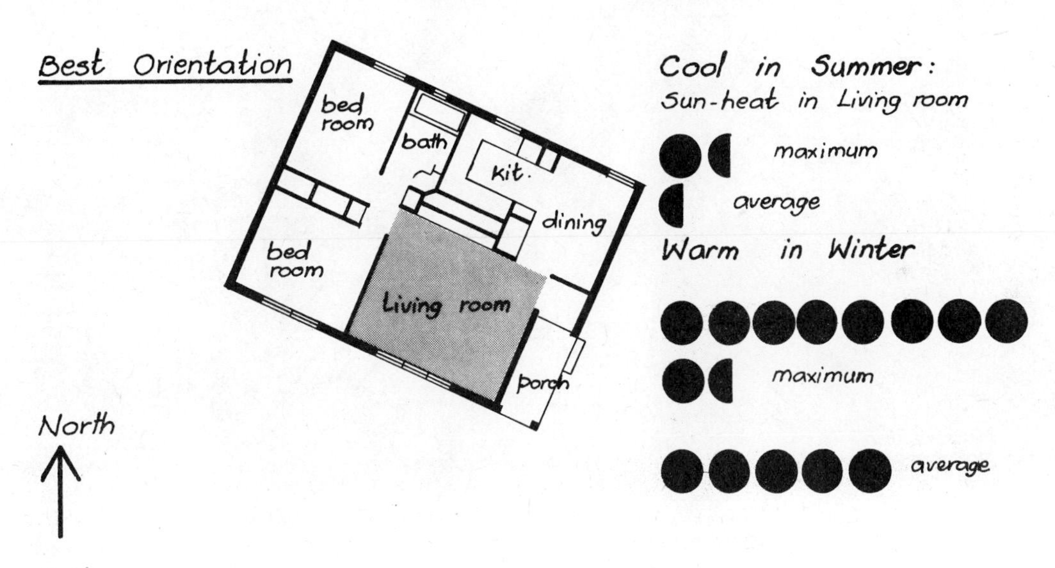

North

Reference 2

COMPARISON OF EFFECTS OF SOLAR RADIATION ON A HOUSE IN NEW YORK FIGURE 7.9

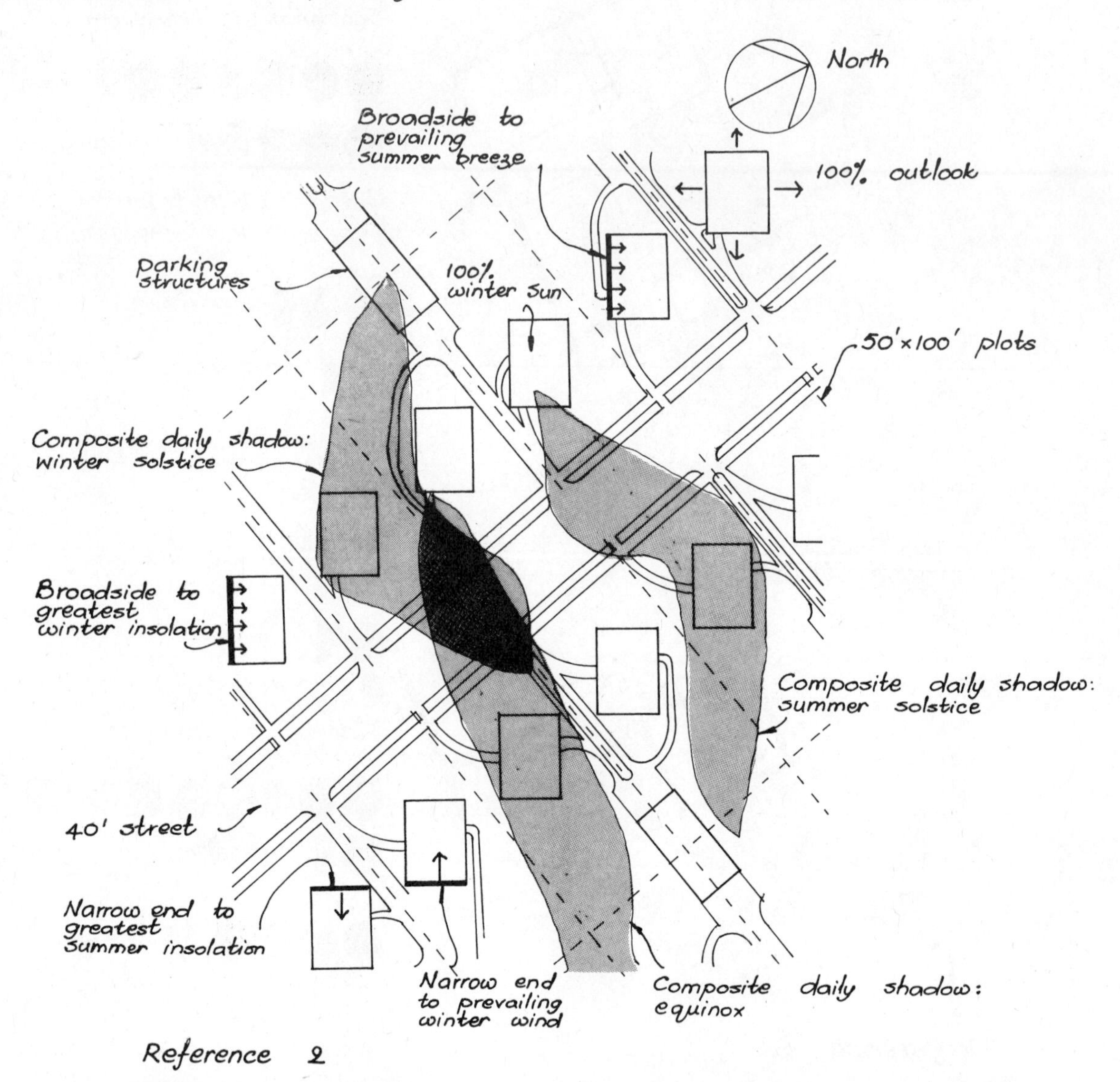
"Heliothermic" site planning for Residences in New York:
North
Broadside to prevailing summer breeze
100% outlook
parking structures
100% winter sun
50'x100' plots
Composite daily shadow: winter solstice
Broadside to greatest winter insolation
Composite daily shadow: summer solstice
40' street
Narrow end to greatest summer insolation
Narrow end to prevailing winter wind
Composite daily shadow: equinox
Reference 2

HELIOTHERMIC SITE PLANNING

FIGURE 7.10

envelope as a whole. Figure 7.11 illustrates the concept that tall buildings have a proportionately smaller roof and are less affected by solar gains on the surface. On the other hand, tall buildings generally are subjected to greater wind velocities which increase infiltration and heat losses. (Wind effects on buildings are discussed later in this chapter.) The chance is rather remote that a tall building will be shaded or protected from winds by surrounding buildings and trees. Generally tall structures demand more mechanical support systems, including elevators, and longer exhaust duct systems. Low buildings may have greater roof area in proportion to wall area and the heating and cooling loads which they generate may, in turn, influence the selection of mechanical systems. In this case, special attention must be given to the thermal characteristics of the roof.

F. Shading by Sun Control Devices

By knowing the latitude of a particular location, it is possible to predict exactly how the sun will impinge on a building at a given time. With this information it is possible to calculate precisely the length of overhang of a sun shade (usually a horizontal building element), the depth of a fin, the location of an inside courtyard, or the size of shadows cast by trees and adjacent structures. Knowledge of the sun's motion is also important in designing a building envelope efficiently. Olgyay and Olgyay did some useful work in designing sun controls.[5,8] The two basic types of sun control devices are overhangs--horizontal projections to protect against vertically incident sun rays, and fins--vertical elements to control the horizontal component of sun rays. Figure 7.12 illustrates the design method for these shading devices. It is important to study the shading effect of sun control devices for winter solstice, equinox and summer solstice.

7.3 THERMO-PHYSICAL PROPERTIES

A. Radiation, Conduction, and Convection

Of all the building elements with energy saving potential, those which offer the greatest array of choices are in the building envelope. To design efficient building envelopes, an understanding of heat transfer mechanisms is important. Heat transfer in buildings takes place in four ways:

- Radiation
- Conduction
- Convection
- Evaporation or Condensation

Radiation is the process by which heat flows from a higher temperature body to a lower temperature body when the bodies are separated in space or when a vacuum exists between them. *Convection* is a process in which the motion of molecules transfers heat from one region to another. *Conduction* is the flow of heat through a material by transfer from warmer to colder molecules in contact with each other. *Evaporation* or *condensation* involves condition changes (liquid to gas or vice versa) which absorb or evolve heat.

Material properties which cause the greatest heat transfer in and out of a building, and consequently affect indoor thermal conditions and occupancy comfort, are: thermal conductivity, resistance, surface characteristic with respect to radiation, surface convective coefficient, and heat capacity of the building material.

The rate of heat flow through a roof, wall, or floor is known as the overall transmission coefficient U. This and some other coefficients that affect it are defined in Table 7.1 for conditions of steady-state flow of heat.[10,11]

In computing U values it is important to note that component heat transmissions are not additive. The value of the overall coefficient is always less than that of any of its parts. Greater thickness, more parts,

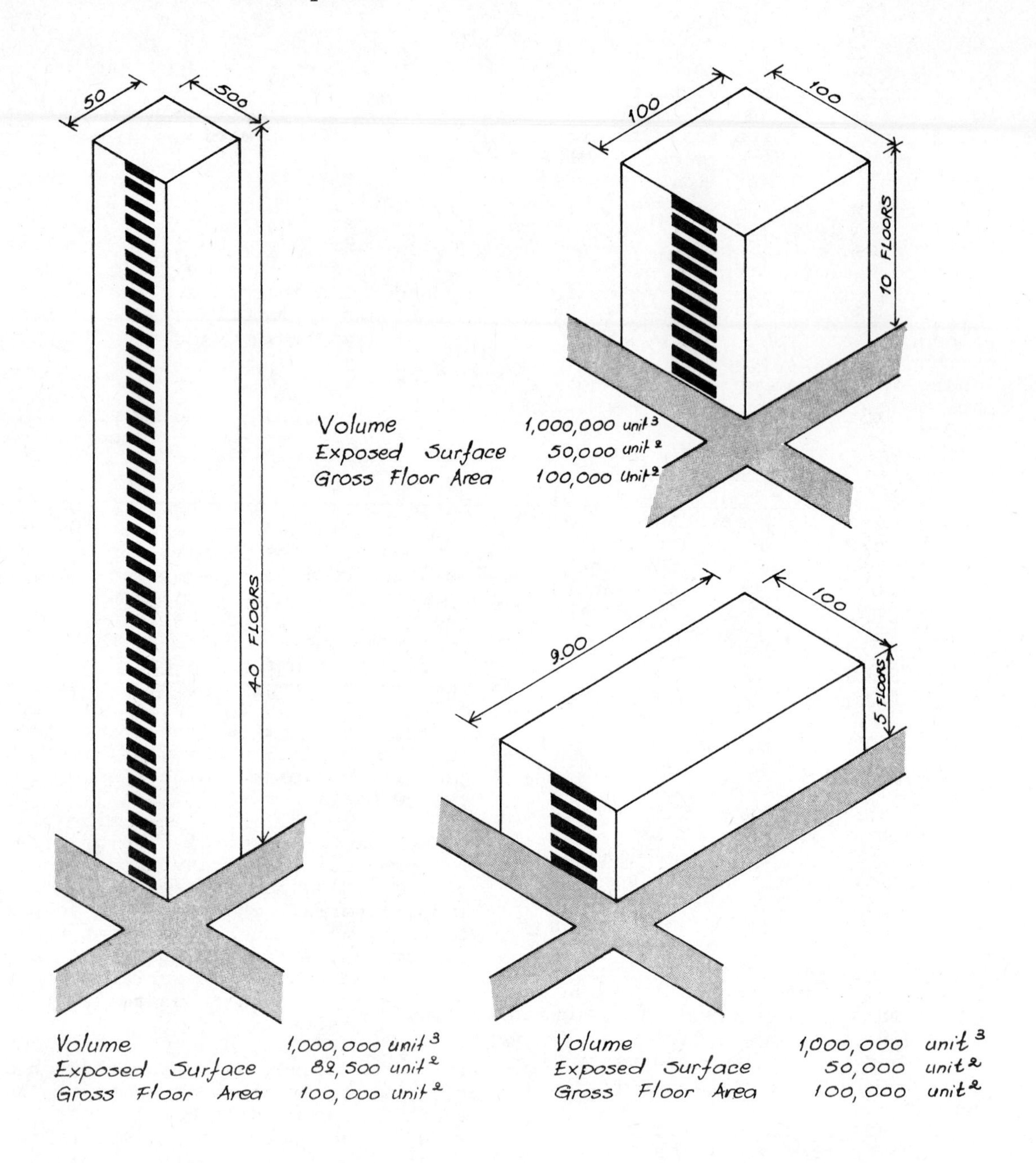

Reference 9

VARIATION OF SURFACE AREA WITH CONFIGURATION **FIGURE 7.11**

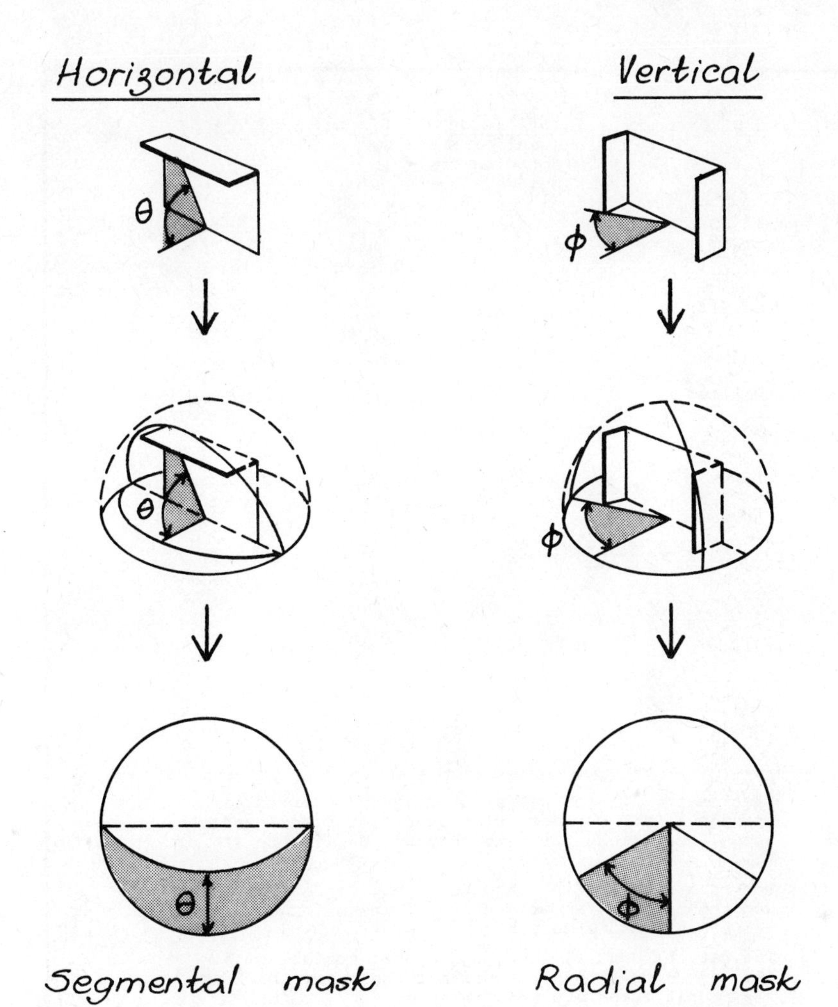

Sun shading masks show the shadows by vertical and horizontal edges

θ = Sun's altitude angle

φ = Sun's horizontal angle

Reference 8

SUN CONTROL ELEMENTS FIGURE 7.12

Symbol	Definition (see corresponding notes below)	SI Unit	B/A Unit
Overall Coefficient of Heat Transmission U	(1)	$\frac{W}{M^2\,°C}$	$\frac{Btu}{hr\ ft^2\,°F}$
Conductivity K	(2)	$\frac{W}{M^2\,°C}$	$\frac{Btu}{hr\ ft^2\,°F}$
Conductance C	(3)	$\frac{W}{M^2\,°C}$	$\frac{Btu}{hr\ ft^2\,°F}$
Air Space Conductance a	(4)	$\frac{W}{M^2\,°C}$	$\frac{Btu}{hr\ ft^2\,°F}$
Surface C Coefficient f	(5)	$\frac{W}{M^2\,°C}$	$\frac{Btu}{hr\ ft^2\,°F}$
Emissivity E	(6)	- dimensionless -	
Thermal Resistance R	(7)	$\frac{M^2\,°C}{W}$	$\frac{hr\ ft^2\,°F}{Btu}$

(1) Heat energy (Btu or watts) flowing from air to air through a unit surface (roof, wall, or floor) under actual conditions for a temperature difference (°F or °C).

(2) The rate of heat flow through a <u>unit</u> surface of a homogenous material of a <u>unit</u> thickness for a <u>unit</u> temperature difference.

(3) The rate of heat flow through a <u>unit</u> surface of a homogenous or combination of materials any thickness for a <u>unit</u> temperature difference.

(4) The rate of heat flow through a <u>unit</u> surface area for a <u>unit</u> temperature difference between bounding surfaces. It is affected by position and emissivity of the surface.

(5) The rate of heat flow through a unit surface due to the motion of air against the surface f, a unit temperature difference. f_1 = inside film coefficient and f_o = outside film coefficient.

(6) The effective thermal emission (or absorption) of the surfaces bounding an air space.

(7) It is reciprocal of heat transfer as expressed by such coefficients as U, C, f or a.

BUILDING HEAT FLOW DEFINITIONS

TABLE 7.1

insulation, and air spaces all help to lower the overall coefficient of heat transmission. To arrive at this U coefficient, it is necessary to add the resistances of the various elements of a wall, floor, roof, glazing and doors, including film coefficients. This results in a total thermal resistance R_T which may be expressed as total seconds (hours) for the passage of 1 Joule (Btu) through the construction. The reciprocal of this total resistance, expressed in Watts (Btu/hr), is the overall U coefficient of transmission.

Figure 7.13 shows a method by which conductance and consequently the U value of a composite material can be calculated.

It is appropriate to discuss a few other aspects of building envelope design at this point. They are:

- effect of wind;
- effect of controlled air spaces in buildings; and,
- effect of glazing.

B. Effects of Wind on Exterior Building Surfaces

Heat loss from a building envelope is also dependent on the outside wind velocity and the nature of inside air in contact with the envelope. In calculating the rate of heat flow between indoor and outdoor air, the thermal resistance of air adjacent to the surfaces should be considered. In winter the inside surfaces of walls are usually cooler than the room temperature. Because of convection warm air collides with the cooler wall surface and increases its temperature. The outside air temperature is less than the temperature of the outside wall's surface. When the wind blows adjacent to the building, the wall is cooled, thereby increasing the rate of heat loss. Thus, inside air and outside air have heat resistance coefficients. These coefficients determine the heat flow from the surface to the ambient air. Still air has higher thermal insolation properties than outside air which is affected by wind speed.

C. Effect of Controlled Air Spaces in Buildings

Properly introduced air spaces in building walls, roofs, and floors can significantly reduce heat transfer from the building. It is important to note that the resistance of an air space is not related to its thickness but to other factors such as its position and the direction of heat flow as shown in Figure 7.14. By lining one side with reflective foils and introducing 1.9 cm (3/4 in.) to 10 cm (4 in.) air spaces between them, it is possible to achieve the equivalent insulating value of 7.5 cm (3 in.) insulating material.[5,10] (See Figure 7.15.)

D. Effect of Glazing

Windows have a major effect on building energy use due to transmission, solar gain and air infiltration. Heat transmission is much greater through glass than through most opaque walls. U values for walls, in the British system of units, can be reduced to 0.04 or less but single glass has a U value of 1.15, double glass 0.55 to 0.69, and triple glass 0.35 to 0.47.

In some modern commercial buildings, window area is in excess of requirements for natural light, ventilation, or view.

Large glass areas can create discomfort for persons who must sit in front of them receiving the sun's heat, radiation, and glare. Elimination of excessive window areas and providing smaller windows which act as horizontal visual panels or picture windows can reduce energy loss brought about by the need for more cooling.* [2,5] The shape of a window can be important even when the window area remains constant. Reducing the area of a window does not necessarily

---

*In the winter this situation may be reversed, i.e., more energy is needed for heating. However, in some buildings the solar heat load is excessive even during winter, requiring cooling!

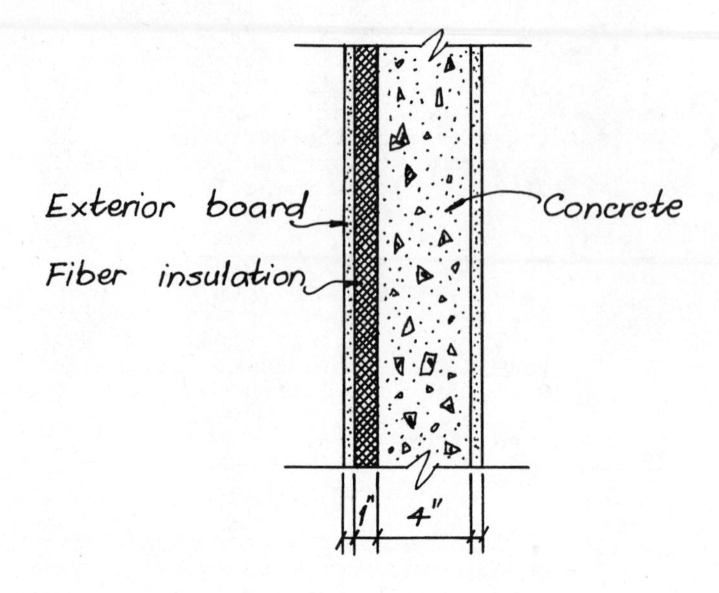

x = Material thickness

k = Conductivity

C = Conductance

R_T = Total Resistance

			R_T
4"	Concrete	$x/k = 4/9$	0.44
1"	Exterior board	$1/0.38$	2.63
	Total		3.07

Conductance, $C = 1/R_T = 0.33$ Btu/h/sq.ft/°F

TYPICAL CONDUCTANCE CALCULATIONS

FIGURE 7.13

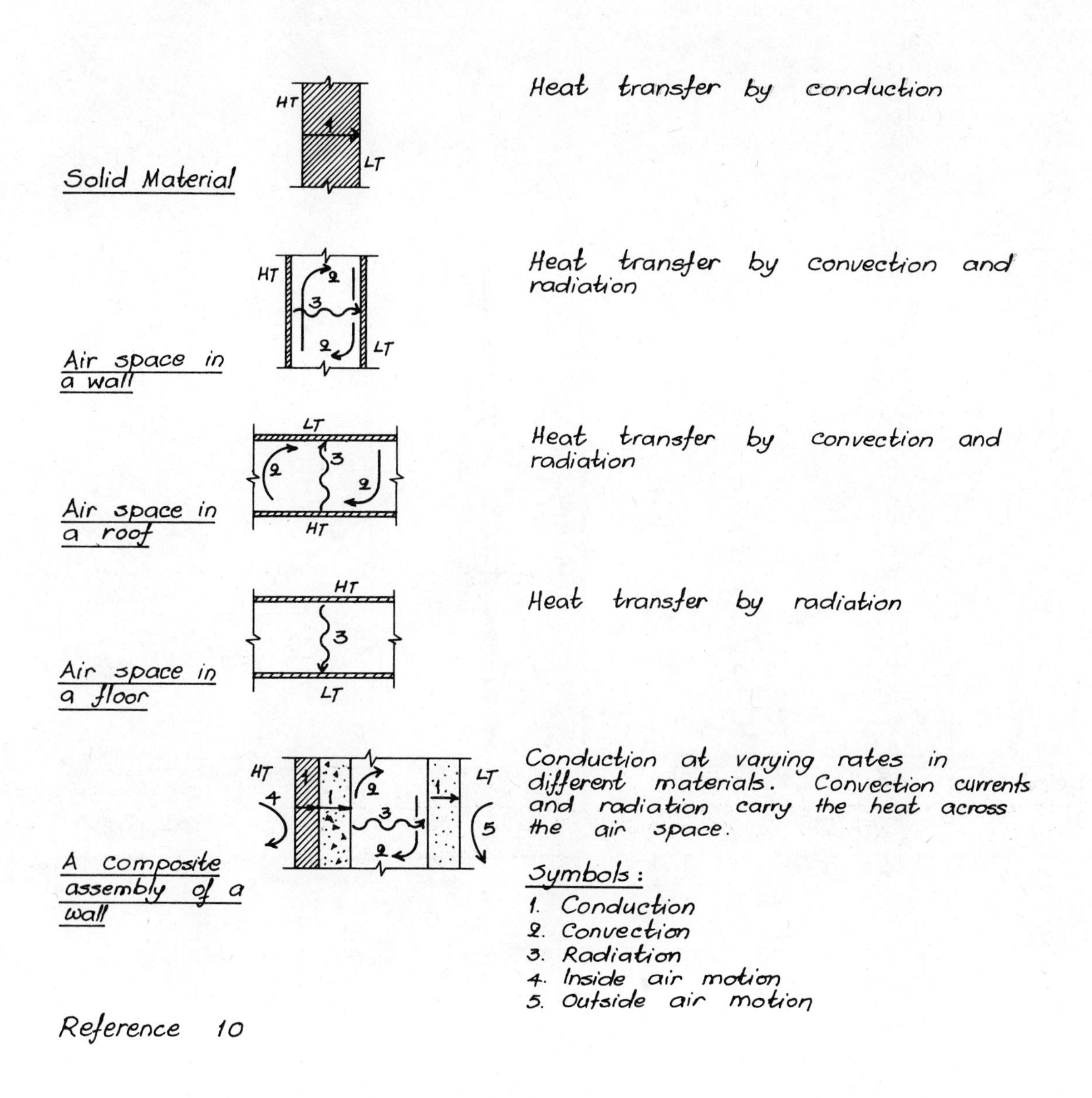
Solid Material
HT
LT
Heat transfer by conduction
Air space in a wall
HT
LT
Heat transfer by convection and radiation
Air space in a roof
LT
HT
Heat transfer by convection and radiation
Air space in a floor
HT
LT
Heat transfer by radiation
A composite assembly of a wall
HT
LT
Conduction at varying rates in different materials. Convection currents and radiation carry the heat across the air space.
Symbols:
1. Conduction
2. Convection
3. Radiation
4. Inside air motion
5. Outside air motion

Reference 10

HEAT TRANSFER MECHANISMS **FIGURE 7.14**

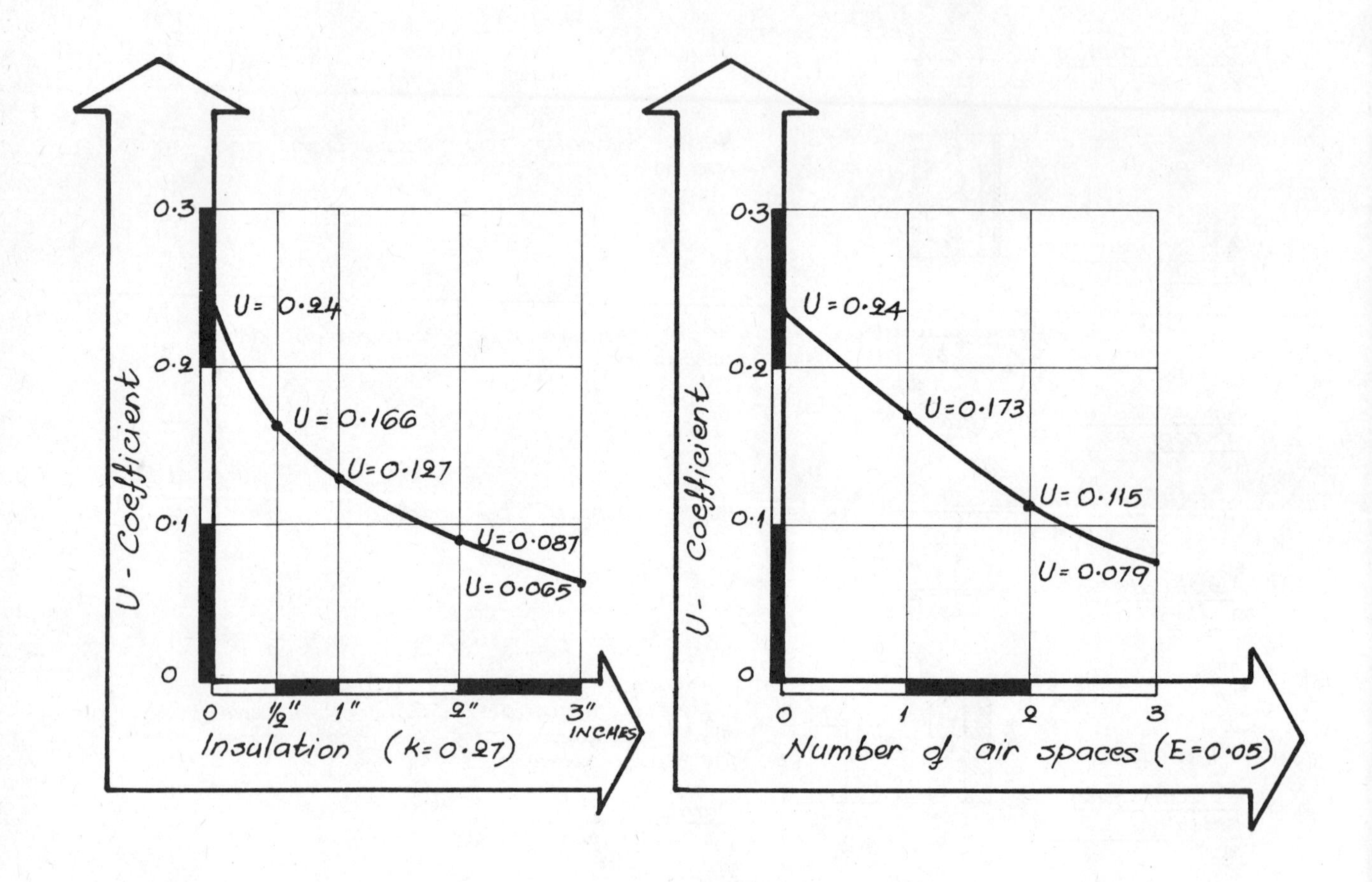

Reference 10

EFFECTIVENESS OF AIR SPACES AND INSULATION

FIGURE 7.15

reduce the amount of natural light. Clearly, proper window design is an important part of building design.

In some buildings where large internal heat gains are caused by operations within the building, an optimum design might require larger window areas and less insulation to allow heat dissipation to the outside. Reflecting or heat absorbing glass intercepts up to 80 percent of radiant energy which is helpful for cooling in summer but results in loss of useful heat in winter. Natural light may be lost as well. The impact on annual energy use and cost must be the criterion for making any choice in this area.

An important relationship exists between natural light and window sizes. Larger windows let in more natural illumination but lose heat when the air temperature is lower outdoors. A thermal barrier installed over windows at night and on weekends reduces heat loss or heat gain when the building is unoccupied and will reduce considerably annual heating and cooling requirements. Solar control to reduce heat gain in the summer is most effective when located on the exterior of the building. Some increase in performance is certainly possible with the use of double and even triple glazing.[12] The rate of heat transfer through single-plate windows and double-plate windows is illustrated in Figure 7.16. When the price of energy exceeds certain levels, double glazing will be very economical in the long run. Case Study 7-1 illustrates the economic benefits of double glazing (see Chapter 15, Case Studies).

Presently some special glazes have heat absorbing and reflective properties. Figure 7.17 illustrates some comparative performances of different combinations of glazing.[2] Solar control is most effective when designed specifically for each facade. As indicated in Figure 7.12, horizontal shading is most effective on the southern exposure; if it is not extended far enough beyond the windows, however, it can permit solar impingement at certain times of the day. On east and west walls a combination of vertical and horizontal sun control baffles is required. In certain cases the building configuration itself can be designed to give the best solar protection while also enhancing the form of the building. In hot climates it is effective to shade walls and roofs as well as windows. It is also possible to use equipment located on the roof to act as a solar shade. Operable windows can be designed to permit the use of natural ventilation so long as weather stripping, gaskets, and locking devices are provided to prevent infiltration loads. Natural ventilation is delightful when the building configuration permits outside air to permeate deeply into the building and conditions are such that the air is clean, pure, and dry enough during the summer to be enjoyed. The number of hours in the year during which natural ventilation can be effective and useful must be evaluated.

Another way in which natural light and building configuration affect energy use is shown in Figure 7.18. Considering installed electrical capacity for lighting only, natural light can reduce electricity needs depending on the building configuration. These savings can not be evaluated alone; consideration must be given to compensating effects such as solar heat loads, building heat losses, etc. These are discussed next.

E. Space Conditioning

Much of the energy presently expended in heating and cooling occupied spaces can be saved by appropriate alterations to the building envelope. A range of savings up to 53 percent (Table 7.2) is reported for homes in three typical cities with a variety of insulation treatments, starting with the FHA-MPS standards and going to the economic optimum.[13]

The effect of variable wall and ceiling insulation for a cited New York residence is shown in Figure 7.19. Here, for instance, as the insulation is increased (upper quadrilateral), ceiling, from 9 to 15 cm (3.5 to 6 in.), wall, from 0 to 9 cm (0 to 3.5 in.), the saving for electric heat and air conditioning rises from $60 to $130 per year, or about 42 percent.

Window heat losses typically are about half the wall heat losses in the

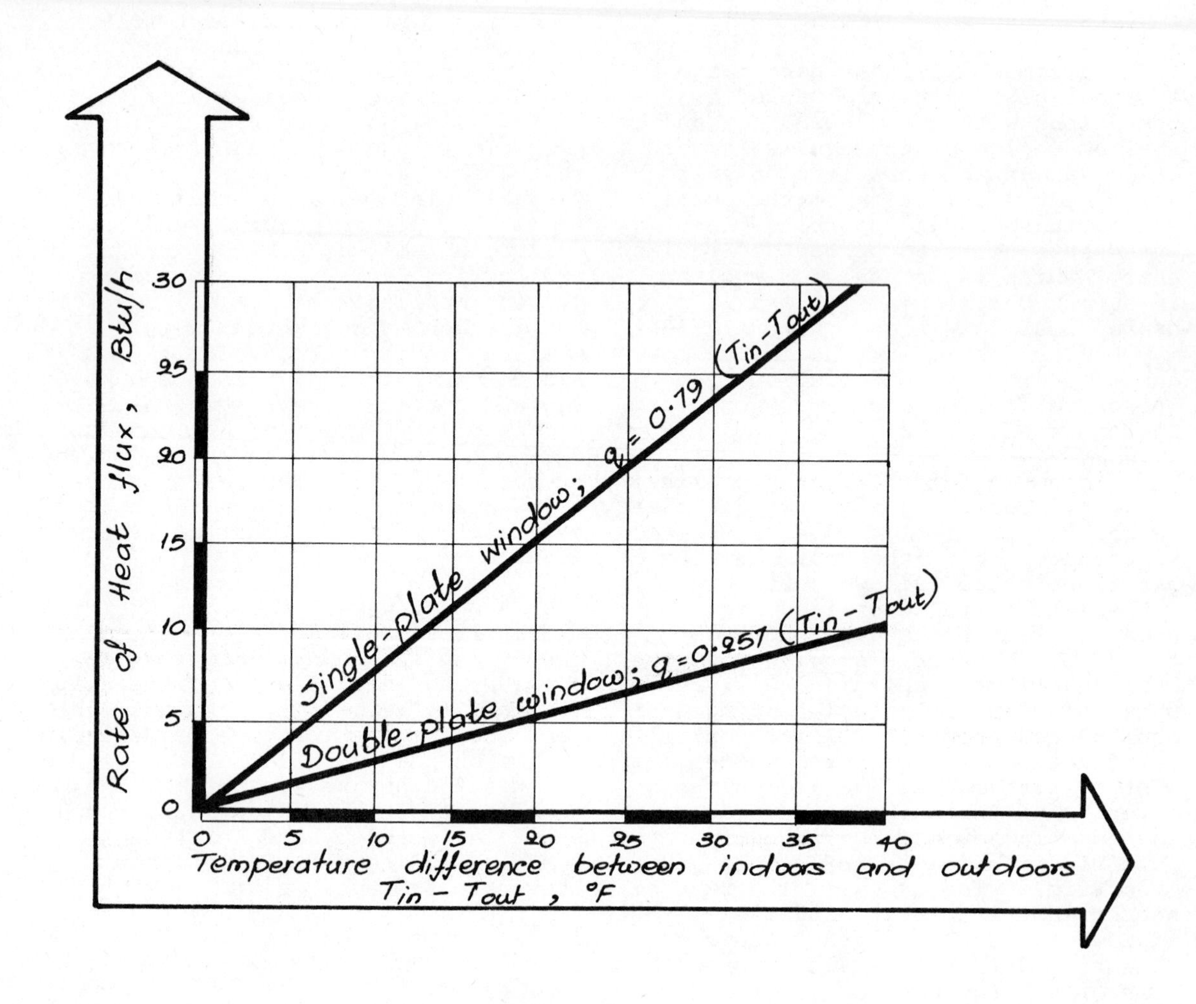

Reference 12

HEAT LOSS FROM TWO TYPES OF WINDOW GLAZING FIGURE 7.16

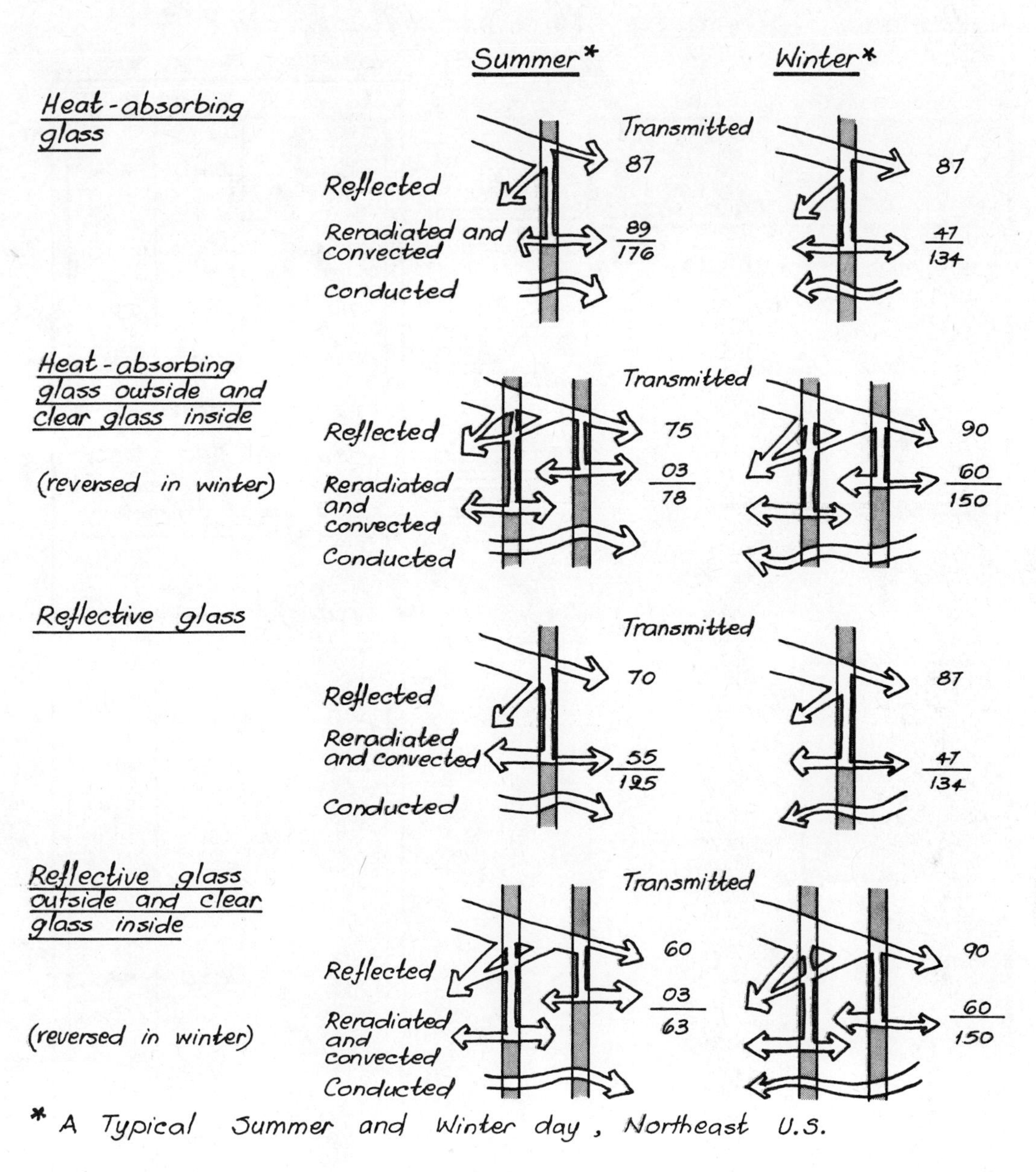

* A Typical Summer and Winter day, Northeast U.S.

Reference 2

REACTIONS OF DIFFERENT GLAZING CONFIGURATIONS **FIGURE 7.17**

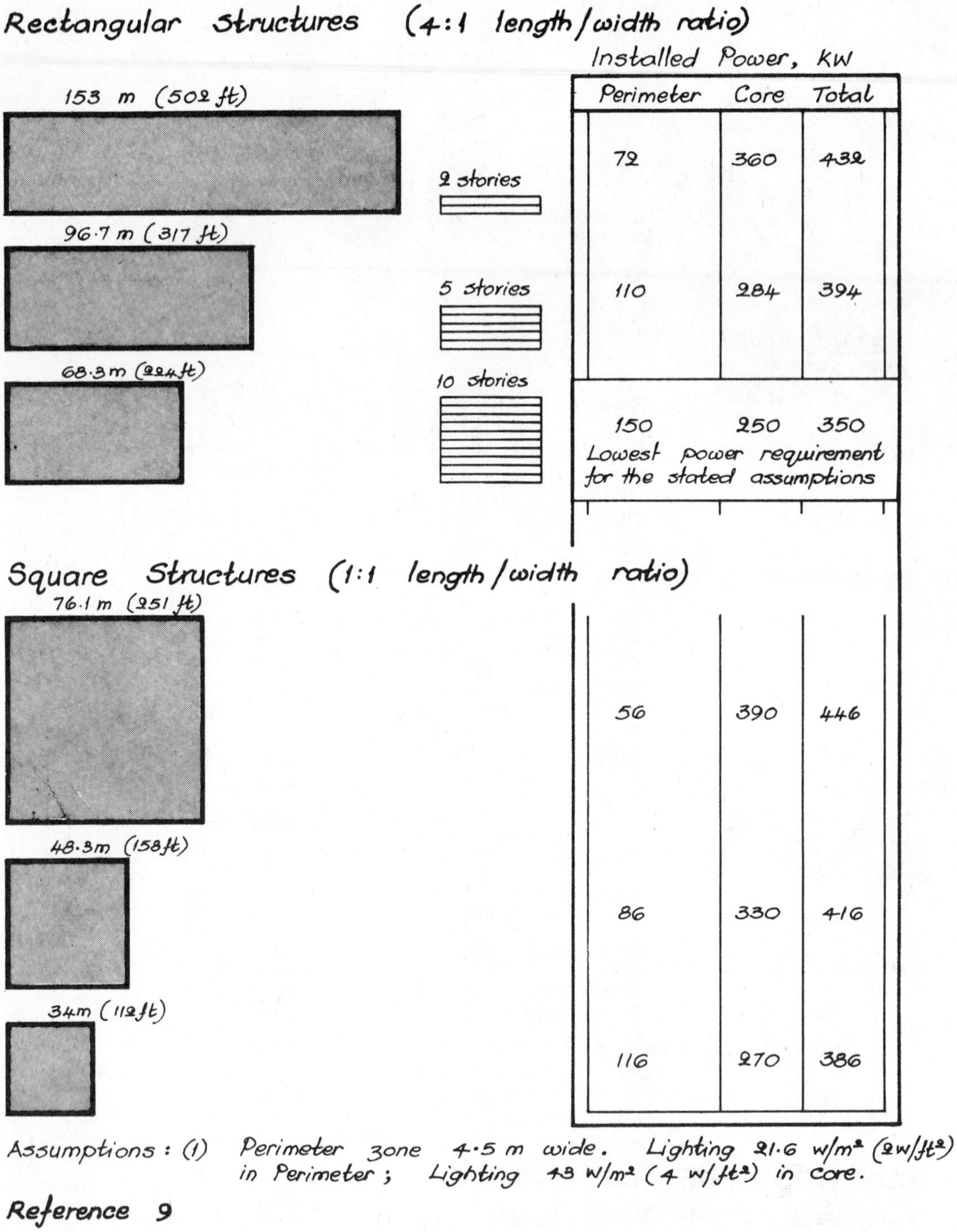

Rectangular Structures (4:1 length/width ratio)

Installed Power, KW

Building	Stories	Perimeter	Core	Total
153 m (502 ft)	2 stories	72	360	432
96.7 m (317 ft)	5 stories	110	284	394
68.3 m (224 ft)	10 stories	150	250	350

Lowest power requirement for the stated assumptions (10 stories)

Square Structures (1:1 length/width ratio)

Building	Perimeter	Core	Total
76.1 m (251 ft)	56	390	446
48.3 m (158 ft)	86	330	416
34 m (112 ft)	116	270	386

Assumptions: (1) Perimeter zone 4.5 m wide. Lighting 21.6 w/m^2 (2 w/ft^2) in Perimeter; Lighting 43 W/m^2 (4 W/ft^2) in core.

Reference 9

EFFECT OF NATURAL LIGHT AND BUILDING CONFIGURATION ON ENERGY USE

FIGURE 7.18

	Revised FHA-MPS savings*			Economically optimum savings*		
	$/year	Gas (%)	Electric (%)	$/year	Gas (%)	Electric (%)
Atlanta						
Gas heat	6	16	--	6	31	--
Gas heat + A-C	3	12	0	6	20	7
Electric heat	36	--	16	87	--	53
Electric heat + A-C	21	--	10	63	--	39
New York						
Gas heat	28	29	--	32	49	--
Gas heat + A-C	28	24	10	37	50	26
Electric heat	75	--	19	155	--	47
Electric heat + A-C	47	--	13	135	--	42
Minneapolis						
Gas heat	37	37	--	42	43	--
Gas heat + A-C	39	37	11	45	43	18
Electric heat	80	--	22	119	--	29
Electric heat +A-C	82	--	22	122	--	29

*Savings are from using revised FHA-MPS standards or the economically optimum amount of insulation, rather than the unrevised (pre-June 1971) FHA-MPS re-requirements.

Reference 13

MONETARY AND ENERGY SAVINGS FROM REVISED FHA-MPS STANDARDS

TABLE 7.2

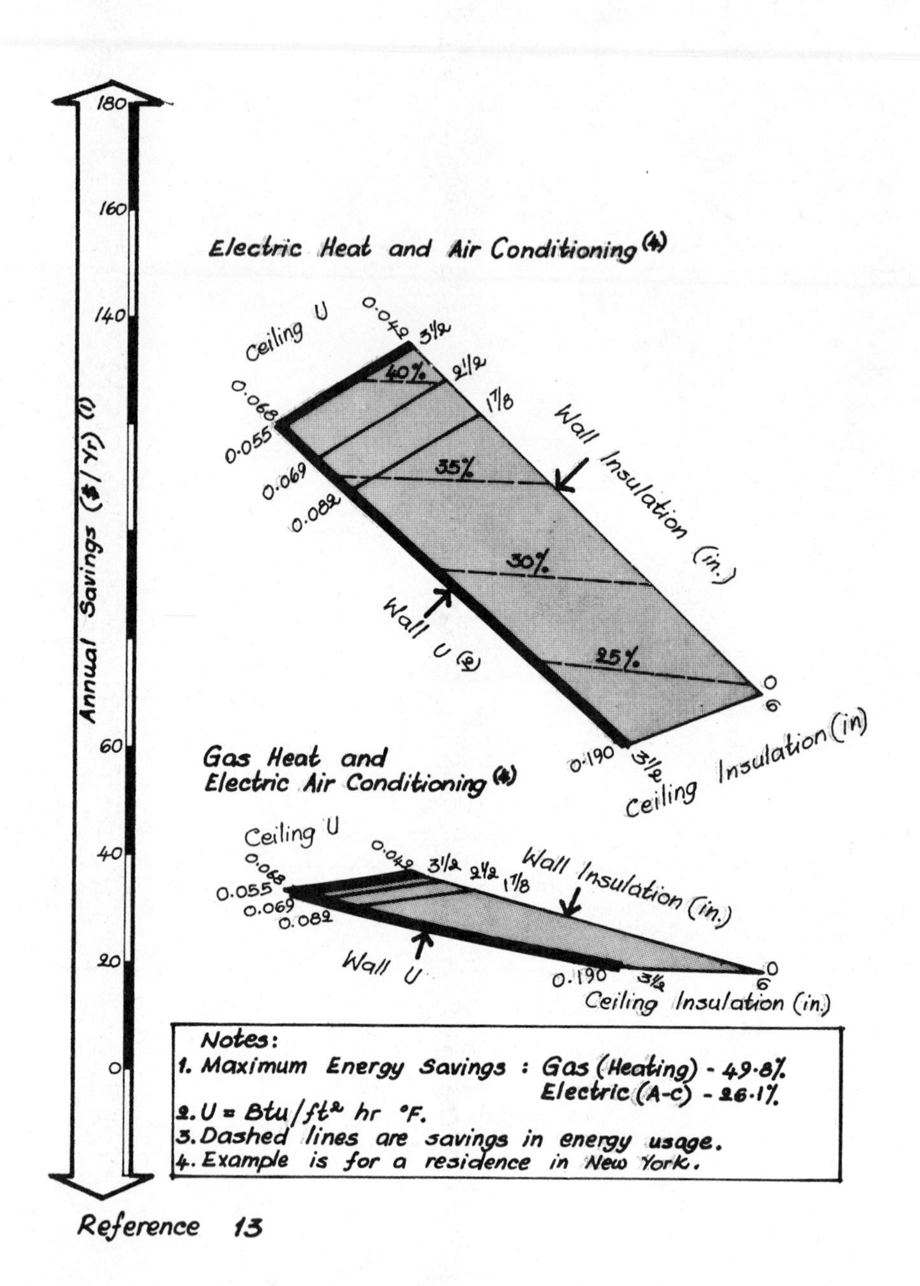

ANNUAL SAVING DUE TO INSULATION AND STORM WINDOWS FIGURE 7.19

two-floor, 100 m^2 (1100 ft^2) home, as shown in Table 7.3a (73 versus 150 W/°C).[14] Figure 7.20 illustrates several ways to use double panes, and Table 7.3b lists the respective thermal transmittances through the glazing. See Table 7.3c for windows and frames. For glazing systems using heat reflective glasses with metallic surface coatings of low emissivity, the metallic coating improves the heat loss (as much insulation improvement as 30 to 40 percent for double glazed units). Aluminum window frames account for about 25 percent of the total thermal losses of such windows. Wood frames, however, account for only 13 percent of window thermal losses. Air leakage (infiltration) is also important and can best be controlled by tightly closing types.

Internal shades, while less effective than external shades, are often less expensive and easier to install. Closed draperies, for example, cut solar gains by 65 percent. Internal shading devices can have the following impact on single-pane windows: medium-colored venetian blinds can cut solar transmission by 36 percent; light-colored versions by 45 percent; dark opaque shades by 75 percent; and light, translucent shades by 61 percent. Vertical baffles can effectively control solar gains on east or west exposures.[15]

Color is an important solar control. Exterior shutters, interior shutters, and both sides of operable shutters should be light in color so they will reflect light, and therefore heat, away from the building interior; dark colors tend to absorb light and heat.

7.4 HEAT LOSS AND HEAT GAIN

Heat flows from one point to another whenever a temperature difference exists between the two points. The direction of flow is towards the lower temperature. Water vapor also flows from one point to another whenever a difference in vapor pressure exists between two points. The direction of flow is toward the point of lower vapor pressure. The rate at which heat will flow varies with the resistance to flow between the two points in the material.

It is clear that a reduction in heat loss can be achieved by the use of insulating materials and by interposition of one or several air spaces in the direction of heat flow. Radiant transfer can be reduced by the use of reflective lining in air spaces. Greater thickness of insulative and reflective materials helps, but there can be little control of convection on the inside surface and conductance losses on the outside.

For the purposes of designing a building envelope, it is recommended that the designer refer to ASHRAE 90-75 and the 1972 ASHRAE Handbook of Fundamentals.[15,16] The intent of this recommendation is not to limit the creativity of any designer nor to suggest that this is the ultimate approach for any building envelope design. However, ASHRAE 90-75 is a reference which can help a designer make an evaluation on any given building envelope.[17,18]

In addition to referring to the criteria given in ASHRAE 90-75, a designer should also increase energy use efficiency by determining proper orientation of the building on its site, the geometric shape of the building, the building aspect ratio, the number of stories given, the floor area requirement, the building's thermal mass, its thermal time constant, and shading or reflection from adjacent structures. ASHRAE 90-75's requirements, in convenient tabular and condensed form, for thermal coefficients, cooling criteria, and air leakage limitations are shown in Table 7.4.

A. Heating and Cooling Degree-Days

An analysis of selected temperature differences between the interior and exterior of a building will establish that building's heating and cooling load. The relationship of the outdoor air temperature to the single family residential building heating load has led to the concept of a "degree-day" for predicting building energy requirements. The degree-day concept is useful for simplified or preliminary calculations. Figure 7.21 shows typical data for the US. Refer to Chapter 8 for additional details.

(A)

	Average U-Value	Area		Rate of Heat Loss
Conduction Heat Loss	$W/m^2 °C$	m^2		W/°C
Roof	0.50	x 50	=	25
Ground Floor	0.76	x 50	=	38
Unglazed walling	1.50	x 100	=	150
Windows (single-glazed)	4.30	x 17	=	73
Total rate of conduction heat loss through fabric				286
Ventilation heat loss (assuming 1 air change/hour)				
$= 240\ m^3 \times \frac{W/m^3\ °C}{3}$				80
		Total		366

(B)

Glazing System	Degree of Exposure:* Sheltered	Normal	Severe
Single	5.0	5.6	6.7
Double			
air space 3 mm wide	3.6	4.0	4.4
6 mm	3.2	3.4	3.8
12 mm	2.8	3.0	3.3
20 mm (or more)	2.8	2.9	3.2
Triple			
each air space 3 mm wide	2.8	3.0	3.3
6 mm	2.3	2.5	2.6
12 mm	2.0	2.1	2.2
20 mm (or more)	1.9	2.0	2.1

(C)

Window Type	% Total Window area occupied by frame	Degree of Exposure: * Sheltered	Normal	Severe
Single-glazed:				
metal casement†	20	5.0	5.6	6.7
wood casement	30	3.8	4.3	4.9
Double-glased:¶				
metal horizontal sliding window with thermal break	20	3.0	3.2	3.5
wood horizontal pivot window	30	2.3	2.5	2.7

*All units are $W/m^2\ °C$ unless indicated.
†Metal frame assumed to have a thermal transmittance similar to that of the glass.
¶With 20 mm airspace.

Reference 14

BUILDING HEAT LOSSES TABLE 7.3

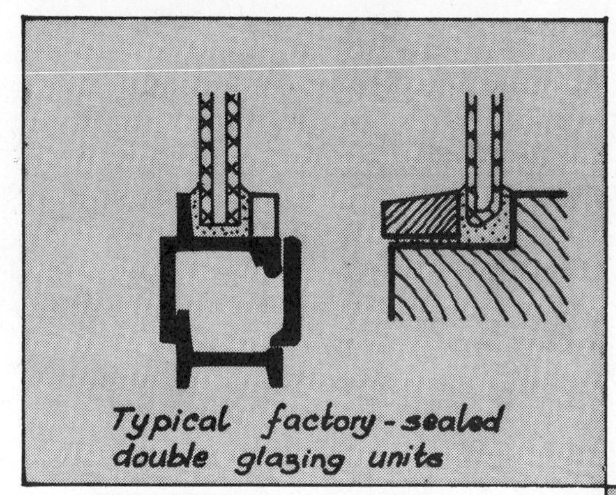

Typical factory-sealed double glazing units

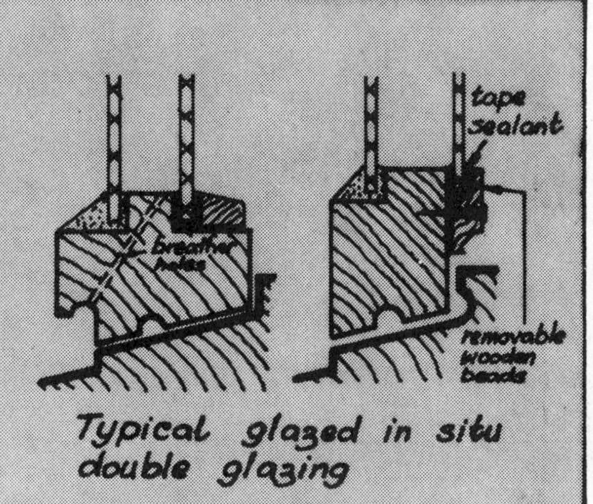

Typical glazed in situ double glazing

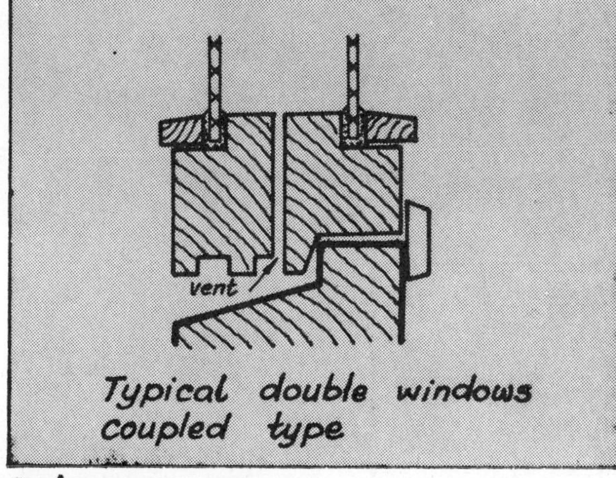

Typical double windows coupled type

Reference 14

REDUCING WINDOW LOSSES FIGURE 7.20

A. THERMAL COEFFICIENTS

Annual Heating Degree Days	Maximum Transmittance $U_O = Btu/h \cdot ft^2 \cdot F$					Min. Insulation†R $(h \cdot ft^2 \cdot F/Btu)$ For Slab-on-Grade	
	Gross Wall			Floor Over Unheated Space	Roof & Ceiling		
	Over 3 Stories	3 Stories & Under	1-2 Family Dwelling			Heated	Unheated
12,000						11.5*	8.8*
11,000		.20	.16			10.7*	8.2*
10,000	.28				.060	10.0	7.5
9,000		.22	.18			9.3	6.8
8,000		.24	.19	.08		8.5	6.2
7,000	.31	.26	.21		.068	7.8	5.5
6,000	.33	.28	.22		.076	7.0	4.8
5,000	.36	.30	.23		.084	6.3	4.2
4,000	.38	.32	.25	.11	.092	5.5	3.5
3,000	.41	.33	.27	.19		4.8	2.8
2,000	.43	.35	.28	.26		4.0	2.2
1,000	.46	.37	.29	.33	.100	3.3	1.5*
500	.47	.38	.30	.36		2.9	0.8*

*Extrapolation from Figure 2 of Standard.
†Total insulation length to be 24 inches minimum.

B. COOLING CRITERIA AND FACTORS

North Latitude	25°	30°	35°	40°	45°	50°	55°
Max. OTTV $(Btu/h \cdot ft^2)$	29.3	30.7	32.1	33.5	34.9	36.3	37.7
SF $(Btu/h \cdot ft^2)$	118	121	124	127	133	138	144

Wall					
Wall	mass/area (lb/ft^2)	0-25	26-40	41-70	>70
	TD_{EQ} (F)	44	37	30	23

C. AIR LEAKAGE LIMITATION

Type of Window/Door	Residential	Non-Residential
Window	0.5 cfm/L.ft	0.5 cfm/L.ft
Sliding Doors	0.5 cfm/ft^2	11.0 cfm/L.ft
Swinging Doors	1.25 cfm/ft^2	11.0 cfm/L.ft
Revolving Doors	---	11.0 cfm/L.ft

ENERGY EFFICIENT BUILDING DESIGN CRITERIA BASED ON ASHRAE 90-75

TABLE 7.4

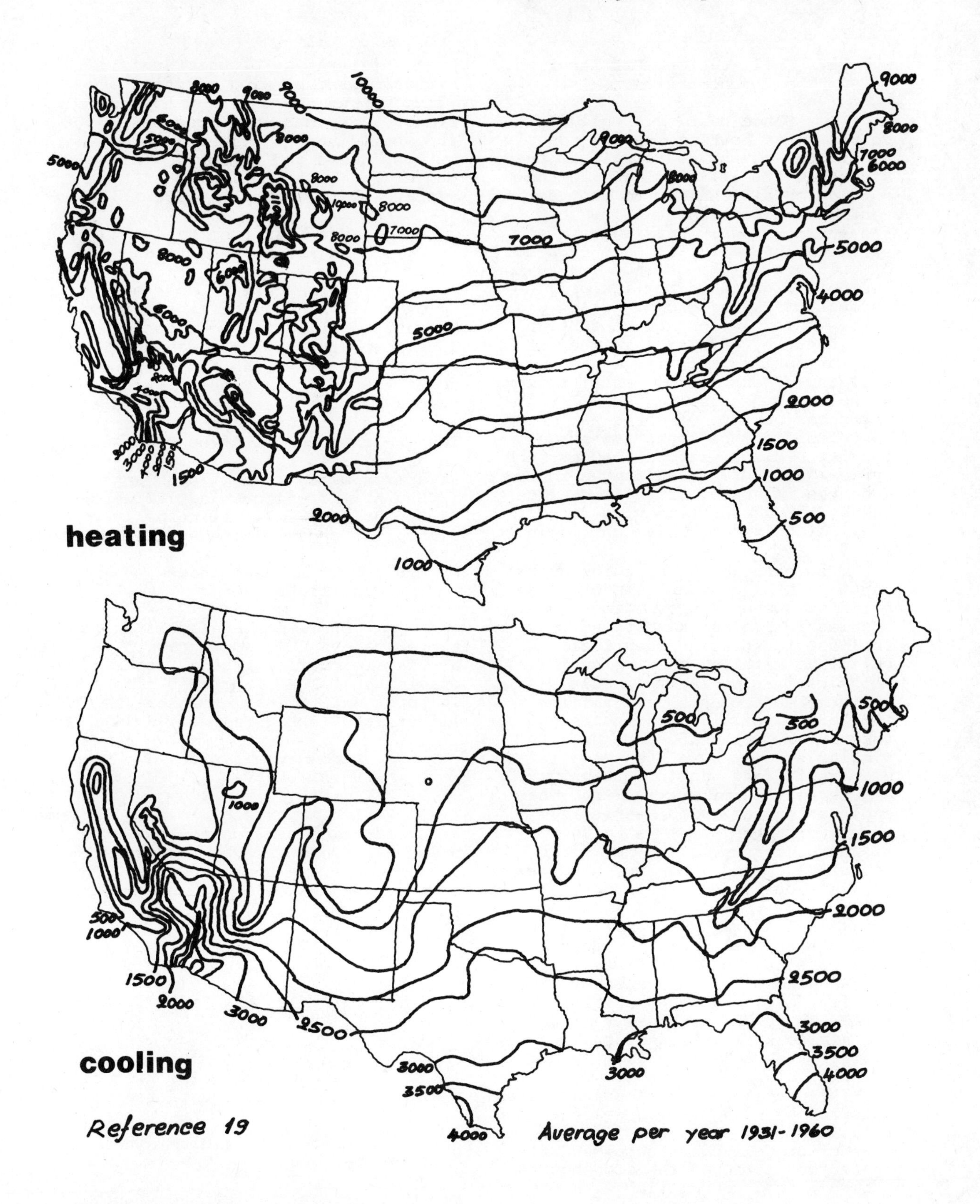

HEATING AND COOLING FIGURE 7.21

B. Infiltration, Exfiltration, and Ventilation

Since heating a building involves first heating the air contained within the structure as well as the structure itself, air leaks to the outside or air changes to provide ventilation cause heat to be lost to the surroundings. Leaking--called exfiltration or infiltration--of hot or cold air to the outside (or to the inside) takes place particularly through cracks around doors and windows. These losses can be substantially cut by effective weather stripping or sealing. In large and in tall buildings, considerable infiltration of cold air occurs at ground level entrances because of the creation of convection air currents ("stack effect"). This effect can be lessened by designing proper entrances to reduce the impact of prevailing winds and by the use of revolving doors, etc., which are effective air locks.

The behavior of building users can have a significant impact on a building's energy use. Some recent studies in England have found that fuel use for heating identical buildings with the same orientation, set side-by-side in the middle of a terrace, and occupied by families of similar composition and income can vary by as much as 50 percent. Similar studies have been made for electricity and gas in US housing tracts, with similar results. The explanation for this interesting observation has to do with the habits of the occupants, e.g., how often doors are opened, how long windows are left open, etc.

Air leakage values for various elements such as doors, windows and wall constructions are given in the ASHRAE Handbook of Fundamentals and air leakage limitations are indicated in ASHRAE 90-75. Table 7.4 lists the air leakage limitations as specified in ASHRAE 90-75.[18]

Air infiltration rates have been measured by using tracer gases such as helium and sulfur hexafluoride. A study on well constructed and tightly weather stripped research houses showed a good correlation with the ASHRAE crack method of air leakage determination.[17]

Infiltration of warm or cold air is caused by wind velocity creating a pressure on the windward side. In summer infiltration into buildings is primarily due to open doors and windows. In winter infiltration is caused by wind velocity and stack effect.[17]

C. Heat Loss from Basement Walls and Floors below Grade

In the case of most basements, no outside air surface is sufficient to carry heat away by convection. Therefore, the loss through the floor is normally small and temperatures under the floor vary only a little throughout the year. The ground acts as a good heat sink and can absorb or lose a large amount of heat without an appreciable change in temperature. An approximate value of 0.6 watts (2 Btu/hr) for floors and 1.2 watts (4 Btu/hr) for walls is normally accepted for construction below grade.[10]

D. Heat Loss at Edges of Slabs

Concrete slab floors exposed above grade and in contact with outside air have significant heat losses. To prevent heat loss from slab floors, it is important to insulate them, as illustrated in Figure 7.22.[9,10]

E. Fundamentals of Heat Transfer

Two major heat loss mechanisms are conduction and convection. The basic relations for heat transfer by conduction state that the rate of heat flow by conduction Q_{cond} in a material is equal to the following three quantities.[5,10,12,17]

1. C, thermal conductivity of the material, in Btu per hour per square foot per °F per foot;
2. A, area of the medium, e.g., the building wall, roof or glazing, in square feet;
3. dt/dx, temperature gradient through the medium, in °F per foot.

In the equation this can be written as

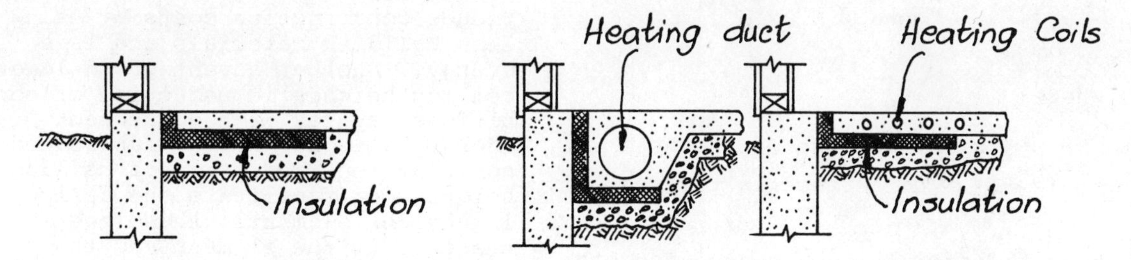

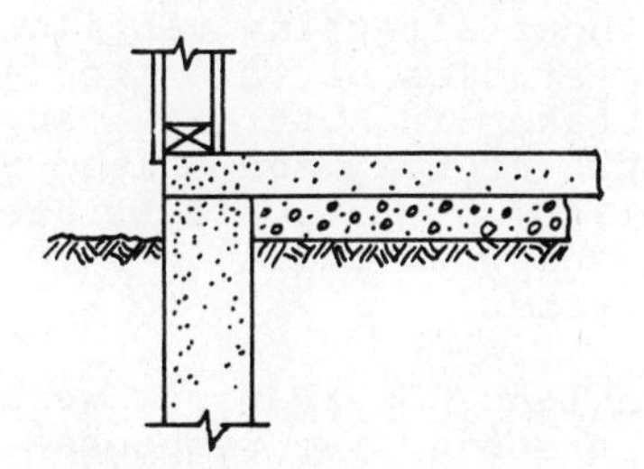

(c) Uninsulated and unheated
(not recommended)

Reference

FIGURE 7.22

PREFERRED EDGE CONDITION TO REDUCE HEAT LOSSES

$$Q_{cond} = CA\frac{dt}{dx}$$

or

$$Q_{cond} = \frac{C}{X}A\ (T_o - T)$$

where:

x = thickness of the material,

$(T_o - T)$ = difference in temperature.

Heat transfer through convection can be represented by an equation similar to conduction:

$$Q_{conv} = C_c A(T_s - T_f)$$

where:

Q_{conv} = rate of heat flow by convection, Btu/hr,

A = base area of heat transfer by convection, ft^2,

T_s = surface temperature, °F,

T_f = fluid temperature, °F,

C_c = convection heat transfer coefficient, Btu/hr ft^2/°F.

Table 7.5 provides U factor values for typical residential wall construction. For the U value of many other building and insulating materials, refer to the ASHRAE guide or Carrier System Manual 1.[16,20]

Case Study 7-3 illustrates typical heat loss calculations for a simple building.

F. Effect of Ceiling Height on Energy Use

The ceiling plenum, floor thickness and floor-to-ceiling height constitute the floor-to-floor height. Floor height is usually determined by construction and structural design. The floor-to-floor dimension most directly influences efficient energy use in that it affects the area of the exterior building that is exposed to weather. The effect of reduced ceiling height on indoor comfort in hot regions has aroused scientific interest in many countries for at least two reasons: the desire to economize without affecting the comfort of the occupants and the desire to reduce construction costs by using less building materials and less energy. Another advantage of lower ceiling heights is reduction of conditioned space, with consequent energy savings. For prefabricated construction, reduction of ceiling heights may facilitate the design of low height elements, thus lowering the cost of the element and the erection cost. In multi-story buildings, the number of stories can be increased by lowering the ceiling height without affecting ventilation. Systematic studies of this problem were undertaken in Australia, England, India, South Africa, the United States and Israel. Table 7.6 illustrates acceptable height limits in a few of these countries.

In England observations were also made of the subjective responses of occupants in apartments with lower ceilings (from 2.4 m to 2.25 m high). The conclusion was that there was no adverse effect either climatically or psychologically from the reduction in ceiling height. Ninety percent of the occupants in the lower ceiling apartments informed the interviewers that the height seemed appropriate to them and a large majority did not notice that the height was lower than usual.

In India, an experimental study was conducted in which indoor and outdoor temperatures were measured in four experimental units, 3.6 m x 3.05 m (12 ft x 10 ft) with a veranda out to the west. The units were identical except for the ceiling height which was 2.4, 2.7, and 3.3 m. The results of this study showed that raising the ceiling height to about 2.7 m had no significant thermal advantage.[5] In the US, at a nationwide conference on apartment design, the question ot ceiling height was also discussed. The majority at the conference stated that the customary ceiling height in the US, 2.4 m, was adequate, while a small number preferred a slightly higher ceiling in living rooms, that is 2.55 m.[5]

U-factor Values for Typical Residential Wall Construction*

Wall Number	Construction Component	Component Thickness, in.	U value Btu/(hr)(ft^2)(°F)
	Face brick	4	
	Block	8	
1	Firring space	3/4	0.09
	Urethane board	1/2	
	Plasterboard	1/2	
	Stucco	1/2-1	
	Block	8-12	
2	Firring space	3/4	0.13
	Urethane board	1/2	
	Plasterboard	1/2	
	Wood siding	1/2-3/4	
	Building paper		
3	Sheathing	5/8	0.13
	Stud/space (with insulation)	3-5/8	
	Plasterboard	1/2	
	Glass, double-plate	1/4	
4	Insulated window (separated by 1/2 in. air space)		0.60
5	Outside air layer, vertical and horizontal		5.9
	Inside air layer, vertical		1.5

* It is assumed the windows are double-glazed, and the glass-to-wall ratio is 20%.

**For more U-value information, refer to References 10,12, and 20.

TYPICAL U-VALUES OF WALLS, ROOF, ETC.

TABLE 7.5

	Ceiling Height
USA	2.4 m to 2.55 m
England	2.25 to 2.4 m
India	2.4 t0 2.7 m
Israel	2.5 m

Reference 5

PREFERRED CEILING HEIGHTS IN DIFFERENT PARTS OF THE WORLD TABLE 7.6

7.5 RETROFITTING BUILDINGS FOR EFFICIENT ENERGY USE

Opportunities to use less energy than currently used while maintaining equipment performance exist in most buildings. The major energy use parameters of existing building have already been established. These include the building's envelope, its construction, its shape and height, its location on a site, and its mechanical systems.

Most of these existing situations are difficult and probably very expensive to modify. In some cases energy efficient changes can be identified, for example, addition of insulation, double glazing, etc.; however, these may be impossible to implement because of high costs. On the other hand, often it is possible to change the operation of the building or the design limits, since the original design probably incorporated large safety factors. A careful study of a building that has been in operation for a period of time permits much more precise definition of the activities which take place there, such as the number of people actually allowed in each space, their behavior in terms of space usage, and their relationship to the building energy use. These might show possibilities for lower ceiling heights or task-oriented space allocation and thus energy savings. An investigation of the design criteria of mechanical systems is also useful in determining energy saving possibilities.

Investigating opportunities to add components which act as a facade can significantly reduce energy usage in any building--sun shades, exterior wall shutters, installation of open windows, modification of the exterior color. Study the possibilities for making changes to the site itself by adding deciduous vegetation, rearranging planter areas near the building, and changing parking lots to make them less reflective. It is useful to establish the energy benefits of day lighting, effects of wind, and heat loss and heat gain from adding insulation to the building. It may be possible to alter the exterior of buildings by changing the glazing types and preventing infiltration or exfiltration.

Table 7.7 is a checklist for identifying energy saving possibilities for an existing building.[21]

A. Performance Codes

One way to encourage efficient energy use in the building envelope is to propose a realistic "energy budget" and allow the designer to select an appropriate approach which would meet performance codes and even to select alternative energy sources. The energy budget would vary from place to place depending on climatic and economic considerations.

An alternative is to force the designer, through legal measures, to incorporate certain design features which would make the building energy efficient.

The General Service Administration published a guide book for new office buildings in 1975.[9] In this guide book, the Federal Government established a target gross energy use of 55,000 Btu per equivalent gross square feet per year. In addition to efforts at the federal level, some states (e.g., Florida, California, Ohio, Minnesota) have taken action to establish their own codes. (See Appendix D.) Another guide is ASHRAE 90-75, discussed previously.

Energy conscious design codes should have a significant impact on future building design. Table 7.8 shows possible impacts on new housing design.

B. A Checklist for Energy Efficient Designs

Table 7.9 provides a checklist of energy management opportunities for designers. These can help the designer to come up with innovative measures for optimal use of and conservation of non-renewable fuels. While using this checklist attention must be paid to the varying climatic conditions of the US.[9,12]

Site	- shade walls and paved areas adjacent to the building
	- plant deciduous and coniferous trees for their shading and windbreak effects
	- plant lawns and shrubs between building and side walks
	- cover roof and walls with plants to limit unnecessary solar gains
	- study the possibility of eliminating solar re-radiation from parking surfaces
Building	- study the benefits of adding roof and wall insulation
	- protect walls from severe wind effect
	- study the benefits and burdens of solar energy on the building
	- look for existing possibilities to lower ceiling heights
	- study the location of thermal insulation
	- determine if the wall colors can be changed to reduce solar heat gain
	- determine if glass areas can be reduced and double or triple glazing added
	- look for areas where infiltration can be controlled
	- determine if daylighting can be utilized to reduce lighting heat load

References 11, 21

CHECKLIST: RETROFITTING BUILDING ENVELOPE **TABLE 7.7**

Housing type

Smaller, higher density, fewer detached houses
Increased shift to townhouse and low-rise from single-family detached
Diminished relative attractiveness of mobile homes in life-cycle cost terms
Improved designs with lower unit demands will help keep energy using conveniences affordable.

Architectural features

Thicker wall (cavity and sandwich) and roof construction for more insulation
Fewer picture windows, more double- and triple-glazed windows, some specially coated glass
More functional windows--designed as passive solar collectors
Tighter, better sealed joints, higher performance sealants, better workmanship
Better control of moisture to protect insulation
Attention to shape, orientation, landscaping in design
Control of air movement between floors--fewer open stairwells and split-level designs
Insulated foundation walls in cold climates
Greater thermal resistance for more expensive fuels, such as electric heating
Better thermal comfort and greater acoustical privacy

Institutional

Life-cycle cost-based performance standards (voluntary or mandatory) for new housing design
Labeling of houses, equipment, and appliances for energy use, cost, and performance
Householders knowledgeable of how to operate homes efficiently

Reference 11

ANTICIPATED IMPACTS OF ENERGY CONSERVATION ON NEW HOUSING DESIGN TABLE 7.8

This checklist of energy-saving opportunities, appended to the guidelines, includes some items that subsume others. Some seem to border on the obvious, yet many contemporary buildings are testimony to the need for even seemingly obvious measures.

The items are ranked in priority and coded to the following climatic features: For winter *A* indicates a heating season of 6,000 degree-days or more; *B* a heating season of 4,000 to 6,000 degree-days, and *C* 4,000 degree-days or less. The numeral *1* following these letters indicates sun 60 percent of daylight time or more and wind nine miles per hour or more; *2* indicates the sun condition but not the wind condition; *3* indicates the wind condition without the sun condition, and *4* the absence of either condition.

For summer, the letter *D* indicates a cooling season of more than 1,500 hours at 80 degrees Fahrenheit; E 600 to 1,500 hours at the same temperature and *F* less than 600 hours. The numeral *1* indicates a dry climate of 60 percent relative humidity or less and *2* indicates 60 percent or more humidity.

Guidelines that are independent of climate are not rated in priority columns and are marked '*'.

SITE	Priority 1	Priority 2	Priority 3	N/A
1. Use deciduous trees for their summer sun shading effects and wind break for buildings up to three stories.	A1 D1	A2 D2	A4 E1	C4 F
2. Use conifer trees for summer and winter sun shading and wind breaks.	C4 D1	C1 D2	C2 E1	A2 F
3. Cover exterior walls and/or roof with earth and planting to reduce heat transmission and solar gain.	A1 D1	A2 D2	A4 E1	C4 F
4. Shade walls and paved areas adjacent to building to reduce indoor/outdoor temperature differential.	C2 D1	C1 D2	C3 E1	C2 F
5. Reduce paved areas and use grass or other vegetation to reduce outdoor temperature buildup.	C2 D1	C1 D2	C3 E1	A2 F

A CHECKLIST OF CONSERVATION OPPORTUNITIES TABLE 7.9

SITE	Priority 1	2	3	N/A
6. Use ponds, water fountains to reduce ambient outdoor air temperature around building.	C2 D1	C1 E1	C3 D2	A4 F
7. Collect rain water for use in building.	*			
8. Locate building on site to induce air flow effects for natural ventilation and cooling.	C2 F	C1 E1	C3 E2	A4 D3
9. Locate buildings to minimize wind effects on exterior surfaces	A4 F	A1 E2	B4 E1	C2 D1
10. Select site with high air quality (least contaminated) to enhance natural ventilation.	C2 F	C1 E1	C3 E2	A4 D2
11. Select a site which has year round ambient wet and dry bulb temperatures close to and somewhat lower than those desired within the occupied spaces.	*			
12. Select a site that has topographical features and adjacent structures that provide breaks.	A4 F	A1 E2	B4 E1	C3 D2
13. Select a site that has topographical features and adjacent structures that provide shading.	C2 D2	C1 D1	B2 E2	A1 F
14. Select site that allows optimum orientation and configuration to minimize yearly energy consumption.	*			
15. Select site to reduce specular heat reflections from water.	C2 D2	C1 D1	B2 E2	A4 F
16. Use sloping site to bury building partially or use earth berms to reduce heat transmission and solar radiation.	A4 D1	A1 D2	A3 E1	C2 F
17. Select site that allows occupants to use public transport systems.	*			

TABLE 7.9

CONT.

BUILDING	Priority 1	2	3	N/A
1. Construct building with minimum exposed surface area to minimize heat transmission for a given enclosed volume.	A4 D1	A1 D2	A3 E1	C2 F
2. Select building configuration to give minimum north wall to reduce heat losses.	A4	A1	A3	C2
3. Select building configuration to give minimum south wall to reduce cooling load.	D1	D2	E1	F
4. Use building configuration and wall arrangement (horizontal and vertical sloping walls) to provide self shading and wind breaks.	A4 D1	A1 D2	B4 E1	C3
5. Locate insulation for walls and roofs and floors over garages at the exterior surface.	A4 D1	A3 D2	A1 E1	C2 F
6. Construct exterior walls, roof and floors with high thermal mass with a goal of 100 pounds per cubic foot.	A4 D1	A1 D2	A3 E1	C3 F
7. Select insulation to give a composite U factor from 0.06 when outdoor winter design temperatures are less than 10 degrees F to 0.15 when outdoor design conditions are above 40 degrees F.				
8. Select U factors from 0.06 where sol-air temperatures are above 144 degrees F up to a U value of 0.3 with sol-air temperatures below 85 degrees F.				
9. Provide vapor barrier on the interior surface of exterior walls and roof of sufficient impermeability to prevent condensation.	*			
10. Use concrete slab-on-grade for ground floors.	A4 D1	A1 D2	A3 E1	C2 F
11. Avoid cracks and joints in building construction to reduce infiltration.	A4 D2	A1 E2	A3 D1	
12. Avoid thermal bridges through the exterior surfaces	A4 D2	A1 D1	A3 E2	C3 F

TABLE 7.9
CONT.

BUILDING	Priority 1	2	3	N/A
13. Provide textured finish to external surfaces to increase external film coefficient.	A4	A1	B4	C2
14. Provide solar control for the walls and roof in the same areas where similar solar control is desirable for glazing.	D2	D1	E2	A
15. Consider length and width aspects for rectangular buildings as well as other geometric forms in relationship to building height and interior and exterior floor areas to optimize energy conservation.	A4 D1	A1 D2	A3 E1	C2 F
16. To minimize heat gain in summer due to solar radiation, finish walls and roofs with a light-colored surface having a high emissivity.	D1	D2	E1	
17. To increase heat gain due to solar radiation on walls and roofs, use a dark-colored finish having a high absorptivity.	A1	A2	A4	C2
18. Reduce heat transmissions through roof by one or more of the following items:				
a. Insulation.	A4 D1	A1 D2	A3 E1	C3 F
b. Reflective surfaces.	C2 D1	C1 D2	C3 E1	A4
c. Roof spray.	D1	E1	F	
d. Roof pond.	D1	E1	F	
e. Sod and planting.	A4 D1	A1 D2	A3 E1	C2 F
f. Equipment and equipment rooms located on the roof.	A4 D1	A1 D2	A3 E1	
g. Provide double roof and ventilate space between.	D1	D2	E1	F
19. Increase roof heat gain when reduction of heat loss in winter exceeds heat gain increase in summer.				
a. Use dark-colored surfaces.	A2	A1	B2	B1
b. Avoid shadows.	A2	A1	B2	B1
20. Insulate slab on grade with both vertical and horizontal perimeter insulation under slab.	A	B	C	

TABLE 7.9
CONT.

BUILDING	Priority 1	2	3	N/A
21. Reduce infiltration quantities by one or more of the following measures:				
a. Reduce building height.				
b. Use impermeable exterior surface materials.	A4	A1	A3	C4
c. Reduce crackage area around doors, windows, etc., to a minimum.	D2	E2	D1	F
d. Provide all external doors with weather stripping.				
e. Where operable windows are used, provide them with sealing gaskets and cam latches.				
f. Locate building entrances on downwind side and provide wind break.				
g. Provide all entrances with vestibules; where vestibules are not used, provide revolving doors.				
h. Provide vestibules with self-closing weather stripped doors to isolate them from the stairwells and elevator shafts.	A4 D2	A1 E2	A3 D1	C4 F
i. Seal all vertical shafts.				
j. Locate ventilation louvers on downwind side of building and provide wind breaks.				
k. Provide break at intermediate points of elevator shafts and stairwells for tall buildings.				
22. Provide wind protection by using fins, recesses, etc., for any exposed surface having a U value greater than 0.5.	A4	A1	B4	C2
23. Do not heat parking garages.	*			
24. Consider the amount of energy required for the protection of materials and their transport on a life-cycle energy usage.	*			
25. Consider the use of the insulation type which can be most efficiently applied to optimize the thermal resistance of the wall or roof; for example, some types of insulation are difficult to install without voids or shrinkage.	*			

TABLE 7.9
CONT.

BUILDING	Priority 1	2	3	N/A
26. Protect insulation from moisture originating outdoors, since volume decreases when wet. Use insulation with low water absorption and one which dries out quickly and regains its original thermal performance after being wet.	*			
27. Where sloping roofs are used, face them to south for greatest heat gain benefit in the wintertime.	A1	A2	B1	C4
28. To reduce heat loss from windows, consider one or more of the following:				
a. Use minimum ratio of window area to wall area.				
b. Use double glazing.				
c. Use triple glazing.				
d. Use double reflective glazing.	A4	B4	C4	
e. Use minimum percentage of the double glazing on the north wall.	A1 A2	B1 B2	C1 C2	
f. Manipulate east and west walls so that windows face south.	A3	B3	C3	
g. Allow direct sun on windows November through March.				
h. Avoid window frames that form a thermal bridge.				
i. Use operable thermal shutters which decrease the composite U value to 0.1.				
29. To reduce heat gains through windows, consider the following:				
a. Use minimum ratio of window areas to wall area.				
b. Use double glazing.				
c. Use triple glazing.	D1	E1	F	
d. Use double reflective glazing.	D2	E2	F	
e. Use minimum percentage of double glazing on the south wall.				
f. Shade windows from direct sun April through October.				
30. To take advantage of natural daylight within the building and reduce electrical energy consumption, consider the following:				

TABLE 7.9

CONT.

BUILDING	Priority 1	2	3	N/A
a. Increase window size but do not exceed the point where yearly energy consumption, due to heat gains and losses, exceeds the saving made by using natural light.				
b. Locate windows high in wall to increase reflection from ceiling, but reduce glare effect on occupants.	C2 C1	B2 B1	A2 A1	
c. Control glare with translucent draperies operated by photo cells.	C3 C4	B3 B4	A3 A4	
d. Provide exterior shades that eliminate direct sunlight, but reflect light into occupied spaces.	F	E	D	
e. Slope vertical wall surfaces so that windows are self-shading and walls below act as light reflectors.				
f. Use clear glazing. Reflective or heat absorbing films reduce the quantity of natural light transmitted through the window.				
31. To allow the use of natural light in cold zones where heat losses are high energy users, consider operable thermal barriers.	A4	A1	B4	C3
32. Use permanently sealed windows to reduce infiltration in climatic zones where this is a large energy user.	A1 D1	A4 D2	B1 E1	C3 F
33. Where codes or regulations require operable windows and infiltration is undesirable, use windows that close against a sealing gasket.	A1 D1	A4 D2	B1 E1	C3 F
34. In climatic zones where outdoor air conditions are suitable for natural ventilation for a major part of the year, provide operable windows.	C2 F	C3 E1	C1 E2	A4 D2
35. In climate zones where outdoor air conditions are close to desired indoor conditions for a major portion of the year, consider the following:				
a. Adjust building orientation and configuration to take advantage of prevailing winds.				
b. Use operable windows to control ingress and egress of air through the building.				

TABLE 7.9

CONT.

BUILDING	Priority 1	2	3	N/A
c. Adjust the configuration of the building to allow natural cross ventilation through occupied spaces.	F	E1	E2	D2
d. Use stack effect in vertical shafts, stairwells, etc., to promote natural air flow through the building.				

Reference 12

TABLE 7.9
CONT.

C. Need for Computer Use

If a designer wants to meet an energy performance standard and desires to study alternatives for the building envelope to achieve energy optimization, a computer can be a useful tool--almost necessary if the study is highly complex and many variations in design are to be analyzed.[22,23,24] At present there are several computer programs which can assist the designer. These include (typical examples; there are others):

- AXCESS, by Edison Electric Institute, is a method for comparing the energy requirements of alternate mechanical/electrical/building systems.
- TRACE, by Trane Company, for performing energy analysis of new building construction in the conceptual stage. Does economic analyses.
- NBSLD, by the Department of Commerce.
- SOLCOST, by ERDA with Martin Marrietta Company, is designed to make solar cost and design calculations.

Refer also to Chapter 5 for additional information.

* * *

REFERENCES

1. McHarg, Ian L., *Design with Nature*, (Garden City, New York: Doubleday and Company, Inc., Natural History Press, 1971).

2. Steadman, Philip, *Energy, Environment and Building*, (Cambridge, England: Cambridge University Press, 1975).

3. US Federal Energy Administration, Research Planning and Systems Studies, Energy Conservation and Environment, *Five Year Program Planning Document for End Use Energy Conservation, Research, Development, and Demonstration*, (Washington, D.C.: June 1974).

4. Fitch, James Marston, *American Building, The Environmental Forces that Shape It*, 2nd edition, (New York: Schocken Books, 1976).

5. Givoni, B., *Man, Climate and Architecture*, (London, England: Applied Science Publishers, Ltd., 1975).

6. The AIA Research Corporation, *Solar Dwelling Design Concepts*, HUD-PDR-156, (Washington,D.C.: US Department of Housing and Urban Development, Office of Policy Development and Research, May 1976).

7. US Federal Energy Administration, *Project Independence Report*, (Washington, D.C.: US Government Printing Office, November 1974), pp. 186-191.

8. Olgyay, Victor and Olgyay, Aladar, *Sun Control and Shading Devices*, (Princeton, New Jersey: Princeton University Press, 1957).

9. Dubin-Mindell-Bloome Associates, P.C. in cooperation with AIA/Research Corporation and Heery and Heery, Architects, *Energy Conservation Design Guidelines for Office Buildings*, (Washington, D.C.: General Services Administration, Public Buildings Service, January 1974).

10. McGuinness, William J. and Stein, Benjamin, *Mechanical and Electrical Equipment for Buildings*, 5th edition, (New York: John Wiley and Sons, Inc., 1971).

11. Snell, Jack E.; Achenbach, Paul R.; Petersen, Stephen R., "Energy Conservation in New Housing Design," *Science* 192(25 June 1976): 1305-1311.

12. Kreider, Jan F. and Kreith, Frank, *Solar Heating and Cooling*, (Washington, D.C.: Hemisphere Publishing Corporation, 1975).

13. Seidel, Marquis R.; Plotkin, Steven E.; and Peck, Robert O., *Energy Conservation Strategies*, Report No. EPA-R5-73-021, (Washington, D.C.: US Environmental Protection Agency, Office of Research and Monitoring, Implementation Research Division, July 1973).

14. Godfrey, J.A., "Double Glazing and Double Windows, *Building Research Digest*, (Great Britain), 140(April 1972).

15. American Society of Heating, Refrigerating and Air Conditioning Engineers, Inc., *ASHRAE Handbook of Fundamentals*, (New York: 1972).

16. American Society of Heating, Refrigerating and Air Conditioning Engineers, Inc., *Energy Conservation in New Building Design*, ASHRAE Standard No. 90-75, (New York: 1975).

17. Gargus, A. Gregory; Shull, H. Eugene; Essenhigh, Robert H., "Estimations of Target Degree-Day Energy Densities for Buildings," Submitted through AICHE for Eleventh Intersociety Energy Conversion Engineering Conference, 1976.

18. Tao, William, "The ASHRAE Energy Standard for New Buildings: A Digest," *Architectural Record*, (July 1976):127-128.

19. McKenzie, J.J. and Osherenko, G., "Regional Patterns and Trends in the Generation and Consumption of Electricity in the U.S.," in *Towards an Energy Policy*, ed: Keith Roberts, (San Francisco, California: The Sierra Club, 1973).

20. Carrier Air Conditioning Company, *Carrier System Design Manual*, 12 Parts, (Syracuse, New York), Part 1, "Load Estimating," (1972).

21. General Services Administration, Public Buildings Services, *Energy Conservation Guidelines for Existing Office Buildings*, (Washington, D.C.: 1975).

22. American Society of Heating, Refrigerating and Air Conditioning Engineers, Inc., *Procedure for Determining Heating and Cooling Loads for Computerizing Energy Calculations*, (New York: 1975).

23. Jones, Charles D. and Sepsy, Charles F. (eds), *Load Profiles and Energy Requirements for Heating and Cooling of Buildings*, ASHRAE Research Project 66-OS, (New York: American Society of Heating, Refrigerating and Air Conditioning Engineers, Inc., 1976).

24. Stoecker, W.F. (ed), *Procedures for Simulating the Performance of Components and Systems for Energy Calculations*, (New York: American Society of Heating, Refrigerating and Air Conditioning Engineers, Inc., 1975).

CHAPTER 8

EFFICIENT HEATING, VENTILATING AND AIR CONDITIONING

Michael K.J. Anderson*

CHAPTER CONTENTS

KEY WORDS

Air Conditioning
Building Codes
Commercial Buildings
Commercial Equipment
Environmental Conditioning
Equipment Modification
Heat Losses
Heat Pumps
Heat Recovery
Refrigeration

SUMMARY

The maintenance of comfortable environmental conditions (heating and cooling) accounts for up to 50 percent of the energy used in many commercial and industrial buildings (see Chapter 3). The amount of energy used for this purpose is increasing, particularly for air conditioning. In fact, roughly 90 percent of all new commercial space is air-conditioned.

The efficiency of many existing heating, ventilating and air-conditioning (HVAC) systems can be improved. Modifications to these systems can decrease energy use by a range of 10 to 50 percent, while causing little or no change in the comfort level of the conditioned space. The most common modifications involve resetting automatic controls, increasing control capabilities, modifying existing equipment, or installing more efficient systems.

More opportunities for efficient energy utilization exist in new buildings because the HVAC systems can be integrated with the total building design. This allows more control of solar load, heat recovery, heat storage, and total installed heating and cooling capacity. Studies indicate that constructing a building according to more efficient standards, particularly ASHRAE 90-75, can greatly reduce the amount of energy used during normal operation without increasing the initial cost of the total facility. [1] For example, the additional costs of improvement in a tighter building (insulation, double glazed windows, shading) are generally more than offset by the savings in installing smaller chillers, boilers, and other HVAC equipment.

*Member of the Technical Staff, Applied Nucleonics Company, Inc.

8.1 INTRODUCTION

This chapter discusses energy efficient practices for HVAC systems. The first section describes the scope of material covered and indicates the type of people who would be most interested in this chapter.

The next section is a description of some of the physical constraints under which HVAC systems operate. Comfort conditions are defined for a variety of occupations in accordance with building codes. The constraints of the outside environment are described in terms of temperature and solar inputs. Psychrometrics, the behavior of air, is also briefly described to illustrate its relation to HVAC processes.

Section 8.4 covers the efficient operation of HVAC equipment by three major approaches:

- Control Modifications--changes which can be made with essentially no initial investment yet which result in money and energy savings from the elimination of unnecessary energy usage.
- Retrofit Modifications--changes in existing equipment or systems which improve efficiency, usually without decreasing overall performance.
- New Design--new methods of system design as well as selection of more efficient equipment.

The manner in which individual components in an active system are combined and operated determines how efficiently they are utilized. This is addressed in the following section on system selection.

A potpourri section describes other methods for improving the operating efficiency of HVAC systems. These include computer control, thermal storage techniques, heat recovery, and the application of solar heating and cooling. Building codes, which constrain the designer's options in certain ways and impact on system efficiency, are also mentioned.

Case histories to illustrate the application of some of these methods are included in Chapter 15. Economic benefit--that is, payback period and return on investment--is a key factor in determining whether or not HVAC modifications should be made.

8.2 OBJECTIVE, SCOPE, ASSUMPTIONS

A. Objective

The objective of this chapter is to present a variety of practical HVAC energy management ideas which can be readily applied to existing or new buildings. Instead of providing a checklist of ideas to consider, the chapter is intended to provide background information which will allow an initial analysis of items which may be found in checklists (such as those in References 2, 3, 4, and 5).

The great variety in modern equipment and systems makes it impossible to cover all of the techniques which could improve efficiency. The information in this chapter is limited, therefore, to ideas which are fairly general in application. Note that for most systems the assistance of specialists will be required.

Every building has its own characteristics and special uses which make its operation unique.[6] Approaches which are appropriate for one situation may not be so for others. Therefore, each building should be analyzed as a separate case. The information in this chapter will provide a starting point and data for that analysis.

B. Scope

Medium to large HVAC equipment and systems are discussed in this chapter. These would be found in almost all commercial buildings and a large proportion of buildings housing industrial operations.

Within the scope of this chapter are the processes of actively adding or removing heat from buildings. For a discussion on the transfer of heat into or out of the building skin, or *building envelope*, refer to Chapter 7 which also provides more details related to outside air effects, heat transfer and infiltration (mass transfer), and energy input from the sun.

C. Assumptions

In this chapter it is assumed that the reader has some familiarity with the operation of industrial or commercial HVAC systems. This familiarity may be obtained through design experience, operation experience, ownership, or schooling. A section of basics is also provided so that others who may be interested can learn some of the characteristics of HVAC systems which relate to their efficient operation.

For a more complete discussion of the subject, the reader should consult design manuals and mechanical engineers experienced in the design of efficient systems.

8.3 USEFUL FACTS AND BACKGROUND

A. The Natural Environment

The natural environment is the set of physical conditions which surrounds a building. Variations in the environment create changes in building energy use. Environmental variables include temperature, relative humidity, solar energy, and wind velocities.

A simplified hand calculation procedure to estimate the effect of outside temperature on building operation is based on the "degree-day" concept. A heating degree-day is defined as the difference betweeen 19 °C (65 °F) and the 24-hour average temperature if it is below 19 °C (65 °F). The temperature 19 °C (65 °F) is chosen because it is roughly the balance temperature when heat generated in a typical building is equal to heat lost to the outside air. Theoretically, no additional heating or cooling should be needed at this temperature. For example, an average temperature of 16 °C (60 °F) on a given day would represent 3 degree C-days (5 degree F-days). An average temperature of 13 °C (55 °F) would represent 6 degree C-days (10 degree F-days), implying roughly twice the heating load.

Individual degree-days, when the daily average temperature is below 19 °C (65 °F), are summed for an entire year to give an indication of how much heating is required. Heating degree C-days in the US range from several hundred in Florida to over 5,500 in Minnesota. Refer to Figure 7.20 for more detail.

The following calculation is shown as an example of how the degree-day would be used in energy calculations. Due to its inaccuracy, this approach is recommended only for rough calculations.* Suppose the outside air introduced into a building which operates 24 hours per day were reduced by 10,000 CFM. The amount of heat required to heat this air in a climate with 5,000 degree F-days is calculated as:

$$10{,}000 \text{ CFM } (5000 \text{ °F days})\left(1.08 \frac{\text{Btu}}{\text{HR CFM °F}}\right)\left(\frac{24 \text{ hr}}{\text{Day}}\right) = 1{,}296{,}000{,}000 \frac{\text{Btu}}{\text{yr}}$$

$$= 1{,}296 \frac{\text{MBtu}}{\text{yr}}$$

The heating requirement of the building would be reduced by this amount as a result of the change. Note that this accounts only for heat in the air. When the losses of the heating system are included, the fuel savings would be greater.

Cooling degree-days, as shown in Figure 7.21 (Chapter 7), are determined in a similar manner for days with an average temperature greater than 19 °C (65 °F). This value can be used in determining the heat input to a building from outside ventilation air. It does not, however, reflect the cooling load caused by solar radiation, lights, and people, so estimates of total energy use require more detailed calculations. Summer climate is more generally expressed in terms of the

*It is described here because it is often used for preliminary calculations. As an example of one of the limitations of this method, it ignores internal heat gains. Thus, for certain types of buildings, the heating requirement will bear no relation to the outside air temperature.

number of hours per year when cooling is required. This is tabulated from four major regions in the US in Table 8.1.

To perform more precise calculations concerning the effects of climate on building energy use, computer simulation can be used. Some programs use hourly weather data for an entire year to give a more precise analysis. These programs are discussed in more detail in Chapter 5.

Wind conditions can significantly affect heat gains or losses through building skins. The effective thermal conductivity of the layer of air surrounding a building increases as the wind velocity increases. Usually, an average wind speed for summer or winter is used in calculations. For more on wind effects, refer to Chapter 7.

The final major climatic factor which affects energy use in buildings is solar radiation. The heat produced can be utilized by an active HVAC system, as discussed in this chapter, or by a passive system involving the architectural layout of the facility. For a further discussion of solar radiation effects, refer to Chapter 7.

B. The Conditioned Environment

Before energy use became a critical issue, HVAC systems in buildings were designed to maintain exact control of temperature and humidity. Designing for tight control led to an increase in energy use compared to earlier systems with less exacting controls.

The range of comfort conditions now accepted is wider than before, however, particularly when appropriate dress is worn. The most generally recognized new standard which accounts for this change is ASHRAE 90-75. It recommends design criteria of 22 °C (72 °F) during winter and 25.5 °C (78 °F) during summer, and requires that humidification be designed to a maximum value of 30 percent relative humidity. If dehumidification controls are installed, they shall be capable of being set to prevent use of new energy to produce a space relative humidity below 60 percent. [8] The standard also recommends that if *lower energy use is the goal*, the temperature and humidity can and should range within the boundaries of the comfort envelope.

ASHRAE 90-75 also states that winter and summer outdoor design temperatures will be chosen so that they are exceeded only 2.5 percent of the time.

Ventilation requirements were originally defined by ASHRAE 62-73 in terms of "minimum" and "recommended" levels. ASHRAE 90-75 now requires the use of the "minimum" levels for design. The volume of outside air which must be introduced to the building is 33 percent of the total ventilation or 15 percent where odor and gas removal equipment are employed. In no case shall the outdoor quantity be less than 5 CFM per person. Some typical ventilation levels can be seen in Table 8.2. In instances where permitted it may be desirable to account for a part of the outside air requirement with infiltration.

C. Psychrometrics

Psychrometrics is the engineer's technique for quantifying the thermodynamic behavior of air. A basic tool used is the psychrometric chart which can be seen in Figure 8.1. The two normal axes are dry-bulb temperature and specific humidity (kilograms of water per kilogram of air). The diagonal lines represent the wet-bulb temperature and the enthalpy (within a small deviation). The curving lines show the relative humidity. These terms are explained in Table 8.3.

The four basic psychrometric processes performed by HVAC systems are heating/cooling, dehumidifying, humidifying, and mixing. Each will be described briefly.

When heating or cooling, the process is traced across a horizontal (constant specific humidity) line. If air starts at 1 (Figure 8.1) and heat is added, it travels along the line to 2. Starting at 2, air can be cooled to pass along the line back to 1.

Region	Pop. 10^6	Number of Households 10^6		Heating °F Degree Days per Year	Cooling Hours per Year
Northeast	49.0	15.3		5470	300
North Central	56.6	17.4		6345	500
South	62.8	19.1		2795	1600
West	34.8	11.1		3515	1600
Totals:	203.2	63.9	Averages:	∿4500	∿1000

Reference 7

CLIMATE IN US BY REGION TABLE 8.1

Function	Total Ventilation in CFM per Occupant	Outside Air Ventilation in CFM per Occupant*
Sales Floors and Showrooms	7	5
Beauty Shops	25	9
Auditoriums (no smoking)	5	5
Auditoriums (smoking permitted)	10	5
General Office Space	15	5
Computer Rooms	5	5
Underground Mine Shafts	20	7
Semiconductor Processing Rooms	10	5
Sawmills	20	7
Flour Milling, Bagging	30	10
Classrooms	10	5

*Assuming no gas and odor control equipment.

Reference 8,9.

VENTILATION REQUIREMENTS FROM ASHRAE 62-73 AS MODIFIED BY 90-75

TABLE 8.2

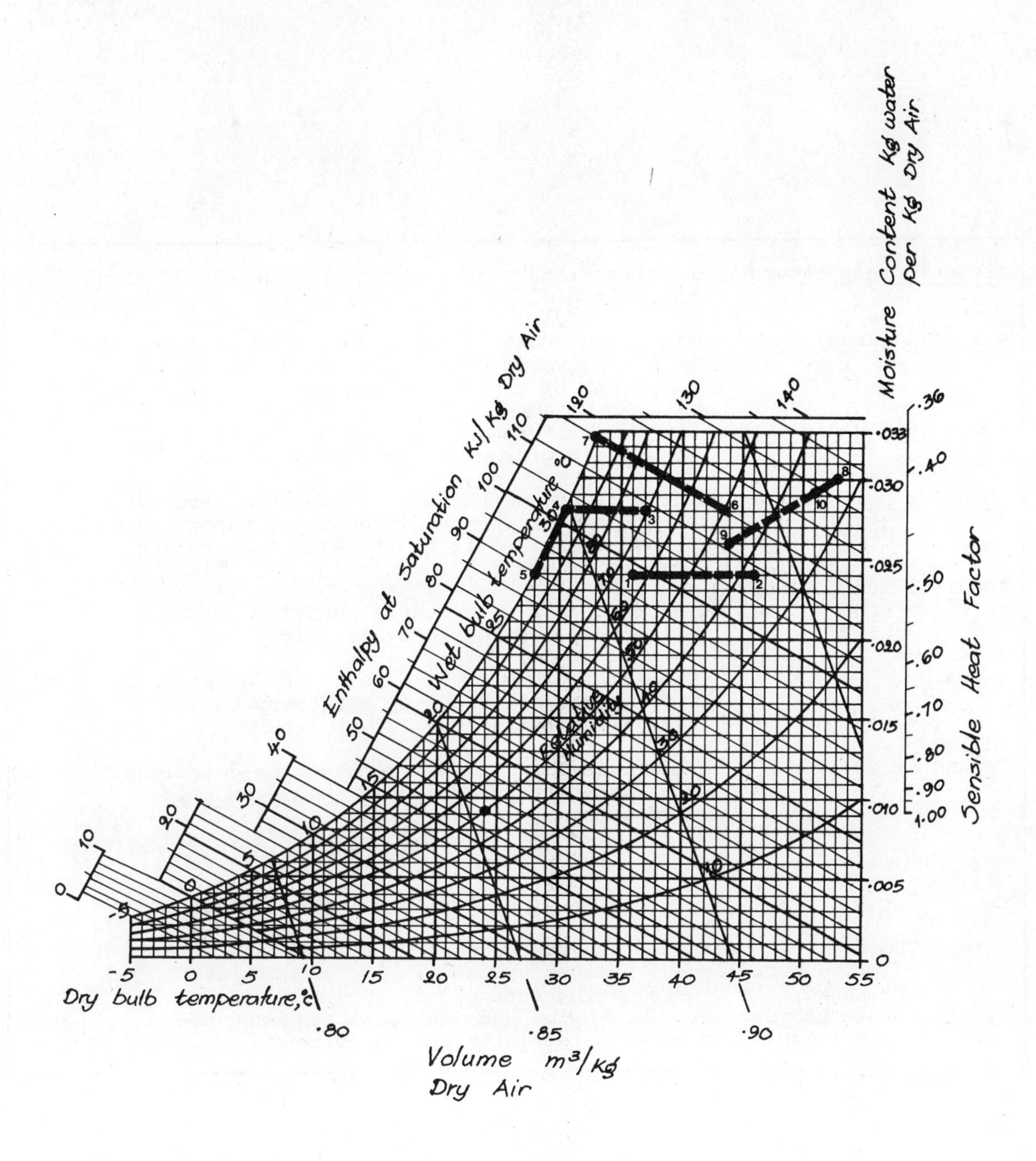

PSYCHROMETRIC CHART FIGURE 8.1

Dry-bulb temperature	--Air temperature registered by a standard thermometer.
Wet-bulb temperature	--Air temperature registered by a thermometer whose bulb is covered by a wetted wick and exposed to a current of rapidly moving air.
Dew point temperature	--The temperature at which condensation of moisture begins when the air is cooled.
Relative humidity	--Ratio of the actual water vapor pressure of the air to the saturated water vapor pressure of the air at the same temperature.
Specific humidity or moisture content	--The weight of water vapor in kilograms of moisture per kilogram of dry air.
Enthalpy	--A thermal property indicating the quantity of heat in the air above an arbitrary datum, in kJ/kg (Btu/lb) of dry air.
Pounds of dry air	--The basis for all psychrometric calculations. Remains constant during all psychrometric processes.

Note: The dry-bulb, wet-bulb and dew point temperatures and relative humidity are related so that if two properties are known, the others may be determined. When air is saturated, dry-bulb, wet-bulb, and dew point temperatures are all equal and relative humidity is 100 percent.

Reference 10

PSYCHROMETRIC DEFINITIONS **TABLE 8.3**

Note that as the dry-bulb temperature changes, the wet-bulb temperature and enthalpy change while the dew point remains the same.

Dehumidifying is performed by cooling the air until the left edge of the chart is reached. This represents the line of saturation, or 100 percent relative humidity. For example, in moving from 3 to 4 (Figure 8.1), this limit is reached. If further cooling is done water will condense out of the air and the temperature will follow the saturation curve towards 5. In actual practice the transition to condensing is a slightly curved path instead of the sudden transition at 4.

Humidification which occurs in an air washer is a constant wet-bulb (isenthalpic) process. This is represented by the line from 6 to 7. As water evaporates into the air, the dry-bulb temperature drops, the relative humidity increases, and the enthalpy remains the same. This is the basis of the operation of evaporative coolers.

Humidification can also occur through the release of dry steam directly into the conditioned air. In this case the humidification is a mixing process and is used to heat the air as well.

Mixing--such as the combining of return and supply air--occurs often in HVAC processes. It is represented on the chart by locations 8 and 9. Air with these initial properties is mixed with a ratio of 75 percent of the air at 8 and 25 percent of the air at 9. The resulting air is located on a line directly between them at 10. This point is 75 percent of the way from 9 to 8 because of the proportion of masses which were mixed.

8.4 EQUIPMENT

This section will briefly discuss some of the modifications which can increase efficiency in typical HVAC equipment. Modifications are considered in three classifications:

- control modifications,
- retrofit modifications, and
- new design.

A. Fans

All HVAC systems involve some motion of air. The energy needed for this motion can make up a large portion of the total system energy used. This is especially true in moderate weather when the heating or cooling load drops off but the distribution systems often operate at the same level.

Control

Simple control changes can save electrical energy in the operation of fans. Examples include turning off large fan systems when relatively few people are in the building or stopping ventilation a half hour before the building closes. The types of changes which can be made will depend upon the specific facility. Some changes involve more sophisticated controls which may already be available in the HVAC system. These will be discussed in the next section.

Retrofit

The capacity of the building ventilation system is usually determined by the maximum cooling or heating load in the building. This load has been changing due to reduced outside air requirements, lower lighting levels and wider acceptable comfort ranges. As a result it is now feasible to decrease air flow in many existing commercial buildings.

This operation is governed by fan laws which will be briefly described. Basic fan laws assume a constant air density, fan size, and distribution system. Rotational speed of the impeller, N, is the main variable in these laws which apply to all types of fans.[10]

The volume rate of air flow through the fan, Q, varies directly with the speed of the impeller's rotation. This is expressed as follows for a fan whose speed is changed from N_1 to N_2.

$$Q_2 = \left[\frac{N_2}{N_1}\right] Q_1$$

The pressure developed by the fan, P, (either static or total) varies as the square of the impeller speed.

$$P_2 = \left[\frac{N_2}{N_1}\right]^2 P_1$$

The power needed to drive the fan, H, varies as the cube of the impeller speed.

$$H_2 = \left[\frac{N_2}{N_1}\right]^3 H_1$$

The result of these laws is that for a given air distribution system (specified ducts, dampers, etc.), if the air flow is to be doubled, eight (2^3) times the power is needed. Conversely, if the air flow is to be cut in half, one-eighth ($1/2^3$) of the power is required. This is useful in HVAC systems because even a small reduction in air flow (say 10 percent) can result in significant energy savings (27 percent).

The manner in which the air flow is reduced is critical in realizing these savings. Maximum savings are achieved by sizing the motor exactly to the requirements. Simply changing pulleys to provide the desired speed will also result in energy reductions according to the cubic law. The efficiency of the existing fan motors tends to drop off below the half load range.

If variable volume air delivery is required, it may be achieved through inlet vane control, outlet dampers, variable speed motors, controlled pitch fans, or cycling. The relative efficiency of these approaches is shown in Figure 8.2. Energy efficiency in a retrofit design is best obtainable with variable speed motors or controlled pitch fans.

New Design

The parameters for new design are similar to those for fan retrofit. It is desirable, when possible, to use a varying ventilation rate which will decrease as the load decreases. A system such as Variable Air Volume incorporates this in the interior zones of a building. In some cases there will be a trade-off between power saved by running the fan slower and the additional power needed to generate colder air (see retrofit discussion above). The choices are complex and must be determined on a case-by-case basis.

B. Pumps

Pumps are found in a variety of HVAC applications such as chilled water, heating hot water, and condenser water loops. They are another piece of peripheral equipment which can use a large portion of HVAC energy, especially at low system loads.

Control

The control of pumps is often neglected in medium and large HVAC systems where it could significantly reduce the demand. A typical system would be a three chiller installation where only one chiller is needed much of the year. Two chilled water pumps in parallel are designed to handle the maximum load through all three chillers. Even when only one chiller is on, both pumps are used to propel water through all chillers. By manual adjustments two chillers could be by-passed and one pump turned off. All systems should be reviewed in this manner to ensure that only the necessary pumps operate under normal load conditions.

Retrofit

Pumps follow laws similar to fan laws, the key being the cubic relationship of power to the volume pumped through a given system. Small decreases in flow rate can save significant portions of energy.

In systems in which cooling or heating requirements have been permanently decreased, flow rates may be reduced also. A simple way to do this is by trimming the pump's impeller. The pump curve must be checked first, however, because pump efficiency is a function of the impeller diameter, flow rate, and pressure rise. It should be ensured that after trimming the pump will still be operating in an

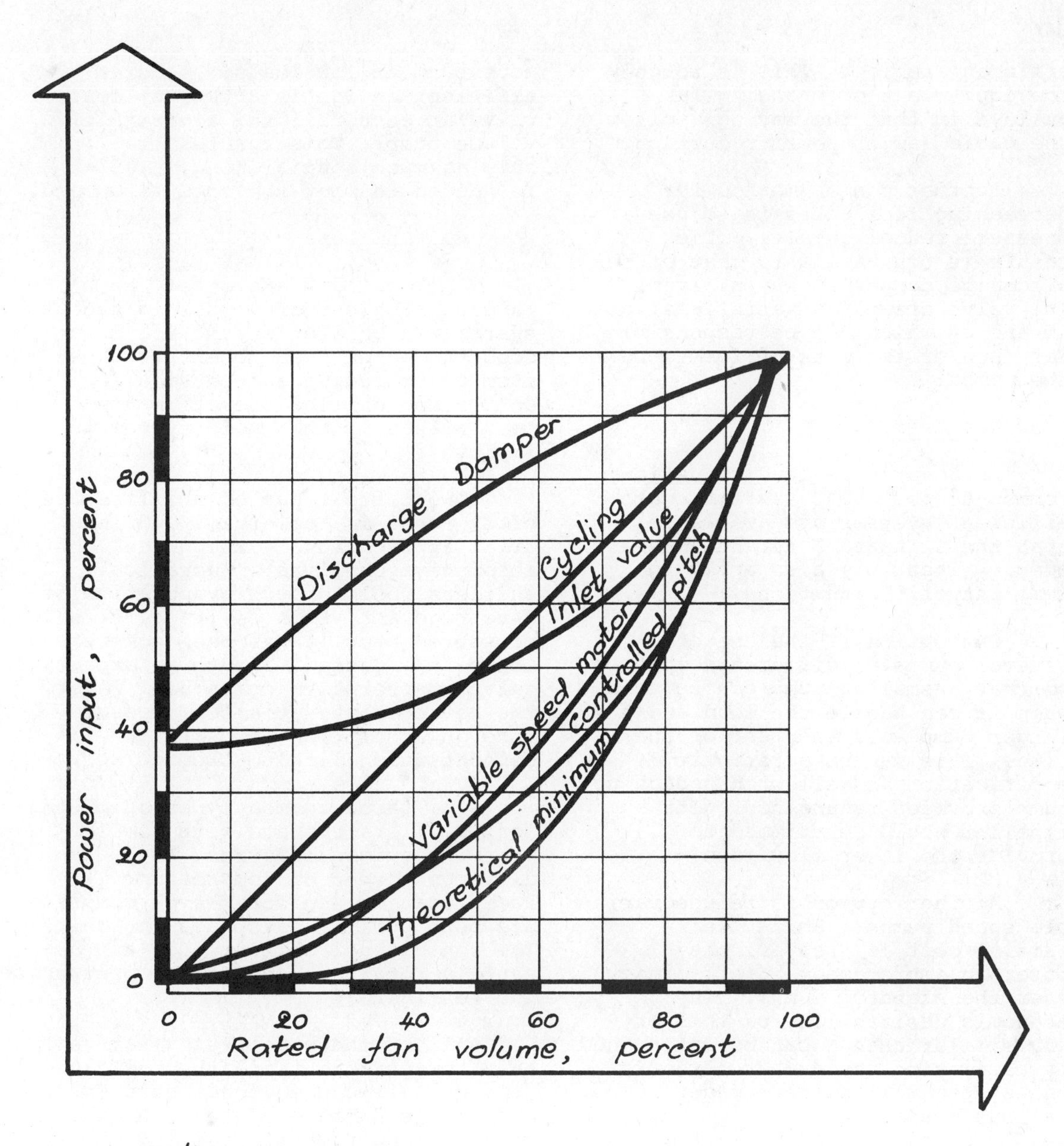

Reference 11

FAN POWER CONSUMPTION FOR VARIOUS TYPES OF PART LOAD CONTROLS

FIGURE 8.2

efficient region. This is roughly the equivalent of changing fan pulleys in that the savings follow the cubic law of power reduction.

Another common method for decreasing flow rates is to use a pressure reducing valve. The result is equivalent to that of the discharge damper shown in Figure 8.2. The valve creates an artificial use of energy which can be responsible for much of the work performed by the pumps.

New Design

In a variable load situation, common to most HVAC systems, more efficient systems are available than the standard constant volume pump. (These may also apply to some retrofit situations.)

One option is the use of several pumps of different capacity so that a smaller pump can be used when it can handle the load and a larger pump used the rest of the time. This can be a retrofit modification as well when a back-up pump provides redundancy. Its impeller would be trimmed to provide the lower flow rate.

Another option is to use variable speed pumps. While their initial cost is greater, they offer an improvement in efficiency over the standard pumps. The economic desirability of this or any similar change can be determined by estimating the number of hours the system will operate under various loads.

The results of combining these two techniques are shown for a hypothetical case in Figure 8.3. By using two variable speed pumps in parallel, a large (up to 50 percent) increase in efficiency is possible. An analysis of this type will reveal the most efficient pumping system for a new or existing HVAC system.

Most existing water systems, such as the chilled water or heating hot water loops, are designed with three-way by-pass valves for control at each coil. This results in a constant volume of water being pumped, most of which is not necessary at low loads. A more efficient design is a two-way control valve at each coil and a variable volume pump. This ensures that only as much conditioning water as is needed to meet the load is pumped.

C. Chillers

Chillers are often the largest single energy user in the HVAC system. A chiller's function is to cool the water used to extract heat from the building and outside air. By optimizing chiller operation the performance of the whole system is improved.

Two basic types of chillers are often found in commercial and industrial applications: mechanical and absorptive chillers. Mechanical chillers cool through evaporation of a refrigerant, such as freon, at a low pressure after it has been compressed, cooled and passed through an expansion valve. Absorptive units boil water, the refrigerant, at a low pressure through absorption into a high concentration lithium bromide solution.

The three common types of mechanical chillers operate with similar thermodynamic properties, but different types of compressors. Reciprocating and screw-type compressors are both positive displacement units. The centrifugal chiller uses a rapidly rotating impeller to pressurize the refrigerant.

All of these chillers must reject heat to a sink outside the building. Some use air-cooled condensers while most large units operate with evaporative cooling towers. Cooling towers have the advantage of rejecting heat to a lower temperature heat sink because the water approaches the ambient wet-bulb temperature while air-cooled units are limited to the dry-bulb temperature. This results in a higher condensing temperature, which lowers the efficiency of the chiller. Air-cooled condensers are used because they require much less maintenance than cooling towers.

Mechanical cooling can also be performed by Direct Expansion (DX) units. These are very similar to chillers except that they cool the air directly instead of using the

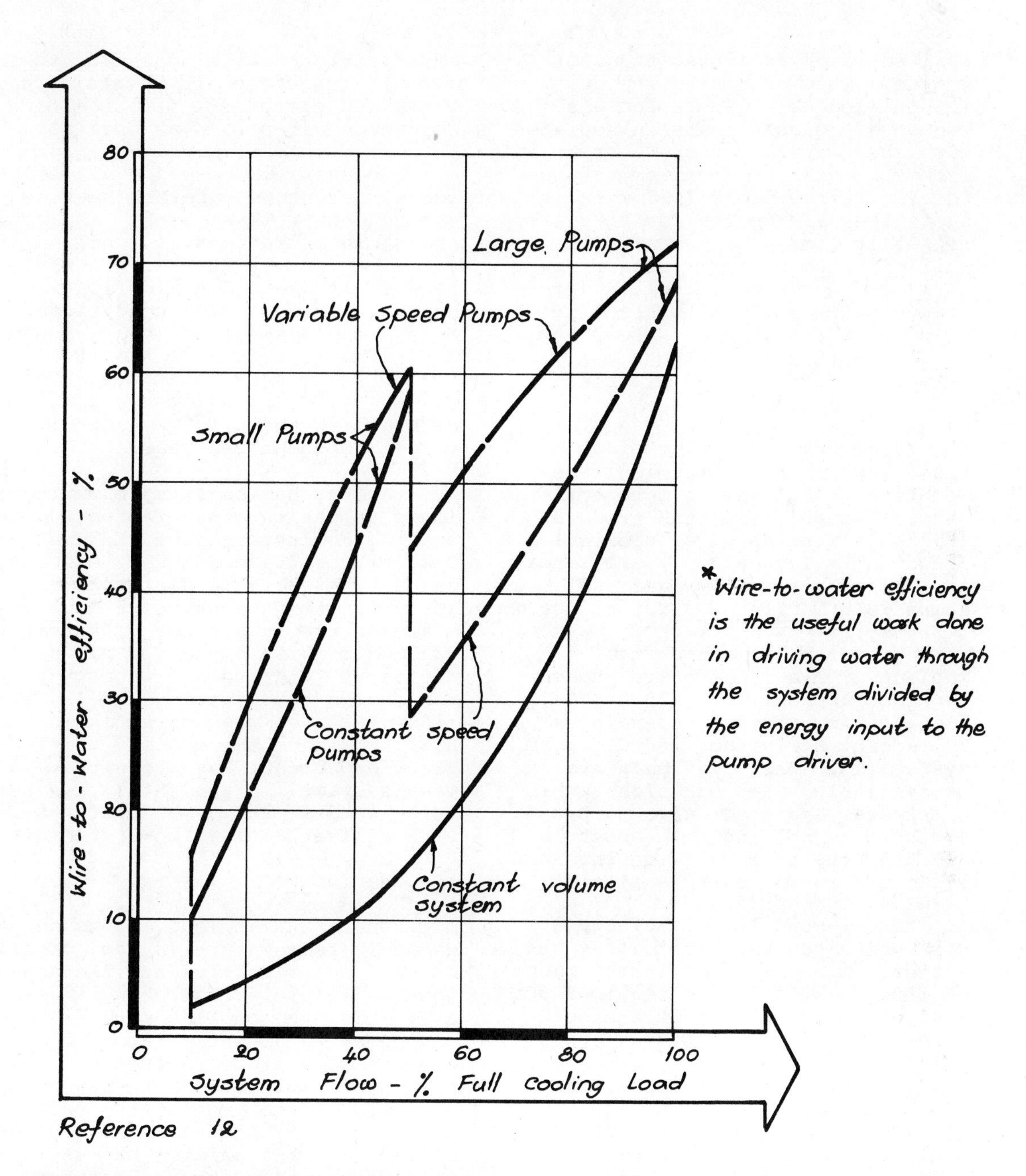

WIRE-TO-WATER EFFICIENCY CURVES FOR SAMPLE CHILLED WATER PUMPING SYSTEMS*

FIGURE 8.3

chilled water as a heat transfer medium. They eliminate the need for chilled water pumps and also reduce efficiency losses associated with the transfer of the heat to and from the water. DX units must be located fairly close (∿30m) to the ducts they are cooling so they are typically limited in size to the cooling required for a single air handler. A single large chiller, however, can serve a number of distributed air handlers. Where the air handlers are located close together, it can be more efficient to use a DX unit.

From the standpoint of the first law of thermodynamics, the compressive chiller is more efficient. Roughly 1.0 kWhe produces 3.5 kWh (12,000 Btu) of cooling. (1 ton hr = 12,000 Btu.) To generate this electricity requires about 10.8 MJ (10,200 Btu) of fuel.

To do the same amount of cooling with an absorptive process requires approximately 30.1 MJ (28,500 Btu) of fuel, assuming a 70 percent efficiency in the steam system.[13] Where 125 psig steam is available, a combination turbine compressor and absorptive unit can do the same cooling with about 18 MJ (17,000 Btu) of fuel, with the same steam generating efficiency.[14]

The second law efficiency of absorptive cooling is a different matter. A heat source in the range of 100 °C (∿200 °F) is required for a single stage unit. If this can be derived from some source other than direct combustion, such as waste heat, there can be an efficiency advantage to using absorption. However, utilizing waste heat sources at this temperature usually involves capital investment and additional operating expense.

Controls

Mechanical chillers operate on a principle similar to the heat pump. The objective is to remove heat from a low temperature building and deposit it in a higher temperature atmosphere. The lower the temperature rise which the chiller has to face, the more efficiently it will operate. It is useful, therefore, to maintain as warm a chilled water loop and as cool a condenser water loop as possible.

Theoretical power requirements for different evaporator (heat source) and condenser (heat sink) temperatures are shown in Table 8.4. Chilled water may be set at 4-7 °C (40-45 °F) to provide adequate dehumidification during extreme summer conditions. During the rest of the year, however, a setting of around 10 °C (50 °F) is adequate for many buildings. The optimal point may be different for each type of building and climate, and will vary throughout the year.

Energy can be saved by using lower temperature water from the cooling tower to reject the heat. However, as the condenser temperature drops, the pressure differential across the expansion valve drops, starving the evaporator of refrigerant. Many units with expansion valves, therefore, operate at a constant condensing temperature, usually 41 °C (105 °F), even when more cooling is available from the cooling tower. Field experience has shown that in many systems, if the chiller is not fully loaded, it can be operated with a lower cooling tower temperature.

Retrofit

Where a heat load exists and the wet-bulb temperature is low, cooling can be done directly with the cooling tower. If proper filtering is available, the cooling tower water can be piped directly into the chilled water loop. Often a direct heat exchanger between the two loops is preferred to protect the coils from fouling. Another technique is to turn off the chiller but use its refrigerant to transfer heat between the two loops. This "thermocycle" uses the same principle as heat pipes, and only works on chillers with the proper configuration.

A low wet-bulb temperature during the night can also be utilized. It requires a chiller which handles low condensing temperatures and a cold storage tank. This technique may become even more desirable as time-of-day or demand pricing for electricity increases.

	°C	°F	Kilowatt per ton of cooling*					
Evaporator Temperature	-12	10	.51	.62	.73	.90	.97	1.12 (3.1)†
	-7	20	.41	.51	.61	.72	.82 (4.3)†	.97
	-1	30	.32	.41	.51	.61 (5.8)†	.72	.82
	4	40	.23	.31	.40 (8.8)†	.50	.60	.71
	10	50	.15	.22 (16)†	.31	.40	.49	.60
	16	60	.07 (50)†	.15	.22	.31	.40	.48
Condenser Temperature		°F	70	80°	90°	100°	110°	120°
		°C	21	27	32	38	43	49

*1 ton of cooling is equivalent to 12,000 Btu/hr = 3.52 kW cooling.

†coefficient of performance = kW (cooling) per kW electrical.

Reference 15

THEORETICAL POWER REQUIREMENTS FOR COMPRESSIVE CHILLERS USING REFRIGERANT R-12

TABLE 8.4

New Design

In the purchase of a new chiller, an important consideration should be the load control feature. Since the chiller will be operating at partial load most of the time, it is important that it can do so efficiently. Figures 8.4, 8.5 and 8.6 show the efficiency of unloading techniques for reciprocating, centrifugal and absorptive chillers, respectively. Note in particular the artificial loading caused by the rather common "hot gas by-pass."

In addition to control of single units, it is sometimes desirable to use multiple compressor reciprocating chillers. This allows some units to be shut down at partial load. The remaining compressors operate near full load, usually more efficiently.

D. Heat Pumps

The mechanical operation of a heat pump is identical to that of a chiller except that the heat pump usually removes heat from the outside to use it inside the building. The coefficient of performance (COP) of typical heat pumps ranges from two to five. That is, for every joule (Btu) of electric energy input, from two to five joules (Btus) of heat are transferred. For this reason heat pumps operate more efficiently than electrical resistance heating.

The efficiency of heat pumps drops off as the temperature differential between the heat source and sink increases. This means that if outside air is used as the heat source, heat is the most difficult to get when it is the most needed. Most systems using heat pumps also use electric resistance heating as a backup. On a season-averaged basis this results in fairly efficient fuel use. However the peak capacity which must be supplied by the utility will be just as great. From a utility viewpoint it might be advantageous to reduce such peaks by load control or customer heat storage which would permit off-peak generation.

In some buildings a simultaneous demand (e.g., at the same time on a given day) for heating and cooling may exist, making it practical to use a heat pump to supply both. Note, however, that this simultaneous demand may be a result of inefficient system design. Before installing a heat pump for this purpose it should be determined that there actually is a need for simultaneous heating and cooling and that this need cannot be met using outside air for cooling.

Commonly, in commercial and industrial buildings, a convenient source of heat for a heat pump is the building exhaust air. This is a constant source of warm air available throughout the heating season. A typical heat pump design could generate hot water for space heating from this source at around 32-35 °C (90 to 95 °F). Heat pumps designed specifically to use building exhaust air can reach 66 °C (150 °F).

Another application of the heat pump is a continuous loop of water traveling throughout the building with small heat pumps located in each zone. Each small pump can both heat and cool, depending upon the needs of the zone. This system can be used to transfer heat from the warm side of a building to the cool side. A supplemental cooling tower and boiler are included in the loop to compensate for net heating or cooling loads.

A double bundle condenser can be used as a retrofit design for a centralized system. (See Figure 8.7.) This creates the option of pumping the heat either to the cooling tower or into the heating system hot duct. Also shown in Figure 8.7 is a circuit for recovering heat from the building exhaust or from the cooling system (cold duct).

Some chillers can be retrofitted to act as heat pumps. Centrifugal chillers will work much more effectively with a heat source warmer than outside air (exhaust air, for example). The compression efficiency of centrifugals falls off as the evaporator temperature drops.

Because they are positive displacement machines, reciprocating and screw-type compressors operate more effectively at lower evaporator

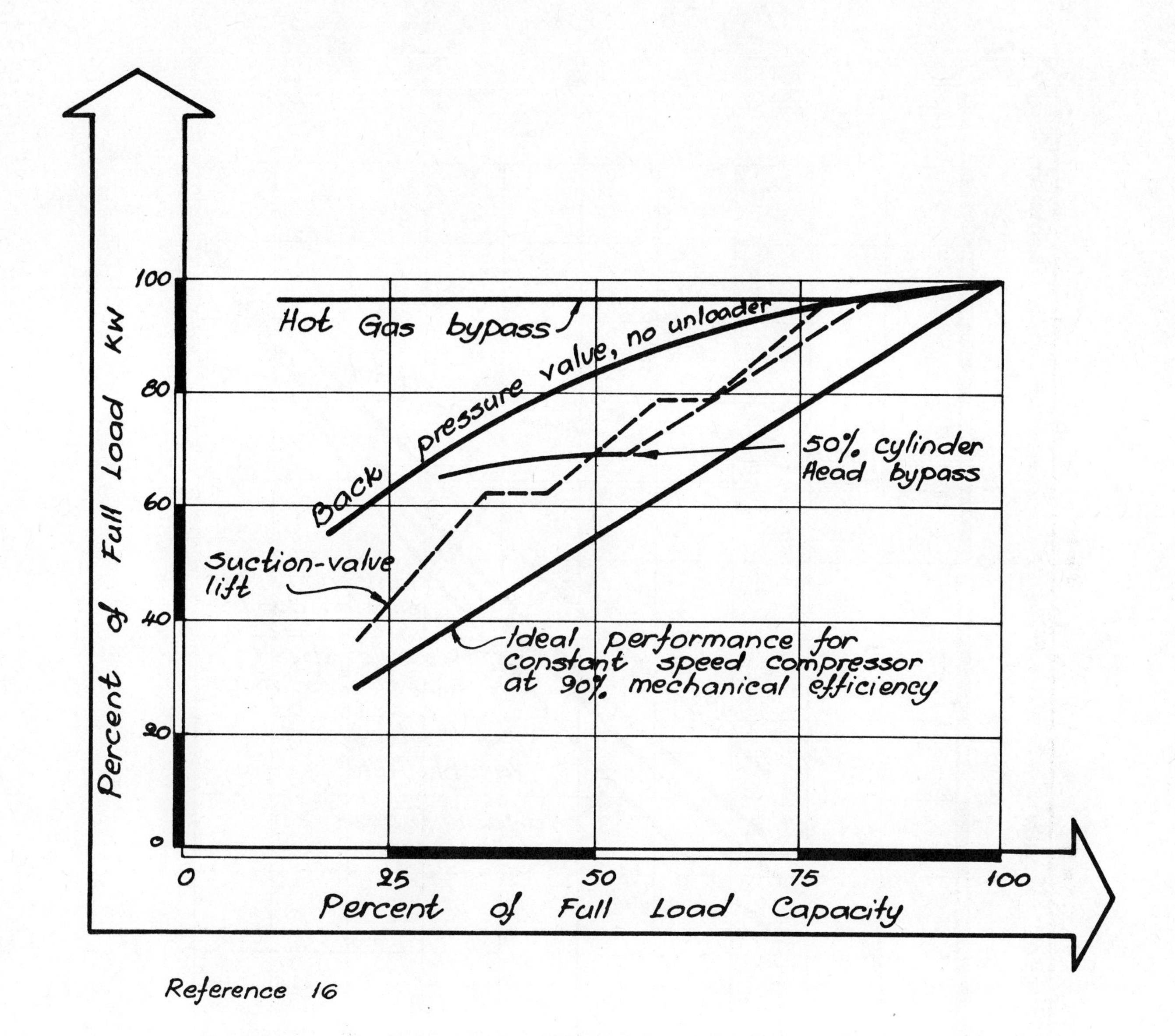

Reference 16

POWER SAVING CHARACTERISTICS OF TYPICAL RECIPROCATING CHILLER UNLOADING DEVICES

FIGURE 8.4

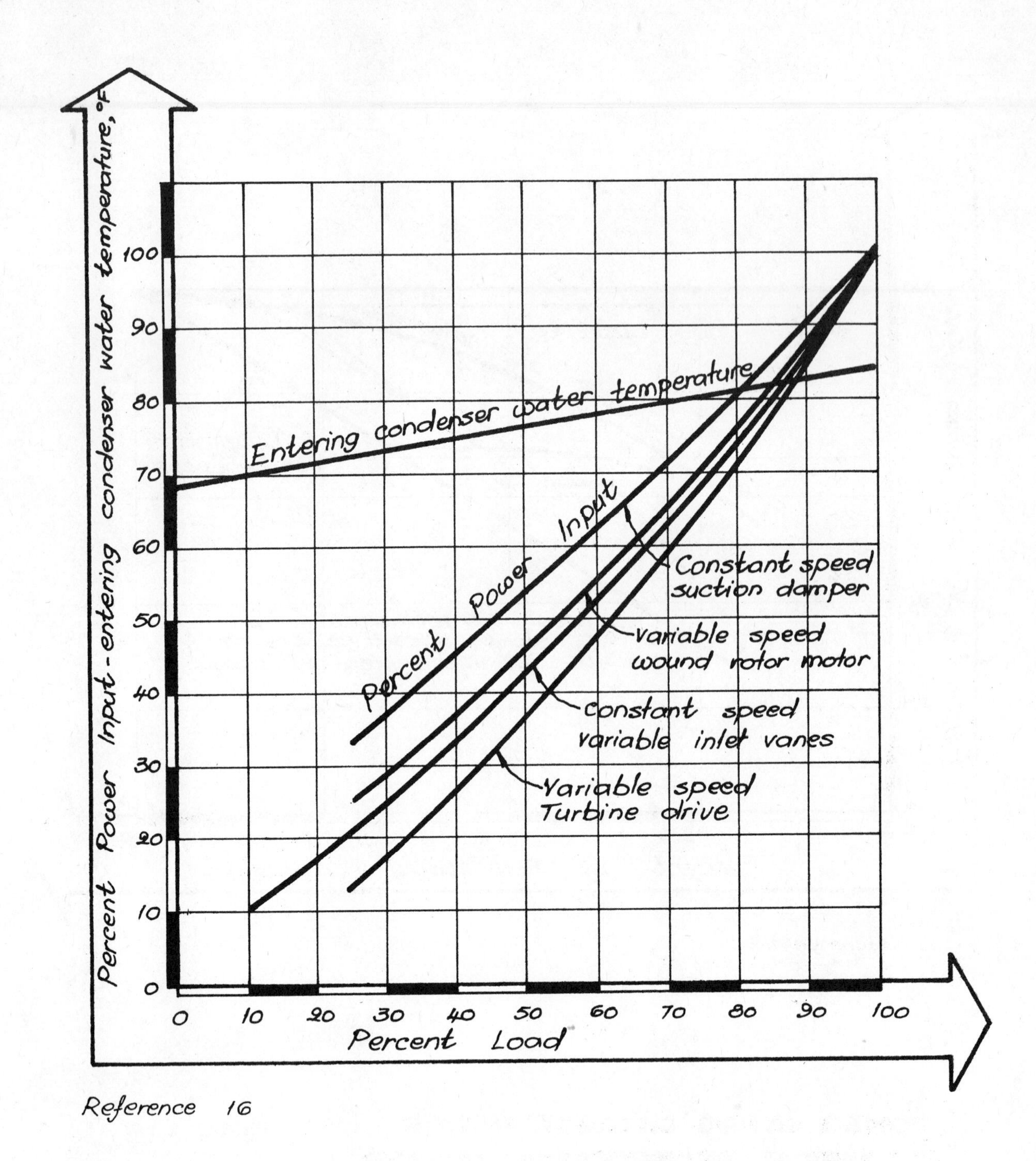

Reference 16

POWER SAVING CHARACTERISTICS OF TYPICAL CENTRIFUGAL COMPRESSOR UNLOADING DEVICES **FIGURE 8.5**

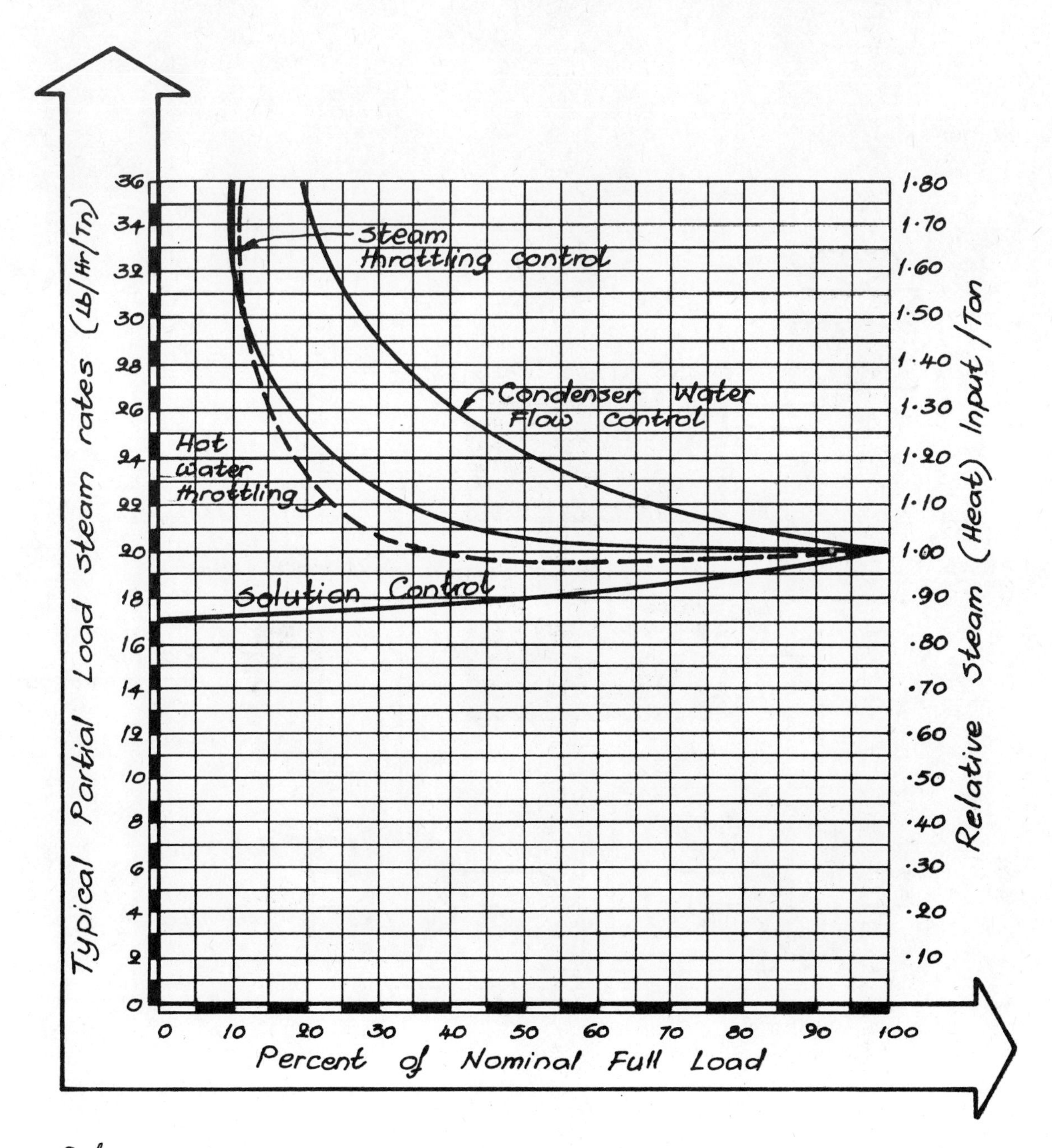

Reference 16

STEAM SAVING CHARACTERISTICS OF TYPICAL ABSORPTIVE CHILLER

FIGURE 8.6

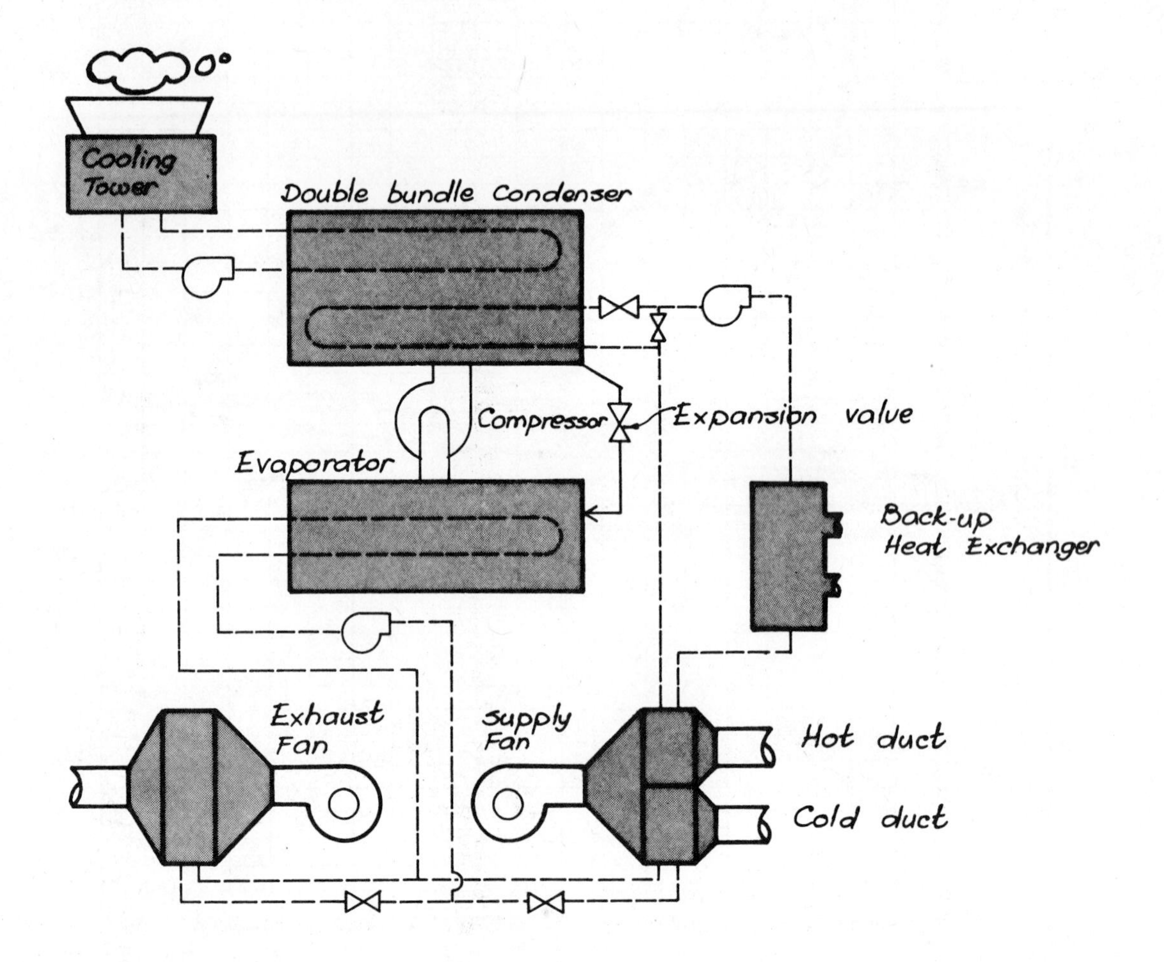

HEAT PUMP SCHEMATIC USING DOUBLE BUNDLE CONDENSER AND EXHAUST AIR HEAT RECOVERY FIGURE 8.7

temperatures. They can be used to transfer heat across a larger temperature differential. Multistage compressors increase this capacity even further.

E. Heating Equipment

Electric space heating is discussed above, under Heat Pumps. For information pertaining to efficient fossil fuel heating (e.g., coal- or oil-fired boilers), refer to Chapter 10 or to the technical literature.

F. Ducting/Dampers

Control

In HVAC systems using dual ducts, static pressure dampers are often placed near the start of the hot or cold plenum run. They control the pressure throughout the entire distribution system and can be indicators of system operation. Often in over-designed systems, the static pressure dampers may never open more than 25 percent. Fan pulleys can be changed to slow the fan and open the dampers fully, eliminating the previous pressure drop. The same volume of air is delivered with a significant drop in fan power.

Retrofit

Other HVAC systems use constant volume mixing boxes for balancing which create their own pressure drops as the static pressure increases. An entire system of these boxes could be overpressurized by several inches of water without affecting the air flow, but the required fan power would increase. (One inch of water pressure is about 250 N/m^2 or 250 Pa.) These systems should be monitored to ensure that static pressure is controlled at the lowest required value. It may also be desirable to replace the constant volume mixing boxes with boxes without volume control to eliminate their minimum pressure drop of approximately one inch of water. In this case, static pressure dampers will be necessary in the ducting.

Leakage in any dampers can cause a loss of hot or cold air. Neoprene seals can be added to blades to slow leakage considerably. If a damper leaks more than 10 percent, it can be less costly to replace the entire damper assembly with effective positive-closing damper blades rather than to tolerate the loss of energy.[11]

New Design

In the past small ducts were installed because of their low initial cost despite the fact that the additional fan power required offset the initial cost on a life cycle basis. ASHRAE 90-75 provides a guideline to reduce this additional use of electricity by defining an air transport factor which may not be less than 4.0.

$$\text{Air Transport Factor} = \frac{\text{Space Sensible Heat Removal}^*}{(\text{Supply + Return Fan(s) Power Input})^*}$$

This sets a maximum limit on the fan power that can be used for a given cooling capacity.[8] As a result the air system pressure drop must be low enough to permit the desired air flow. In small buildings this pressure drop is often largest across filters, coils and registers. In large buildings the duct runs may be responsible for a significant fraction of the total static pressure drop, particularly in high velocity systems.

8.5 SYSTEMS

The use of efficient equipment is only the first step in the optimum operation of a building. Equal emphasis should be placed upon the combination of elements in a system and the control of those elements.

A. Control

The HVAC control point most people are familiar with is the room thermostat. In small systems the room thermostat may exert sufficient

*Expressed in either Btu/h or W.

control over the energy use of the components. In medium and large systems, however, the effect of room thermostats decreases and if they are not coordinated with control of the actual system, they may have only a small effect on efficient energy use. In systems using terminal reheat, it is possible that adjusting the thermostat to a warmer setting during the summer will result in a higher use of energy. In general, however, higher thermostat settings in the summer and lower settings in winter do reduce energy use.

A more critical control point is in the mechanical room where the operation of chillers, boilers, pumps and fans is controlled. The type of changes necessary for efficient operation is similar for most systems. The major objective is to do as little mechanical heating and cooling as possible.

Many systems use a combination of hot and cold to achieve moderate temperatures. Included are dual duct, multizone, and terminal reheat systems, and some induction, variable air volume and fan coil units. Whenever combined heating and cooling occurs, the temperatures of the hot and cold ducts or water loops should be brought as close together as possible, while still maintaining building comfort.

This can be accomplished in a number of ways. Hot and cold duct temperatures are often reset on the basis of the temperature of the outside air or the return air. A more complex approach is to monitor the demand for heating and cooling in each zone. For example, in a multizone building, the demand of each zone is transferred back to the supply unit by electric or pneumatic signals. At the supply end hot and cold air are mixed in proportion to this demand. The cold air temperature should be just low enough to cool the zone calling for the most cooling. If the cold air were any colder, it would be mixed with hot air to achieve the right temperature. This creates an overlap in heating and cooling not only for that zone but for all the zones because they would all be mixing in the colder air.

If no zone calls for total cooling, then the cold air temperature can be increased gradually until the first zone requires full cooling. At this point the minimum cooling necessary for that multizone configuration is performed. The same operation can be performed with the hot air temperature until the first zone is calling for heating only.

Note that simultaneous heating and cooling is still occurring in the rest of the zones. This is not an ideal system but it is a first step in improving operating efficiency.

The technique for resetting hot and cold duct temperatures can be extended to the other systems which have been mentioned. It may be performed automatically with pneumatic or electric controls, or manually. In some buildings it will require the installation of more monitoring equipment (usually only in the zones of greatest demand) but the expense should be relatively small and the payback period short.

Nighttime temperature setback is another control option which can save energy without significantly affecting the comfort level. Energy is saved by shutting off or cycling fans. Building heat loss may also be reduced because the building is cooler and no longer pressurized.

In moderate climates complete night shutdown can be used with a morning warm-up period. In colder areas where the average night temperature is below 4 °C (40 °F), it is usually necessary to provide some heat during the night. Building setback temperature is partially dictated by the capacity of the heating system to warm the building in the morning. In some cases it may be the mean radiant temperature of the building rather than air temperature which determines occupant comfort.

Some warm-up designs use "free" heating from people and lights to help attain the last few degrees of heat. This also provides a transition period for the occupants to adjust from the colder outdoor temperatures.

In some locations during the summer, it is desirable to use night air for a cool-down period. This "free cooling" can decrease the temperature of the building mass which accumulates heat during the day. In some buildings with high heat content (such as libraries or buildings with thick walls), a long period of night cooling may decrease the building mass temperature by a degree or two. This represents a large amount of cooling that the chiller will not have to perform the following day.

B. Retrofit

Retrofitting HVAC systems may be an easy or difficult task depending upon the possibility of using existing equipment in a more efficient manner. Often retrofitting involves control or ducting changes which appear relatively minor but will greatly increase the efficiency of the system. Some of these common changes, such as decreasing air flow, are discussed elsewhere in this chapter. This section will describe a few changes appropriate to particular systems.

Both dual duct and multizone systems mix hot and cold air to achieve the proper degree of heating or cooling. In most large buildings the need for heating interior areas is essentially nonexistent, due to internal heat generation. A modification which adjusts for this is simply shutting off air to the hot duct. The mixing box then acts as a variable air volume box, modulating cold air according to room demand as relayed by the existing thermostat. (It should be confirmed that the low volume from a particular box meets minimum air requirements.)

Savings from this modification come mostly from the elimination of simultaneous heating and cooling. Since fans in these systems are likely to be controlled by static pressure dampers in the duct after the fan, they do not unload very efficiently and represent only a small portion of the savings.

<u>Economizer System/Enthalpy Control</u>

The economizer cycle is a technique for introducing varying amounts of outside air to the mixed air duct. Basically it permits mixing warm return air at ∿24 °C(75 °F) with cold outside air to maintain a preset temperature in the mixed air plenum (typically 10-15 °C, 50-60 °F). When the outside temperature is slightly above this setpoint, 100 percent outside air is used to provide as much of the cooling as possible. During very hot outside weather, minimum outside air will be added to the system. A schematic of the system is shown in Figure 8.8. Recently, improvements have been introduced for economizer systems in dual fan plenums. For details refer to Case Study 8.2.

A major downfall of economizer systems is poor maintenance. The failure of the motor or dampers may not cause a noticeable comfort change in the building because the system is often capable of handling the additional load. Since the problem is not readily apparent, corrective maintenance may be put off indefinitely. In the meantime the HVAC system will be working harder than necessary, wasting energy and money. A continual maintenance program is necessary for any economizer installation.

Typically economizers are controlled by the dry-bulb temperature of the outside air rather than its enthalpy (actual heat content). This is adequate most of the time, but can lead to unnecessary cooling of air. When enthalpy controls are used to measure wet-bulb temperatures, this cooling can be reduced. However, enthalpy controllers are more expensive and less reliable.

The rules which govern the more complex enthalpy controls for cooling-only applications are as follows:[17]

- When outside air enthalpy is greater than that of the return air or when outside air dry-bulb temperature is greater than that of the return air (Region III in Figure 8.9), use minimum outside air.

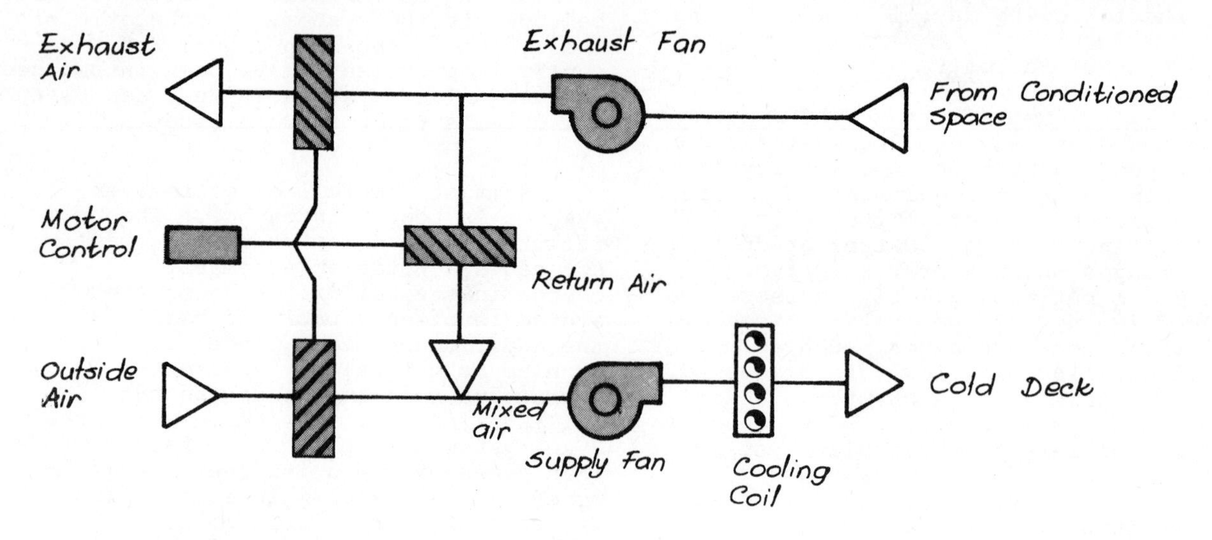

SCHEMATIC OF THE ECONOMIZER SYSTEM

FIGURE 8.8

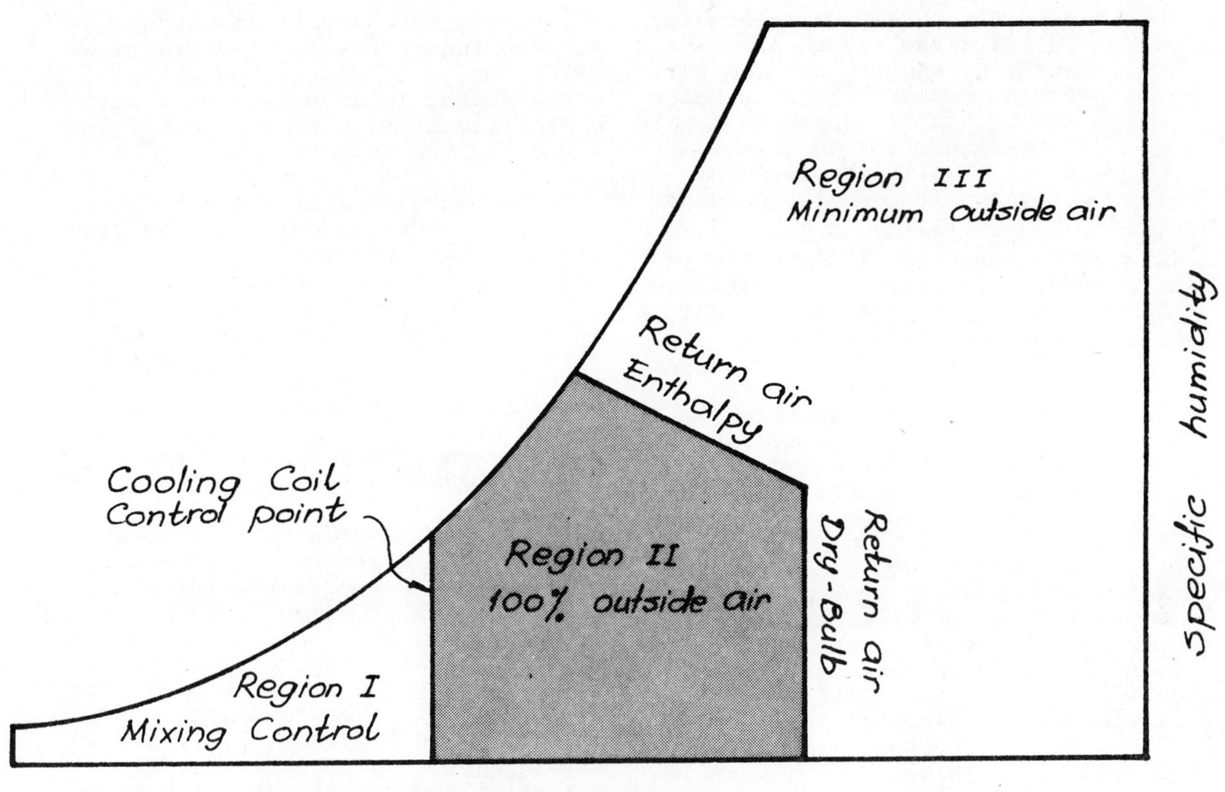

Reference 17

ENTHALPY OPTIMIZATION LOGIC FOR OUTSIDE AIR FIGURE 8.9

- When the outside air enthalpy is below the return air enthalpy and the outside air dry-bulb temperature is below the return air dry-bulb temperature but above the cooling coil control point (Region II), use 100 percent outside air.
- When outside air enthalpy is below the return air enthalpy and the outside air dry-bulb temperature is below the return air dry-bulb temperature and below the cooling coil controller setting (Region I), the return and outside air are mixed by modulating dampers according to the cooling set point.

These points are valid for the majority of cases. When mixed air is to be used for heating and cooling, a more intricate optimization plan will be necessary, based on the value of the fuels used for heating and cooling.

C. New Design

In the selection of systems for new buildings, the foundation has been laid by ASHRAE Standard 90-75, *Energy Conservation in New Building Design*. With the assistance of computer simulation the engineer can go much further in fine tuning the design of a building for efficient energy use (refer to Chapter 5). Some of the concepts which the designer should consider in a new system are the following:

Variable Speed Fans
Exhaust Air Heat Recovery
Solar Heating/Cooling
Split Duct Economizer
Heat/Cool Storage
Early Morning Warm-Up/Cool-Down
Evaporative Cooling
Natural Ventilation
Variable Volume Air Flow
Variable Speed Pumps
Through the Lights Return Air

8.6 POTPOURRI

A. Solar Heating/Cooling

Active use of the sun's energy for space heating is a relatively new concept in the US but is becoming increasingly attractive as the costs of fossil fuels rise. It appears that solar energy will be a significant supplemental energy source in the future.

The actual collection of solar energy is performed with either flat plate (lower temperatures) or concentrating (higher temperatures) collectors. The heat transfer medium is commonly air, water or a water-glycol solution. Storage is performed in rock beds, aluminum oxide, water tanks or beds of eutectic salts (phase change material). A comparison of the storage volume required for these systems is shown in Figure 8.10.

The most common type of solar heating is a flat-plate warm-water system using temperatures in the range of 38-93 °C (100 to 200 °F). A typical system schematic is shown in Figure 8.11. Another concept is the use of stored solar heat in the evaporator of a heat pump, as diagrammed in Figure 8.12. This allows very high COPs for normal operation and permits the heat storage medium, which ordinarily might not be warm enough, to act as a heat source for the building.

Solar cooling can be assigned to three major categories: night radiation, compressive chilling and absorptive chilling. Night radiation simply involves operating the solar collectors at night so that their warmth is radiated to the sky. The resulting "cool" can be stored in one of the systems already described. (Note that phase change material can be used for storage at only one temperature so that the same material cannot be used for both heat and cool storage.)

Most compressive chilling systems are in the development stage. They operate by using the expansion of a refrigerant to operate a turbine or positive displacement drive for a compressor. The compressor is then used with another refrigerant as in a normal air-conditioning system.

Absorptive chilling has been used for many years where a source of heat was inexpensive, or where steam or another moderately high temperature (>93 °C,>200 °F) source was available

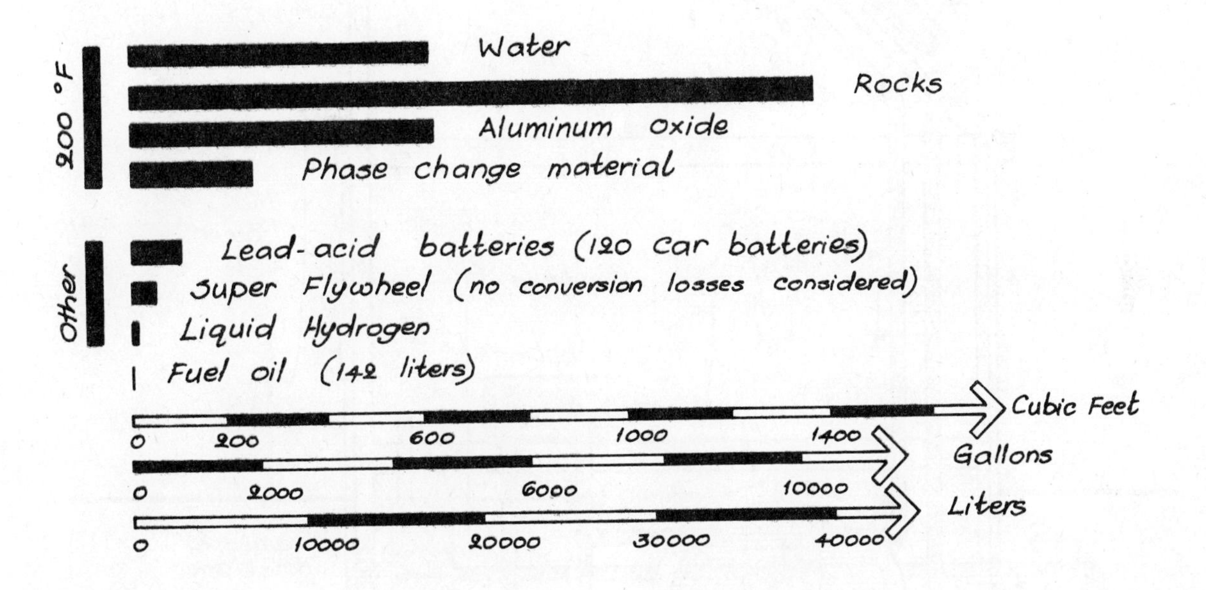

Reference 18

ENERGY STORAGE VOLUME REQUIRED BY A TYPICAL HOUSE FOR THREE DAYS AT −7°C (20°F) (OR 5.3 GJ, 5 MBTu)

FIGURE 8.10

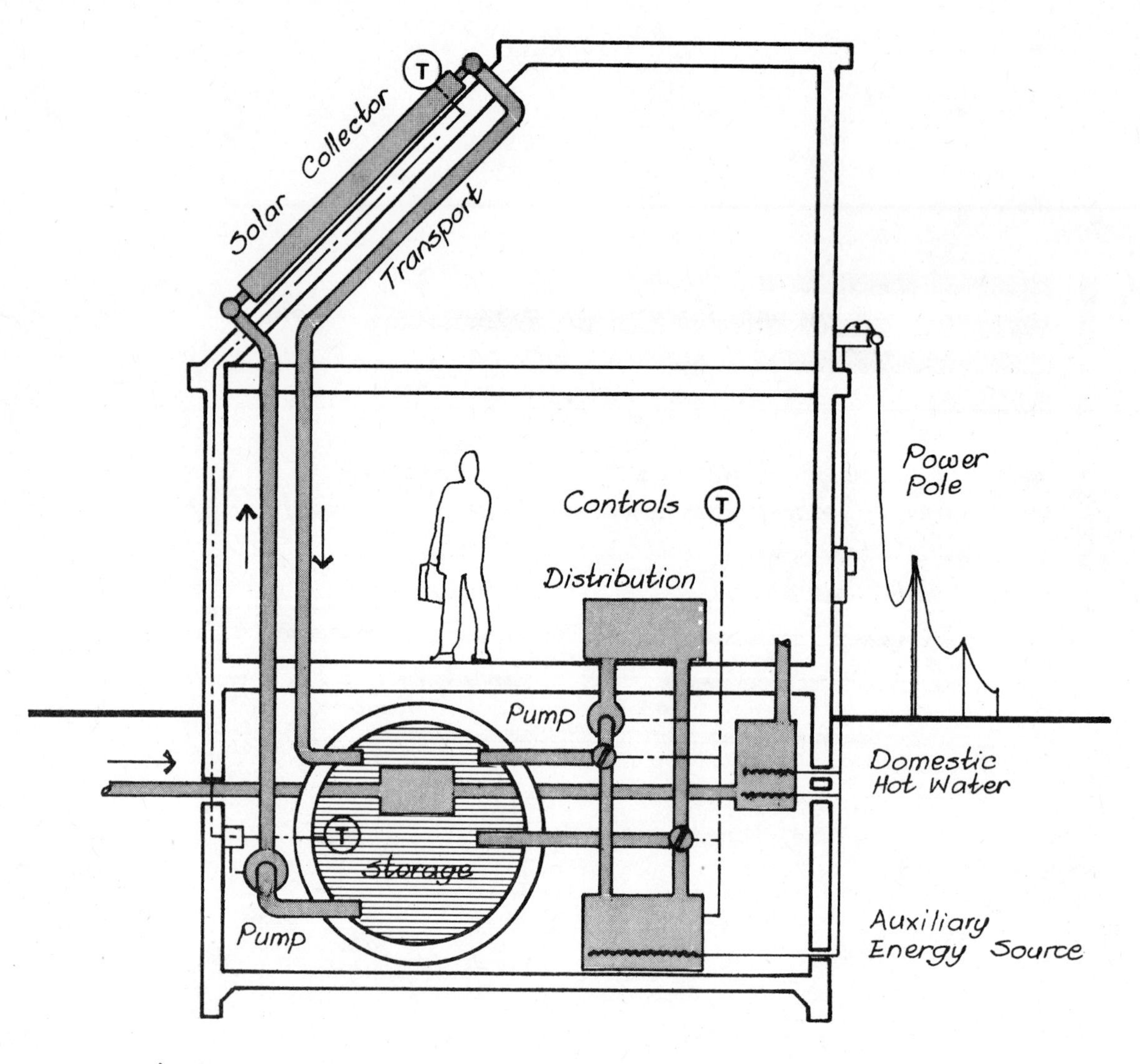

Reference 18

SCHEMATIC DIAGRAM OF WARM-WATER FLAT-PLATE SYSTEM FIGURE 8.11

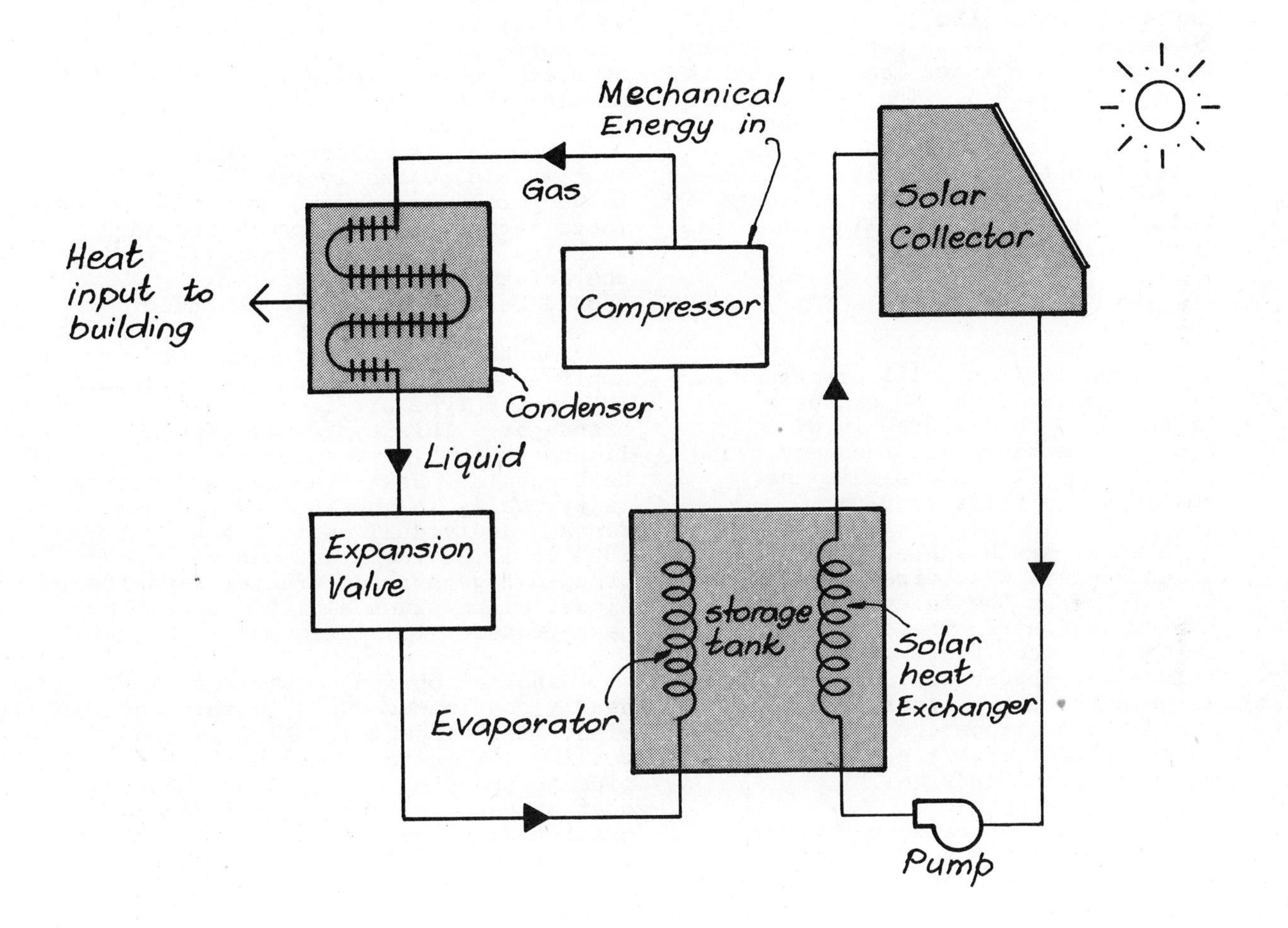

SOLAR ASSISTED HEAT PUMP SCHEMATIC

FIGURE 8.12

to heat the generator. In solar applications the temperature of collected heat is usually less than this. As a result, if an existing absorptive unit is solar powered, it will supply roughly 40 percent of the cooling capacity of its original rating.

Manufacturers also build absorptive units designed specifically to handle the lower input temperatures. Their capacity is very dependent upon inlet temperature, as seen in Table 8.5. The difficulty in collecting these higher temperatures efficiently and the higher initial cost of absorptive chillers imply that it may be several years before active solar cooling is economically attractive.

B. Heat Recovery

Heat recovery is often practiced in industrial processes which involve high temperatures. It can also be employed in HVAC systems. Active heat recovery using a heat pump is discussed in the heat pump retrofit section.

Other systems are available which operate with direct heat transfer from the inlet air to the exhaust air. These are most reasonable when there is a large volume of exhaust air, for example in once-through systems, and when weather conditions are not moderate. An example of such a system can be found in Case Study 8-1.

Common heat recovery systems are broken down into two types, regenerative and recuperative. Regenerative units use alternating air flow from the hot and cold stream over the same heat storage/transfer medium. This flow may be reversed by dampers or the whole heat exchanger may rotate between streams. Recuperative units involve continuous flow; the emphasis is upon heat transfer through a medium with little storage.

The rotary regenerative unit, or heat wheel, is one of the most common heat recovery devices. It contains a corrugated or woven heat storage material which gains heat in the hot stream, as can be seen in Figure 2.13. (See Chapter 2.) This material is then rotated into the cold stream where the heat is given off again. The wheels can be impregnated with a dessicant to transfer latent as well as sensible heat. Purge sections for HVAC applications can reduce carry-over from the exhaust stream to acceptable limits for most installations, as little as 0.04 percent in some units.[20] For more detail on the various types of heat wheels, please refer to Chapter 10.

The heat transfer efficiency of heat wheels generally ranges from 60 to 85 percent depending upon the installation, type of media and air velocity. For easiest installation the intake and exhaust ducts should be located near each other.

Another system which can be employed with convenient duct location is a plate type air-to-air heat exchanger. This system is usually lighter though more voluminous than heat wheels. Heat transfer efficiency is typically in the 60 to 75 percent range. Individual units range from 1000 to 11,000 SCFM and can be grouped together for greater capacity. Almost all designs employ counterflow heat transfer for maximum efficiency.

Another option to consider for nearly contiguous ducts is the heat pipe. This is a unit which uses a boiling refrigerant within a closed pipe to transfer heat. Since the heat of vaporization is utilized, a great deal of heat transfer can take place in a small space. The heat pipe is illustrated in Figure 2.15 in Chapter 2.

Heat pipes are often used in double wide coils which look very much like two steam coils fastened together. The amount of heat transferred can be varied by tilting the tubes to increase or decrease the flow of liquid through the capillary action. Heat pipes can not be "turned off" so by-pass ducting is often desirable. The efficiency of heat transfer ranges from 55 to 75 percent depending upon the number of pipes, fins per inch, air face velocity, etc.

Hot Water Flow . . . . 41.6 L/min (11.0 GPM)
Condensing Water Flow . . . . 45.4 L/min (12.0 GPM)
Chilled Water Flow . . . . 27.3 L/min (7.2 GPM)
Chilled Water Leaving Temperature . . . 7.2 °C (45 °F)

Inlet Temp.		Outlet Temp.		Energy Input		Inlet Cond. Water		Delivered Capacity			Rejected Heat	
°C	°F	°C	°F	kW	Btu/hr	°C	°F	kW	Btu/hr	Tons	kW	Btu/hr
		75	167.0	4.80	16,400	26.7	80	2.84	9,700	0.81	7.64	26,100
76.7	170	75.2	167.4	4.25	14,500	29.4	85	1.87	6,400	0.53	6.12	20,900
			*		*	32.2	90		*	*		*
		77.1	170.7	6.97	23,800	26.7	80	5.07	17,300	1.44	12.04	41,100
79.4	175	77.3	171.1	6.33	21,600	29.4	85	3.84	13.100	1.09	10.16	34,700
			*		*	32.2	90		*	*		*
		79.1	174.3	9.14	31,200	26.7	80	7.15	24,400	2.03	16.28	55,600
82.2	180	79.3	174.8	8.43	28,800	29.4	85	5.68	19,400	1.62	14.12	48,200
		79.8	175.7	6.97	23,800	32.2	90	4.16	14,200	1.18	11.13	38,000
		81.1	178.0	11.25	38,400	26.7	80	9.11	31,100	2.59	20.35	69,500
85.0	185	81.4	178.5	10.51	35,900	29.4	85	7.50	25,600	2.13	18.01	61,500
		81.9	179.4	8.96	30,600	32.2	90	5.65	19,300	1.61	14.61	49,900
		83.2	181.7	13.41	45,800	26.7	80	10.78	36,800	3.07	24.19	82,600
87.8	190	83.4	182.2	12.56	42,900	29.4	85	9.17	31,300	2.61	21.73	74,200
		84.0	183.2	10.98	37,500	32.2	90	6.97	23,800	1.98	17.95	61,300
		85.2	185.3	15.55	53,100	26.7	80	11.89	40,600	3.38	27.44	93,700
90.6	195	85.5	185.9	14.64	50,000	29.4	85	10.54	36,000	3.00	25.19	86,000
		86.1	186.9	12.07	44,300	32.2	90	8.08	27,600	2.30	21.06	71,900
		87.4	189.3	17.22	58,800	26.7	80	12.24	41,800	3.48	29.46	100,600
93.3	200	87.7	189.8	16.40	56,000	29.4	85	11.77	40,200	3.35	28.17	96,200
		88.2	190.7	14.94	51,000	32.2	90	8.93	30,500	2.54	23.58	80,500
		89.7	193.4	18.69	63,800	26.7	80	12.30	42,000	3.50	30.99	105,800
96.1	205	89.9	193.9	17.81	60,800	29.4	85	12.30	42,000	3.50	30.11	102,800
		90.4	194.8	16.46	56,200	32.2	90	9.52	32,500	2.71	25.98	88,700

*Unit operation is unstable in these areas.

Reference 19

CAPACITY OF SOLAR HEATED ABSORPTIVE CHILLER VERSUS INPUT TEMPERATURES

TABLE 8.5

Runaround systems are also popular for HVAC applications, particularly when the supply and exhaust plenums are not physically close. Runaround systems involve two coils (air-to-water heat exchangers) connected by a piping loop of water or glycol solution and a small pump. (See Figure 8.13.) The glycol solution is necessary if the air temperatures in the inlet coils are below freezing. The exact solution may vary, as seen in Figure 8.14.

Standard air-conditioning coils can be used for the runaround system and some equipment manufacturers supply computer programs for size optimization. Precaution should be used when the exhaust air temperature drops below 0 °C (32 °F) which would cause freezing of the condensed water on its fins. A three-way by-pass valve will maintain the temperature of the solution entering the coil at just above 0 °C (32 °F).

The heat transfer efficiency of this system ranges from 60 to 75 percent depending upon the installation. An example of runaround coil design can be seen in Case Study 8-1.

Another system similar to the runaround in layout is the dessicant spray system. Instead of using coils in the air plenums, it uses spray towers. The heat transfer fluid is a dessicant (lithium chloride) which transfers both latent and sensible heat -- desirable in many applications. Tower capacities range from 7700 to 92,000 SCFM; multiple units can be used in large installations. The enthalpy recovery efficiency is in the range of 60 to 65 percent.[20]

C. Computer Control of HVAC

Digital computers are becoming increasingly popular in the control of HVAC systems. They are very useful in some control aspects while in others they do not match the energy saving capabilities of more conventional controls.

Computer systems can be very effective, for example, in optimizing the start/stop timing of equipment. They also interface well with other functions such as building fire safety, security, and equipment maintenance.

A drawback to computer control is that the partial load operation of fans and pumps is usually controlled by cycling. This has been known to burn out large motors which were not designed to handle frequent starts. In addition, cycling does not result in the same energy savings that can be achieved by slowing fans down by changing pulleys or typical partial controls (see Figure 8.2).

Many HVAC changes can be made at low cost with existing controls such as time clocks and thermostats, as described in this chapter. Once these have been accomplished, a new energy use level will result. Only then should automated controls be considered, on the basis of what they can do to lower this energy use, after all of the basic changes have been made.

* * *

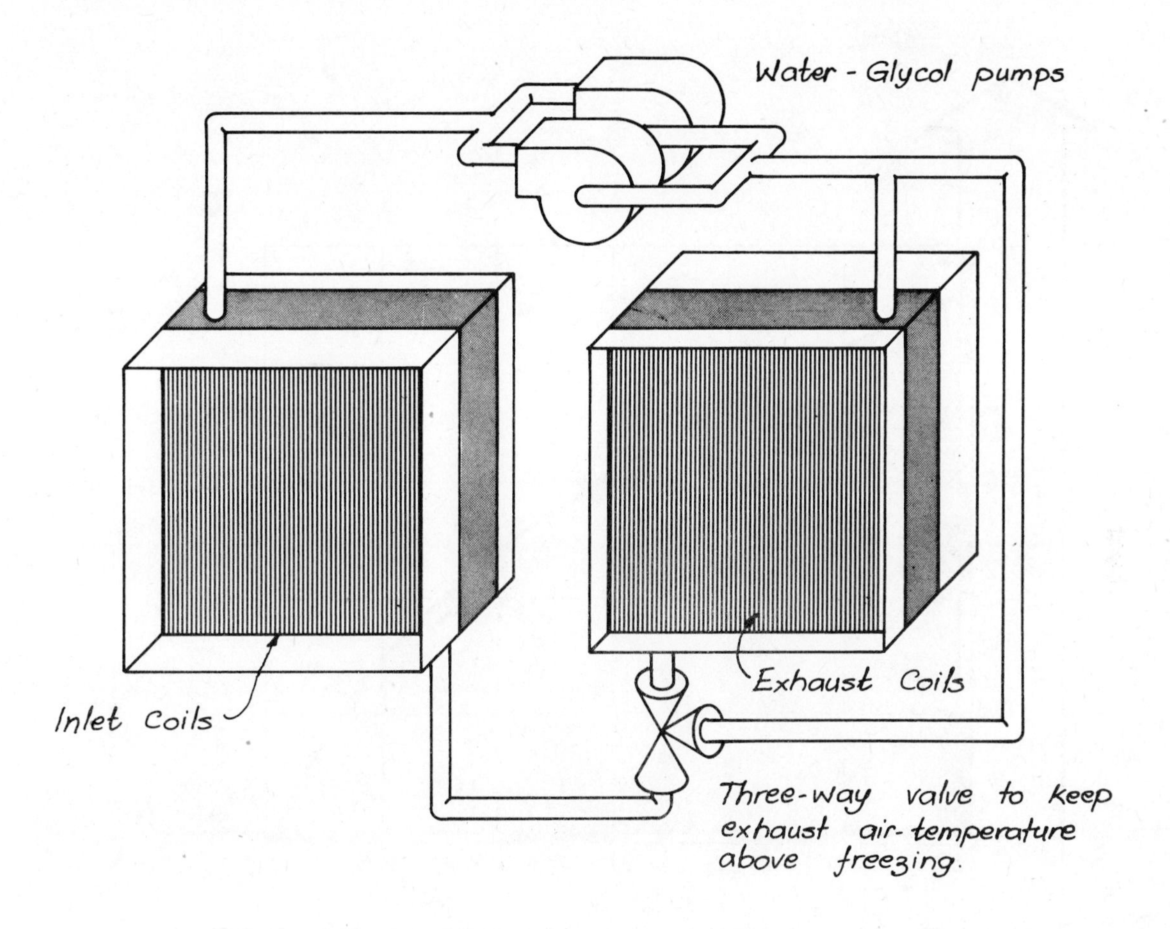

RUNAROUND COILS FOR HEAT RECOVERY

FIGURE 8.13

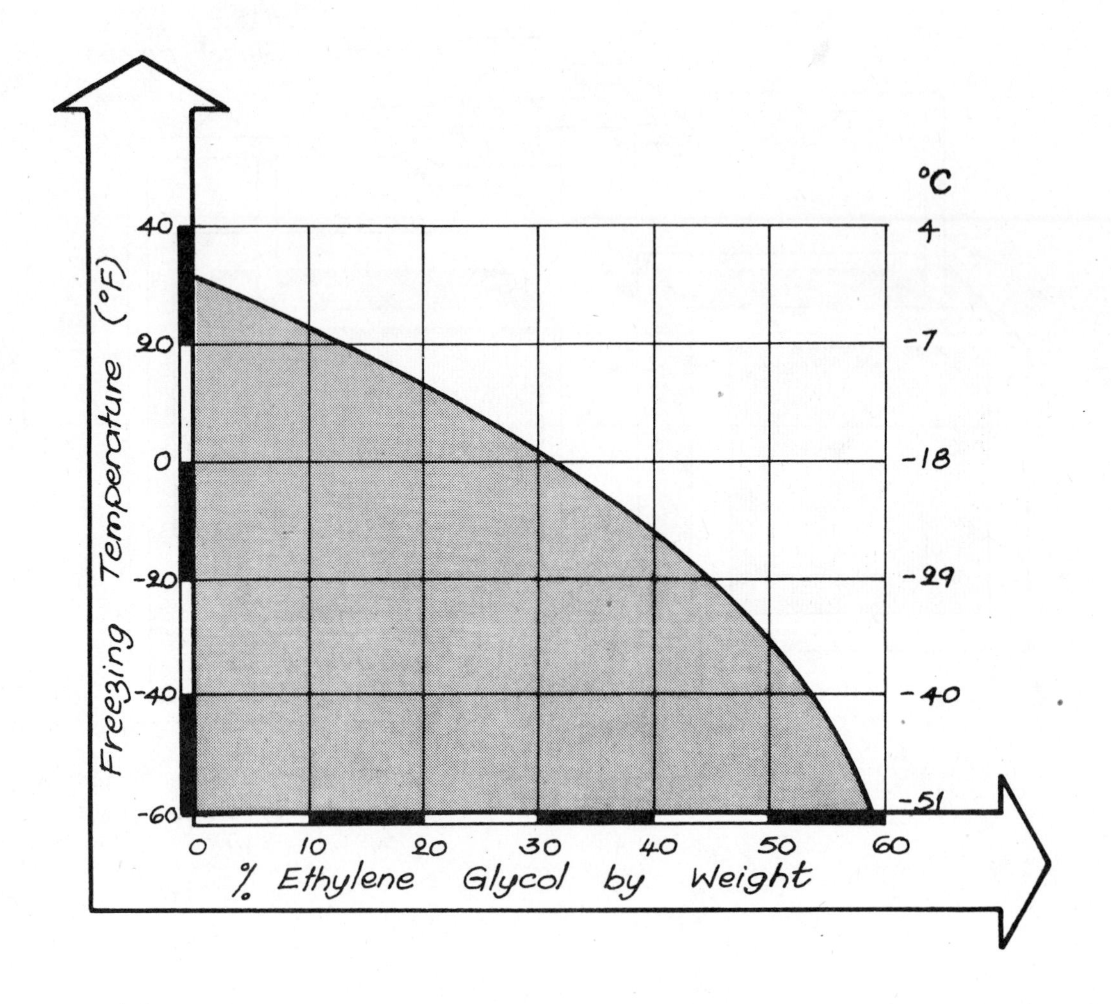

Reference 21

FREEZING TEMPERATURE OF AQUEOUS ETHYLENE GLYCOL SOLUTION

FIGURE 8.14

REFERENCES

1. "ASHRAE 90-75 Seen to Affect Design Engineer, Building Product Industry," *Professional Engineer*, 46(February 1976): 35-37.

2. National Electrical Contractors Association and National Electrical Manufacturers Association, *Total Energy Management: A Practical Handbook on Energy Conservation and Management*, (Washington, D.C.: 1976).

3. Gatts, Robert R.; Massey, Robert G.; Robertson, John C., *Energy Conservation Program Guide for Industry and Commerce (EPIC)*, NBS Handbook 115, (Washington, D.C.: US Department of Commerce, National Bureau of Standards, September 1974).

4. Massey, Robert G. (ed), *Energy Conservation Program Guide for Industry and Commerce (EPIC)*, NBS Handbook 115, Supplement 1, (Washington, D.C.: US Government Printing Office, December 1975).

5. General Services Administration, Public Buildings Services, *Energy Conservation Guidelines for Existing Office Buildings*, (Washington, D.C.: 1975).

6. Spielvogel, Lawrence G., "Exploding Some Myths about Building Energy Use," *Architectural Record* 159(February 1976): 125-128.

7. Rothberg, Joseph E., "Energy Consumption and Conservation in the U.S." Paper presented at the U.K. Science Research Council Summer School on "Aspects of Energy Conservation," Lincoln College, Oxford, England, 14 July 1975, p.13.

8. American Society of Heating, Refrigerating and Air-Conditioning Engineers, Inc., *Energy Conservation in New Building Design*, ASHRAE Standard No. 90-75, (New York: 1975).

9. American Society of Heating, Refrigerating and Air-Conditioning Engineers, Inc., *Standards for Natural and Mechanical Ventilation*, ASHRAE Standard No. 62-73, (New York: 1973).

10. Carrier Air Conditioning Company, *Carrier System Design Manual*, 12 Parts, (Syracuse, New York), Part 1, "Load Estimating," (1972).

11. "Application and Adjustment of Controls for Energy Economy." *Heating/Piping/Air Conditioning*, (February 1975): 36-43.

12. Rishel, James B., "Wire-To-Water Efficiency of Central Station Chilled Water Pumping Systems," in *Central Chilled Water Conference Proceedings - 1976*, edited by J.T. Pearson and D.L. Weast, (Lafayette, Indiana: Purdue Research Foundation, 1976), pp. 85-95.

13. The Energy Task Force of the American Council on Education, American Physical Plant Administrators, National Association of College and University Business Officers, *Energy Management: A Manual for Use in Workshops*, (Washington, D.C.: National Association of College and University Business Officers, 1976), p. II-9.

14. Carrier Air Conditioning Company, *Absorption Refrigerating Equipment*, T-200-30C, (Syracuse, New York: 1974).

15. Trane Air Conditioning Corporation, *Trane Air Conditioning Manual*, (Lacrosse, Wisconsin: 1961), pp. 344-345.

16. Carrier Air Conditioning Company, *Carrier System Design Manual*, 12 Parts, (Syracuse, New York), Part 7, "Refrigeration Equipment," (1969).

17. Shih, James Y., "Energy Conservation Methods Through the Use of Building Automation Systems," in *Proceedings of the Conference on Improving Efficiency and Performance of HVAC Equipment and Systems for Commercial and Industrial Buildings*, Volume 1, edited by Victor W. Goldschmidt, David Didion, and Alwin B. Newton, (Lafayette, Indiana: Purdue Research Foundation, 1976), pp. 61-69.

18. The AIA Research Corporation, *Solar Dwelling Design Concepts*, HUD-PDR-156, (Washington, D.C.: US Department of Housing and Urban Development, Office of Policy Development and Research, May 1976).

19. Arkla Industries, Inc., "Specifications, Model WF 36," (Evansville, Indiana: November 1976).

20. Pannkoke, Ted, "Air to Air Energy Recovery," *Heating/Piping/Air Conditioning*, (March 1976): 37-44.

21. Trane Company, "Trane Coil Heat Recovery Loop," (LaCrosse, Wisconsin: 10 January 1975).

CHAPTER 9

LIGHT ENERGY

C.B. Smith*

CHAPTER CONTENTS

KEY WORDS

Architect-Engineer Services
Efficacy
Efficient Energy Use
Electroluminescence
Energy Management
Gas Discharge Lamps
Illumination Practices
Illumination Standards
Incandescent Lamps
Lamps
Light Sources
Sodium Vapor Lamps

SUMMARY

In 1973 lighting used about 20 percent of US electric energy and 5 percent of all energy. Of all lighting energy, about 20 percent is residential, 40 percent commercial (including schools, public buildings, etc.), 20 percent industrial, and 20 percent miscellaneous. Split another way, about two-thirds is fluorescent and one-third is incandescent.

Currently lighting guidelines are specified in terms of worker performance. A review of the literature shows that controversy exists concerning these guidelines. For example, there are disagreements among individuals and countries on the relationship between visibility and human performance and on the level of illumination required for tasks.

The major use of lighting is in buildings. Improved efficiency in this usage can result from the following technological changes, all of which can be implemented in the near-term:

- increased use of natural light;
- non-uniform illumination practices;
- revised illumination guidelines; and,
- use of more efficient lamps and devices.

*Principal, Applied Nucleonics Company, Inc.

With currently available technology, it is estimated that energy use in lighting could be reduced by 60 percent: (1) 10 percent by better controls, e.g., turning lights off when not in use; (2) 30 percent by widespread substitution of fluorescent lamps for incandescent in residential, commercial, and industrial applications; (3) 20 percent by using high intensity discharge lamps in industrial applications; and (4) reduced lighting where the task does not require present standards.*

Note that in the US an average of 20-30 We/m^2 (2-3 We/ft^2) is used for lighting in the existing inventory of office buildings. With commercial building space of 25 x 10^8 m^2(25 x $10^9 ft^2$), a reduction of 40-60 percent would reduce electricity use by 2.4 x 10^{11} kWh. For illustrative purposes, if this electric energy were distributed on the peak as is the average of all other electrical energy, then the saving could reduce capacity requirements by ∿30,000 MWe or about 7 percent of 1973 installed capacity.

Several case studies are presented in Chapter 15 to: 1) support the conclusions reached in this chapter; 2) demonstrate typical applications in industry and commerce of the concepts presented in the chapter; and 3) show economic savings which can result. In a typical example, the use of high-pressure sodium lamps in an industrial building reduced electricity use by 50 percent and the added investment was repaid in less than three years. Improvements in maintenance cost and lamp-life also resulted.

New technological developments, such as light emitting diodes (LED's), have potential for providing small capacity light sources in the near future, although they will have to be improved at least two orders of magnitude in output and performance before they will be useful for general lighting. In addition to saving lighting energy, improved lighting efficiency can reduce air-conditioning energy use but might increase energy use for heating.

*To the extent that lights add useful heat, such a reduction would require additional heating energy. To the extent that reduced lighting decreases air-conditioning requirements, there would be further energy savings.

9.1 INTRODUCTION

Lighting is an important use of electrical energy, accounting for 20 percent of all the electricity sold in the United States.[1] Electric light is a convenient supplement to natural light, and its use is necessary in our societies. An ultimate goal is to design buildings which combine electric and natural light to realize the best advantages of each.

This chapter has been organized to present information on efficient energy use in lighting in five broad areas:

- OPERATING STRATEGIES which the *user* can apply, some immediately and at no expense and others which require some effort and expenditure;
- FACILITY MODIFICATIONS which the *owner* or *lessor* may undertake to improve efficiency;
- NEW DESIGN OR REDESIGN OF EQUIPMENT AND SYSTEMS which the *architect, engineer, planner, or manufacturer* may develop, leading to reductions in energy use and savings in initial cost, operating cost, or life cycle cost;
- REVIEW OF ILLUMINATION PRACTICES to provide more efficient energy utilization without sacrificing work performance or health;
- POTENTIAL TECHNOLOGICAL IMPROVEMENTS which include longer term developments aimed at improving efficiency of light sources and devices.

In this chapter an attempt has been made to provide some basic data and to outline some of the methodologies necessary to improve lighting energy use efficiency. However, this chapter supplements rather than replaces industry handbooks on illumination practices.[2] The chapter provides an indication of techniques which can be used to improve lighting efficiency; it does not presume to be a complete treatise on lighting system design.

9.2 OBJECTIVE, SCOPE, ASSUMPTIONS

A. Objective

The objectives of this chapter are:

- to define light units;
- to present useful facts and background data;
- to present specific suggestions for more efficient use of light energy;
- to present benefit-cost data for decision making;
- to give an indication of areas where future research might be beneficial; and,
- to estimate potential savings with improved lighting efficiency.

B. Scope

The scope of the material in this chapter deals almost solely with the conversion of electrical energy into light. Light sources, devices, types of lamps, and illumination practices and standards are discussed.

C. Assumptions

It is assumed that the reader is an architect or engineer generally familiar with lighting systems and devices. This chapter is not a substitute for the detailed information obtainable from manufacturers or lighting consultants but can be used as a guide when efficient use of energy is one of the constraints on the lighting system.

9.3 USEFUL FACTS AND BACKGROUND

Luminous flux (light) is measured in lumens (lm); a 100-watt incandescent bulb, as an example, yields 1750 lm. The luminuous flux per unit area (e.g., the light on this page) is called the "illuminance" and is measured in lm/m^2, also called the *lux*. In the British system of units, 1 lm/ft^2 is also widely called a footcandle. Table 9.1 shows other units sometimes used and conversion factors.

Figure 9.1 shows the relative visual response of the cones of the human eye to unit incident energy, as a function of wavelength. A monochromatic light source at $\lambda = 555$ nm has the maximum possible efficacy,* 680 lm/watt for the internationally defined standard observer. For white light, the average efficacy (between 400 and 700 nm) is much less, being 250 lm/We. Phosphors can be optimized to emit "acceptable" white light with an average efficacy of 400 lm/We, as in the case of the yellowish green light of "cool white" fluorescent bulbs.

Specifications for the illumination of a given task are usually given in lm/m^2 (or footcandle), but for a quantitative understanding of a curve of visual acuity (Figure 9.2), we must distinguish between illuminance (lm/m^2) and the brightness of the light reflected from an object. This is called "luminance" and is commonly measured (as in Figure 9.2) in cd/m^2, (foot lamberts [fL]). Conveniently, a fL is defined such that a white card (perfect uniform diffuser) illuminated by 1 lm/ft^2 has a luminance of 1 fL.

In Figure 9.2, the ordinate, called "visual acuity," is just $1/\theta$ where θ, measured in minutes of arc, is the angle subtended by the gap in a broken printed circle. Thus, an acuity of 2 (at 10 fL) means one can see any gap larger than 0.043 mm (0.0017 in.) at a viewing distance of 30.5 cm (12 in.).

---

*Lighting engineers use the term "efficiency" for the dimensionless quantity (energy out/energy in), and use "efficacy" for light output (lumens)/power in (watts).

Recommended Units

	SI Units	American/British Units
Illumination (illuminance)	lux = 1 lm/m^2	Footcandle fc = 1 lm/ft^2
Luminous flux	lumen = cd·sr	lumen = cd·sr
Luminous intensity	cd = lm/sr	cd = lm/sr
Luminance (photometric brightness)	cd/m^2	fL

Conversion to Other Units

To Convert	Into	Multiply By
Footcandle (fc)	lux (lx)	10.76
Lux (lx = lm/m^2)	Footcandle (fc)	0.0929
Footlambert (fL)	cd/m^2	3.43
$Candela/m^2$	Footlambert (fL)	0.292
Lumen	light watt	0.001496
Lumen·hr	Joules	5.386

Definitions

Candela (cd): The luminous intensity, in the perpendicular direction, of a surface of 1/600,000 m^2 of a black body at the temperature of freezing platinum under a pressure of 101,325 N/m^2.

Lumen (lm): The flux emitted within a unit solid angle of one steradian by a point source having a uniform intensity of one candela.

Lambert (L): A unit of luminance (photometric brightness) equal to $1/\pi$ $candela/cm^2$, and, therefore, equal to the uniform luminance of a perfectly diffusing surface emitting or reflecting light at the rate of 1 lm/cm^2.

Footcandle (fc): Same as a Lambert, except equal to $1/\pi$ $candela/ft^2$.

SELECTED LIGHT UNITS & CONVERSION FACTORS

TABLE 9.1

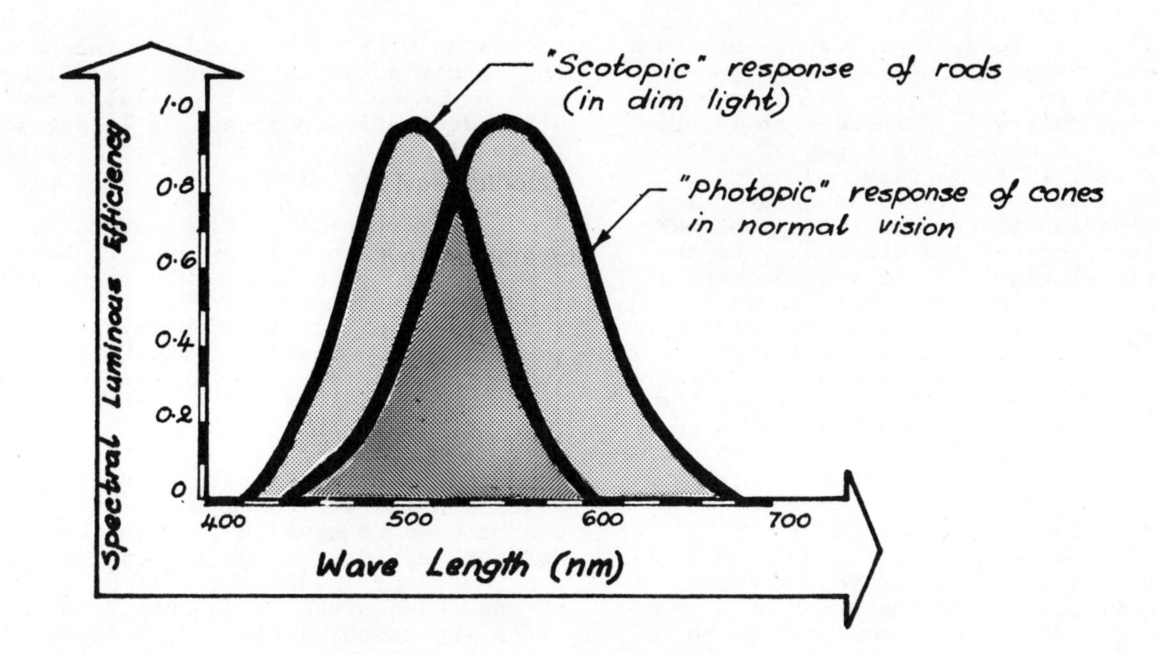

SPECTRAL LUMINOUS EFFICIENCY FIGURE 9.1

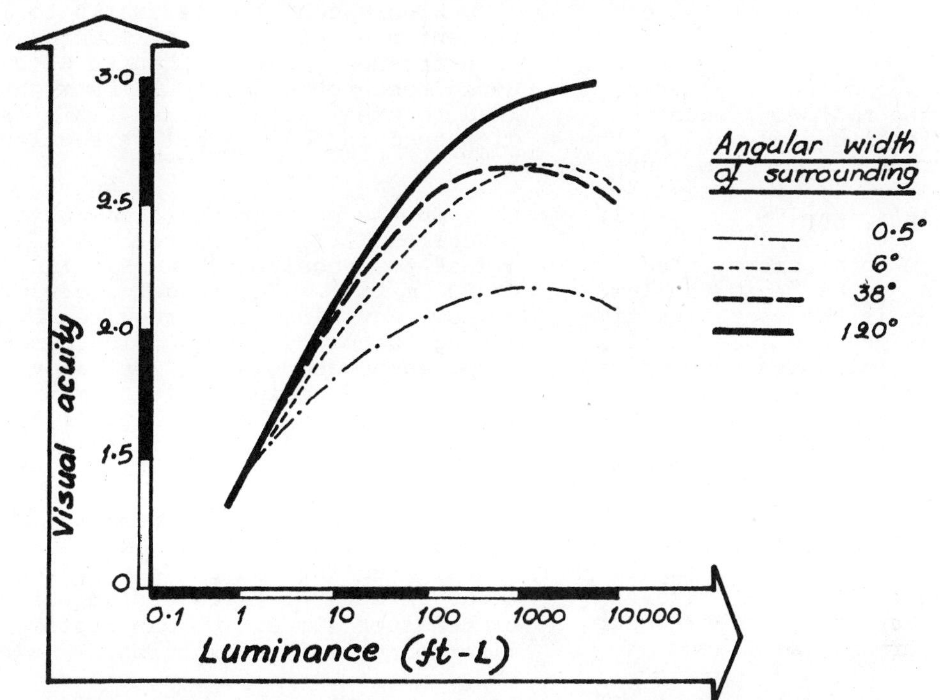

For white paper, luminance in fL is about the same as luminance in lm/ft². The various curves are for different sizes of surrounding, illuminated the same as the central field.

Reference 3

VISUAL ACUITY AS A FUNCTION OF LUMINANCE FIGURE 9.2

Note the logarithmic nature of the acuity curve; one can see a gap of 0.075 mm (0.003 in.) (acuity = 1.2) with only 0.5 fL instead of 10 fL, and one cannot see a gap of 0.025 mm (0.001 in.) (acuity = 3.5) at all. To some extent this is an oversimplification since acuity does not take into account the difficulty of the visual task. As to the horizontal scale, a luminance of 1 fL is roughly the brightness of a white card exposed to an illuminance of 1 lm/ft^2. A horizontal white surface outdoors by day will generally lie between 500 and 5000 fL. The surface of a common fluorescent lamp is about 2500 fL.

9.4 DEVICES AND LOSSES [3,4]

The best light sources have efficiencies on the order of 30 percent.[4] There appears to be no fundamental reason why the theoretical limit of 100 percent cannot be approached more closely with technological improvement. Table 9.2 shows the energy distribution in typical lamps.

For the purpose of evaluating efficiency, the following equation relates the total energy used by the light source and its luminous output:

$$E \propto \phi h \{\eta_l \eta_b \eta_f\}^{-1} \qquad (9.1)$$

where E is the total energy used by the device in joules, ϕ is the luminous output, h is the operation time, and η_l, η_b, η_f are the lamp, ballast, and fixture efficiencies respectively. The efficacies for several common sources of light are shown in Table 9.3. A more complete listing of data for Equation 9.1 is shown in Table 9.4.

From the data it is apparent that fluorescent lamps are four to five times more "efficient" (referring to efficacy) than incandescent lamps. Furthermore, efficacy increases with power for both incandescent and fluorescent types. Gas flames, used occasionally for lawn lights and other decorative purposes or in areas where electricity is not available, are extremely inefficient.

The custom of putting several small incandescent lamps in a fixture instead of one large one may be decorative, but uses more energy. The same remark applies to fixtures which use four 20 We fluorescent lamps to provide two-thirds of the light available from two 40 We tubes, or four-sevenths that from one 75 We tube.

Solid-state "dimmers," inexpensive devices using semiconductor rectifiers, can be used to save on lighting energy. However, besides dimmer losses, it is a property of incandescent lamps that if one reduces the lumen output to one-fifth (by reducing the line voltage to 67 percent), the power consumed by the lamp drops only by 50 percent (see Reference 5, Figure 3.3).

Fluorescent lights are 10 to 15 percent more efficient if powered by high-frequency rather than 60 Hz current. Basic research with kHz lamps might lead to even larger gains. This is discussed in more detail in section 9.6.

In the section on illumination practices (9.5), it is recommended that a desk top receive 540 to 860 lux (50 to 80 lm/ft^2). We can get a crude idea of how many lamps this requires by noting that a bare 40 We fluorescent lamp, suspended 0.6 m (2 ft) above a desk top and equipped with a mirror so that the upwards half of its light is directed back down, provides 1300 lux (120 lm/ft^2) at the desk top. This would be an unattractive and unacceptable way to illuminate a one-desk office due to the glare but the example provides a feeling for the energy requirement. A modern office contains five or ten such lamps. We can restate the discrepancy in energy terms. A single

Lamp Type	Incandescent	Incandescent halogen	Fluorescent	Sodium Low-Pressure	Sodium High-Pressure	Mercury	Metal-Halide
Power (We)	100	100	40	180	400	400	400
Electrode losses	---	---	15	10	6	8	9
Arc power	---	---	85	90	94	92	91
Non-radiative loss	34	16	25(63)*	59	44	44(63)*	40
Radiative	66	84	60(22)*	31	50	48(29)*	51
UV	0	0	58	0	1	18(2)*	4
Visible	5	13	2(22)*	27	29	15(15)*	23
Infrared	61	71	0	4	20	15(12)*	24
lm/We	15	30	80	180	120	57	80
$(lm/We)_{max}$	260	310	(330)	530	(390)	250	(325)

*Energy conversion fluorescent layer included.

Reference 4

ENERGY DISTRIBUTION (PERCENTAGE) FOR VARIOUS LAMPS

TABLE 9.2

Light Source*	Lamp Alone	Lamp plus Ballast[a]
Open gas flame, in lm/W [b]	0.2	
Gas mantle, in lm/W	1 to 2	
Incandescent lamps[c]		
40 We	12	
100 We	18	
Fluorescent lamps,[c] plus ballast		
20 We, 24-in T12, plus 13 We[d]	(65) [e]	46
40 We, 48-in T12, plus 13.5 We[d]	(79)	68
75 We, 96-in T12, plus 11 We[d]	(84)	72
Metal halide,[a] 400 We plus 26 We[d]	(85)	75
Sodium:[a]		
400 We high pressure, plus 39 We[d]	(120)	108
180 We low pressure, plus 30 We [d]	(180)	155

*(All entries except for flames and mantles are in lm/W_e)

Notes:

(a) Source: Reference (4)
(b) Source: Reference (6)
(c) Source: Reference (7), p. E-185
(d) These are typical ballasts; however, 0.5 W ballasts are available for an extra 10-20% in price.
(e) The numbers in parentheses are for a hypothetical lamp without ballast losses.

EFFICACIES OF SELECTED FLAMES AND LAMPS

TABLE 9.3

Lamp Type	Power Lamp Efficacy (We)	lm/We	%	Ballast Power (We)	Ballast + Lamp Efficacy lm/We	%	Fitting Loss	Overall Efficacy lm/We	%
Incandescent	100	15	5	--	---	--	10-50 (in-door)	7-13	2.5-4.5
Incandescent halogen	100	20-30	13	--	---	--		15-27	6.5-12
Fluorescent	40 65	80 85	22 --	13.5 11	60 73	16.5 --		30-54 36-66	8.5-15
Low-pressure sodium	90 180	140 180	-- 27	35 30	100 155	-- 23	20-45 (out-door)	55-80 85-125	--- 13-18.5
High-pressure sodium	250 400	100 120	-- 29	33 39	--- 108	-- 24		60-86 72-103	--- 14-20
High-pressure mercury	80 250 400	44 54 57	-- -- 15	8.5 18.5 26	39 50 53.5	-- -- 14		22-31 28-40 30-43	--- --- 8-11
High-pressure metal-halide	400 1000 2000	85 90 100	23 -- --	26 43 68	75 86 97	21.5 -- --	30-40 (flood light)	45-52 51-60 58-68	13-15 --- ---

Reference 4

PROPERTIES OF INCANDESCENT AND ELECTRICAL DISCHARGE SYSTEMS FOR ILLUMINATION PURPOSES

TABLE 9.4

40 We lamp (with 10 We of ballast) in a 10 m^2 (100 ft^2) office (a 2.5 m x 4 m module) uses 5 We/m^2 (0.5 We/ft^2). Clearly, most of the light is not reaching work surfaces.

What happened to all the light from the 40 We lamp? First, the lamp is not at desk-lamp height but in the ceiling and shielded from direct view by louvers or plastic to prevent glare. As was mentioned earlier, the surface brightness of a standard 40 We cool-white fluorescent lamp is $\sim$2500 fL, which is no brighter than the surface would look if we turned it off and carried it outside on an average day. Yet most people find it too bright to look at directly when inside. The plastic prismatic lenses used to shield lamps actually absorb about one-fourth of the light, even when the plastic is new, clear, and clean.

The familiar "egg crate" louver is inefficient; more efficient types can be found both in Europe and in the United States. In selecting lamp shields, efficiency is only one of several criteria normally considered. Table 9.5 indicates values for several types of troffers. Note that a clear lens can have efficiencies as low as 45 percent. The parabolic louver provides the best brightness control.

Other lighting losses are caused by deterioration of lamps (fluorescent lamps drop slowly in light output to a plateau of 80 percent of their original output) and dirt accumulation on lamps and fixtures.

To summarize, a checklist to reduce loss from lamps and lighting devices would include these items:

- Is the highest efficacy lamp (suitable for the application) being used?
- Is the highest powered lamp available (suitable for the application) being used?
- Is the most efficient voltage, frequency, ballast being used?
- Is the most efficient fixture (luminaire) consistent with good glare control being used?
- Can lamps or luminaires be eliminated or combined to in-increase efficiency?
- Have adequate provisions been taken for maintaining and cleaning lighting equipment and lamps?

9.5 ILLUMINATION PRACTICES

A. General Discussion

The current approach to task lighting standards is to specify illumination levels in terms of worker performance. In the broadest sense this includes consideration of such factors as age, fatigue, physiological and psychological effects, impairment of vision or health, economics, energy use and availability, and even cultural or emotional effects of light. A review of the literature shows that controversy exists concerning illumination standards; for example there are disagreements among individuals and countries on the relationship between visibility and human performance.[6] Since the ultimate criterion in establishing lighting standards is human performance, which in turn involves subjective judgments, a synopsis of several points of view is presented below.

Engineers usually adopt conservative approaches when there is uncertainty in their knowledge. Thus, lighting levels are sometimes set above the minimum recommended task lighting guidelines of the Illuminating Engineering Society (IES).[7] In addition, the IES recommendations have been incorporated into various school codes, state codes, and building codes. In the absence of other data or experience, they are used by architects and design engineers in order to have guidelines for recommended illumination levels.[7]

It is interesting to examine US task lighting guidelines as a function of time and to compare them with standards in other countries. [2,3,7]

VISUAL COMFORT	(VCP)	EFFICIENCY	(%)	TYPICAL COST[*]	($)
Parabolic Louver	99	Clear Lens	45-70	Parabolic Louver	20-25
Dark Metal Louver	70-90	Polarizer	55-60	Polarizer	15-24
Toned Lens	65-95	Diffuser	40-60	Toned Lens	10-25
White Metal Louver	65-85	Plastic Louver	45-55	Dark Metal Louver	10-20
Polarizer	60-70	White Metal Louver	35-45	White Metal Louver	8-15
Clear Lens	50-85	Parabolic Louver	35-45	Plastic Louver	8-10
Plastic Louver	50-70	Toned Lens	30-65	Diffuser	6-15
Diffuser	40-50	Dark Metal Louver	25-40	Clear Lens	5-20

* Contractor net prices for 2' x 4' panels. (US, c. 1970)

Notes:

a) Shown are the three most often considered trade-offs in selecting shielding. Other considerations might include: fire rating, cleanability, color stability, local codes, reflected glare, hiding power, module size, type distribution required, etc. All figures shown are approximate. Source: Reference 1.

b) A troffer is a long lighting unit recessed in the ceiling.

SHIELDING MEDIA[a]
Comparison of Some Characteristics for Typical 2ft x 4ft Troffers[b]

TABLE 9.5

In the US, guidelines have been changed periodically, most recently in 1958. The IES recommendations have not been changed since then.[1] For example, in 1952, the New York City Board of Education *Manual of School Planning* called for 20 lm/ft^2; but this was raised to 30 in 1957, and to 60 in 1971. Similar increases have been registered for libraries, shops, and drafting rooms. Until recently modern office buildings have been designed for 80 to 100 lm/ft^2 and frequently have had uniform illumination even in corridors and stairways. Yet the IES recommendations specifically draw a distinction between recommended levels for *work spaces* versus *non-work spaces*. This deviation, then, may not be so much a problem of the guidelines as of practice, which takes the line of least resistance and designs for uniform illumination.

US lighting guidelines are said to be above those of Western Europe. [5,8] Stevens gives several comparisons and then states: "These very high values may be a reflection of a higher standard of living rather than being based on convincing proof that they are economically justified." [5]

In rebuttal, Robert Dorsey points out that the US values are recommendations only, that the British standard has been revised and has many recommendations identical to the US, and that higher US levels quoted are intended to be provided by local lighting. [1]

Energy standards for building lighting are sometimes expressed in terms of installed power per m^2 (ft^2) of usable building floor area. Modern office buildings typically install 2.5 to 5.0 We/ft^2, whereas the General Services Administration recently recommended 1 to 2 We/ft^2. The energy efficient Federal Office Building in Manchester, New Hampshire, actually uses 0.7 We/ft^2. [3] The average in existing office buildings is 2 We/ft^2. [1]

If we combine lighting power in We/ft^2 of floor area with illumination standards in lm/ft^2 of task surface, and add the assumption that interior lighting is uniform, we can estimate the efficiency with which light is conveyed to any horizontal surfaces, as opposed to being absorbed on fixtures and walls. A building equipped with 2.5 We/ft^2 in the form of 40 We fluorescent lamps will emit 197 lm/ft^2 of floor area, at least double the 80 to 100 lm/ft^2 figure quoted earlier. Thus, about half of the light does not reach the task surface either directly or through reflection.

Patterns of electricity use and lighting use in the US make clear that average illumination in homes, where lighting is governed by personal choice, is well below the average level in commerce and industry, where lighting is governed principally by established standards. On the average, each person encounters ten times as much artificial light (in lm/hr) outside the home as inside the home (even though most people work during the day and spend their nights at home!). At home, moderate levels of general lighting are used and desk lights and reading lights are used for special tasks.

Electric lighting used about 1.8 GJe per household in 1950 and 2.7 GJe (750 kWh) per household in 1970. [8] Almost all residential lighting is incandescent, although more than 75 percent of industrial, commercial, and institutional lighting is fluorescent. As an example of the potential energy savings resulting from the use of more efficient lighting, if all households in 1970 had fluorescent lamps rather than incandescent, residential electricity use for lighting (at equivalent levels of illumination) would decrease by about 75 percent and total residential electricity use by 8 percent. [8]

Concerning lighting standards, several authors have questioned whether further increases in the present levels are needed and in some cases have questioned the current levels. [7,9] Suffice it to say that the recommendations and standards have been prepared by experts after considerable study, and it is not the purpose of this chapter to refute them. As energy cost and availability change, however, the cost-benefit ratio of increased illumination may be expected to change and consequently users are advised to watch for changes in recommended illumination levels. Perhaps most critical is the need for better understanding of the relationship between illumination, human performance, and variables such as age, visual task difficulty, and contrast.

A review and critique of research studying the relation between illumination and performance has been made. [10] Other studies have been carried out by the lighting industry which indicate that productivity increases with increased illumination to a certain extent. Additional research work is needed to establish a definitive physiological link between human performance and comfort and light.

B. US Federal Energy Administration (FEA) Lighting Guidelines [11]

The following discussion summarizes FEA suggested approaches for energy conservation in lighting systems design and operation, particularly in public buildings. Design, installation, and operation of effective lighting systems have complex scientific, management, engineering, and architectural components. Of the many elements that must be considered in providing an adequate visual environment at acceptable cost, energy conservation is only one, but recent events and future prospects for the demand and supply of energy have underscored the necessity of giving conservation greater weight. Other elements that must be taken into account are the visual tasks to be performed, the physiological state of the observer's eyes, the psychological state and perceptual skill of the observer, the design of task and surrounding areas, the availability of daylight, the level of illumination, and the lighting system quality with regard to spectral characteristics, glare, veiling reflections, and geometrical factors.

These complexities limit the degree to which simple guidelines for energy conservation in lighting can be applied in all cases. In most situations, however, they are useful for beginning the analysis to identify possible areas for saving lighting energy while also providing an adequate visual environment.

Variation in visual requirements for different tasks, for different observers, and other considerations make necessary additional analyses. The objective of the guidelines is to provide useful assistance in the design and operation of lighting systems to minimize energy usage, both directly and indirectly. There are some situations in which measures beyond those specified in the guidelines may result in additional energy savings. The expert assistance of architects and lighting engineers can provide additional guidance.

Illumination Level Guidelines

The following illumination levels were suggested by FEA as targets for public buildings. Whether these guidelines are sufficient or acceptable for other applications is under debate by experts.

Commercial Buildings

Commercial buildings include office buildings, administrative spaces, retail establishments, schools, and warehouses. During working hours illumination levels of 50 fc at occupied work stations, 30 fc in work areas, and 10 fc in areas that are seldom occupied or which have minimal visual requirements such as hallways and corridors, may be sufficient. Where needed, because of exceptional individual requirements or because of the difficult nature of a specific task, non-uniform supplemental lighting should be provided for the task duration. Individual switches should be provided to permit maximum control over both standard and supplemental lighting when not needed. Lights should be switched off whenever daylight can be used.

Industrial Buildings

Industrial buildings include factories and plants. For industrial lighting, levels at the work station should be those recommended by the American National Standards Institute, *Practice for Industrial Lighting*, ANSI-A11.1-1973 (June 1973)--that is, 30 fc in work areas, and 10 fc in nonworking areas, except in a few special cases specified by the Occupational Safety and Health Administration. Daylight, when available, and switching should be used to the greatest degree possible to reduce energy use.

Hospitals

Illumination levels at the task should be no greater than those recommended in *Illuminating Engineering Society Lighting Handbook* (Reference 2, pp. 84-85) and no greater than 10 fc in non-working areas such as hallways and corridors.

Discussion

No conclusive evidence is available to show harmful effects from either too much or too little light within the range found in commercial and industrial buildings. There are differences however in lighting requirements for individuals; for example, older people generally need more illumination because of degenerative effects on pupil size, corneal transmission, visual acuity, scattered light, and muscular response. In addition, both level and quality of illumination are important, since glare and other factors can make lighting offensive at high or low levels.

Placement and orientation of luminaires and work stations with respect to each other are important in realizing energy savings. Work stations, places where the principal visual tasks are performed, are to be distinguished from general work areas which surround work stations and which usually have lower illumination level requirements. Wherever possible non-uniform lighting that is task-oriented, with respect to both placement and illumination level, should be used. Uniform lighting systems which light large general areas independently of the task locations within them do not usually make the most effective use of energy. Coincidentally, the esthetic appearance of indoor space can often be improved by following non-uniform lighting practice.

Guidelines for Efficiency in Lighting

Selection of Efficient Lighting Equipment

In the design of new lighting systems and in modifying existing ones, the most efficient light sources that can provide the illumination required should be selected. As a general rule, the efficiencies of some available lamp types rank according to the following list, with the most efficient first: low pressure sodium, high pressure sodium vapor, metal halide, fluorescent, mercury, and incandescent. Use of low pressure sodium is limited to applications where a monochromatic light source is acceptable. In many cases replacement of existing low-efficiency lamp types with lower wattage more-efficient types will result in reduced total costs and improved lighting. See Table 9.6 for detailed examples.

Control and Scheduling

Maximum control over lighting systems can be accomplished by having switches to permit turning off unnecessary lighting. Large general areas should not be under the exclusive control of a single switch if turning off small portions would permit substantial energy savings when they are not occupied. Lights should be turned off as a regular practice when buildings are not occupied, such as after working hours or on weekends and holidays. When opportunities for using daylight exist, artificial light should be turned off. Occupants of buildings should be educated and periodically reminded to adopt practices which will save lighting energy, such as turning off lights when leaving a room.

Caution: some buildings rely on lighting systems as a source of heat for space heating.

Frequent on-off use shortens lamp life. Therefore, there is an optimum point between energy cost savings and the costs of lamps and replacement labor. Variations in energy prices, labor costs, and facility convenience and schedules influence the decision. Under typical conditions, however, the breakeven point for fluorescent lamps is reached in five to ten minutes when replacement costs are low and in 20 to 30 minutes when costs are high. It certainly pays to turn off fluorescent lights for any period greater than 30 minutes. Note that certain lamp types have a warm-up period, which must be considered in a decision to turn lamps off.

Proper luminaire placement in the design of new lighting systems and the removal of unnecessary lamps in

CHANGE OFFICE LAMPS (2700 hours per year)		ENERGY SAVINGS/COST SAVINGS kWh	GJ	3¢/kWh	5¢/kWh
from	*to*	*to save annually*			
1 300-watt incandescent	1 100-watt mercury vapor	486	5.25	$14.58	$24.30
2 100-watt incandescent	1 40-watt fluorescent	400	4.32	12.00	20.00
7 150-watt incandescent	1 150-watt sodium vapor	2360	25.5	70.80	118.00
CHANGE INDUSTRIAL LAMPS (3000 hours per year)					
from	*to*	*to save annually*			
1 300-watt incandescent	2 40-watt fluorescent	623	6.73	18.69	31.15
1 1000-watt incandescent	2 215-watt fluorescent	1617	17.5	48.51	80.85
3 300-watt incandescent	1 250-watt sodium vapor	1806	19.5	54.18	90.30
CHANGE STORE LAMPS (3300 hours per year)					
from	*to*	*to save annually*			
1 300-watt incandescent	2 40-watt fluorescent	685	7.40	20.55	34.25
1 200-watt incandescent	1 100-watt mercury vapor	264	2.85	7.92	13.20
2 200-watt incandescent	1 175-watt mercury vapor	670	7.24	20.10	33.50

RELAMPING OPPORTUNITIES **TABLE 9.6**

existing installations are examples of energy saving measures. Luminaires should be positioned to minimize glare and veiling reflections, and work stations should be oriented and grouped to utilize light most effectively. Daylight should be used when available, maximum switching control should be provided to the user, and light colors should be used on walls, ceilings, and floors. Tasks should be designed to present high contrast to the observer.

Deterioration in illumination level due to dirt accumulation on lighting equipment should be prevented by adequate maintenance programs, cleaning lamps and luminaires, and replacement of lamps. As a part of maintenance programs, periodic surveys of installed lighting with respect to lamp positioning and illumination level should be conducted to take advantage of energy conservation opportunities as user requirements change.

Indirect Impact of Lighting Energy on Heating and Cooling of Buildings

The adoption of lighting energy conservation methods should be considered in conjunction with the operation of the heating and cooling systems. As a rule of thumb, when air-conditioning equipment is operating, each watt of lighting causes the expenditure of about one-third watt of air-conditioning power. Substantial cooling energy can be saved by reducing electrical lighting loads to a minimum. Moreover, substantial savings in initial cost may be realized by reflecting the reduced heat load from an energy-conservative lighting system in the design of the air-conditioning system. When possible, heat removal techniques should be considered to conduct waste heat from lighting sytems out of the building without imposing an additional load on air-conditioning equipment.

Heat gain from lights should be included in calculating heat load; in addition, schemes utilizing waste heat from lighting are encouraged. Reliance on heat produced by lighting systems, beyond that produced by systems operated in normal energy-efficient ways, is not encouraged.

Measurement of Recommended Lighting Levels

Light levels can be determined with portable illumination meters such as a photovoltaic cell connected to a meter calibrated in footcandles. The light meter should be calibrated to a basic accuracy of ±15 percent over a range of 30 to 500 fc and ±20 percent from 15 to 30 fc. The meter should be color corrected (according to the CIE Spectral Luminous Efficiency Curve) and cosine corrected. Measurements refer to average, maintained, horizontal footcandles at the task or in a horizontal plane 30 inches above the floor.

Measurements of work areas and nonworking areas should be made at representative points between fixtures in halls, corridors, and circulation areas. An average of several readings may be necessary. Daylight should be excluded or corrected for during illumination level readings in order to determine minimum lighting level when the system is operated without available daylight.

Recommended illumination levels should be used when work stations are occupied; otherwise, consideration should be given to turning lights off or to switching to 30 fc if other workers remain nearby. For tasks requiring levels higher than 50 fc, switching to lower levels is desirable if the work changes to less critical tasks. Illumination at the task should be reasonably free of veiling reflections and body shadows. Levels for industrial work are from the American National Standards Institute, *Practice for Industrial Lighting*, ANSI-A11.1-1973 (June 1973).

Equivalent sphere illumination (ESI) is a relatively new concept that is used to evaluate lighting systems for both quantity and quality of illumination. For the purpose of this brief discussion, quality lighting can be evaluated in terms of contrast. The Contrast Rendition Factor (CRF) is a measure of the actual contrast to the contrast on a task located within a hypothetical, uniformly illuminated, spherical surface. Finally, the effectiveness of a lighting system (from a visual point of view) is given by:

Lighting Effectiveness Factor (LEF)=

$$\frac{\text{ESI lux (footcandles)}}{\text{Raw lux (footcandles)}}$$

In this expression "raw" footcandles are what would be measured conventionally as described above. Thus, the LEF is always less than or equal to 1.0.

ESI calculations will increasingly be made in the future as an attempt to squeeze better performance out of lighting installations. To date, ESI measurement techniques are still in the process of exploration and development. For additional ESI details, consult the technical literature, particularly reports by the IES "Committee on Recommendations for Quantity and Quality of Illumination."

Relamping Opportunities

Relamping to a lower wattage can save substantial amounts of energy. For example, relamping from a 150 watt to a 75 watt bulb saves 50 percent of previous use, or relamping fluorescents to smaller wattages will save energy. Relamping two lamps with one can save also. For example, replacing two 60 watt lamps with one 100 watt will save 12 percent of previous usage and will normally provide the same amount of light as before.

Table 9.6 lists some illustrative examples of changes that can be made in installed lighting systems. They result in roughly the same illumination levels, but reduce energy usage. Also, see the Case Studies.

C. Example:

Governmental Actions During an Energy Emergency

During the 1973 energy crisis, the Federal Government formed a Task Force to improve energy use efficiency in federally occupied buildings. One of the measures enacted was to require an "Energy Usage Analysis" report to be completed each year showing energy usage by month. Among other reductions, lighting energy use was to be reduced by removing lamps and fixtures and by applying non-uniform lighting standards to the existing systems (see Table 9.7). [12]

D. Other Suggestions

Air-cooled or water-cooled heat-of-light systems can be used to prolong lamp life, provide higher efficacy, and reduce the load on heating and air-conditioning systems. The following description is from Reference 13:

An air-cooled heat-of-light system works as follows: Dry air in a room is extracted from a light fixture so that heat from the light system, which would normally go into space, is picked up by the air passing through the lighting fixture. A water-cooled system consists of special light fixtures equipped with passages through which water flows and the water takes up the heat from the lights. The water then flows to a cooling tower, is cooled, and then is recirculated to the light. Normally, the water-cooled system is used together with an air-cooled system. Air-cooled systems alone are much more common and have fewer mechanical problems, but are not as efficient.

The instantaneous rate of heat gain (thermal power) due to lighting can be calculated from: [2]

$$q_{el} = \frac{A \cdot E}{\phi \cdot CU \cdot LLF} W \cdot UF \cdot EDF \qquad (9.2)$$

where:

q_{el} = sensible lighting heat load, W

A = total area, m^2

E = average illumination, lx

ϕ = total lumens per luminaire

CU = coefficient of utilization

LLF = light loss factor

W = actual power per luminaire in service, We

UF = use factor, ratio of wattage in use for the conditions under which the load estimate is being made to the total installed wattage

1. The full potential for energy savings which may be achieved from reduced levels of illumination in office areas has not yet been achieved. Under the nonuniform concept, where lighting is concentrated at the work stations (desk tops), care must be exercised in selecting the fixtures which should have their lamps removed. Lighting levels at work stations for general office work under the nonuniform lighting concept should be reduced to approximately 50 ft-candles. As a general guide, one watt per square foot for lighting should be the maximum allowed. Lighting levels in warehouse and records space should be reduced to the maximum extent and utilized only in those areas in which people are working and while they are working.

2. Fixture location in relation to the fixed desk location is the major consideration. Therefore, the most desirable location for fixtures or lamps remaining energized should be directly over the desk or the sides of the desk where they will not create reflections. Care should be taken to ensure that surroundings seen by the worker do not become dark, producing visual discomfort.

3. In removing lamps from existing fixtures, always remove all lamps that are operated by the same ballast. When these tubes are removed from fixtures without disconnecting the ballast, the ballast energizing current is very small, particularly in the case of the rapid start series-lead type that have been specified for the past 10 to 12 years. During normal operation of a fixture with four 40 watt lamps, the two ballasts and four lamps will use a total of about 185 watts. With the removal of two tubes leaving both ballasts in place, the wattage is reduced to about 100 watts per fixture. There are, however, some older type fixtures in our buildings such as lead-lag, preheat, and instant start that may not have a similar wattage reduction with the tubes removed. Consequently, it is necessary that a check be made of the type of fixtures used in our buildings and where the use of a single phase wattmeter indicates the ballast consumes 15 watts or more, the ballast should be disconnected and the tubes removed. Do not use a ammeter and voltmeter to obtain wattage readings because the results will be incorrect.

NONUNIFORM LIGHTING LEVELS **TABLE 9.7**

4. Certain building areas should not receive the nonuniform treatment. Where more difficult seeing tasks require a higher level of illumination, reductions should be made only to the level appropriate for the task being accomplished. These areas include drafting rooms, computer rooms, and areas containing printing presses or rotating machinery where lower illumination might result in a hazardous condition.

5. Lamps in the corridors, lobbies, toilet rooms, and other public spaces should be removed to the maximum extent possible without creating a hazardous condition. In these nonwork areas and other similar space, lighting levels of between 25 to 30 ft-candles are specified. All outside architectural lighting (except security lighting) should be extinguished.

6. To the extent possible, at the time that this nonuniform conversion is made, all fixtures should be cleaned and those to remain in operation provided with new lamps. This will increase the light output of the fixtures remaining approximately 15% or 20% initially, and provide a psychological boost toward the acceptance of this new concept by the employees occupying the space. The fixtures to be left without lamps should be marked to prevent inadvertent relamping.

7. Occupants of the space affected should be made aware of the purpose and intent of the lowering of lighting levels well in advance of implementation to gain their acceptance of the plan.

8. Special consideration should be given to those employees who have particular problems relating to vision or unusually ardous visual tasks.

Reference 12

Note: Certain items (e.g., 1, 5) in this table are controversial. It should be noted that the table reflects what the government did to public buildings during an emergency, and these measures are not necessarily recommendations for other buildings during normal times.

TABLE 9.7 CONT'D

EDF = energy distribution factor, to account for fluorescent and other luminaires which are either ventilated or installed so that only part of their heat goes to the conditioned space.

The lighting heat load is generally significant and efficient design requires that heating, air conditioning, and lighting be considered as a unit, rather than isolated in unrelated design tasks. As mentioned above, proper consideration of lighting system temperatures has the additional advantage of improving efficiency and prolonging the life of lamps and equipment.

Lighting energy should be considered as follows:

$$E_{net} = E_L + E_R - E_H \qquad (9.3)$$

where:

E_{net} = net lighting energy impact on building

E_L = energy to operate lighting system

E_R = energy to refrigerate lighting system on cooling cycle

E_H = contribution of lighting heat toward heating the building.

Typical numbers for US commercial buildings are:

	MJ/m^2yr	Btu/ft^2yr
E_{net}	81.8	7,200
E_L	250.0	22,000
E_R	43.2	3,800
E_H	125.0	11,000

These can be contrasted with total building energy use, which ranges from 0.85 to 2.27 GJ/m^2yr (75,000 to 200,000 Btu/ft^2yr) of gross space. One US government objective is to bring the total down to less than about 0.6 GJ/m^2yr (∿50-55,000 Btu/ft^2yr) in new government owned or leased buildings.

One final point should be made. Significant indirect savings can result from fewer lamps, fewer fixtures, smaller wiring loads, and smaller switch gear. Of roughly 5.4 x 10^9 kWh used in the building construction industry for the manufacture of electric lighting and wiring, at least 25 percent, or 1.35 x 10^9 kWh, could be saved. [7]

Apart from basic research in the conversion of electric energy to light and applied research in human response to lighting patterns and intensity, there are obvious near-term steps for more efficient energy use connected with lighting. The most dramatic way to save power in residential lighting would be to switch from incandescent to fluorescent lamps; the biggest commercial saving would be to use recommended illumination levels only where needed rather than to illuminate uniformly a whole office or a whole floor, to switch more completely to fluorescent lamps, and to use timed controls. In industry, changing to high pressure sodium lamps could reduce electricity use by 50 percent. Finally, a reexamination of industrial and commercial lighting standards and practices is in order. [14]

9.6 POTENTIAL TECHNOLOGICAL IMPROVEMENTS[4]

Light sources can be classified according to the energy form which generates the light:

Energy Form	(Phenomenon)
Electrical	- (Incandescence, Gas discharge, Electroluminescence, p-n luminescence, Phosphorescence/fluorescence)
Chemical	- Flame, Flame + selective radiator, Chemiluminescence, Candoluminescence)
Radiative	- (Photoluminescence)

A. Electrical Energy

Incandescent Lamps

Because of its low efficiency, the incandescent lamp has been replaced by other light sources in many applications. In theory, and neglecting losses, a considerable improvement could be obtained if the following conditions for the temperature dependent spectral emissivity E(λ) of the electrically heated radiator could be satisfied:

$$E \simeq 1 \text{ for } \lambda<0.7\ \mu m,\ T = 0,\ R = 0 \tag{9.4}$$

$$E \simeq 0 \text{ for } \lambda>0.7\ \mu m,\ T = 1 \text{ or } R \simeq 1$$

T = transmission, R = reflection

The efficiency of incandescent lamps could also be increased by the application of heat filters. Maximum efficacy of approximately 250 lm/We for white light could theoretically be obtained with a filter that transmits only radiation of a wave-length from 0.4 μm to 0.7 μm (maximum of eye-sensitivity), the light source then emitting monochromatic green light.

Increasing the temperature of the thermal reactor is another method for improving incandescent lamp efficiency. This could be achieved through a regenerative cycle using fluorine compounds.

It is recommended, therefore, that research be aimed at:

- improved optical filters,
- selective radiators, and
- fluorine chemistry.

The possibilities of using multiphoton phosphors to improve the efficiency of the incandescent lamp should be investigated more closely. Finally, the development of halogen-tungsten lamps of lower wattage for residential lighting should be encouraged.

Gas Discharge Lamps

Fluorescent Lamps

The size and shape of the fluorescent lamp have generally hindered its application in residential lighting. Yet, the fluorescent lamp would be preferable for residential lighting because its efficiency is higher than that of the incandescent lamp. Therefore, research and development should be carried out or expanded:

- to improve the efficiency at acceptable values of the color temperature; and,
- to reduce the size without loss of efficiency.

Since the operation of fluorescent lamps at higher frequencies (1-3 kHz) improves the lamp efficiency (10 to 15 percent) and reduces ballast losses (about 50 percent), higher frequency operation should be introduced, especially in large scale applications.

Improvement in the efficiency of the supply units (inverters) should also be developed, and as a general recommendation, further investigations for better and new phosphors should continue to be given attention.

High-Pressure Lamps (General)

A variety of so-called high-pressure lamps are very efficient and, in particular cases, also provide good color and good color rendering properties. In principle, the relatively small physical size of these lamps makes them suitable for indoor lighting.

It is recommended that further research be aimed at the development of these light sources for many lighting applications (industrial, commercial, and residential) by:

- adaptation of the properties to the specific applications; and,
- studying in greater detail the power/efficiency relationships.

High-Pressure Sodium Lamps

The high-pressure sodium lamp has an *excellent* overall efficiency and is suited for applications ranging from street lighting to industrial use. Questions have been raised about botanical side effects which have been attributed to emission in the red part of the spectrum (around 680 nm). These have been investigated and evidence to date shows no deleterious effects attributable to the lamps when used in typical municipal installations.[1]

Low-Pressure Sodium Lamps

The low-pressure sodium lamp is the *most efficient* light source known to date. However, the light source is so large that optical control is not as great as for smaller sources such as high-pressure sodium lamps. Furthermore, the light is monochromatic (about 590 nm), which is a disadvantage for applications where it is important to distinguish colors.

Electroluminescence and p-n Luminescence

Despite the effort that has been devoted over the past 25 years to study of these effects, relatively little progress has been achieved. The prospects for further improvements are not rated highly. On the other hand, there has been a rapid development in the field of p-n or injection luminescence with a variety of materials. High quantum efficiencies have been obtained in the red and infrared. Injection diodes for general lighting purposes would, however, require much larger radiating areas.

The extremely high theoretical internal conversion efficiency warrants further investigation of semiconductor materials with regard to:

- spectral output, and,
- technology of large-size devices.

The field of p-n luminescence is of great interest for long-term development of high-efficiency "cold light" radiation sources.

B. Chemical Energy

There is only a limited application for direct chemical energy conversion into light, largely because high conversion efficiency is rarely obtained. Examples are known, however, when efficiencies on the order of 10 percent can be achieved (e.g., flash lamp). In addition, only about half a century ago artificial light was produced predominantly by direct conversion of chemical (free) energy (gas, candle, petroleum lamp).

In developing countries or in gas/H_2 economies, a potential for the flame/selective radiator (e.g., Auermantle) is recognized.

Therefore, the problems related to the properties of refractory materials at high temperatures and in chemically reactive atmospheres should be reconsidered. Special attention should be given to vapor pressures, absorbing centers, and chemical and mechanical stability.

Another possibility, namely production of selective radiation by seeding flames with small particles, should also be investigated.

Bioluminescence is a special case of chemiluminescence and might be useful in specialized circumstances, such as camping and blackouts. The phenomenon, however, is not considered as energy saving in general.

C. General Conclusions

The efficiency of ballasts for gas discharge light sources should be improved by using solid-state devices or other components. Furthermore, additional efforts should be initiated to develop new high efficiency lamps which are directly interchangeable with older types. Some work along these lines has been done, such as the "watt-miser" and other types mentioned previously.

Recommendations for improving the efficiency of light sources should be brought to the attention of international bodies, especially the Commission Intérnationale de l'Eclairage (CIE). The possibilities for developing self-cleaning light source/fixture combinations should be investigated further. In one design, slots near the lamp socket allow convection air currents to flow past the reflector and lamp, carrying much of the dirt out through the slots. This can reduce maintenance costs and light losses.

The optimization of lighting systems should take into account the fact that a wide variability in electrical usage increases energy loss in the generation part of the system.

For the purpose of including lasers in this discussion, "light" was defined to cover the range of electromagnetic radiation between 10^2 and 10^4 nm. Recent developments in thermonuclear fusion research emphasize the importance of the efficient production of radiation in wavelength regions shorter and longer than that of "visible light" (400-700 nm). Furthermore, there is a tendency to increase the application of "light" in industrial processes. Research and development for the further improvement of light sources should be given the proper attention.

Lasers may have some long-term potential as efficient light sources, either alone or in specific applications. Primary energy sources for lasers can be electrical, optical, chemical, thermal, or combinations of these.

All aspects of lighting, including the resultant heat load, should be taken into account in the design consideration of buildings and houses. For example, the optimal requirements for outdoor lighting can be entirely different from those to be met for indoor lighting applications.

Investigations should be initiated to explore the potential for "central lighting" systems in which a central light source is used in conjunction with a light distribution system. In view of this general recommendation, the application possibilities of lasers and light pipes should be considered in detail.

Research should also be carried out to adapt the properties of "central lighting" systems to an integrated sunlight electric light system.

9.7 POTENTIAL ENERGY SAVINGS

Based on the considerations discussed in this chapter, it is reasonable to assume lighting could be reduced 10 to 15 percent by better controls, e.g., turning lights off when not in use and by delamping and relamping. More important would be the possibility of widespread substitution of fluorescent lamps for incandescent lamps in homes, commerce, and industry. This alone could reduce light enegy use by 10 to 30 percent, and still provide equivalent (or near-equivalent) illumination levels (see Table 9.2). In industry, widespread substitution of high-pressure sodium lamps for fluorescent could reduce electricity use by 20 to 50 percent. To stay on the conservative side, values of 10 to 15 percent, 10 to 15 percent, and, 20 to 30 percent have been assumed for *immediate*, *near-term*, and *long-term* savings respectively. This gives total savings of 40 to 60 percent of electricity use for lighting.

The potential savings are summarized in Table 9.8. This estimate is expressed in terms of an equivalent number of 1000 MWe power plants operating with a 60 percent capacity factor. (See Chapter 1 for additional details concerning the methodology and assumptions for these estimates.)

The implementation of these changes would be spread over a 10 to 20 year period. A substantial immediate improvement (at least 10 percent) could be made by better controls, delamping, and relamping. Generally this would cost very little and would yield a significant economic savings. An intermediate step would be relamping to more efficient lamps where different lamps and ballasts would be required. This would take some capital expense and one to two years to accomplish nationwide. The final step of conversion to more efficient lamps on a widespread scale would require a substantial investment and a 10 to 20 year period.

An important non-technological factor is related to observing (not exceeding) illumination guides. Careful differentiation between working and non-working areas is of particular importance in design; appropriate light levels should be provided for each. Finally, innovative designs will consider maximum use of natural light, such as is described in the Case Studies.

* * *

Energy Input for Lighting, Year 2000(1) (10^9 GJ)	Potential Savings, Year 2000(2) (%)	Number of 1000 MWe Power Plant Equivalents(3)
21.6	40-60	150-225

Notes

(1) Lighting is assumed to be 20% of US electricity use. Fuel for electricity is assumed to be 60% of total energy in the year 2000, or 180 x 10^9 GJ/yr x 0.60 gives 108 x 10^9 GJ/yr for all electricity uses, 22 x 10^9 GJ/yr for lighting.

(2) See text for source of estimates.

(3) Assumes 60% system capacity for lighting load, one 1000 MWe plant produces 1.9 x 10^7 GJe/yr = 5.7 x 10^7 GJ/yr.

POTENTIAL ENERGY SAVINGS IN US LIGHTING **TABLE 9.8**

REFERENCES

1. Personal communication, R.T. Dorsey, Manager, Lighting Development, General Electric Company, 13 May and 6 June 1975.

2. Kaufman, J.E., ed., *IES Lighting Handbook*, 5th edition, (New York: Illuminating Engineering Society, 1972).

3. Part of the material in this chapter has been excerpted from a study made by the American Physical Society under a subcontract to Applied Nucleonics Company. We acknowledge the contributions of Professor Marc Ross and the other APS members who contributed to this work:

   American Physical Society, *Efficient Use of Energy: A Physics Perspective*, (New York: January 1975).

4. Kovach, Eugene G., ed., *Technology of Efficient Energy Utilization*, Report of a NATO Science Committee Conference, Les Arcs, France, 8-12 October 1973, (Oxford: Pergamon Press, 1974).

5. Stevens, W.R. *Building Physics: Lighting*, (New York: Pergamon Press, Inc., 1969).

6. Henderson, R.L.; McNelis, J.F.; and Williams H.G., "A Survey and Analysis of Important Visual Tasks in Offices," presented at the Annual Conference of Illuminating Engineering Society, New Orleans, Louisiana, July 1974.

7. Stein, Richard G., "A Matter of Design," *Environment* 14(October 1972): 17-29.

8. Tansil, John, *Residential Consumption of Electricity, 1950-1970*, Report No. ORNL-NSF-EP-51, (Oak Ridge, Tennessee: Oak Ridge National Laboratory, July 1973).

9. Berg, Charles A., "Energy Conservation Through Effective Utilization," *Science* 181(13 July 1973): 128-138.

10. Taussig, Robert T., ed., *Energy Conservation in High Density Urban Areas: A Study of the Energy Crisis in Metropolitan New York City*, (New York: Columbia University, School of Engineering and Applied Science, 20 September 1974): 30-43.

11. US Federal Energy Administration, Office of Conservation and Environment, *Lighting and Thermal Operations--Guidelines*, (Washington, D.C.: 1974).

12. General Services Administration, *Conservation of Energy, Particularly Fuel Oil, in GSA-Operated Buildings*, Order PBS P 5800.34(Washington, D.C.: 28 November 1973).

13. Dubin, Fred (as reported by Margot Villecco), "Energy for Architects," *Architecture Plus*, (July 1973).

14. Appel, J. and MacKenzie, J.J., "How Much Light Do We Really Need," *Bulletin of the Atomic Scientists*, (December 1974).

CHAPTER 10

PROCESS HEAT AND HEAT RECOVERY

G.E. Howard*

CHAPTER CONTENTS

KEY WORDS

Combustion
Energy Conversion
Equipment Modification
Heat Exchangers
Heat Losses
Heat Recovery
Heat Sources
Heating Technique Modification
Insulation
Integrated Utility Systems

SUMMARY

This chapter describes more efficient methods for the use of heat sources and process heat systems. The emphasis is on electric heat, but some discussion on direct-fired systems is also included where the technology is applicable to a variety of equipment types. This has been a significant impact on electricity, since many components--pumps, fans, controls--use electricity.

For the *industrial sector*, generally applicable methods include:

- diligent management of heat sources (operating and maintenance practices);
- the use of on-line computer controls for operation of large thermal processing plants;
- improved electric furnace and oven designs and operating schedules;
- improved utilization of low temperature recovered process heat for lower level process use, building heating, or water heating;
- heat conservation by recycling of waste materials; and,
- possible use of high temperature recovered heat for process use or to generate electricity in conjunction with process steam systems.

Methods specific to certain industries are also discussed.

*Principal, Applied Nucleonics Company, Inc.

For the *commercial sector*, generally suitable methods include:

- careful attention to heating and air-conditioning system designs which minimize power and thermal requirements and maximize opportunities for heat recovery;
- more efficient water heating systems;
- more efficient refrigeration systems; and,
- improved equipment designs (e.g., ranges, warming ovens, etc.).

For the *residential sector*, methods similar to those applicable to commercial systems are appropriate, including:

- minimum power consumption systems;
- lower loss water heating systems and refrigeration systems; and,
- improved equipment designs.

10.1 THE POTENTIAL FOR IMPROVED PROCESS HEATING: A BRIEF SUMMARY

It has been estimated that six industries (iron and steel, petroleum refining, paper, aluminum, copper, and cement) account for approximately 40 percent of US industrial fuel consumption. With existing technology, it appears *possible* to reduce specific fuel consumption (energy utilized per unit of output product) in these industries by one-third; however, due to the capital investment required, actual savings "...will be somewhat lower."[1] This potential saving of one-third is quite significant, accounting for about 5 percent of total US fuel consumption.

Parallel studies have been conducted at the state and local level to evaluate fuel utilization and conservation opportunities. A study of industrial consumption in Texas focused on that state's petroleum refining, chemical manufacturing, pulp and paper, and metals industries.[2] The study concluded that an "aggressive energy conservation program" could achieve energy usage reductions of the following percentages over a five to seven year period:

Sector	Usage Reduction
Petroleum	30%
Chemical Manufacturing	31%
Paper and Pulp	20%
Metals	11%

It was estimated that the average medium-to-large firm conservation investment of $7x10^6$$ would have a payout period of one to three years.

A major fraction of US energy use is allocated to process heating end uses:

End Use (Amount of US Energy)

Industrial Process Steam (16.7%)
Industrial Direct Heat (11.5%)
Residential Space Heating (11.0%)
Commercial Space Heating (6.9%)
Residential and Commercial
  Water Heating (4.0%)
Total (50.1%)

Thus, approximately one-half of the total annual energy use of the United States may be attributed to heating applications. Direct heat and process steam raising in the industrial sector alone account for somewhat more fuel consumption than required for all forms of transportation. The fuel required to operate industrial direct heat apparatus is comparable to that consumed by automobiles.[3]

The steady-state heat transfer efficiency of individual items of direct heat plant equipment is often on the order of 20 to 30 percent (efficiencies as low as 5 percent have been reported). Hot water heating, which represents about 4 percent of total annual US energy utilization, is an example where improved efficiency could have national significance. Although under ideal conditions efficiency for water heating is in the range of 60-80 percent, dirt accumulations on heat exchange surfaces and the effect of transient operations can reduce practical efficiencies 35 to 50 percent. Residential and commercial cooking represents 1.3 percent of total energy utilization, although the heat transfer efficiency may be quite low and redesign could increase this efficiency substantially. For example, Berg estimates that the heat transfer efficiency of the common tea kettle is currently 15 percent, but could be 60 percent if properly redesigned.[3] Proper design of heat transfer surfaces is particularly important for electric heating applications which may depend primarily on conduction.

In view of the large energy utilization devoted to heating applications, and the performance of typical equipment and systems for heating, the application of technology to improvement in these operations can yield nationally significant reductions in energy utilization.

Large amounts of energy are required to provide low temperature heat for such purposes as hot water heating, space heating, cooking, and drying. The burning of high quality fuels ("high" temperatures are >1000°C) to obtain low temperature heat is fundamentally wasteful, even if all the heat could be transferred without losses. The determination of efficiency should be based upon a comparison of the amount of fuel actually consumed in a process with the minimum amount of fuel that might be used in the process. As an illustration, let fuel be burned at 1000°C (T_c = 1273°K) and the high temperature heat of combustion (Q_c) be used to drive a power plant rejecting heat to the atmosphere ($T_o \sim$ 20°C, or 293°K). If the work obtained is used to drive a heat pump to extract heat from the atmosphere and to provide heat (Q_p) to a low temperature process at T_p (say, hot water of 50°C or 323°K), the ratio of the heat of combustion (Q_c) to the heat delivered to process (Q_p) is

$$Q_c/Q_p \geq \frac{1 - T_o/T_p}{1 - T_o/T_c}$$

$$\geq \frac{1 - \frac{293°K}{323}}{1 - \frac{293}{1273}} = 0.12$$

The ratio (Q_c/Q_p) represents the ratio of the fuel required to operate a process by the heat engine (heat pump) to that required by direct combustion with *100 percent efficient heat transfer*. This sample calculation shows that direct combustion in hot water heating requires many more times the fuel energy than would be needed with a heat pump system. Since the best heaters actually have an efficiency of less than 70 percent, their ratio is at best 0.12 x .7 $\sim$ 0.08.

The example illustrates that *work* rather than heat is the valuable energy form. Once work is extracted, it may be used with high efficiency to provide low temperature heat or mechanical power. The extent to which the energy of a supply may be made to yield useful work brings one to the concept of *thermodynamic availability*.[4] (See Appendix C.) The essence of this concept is that work, rather than heat, is the intrinsically valuable energy form and that the extent to which a given fuel can be made to do work is the proper measure of value for that fuel.

The most significant losses in thermodynamic availability, and the subsequent wastage of fuel, occur in two areas: (1) in the combustion process because of irreversible high temperature heat transfer phenomena; and (2) at the point of consumption (where the efficiency of energy utilization may be based upon a comparison between the minimum thermodynamic availability required for a process and the actual availability consumed in the process). [1]

Point of consumption processes for low temperature operations are extremely costly in terms of availability (e.g., 15 times the theoretical minimum is consumed in home heating with gas and oil, and ten or more times the theoretical minimum for water heating). [1] Similar numbers hold for electric resistance heating.

In high temperature operations the efficiency of availability usage is about the same as the heat transfer efficiency for the process. Thus, reductions in fuel consumption in high temperature processes appear obtainable through heat transfer and control improvement; low temperature process improvements will require the third step of process modification--e.g., the use of electric heat pumps rather than direct combustion for space heating.

Availability provides a measurement of the quality *and* quantity of an energy form. It is in the provision of relatively low temperature heat (space heating and cooling, industrial process steam, etc.) that the most wasteful consumption of available work occurs. The example of a furnace is instructive. Its first law efficiency is 60 percent; its second law efficiency, where it provides air at 110°F (316°K) and outside air is at 32°F (273°K), is only 8.2 percent. (See Appendix C, Section C.)

Perfectly reversible equipment operation (and thus negligible losses in availability) would require infinite investment in heat transfer apparatus, controls, etc. This is not a realistic goal. Nevertheless, there are wastes which are not justifiable on a rational economic basis, and improvements are possible.

10.2 PROCESS HEAT SOURCES

Electric Heat Sources

Electricity is widely used as a source of process heat due to ease of control, cleanliness, wide range in unit capacities (watts to megawatts), safety, and low initial cost. Several different physical principles are employed. Resistance heating is most common, but systems using induction heating, dielectric heating, electric arcs, or high frequency electromagnetic energy (microwaves) are also found. Table 10.1 summarizes typical types of electric heaters and their applications.

Process heaters are of the electrode type for direct immersion in the processed material inside process vessels, for direct fluid immersion in pipeline type heaters, and for radiant heating applications in industrical equipment, commercial equipment, and residential appliances. Control is generally achieved by automatically monitoring process material or fluid temperature and adjusting control system input current or voltage. Many commercial and residential units are operated on simple on-off controls around a preselected temperature set point.

For direct immersion elements operated on a continuous basis at high temperatures, the efficiency at the point of use approaches 100 percent. If the process is batch-type or operated intermittently, then the system thermal mass becomes a factor. In the case of melting processes for solids, until melting occurs radiation and convection of heat will cause losses through the material out the top of the vessel. After melting there will be conduction, convection, and radiation losses as molten materials are poured and processed. Vessels are generally well-insulated.

Electric arc, steel making furnaces are an example. Energy management options in arc furnaces include better maintenance of refractories to minimize heat loss; maintenance of roof and doors to reduce heat loss; careful control of electrode feed rates; and control of furnace cycles to avoid peak loads.

Typical energy use in a direct arc furnace is:[5]

	kWh/ton	%
Heat content of steel	370	63.9
Heat content of slag	22	3.8
Exothermic reaction	-31	-5.3
Heat in waste gases	15	2.6
Heat in cooling water	25	4.3
Electrical losses	45	7.7
Radiation and misc. losses	134	23.0
TOTALS	580	100

Thus, about 38 percent of the input energy is dissipated and potentially could be saved by improved equipment design or heat recovery.

Electric furnace or oven design offers several opportunities for improved efficiency. Improved insulation is a first step, as is the use of better seals to reduce heat loss while charging. Waste heat can be directed to preheat feed materials. Rather than rely solely on radiative or conductive heat transfer, oven design should incorporate fans for improved performance using convective heat transfer.

Type	Applications	Electrical Characteristics
1. Resistance Heaters 1.1 Metal Sheath Heaters (strip, finned, cartridge, tubular, immersion, and cable heaters)	Temperatures: 40-650°C (100-1200°F) Surfaces (platens, dies, hot plates) Liquids (chemicals, electroplating) Air (ovens, comfort heating) Metal Melting (solder, lead) Pipes (winterization processes) Radiant (paint, ink drying)	120-480 Volts, 0.2-2.0 kW, 50-60 Hz, Typical installations use multiple heaters to achieve 1-10 W/cm^2 (6.5-65 W/in.^{2})
1.2 Resistance Ovens (batch type ovens, including cabinet or bench, truck-loaded, elevator loaded, etc.; and conveyor types, both horizontal and vertical)	Temperatures: 40-540°C (100-1000°F) Baking (paints, abrasive wheels, armatures, chemicals, foods) Annealing (aluminum, glass) Aging (metals and finishes) Curing (textiles, plastics) Drying (washed parts, ink) Dehumidifying and Moisture Control	120-480 Volts, 1-100 kW, 50-60 Hz
1.3 Resistance Furnaces (batch type furnaces, including box, car-bottom, elevator, bell, and vertical types; and continuous types: rotary hearth, pusher, roller, walking beam, etc.)	Temperatures: 540-1370°C (1000-2500°F) (Natural or artificial atmospheres are used) Heat Treating (annealing, tempering, hardening, nitriding) Melting Metal Working (forging, rolling, hot extrusion, sintering, welding, etc.) Other Processes (calcining, flux drying, tinning, etc.)	1-1000 kW
1.4 Electric Salt Bath Furnaces (externally heated types, generally limited to 900°C [1650°F]; and internally heated, with immersed heaters, immersed electrodes, or liquid electrodes)	Temperatures: 175-1315 °C (350-2400°F) Carburizing, Cyanide Hardening Nitriding (tool steel) Brazing (steel, copper, aluminum) Quenching Heat Treating, Tempering Descaling	120-480 Volts, 1-100 kW, 50-60 Hz

Reference 5

ELECTRIC HEAT APPLICATIONS **TABLE 10.1**

Type	Applications	Electrical Characteristics
1.5 Infrared Heaters (glass heat lamps, tubular quartz lamps and heaters, metal sheath heaters, panel and ribbon heaters)	Temperatures: 40-1650°C (100-3000°F) Product Heating (heat treating metals, brazing, soldering, bonding, laminating, cooking, roasting, baking, etc.) Drying, Dehydrating (metals after cleaning, foundry cores, textiles, sterilizing bottles, leather, etc.) Baking or Curing (paint, enamel, resins, plastics, papers, wood) Space Heating (airport terminals, warehouses, gymnasiums, foreman's booth, snow melting on sidewalks)	120-960 Volts, 0.1-5.0 kW, 50-60 Hz
2. Induction and High Frequency Resistance Heating (high frequency, low frequency, and dual frequency units)	Temperatures: 150-1650°C (300-3000°F) Melting (core and coreless-type furnaces, vacuum melting) Centrifugal Casting Tin-Flowing on Steel Continuous Casting of Steel Slabs Hardening and Quenching Stress Relieving, Normalizing, and Preheating Shrink Fitting, Tube Welding, Hot Forming, Forging, Extruding	120-13,200 Volts, $10\text{-}10^6$ Hz 1-6,000 kW
3. Dielectric Heating	Temperatures: 100-350°C (212-660°F) Moisture Removal Plastics (thermoplastic and thermosetting resins, molding, forming) Wood (drying, curing, gluing, bonding, laminating) Foundry Uses (core baking) Food (cooking, thawing)	120-960 Volts 1-500 Megahertz 0.2-400 kW
4. Direct-Arc Electric Furnace	Temperatures: 1000-1650°C (1800-3000°F) Steel Making (specialty steels, low carbon steels) Foundries (iron and steel) Smelting (copper, silver, etc.)	480-138,000 Volts 50-60 Hz 500-150,000 kW

TABLE 10.1 CONT'D

Most commercial and residential heaters are radiant units provided only with minimal insulation, since all heat presumably goes into occupied spaces. Because of the relatively coarse controls, precise temperature control is uncommon. Ceiling-type electric residential heating units are efficient in rejecting all their heat to the ceiling material but, unless suitably backed with insulation, much of the heat can find its way to attic spaces where it is of little value and is eventually lost. As discussed earlier, low temperature applications of heat for space heating have a large cost in terms of thermodynamic availability.

Direct-Fired Heaters

Direct-fired heaters depend on the heat of combustion derived from burning coal, various grades of oil, other liquid fuels (propane, for example), or gas (natural or manufactured) in the heater. Heat transfer is usually achieved within the heater by indirect means, with the products-of-combustion gases flowing around tubes containing the material or fluid to be heated, or vice versa. In cases where the products of combustion are not detrimental to the process, direct mixing of the heating and heated fluids can be employed.

Direct-fired heaters are common to all sectors (industrial, commercial, residential, and agricultural). The application of a heater to a particular process can be as critical to its efficiency as the design of the equipment. For example, the need to leave a heater in operation for extended periods with no useful work being done on the process is fairly common. This may be done to avoid time delays due to shutdown and startup cycles, extra labor charges associated with shutdown and startup operations, or thermal cycle limitations on the equipment. The energy required during such periods can be a significant fraction of the total.

Inefficiencies in direct-fired heater operations may result from the following:

- internal losses associated with the physics of combustion (thermodynamic irreversibility);
- improper fuel-air ratios or combustion processes due to faulty or improperly adjusted or maintained combustion equipment;
- heat lost in stack gases exhausted from the heater;
- heat rejected to the surroundings from the heater casing by conduction, convection, and radiation;
- heat lost to the surroundings due to casing leaks and openings;
- heat lost due to process requirements to open the heater to insert and withdraw process materials; and,
- heat required to facilitate the operation of pollution control devices.

Each particular heater design is not subject to all of the above losses but will generally experience most of those listed to a greater or lesser degree. Heater efficiencies are frequently in the range of 30 to 40 percent; substantial improvement is often possible through the use of waste heat recovery devices. All of the above losses relate in a major way to the design and construction of the heater or to the application of a particular heater design to a specific process.

Figure 10.1 illustrates fuel efficiencies as a function of stack gas temperature for a variety of fuels in large boilers. Factors affecting efficiency are stack gas temperature, fuel analysis and fuel heating value, excess air, unburned carbon, and radiation losses. Efficiency here refers to that percentage of the fuel's gross energy, or "higher heating value" (HHV), that is recovered by the boiler and associated equipment; heat input preheating fuel and air is not considered

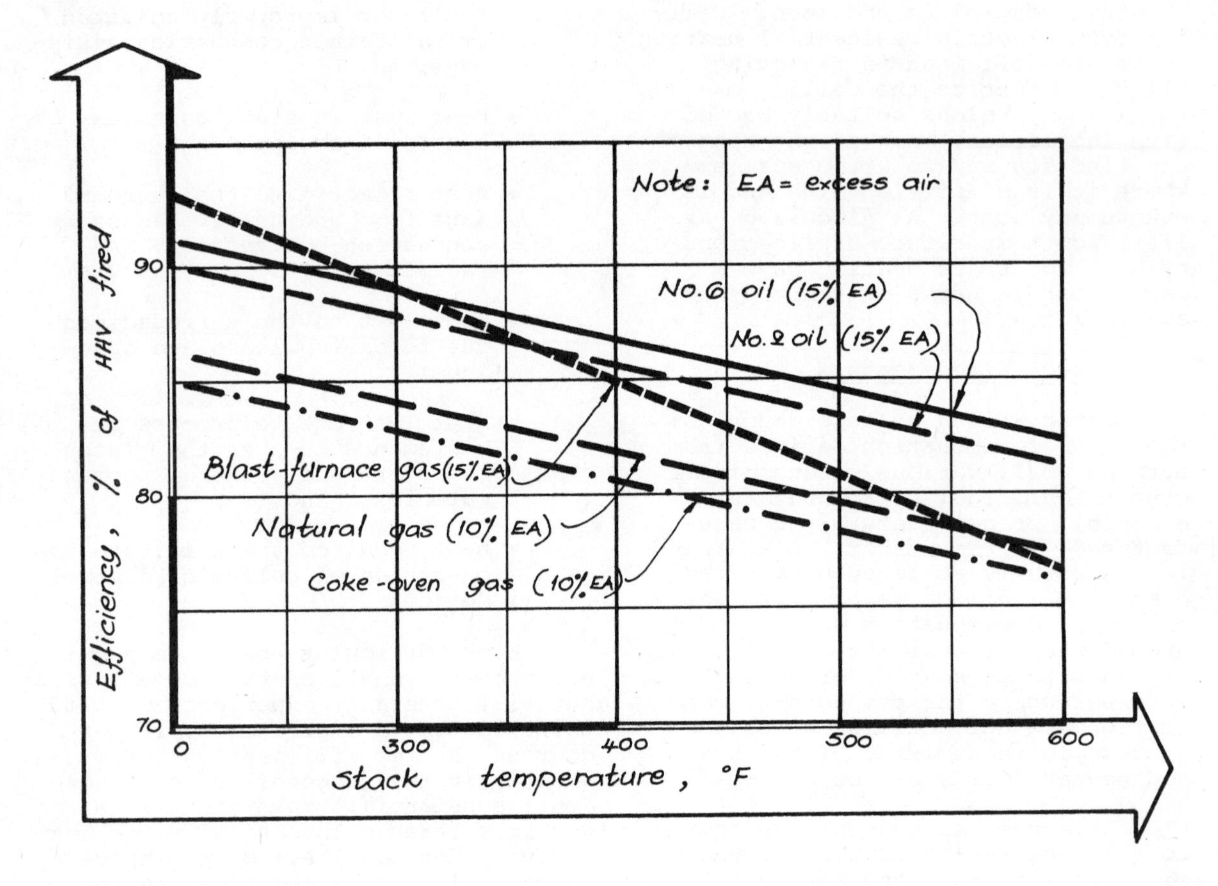

Reference 6

TYPICAL FUEL EFFICIENCIES FIGURE 10.1

Combustion control is particularly important to direct-fired heater efficiency. Fuel consumption may be reduced up to 5 percent by careful control. Approaches currently employed include:

- positioning controls to maintain predetermined fuel and air flow area ratios;
- pressure controls to maintain a desired pressure ratio between fuel and air fluid supplies;
- metering controls to maintain preset volume or weight flow ratios of fuel and air; and,
- simple on-off firing controls with a constant built-in fuel-air ratio (usually on small units).

The impact of close combustion control is illustrated in Figure 10.2 from which one may perform a quick economic impact evaluation.

Indirect-Fired Heaters

Indirect-fired heaters (heat exchangers) are widely used in industrial and large commercial applications and, to a lesser degree, in small commercial and residential applications. All involve the exchange of heat for heating (or cooling) applications, usually between two fluids, where one of the fluids has been previously supplied with heat in another process unit (usually a direct-fired heater). The geometry of these heaters varies considerably to suit the specific application but for purposes of discussion the shell and tube type is commonly used. The higher pressure fluid is generally routed through the tubes which are surrounded by the lower pressure fluid. Indirect-fired heaters are used in lieu of direct-fired units for a variety of reasons including local space limitations, economy of size in using a larger central-fired unit to serve several indirect units, better controllability for certain process applications, and safety (no fire or flame).

Control of the heater (exchanger) typically is achieved by monitoring the exit temperature of the fluid being heated and adjusting the heating fluid accordingly. Occasionally this is done manually, but more often, especially on larger installations, it is performed by an automatic temperature controller. The heat energy of the heating fluid can be adjusted by varying the flow rate, by mixing with recycled fluid for temperature control, or by adjusting the temperature of the heating fluid at the remote heating source.

Most exchangers are used in a closed loop system so that all heating fluid (and residual heat) is returned to the heat source and recycled. Heat losses through the insulation on piping to and from the exchanger and on the exchanger itself, heat losses associated with fluid leaks in the system, and losses in the heat source all contribute to the decrease in efficiency of the overall system. If the system is not operated on a continuous process basis, then the thermal mass (capacitance effect) of the material in the system becomes a factor. Because of design and construction differences, it is not possible to generalize on system efficiencies. However, heat losses are controllable to a large degree with appropriate selection of insulation, and piping and insulation efficiencies are primarily a function of economics (energy costs and insulation costs).

For those installations where indirect equipment must be used, the only latitude is in choice of operating temperature and insulation. The system must be maintained leak-free as well. For other installations, the overall system (including the remote source) must be evaluated.

10.3 GENERAL ENERGY MANAGEMENT TECHNIQUES APPLICABLE TO PROCESS HEAT

General techniques available for more efficient management of process heat fall into three categories:

- reduction of heat losses,
- more efficient equipment or processes, and
- heat recovery.

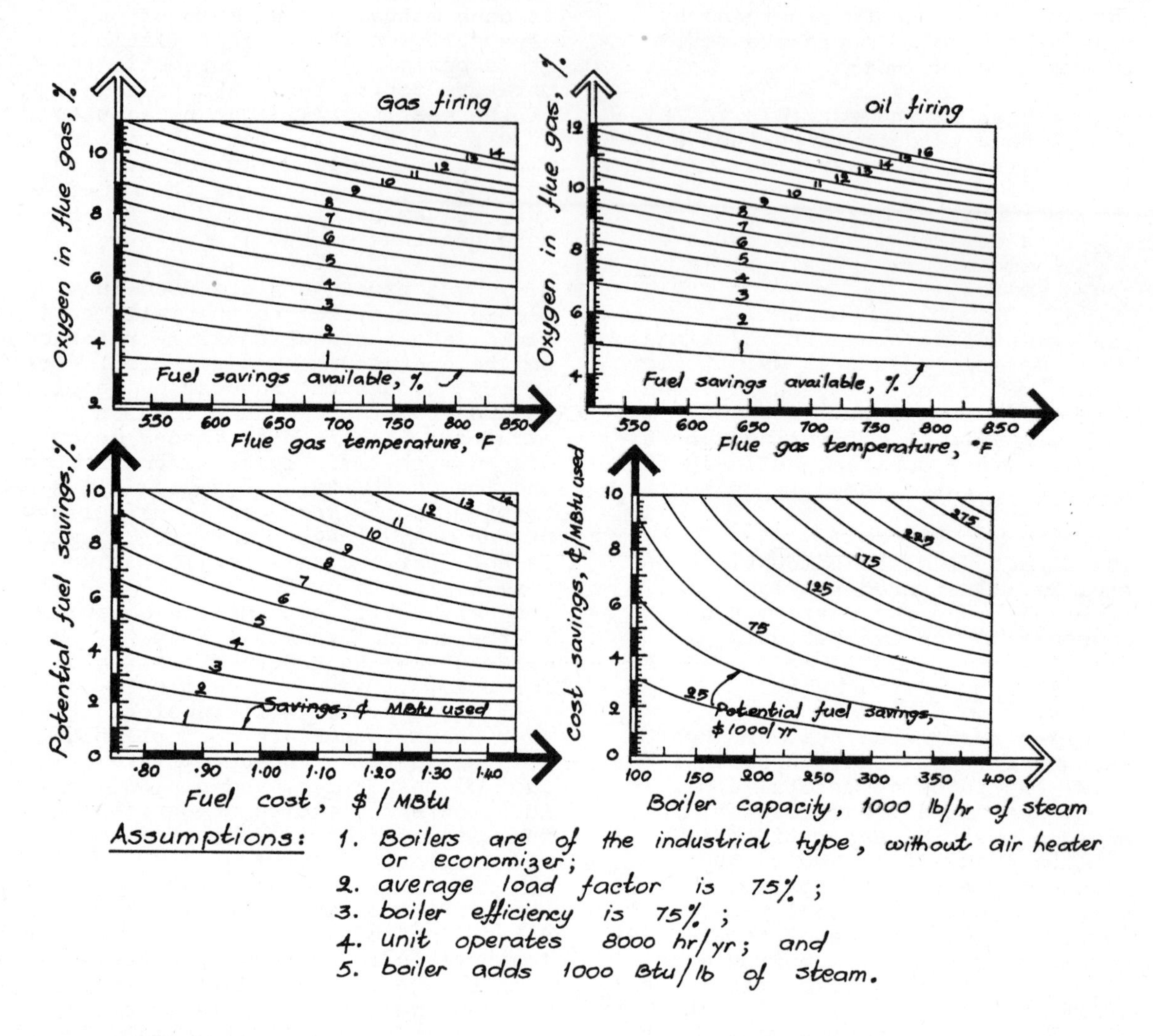

Assumptions:
1. Boilers are of the industrial type, without air heater or economizer;
2. average load factor is 75%;
3. boiler efficiency is 75%;
4. unit operates 8000 hr/yr; and
5. boiler adds 1000 Btu/lb of steam.

Reference 6

POTENTIAL FUEL SAVINGS FOR BOILERS FROM FLUE-GAS ANALYSIS FIGURE 10.2

These apply for both electric and non-electric heat sources. The first two categories are discussed next. Heat recovery is of such importance as an energy management technique that it is discussed separately.

Table 10.2 presents general ideas and concepts one might consider when reviewing opportunities for improving energy utilization efficiencies in any process heating situation.

Heat Loss Reduction Through Insulation

Traditionally, electric heating systems have had less heat loss than fuel-fired systems due to higher cost (and the need to keep operating costs down to remain competitive), better heat transfer (application of heat directly to the work), and better controls. Still, reduction of losses from electric heating systems is an important concept.

Insulation is a significant variable in the design of electric process heat systems. While wall and ceiling areas are obvious candidates in buildings, mechanical equipment insulation can be very important also. Equipment insulation is of particular relevance in the industrial and commercial sectors.

Improved piping, furnace, and other relatively high temperature applications can provide very attractive investment opportunities. Energy utilization per unit of product can be reduced 10 percent to 20 percent in particular applications, thus permitting increased production for a constant energy input or (of growing significance) permitting continued production in the face of escalating energy costs.

As an example, consider the following case:

A glass company's bending furnace was modified--reinsulated with low density ceramic felt material at a cost of $1750. The 45-foot (15 m) long by 6-foot (2 m) wide furnace operated 16 hours per day with one cycle per day. Following modification, energy usage dropped by 19 percent, and an annual cost saving of $924 was achieved, for a payback of less than two years. In addition to savings which will grow with increased energy costs, other advantages of the increased insulation included: a 10 percent production increase due to faster heat-up; improved working environment due to lower heat leakage; more uniform furnace conditions across furnace width.[7]

The details of evaluating heat losses per unit area may be found in any standard engineering reference or in manufacturers' catalogs. From knowledge of these heat losses, the cost of energy, thermal efficiency of the heating process, and hours of operation, the cost of heat losses per unit area may be calculated. By estimating the capital cost of insulation, the annual cost of insulating may likewise be determined (it should include a provision for maintenance costs). The economic thickness of insulation is that thickness for which the sum of the cost of the heat loss plus the cost of insulation is a minimum.

As an example, Table 10.3 illustrates the magnitude of losses of bare and calcium silicated insulated piping for a specific case of a 260°C (500°F) pipe temperature, 26.5°C (80°F) ambient temperature. Table 10.4 illustrates the next step, a comparison of annual heat loss costs and savings for bare and insulated pipe for an energy cost of $2.50 per 1000 pounds of steam. For example, if one examines the case of a six-inch pipe under the given conditions, the first inch of insulation yields a saving of $81.66 per foot per year and a remaining heat loss cost of $9.49 per foot per year. Specifying two inches saves an additional $3.84 for a remaining loss of $5.65 per foot per year.

Heat input into electric ovens and furnaces includes:

- heat absorbed by the work;
- heat absorbed by auxiliary equipment (trays, conveyors, etc.);
- heat absorbed by the oven or furnace;
- standby losses through walls and doors;
- open door losses; and,

- Preheat feedstocks or combustion air with waste heat from electric or other heaters.
- Reduce disaggregation of process requiring heat/cooling/heating cycles; use continuous processes where possible (continuous digesters in paper industry reduce fuel consumption by 1.3 x 10^6 Btu/ton).
- Convert processes requiring scarce fuels, e.g., use solar or microwave in pulp drying rather than natural gas; use mechanical pressing to reduce drying energy demand.
- Reduce moisture content in process operations before heating or convert to dry processes--for example, in cement, water evaporation in heating requires $\sim$1 x 10^6 Btu/ton, or $\simeq$ 12.6% of total fuel used. Or, use screw, centrifugal, or pneumatic presses to dewater products before drying.
- Examine process modification (temperature, pressures)--e.g., Japan coke consumption is 20 percent less than US in steel making because of higher air blast temperatures and higher top gas pressures; quench coking with recycled inert gas through heat exchangers (Europe) rather than water.
- Generate electricity or mechanical work while raising process steam.
- Increase recuperation in all combustion systems used in direct heating.
- Improve efficiency of steam boilers; reduce steam losses in standby equipment.
- Improve combustion efficiency in all direct-fired heaters (excess air).

GENERAL ENERGY MANAGEMENT TECHNIQUES FOR PROCESS HEATING **TABLE 10.2**

- Increase heat interchange between process streams (recuperators, run-around systems).
- Use hydraulic turbines to receive mechanical power from the depressurization of high pressure or duct streams.
- Reduce convective, radiation losses in heat applications--US cement kilns have losses of ∿17 percent of total fuel input.
- Use more efficient ovens, kilns, heat transfer devices, boilers, etc.
- Use mechanical vapor recompression in evaporation processes.
- Use microwave drying and cooking rather than steam.
- Shut down process heating equipment when not in use, or at least reduce temperature.
- Convert liquid heaters from underfiring to immersion or submersion heating.
- Replace steam use by high temperature water -- eliminate steam losses.
- Use infrared dryers where applicable in textile operations, induction heating of steel slabs, etc.
- Use waste heat boilers, waste heat drying, waste heat to preheat furnace charges, etc. Convert rotary kilns for calcining alumina to fluidized bed calciners.
- Use water based solvent systems to eliminate need for afterburners on coating ovens, or use coatings that do not require firing.
- Use cold wash processes or lower temperature of hot wash processes.

TABLE 10.2 CONT'D

Pipe Size In.	Bare Pipe, Btu/ft-hr	Insulated pipe, Btu/ft-hr Thickness of insulation, In.					
		1	2	3	4	5	6
1	647	87	63	53	47	43	41
2	1127	131	88	71	62	56	52
3	1623	179	114	90	78	69	63
4	2056	214	136	105	89	80	72
6	2969	309	184	138	114	99	87
8	3810	389	228	170	136	118	107
10	4696	475	267	198	162	139	123
12	5546	556	309	230	184	158	140

Note: Conditions 260°C (500°F) pipe and 26.5°C (80°F) ambient temperatures; calcium silicate insulation

*Reference 6

HEAT LOSS FROM BARE AND INSULATED PIPE

TABLE 10.3

Pipe Size In.	Bare Pipe Cost, $/ft-yr	Insulated pipe, $/ft-yr Thickness of Insulation, In.											
		1		2		3		4		5		6	
		Cost	(Saving)	Cost	(Saving)	Cost	(Saving)	Cost	(Saving)	Cost	(Saving)	Cost	(Saving)
1	19.86	2.67	17.19	1.93	0.74	1.63	0.30	1.44	0.19	1.32	0.12	1.26	0.06
2	34.48	4.02	30.46	2.70	1.32	2.18	0.52	1.90	0.28	1.72	0.18	1.60	0.12
3	49.83	5.50	44.33	3.50	2.00	2.76	0.74	2.39	0.37	2.12	0.27	1.93	0.19
4	63.12	6.57	56.55	4.18	2.39	3.27	0.91	2.73	0.54	2.45	0.28	2.71	0.24
6	91.15	9.49	81.66	5.65	3.84	4.23	1.42	3.50	0.73	3.04	0.46	2.67	0.37
8	116.97	11.91	105.03	7.00	4.94	5.22	1.78	4.18	1.04	3.70	0.48	3.28	0.42
10	144.97	14.58	130.39	8.20	6.38	6.08	2.12	4.97	1.11	4.27	0.70	3.77	0.50
12	170.26	17.07	153.19	9.49	7.58	7.06	2.43	5.64	1.42	4.85	0.79	4.30	0.55

Note: Conditions 260°C(500°F) pipe and 26.5°C(80°F) ambient temperatures; steam, $2,50/1,000 lb; calcium silicate insulation.

ANNUAL HEAT LOSS AND SAVING FROM BARE AND INSULATED PIPE

TABLE 10.4

ventilation losses.

This breakdown suggests opportunities for energy management. The energy input to the oven and the work can be calculated from the mass, specific heat, and temperature difference. Standby losses depend on specific equipment designs, but Tables 10.5 and 10.6 show typical numbers.

To illustrate the importance of standby losses, consider this example for an electric salt bath heater operating at approximately 850°C:

	kW	%
Heat in product	13.8	43
Top losses (no cover)	7.8	57
Side wall losses	6.7	
Electrode and transformer losses	3.7	
Total	32	100

With the furnace in production, but the cover closed between loading and unloading, the losses are reduced from 18.2 kW to 10.7 kW (57 percent to 43 percent).

As a rule of thumb, low temperature applications should have at least 2.5 cm of insulation for each 50°C increase in temperature (1 inch per 100°F).[5]

Adding insulation to pipelines and equipment in industrial plants can be economically attractive as well as feasible since equipment is usually accessible. Retrofit of mechanical installations in commercial facilities is usually more difficult to justify. It has been estimated that about 80 percent of mechanical systems are concealed, with attendant high costs in gaining access; the remaining 20 percent is typically in equipment room areas and can be economic to retrofit.[6] Insulation calculations can indicate the benefits.

More Efficient Equipment or Processes

Energy management opportunities in this category vary widely depending on the specific process. While a complete delineation of the possibilities is outside the scope of this chapter, several examples can be given to illustrate the possibilities.

In the industrial sector, use of microwave drying as an energy efficient replacement for scarce fuels is an attractive possibility. Potential applications include drying of agricultural crops, paper production, paint and finishes, and water removal from chemical production.

Additional conversions can and are being made from open hearth to electric furnace operations in the iron and steel industry. Since the electric furnace uses such a high percentage scrap charge, the extent to which it can be employed will be limited by available in-plant generated scrap and the availability of purchased scrap. Increasing use of the electric arc furnace is to be expected in the future.

Power requirements for industrial heaters are dictated by factors such as heat-up times, throughput, and unit capacity. Consideration should be given to scheduling units for off-peak operation where possible. Watt density ratings (W/cm^2) must be selected to be appropriate for the job and not require excessive capacity. In water, relatively high densities are possible (6-8 W/cm^2), but in oil lower ratings (2-3 W/cm^2) must be used to avoid carbon encrustation on heating elements. Choice of heater type, sheath material, etc., depends on the temperature range and compatibility with the product.

Besides choice of insulation and heating element size, the designer can select either forced or natural circulation. In most cases forced circulation provides better heat transfer, more uniform temperatures, and more efficient designs.

A unique application of energy management is found in the continuous casting electric arc furnace.[5] Molten metal is poured from the furnace or from a heater reservoir through a downspout. The molten metal is formed up in a cooled reciprocating mold. By the time the metal leaves the mold, it has solidified in slab form and is ready to be rolled. As the bar leaves the roll, it is sawed off in convenient

Insulation Thickness*	Temperature Difference °C			
	100	200	300	400
2.5 cm	700	1200	1780	2310
5.0 cm	350	600	890	1155
7.5 cm	230	400	590	770
10.0 cm	175	300	445	580

* Insulation assumed to be 85 percent magnesia, Rockwool, etc.

Reference 5

APPROXIMATE HEAT LOSSES FROM INSULATED OVEN OR TANK WALLS (in W/m²)

TABLE 10.5

Liquid	Temperature, °C (Air at 27°C)						
	50	75	100	200	300	400	500
Water	1.3	5.0	10.1*	--	--	--	--
Oil & Paraffin	--	--	1.5	4.5	9.3	--	--
Metals	--	--	--	--	4.5	9.5	15
Molten Salts	--	--	2.5	3.7	5.0	11.3	21

*Actually 93°C. Air relative humidity assumed to be 70%. Note that air movement significantly increases these values. For water, with air velocity of 3 m/sec, heat losses are three times as great; at 15 m/sec, about 12 times greater.

Reference 5

APPROXIMATE HEAT LOSSES IN kW/m² FROM LIQUID SURFACES IN STILL AIR

TABLE 10.6

lengths. This process avoids the heat losses in a conventional operation which requires reheating of the metal.

Other major savings opportunities in the industrial sector may be obtained by:

- large-scale use of recuperators,* regenerators, and low temperature heat engines for the recovery of waste heat;
- use of unfired and supplementary fired waste heat for heating of water and other fluids; and,
- industrial cogeneration of electricity (where practical).

Use of submerged conduction type direct-fired heat exchangers with efficiencies of 80 to 90 percent should be considered in lieu of more commonly used indirect-fired heat exchange cycles with overall efficiencies of 40 to 60 percent.

Development of improved industrial heaters and furnaces is needed, both fossil fuel-fired and electric, which could have efficiencies much higher than the now prevalent 30 to 40 percent.

Another general approach for reducing industrial process heat involves a change of process or substitution of materials.

Listed below are examples of techniques for saving both direct and indirect energy by reducing material consumption and waste through material changes, either in equipment and systems, or in basic processes.

- use materials that are less energy intensive to process (e.g., powder metallurgy vs foundry operations);
- use materials with lower thermal capacity where thermal cycling is required;
- use high solids type paints with ultraviolet curing in lieu of conventional solvent-based paints requiring high temperature bake curing;
- use higher yield ore in raw material feed to blast furnaces for iron and steel manufacture;
- switch some raw materials to natural, replenishable materials instead of using petroleum-derivatives from petroleum refining;
- for foods and kindred products, emphasize easier (lower energy) foods to process; and,
- use a higher percentage of natural fibers in fiber blends with synthetics for impact on textile use.

In the commercial sector, gains can be realized from the use of more efficient heat sources and heat systems such as heat pumps. Furthermore, existing systems can be improved by reducing losses (for example rotating equipment seals) and by proper equipment sizing (see Chapter 3), where careful matching of heating equipment and load leads to greater efficiency. Reduction of heat losses in buildings--either by greater efficiency or by heat recovery--has a dual advantage, since fuel consumption is reduced at the same time the air-conditioning load is decreased.

More efficient, small space heating and air-conditioning units should be developed. In many instances these will use heat pumps. Microwave techniques should be considered increasingly for commercial cooking, drying, and food processing.

For water heating greater use of instantaneous, local electric heating units for small users, instead of larger remote electric or gas-fired units, will save energy, particularly standby and piping losses. The use of electric-start intermittent gas pilots ("electric

*Recuperative air heaters transfer heat directly across an interface between two gas flow streams; regenerative devices use an intermediate medium such as rotary heat exchangers (heat wheel) and pebble heaters.

igniters") instead of continuous pilots also saves energy. They are available on industrial equipment but typically are not used on small commercial units. More efficiently located hot water storage tanks eliminate long runs of high thermal mass, high heat loss pipe.

In the residential sector, more efficient space heating and air-conditioning units, similar to those used by the commercial sector, are required. For water heating, observations similar to those made for the commercial sector apply. For clothes drying, advantages could be obtained with better insulated recirculation units. Currently available units are generally uninsulated. The use of recirculation of hot air through a moisture "getter" or alternative hot air from air cycles could save heating energy. For refrigerators and freezers, better interior access units would save energy. They would allow contents to be removed more quickly and shorten the "door open" time. Additional insulation could be used and more efficient compressors could be developed.

Here also the use of more efficient heat sources is important. These can range from microwave ovens for cooking to heat pumps and solar-assisted space and water heating. (See Chapter 4 for discussion and examples.) The discussion (above) for the commercial sector is also relevant.

Use of internal refrigerator compartment doors to reduce heat gains with the door open is beneficial, as is the use of on-off controls on mullion heaters. Such a simple step as including a switch to turn off the heaters has resulted in reductions of 15 percent in power consumption. The case studies detail these methods of improving residential refrigerator efficiency.

10.4 HEAT RECOVERY TECHNOLOGIES

Heat recovery is an energy management tool of fundamental importance. As broadly considered herein, heat recovery refers to any of the following techniques:

- heat recovery, regeneration, or recuperation in various industrial operations including electric ovens or kilns, air compressors, chillers, process equipment, or boilers. Heat can also be recovered from process streams, waste service hot water, etc.;
- heat recovery from building heating, ventilating, and air-conditioning systems;
- heat recovery by use of wastes or by-products (waste energy reclamation); and,
- heat production made possible by better fuel utilization through combined cycles.

Heat recovery can reduce energy utilization, permit increased production for a given utilization, or effectively extend available energy supplies.

Economically attractive applications of heat recovery under diverse conditions support the case for heat recovery and include the following typical examples from Reference [7]:

- By recovering heat from the exhaust of a drying oven, space heating for a midwestern (6588 degree days) plant was supplied. A $5600 heat wheel installation produced annual cost savings of about $4100/year; payback was 1.4 years.
- A Kaolin powder manufacturer used a recovery system to preheat combustion air from an air-hot powder heat exchanger. Fuel savings of $67,000 per year (at $1.2 per million Btu) were realized for an investment of $90,000, for a payback of less than two years.
- A textile manufacturer identified an opportunity to save $34,000 (net) per year through recovery of 10 million Btu/hr from hot waste water at 65°C (150°F). For a five-year capital recovery period, the estimated annual amortization cost of the installation was $6000/yr providing a fuel saving of $40,000 per year at 0.85 $/GJ ($0.90

per million Btu) while preheating incoming process water.

- A metal finishing plant added an incineration unit to a large paint curing oven in order to discharge only clean air to the atmosphere. Heat recovery from the 760°C (1400 °F) incinerator exhaust was used to preheat the incinerator inlet stream and, upon leaving the heat exchanger, sent to a waste heat boiler to generate 50 psi plant steam. Total energy recovered was about 50 percent of the heat generated in the incinerator, for an annual fuel saving of $120,000 at 1.89 $/GJ ($2.00 per million Btu).

In reviewing opportunities for heat recovery in any facility, one might consider the following applications of heat recovery, most of which have potential for conserving electricity:

- generation of hot water for space heating and process heating--e.g., preheating incoming feed water for boilers can lead to fuel savings in range of 5 to 10 percent;
- generating hot air for both space and process heating--e.g., preheating combustion air can reduce fuel consumption 10 percent or more;
- preheating effluents entering fume incinerators to reduce fuel required for combustion;
- heating of process fluids such as organic liquids for process applications, or preheating process materials;
- generation of process steam; and,
- generation of space conditioning steam for both heating and cooling (comfort).

The actual economic advantage of various heat recovery forms depends upon such factors as the availability and cost of fuels, required capital investment, amortization, maintenance, operating cost, taxes, insurance, and other cost factors associated with owning and operating equipment. As of early 1976, the following payback periods were typical:[8]

- boiler or incinerator air-to-air heat recovery - 2-3 years;
- heat recovery liquid heater on an incinerator or engine exhaust - 3-4 years;
- low pressure heat recovery boiler on gas turbine or incinerator exhaust - 4-5 years; and,
- high pressure heat recovery boiler on gas turbine or incinerator exhaust - 5-7 years.

Refer to the Case Studies in Chapter 15 for examples.

Heat recovery can be considered for any electricity-powered or fuel-fired process from which a hot product, off-gas, or liquid is extracted. It is beyond the scope of this book to provide detailed design information. However, some of the salient features are summarized below.

The first step is to determine the amounts of heat available, the temperature ranges, and the load curves. The economically recoverable fraction can be estimated. Then the availability of recovered heat must be compared with heat demands to establish to what extent the available supply can satisfy the demand.

In some cases heat will be available at precisely the time when it is least needed. In other cases, where heat is currently being discharged through a cooling tower, heat recovery can lead to double savings--first in terms of the electricity or fuel saved, and second in terms of the electricity no longer needed for the cooling tower.

From electric installations heat is usually recovered as hot air, hot water or steam, or occasionally as process off-gas. Today, economics are such that it is usually a good investment to add a heat wheel or other heat exchanger and use oven exhaust heat to preheat supply air.[5] Heat can also be recovered from electric furnaces with artificial atmospheres.

In fuel-fired systems, heat recovery can take any of a number of forms as are appropriate to the process requirements, including air-to-gas, gas-to-water, gas-to-organic fluids, gas-to-water, and gas-to-steam. Gas-to-steam systems, in particular, have a broad range of applications: steam can be generated at low pressures (1 atm or 15 psig) for absorption air conditioning and heating; at medium pressure (100 to 150 psig) for process applications; and at pressures of 250 psig to 600 psig[2] with or without superheaters for supplementary power generation. Refer to Reference [8] for a number of examples.

When heat recovery is applied to combustion processes, the difficulties are related to fuel composition. Turbine exhaust (from natural gas or light oil) is relatively clean and typically is in the temperature range of 385°C (725°F) to 590°C (1100°F). Reciprocating engine exhaust is not as clean and varies from 425°C (800°F) to 730°C (1350°F), depending upon engine size and design. Incinerator exhaust gases are highly variable and fouling or corrosion of heat exchange surfaces can be a significant problem.

Table 10.7 summarizes important design parameters. Reference [9] is a particularly useful compendium of design information.

An important collateral benefit of heat recovery systems that will not be examined in this chapter is that of reduced plant emissions per unit of final product. Heat recovery can improve the economics of processes *and* reduce environmental consequences of production since less fuel consumption is required. In addition, heat recovery systems have been applied successfully to the problems of conditioning exhaust to meet regulatory requirements, e.g., removing sulfur from off-gas streams while extracting useful energy.[10]

Heat recovery from building HVAC systems is discussed in Chapter 8.

Methods include use of heat wheels, run-around systems, air-to-air heat exchangers, and heat pipe heat exchangers. Such systems are of particular importance to facilities with large heating loads, since up to 50 percent of the exhaust heat can be recovered. There is a small increase in electricity used to overcome system pressure drops in such installations.

As an illustration of the range of equipment available, Table 10.8 summarizes the characteristics of typical heat wheels. Heat pipes are available for a wide range of applications. Typical materials would be aluminum below 200°C (400°F), copper below 425°C (800°F), and steel below 700°C (1300°F).

Waste energy reclamation systems have been employed actively for many years. Waste heat boilers, turbines, and recuperators have almost always been employed in industrial applications to some degree and have depended largely on the economics of the devices at the time they were under consideration. Current examples are treated in the Case Studies in Chapter 15. Alternate use of waste products found in landfills has been common in the past. Recently, emphasis has been on a modified approach, e.g., methane generation from solid wastes or their direct combustion as fuels. Table 10.9 lists fuel values for typical industrial wastes.

Heat production is also possible by various combined cycles, either with or without electricity generation. These can take the form of either "topping" or "bottoming" cycles. Some industrial processes have high temperature exhausts:

<u>Source - Temperature °C (°F)</u>

Annealing furnace
590-1090°C (1100-2000°F)

Cement kiln (dry process)
620-815°C (1150-1500°F)

Cement kiln (wet process)
425-590°C (800-1100°F)

Diesel engine exhaust
535-650°C (1000-1200°F)

1. Materials in air heat exchangers require tubes for pressure applications significantly above atmospheric. The appropriate metals are a function of corrosives in the gases and a strong function of temperature. Copper is satisfactory at or below maximum metal temperatures of 205°C (400°F); transition to brass, copper-nickel and carbon steel is required as temperatures go to 290°C (550°F), 370°C (700°F), and 400°C (750°F). Low alloy steels are needed at about 510°C (900 to 1000°F), high alloy steels above 790°C (1450°F), and inconel X and hasteloy-C above 980°C (1800°F).

2. For corrosive control in gas systems, it is important that the temperature of the heat exchanger be everywhere well above the dewpoint temperature of the gases. Sulfur content of gases raises the minimum allowable tube temperatures. By-pass dampers can control tube temperatures under low load conditions.

3. Heat recovery heaters and boilers are frequently designed for a gas pressure drop of 4 inches to 6 inches of water (0.010-0.015 atmospheres, or $\sim$1250 N/m^2). Adequate pressure should be maintained for good water distribution in parallel tubes and coils.

4. Steam carrying systems require large pipelines because specific volume is high. A 100,000 lb/hr saturated steam system at 100 psig can release 88 million Btu/hr by condensation. A 550,000 lb/hr water system can release 88 million Btu/hr by cooling from 232°C (450°F) to 149°C (300°F). The steam system requires a 30 cm (12 in.) pipeline, while the water system requires a 20 cm (8 in.) pipeline.

5. Water-tube type high temperature boilers are usually more economical than finned tube types.

6. When the convective thermal resistance on one side of a conductive tube is significantly higher than on the other side, the exchanger performance can be greatly improved by fins on the high resistance side. Best heat performance is obtained for gas flow normal to cylindrical fins.

7. Sequential or serrated fins improve transfer by increasing gas turbulence; increases by a factor of two or more over continuous fins are feasible for tolerable pressure drops. Sequential fins are easier to clean.

HEAT RECOVERY SYSTEM DESIGN NOTES — **TABLE 10.7**

8. Clean exhaust (natural gas) can tolerate fin spacing of 4 fins per centimeter (10 fins per inch). Light oil-fueled systems: 3 fins per centimeter (8 fins per inch) maximum. Heavier fuel oils: 2.4 fins per centimeter (6 fins per inch) or less, with several leading unfinned rows. Municipal waste: unfinned to avoid clogging.

9. Organic heat transfer fluid systems for gas-to-organic fluid heat recovery offer very attractive high temperature - low pressure characteristics. An organic system at 260°C (500°F) can operate at atmospheric pressure; a steam system at 260°C (500°F) requires a pressure of about 45 atmospheres. The major economic advantage is in the substantial capital cost reduction possible in going from pressurized to unpressurized or low pressure piping systems. Operational savings are also possible. Special heat transfer fluids can provide advantages over oils, including operating temperature range and chemical stability. Fluids should be selected on the basis of maximum film temperature with a margin of longer service; temperatures of 315°C (600°F) are common and 426°C (800°F) possible.

10. Fluid-side pressure drop is an important factor in design and involves a trade-off between pumping power (proportional to the flow velocity cubed) and heat exchanger performance. Velocity through the exchanger is the single most important factor since it affects the film temperature and the convective heat transfer coefficient. At the cost of larger piping and valves, flow through piping can be reduced to less than 1.2 meters/second (4 feet/second) for power conservation.

11. Common flow velocity practice is 1.5 - 2.4 m/second (5 - 8 feet/second) through heaters, 1.2 meters/second (4 feet/second) in piping and 1.2 - 1.8 m/second (4 - 6 feet/second) through user equipment.

TABLE 10.7 CONT'D

- Dessicant impregnated: 1000 to 75,000 SCFM*
  maximum temperature 50 to 200°C (120 to 400°F)
- Woven aluminum mesh: 750 to 72,000 SCFM
  65°C (150°F) for aluminum, 150°C (300°F) for aluminum oxide surface
- Stainless steel mesh: 750 to 72,000 SCFM, 200 to 315°C (400-600 °F)
- Corrugated aluminum: 750 to 72,000 SCFM, 150°C (300 °F)
- Corrugated aluminized steel: 750 to 72,000 SCFM, 315°C (600°F)
- Corrugated alloy steel: 400 to 43,400 SCFM, 400°C (750°F)
- Corrugated stainless steel: 400 to 18,400 SCFM, 815°C (1500°F)
- Honeycomb ceramic: 1000 to 15,000 SCFM, 815°C (1500°F)

*Note: Multiply SCFM by 2.84×10^{-3} to get SCMM, e.g., 1000 standard ft^3/min = 2.84 standard m^3/min.

OPERATING CHARACTERISTICS OF HEAT WHEELS

TABLE 10.8

Waste	Average Heating Value (as fired) Btu/lb	MJ/kg
Gases:		
Coke-oven	19,700	45.8
Blast-furnace	1,139	2.6
Carbon monoxide	575	1.3
Refinery	21,800	50.6
Liquids:		
Industrial sludge	3,700-4,200	8.6 -9.8
Black liquor	4,400	10.2
Sulfite liquor	4,200	9.8
Dirty solvents	10,000-16,000	23.2-37.2
Spent lubricants	10,000-14,000	23.2-32.5
Paints and resins	6,000-10,000	13.9-23.2
Oily waste and residue	18,000	41.8
Solids:		
Bagasse	3,600- 6,500	8.4-15.1
Bark	4,500- 5,200	10.4-12.1
General wood wastes	4,500- 6,500	10.4-15.1
Sawdust and shavings	4,500- 7,500	10.4-17.4
Coffee grounds	4,900- 6,500	11.4-15.1
Nut hulls	7,700	17.9
Rice hulls	5,200- 6,500	12.1-15.1
Corn cobs	8,000- 8,300	18.6-19.3

TYPICAL INDUSTRIAL WASTES WITH SIGNIFICANT FUEL VALUE

TABLE 10.9

Gas turbine exhaust
425-535°C (800-1000°F)

Garbage incinerator
840-1090°C (1550-2000°F)

Sewage sludge incinerator
535-760°C (1000-1400°F)

Glass melting furnace
650-870°C (1200-1600°F)

At these high temperatures recovered heat can be utilized in a variety of ways.

More often, however, industrial heat is recovered in lower temperature ranges, e.g., 150-370°C (300-700°F). At the higher limit of the low range, heat can be returned to the process through recuperators, by heating feedstocks, or raising steam. Near the lower limit heat can be extracted for useful work by heat pumps or an organic Rankine cycle.

In an organic Rankine cycle, superheated vapor is expanded through a turbine or a reciprocating engine. The low pressure vapor then passes through the gas side of a regenerator where it transfers heat to the boiler feed stream. The vapor is then condensed, pumped up to boiler pressure, preheated by the regenerator and returned to the boiler. The organic working fluid is selected to have a low heat of vaporization relative to water (much smaller ratios of latent heat of vaporization to sensible heat than water), which greatly reduces the temperature differential in the boiler needed to drive the heat transfer. This in turn permits the organic cycle to approach thermodynamic reversibility and avoid a major loss term discussed previously. Researchers report that, depending upon gas temperature, organic cycles can produce 25 percent to 50 percent more work than produced in the equivalent steam bottoming cycle.[9]

Figure 10.3 illustrates the Carnot efficiency of various thermodynamic cycles, and how the low temperature organic Rankine cycle can approach Carnot efficiency at moderate to low temperatures.

* * *

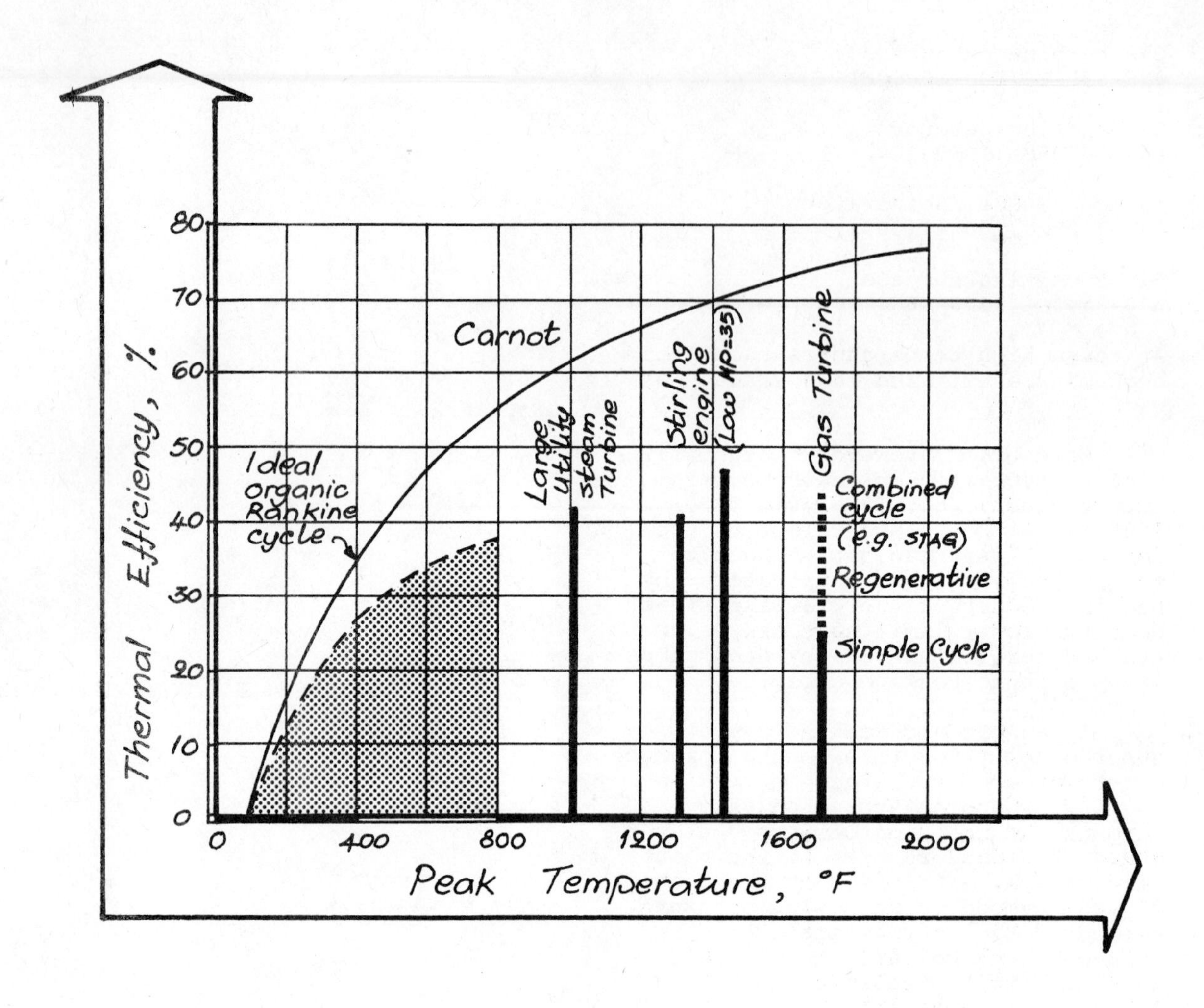

Reference 11

CARNOT EFFICIENCY AS A FUNCTION OF PEAK CYCLE TEMPERATURE

FIGURE 10.3

REFERENCES

1. Gyftopoulos, Elias P.; Lazaridis, Lazaros J.; and Widmer, Thomas F., *Potential Fuel Effectiveness in Industry*, (Cambridge, Massachusetts: Ballinger Publishing Company, 1974).

2. Prengle, H. William Jr.; Crump, Joseph R.; Fang, C.S.; Grupa, M.; Henley, D.; and Wooley, T., *Potential for Energy Conservation in Industrial Operations in Texas*, Report No. S/D-10, (Houston, Texas: University of Houston, Department of Chemical Engineering, Cullen College of Engineering, November 1974).

3. Berg, Charles A., "Conservation in Industry," *Science*, 184(19 April 1974): 264-270.

4. Keenan, J.H., "Availability and Irreversibility in Thermodynamics," *British Journal of Applied Physics*, 2(July 1951): 183-192.

5. Edison Electric Institute, *Power Sales Manual*, (New York: 1970).

6. "The 1975 Energy Management Guidebook," *Power*, Special Issue, (1975).

7. Massey, Robert G., ed., *Energy Conservation Program Guide for Industry and Commerce (EPIC)*, NBS Handbook 115, Supplement 1, (Washington, D.C.: US Government Printing Office, December 1975).

8. Boyen, J.L., "Practical Heat Recovery," Presented at "Energy Management for Industry and Commerce," Short Course, University of Arizona, Tucson, Arizona, 5-9 April 1976.

9. Boyen, J.L., *Practical Heat Recovery*, (New York: John Wiley and Sons, 1975).

10. Csathy, Denis, "Energy Conservation by Heat Recovery," Presented at the Ninth Intersociety Energy Conversion Engineering Conference, San Francisco, California, 26-30 August 1974.

11. Sternlicht, Beno, "Low-Level Heat Recovery Takes on Added Meaning as Fuel Costs Justify Investment," *Power*, 119(April 1975): 84-87.

CHAPTER 11

ELECTROMECHANICAL ENERGY

P. Ibáñez*

CHAPTER CONTENTS

KEY WORDS

Design Improvement
Energy Conversion
Equipment Modification
Equipment Substitution
Industrial Use
Manufacturing Industries
Material Substitution
Mechanical Energy Sources
Motors
Prime Movers
Process Modification

SUMMARY

In the United States today, the residential, commercial, and industrial sectors use a considerable quantity of electrical energy, much of it in the form of electromechanical energy. For example, of total electrical energy use, approximately 60 percent is for electrical drives. Thus, improvements in electrical drive efficiency would have a significant effect on reducing industrial electrical energy requirements.

Improvements in the utilization efficiency of electrical prime movers can be achieved by good maintenance practices. The efficiency of electric motors can also be significantly increased (particularly for small motors rated less than 1 kW) by improved motor design and manufacturing practices.[1] In recent years the older practices have been discarded to achieve less expensive motors, but the result has been lower efficiencies.[2] With current energy price trends, higher initial cost motors will be cheaper in the long-term when energy use, maintenance requirements, and usable life are considered. Recent technological developments can also provide greater electrical use efficiency (e.g., utilization of pulse-width modulation techniques for the control of motor speed).

Material shaping and forming operations have potential for improved electrical use efficiency. Some examples

*Principal, Applied Nucleonics Company, Inc.

are the use of cold forgings from easily wrought alloys to replace hot forgings; use of fine blanking (a shaping operation that uses close-fitting dies) to produce precision stampings requiring no subsequent finishing, thereby lowering the energy content of the part; and the use of stretch forming (pre-stretching the material 2 to 3 percent prior to forming) which reduces the energy used to form the part and increases the strength of the finished part.

The transport of material and products within a given industrial facility presents special requirements for electrical energy use. For any given process a comparison can be made of the advantages and disadvantages of providing continuous versus a discontinuous (batch) mode of transport and product movement. Energy requirements and costs for the two types can have an effect on the outcome of the decision.

Based on estimates of potential improvements in motor efficiency and projected electricity demand, there is a potential for saving electricity equivalent to that generated by 100 to 200 1000 MWe power plants over the period 1975-2000.

11.1 INTRODUCTION

In industry, approximately 80 percent of electrical energy use is for electrical drives which form part of electromechanical systems. Thus, even small improvements in the utilization efficiency of these systems could have a significant effect on reducing industrial electricity requirements.

Electromechanical systems generate mechanical forces which move and form products within the manufacturing operation. *Motion* refers to the transportation of material from one point to another within the facility, while *formation* refers to changes in the material's physical dimensions and characteristics. Both motion and formation of products normally require conversion of electrical energy (EE) to mechanical energy (ME) and are traditionally performed by means of electrical motors.

Figure 11.1 shows schematically the process by which this energy conversion takes place. The equation governing this process is:

$$ME_{end\ use} \propto EE \cdot \eta_p \cdot \eta_c \cdot \eta_t \cdot \eta_d \qquad (11.1)$$

where ME = Output mechanical energy, joules (ft·lbf);

EE = Input electrical energy, joules (kWh);

η_p = Prime mover efficiency, energy out/energy in;

η_c = Clutch efficiency, energy out/energy in;

η_t = Transmission efficiency, energy out/energy in; and,

η_d = Mechanical device efficiency, energy out/energy in.

In this approach, efficiency can be improved in four principal areas:

- prime mover,
- clutch,
- transmission, and
- device.

This chapter presents information on improving the efficiency of energy use in the conversion from EE to ME (i.e., by an electric motor) and in the employment of the required ME, thus reducing the need for ME and, in turn, reducing EE requirements. There are three methods for improving energy efficiency:

- modification of user operating procedures;
- modification of equipment at the user level; and,
- redesign at the electric motor manufacturer level.

Improvements in efficiency by the first method will result in less energy demand and subsequent dollar savings. Efficiency improvements by the second method will result in less net use of energy for the same requirements. Improvements by the third method will result in reduced operating costs and reduced total life cycle costs for a given electric motor.

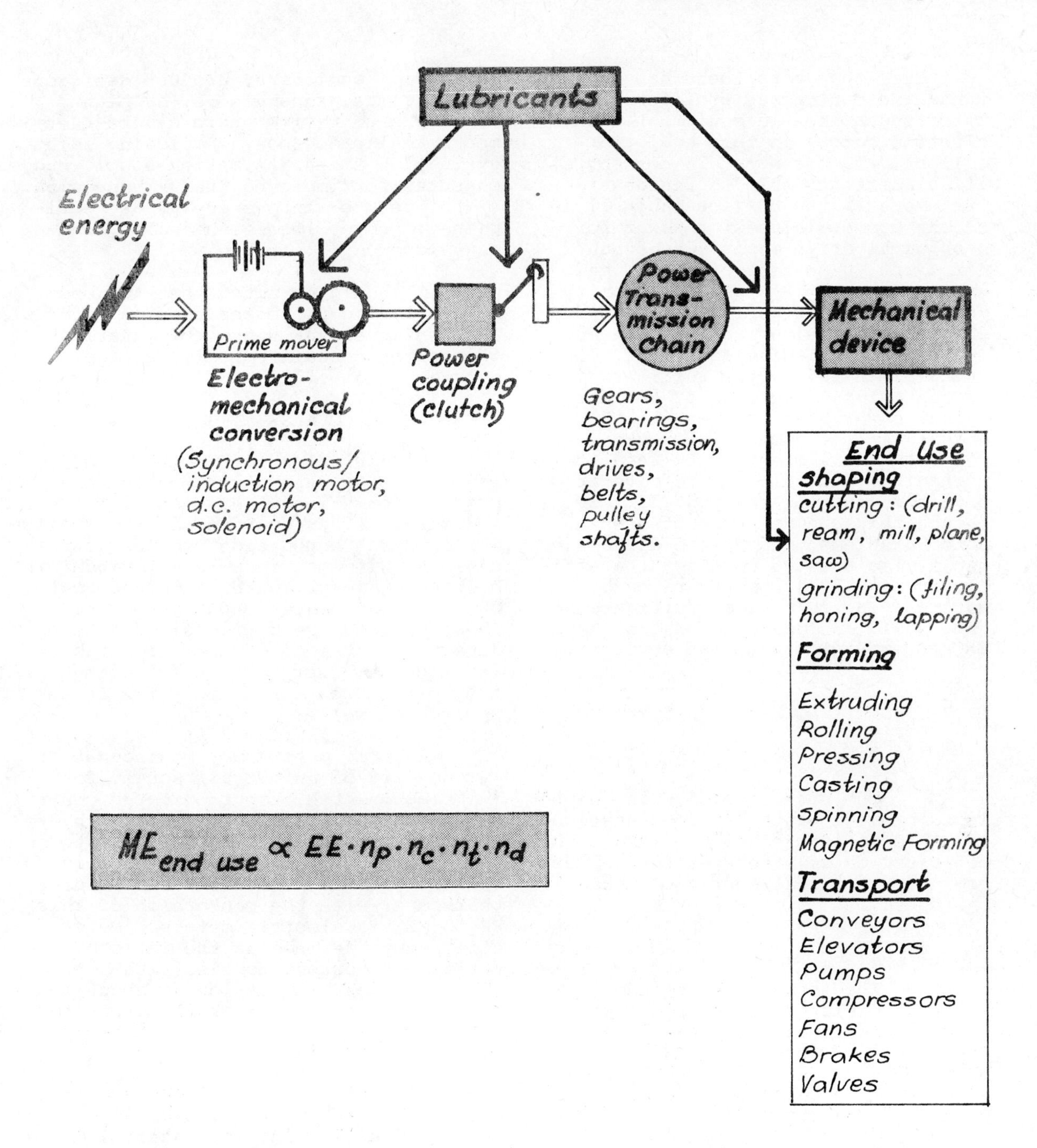

ELECTROMECHANICAL ENERGY CONVERSION

FIGURE 11.1

Over the years there has been a rapid and continuous proliferation of motorized household equipment. Today electric motors in this equipment collectively use about 38 percent of all electricity sold to residential customers in the US (see Table 11.1). In the commercial sector, power for mechanical drive apparatus accounts for about 70 percent of electrical energy demand. By far the majority of this motive power is used for running refrigeration machines, including air-conditioning equipment. In the residential sector, refrigeration accounts for 88 percent of the electrical power used for mechanical drive. In the commercial sector, 62 percent of the electricity used for mechanical drive is for refrigeration or air motion for cooling.[3] Although this chapter concentrates on motor problems, a complete discussion would also have to consider the efficiency of the equipment driven by motors. In the case of refrigeration, this would include compressors, heat exchangers, and insulated enclosures.

11.2 OBJECTIVE, SCOPE, ASSUMPTIONS

A. Objective

The objective of this chapter is to provide basic data, suggestions, and information needed to improve the efficiencies of processes for conversion of electrical energy to mechanical energy.

B. Scope

The scope of this chapter is limited to electromechanical energy conversion processes, devices, and equipment. Emphasis has been placed on manufacturing applications, which account for a major part of electromechanical energy use.

C. Assumptions

This book is not a reference work on mechanical design; therefore, the reader, assumed to have an engineering background, is considered to be familiar with appropriate references and handbooks for this purpose. An attempt has been made, however, to draw together sufficient information to permit mechanical engineers, electrical engineers, industrial engineers, product designers, shop foremen, supervisors, or plant engineers to review and evaluate electromechanical processes. Following such an evaluation this information should suggest measures for improved energy management: techniques to improve energy use efficiency, reduce losses and waste, and save money.

It should be noted that for most manufacturing processes, saving materials also saves energy since the materials themselves contain embodied energy.

11.3 PRIME MOVERS

Electrical prime movers form the basis for the majority of motion and product formation within most of US industry. Other sources of prime motion are internal combustion engines, steam engines, wind machines (e.g., windmills), hydropower (e.g., water turbines, water wheels), and gravity-powered devices. These sources are not considered here directly, but some of the suggestions for reducing energy use by changing operating procedures would apply for these sources as well.

Electric motors may be thought of as one part of an overall scheme for distributing the output of large central mechanical devices (heat engines, hydroturbines) to machines too distant to be driven by direct mechanical transmissions. The end user is primarily concerned, therefore, with the conversion of this EE to ME via electric motors. Motor efficiency, defined as the ratio of useful work output to electrical energy input, varies from as low as about 15 percent for small universal motors to about 95 percent for 500 kW, three-phase machines (see Figure 11.2).

Currently, modern motors are less efficient than older units because current designs minimize material costs and labor. Larger motors are more efficient than smaller motors. It has not been considered important, for example, if a small room exhaust fan uses 100 as opposed to 50 watts, while a factor of two in an air-conditioning system using a several horsepower motor does warrant more careful design and higher initial costs.

Most residential motors are rated at less than 0.75 kW (1 hp), although

Sector	Total Use		% of	Electric Motor Use	
	10^9 kWh	10^6 GJe	Sector	10^9 kWh	10^6 GJe
Residential	373	1,343	38.6	144	518
Commercial	290	1,044	68.9	200	720
Sub-total	663	2,387	--	344	1,238
Industrial	549	1,976	79.8	438	1,576
Transportation	4.8	17.3	--	--	--
	1,217	4,381	∿64.2	∿782	∿2,814

Reference 3

ELECTRICITY USED FOR ELECTRIC MOTORS

TABLE 11.1

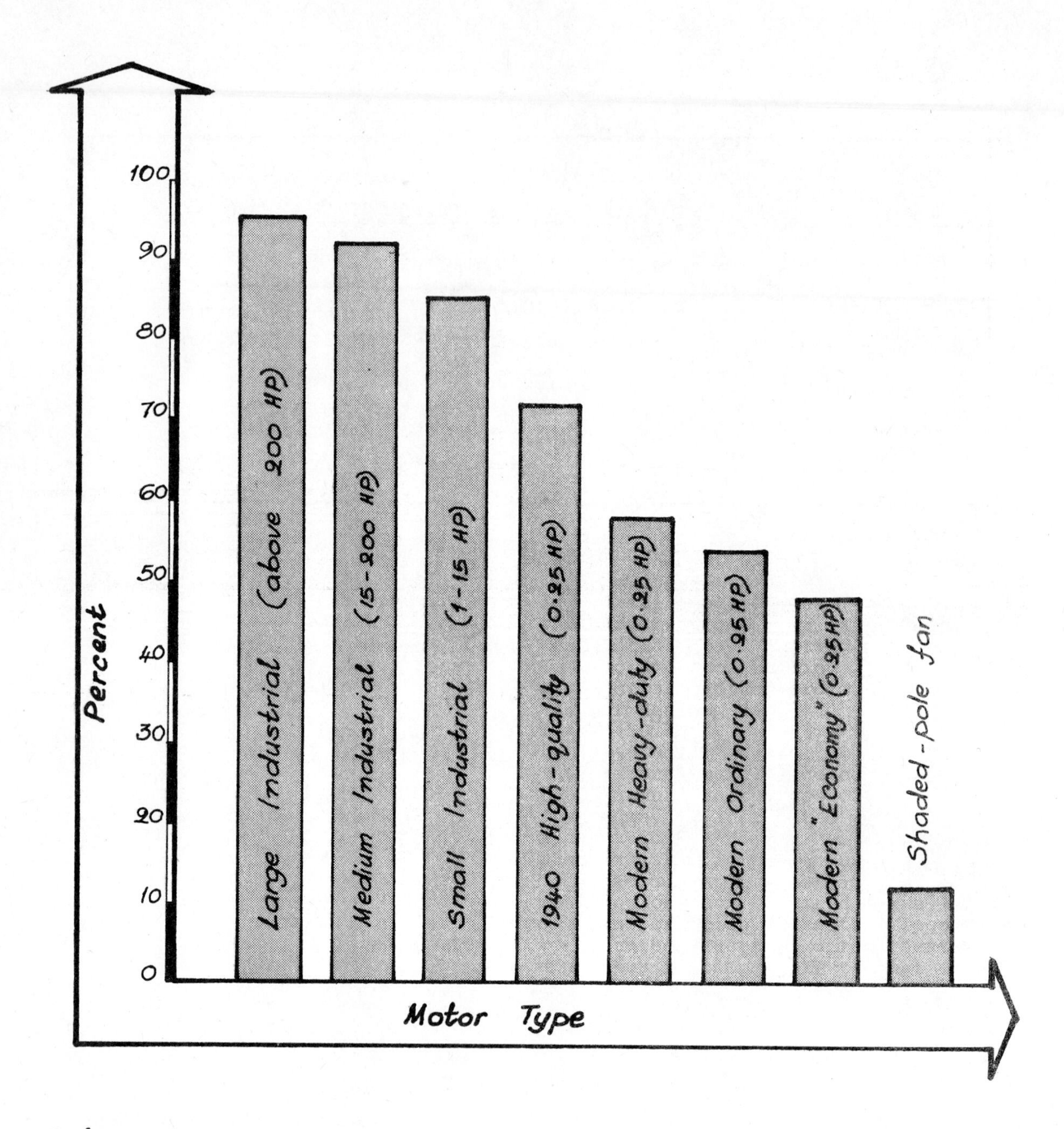

References 2 and 4

EFFICIENCY OF REPRESENTATIVE ELECTRIC MOTORS FIGURE 11.2

a few as large as 3 to 4 kW (4 to 5 hp) are installed in residential central air conditioners. Almost all of these motors operate on single-phase, alternating current, which is the service utilities normally sell to households. Commercial establishments, however, use a greater variety of motors, some rated up to thousands of kilowatts (thousands of horsepower). Unlike most residential customers, commercial customers can obtain three-phase, alternating current. This electrical service offers large businesses the better economy of purchase, operation, and maintenance associated with polyphase motors.

A. Operational Procedure Improvements

Certain improvements in the utilization efficiency of electrical motors can be obtained by modifications in the user's operational procedures. This section assumes that the user has already purchased a motor for his particular needs and has made whatever modifications to the equipment that he can in order to improve the efficiency. The next section presents several suggested methods of improving electricity utilization efficiency by means of modifying user equipment. (See Chapter 4 for specific recommendations related to home appliances and their associated motors.)

1. Optimum Power

Electric motors operate most efficiently at rated voltage. The supplying utility will check the service voltage upon a customer's request and, if the voltage is outside the permitted range, will correct it. Figures 11.3 and 11.4 show typical torque, speed, and operating curves for typical induction and single-phase motors. Voltage balance of three-phase power supply to motors, for example, is important.[5] An unbalance of 3 percent can increase losses 25 percent.

2. Equipment Scheduling

All electric motorized equipment should be turned off when not in use. Peak demand should be controlled by rescheduling operations to level off-peak demand periods. For example, test run large motors on fire pumps at off-peak hours. While this does not affect user efficiency, the generating efficiency is improved, and some utilities give reduced rates for off-peak hours. At lower load levels fewer motors should be used, and if possible, off-shift operation could be performed by a small motor selected just for the application. (See also Chapters 2 and 3.) Note that very large motors can not be stopped and started as often as smaller motors without decrease of life and possible impact on electrical distribution systems.

3. Equipment Maintenance

Regular maintenance of electric motors will help retain their original efficiency and increase their life. Cleaning air filters will reduce heating (lowering resistance) of motors. Regular lubrication of bearings will minimize frictional losses and heating of motor materials (see the Case Studies). Connections should be checked periodically to insure that arcing or corrosion is not taking place. The efficiencies of mechanical equipment in general can be increased typically 10 to 15 percent by proper maintenance.

4. Continuous versus Batch Processes

Depending on the type and volume of production, considerable energy can be saved by considering the trade-offs between continuous and batch-type processes. For example, for low production processes requiring equipment with little warmup time, batch processes can be employed most economically while shutting down operations completely between batches. For high volume and high-peak production work, a more moderate and continuous process will conserve energy by lowering peak-load demands on equipment.

5. Power Factor Correction

In some cases the use of power factor correction may be justified to reduce energy loss in feeder and branch circuit equipment. Such correction is also helpful to the utility supplying the power and results in reduced transmission losses.

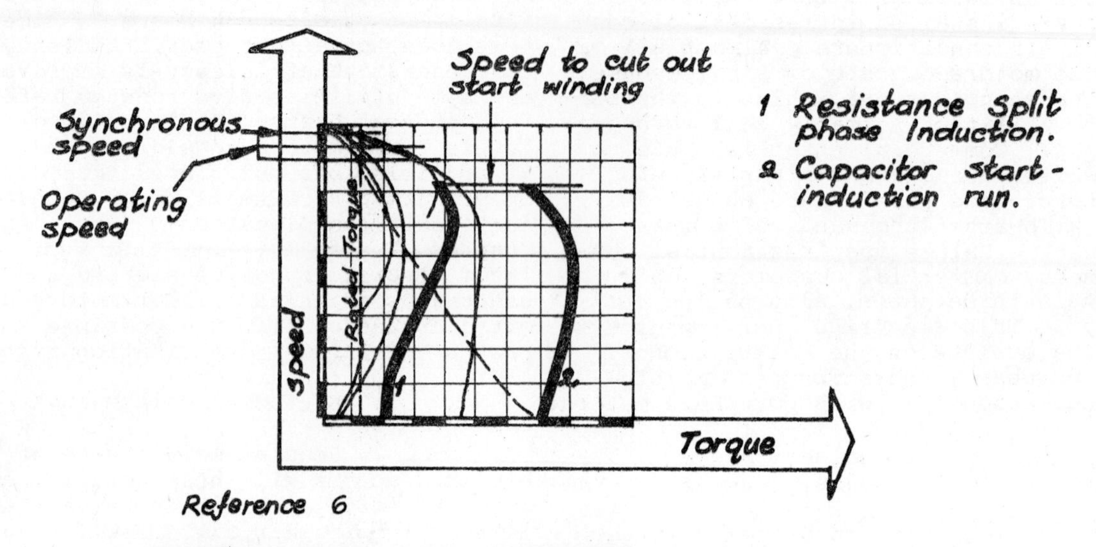

Reference 6

COMPARISON OF TORQUE VERSUS SPEED CURVES FOR INDUCTION MOTORS

FIGURE 11.3

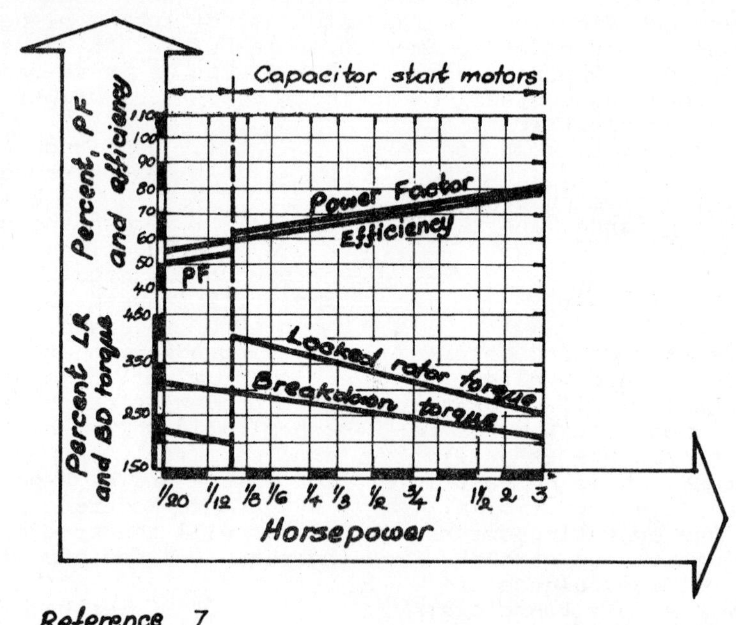

Reference 7

FIGURE 11.4

TYPICAL OPERATING CURVES, SINGLE PHASE MOTORS

B. User Equipment Improvements

Improvements in the utilization efficiency of electrical motors can also be obtained by changes in the equipment. Following are several recommendations.

1. Improve Heat Removal

Cooler motors operate slightly more efficiently. If a motor is tending to heat up, it may be relocated to a cooler area or some form of cooling may be used to remove heat. An improved thermal heat sink for the motor frame and supports will aid in heat removal. Any unnecessary restrictions to air flow or thermal insulation should be removed from around the motor. For example, resistance of the copper windings in a motor increases roughly linearly with absolute temperature. A rise in temperature from 27°C (80°F) to 32°C (90°F) results in a 2 percent increase in resistance and energy loss due to resistive heating. Efforts spent in cooling motors must be consistent with the cost of the cooling system and benefits achieved.

2. Equipment replacement

Replacement of old, inefficient equipment with newer, more efficient equipment will improve the electrical utilization efficiency of a given facility. Grossly oversized motors should be replaced since motors operate more efficiently near rated capacity and develop a better power factor. (See Table 11.2.)

3. Equipment Selection

Avoid oversize motors, pumps, and compressors. Smaller pumps and compressors with receivers or storage tanks should be used for sustained periods of low demand operation. When ordering electrical equipment, demand rigid compliance with specifications and verify the efficiency of the equipment once it is installed.

Figure 11.5 indicates the various types of electric motors available for different applications. [8,9] Table 11.2 shows the power factor and efficiency of typical induction motors, demonstrating the increase in efficiency with motor size.

Reference [5] points out that electric motors are relatively efficient machines well suited as a prime mover where loads are highly variable. It stresses the following points for proper selection:

"A. Determine the optimum motor rating to handle the load. Where the load is constant, the appropriate motor rating is readily indicated. Since the motor efficiency is nearly constant in its normal load range, the exact matching of a motor to its load is not necessarily required. Adequate matching of the motor rating to the load is important to avoid overheating of the motor. The use of motors having an output rating greater than the load (oversizing) causes a reduction in the system power factor with resultant added losses in the distribution system.

"B. With a widely varying load involving a number of stops and restarts necessary to perform the function of the driven machine, a careful analysis of the application can result in energy savings. Operating conditions, such as starts, plug stops, reversals, speed changes, dynamic braking, etc. all consume energy at rates much higher than that when the motor is operating continuously at a constant load. When variable duty cycles are encountered, two actions can be taken to minimize energy usage. The first is to reduce the mass of moving parts wherever possible, because the energy used to accelerate these parts is proportional to the mass. Secondly, all aspects of the load should be carefully analyzed which should involve consultation with the motor and control manufacturer for recommendations as to the motor and control best suited to the application. Motors designed for optimum full load efficiency would in a number of instances be totally unsuitable for many applications. Applications often have characteristics such as frequent starting, duty cycle operation, repetitive shock loading where motor torques and motor slip characteristics are the more important factors.

"C. Select the most efficient process and machinery. Often alternate means are available for doing a job and a variety of machines often exist that are capable of performing a task. Once these determinations are made, the lowest motor rating consistent with system economics can be specified.

Power kW	Weight kg	Horse-power	Weight lb.	Amp.	Power factor, percent 1/2 load	3/4 load	4/4 load	Efficiency, percent 1/2 load	3/4 load	4/4 load
Squirrel-Cage Type										
.75	30	1	65	3.31	60	71	78	71	76	76
1.5	45	2	100	5.70	71	81	86	78	80	80
3.7	72	5	159	13.4	76	84	87	83	84	84
7.5	116	10	255	26.2	81	87	88	85	86	85
14.9	190	20	418	52.2	85	88	89	84	85	84
29.8	365	40	804	98	86	88.5	89.5	89.5	90	89.5
74.6	802	100	1,769	238	85	89.5	90.4	89	90.5	91
149.1	1,463	200	3,225	463	91	93	94	87	90	90
Wound-Rotor Type										
3.7	100	5	220	14.3	72.5	80	82.5	78	79	79.5
7.5	152	10	336	26.6	69	79	83	83	84.5	85
18.6	262	25	578	62.9	75	83.5	87	84	86	86.5
37.2	450	50	991	118.4	84	89	90	86	88	88
74.6	1,187	100	2,618	233	88	90.5	89.5	86	88	88
149.1	1,770	200	3,902	473	89	91	92	87	89	90
Three phase, 2,300 volts, 60 cycle, 1,1775 rpm Squirrel-Cage Type										
224	1,451	300	3,200	67	87.5	89.3	90.6	90.0	91.8	92.7
522	2,359	700	5,200	151	90.2	92.0	92.9	91.6	93.0	93.6
745	3,493	1,000	7,700	212	91.2	92.8	93.7	92.2	93.4	94.0
Wound-Rotor Type										
224	1,769	300	3,900	67	84.7	89.0	90.0	90.0	91.8	92.7
522	2,608	700	5,750	151	88.5	91.8	92.6	91.6	93.0	93.6
745	3,833	1,000	8,450	212	90.0	92.8	93.5	92.2	93.4	94.0

Reference 10

TYPICAL INDUCTION MOTOR DATA **TABLE 11.2**

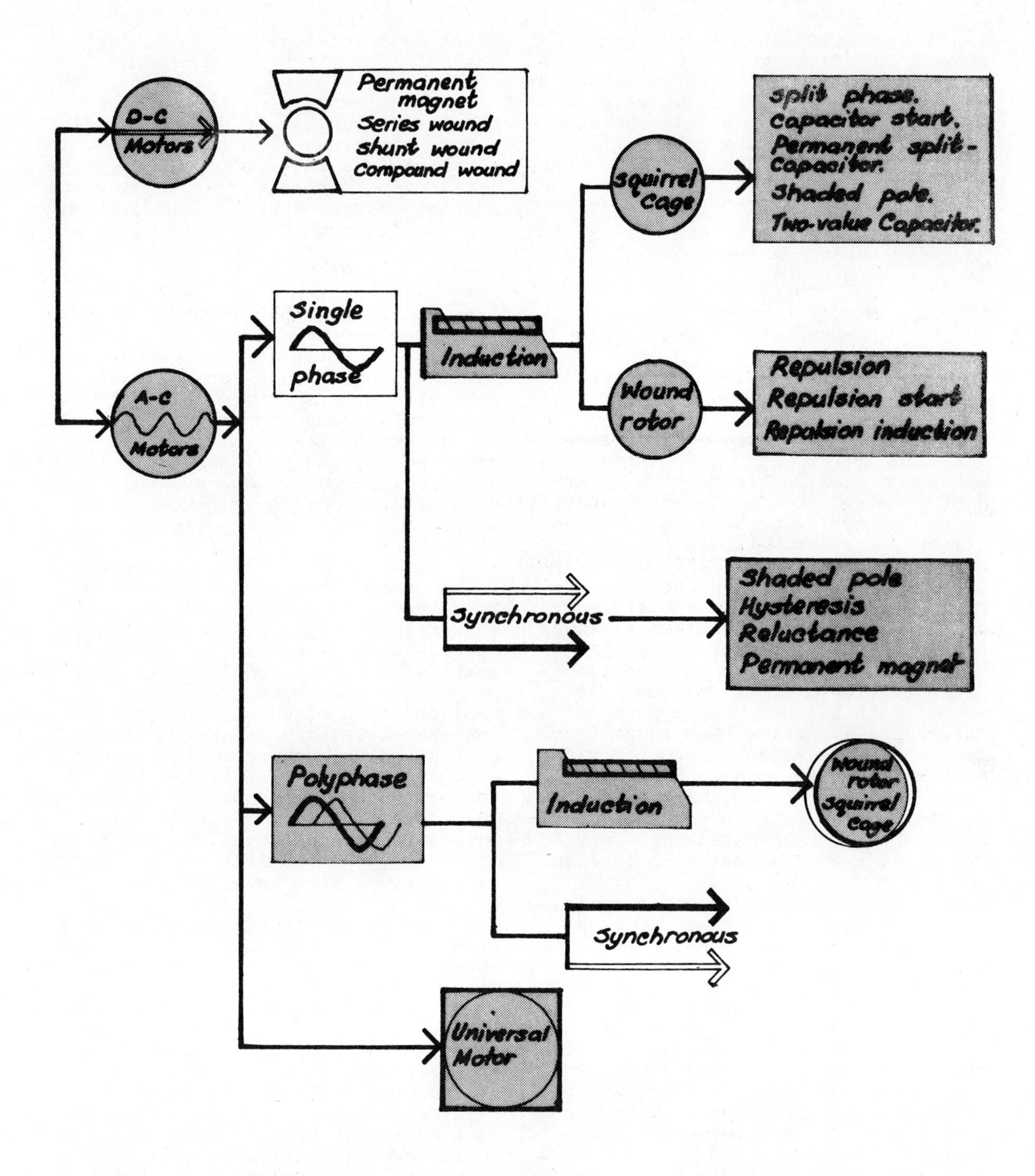

ELECTRIC MOTOR CLASSIFICATION FIGURE 11.5

"D. For variable or multi-speed drives, the first cost and long-range energy costs should be carefully evaluated because drive systems vary widely in first cost and in operating efficiency, i.e. - the choice of multi-speed or adjustable speed motors for process control as compared to throttling control; or the choice of high-speed motors with speed reduction as compared to low-speed motors."

C. Motor Equipment Manufacturer Improvements

1. Small Electric Motor Efficiency Improvements

Over the last 30 years electric motors which power household and commercial machinery have undergone a marked decline in efficiency.[1] Modern motors, especially small ones (<1 kW), are less efficient than earlier models, in part, because manufacturers have designed them for minimum initial cost to satisfy purchaser demands. In addition to lower cost these modern motors have also been more compact than the massive, more efficient, older designs. Because of this trend, the amounts of copper and iron have been reduced and design factors essential to efficient machines have been de-emphasized. Thus, while new motors may be cheaper initially, the extra power they dissipate as waste heat makes them less efficient and more costly to operate.

As one solution the decline in motor efficiency could be reversed by engineering design that combines a carefully analyzed structure, new tooling methods, and quality control with the best modern insulating and magnetic materials. Such an approach would create a new generation of motors that would, with occasional maintenance, last considerably longer than standard electric motors. This new generation of electric motors would probably be larger, heavier, and more expensive. With rapidly rising energy costs, the higher initial cost would be justified through longer life and improved operating economy. A complete conversion to more efficient motors throughout the commercial and residential sectors would reduce total US electrical usage by nearly 6 percent.[2,11]

As an example of the economic benefits which can be gained by using a more efficient electrical motor, the total savings for a motor can be calculated using variable efficiencies and initial costs. Table 11.3 illustrates the significance of efficiency in a dramatic way. The life cycle cost of providing 3600 GJ (10^6 kWh) of mechanical energy with motors ranging in size from 0.1875 to 75 kW (0.25 - 100 hp) was selected to compare the effects of efficiency and of using numerous small motors rather than a central power source. For simplicity this calculation does not consider several factors such as investment tax credit, financing, maintenance, and regional variation of electricity costs. For a more detailed consideration of these factors, see Appendix E.

Note that the life cycle cost is up to ten times greater for smaller motors. This is due to greater initial cost and greater energy cost. The greater initial cost is due to the fact that small size motors require more material per unit power capacity and have to be replaced more often. (See Table 11.2, for example, which indicates that a 0.75 kW [1 hp] squirrel cage motor requires about 50 kg/kW, while the 150 kW [200 hp] motor requires about 10 kg/kW.) Over the full range of motor powers, from fractional to 1000 kW (1340 hp), the specific weight (kg/kW) varies by as much as a 10:1 ratio.

The greater energy cost results from the fact that the less efficient motors require more kWhe to deliver a specified amount of work. This has an important effect on life cycle cost. For example, consider Table 11.4 which compares two 1 hp motors with different efficiencies. The greater first cost of the more efficient motor is more than compensated for by the life cycle cost savings. Additional benefits would result as the more efficient motor will run cooler and probably have a longer life. Escalating energy costs further enhance the benefit of the more efficient unit. Note that for motors which operate infrequently, the extra cost of an efficient unit may not be warranted.

	Motor 1	Motor 2	Motor 3	Motor 4	Motor 5
Motor Size, kW	0.1875	0.375	0.75	7.5	75
Motor Size, hp	0.25	0.5	1.0	10	100
Initial Cost $(b)	25	60	110	420	2,200
Assumed Motor Life (Years)	5	5	10	20	50
Number of Motors	3,995	1,998	499	25	1
Efficiency (%)	50	60	80	85	95
Cost of Motors $	99,875	119,880	54,890	10,500	2,200
Annual Energy Cost(c) $	2,002	1,669	1,252	1,178	1,054
Present Value of AEC $(d)	19,855	16,548	12,414	11,680	10,450
Life Cycle Cost(e)	119,730	136,428	67,304	22,180	12,650
Cost Relative to Motor 5	9.5	10.8	5.3	1.8	1.0

Notes: (a) Assume 267 hrs per year operation, 50 year life, 10^6 kWh total delivered
(b) Source: References [11,12]
(c) At 0.05 $/kWhe
(d) At 50 years at i = 10%, PWF = 9.915
(e) Equals initial cost plus present worth of annual energy cost (AEC)

COMPARISON OF MOTOR LIFE CYCLE COSTS[a]

TABLE 11.3

	Typical Values	
	Motor 1	Motor 2
kW	.75	.75
hp	1.0	1.0
Initial Cost[b] $	100	120
Life, years	10	10
Efficiency %	75	85
Annual Energy Cost[c] $	66.67	58.82
Present Value of AEC[d] $	409.60	361.41
Life Cycle Cost[e]	509.60	481.41
% Savings Compared to Base	base	6

Notes:
(a) Assume 1333 hr/yr operation, 10 year life, delivering 10^4 kWh
(b) Source: Reference[11]
(c) At 0.05 $/kWh
(d) At 10 yr and i = 10%; PWF = 6.144
(e) Equals motor cost plus present worth of energy cost

COMPARISON OF MOTOR EFFICIENCIES AND LIFE CYCLE COSTS[a]

TABLE 11.4

Finally, more energy is required to manufacture a more efficient motor. Is it justified? Consider two 0.75 kW (1 hp) motors selected for air-conditioner service.

Motor #1 = 0.75 kW, 24 kg, 60 percent efficient, 10 yr, 1000 hr/yr

Motor #2 = 0.75 kW, 30 kg, 75 percent efficient, 10 yr, 1000 hr/yr

At roughly 100 MJ/kg to manufacture (Appendix B), the more efficient motor requires 3 GJ versus 2.4 GJ for the less efficient one.

The 10 year output of both motors is 27 GJ (7.5×10^3 kWh). The 60 percent efficient unit requires an input of 27 GJ ÷ 0.6 = 45 GJe, against 36 GJe for the 75 percent efficient unit. Thus, the added energy investment (in materials and manufacturing) of 0.6 GJ to produce the more efficient unit saves 45 - 36 = 9 GJe.

For small electric motors, one trend in the direction of increasing efficiency has been the increased popularity of the capacitor-run (or permanent-split capacitor) motor which tends to be more efficient under constant load conditions than other types of single-phase motors.[2] Yet within the class of capacitor-run motors, efficiency is still dependent on the same rules of good design and construction as in other classes.

Modern motors are, of course, smaller and lighter than their predecessors. They weigh about one-third to two-thirds as much; and a 0.5 kW motor of today is actually smaller in size than the 0.25 kW motor of 30 years ago.[2] Clearly, modern motors save on materials (representing some energy savings) and labor, but do so at the sacrifice of efficiency (see Figure 11.2).

A possible alternative would be to construct motors with designs similar to those of 30 to 40 years ago, but with the best modern insulating materials.[2] Magnetic materials would be selected for optimum performance with minimum losses, and, finally, the bearings would be of the best quality and easily replaceable with ordinary tools as is now done for better quality motors. The result would be an extremely long-lived motor. The high initial cost (perhaps twice that of a modern motor or more) could be partially offset by complete standardization of frames and mounts, an arrangement which already exists for large industrial motors.[2] When the machine containing the motor is eventually scrapped or becomes obsolete, its motor could be saved and transferred into some new apparatus. (A few commercial refrigeration companies have actually been applying this concept on a very limited basis by rebuilding old motors with modern insulation and supplying them to special customers.)

Efficiently designed motors (1 to 200 hp) are beginning to appear in the marketplace.[11] Motors with efficiencies 2 to 15 percent greater than standard units are available at cost increases up to 25 percent. The reduced cooling requirement for these new motors is also a significant advantage. Note that proper motor selection, operating strategy, maintenance, and driven equipment design may offer opportunities for efficient energy use far greater than fundamental motor efficiency.

It was pointed out earlier that most of the output from motors in homes and businesses is used to run refrigerator and air-conditioner compressors. Virtually all refrigerators and air conditioners made for residential use, and the majority of those made for commercial use, employ what is called the hermetically sealed motor compressor unit.[2] That means the motor and compressor are directly coupled and sealed inside a welded steel case. The advantages here are that the coupled system runs more quietly, occupies less space, and requires less maintenance. It would clearly be wasteful, however, to attach a non-depreciating motor inseparably to a compressor destined to wear out within a few years, and then add a housing which makes repair impossible and salvage for copper difficult. Thus, the proposal to return to efficient motors would also require a return to an open-type, coupled (possibly belt-driven) compressor--a change which also would be for the better. Belt-driven compressors run at lower rotational speeds, thus having a potential for longer life. Even more important, with proper design less waste heat from the motor will be added to the working fluid than in the hermetic machines. Thermodynamically, this is advantageous for it permits one to get more refrigerating effect per motor unit power. There would be, therefore, a further gain in conservation of energy resources.

If such motors were to be adopted generally, the saving in electric power would be sizable. Suppose motor efficiency were raised to the levels typical of the better units of the 1930s. For a low cost 0.25 kW motor, this would mean more than a 20 percent reduction in the power use of motorized appliances. (Larger motors would not have such a drastic reduction of power losses.) The residential sector would save 20 percent of 518 x 10^6 GJe (144 x 10^9 kWh) total annual residential motor energy use, or 104 x 10^6 GJe (28.8 x 10^9 kWh) per year, and the commercial sector would save 20 percent of 720 x 10^6 GJe (200 x 10^9 kWh) total annual commercial motor energy use, or 144 x 10^6 GJ (40 x 10^9 kWh) per year. The total savings of 248 x 10^6 GJ (69 x 10^9 kWh) would amount to 5.7 percent per year of the total 1968 US electrical use of 4381 x 10^6 GJe (1,217 x 10^9 kWh) (see Table 11.1).[11,13]

In addition to the energy savings, there also would be a commensurate long-term cost savings. The 70 percent efficient, 0.25 kW motor demands ~100 watts less than the 55 percent efficient one. If electricity sells for 0.05 $/kWh, the motor would save $5 for every 1000 hours of operation. Thus, if the efficient motor costs $30 extra, this investment would be paid back in about 1-1/2 years if the motor ran continuously.

In a typical refrigerating machine, the motor operates about one-third to one-half of the time (33-50 percent duty cycle). In this case the repayment time would be about three or four years. These repayment times are computed on the basis of the present price of energy. If that price continues to rise as it has done recently, the efficient motor would become that much more attractive.

One exception to these long-range savings would be the case of a motor which is used infrequently, as in a power tool. There would be no compelling economic reason to install the more efficient motor in such a tool, except perhaps to increase reliability and to improve resistance to periodic overloads during operation.

The possibility of major improvements in motor design should not be ruled out. Recent use of rare earth magnets in DC motors results in motors with more power, faster response, lighter weight, and interesting new designs (for example with the magnets on the rotor and windings on the stator).[14] Recently the development of a "controlled torque motor" was reported with improvements in efficiency claimed to be as large as 30 percent. [15,16] (See the Case Studies.)

2. Large Electric Motor Efficiency Improvements[4]

In the following discussion, large motors are considered to be those in excess of 37.5 to 45 kW (50 to 60 hp). Such motors would normally be NEMA frame 360 and larger. Motors of this size are normally polyphase (usually three-phase), either synchronous or squirrel cage induction. These motors operate on voltages ranging from 200 V to 13 to 15 kV.

Large motors are usually custom designed to meet specific customer application requirements. This custom design must account for all the requirements of the driven load and of the electrical system supplying power to the motor.

Such factors as load speed-torque characteristics, load inertia, duty cycles, impact loading, and voltage drops while starting and running must be considered by the designer. Customer specifications on power factor and efficiency are also considerations. In addition, motor design is affected by noise level requirements.

Motor purchasers have tended to emphasize low initial cost and, therefore, manufacturers have supplied motors which are competitive on this basis. As energy costs continue to escalate rapidly, it becomes profitable for users to evaluate motors on a life cycle cost basis including initial costs, maintenance, and energy costs.

With energy costs high, it may well be more profitable to pay a higher initial cost for a premium efficiency motor and reap the savings in energy cost differences during the life of the motor.

The general procedure for such a comparison among several bids is to normalize all bids to a given efficiency. For example, all bids would be normalized to 95 percent efficiency by applying a penalty of X dollars for each percentage point below the 95 percent, and a similar credit for each point above 95 percent. The X dollars represents the present worth of energy costs over the lifetime of the motor. In this way the user recognizes which motor will be most economical to use.

Motor manufacturers are subject to the demands of motor users. If the user will make such a comparison of actual costs rather than only initial costs, manufacturers will respond by providing higher efficiency motors.

The user must realize, however, that there are many application requirements which must take precedence over motor efficiency. Safe acceleration of the load without injurious thermal and mechanical strain on the motor, under all expected conditions, is a primary concern. The environment in which the motor will operate dictates the minimum requirements of motor enclosure to protect electrical parts. Maximum efficiency is obtained with an open motor, but a protective enclosure is required in many applications, such as some outdoor, chemically corrosive, or explosive atmospheres.

Many of the steps which can be taken to improve efficiency are in conflict with other desirable motor performance characteristics, and, therefore, proper application of motors designed for premium efficiency will require careful analysis of the load requirements and of the electrical power system. The user must be responsible for this careful analysis and for using this analysis as part of his purchase specifications.

The design and construction of a premium efficiency motor will definitely be more expensive in the initial cost stage. Furthermore, motor manufacturers will require several years to complete development and evaluation of the design concepts involved. This trend will be helped if users begin now to evaluate motors based on life cycle costs.

In order to explore the concept of a premium efficiency motor, let us first consider the various factors affecting motor efficiency. Equation (11.2) defines the efficiency of an electric motor:

$$\text{EFFICIENCY} = \frac{\text{POWER OUT}}{\text{POWER IN}} = \frac{\text{POWER IN - POWER LOSS}}{\text{POWER IN}}. \quad (11.2)$$

The "power loss" term is composed of the following elements:

1. STATOR LOSS (WS). This is the I^2R loss in the stator winding.
2. ROTOR LOSS (WR). This is the I^2R loss in the rotor winding.
3. CORE LOSS (WC). This is the hysteresis and eddy current loss of the laminated stator and rotor core.
4. FRICTION AND WINDAGE LOSS (FW). This is the loss due to fans and the bearing friction.
5. STRAY LOSS (WL). This is the lump sum of all losses in the motor which cannot be attributed to one of the other four components. It is principally due to electrical harmonics and stray currents in the motor.

Various design steps can be taken to reduce each of the above components. Some of these steps, and associated problems, are as follows:

- *Increasing the amount of copper in the stator will lower stator resistance and thus lower stator loss.* The major penalty for this is increased motor size to accommodate the larger coils.
- *Reducing the number of turns in the stator cells reduces stator resistance and stator loss.* The penalties are higher magnetic density and higher starting current. The higher magnetic density causes lower motor power factor and higher core loss. In an induction motor the higher magnetic strength also reduces rotor loss. The net result is usually improved efficiency, unless carried to extremes.
- *By increasing the air gap (the clearance between rotor and stator), stray loss can be reduced because the strength of the harmonics is reduced.* The penalty is a reduction in motor power factor.
- *Use of high-silicon laminated steel can reduce core loss due to decreased hysteresis loss.* High-silicon steel has more reluctance than carbon steel, so the penalty is a slight reduction in motor power factor.
- *Use of thinner lamination steel can reduce core loss by reducing eddy current loss.*
- *In squirrel cage induction motors, use of large, high-conductivity rotor bars and end rings will reduce rotor loss.* The penalty may be very severe since rotor cage resistance greatly affects motor starting torque and current. The end result may be such a reduction in torque and severe starting voltage reduction, due to high starting current, that the motor will not accelerate to full speed.
- *Stray loss can be reduced by eliminating the skew in the rotor.* This skew is normally used to reduce or eliminate certain harmonics. Without the rotor skew, the motor noise level may increase by 2 to 5 db.
- *Insulating rotor bars from the laminations reduces stray loss due to stray rotor currents.* On aluminum rotors this is accomplished by anodizing the rotor bars before they are inserted in the core.

There are many other steps which the designer can take in some instances. It should be obvious that for a designer to establish maximum efficiency for a given application, he must have complete data on the load characteristics and expected maximum voltage drop during starting. Environmental requirements, such as noise level, also must be known.

The preceding discussion centered on the efficiency of the motor itself. The user has the additional responsibility of optimizing the overall system efficiency and his energy costs by means of a system analysis. This optimization includes motor losses, distribution cable losses, transformer losses, and utility charges based on system power factor. The system analysis will determine which of the following systems to use:

- high voltage distribution system with high voltage motor;
- low voltage distribution system with low voltage motor; or,
- high voltage distribution system with step down transformer and low voltage motors.

The decision should be based on energy costs as well as initial costs of motor, switchgear, and distribution cable costs.

Specifying high voltage for low power motors usually results in a motor which costs more and has a lower efficiency and decreased power factor. Logical horsepower and voltage relationships are shown in Table 11.5.

Voltage Class	Power	
200-600	Up through 500 hp	Up through 375 kW
2300-4000	300 through 5000 hp	225 kW through 3750 kW
6000-6900	1500 through 10,000 hp	1025 kW through 7500 kW
13-15 kV	3000 hp and up	2250 kW and up

LOGICAL VOLTAGE AND MOTOR POWER RELATIONSHIPS **TABLE 11.5**

Because many of the steps which can be taken to improve motor efficiency also cause a decrease in power factor, it is worth noting that decreased power factor causes higher distribution losses. This occurs because the motor draws more electrical current despite the improved efficiency. An effective method of overcoming this is to install power factor correction capacitors as near the motor terminal as possible.

Warning: Improper sizing of power factor correction capacitors can result in extremely high voltages which are dangerous to personnel and equipment. Consult with your motor manufacturer to verify specific requirements.

An indirect effect on overall efficiency is the selection of motor space heaters. Two basic types of heaters are used: the cartridge heater and the belt heater. The cartridge heater mounts in the motor frame; the belt heater is wrapped around the stator coil end turns. The purpose of space heaters is to prevent moisture condensation on coils when the motor is not running. Because belt heaters apply the heat directly on the coils, they can operate on less power than cartridge heaters. In addition, because belt heaters operate at a lower power density, they usually have a longer life, comparable to the life of the motor winding.

In summary, the following important considerations should be reiterated:

- responsibility for the selection and utilization of premium efficiency motors rests with the motor user;
- motor selection should be based on life cycle costs and not on initial cost;
- closer liaison between manufacturer and user should be established; and,
- user specifications should be written for the particular application rather than for all motors in general.

3. Control Systems

Motor controllers are not power utilizing equipment, but only transmit and control power. For example, the losses in an across-the-line controller at full load are small compared to the total load.

While controllers themselves do not use much energy, their design and operating strategy can greatly affect the efficiency and life cycle costs of the motor/equipment/controller system. In the control of DC motors, for example, recent advances in solid state electronics have allowed efficient control of motors ranging from fractional to in excess of 100 hp by pulse-width modulation (PWM).[17] These systems, which include conventional transistors (low power applications), high current transistors (moderate power), and SCR units (high power), are a marked improvement over earlier control systems which, for example, simply routed a portion of the energy through a resistor bank to achieve speed control.

Motor controller choice is primarily governed by specific operational requirements and motor type. More modern solid state control systems and "intelligent" circuitry often offer both improved control and energy savings.[18] Microprocessors and minicomputers allow the control system and operating strategy of the motor to be merged into one efficient unit.[19] (Also see Chapter 5.)

11.4 MECHANICAL ENERGY TRANSFER

The ultimate need for electrical energy is dependent upon the efficiency of mechanical energy use once electromechanical conversion has taken place. The purpose of this section is to consider a few of the concepts related to the transfer of mechanical energy and the important factors affecting system efficiency.

An excellent review of mechanical drives is given in Reference [20]. The reader is directed to this source for detailed information on the types, characteristics, and operation of mechanical drive systems, bearings, and seals. Review of the information will provide guidance for deciding which type of mechanism would be the most efficient for the system being used.

The information presented here deals primarily with the efficiencies of a few examples of standard methods and mechanisms for transferring mechanical energy.

Table 11.6 shows the more commom mechanisms for transferring mechanical energy. Each of the mechanisms has an inherent efficiency and each is suited for certain types of operations. For example, for high precision speed control and ratio of drive rotation as used in precision threading, conventional or worm gearing is preferable to belts which can slip under certain load conditions. The following paragraphs briefly describe the important aspects of the efficiencies of a few typical mechanical energy transfer systems.

The approximate dollar sales breakdown of the product groups making up mechanical systems is shown in Figure 11.6. The dominant product group among components is mechanical drives; and nearly 50 percent of these drives are gear type.[21] Over the next ten years, industrial purchases of all mechanical drive components are expected to increase by $1 billion.[21] The single most important factor in future applications is expected to be the introduction of new materials, principally plastics, high strength metals, and powdered metal parts.[21] Improved design of mechanical energy transfer mechanisms utilizing new materials can have an effect on increasing the efficiency of systems. The Case Studies illustrate several examples of recent design improvements aimed at improving energy transfer efficiency.

A. Effect of Friction

Friction is the resistance to motion which takes place when one body is moved upon another.[22] The force of friction, F, bears--according to the conditions under which sliding occurs--a certain relation to the normal force between the sliding surfaces. The relation between force of friction and the normal force N is given by the "coefficient of friction":

$$\text{coefficient of friction} = \mu = F/N. \quad (11.3)$$

For well lubricated surfaces, the laws of friction are considerably different from those governing dry or poorly lubricated surfaces. If the surfaces are flooded with oil, the frictional resistance is almost independent of the pressure per unit area and the nature of the surface materials in contact.[22] For well lubricated surfaces, friction varies directly as the speed, at low pressures; but for high pressures friction is very great at low velocities, approaching a minimum at about 60 cm/sec (2 ft/sec) linear velocity, and afterwards increasing approximately as the square root of the speed.[22] With less lubrication the coefficient of friction becomes more dependent upon the material of the surfaces. Typically, coefficients of friction for various materials (sliding friction) range from 0.05 (continually greased smooth surfaces) to 0.56 (leather on dry metal).[22] Friction between machine parts lowers the efficiency of machine elements. Table 11.7 shows typical values of efficiency for common machine elements when carefully made.

When a body rolls on a surface, the force resisting the motion is termed "rolling friction". This has a different value from that of the ordinary, or sliding friction:

$$\text{Resistance to Rolling, newtons} = \frac{Wf}{r} \quad (11.4)$$

where W = total weight of rolling body or load on wheel, in newtons;

r = radius of wheel, in meters; and,

f = coefficient of rolling friction, meters.

The coefficient of rolling friction varies with the conditions. Typical values range from 0.5×10^{-3} m (iron on iron) to 5×10^{-3} m (iron on wood).

The value of an oil as a lubricant depends mainly upon its film-forming capacity; that is, its capability of maintaining a film of oil between the bearing surfaces. Film-forming capacity depends to a large extent on the

Adjustable Speed Drives:

- Gear drives
- Belt drives
- Chain drives
- Hydroviscous drives
- Variable-stroke drives
- Traction drives

Gears:

- Spur gears
- Helical gears
- Worm gears
- Bevel gears
- Celestial gears
- Combination of above

Chains:

- Bead chains
- Detachable chains
- Pintle and welded steel chains
- Offset sidebar chains
- Roller chains
- Double pitch chains
- Inverted tooth silent chains

Belts:

- Film belt
- Flat belt
- V-belt
- Synchronous drive belt

Mechanical Clutches:

- Positive clutches
- Friction clutches
- Overrunning clutches
- Overload clutches
- Centrifugal clutches

Electric Clutches:

- Friction clutches
- Tooth clutches
- Magnetic spring clutches
- Hysteresis clutches
- Magnetic-particle clutches
- Eddy-current clutches

Fluid Couplings:

Mechanical Brakes:

- Band brakes
- Drum brakes
- Disk brakes

Electric Brakes:

- Friction brakes
- Hysteresis brakes
- Eddy-current brakes
- Magnetic particle brakes

Couplings:

- Rigid couplings
- Flexible couplings

Universal Joints:

- Cardan joints
- Constant velocity joints

MECHANICAL ENERGY TRANSFER MECHANISMS

TABLE 11.6

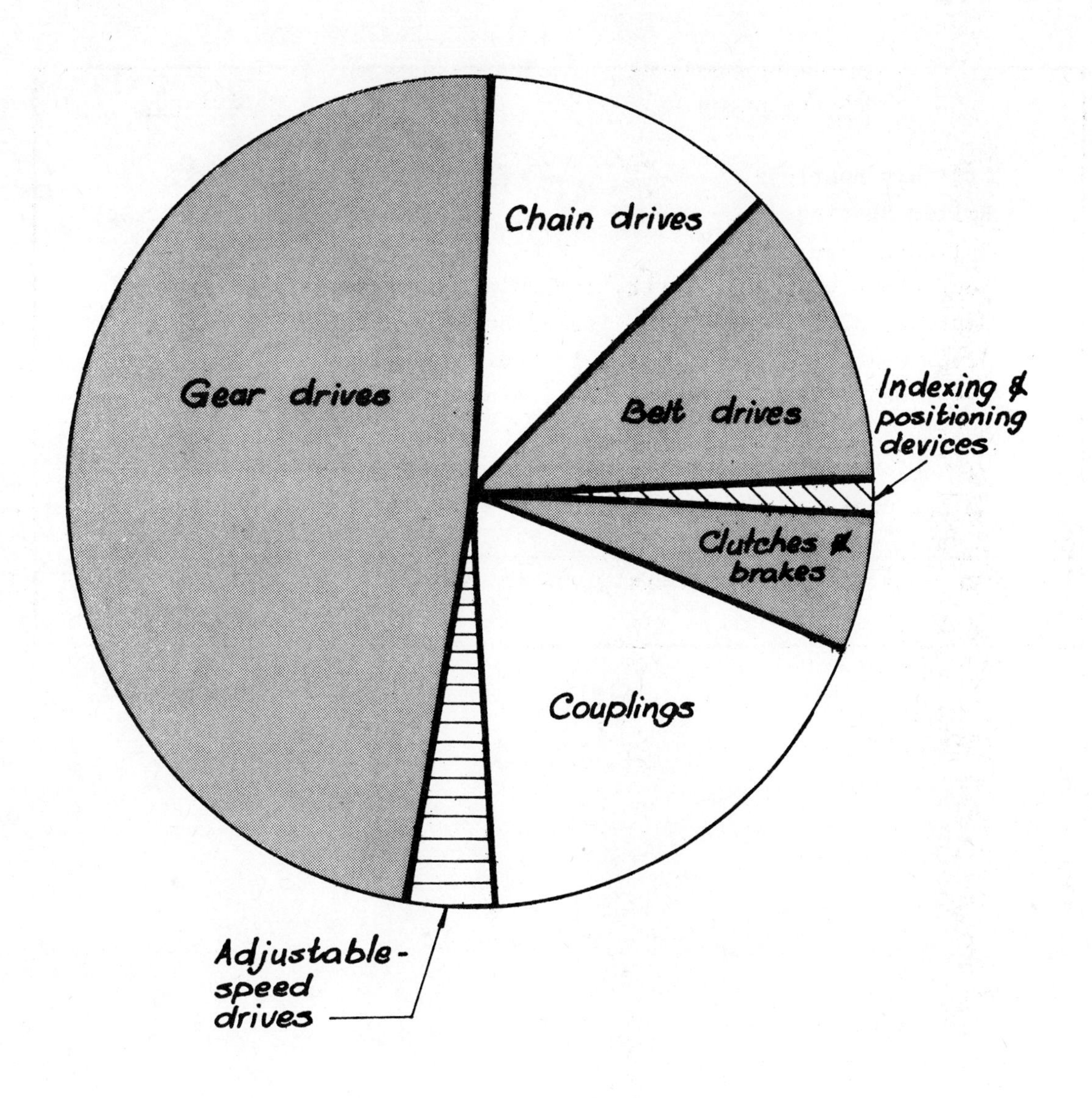

Reference 21

INDUSTRIAL USE OF MECHANICAL COMPONENTS

FIGURE 11.6

Description	% Efficiency
Ordinary Bearings	95-98
Roller Bearings	98
Ball Bearings	99
Spur Gears with Cast Teeth, Including Bearings	93
Spur Gears with Cut Teeth, Including Bearings	96
Bevel Gears with Cast Teeth, Including Bearings	92
Bevel Gears with Cut Teeth, Including Bearings	95
Belting	96-98
High-class Silent Power Transmission Chain	97-99
Roller Chains	95-97

* Table generated from numbers given in text of Ref. [22]

EFFICIENCY OF SMALL MACHINE ELEMENTS*

TABLE 11.7

viscosity of the oil, but this should not be understood to mean that the oil of the highest viscosity is in every case the most suitable lubricant. An oil of the lowest viscosity which will retain an unbroken oil film between bearing surfaces is the most suitable for purposes of lubrication because a higher viscosity than that necessary to retain the oil film results in a waste of power, due to the expenditure of energy necessary to overcome the internal friction of the oil itself.

The use of bearings permits smooth low-friction movement between two surfaces. The movement can be either rotary (a shaft rotating within a mount) or linear (one surface moving along another).[21] Bearings can employ either a sliding or rolling action (governed by the laws of sliding and rolling friction, respectively). For both cases, a strong attempt is made to provide enough lubrication to separate the bearing surfaces by a film of lubricant. The long service lives of most bearings are attributable to the absense of contact provided by the lubrication.[21]

In order to provide additional protection under conditions of heavy shock or cyclic loading, as well as during periods of abnormally high temperatures, lubricant suppliers make use of both oil-soluble chemical additives and sub-micron sized particles of solid additives in suspension as lubricant supplements. These additive components are carefully selected to fit specific application requirements and to protect working surfaces when sudden shock loading may rupture the oil film or when extreme temperatures severely reduce oil film strength due to lowered oil viscosity. Such materials also serve to extend the useful service life of the lubricant. Effective lubrication reduces power losses in a mechanical system and the improved efficiency can appear as a measurable energy saving.

Effective lubrication deserves plant management consideration as an opportunity to save energy and improve the cost effectiveness of mechanical systems. Savings of 9 percent on energy usage have been reported with improved lubrication programs. (See the Case Studies.)

B. Flywheels

Flywheels may be classified as either balance wheels or flywheel pulleys. The object of all flywheels is to equalize energy exerted and work done over a cycle and thereby prevent excessive or sudden changes of speed. The permissible speed variation is an important factor in all flywheel designs. The allowable speed change varies considerably for different classes of machinery; for instance, it is about 1 or 2 percent for certain machinery, while in punching and shearing machinery a speed variation of 20 percent may be allowed.[22]

When a flywheel absorbs energy from a variable driving force, the velocity increases; when this stored energy is given out, the velocity diminishes. When the driven member of a machine encounters a variable resistance in performing its work--as when the punch of a punching machine is passing through a steel plate--the flywheel gives up energy while the punch is at work and, consequently, the speed of the flywheel is reduced. The total energy that a flywheel would give out if brought to a standstill is given by:[22]

$$E = \frac{mv^2}{2} = \frac{Wv^2}{2g} \qquad (11.5)$$

in which
- E = total energy of flywheel, joules;
- W = weight of flywheel rim, newtons;
- m = mass of flywheel rim, in kg;
- v = velocity at mean radius of flywheel rim, m/sec; and,
- g = acceleration due to gravity = 9.8 m/sec^2.

If the velocity of a flywheel changes, the energy it will absorb or give up is proportional to the difference between the squares of its initial (v_1) and final (v_2) speeds, and is equal to the difference between the energy which it would give out if brought to a full stop and that which is still stored in it at the reduced velocity.[22] Hence:

$$E_1 = \frac{Wv_1^2}{2g} - \frac{Wv_2^2}{2g} = \frac{W(v_1^2-v_2^2)}{2g} \quad (11.6)$$

in which E_1 = energy in joules which a flywheel will give out while the speed is reduced from v_1 to v_2; other symbols are defined in equation (11.5)

The general method of designing a flywheel is to determine first the value of E_1 or the energy the flywheel must either supply or absorb for a given change in velocity which, in turn, varies for different classes of service. The mean diameter of the flywheel may be assumed, or it may be fixed within certain limits by the general design of the machine. Ordinarily the speed of the flywheel shaft is known, at least approximately; the values of v_1 and v_2 can then be determined, the latter depending upon the allowable percentage of speed variation. When these values are known, the weight of the rim and the cross-sectional area required to obtain this weight may be computed.[22]

C. Clutches

When the driving and driven members of a clutch are connected by the engagement of interlocking teeth or projecting lugs, the clutch is said to be "positive" in order to distinguish it from the type in which the power is transmitted by frictional contact fluids or magnetic fields.[22] A positive clutch is employed when a sudden starting action is not objectionable and when the inertia of the driven parts is relatively small. The various forms of positive clutches differ merely in the angle or shape of the engaging surfaces. The least positive form is one having planes of engagement which incline backward, with respect to the direction of motion. The tendency of such a clutch is to disengage under load, in which case it must be held in position by axial pressure. This pressure may be regulated to perform normal duty, permitting the clutch to slip and disengage when overloaded. Positive clutches, with the engaging planes parallel to the axis of rotation, are held together to obviate the tendency to jar out of engagement, but they provide no safety feature against overload. So-called "undercut" clutches engage more tightly as the loads become heavier and are designed to be disengaged only when free from load.[22]

When selecting a clutch for a given class of service, it is advisable to consider any overloads that may be encountered and base the power transmitting capacity of the clutch upon such overloads. When the load varies or is subject to frequent release or engagement, the clutch capacity should be greater than the actual amount of power transmitted. Design equations are tabulated in the literature.[22]

The approximate amount of power that a disk clutch will transmit may be determined from the following formula:

$$H = \mu RFnS \quad (11.7)$$

where H = power transmitted by the clutch, watts;

μ = coefficient of friction, dimensionless;

R = mean radius of engaging surfaces, meters;

F = axial force (spring pressure) holding disks in contact, newtons;

n = number of frictional surfaces; and,

S = Speed of shaft, hertz.

While the frictional coefficients used by clutch designers differ somewhat and depend on variable factors, typical values range from 0.1 (lubricated disk clutches) to 0.35 (metal and cork on dry metal).

Clutches of the electric type, like other electrical apparatus, are adapted to remote and automatic control. They are especially applicable for high-speed drives; for heavy duty (for use with motors that cannot start heavy loads); and for stopping machinery quickly, in which case a brake is used in combination with the other clutch. Clutches and brakes have similar operating principles: a brake is basically a clutch with one member held stationary.[21]

Electric clutches use two different operating principles. One class consists of clutches that are essentially mechanical but are electrically actuated (all of the friction and tooth-type electric clutches). The other class uses electrical forces to engage input and output without direct mechanical connection (hysteresis, eddy-current, and magnetic-particle clutches).

One type of the electrically actuated clutch, the Cutler-Hammer magnetic clutch, has a field or driving member and an armature or driven member.[22] Each of these parts is carried by a flexible spring steel plate so that when current passes through the winding of the field, the armature is attracted to it and the friction surfaces come into engagement. The turning power of the clutch depends entirely upon the friction surfaces which are held together by magnetic attraction. Current is conducted to the magnetizing winding of the field through two collector rings and graphite brushes. These clutches are operated by direct current. The ratings of two of the different sizes are given in Table 11.8 [22]

D. Worm Gearing

Primary considerations in industrial worm gearing are usually:

- to transmit power efficiently;
- to transmit power at a considerable reduction in velocity; and,
- to provide considerable "mechanical advantage" when a given applied force must overcome a comparatively high resisting force.

Worm gearing for use in such applications is usually a relatively coarse pitch.

Efficiency of worm gearing at a given speed depends on the worm lead angle, the workmanship, lubrication, and the general design of the transmission. When worm gearing consists of a hardened and ground worm running with an accurately hobbed wheel properly lubricated, the efficiency depends chiefly upon the lead angle and coefficient of friction between the worm and wheel. In the lower range of lead angles, the efficiency increases considerably as the lead angle increases. This increase in efficiency remains practically constant for lead angles between 30° and 45°. Handbooks provide equations for calculating gearing efficiency. Table 11.9 shows typical values.

Although not discussed herein, similar efficiency considerations apply to other elements of the power transmission chain, such as belts, flexible shafts, and other gear types (see Figure 11.1). Also see the Case Studies.

11.5 MATERIAL SHAPING AND FORMING

The shaping and forming of materials to produce commercial products is one of the main functions of industry. Some production processes have a considerable technological base; the metal working industries, woodworking industries, plastics, and chemicals are representative of these. Metal working processes have been chosen to illustrate energy use efficiency principles because they are widespread and also because the metals'cutting and forming processes have their counterparts in the processing of other materials, such as wood and plastics.

Metal production processes can be classified into four groups, having the following basic purposes:[23]

- processes used primarily to shape or form metals, such as casting, forging, extrusion, stamping, bending, and spinning;
- processes used for machining parts to specified dimensions, such as turning, planing, drilling, and milling, as well as those used primarily to produce a surface finish, such as polishing, tumbling, electroplating, honing, and shot blasting;
- heat treatment to change physical properties; and,
- processes used to join parts and materials, such as welding, soldering, riveting, screw fastening, and adhesive joining.

		Ratings Type H-30 Clutches			Ratings Type H-60 Clutches		
Nominal Size, Inches	Maximum Speed, rpm	Maximum Torque, Lbs. at 1 Ft. Radius	Safe hp at 100 rpm	Current Consumption, Watts	Maximum Torque, Lbs. at 1 Ft. Radius	Safe hp at 100 rpm	Current Consumption, Watts
10	2000	89	1.1	78	......	......	......
12	1680	154	2.0	93	......	......	......
14	1440	245	3.0	115	490	6	130
16	1260	366	4.5	133	732	9	160
20	1000	714	9.0	177	1,428	18	200
24	840	1233	15.5	260	2,466	31	247
28	725	1960	25.0	280	3,920	49	253
32	635	2920	37.0	315	5,840	74	250
40	500	5710	72.0	380	10,420	132	341
48	420	9860	124.0	460	19,720	250	400
60	340	......	......	......	38,600	485	645

Source; Cutler-Hammer Mfg. Co.

MAGNETIC CLUTCH RATINGS **TABLE 11.8**

Coefficient of Friction	Lead Angle of Worm								
	5 Deg.	10 Deg.	15 Deg.	20 Deg.	25 Deg.	30 Deg.	35 Deg.	40 Deg.	45 Deg.
0.01	89.7	94.5	96.1	97.0	97.4	97.7	97.9	98.0	98.0
0.02	81.3	89.5	92.6	94.2	95.0	95.5	95.9	96.0	96.1
0.03	74.3	85.0	89.2	91.4	92.7	93.4	93.9	94.1	94.2
0.04	68.4	80.9	86.1	88.8	90.4	91.4	92.0	92.2	92.3
0.05	63.4	77.2	83.1	86.3	88.2	89.4	90.1	90.4	90.5
0.06	59.0	73.8	80.4	84.0	86.1	87.5	88.2	88.6	88.7
0.07	55.2	70.7	77.8	81.7	84.1	85.6	86.4	86.9	86.9
0.08	51.9	67.8	75.4	79.6	82.2	83.8	84.7	85.2	85.2
0.09	48.9	65.2	73.1	77.6	80.3	82.0	83.0	83.5	83.5
0.10	46.3	62.7	70.9	75.6	78.5	80.3	81.4	81.9	81.8

EFFICIENCY OF WORM GEARING FOR DIFFERENT LEAD ANGLES AND FRICTIONAL COEFFICIENTS

TABLE 11.9

The major metal production processes are listed in Table 11.10.

Forming processes are often done by specialized plants that produce semifinished shapes which serve as raw material for other plants where the finishing operations take place. Examples are steel and aluminum rolling mills, which produce various standard and nonstandard shapes such as angles, I-beams, and tubes. Foundry operations also are often specialized and produce rough castings to order according to specified shapes and dimensions. Other forming operations such as forging, drawing, bending, and shearing are more often integrated within a manufacturing plant which performs finishing operations.

In all machinery operations, metal is removed from the part in small chips by the cutting action of a tool. The cutting action is accomplished by either a rotating or reciprocating action of the tool relative to the part. In combination with this motion, either the tool or the work must "feed" to produce a continuous cutting action over an entire surface. Figure 11.7 shows a summary of tool and work piece motion for the main machine tools.[24]

The physical properties of the metal specified by the design engineer are not always consistent with the ability to process it; that is, the functional demands of the part may require a hard material which would be difficult or impossible to cut. Fortunately, by means of heat treatment, the physical properties of metal can be altered at almost any point during fabrication so that machining can take place while the metal is in the most machinable state, the final properties being produced at the time desired.

The joining of material is related to the fabrication of products. Some methods require the direct use of energy (e.g., welding) whereas others use a bond (fastener, glue). Energy needed for the various joining operations depends on the type of process and the quantity of material.

Power requirements for performing various metal cutting operations can be calculated from consideration of the cutting forces. For a lathe, for example, the power at the cutting tool is defined by equation (11.8):

$$P = F_t v \qquad (11.8)$$

where P = power of the cutting tool, watts;

v = cutting speed, m/sec; and,

F_t = tangential cutting force component, newtons.

The radial cutting force does not contribute to the power. Although the feed force can be considerable in magnitude, feeding velocity is generally so low that the power required to feed the tool can be neglected.

Although some variations of tangential cutting force with respect to changes in speed may occur at low cutting speeds, the cutting force can be considered to be independent of cutting speed within the practical ranges of normally used cutting speed. The effect of feed and the depth of cut on the cutting force is illustrated in Figure 11.8.[25] The figure indicates that the cutting force is proportional to the feed and depth of cut, each raised to some power. Equations relating power requirements to the cutting rate (amount of material removed per unit time) are available in the literature.[25] Depending on the material, tool shape, and material hardness, the power per cutting rate is in the range of 4.7 to 94 W per cm^3/min (0.1 to 2.0 hp per in^3/min) when the materials range from aluminum to alloy steel.

Force and power predictions for multiple-point tools are more complex than those for a single-point tool. Varying numbers of teeth may be in contact with the work, the chip size can vary in different parts of the cut, and the orientation of the cutting teeth may not be constant with respect to the work piece. The best method to determine forces on such tools is by actual measurement on the machine to be used or on a simulated setup in a

Metal Forming

Casting:	Tolerance
Sand casting	1/32 in. (0.03 n.)
Permanent mold casting	0.0025 to 0.010
Die casting	0.002 to 0.010
Centrifugal casting	±1/64 in.
"Lost wax" precision casting	±0.2 %/in.
Continuous casting	
Power metallurgy	±0.001 in.
Hot Forming:	
Hot rolling	
Hot drop forging	
Press forging	+0.09/-0.03 in.
Extrusion	
Cold Forming:	
Cold rolling	
Roll forming	
Cold forming with presses (stamping)	
Gruerin process	
Spinning	

Metal Machining

Metal Cutting:	Metal Finishing:	
Turning	Grinding (fine)	Electric discharge
Boring	Tumbling	Electroarc
Grinding (coarse)	Shot blasting	Electrolytic
Milling	Polishing (buffing)	Laser cutting
Shaping	Honing	Chem-milling
Planing	Electroplating	Ultrasonic
Broaching		
Drilling		

Heat Treatment

Hardening	Induction hardening
Tempering	Flame hardening
Case hardening	Annealing
Carburizing	Normalizing
Nitriding	Spheriodizing
Cyaniding	

Joining

Welding:	Fastening:	Adhesive Joining:
Acetylene torch	Riveting	Bonding
Heliarc	Screwing	Gluing
Brazing	Bolting	Tape
Soldering		

METAL PRODUCTION PROCESSES TABLE 11.10

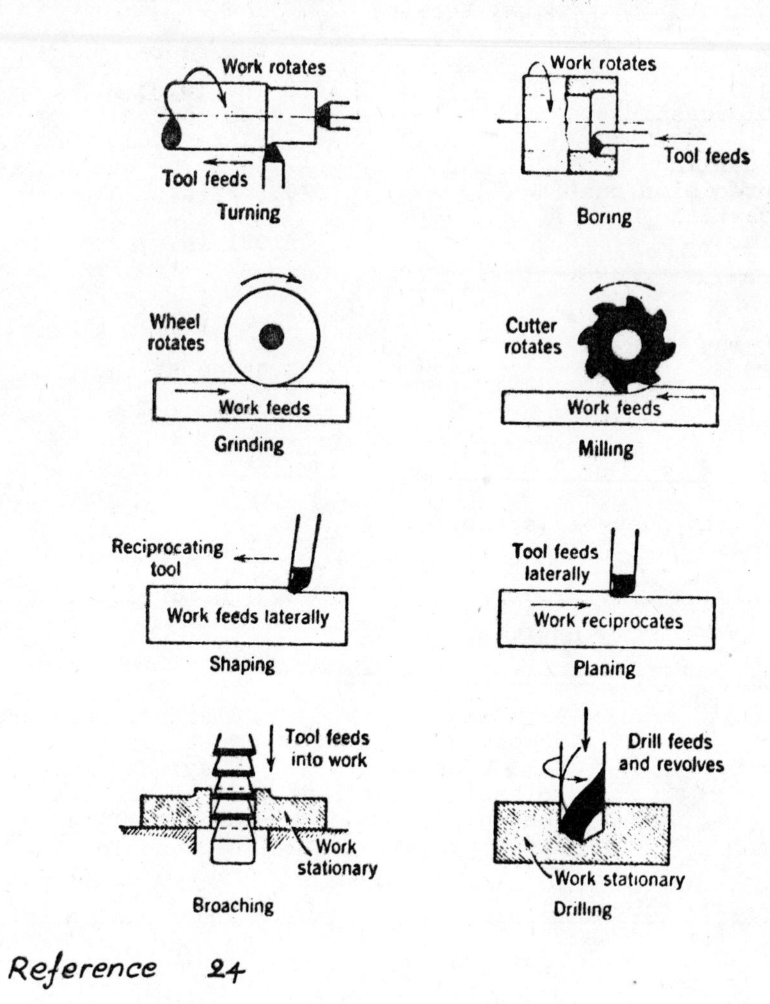

Reference 24

MACHINING OPERATIONS **FIGURE 11.7**

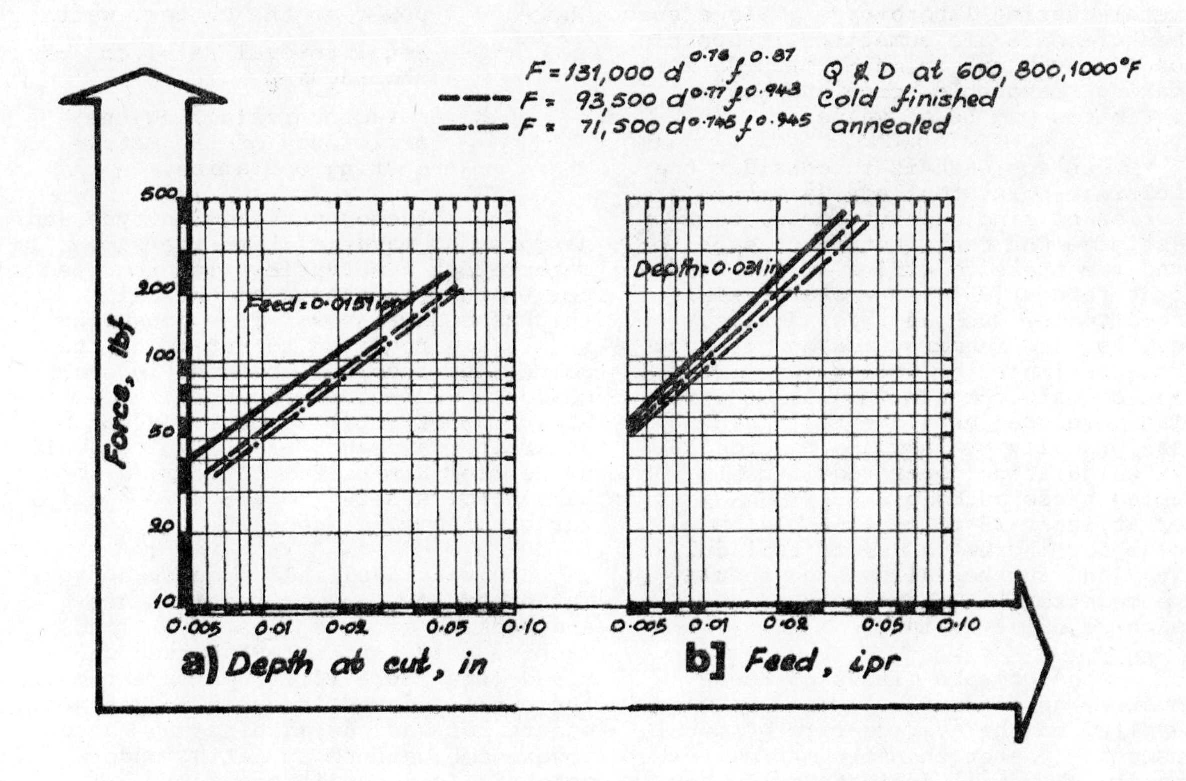

a) Cutting force vs depth of cut.

b] Cutting force vs feed.

Tool material - HSS
Tool geometry, 6, 11, 6, 6, 6, 15, 0.01
Work material - SAE 1045 steel

Depth - 0.031 inches
ipr - inches per reduction

Reference 25

EFFECT OF DEPTH AND FEED ON CUTTING FORCE FIGURE 11.8

metal-cutting laboratory. Since such measurements are sometimes impossible or inconvenient to make, methods for making reasonable force and power estimates may be of value.

One approach is to consider the multiple-point tool equivalent to a series of single-point tools, then estimate the contribution of each tool, and sum these to arrive at the resultant forces.[25] For rotary axial-feed tools, such as twist drills, core drills, and reamers, reasonably accurate estimates of forces and power can be made through the use of formulae developed experimentally and analytically by Shaw and Oxford.[26] In calculating power requirements using these methods, an allowance of at least 25 percent should be made for increases due to tool dulling, and further allowances should be made for the efficiency of the machine drive train.

The concepts discussed above must be applied with caution. If applied to the average rate of metal removal (rather than the maximum rate), they will indicate only average torque or average force, and this could be considerably below the peak forces in the case of intermittent cutting as might be encountered in milling or broaching. Furthermore, many of the equations presented in the references ignore the even higher peak forces resulting from impact or vibration during cutting.

Milling machines, like other machine tools, dissipate part of the input power. Some of the reasons are frictional losses, gear-train inefficiencies, spindle speeds that are too high for the particular machine, and mechanical condition. Consequently, the power for milling must include the machine power losses and the power actually used at the cutter. Efficient use of power at the cutter is influenced by cutter speed, design, and material, and by work piece material. [27]

The total power required at the cutter is given by equation (11.9):

$$P = cmm/K \qquad (11.9)$$

where P = power at the cutter, watts;

cmm = metal removal rate, cm^3 per second; and,

K = a factor reflecting the efficiency of the metal-cutting operation.

The K factor varies with type and hardness of material; for the same material it also varies with the feed per tooth, increasing as the chip thickness increases. Time consuming trials are required to determine the quantities involved, because in each case, the K factor represents a particular rate of metal removal and not a general or average rate. Typical values range from $K = 0.18 \times 10^{-3}$ cm^3/J for hard alloy steels, to 1.5×10^{-3} cm^3/J for aluminum and magnesium.*

To make available a quick approximation of total power requirements and machine efficiencies, a milling machine selector table has been devised (see Table 11.11) which estimates the metal removed in cubic inches per minute for various machine and horsepower combinations operating under constant load conditions.[28]

Methods similar to those described above can be developed for other types of milling operations and for other material shaping and forming operations.[29,30] By comparing the power requirements for various operations, a manufacturer will be able to decide on the importance of the relative energy utilization of various processes. As energy costs increase, efficient energy use will have a greater influence on decisions related to process selection, particularly for high production processes.

Opportunities for Improved Energy Utilization

In the field of material shaping and forming, important energy savings can be obtained through better organization of production, standardization of products, and avoidance of over-specification of material quality.

*In American usage cmm (equation 11.9) is expressed in cubic inches per minute, P in hp, and K has units of in^3/min hp. Representative values for K are 0.05 (alloy steels) and 4.0 (aluminum and magnesium).

Rated hp of machine	3	5	7.5	10	15	20	25	30	40	50
Overall machine efficiency, percent	40	48	52	52	52	60	65	70	75	80
Material	Max metal removal (cu in/min)									
Aluminum	2.7	5.5	8.7	12	18	27	37	48	69	91
Brass, soft	2.4	4.7	7.5	10	16	24	32	41	60	79
Bronze, hard	1.7	3.3	5.3	7.3	11	17	23	30	43	56
Bronze, very hard	0.78	1.6	2.5	3.4	5.3	7.8	11	15	20	26
Cast iron, soft	1.6	3.2	5.2	7.1	11	16	22	28	41	54
Cast iron, hard	1	2	3.3	4.6	7	10	14	18	26	35
Cast iron, chilled	0.78	1.6	2.5	3.4	5.3	7.8	10	13	19	26
Malleable iron	1	2.1	3.4	4.7	7.3	11	14	18	26	36
Steel, soft	1	2	3.3	4.6	7	10	14	18	26	35
Steel, medium	0.78	1.6	2.5	3.4	5.3	7.8	10	13	19	26
Steel, hard	0.56	1.1	1.8	2.5	3.9	5.7	7.7	10	14	19

Reference 28

MILLING MACHINE SELECTION

TABLE 11.11

These savings can be obtained without technical research or development effort.

Inasmuch as material shaping processes absorb an appreciable portion of total energy use in industrialized countries, shaped products should be made to last longer. Life can often be prolonged with negligible extra energy outlay by improving the quality of the material, e.g., by protection against corrosion and wear.[31] (See Chapter 12.)

The primary material processes for conversion of metal ores into the semi-finished materials which constitute the starting point for the shaping processes use as much and frequently even more energy than the latter. Furthermore, shaping is mostly a multi-stage operation, much of it carried out at elevated temperatures, involving repeated intermediate reheating and final heat treatments. Heating uses several times more energy than shaping operations. Additional energy is required in ancillary treatments., e.g., pickling and coating. These facts lead to the following general conclusions about achieving energy economies.[30]

- As far as practicable, replace shaping processes having a low metal yield--i.e., those processes whose end products contain only a small fraction of the metal used--(e.g., machining) by high yield processes, notably casting and plastic forming (e.g., extrusion, spinning). Use fabrication extensively as a means of producing complex shapes.
- Favor processes in which the final shape is achieved in the least number of intermediate stages, e.g., casting, extrusion, powder compacting.
- Favor shaping at lower temerature, even though the mechanical forces required are greater in this case. Where appropriate, electroforming should be developed further.
- Develop scrap recycling processes which bypass the primary stage, e.g., direct compaction of turnings into reinforcing bar.

Much energy is lost between the points of input into the machine and application to the material which is being shaped. For example, in machining a transmission or coupling, efficiency of less than 50 percent is common. There is room for considerable improvement by engineering of more efficient shaping machinery, and even more readily by appropriate operating practices and adequate machine maintenance.

The utilization of shaping machinery is generally low, and energy is wasted by idling. The most obvious way of reducing idling losses is by employing continuous processes. Not all the technology is available; however, even where it is applied, there is room for further development, e.g., in continuous casting of tubular and other slender shapes.

The potential of casting needs to be explored further with the aim of achieving greater complexity of slender shapes in the more "difficult" metals, coupled with the right mechanical properties of the finished product. Powder technology is probably the most promising of the shaping processes with reference to future energy economy.

Processes which use high pressures to force cold metal through or against dies have a triple energy savings in that heating the material is not required, energy is not spent in cutting or removing the material, and less energy intensive material is used. Reference [32] reports a high-pressure process that cut electricity usage 93 percent in the manufacture of coaxial cable connectors (compared to machining), 87.5 percent in production of pinion rods, 60 percent in hollow cathodes, and 50 percent in plumbing equipment.

Conventional heating and heat treatment processes are known to be very inefficient. Processes generating heat *in situ* require further development, e.g., induction heating of conducting materials and perhaps microwave heating of nonconducting materials (see also Chapter 10). Development efforts should concentrate on cheaper sources of energy for these processes (e.g., low vs high frequency, chemical maser, and induction coils operating in the superconducting state).[31]

Where multistage (or even single step) operations are unavoidable, possibilities for recovery of the energy lost by cooling should be considered. In cutting and welding there are opportunities for energy economy by the use of more concentrated heat generation (e.g., by electron beam and laser), but at present the cost is prohibitive. Hence, research and development should be devoted to the cost reduction of these systems. Also, in view of the production and transmission inefficiencies of electricity, non-electrical methods (e.g., chemical maser, thermite welding) and cold methods (gluing) should be developed.

11.6 MATERIAL TRANSPORT

The transport of material and products within industrial facilities presents special requirements for electrical energy use. Material transport as discussed in this section is to be differentiated from material transport between facilities or over long distances (as discussed in Chapter 5).

The most common method of transporting material within industrial facilities is by means of conveying machinery. Conveyors are devices for moving material from one point to another at the same or at a different elevation. In intermittent conveying, the material is moved in a succession of separate loads; in continuous conveying, the material is delivered in a steady stream.

In the transportation of package material in industrial plants, freight houses, stations, etc. electric baggage, freight, and fork-lift trucks are used extensively. Storage-battery trucks with lifting platforms are built in two types: a low-lift truck that lifts ∿10 cm (∿4 in.) and a high-lift truck that lifts the platform up to a height of 2.4 m (8 ft). The low-lift is a single-purpose truck used in factories and warehouses where material can be stored on skid platforms and transported on the platforms by truck between manufacturing processes and to and from cars. The high-lift truck is a general purpose truck used to convey skid platforms; in addition, the high-lift feature is used for piling cases, barrels, and similar packages in warehouses, lifting the loads from ground level to autotrucks, and to car level. Lift trucks are recommended for hauls up to 360 m (1200 ft) and are built from 1 to 10 mt capacity. Trucks of 2 and 3 mt capacity are generally recommended.

Electric tractors and trailers are recommended for hauls from 100 to 450 m (350 to 1500 ft) when more material must be moved per trip than is possible with lift trucks. The trailer load for an electric tractor is 10 to 20 mt.

Overhead trackage is another means of conveying material. Light rigid trackage consists of tracks of various forms suspended from overhead structures for carrying trolleys to which loads are attached by hooks or chain blocks. The four main types of track are the bar type, the Coburn type, the single I-beam, and the double I-beam. The first two types can accommodate loads up to ∿2 mt. The single I-beam is suitable for loads in the range of 1 to 10 mt; the double, in the range of 10 to 20 mt. The trolleys may be moved either by hand or by chain wheel, or a series of trolleys may be attached at suitable intervals to an endless power driven chain (which is called a trolley conveyor). A trolley conveyor may run through several rooms or departments of a factory and up and down between floors. Hooks, trays, or baskets attached to each trolley provide means for carrying the loads. They have their widest application in carrying parts between machines that perform succeeding operations in the manufacture of finished parts and delivering them to the parts storeroom or assembly floor.

Monorails are a form of electric hoist which not only lift the load but transport it along an overhead track from one portion of a plant to another; the operator rides with the machine and controls all movements. Monorails are used to handle any materials that may be suspended from a hook either as units or in quantities in boxes or tubs.

Cableways are aerial hoisting and conveying devices using suspended steel cable for track, the loads being suspended from carriages and moved by gravity or power. The maximum clear span is 600 to 900 m (2000 to 3000 ft). The gravity type is limited to conditions where at least a 20 percent grade is obtainable on the track cable.

Screw or spiral conveyors are used for horizontally conveying dry nonabrasive materials, such as grain, flour, seeds, cement, and fine coal, and also for sand, gravel, fine ashes, etc., although the wear from sand, gravel and ashes is rapid and maintenance costs may be very high. The maximum length when handling grain or light pulverized material is 120 to 150 m (400 to 500 ft), for more abrasive materials, 30 to 45 m (100 to 150 ft). These conveyors may be used as feeders in single sections, on inclines up to 15°.

The power required to drive a typical screw conveyor is given by:[10]

$$P = KCWL/2 \times 10^6 \qquad (11.10)$$

where P = power, hp;

C = capacity, ft^3/hr;

W = weight of material, lb/ft^3;

L = length, ft; and,

K = material constant (1.2 for grain, 2.5 for coal and cement, 4.0 for sand, gravel, and ashes).

Another category of continuous conveyor is the chain conveyor type, including scraper conveyors, apron conveyors, and open top carriers. For these, the power required can be calculated using empirical relations given in the literature.[10] In general the power needed depends on weight, length of the conveyor, speed, and the material being transported. Similar approaches are used to determine power requirements for V-bracket and pivoted-bucket carriers.[10]

If the conveyor is composed of portions on different inclines, the various portions should be considered independently and the results added. Ten percent additional power should be added for each change in direction.

Belt conveyors are used for transporting bulk material in large quantities and will handle practically anything that can be properly fed to the conveyor provided that it will not adhere strongly to, or burn, the belt. The advantages of the belt conveyor are: large capacity (10,000 mt/hr of material can be carried), low power requirement, and low attendance and maintenance costs.

From the standpoint of the application of power to the belt, a conveyor is identical to a power belt; the determining factors are the coefficient of friction between the driver pulley and the belt, tension in the belt, and arc of contact between the pulley and belt.

Electric vibrating feeders are in wide use in industry. They often operate magnetically with a large number of short strokes. They are built to feed from a few kg/min to 1000 mt/hr and will handle any material that does not adhere to the vibrating pan. Other types of feeders in general use are short, slow-moving, belt- and apron-conveyors.

Another type of conveyor is the gravity roller conveyor. The principle involved in the gravity roller conveyor is the control of motion due to gravity by interposing an antifriction trackage set at a definite grade. It finds application in the movement of various types of package goods having a smooth surface sufficiently rigid to prevent sagging between rollers. The rollers vary in diameter and strength from ∿2 cm with a capacity of 20 N per roller up to 10 cm with a capacity of 8000 N per roller. Spacing of the rollers in the frames varies with size and weight of the object to be moved. Three rollers should be in contact with the package to prevent hobbling. The grade of fall required to move the object varies from 1-1/2 to 7 percent, depending on the weight and character of the material in contact with the rollers.

Pneumatic conveyors are used for handling: (1) material in bulk such as pulverized coal, grain, wood waste; (2) dust from grinding wheels, sand-blast equipment, and other industrial processes; and (3) ashes in boiler plants. They are also used for handling small packages by placing the material in containers that fit the carrying tube and act as pistons. Pneumatic conveyors for bulk materials may be blower type, exhaust or suction type, or a combination of exhaust and blower type. The majority of installations are probably of the latter type; the material entering under suction passes through the fan and is blown to its destination, the escape of air being accomplished through a dust collector.

For pneumatic conveyors, all pipes should be as straight and short as possible and bends, if necessary, should have a radius of at least three diameters of the pipe. Pipes should be proportioned to keep down friction losses yet maintain air velocities that will prevent material settling. Sudden changes in diameter should be avoided to prevent eddy losses.

Selection of the best material transport mechanism to use for a particular operation should depend on the type of product and the volume of material which must be moved per unit time. The physical limitations of the facility as well as the need for continuous versus batch type processing will affect the final selection. Total energy use for each candidate transport method could affect the final selection. In fact, energy utilization considerations may even affect the decision to use a batch or continuous process.

Efficiency Improvement Opportunities

Several obvious courses of action can provide an improvement in the electrical energy utilization efficiency of material transport systems. They include the following;

- Survey all possible alternative transport system equipment to determine the most suitable for the specific needs of the facility and product operation.
- Determine energy requirements for the operation of the system and factor this quantity into the selection decision.
- Investigate the energy requirement trade-off for intermittent versus continual material movement.
- Use proven design methods and engineering formulations in sizing and constructing material transport mechanisms.
- During operation, do not overload the transport mechanism to cause overworking of the drive motor.
- Provide for regular maintenance and lubrication of all moving equipment parts to keep unnecessary friction to a minimum.
- Make random checks to ensure that equipment is being used at near maximum utilization. Investigate the possibility that transfer equipment can be run for shorter periods of time and during non-peak work times.
- Turn off conveyors, lift trucks, etc. when not in use.
- Recharge batteries on material handling equipment during off-peak demand periods.
- Adjust and maintain fork lift trucks for most efficient operation.
- Use gravity feeds wherever possible.

11.7 POTENTIAL ENERGY SAVINGS

As indicated in Table 11.1, energy used by electrical prime movers for the residential and commercial sectors is significant. The numbers shown in the the table illustrate that 2814×10^6 GJe out of a total of 4381×10^6 GJe (64 percent) for combined residential, commercial, and industrial use is required for electric motors. Extrapolating this number to include transportation and other uses as well, a reasonable assumption is that electric motors require more than 65 percent of the total electrical energy used. Thus, relatively small changes in efficiency potentially can accrue to significant energy savings.

The estimated savings which may be effected are classified as:

- *Intermediate--2 to 5 percent*

  Operational changes for motor use and scheduling, improved maintenance practices, reduced overloading of motors.

- *Near-Term--5 to 10 percent*

  Replacement of multiple small motor applications to improve cooling, replacment of old motors with newer more efficient motors.

- *Long-Term--5 to 10 percent*

  Introduction on market of premium efficiency motors, increased use of new materials, super-conducting motor applications for large engines.

The estimated annual savings (year 2000) due to increased efficiency is 12 to 25 percent. Using the projected year 2000 US energy use for electricity generation of 105×10^9 GJ/yr (see Chapter 1), assuming 50 percent is used by electric drives, and applying the potential range of savings (12 to 25 percent), the fuel savings can be calculated as:

$$(105 \times 10^9 \text{ GJ/yr})(50\%)(12\text{-}25\%) = 6.3 - 13 \times 10^9 \text{ GJ/yr}.$$

Since a 1000 MWe power plant operating with a 60 percent capacity factor requires 5.7×10^7 GJ/yr, this saving is equivalent to the energy use of 110-220 power plants.

In round numbers, improved efficiency in electric drives could eliminate the need for 100 to 200 1000 MWe power plants by the year 2000. This large potential saving warrants a significant expenditure of research funds to improve the efficiency of electric motors and motor driven systems.

* * *

REFERENCES

1. Arthur D. Little, Inc., *Energy Efficiency in Electric Motors*, (Cambridge, Massachusetts: May 1976).

2. Allen, Jonathan, "The Craft of Electric Motors," *Environment* 16(October 1974): 36-39.

3. Stanford Research Institute, *Patterns of Energy Consumption in the United States*, Report prepared for the Office of Science and Technology, (Menlo Park, California: January 1972).

4. Material provided through the courtesy of Mr. Howard E. Barr, Senior Design Engineer, U.S. Electrical Motors, Prescott, Arizona.

5. The Electrification Council, *Motors and Motor Controls*, (New York: 1975).

6. Woodson, Thomas T., "Motors for Integral Mechanisms," *Mechanical Engineering* (August 1950): 615-628.

7. Fink, D.G. and Carroll, J.M., eds., *Standard Handbook for Electrical Engineers*, 10th edition, (New York: McGraw-Hill Book Co., 1969).

8. "Electric Motors and Controls," *Machine Design* 1975 Reference Issue, 47(24 April 1975).

9. "Motors" *Machine Design 1977* Reference Issue, 49(19 May 1977).

10. Baumeister, Theodore, ed., *Marks' Standard Handbook for Mechanical Engineers*, 7th edition, (New York: McGraw-Hill Book Company, 1967).

11. "Motor Market Still Slow for Efficient Units," *Energy User News*, 14 March 1977.

12. Woodson, T.T., *Introduction to Engineering Design*, (New York: McGraw-Hill Book Company, 1966).

13. "Motors, A Special Report," *Power* 113(June 1969).

14. Rashidi, Abdul S., "Better Motors with Rare-Earth Magnets," *Machine Design* 48(8 July 1976): 70-73.

15. "New Electric Motor May be Huge Energy Saver," *Southern California Industry News*, 9 May 1977.

16. "New Electric Motor Idea Hailed as Energy Saver," *Los Angeles Times*, 26 April 1977.

17. Krouse, John K., "Energy Miser Now Controls Any DC Motors," *Machine Design* 49(6 January 1977):79-82.

18. "Motor Controls & Protection," *Machine Design* 49(19 May 1977).

19. "Solid State Switching Devices," *Machine Design* 49(19 May 1977).

20. "Mechanical Drives," *Machine Design* 1977 Reference Issue, 50(30 June 1977).

21. "Mechanical Drives," *Machine Design* 1975 Reference Issue, 47(19 June 1975).

22. Oberg, Erik, ed., *Machinery's Handbook*, 17th edition, (New York; The Industrial Press, Inc., 1964).

23. Buffa, E.S., *Modern Production Management*, (New York; John Wiley & Sons, 1962).

24. Begeman, M.L., *Manufacturing Processes*, 4th edition, (New York: John Wiley & Sons, 1957).

25. Wilson, F.W., ed., *Fundamentals of Tool Design*, (Englewood Cliffs, New Jersey: Prentice-Hall, Inc., 1962).

26. Shaw, M.C. and Oxford, C.J., "On the Drilling of Metals," *Transactions ASME* 79(January 1957).

27. Idem, "The Torque and Thrust in Milling," *Transactions ASME* 79(January 1957).

28. American Standards Association, *Milling Cutters, Nomenclature, Principal Dimensions, Etc.*, American Standard ASA B5.3-1959, (New York: 1959).

29. American Society of Tool and Manufacturing Engineers, *Tool Engineers Handbook*, 2nd edition, (New York: McGraw-Hill Book Company, 1959).

30. The Cincinnati Milling Machine Company, *A Treatise on Milling and Milling Machines*, 3rd edition, (Ohio: 1951).

31. Kovach, Eugene G., ed., *Technology of Efficient Energy Utilization*, The Report of A NATO Science Committee Conference held in Les Arcs, France, 8-12 October 1973, (Oxford: Pergamon Press, 1974).

32. "Metalworking Process Gets Few Buyers," *Energy User News*, 21 March 1977.

CHAPTER 12

ELECTROLYTIC AND ELECTRONIC PROCESSES

G.B. Taylor

CHAPTER CONTENTS

KEY WORDS

Alternate Sources
Automobiles
Corrosion
Efficient Energy Use
Electrolytic and Electronic Process Use
Energy Conversion
Energy Sources
Energy Storage
Equipment Maintenance
Primary Metals
Process Modifications
Research Recommendations

SUMMARY

The electroplating and electrowinning industry is one of the largest industrial users of electricity. Small improvements in mass transfer and plating distribution efficiencies can result in large savings in the electrical energy used for these purposes.

In spite of their direct usage of large amounts of electricity, electrochemical processes such as electrosynthesis, electrowinning, electrochemical recycling, and electrochemical matching should be considered as alternatives to conventional processes where improved energy efficiency is possible.

Direct energy conversion with fuel cells could play an important role in future electricity production. With improvements in ion conduction, electrocatalysis, and oxygen reduction, for example, each residence or business could produce electricity on demand, using fuels ranging from coal gas to hydrogen with greater efficiency.

Production of more efficient batteries for electrochemical storage of electrical energy will serve to provide two greatly needed capabilities: use of electrical power for transportation and off-peak storage of electricity produced by solar and nuclear generation plants. Basic research and development of electrode mechanisms, materials, configuration, and charging control are needed. The use of hydrogen produced by high temperature electrolysis of steam to be used in both fuel cells and energy storage must continue to be investigated.

Further development of fuel cells and electrochemical storage technology is needed in order to produce the high efficiencies necessary for their successful application to energy production on a large scale.

Corrosion damage to parts, equipment, vehicles, and structures accounts for approximately 1 percent of the United States GNP. It is corrosion which creates loss in fuel cells and batteries and it is one of the factors which limits the life of these systems. Basic research in corrosion prevention will play an important part in preserving natural resources and energy reserves.

12.1 INTRODUCTION/OVERVIEW

The uses of electrical energy in industry are complex and varied. In order to simplify the overall picture so that it may be studied more easily, it has been found convenient to categorize the usage by the process in which this energy is expended.

The use of electrical energy for electrolysis, electrolytic, and electronic processes occurs in residential, commercial, and industrial applications. In the US, the major use is in industry. This chapter will address energy saving procedures in some of these major industry groups: primary metal industries and fabricated metal products, which jointly accounted for 22 percent of the US energy use in 1971; reduction of corrosion, which accounts for approximately 1 percent of the United States GNP; and fuel cells and batteries.

Specifically, this chapter deals with the forming of primary nonferrous metals (e.g., copper, lead, zinc, and aluminum), the problems of attendant corrosion, and corrosion-prevention processes for these metals. It is interesting to note that, in 1971, $20 million were spent in the prevention of corrosion which, during that year, caused over $10 billion damage to equipment (primarily transportation related) in the US alone. The world total is undoubtedly much larger.

To a lesser extent, this chapter also discusses electrical energy use in electronic processes such as batteries, fuel cells, electronics, and the electronics industry.

12.2 OBJECTIVE, SCOPE, ASSUMPTIONS

A. Objective

The primary objective of this chapter is to point out means by which energy, electrical energy in particular, can be saved in electronic and electrochemical processes. This energy can be saved in the *immediate* future through better housekeeping measures; in the *near-term* future by minor modifications of industrial processes and improvements in organization and procedures; and finally, in the *long-term* future, through major changes in industrial processes requiring large capital investments, utilization of future technology, and new breakthroughs which hopefully will occur.

B. Scope

As a result of the large amount of information available in the electronic and electrolytic processing industry, it is necessary to concentrate on what are considered to be the most important areas of this industry in order to avoid a lengthy document which goes beyond the intended scope of this book. Therefore, this chapter will concentrate on the primary metals industries which require electrolytic purification for metals production. Due to the extreme importance of corrosion prevention and the potential energy savings in all sectors of the economy, electrochemical processes presently used to reduce corrosion are investigated. These include energy saving practices in present processes and a look at possible future processes which have the potential of further reduction of energy and material waste through corrosion. In addition, fuel cells and batteries are discussed, since improvements in electrical energy storage will have far-reaching effects on the power production industry.

C. Assumptions

This chapter is not intended to be a detailed manual for use in electrolytic and electronic processes. It is, on the other hand, intended to provide some guidelines through which energy savings can be realized by modifications of processes and procedures, improvement in energy saving practices, and, in the long-term, major process changes which have the potential of saving energy, primarily through corrosion prevention but also in the process of corrosion prevention itself.

12.3 USEFUL FACTS AND BACKGROUND

Measurement of power use, especially electrical power use, is an index which may be used to compare the efficiencies of industrial processes. No such data on efficiencies are regularly collected by government or civilian agencies. Some companies gather such information and use the results of their studies to improve processes, but due to the competitive atmosphere in industry, this information is normally confidential.

The measurement of direct energy used per unit of output does not provide a true measure of efficiency. Indirect energy use must also be measured and added to direct energy use. Table 12.1 lists the direct fuel and electrical energy required to produce various important industrial materials.[1]

Energy use efficiency in electrolytic processes depends heavily on the chemicals being produced and is relatively low when compared with other electrical processes. Primary parameters which affect the process efficiencies are circuitry and electrode design, container heating and heat loss, electrode consumption, and chemical reactions with contaminants. Electrolytic processes are used in the refining of many metals including aluminum, zinc, and copper. Aluminum, for example, is produced principally by electrolysis of alumina in a molten cryolite bath. An anode of carbon, a cathode consisting of a carbon-lined steel shell containing a pad of molten aluminum, and an electrolyte of molten cryolite (in which the alumina is dissolved) form the reduction cell. The alumina is reduced to aluminum at the cathode and carbon is oxidized to carbon dioxide at the anode. Typical power requirements for these cells range from 80,000 to 100,000 amperes with a cell voltage of 4.5 to 5.0 volts. This represents a range of 360 to 500 kWe per cell.[2] The large amount of electrical energy used to refine these metals, aluminum being the greatest user, justifies research on more efficient electrolytic processes due to the large potential savings of electrical energy. Reference 3 is an example.

Corrosion

Corrosion, an electrochemical process, produces losses approaching 1 percent of the Gross National Product of the United States. Approximately 25 percent of this destruction could be saved through present technology, the application of short-term good housekeeping practices, and modification of present processes. Up to 50 to 75 percent reduction in corrosion processes could be achieved in the future by utilizing advanced technical concepts available now and anticipated technical breakthroughs in the future. The greatest impact of corrosion reduction would be felt in the transportation sector through increased vehicle life and the resulting decrease in overall energy cost.[4]

Corrosion of metals occurs as a result of oxidation-reduction reactions between a pure metal or alloy and chemical agents (corroding agents) in the local environmental media. These reactions may be the result of two related processes: (1) direct chemical reactions which are limited to those situations involving a highly corrosive environment, high temperature, or both; (2) electrochemical reactions which take place in the presence of an electrolyte media in contact with a metallic surface, or in the presence of external electrical currents.

Most corrosion reactions of concern are produced by electrochemical reactions. The most common electrolyte is ordinary water containing trace amounts of dissolved salts, acids, or alkalis, which convert it to a conductor. The rates of

Material	Energy Required MJe/kg
Magnesium	107.7
Aluminum	70.7
Zinc	53.4
Manganese	53.3
Copper	32.3
Paper	31.4
Steel	30.2
Tin	23.2
Glass	20.4
Ferroalloys	20.1
Lead	13.0
Nickel	7.4
Lime	6.5

Reference 1

DIRECT ENERGY REQUIRED TO PRODUCE VARIOUS PRIMARY INDUSTRIAL MATERIALS

TABLE 12.1

electrochemical corrosion reactions are dependent on the impurity content of the gas envelope or dissolved gases (O_2, CO_2, H_2S, NO_x, H_3O, and the halogens Cl, Fl, Br, I being most predominant); the dissolved salts, acids or alkalis in electrolytes; and the temperature, surface texture, and chemical constituents of the metal.

In electrochemical reactions, the anodes and cathodes may be at geometrically displaced positions, in which case a net current will flow. The electrochemical potential of each cell constructed may be determined with a volt meter. In this fashion the electrochemical series of the most common metals and alloys has been evaluated and a listing appears in Table 12.2. Those metals appearing at the head of the table (left column) have a greater potential for the anodic oxidation reaction which preferentially places them in higher concentrations in the electrolyte as ions; those at the base of the table (right column) undergo an accompanying cathodic reduction reaction with the ions of these materials plating out on a conducting surface.

Frequently, electrochemical reactions occur in which the anodic and cathodic electrodes are microscopic points, located randomly across the surface of the metallic object. These points of varying potential are a result of differences in surface texture, composition, or microscopic inhomogeneities such as cracks, inclusions, voids, and films.

Factors Which Promote Corrosion

Ordinary water, the most common electrolyte, is always in equilibrium with its dissociation products, H^+ and OH^-. Salts present in the water which tend to increase the H^+ concentration (acids) increase the corrosion potential due to hydrogen gas (H_2) formation at a cathodic surface. Those salts which would tend to increase the hydroxyl ion (OH^-) concentration conversely will reduce the corrosion rate of most metals.*

*Aluminum and Zinc are notable exceptions to this general rule, as are the alkali metals.

Gaseous oxygen dissolved in water reacts with the hydrogen ions on cathodic surfaces and thus destroys this protective film. Zinc and aluminum provide a poorer catalytic surface for this reaction than iron. The rate of corrosion of iron or steel is limited by the diffusion of dissolved oxygen to the metal surface. In the absence of other factors (dissolved salts, external electric currents, or the direct contact of dissimilar metals), the corrosion of metals can be reduced through the deaeration of water.

Factors which stimulate atmospheric corrosion include: high temperature, high humidity, acids produced by dissolved salts or airborne pollutants such as NO_x, SO_2, soot, etc. dissolved oxygen in the moisture film, conditions of nonuniformity in surface properties, concentrations of corrosives, oxygen, or surface films, and the presence of external electric fields or currents.

Factors which tend to inhibit corrosion are: the absence of previously stated stimulating conditions, alkalis, electric fields which promote cathodic actions, and the electrical isolation of the metal from others which would tend to make the former anodic.

The same factors which either stimulate or inhibit atmospheric corrosion apply in the corrosion of metals immersed in liquids or buried underground. Anaerobic bacteria can also contribute.

Corrosion Prevention

The rates of corrosion reactions may be reduced or prevented by a variety of means. Corrosion as a result of direct chemical reactions may be inhibited by placing a barrier between the base metal and the corrosive media. This barrier may be a film which is applied such as a paint or metallic plating, or the outer surface of the base metal may be transformed from a chemically reactive to a non-reactive state through a passivation technique.

Electrochemical corrosion reactions may be inhibited by placing a barrier between the base metal and the corrosive media, the electrolyte, as in chemical corrosion reactions. Another method is to reduce or reverse the flow of current

Magnesium and its Alloys	Cadmium
CB-75 Aluminum Anode Alloy	Nickel
B-605 Aluminum Alloy	Iconel
Aluminum 7072	Lead-tin Solder (50-50)
5456	Tin
5086	Lead
5052	Manganese Bronze
3003, 1100, 6061, 356	Aluminum Bronze
2117 Aluminum Rivet Alloy	Silicon Bronze
Titanium	Naval Brass
Zinc	Yellow Brass
Galvanized Steel	Red Brass
Wrought Iron	Bronze (75% cu, 20% ni, 5% zn)
Cast Iron	Bronze (90% cu, 10% ni)
Mold Steel	Bronze G
Stainless Steel 410	Bronze M
Stainless Steel 304 (active)	Copper
Stainless Steel 316 (active)	Platinum

THE ELECTROCHEMICAL SERIES OF COMMON METALS AND ALLOYS

TABLE 12.2

which accompanies the reaction. Still another is to chemically alter the reactive potential of either the base metal through a passivation technique or the electrolyte through the addition of chemical additives. These would either render the metallic atom at the metal surface insoluble in the media, or prevent the precipitation of those metallic compounds in the corrosive media.

Use of Protective Film

Paints, epoxies, or passive metallic liners may be used to prevent the flow of current or to act as a barrier between the current and base metal. To be effective, this film must not contain corrosive or incompatible substances. The main shortcoming in the use of films is that it is difficult to prevent microscopic holes which form from the actions of abrasives and temperature variations with time. Such holes lead to pitting corrosion which may be a situation worse than having no coating at all. Paints, in particular, require periodic replacement to sustain their anticorrosion capability.

Passivation techniques are used to alter the metallic surface chemistry. Anodizing techniques minimize corrosion by the production of oxide, phosphate, or similar coatings on iron or steel. Metallic coatings of zinc, tin, lead, nickel, copper, or chromium applied in galvanizing techniques may be used to inhibit the corrosion of iron and steel.

Cathodic Protection with Sacrificial Anodes

By using the electrochemical series, metal combinations can be chosen that will control or reverse the flow of current. By this means, some selected metal becomes the cathode and is held secure from deterioration while another metal, the anode, purposely chosen to be sacrificed, is destroyed. Current flows from the anode into the electrolyte and from there into the cathode. The metal atoms on the surface of the cathode are repelled by the electrolyte and protected, while the atoms on the anode surface are ionized and go into solution.

The sacrificial metal is more active and comes from the upper or anodic end of the list. Magnesium would be a first choice, but it is expensive and for some purposes too active. Zinc and its alloys usually substitute with adequate protection. In order for these sacrificial anodes to be effective, they must be connected electrically to the metal they are to protect, and they must be exposed to the electrolyte which is common to both the sacrificial anode and the metallic surface to be protected.

Zinc is extensively used in all types of ocean-going craft throughout the world as anodes to protect exposed metals such as propellers, shafts, structures, and through-hull fittings.

Cathodic Protection with Applied Electromotive Force

The cathodic system differs from the sacrificial anodic system in that the required current of electricity is impressed from external sources and is not self-generated. This method is more flexible although occasionally more costly than sacrificial anode systems. Cathodic protection is realized by establishing a negative electrostatic potential on the metallic surface which both attracts positively charged metallic and hydrogen ions and repels negatively charged anions. This method, for example, is sometimes used to protect aluminum boat hulls in sea water.

Electroplating and Anodizing

Electroplating is one of the electrolytic processes currently used for corrosion protection, and is also a large user of electrical energy. Electroplating is basically the electrodeposition of an adherent coating upon a base metal. Among the non-ferrous metals, plating is applied commercially to copper, brass, nickel-brass, zinc, and aluminum, using tin, cadmium, chromium, copper, gold, platinum, silver, and zinc as plating materials.

Anodizing is another important corrosion-inhibiting electrolytic process and is widely used to provide corrosion resistant and decorative finishes on aluminum. This process is somewhat the reverse of electroplating in that the workpiece is the anode in the electrolytic circuit and, instead of layers of material being added to the surface, the reaction progresses inward from the surface of the workpiece, forming a protective film of aluminum oxide upon the surface.

In summary, electrolytic processes utilize electrical energy as direct energy, with a few exceptions. In the primary metals industry these processes utilize large quantities of electrical energy and are thus of primary importance when considering future electrical energy demands. Corrosion, on the other hand, destroys capital investments (buildings and equipment) in all sectors of the economy. As a result, control of corrosion can have a profound effect on the indirect utilization of energy due to the vast amount of damage produced by corrosion (control of this process has the potential of saving large amounts of capital--estimated at possibly 0.5 to 1.0 percent of the United States Gross National Product). Not all of this, of course, will be electrical energy but the savings will certainly be felt in the primary metal industries in which electricity is utilized primarily in separation and refining processes. In addition, savings will be realized in all subsequent processes of forming, assembly, transportation, construction, and processing of scrap material. When all savings are added, the reduction of corrosion would have a profound effect on the overall energy use of the world. A long and careful study of corrosion-preventive processes is well-justified when the potential savings of such efforts are considered.

Fuel Cells

Of primary importance to the electric power industry is a means of leveling the power demand. If the power demand is leveled as a function of time, the required capital investment for given electric power output is reduced. In order to achieve this, various methods have been utilized for power-demand leveling.

The primary method used to level the power demand is to schedule power consuming processes and activities during the valleys of the demand schedule instead of during the peak periods. This method has been relatively effective. Many processes, however, especially in residential communities (e.g., air conditioning), produce a high power demand at particular times of day which cannot be offset by regulating controllable processes.

Another method is to store electrical energy in another form and reproduce it when needed. One such process is the utilization of the large reversible motor generator systems in storing water. Water is pumped into a reservoir utilizing electricity during low demand and reversed to produce electricity during high demand.

The storage of large amounts of electrical energy utilizing electrochemical means has not found acceptance. The energy density in electrochemical systems such as batteries and fuel cells is not sufficiently high or the system sufficiently low in cost to warrant storing of large quantities of electrical power. In addition, the raw materials required to produce batteries and fuel cells place an additional burden on such industries as the primary metals industry. Recent developments in battery construction, such as the use of heated sodium sulphur cells, have greatly increased the energy density of batteries and thus made possible the use of stored electrical energy in batteries to propel, for example, automobiles.[5] Similar work in battery technology, such as the lithium sulfide, zinc oxide, lithium chloride, and many others, is also progressing.[6] It is anticipated that future development of high-energy density batteries will make possible the practical application of storage batteries in the transportation industry and conceivably batteries could be used to store electrical power, thus allowing further smoothing of the electrical demand.

Fuel cells also can be used reversibly to convert electricity into chemical energy for storage and to reconvert chemical energy back to electrical energy to be utilized during times of peak demand. Figure 12.1 is a schematic diagram of a fuel cell utilizing hydrogen as the fuel. Table 12.3 is a listing of the efficiencies of the various processes used in converting electrical energy to chemical energy and then reconverting the chemical energy back to electrical energy.[7] As the table indicates, the cost of such a converter scaled down to a commercial unit is still quite high and, at the present time, is not competitive with alternative electrical storage methods. This may change in the future, however, due to improvements in fuel cell technology and to the increased cost of electrical energy.

Electro-Forming and Etching Processes

The use of chemical etching and electro-forming has found many applications in modern industry and has become an invaluable tool in the production of micro-circuits and related products requiring precise dimensioning on a small scale. Sensitizing surfaces using photochemical reactions play an important part in the process. Present technical advances indicate that chemical etching of small parts is entirely feasible and has been used in special cases. For example, the utilization of chemical etching and chemical machining to produce small parts could reduce the demand for large investments in highly complex and massive machine tools.

Extensions of present methods in photochemical etching can be used to form intricate shapes which presently are being machined in the conventional way. It is anticipated that breakthroughs in this field will provide much more efficient processes utilizing electrochemical energy which will make the conventional machining processes presently being used obsolete in the production of a large variety of machined products. This conversion will result in the decommissioning of many large machining tools presently in use and will slowly replace present methods as machine tools reach the end of their useful life, thus minimizing changeover costs and allowing time for retraining personnel. This changeover will have greatest impact in the near-term future.

Batteries

Research in new and more efficient batteries is currently being conducted by a number of sponsors in many countries.[8,9,10,11,12] The research covers concepts ranging from efficient, electrically powered vehicles to "super batteries" used for storage of off-peak energy for delivery during daily peak electrical demand periods.

A number of electrically powered vehicles are currently on the market. However, they all suffer from low horsepower limitations and short ranges, clearly demonstrating the inadequate electrical storage per unit weight of the lead-acid battery. The need for more advanced batteries which will satisfy present and future electrical storage requirements is becoming critically important.

Current research in the development of batteries for utility systems is being directed by the Electric Power Research Institute (EPRI).[13] The development of a demonstration plant or a national Battery Energy Storage Test (BEST) facility has been studied by personnel at Argonne National Laboratory and EPRI. A facility design study by Bechtel, Inc. identified the key requirements and concluded in 1975 that establishment of a BEST facility was feasible. Since then, EPRI and ERDA have agreed to implement the concept jointly.[14]

If schedules mesh, the testing of prototype batteries capable of storing up to 10 megawatt hours of energy and delivering several megawatts to the electrical grid should begin in 1980.

Welding

Material design, technique, and equipment used in different kinds of welding vary considerably. Each specific requirement must be considered separately to produce the best possible welds:

Material--The material which must be welded or, more ideally, can be chosen for a particular job, must be adequate for the intended use both structurally and chemically.[15] For example, as a

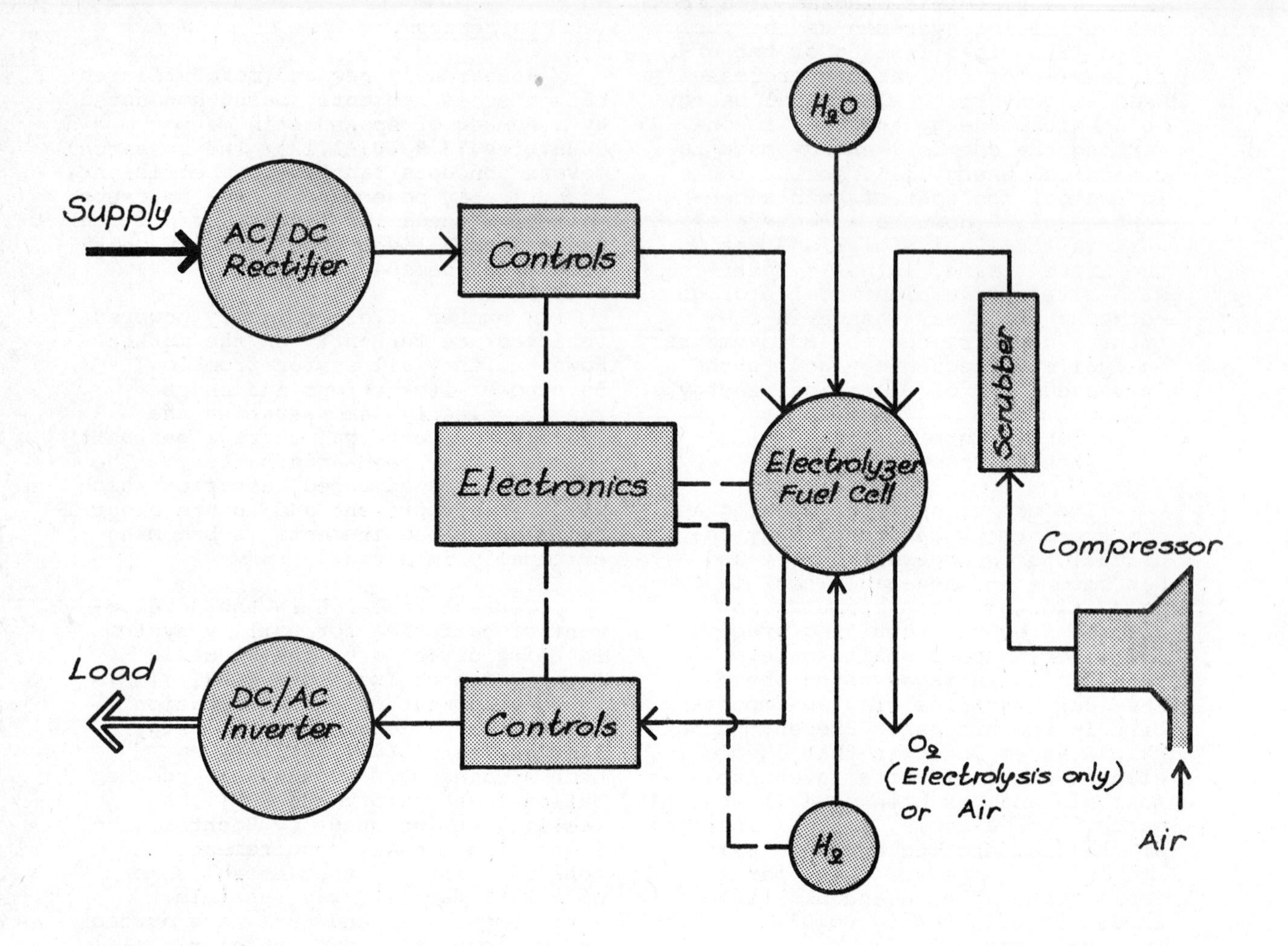

Reference 7

SCHEMATIC DIAGRAM OF ELECTROLYZER / FUEL CELL CONCEPT

FIGURE 12.1

Parameter Efficiency	Value
Electrolyzer	.68-.95
Fuel Cell	.50-.70
dc/ac Inverter	.83-.95
ac/dc Rectifier	.83-.95
Capital Costs	
Electrolyzer/Fuel Cell System	$50/kWe
Storage Tanks (H_2 Gas)	
dc/ac Inverter	$6.7/kWe
ac/dc Rectifier	$4.0/kWe

Reference 7

ELECTROLYZER/FUEL CELL SYSTEM DATA

TABLE 12.3

structural member, it must be sized to support design loads without excessive flexure. Minimization of weld flexure must be assured, since fatigue is a common cause of weld failure.

Technique--The welder must be properly trained to satisfy the particular welding requirement. He must be properly briefed for each job and be thoroughly acquainted with the preheating and heat treating requirements as well as be able to prepare a weld which is both strong and professional in appearance.

Equipment--One of the greatest contributors to inferior welds is the lack of equipment required to properly preheat, deoxidize metal surfaces, and heat treat after welding. For example, stainless steel and aluminum must be welded in an oxygen free atmosphere in order to obtain strong welds. Without a heliarc system, welding these metals is extremely difficult. Cast iron can be welded if the temperature is carefully controlled and proper heat treatment and cooling down periods are employed. Clamps and braces must be used to minimize warping.

Design--In addition to the above, good welds involve careful planning, beginning with initial design of the part or structure and following through to the intended environment, required paints or corrosion inhibitors , expected life, and anticipated performance.

In summary, one final but very important function should be included on all jobs. A qualified independent weld inspector should be called upon to evaluate and pass all welds. This is especially vital for all critically important structural welds.

Careful welds save energy, since reworking is minimized. Selection of the appropriate technique can also reduce energy use for welds. Alternating current welders should be used whenever possible since they have a better power factor and a reduced demand for the same class of service.[16] In addition, ac welders use less energy, principally because of reduced electrical losses under both loaded and no-load conditions.

For example, Table 12.4 compares an ac welder with a dc rectifier welder and a motor-generator welder, each rated at 300 amps. Losses for the ac welder are about 10 percent of the motor-generator welder under similar operating conditions.

12.4 ENERGY-SAVING PRACTICES

A. Electrolysis and Electrochemistry

Electrolysis and electrochemistry processes cover a variety of industrial practices ranging from the use of storage batteries to the control of chemical corrosion. Due to the large variety of processes involved under this general heading, only the more important processes will be enumerated here.

The following is a list of energy saving practices and tips which, in some cases, have been utilized for many years. Others are a result of more recent developments and still others are anticipated through research and development presently going on and to be carried out in the future. Again, it should be emphasized that any improvement in corrosion resistance generally can be applied to all structures. This list indicates practices and procedures which will reduce corrosion. Although not complete, it does cover the more important practices which, if applied, could save billions of dollars per year in production costs and large amounts of energy. The numbers in parentheses indicate the estimated reduction in loss achievable by each process.

Corrosion Reduction Processes

- Reduce undesirable dissolved salts, acids, or alkalis in aqueous solutions, i.e., use a demineralizer in feed water to all closed process systems using water such as boilers, heat exchangers, and related equipment. Inject standard

	ac Welder	Rectifier Welder	Motor Generator Welder
kW input idling	0.19	0.57	2.82
kVa input idling	5.8 (leading)	0.85 (lagging)	7.37 (lagging)
kVa demand	16.0	21.4	25.3
Idling hours per year	1,248	1,248	1,248
Total hours per year	2,080	2,080	2,080
Loaded hours per year	832	832	832
kWh loss idling	237	711	3,520
kWh loss under load	815	5,750	8,500
kWh loss total	1,052	5,861	12,020
Cost of loss per year at 3¢ per kWh	$31.56	$175.83	$360.60
Eff. at rated load, percent	92.5	66.0	54.0
Power factor at rated load, percent	81	85	88

*Assumptions: 300-Amp Arc Welders Operating 8 Hr Per Day, 5 Days Per Week, 52 Weeks Per Yr at 40 Percent Duty Cycle

Reference 16

COMPARISON OF WELDER EFFICIENCIES TABLE 12.4

boiler water treatment chemicals into the system. (10-15%)

- Remove excess salts from all water supplies which involve the use of metal pipes and equipment; building water supplies. (10-15%)
- Reduce humidity wherever possible to reduce corrosion of exposed equipment. (8-12%)
- Filter and cleanse air to reduce airborne corrosives such as NO_x, SO_2, CO_2, etc. (12-18%)
- Avoid use of dissimilar metal in liquid conduits such as water pipes without proper electrical separators. (15-25%)
- Use material with uniform surface properties wherever possible. (5-10%)
- When metals are placed in salt solutions such as sea water, use zinc plates except when protecting aluminum (Cathodic protection). This is the technique used to protect boilers and hot water heaters. (15-20%)
- When using aluminum in corrosive solutions, use an external electrical source to protect metals (Cathodic protection). (30-50%)
- Purge coolants and circulating fluids of dissolved oxygen. (10-15%)
- Check for externally created electrical currents or fields around surfaces subject to corrosion. (5-20%)
- Consider the possibility of electrical isolation of dissimilar metals where electrical contact is possible. (5-15%)

(Estimated percentage savings are based on net savings in expenditure of energy. Actual savings will vary depending on specific applications, but should fall in these ranges. These savings are not necessarily additive.)

Added Use of Surface Treatment (Protective Films)

In addition to the practices listed above, an effective practice which greatly reduces corrosion involves the treatment of surfaces subject to corrosion.

- Use paints, epoxies, oils, or other coatings to protect surfaces. Consult with suppliers to obtain best products. (5-40%)
- Use metallic coatings such as zinc and cadmium plate where feasible. (20-30%)
- When purchasing parts and equipment, inquire about methods used to apply protective metal. Pin holes in galvanized coatings, for example, can spell disaster over long periods. Were assemblies dipped before or after parts were assembled? (5-30%)
- Perform tests under actual environmental conditions whenever possible or use test chambers to accelerate corrosion tests, if necessary. (5-15%)
- Periodically check coatings, especially in hard-to-reach areas where corrosion is most likely to occur. Replace coatings when necessary. (15-20%).
- Use cathodic protection to augment the steps listed above. Due to abrasions and wearing of surfaces, it is virtually impossible to maintain 100 percent coverage of surfaces. Small exposed areas will promote accelerated corrosion due to higher current densities. Therefore, use cathodic protective measures *only* after consultation with an expert. An error in proper use can be costly. (5-15%)

(Estimated percentage savings are based on net savings in expenditure of energy. Actual savings will vary depending on specific applications, but should fall in these ranges. These savings are not necessarily additive.)

B. Storage Batteries

Improper use and maintenance of storage batteries, especially in transportation vehicles, have resulted in unnecessary expenditure of energy required to produce the batteries, not to mention the inconvenience and replacement costs. Also, some basic principles, if properly followed in the construction of batteries, could greatly increase the battery life. Following are some procedures which will increase battery life and minimize energy use in production and maintenance.

- Provide sufficient distance between the bottom of the battery case and the bottom of the plates to allow adequate space for buildup of insoluble products which are generated by corrosion of the battery plates as well as addition of impurities in the electrolyte. If this insoluble material is allowed to reach the bottom of the battery plates, the internal resistance of the battery is greatly reduced and its usefulness, therefore, will be limited. Unfortunately, this practice is not followed in the construction of many commercial batteries and results in a shortened battery life, even though the plates of these batteries have a great deal of life remaining. (20-75%)
- Another practice which greatly reduces battery life is the use of tap water in replacing electrolytes in wet storage batteries. In certain regions, this practice introduces large quantities of impurities in the form of salts, alkalis, and acids which combine with the chemicals in the electrolyte and accelerate the production of insoluble impurities, thus reducing the battery life. In such cases the use of distilled water to replace the electrolyte will greatly increase battery life. (30-60%)
- Proper design of structural supports and spacing of the cathodes and anodes is important. In many cases, unfortunately, insufficient structural support is used for the relatively heavy cathodes and anodes. As a result, small accelerations will often produce distortions of the structural support and can cause shorting of the cathodes to the anodes, rendering the battery useless. (5-75%)
- Sufficient material should be utilized in the production of the cathodes and anodes to ensure that premature loss of cathode and anode material does not occur. (20-30%)
- Adequate structural support of the battery terminals is also extremely important since these are often subjected to large stresses. Inadequate support can destroy the usefulness of the battery. (10-80%).
- Placement of the battery in the proper environmental condition is extremely important. The temperature range should be controlled within battery specifications. Overheating could result in deformation of the battery case and possible leakage of the electrolyte, which can cause extensive damage due to its high corrosiveness. (20-50%)
- Proper venting and choice of materials for battery cases are extremely important in terms of safety considerations since a pressurized battery could explode, spreading electrolyte over a large area and possibly injuring personnel. The enclosure in which the battery is placed should be well ventilated to reduce the possibility of hydrogen buildup which could result in a chemical explosion.

(Estimated percentage savings are based on net savings in expenditure of energy. Actual savings will vary depending on specific applications, but should fall in these ranges. These savings are not necessarily additive.)

C. Battery Use and Maintenance

Before purchasing a battery:

- obtain structural specifications and drawings;
- be certain it is properly constructed;
- be sure that the battery has the proper amount of electrolyte in each cell;
- be sure the battery is not placed on any sharp object which would pierce the case, allowing the electrolyte to drain;
- check the circuit to which the battery is attached to be sure that no current leaks or short circuits appear;
- use only distilled water when replacing electrolyte in wet cell batteries;
- do not charge batteries at a rate greater than the specifications as overheating and possible explosions may occur; and,
- never expose the battery to an open flame due to hydrogen explosion danger.

The above energy saving tips are directed to the private consumer to a greater degree than to the industrial technician. These considerations are important, since vast numbers of batteries are in the care and control of individuals in the private sector.

The following practices are directed more to the technical person who may be involved in the design, manufacture, testing, distribution, and sale of batteries.

- Follow the basic design requirements listed above to produce batteries of long lifetime, thus avoiding unnecessarily rapid recycling of the many energy intensive materials that go into the production of a battery. (10-30%)
- Provide a complete set of instructions on the use and care of batteries to allow the maximum possible useful life. (10-30%)
- Reexamine production methods to determine whether or not the most efficient processes are being used. (20-30%)
- Increase the distance between the top of the battery case and the electrolyte level in order to minimize the escape of the electrolyte through battery vents, thus increasing the life of the battery and minimizing the damage that can be caused by escaping electrolyte. (20-40%)
- Encourage manufacturers of equipment requiring the use of batteries, especially dry cells, to include in their instructions the removal of these batteries periodically for inspection of any loss of electrolyte and possible destruction of equipment. (10-20%)
- Encourage the use of rechargeable batteries when appropriate in order to avoid the high turnover of non-rechargeable batteries, resulting in extensive use of energy intensive materials and natural resources. (10-15%)

(Estimated percentage savings are based on net savings in expenditure of energy. Actual savings will vary depending on specific applications, but should fall in these ranges. These savings are not necessarily additive.)

In summary, large reductions in energy needed in the production, distribution, and use of batteries can be gained by adhering to relatively straightforward procedures. As a result, there is potential for large energy savings.

D. Primary Metals Production

The electrolytic smelting processes required to produce the more reactive primary metals such as aluminum and magnesium require large quantities of electrical energy.

As a result, a great deal of attention is being focused on more efficient methods which can be applied to the smelting process. The Aluminum Corporation of America (ALCOA) has been working for the past 15 years on a research program to reduce power used in smelting to 10 kWh/kg (4.5 kWh/lb) of metal produced. This is a 30 percent reduction below the most efficient units of the Hall process presently used worldwide. ALCOA began operation of this new process in Texas early in 1976.

Developments such as this could represent a major breakthrough in energy savings for electrolytic smelting and might be applied not only to aluminum but to other metals such as magnesium and zinc.[17]

ALCOA has also produced a significant energy saving by replacement of the inert atmosphere produced by burning natural gas with a cryogenic nitrogen plant. Although the initial investment is high, the savings in cost and energy realized over a few years will produce significant and overall savings.

Another significant improvement in smelting processes has been realized by the replacement of conventional mercury arc rectifiers with silicon rectifiers, resulting in an important energy saving. For example, this has provided an additional 3 percent reduction in energy use for the production of aluminum. The use of solid-state rectifiers reduces the amount of waste heat and lowers the power required, thus reducing the need for cooling water.

Since the electrolytic process is highly energy intensive, a great deal of waste heat is produced. Some of this heat is now used to melt cold metal for fabrication processes, resulting in a 20 to 25 percent reduction in the energy normally needed. Additional energy conservation methods used in the primary metals industries include:

- scheduling longer production runs;
- rolling of larger ingots to improve recovery;
- employing advanced instrumentation;
- improving installation of pipes and equipment;
- installing more efficient space heaters and using waste heat for space heating;
- installation of exacting controls on existing equipment;
- modifying existing equipment and processes; and,
- constructing new plants of superior design.

These energy saving practices are being implemented in the aluminum and related industries. Other longer-term energy saving practices which will require more study involve the modification of electrolytic cells. Some areas of research which should be fruitful in improving cell efficiency are:

- fundamental research on electrode processes including electrode configuration, mass transfer, and electrocatalysis;
- improvements in material science technology; and,
- in-depth study of ion conduction in solids, configuration of electrode surfaces, such as control of pore size distributions, porosities, pore tortuosity, and surface tension effects.

Extensive basic research is being done on these processes and, being fundamental, will not only be applicable to the primary metals industry but hopefully will also extend to all electrochemical processes such as plating, surface conditioning, etching, chemical machining, and related processes. The potential energy savings as a result of such fundamental research should not be underestimated. For insurance for the future, when energy costs will be much higher, it is imperative that this work be continued and expanded.

E. Fuel Cells

The economic and logistic impact of fuel cells over the more conventional fossil fuel systems would allow substantial energy savings in many sectors, especially in the transportation sector (see also Chapter 5). This fact has been recognized and discussed with reference to military and defense considerations as well as civilian expenditures.[18] The higher efficiency of the fuel cell relative to burning fossil fuels could make an important contribution to decreasing energy use in vehicles and toward conservation of petrochemical resources. Great reductions in the logistic burden imposed by transportation required for fuel for military as well as civilian establishments could also be realized. Performance parameters such as torque and horsepower as a function of velocity make the fuel cell much more appealing as a motive power source when compared with the conventional gasoline engine.

Figures 12.2 and 12.3 show a comparison of the typical power requirements and torque characteristics of a fuel cell-powered automobile equipped with a dc motor and the conventional gasoline engine-powered vehicle.

Utilization of the fuel cell on a large scale, due to its higher efficiency, could play an important part in reducing energy use in the residential, industrial, and transportation sectors. Stationary cells could be utilized as electrical energy converters and storage units to supply peak demand electricity.

A number of research projects are currently underway with sponsorship by both government and private industry. For example, an ERDA/EPRI Fuel Cell Seminar was held in June, 1976.[19] Research objectives included four critical issues for a five-year fuel cell plan:

- Success of the first generation fuel cell (FCG-1) is a prerequisite for a second generation fuel cell program.
- The cost/availability of fuel cell fuels in the near-to-intermediate future is of major importance in determining the role of fuel cells in the utility network.
- Technical and economic performance of fuel cells in a utility network require delineation. Of primary importance is quantification of the benefits as well as a definition of the potential fuel cell market as a function of capital cost.
- A matrix of technology programs was necessary to maximize the probability of achieving second generation technology goals. The more viable second generation fuel cell concepts include the phosphoric acid, molten carbonate, and alkaline technologies.

Tables 12.5 and 12.6 depict the EPRI program and the relationship of the projects to what are considered to be critical issues.

Many fundamental questions must be answered through basic research and development of the fuel cell before it can be used on a large scale. However, the greater efficiency of the fuel cell warrants exacting research to reduce the uncertainties in its operation. In a conceptual system the fuel cell could be powered by natural gas, methanol, or hydrogen. In the latter case the only combustion product would be water; in addition, high efficiencies are obtainable. An advantage of such a total energy system is its flexibility.[20] It could be applied at different scales for individual buildings, neighborhoods, or central systems. It could also be integrated into the utility grid system as previously mentioned. (Since the hydrogen fuel is not used in conventional operating systems but is still in the research and development stage, energy saving practices are not discussed.)

12.5 CONCLUSIONS

In conclusion, many additional energy saving processes involving electrolytic and electronic processes are being investigated, several of which promise to be important in the reduction of energy use and in the processing of primary metals through recycling or reclaiming used electronic and electrolytic material. For example, selective recovery of primary metals

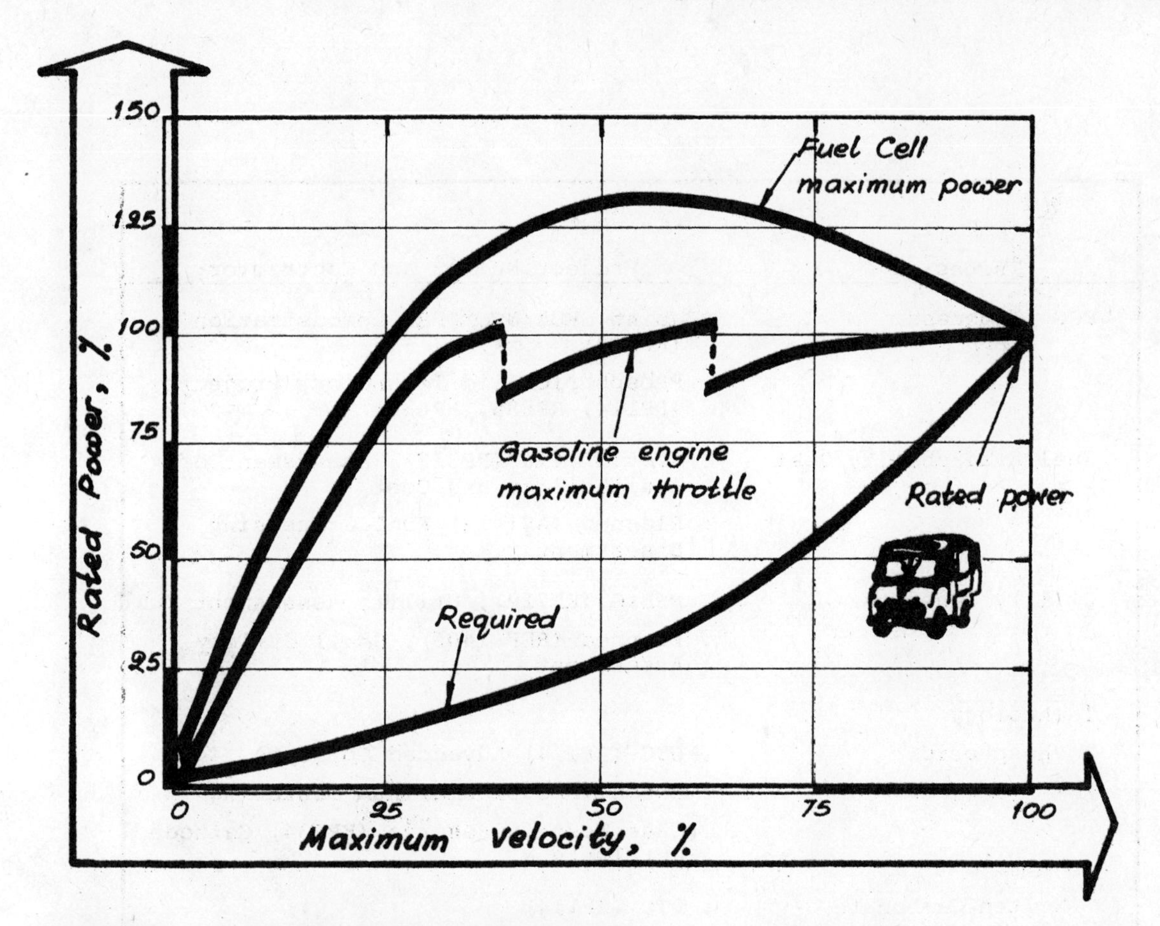

TYPICAL POWER REQUIREMENTS FIGURE 12.2

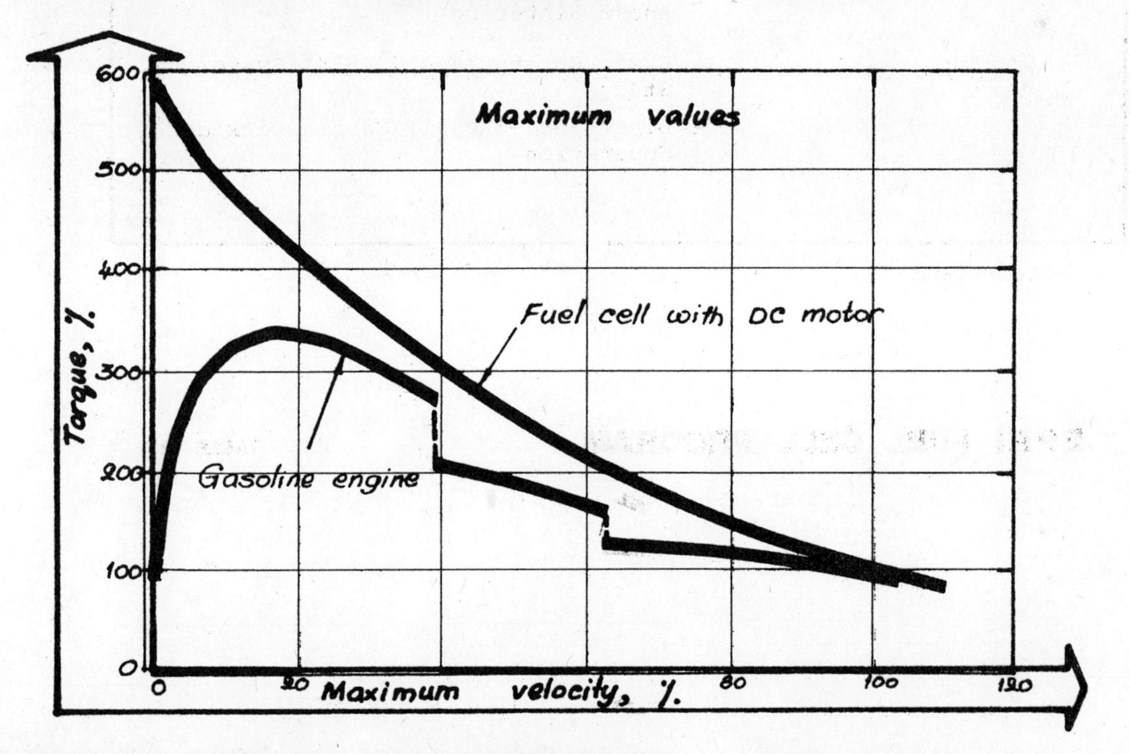

TYPICAL TORQUE CHARACTERISTICS FIGURE 12.3

Issues	Project Number and Contractor
FCG-1 Success	. Joint ERDA/UTC/EPRI Demonstration (RP842)
	. Phosphoric Acid Technology Projects (RP114, RP583, RP634)
Fuel Availability/Cost	. A.D. Little (RP318), Assessment of Availability and Cost
	. Planned (AS1412) Fuel Processing Assessment
Utility Benefits	. PSE&G (RP729), Utility Assessment
	. Planned (RFP 1406), Small Utility Assessment
Technology	
Phosphoric	. UTC (RP114) Advanced Concepts
	. Exxon (RP583) Platinum Sintering
	. Case Western Reserve (RP634) Cathode Catalysts
Molten Carbonate	. UTC (RP114)
	. Northwestern University (RP371) Anode Sintering
Alkaline	. Exxon Enterprises (RP584) System Study
	. Giner Inc. (RP391) Carbon Dioxide Separation

Reference 19

EPRI FUEL CELL PROGRAM **TABLE 12.5**

Characteristic	Goal: First Generation	Goal: Second Generation
Commercial Introduction	1980	1985
Capital Cost (1975 dollars)	$250/kW	$200/kW
Life	20 years	20 years
Stack Refurbishment	40,000 hours	40,000 hours
Thermal Efficiency, percent	37 - 38	45 - 47
Fuel	Naphtha Natural Gas Clean Coal Fuels	Distillate Naphtha Natural Gas Clean Coal Fuels

Reference 19

GOALS OF ELECTRIC UTILITY FUEL CELL PROGRAMS

TABLE 12.6

from used equipment through control of recovery cell voltages would result in the recovery of precious metals which, in many cases, have been totally lost.

These processes should receive increased attention and funds for research and development to improve present concepts such as fuel cells and to introduce new applications of these concepts which will result in significant energy savings throughout the economy. Only thus will the full potential of electrolytic and electronic processes be realized.

* * *

REFERENCES

1. Stanford Research Institute, *Patterns of Energy Consumption in the United States*, Report prepared for the Office of Science and Technology, (Menlo Park, California: January, 1972), p. 152.

2. Prengle, H. William Jr.; Crump, Joseph R.; Fang, C.S.; Grupa, M.; Henley, D.; and Wooley, T., *Potential for Energy Conservation in Industrial Operations in Texas*, Report No. S/D-10, (Houston, Texas: University of Houston, Department of Chemical Engineering, Cullen College of Engineering, November 1974), p. 7-6.

3. Argonne National Laboratory, *Final Report on Improvements in Energy Efficiencies of Industrial Electrolytic Processes*, ANL/OEPM-77-2, (Argonne, Illinois: 1977).

4. Kovach, Eugene G., ed., *Technolog of Efficient Energy Utilization*, The Report of a NATO Science Committee Conference, Les Arcs, France, 8-12 October 1973, (Oxford: Pergamon Press, 1974), p. 42.

5. The Electricity Council Research Centre, *Sodium Batteries for Electric Vehicles*, (Capenhurst, Chester, Great Britain: 1974).

6. Kovach, p. 39.

7. Szego, G.C., *The U.S. Energy Problem*, Vol. 2: *Appendices--Part B.*, (Warrenton, Virginia: Intertechnology Corporation, November 1971), p. I-12.

8. Hogan, B.J., "Lead-Calcium Grids Reduce Battery Gassing," *Design News*, 31(11 October 1976): 70-78.

9. Ikeda, H.; Kato, H.; and Ezoki, T., "A Fast Rechargeable Sealed Nickel-Cadmium Battery," *Journal of the Electrochemical Society of Japan*, 44(November 1976): 747-783.

10. Hattori, S.; Tosano, S.; and Kuzuoka, O., "Temperature Dependence of Capacity, Self-Discharge and Life of Lead-Acid Batteries in High-Concentration Electrolytes," *Journal of the Electrochemical Society of Japan*, 44(December 1976): 792-803.

11. Stefanides, E.J., "Welding Power Supply Gives Linear Control of Output," *Design News*, 31(11 October 1976): 57-70.

12. Roddam, T.R., "Designing Battery Chargers; Difficult-to-Find Curves and an Inexpensive Over-Charging Protection Circuit," *Wireless World*, 82(December 1976): 37.

13. "Storage Batteries: The Case and the Candidates," *EPRI Journal*, 1(October 1976): 6-13.

14. Public Service Electric and Gas Company, *An Assessment of Energy Storage Systems Suitable for Use by Electric Utilities*, EPRI EM-264, (Palo Alto, California: Electric Power Research Institute, July 1976).

15. Jones, M.M., "Cracking of 300 Series Stainless Steel Overlay Deposited by the Metallic Inert Gas Welding Process," in *Proceedings of the American Petroleum Institute, Refining Department*, 41st Mid-Year Meeting, Los Angeles, California, 13 May 1976.

16. Edison Electric Institute, *Power Sales Manual*, (New York: 1970).

17. Remarks by A.C. Sheldon, Vice President of Energy Resources, Aluminum Company of America, at the University of Arizona's Industrial Energy Management Workshop, Tucson, Arizona, 13 May 1975.

18. Szego, p. K-3.

19. The Mitre Corporation, *ERDA/EPRI Fuel Cell Seminar, June 1976*, (Palo Alto, California: Electric Power Research Institute, 1976).

20. Kovach, p. 37.

PART IV

ENERGY MANAGEMENT AND THE FUTURE

IN THE DESIGN OF BUILDINGS AND CITIES, IT HAS BEEN ASSUMED THAT THE PIPE WOULD BE THERE IN THE GROUND OR THE POWER WOULD BE THERE ON THE POLE (OR PREFERABLY UNDERGROUND) AT LITTLE OR NO COST. EXCEPT IN THE DEVELOPMENT OF ALTERNATIVE LIFE-STYLE COMMUNITIES, THE ACTUAL AVAILABILITY OF ENERGY IN ALL NECESSARY FORMS HAS BEEN TAKEN FOR GRANTED.

--R. Schoen, 1975

CHAPTER 13

ENERGY MANAGEMENT IN CITIES

R. Schoen, AIA*

CHAPTER CONTENTS

KEY WORDS

Alternate Energy Sources
Architect-Engineer Services
Communication Use
Electric Vehicles
Energy Conscious Design
Energy Conversion
Energy Management
Energy Storage
Environmental Impact
Freight Transport
Government Policy
Institutional Barriers
Integrated Utility Systems
Municipal Energy Use
Solar Energy
Solid Waste Utilization
Total Energy Systems
Transportation Use
Urban Planning
Waste Water Reclamation

SUMMARY

Today, much more than in the past when energy availability was generally assumed, energy use has become a dominant factor in the planning and design of cities. This chapter considers energy use in cities in an integrated or *systemic* manner. Emphasis has been placed on actions which could improve energy use efficiency in the *near-term* (2-5 years) and *long-term* (5-25 years).

Basic human energy needs are timeless; it is energy forms and resources used to meet these needs which change. In reviewing the historical evolution of city design from prehistoric to modern times, it becomes apparent that general principles leading to efficient energy use have evolved. In ancient times these principles influenced the utilization of natural energy sources provided by wind, sun, and site. In modern times, the same principles or their variations lead to effective use of manmade energy sources such as electricity.

In the design of current or future cities, there are near-term technological options for improving energy use efficiency. Major options include those related to utility systems (combined power cycles, energy corridors,

*Lecturer, School of Architecture and Urban Planning, UCLA

district heating, integrated utility systems); solid waste recovery (reduction of volume to be disposed, heat recovery during incineration, conversion to fuels, energy intensive by-product recovery--metals, glass); energy efficient urban transportation (vehicle type selection, power sources, mode of operation); and, undergrounding of selected major buildings within cities (reduction of heat losses, reduced infiltration, energy and cost savings, improved land utilization).

These and other concepts are discussed in the text; a series of Case Studies in Chapter 15 illustrate the principles and concepts developed herein.

13.1 INTRODUCTION: ENERGY AND CITIES

Much of the energy budget of industrialized nations is devoted to use within urban areas (including non-agricultural transportation, industrial, commercial, and residential energy use).

The city has a history which predates history. It has traditionally been characterized as a focus of human energy use--the gathering place where people create, compete, learn, work, and play. More than anything, the city has been viewed as the ultimate vehicle for face-to-face interaction: "the web, the matrix that interlocks man and his fellow man." [1] Moreover, it has been asserted that

> . . . the ultimate purpose of a city in our times is to provide a creative environment for people to live in. Creative in that a city which has great diversity allows for freedom of choice: it generates the maximum of interaction between people and their urban surroundings.[2]

The degree to which cities of today are able to meet those objectives is in great dispute. Historical and contemporary perceptions as to what the city has been, is, or should be, vary enormously. Yet, it is clear that there are no viable alternatives--cities cannot be eliminated.[3] If anything, emphasis upon cities will continue, as they grow in size and, with less certainty, in number.

By 1920, "the US population had become 50 percent urban."[4] By 1960, "70 percent of the US population had come to live on 1 percent of the land, called cities" (excluding cities with less than 50,000 population).[5] Growth projections for the year 2000 similarly vary, depending on how metropolitan regions are defined and whether or not conservative projections of population growth are used. From a US population of 203×10^6 in 1970, estimates for the year 2000 range from 282 to 361×10^6 people.[6] Based upon these estimates, the size of urban areas will increase by at least 40 percent in the period between 1970 and 2000.[7] Estimates of resulting impact range from 70 percent of US population on 10 percent of the land to 83 percent on 16 percent of the land in the year 2000.[8,9]

At that time, as much as 60 percent of the population may be living in but four regions of the country, covering 7-1/2 percent of the land.[10] These "megapolitan regions" include: 96 km (60 mi) north of San Francisco to the Mexican border (960 km [600 mi] long, 13×10^6 ha [50×10^3 mi^2]); Florida; the Atlantic Seaboard, or in more limited terms, the Boston-Washington Corridor (644 km [400 mi] long, 161 km [100 mi] wide); and the Lower Great Lakes Region, with the possibility that the latter two may merge into a single transregional belt.[11]

Few would quarrel with the conclusion that "we can expect the metropolis to be the normal environment of the future."[12]

Architect William Caudill has stated:

> The energy crisis together with the conservation movement could have much more impact upon building design (in the next decade) than the great "form-givers" of the last three decades: Frank Lloyd Wright, Le Corbusier, Mies Van der Rohe, and Louis Kahn. [13]

This statement could be applied with equal validity to the organization, design, and physical form of future and even, to some degree, present-day cities. As shown in Figure 13.1, the multiple forces of energy, environment, cost, construction technology, architecture and design will increasingly impact and complicate planning and use of cities by their often conflicting demands.[14] Limited availability of energy is but a new, albeit serious, addition to the set of problems which have increasingly plagued the city in recent years--housing, education, transportation, employment, and decay. Over the same period, a number of schemes have been proposed to solve these problems. Although increasing attention has been paid to environmental and ecological concerns at the proposed place of city-building, few, if any, of these proposals have considered the implications concerning energy use. In almost all such concepts, it was simply assumed that the pipe would be there in the ground or the power would be there on the pole (or preferably underground) at no or minor cost. Except in the development of alternative life-style communities, the actual availability of energy in all of the forms necessary for city-building and operation has simply been assumed.

13.2 OBJECTIVE, SCOPE, ASSUMPTIONS

A. Objective

The objective of this chapter is to establish an appropriate perspective for energy efficiency innovations at the urban/municipal scale. In part, it draws upon historical example in order to establish the required context of man in balance with his available resources including energy. The approach adopted is that of an architect or urban planner evaluating the technology of efficient energy use in cities.

B. Scope

The chapter considers energy use in cities in an integrated or *systemic* manner.[15] Thus, while individual energy-using systems e.g., a lighting system, commercial building, or transportation system) have been considered elsewhere in this book, here an effort shall be made to consider their interactive effect on energy use efficiency. The city has been reviewed as a *human-centered* system. Selected examples from urban history, planning, and design have been used to develop principles and illustrate concepts for efficient use of energy and resources in cities. While the primary thrust involves electrical energy, the complexity of urban systems precludes any actual exclusivity in the energy forms discussed.

C. Assumptions

Emphasis has been placed on actions which could improve the efficiency with which energy is currently used in cities. A *near-term* time frame of two to five years has been the criterion for selection of most of the options discussed. While the criterion applies to options for altering *existing* ways in which energy is used at the urban scale, this time frame is simply not practicable for most, if not all, of the *new*, innovative approaches to urban-scale energy use. For example, the realization of a comprehensive urban subway system can take between 10 and 20 years from initial feasibility investigation to installation and operation. Even in the case of a new community or town being created by a single developer, a minimum of three to five years will elapse from first acquisition of land to initial installation of services. At that, another 10, or more likely, 20 years will elapse before such systems are fully operative at their design capacity. Individual buildings have useful lives of 30 to 80 years. The lifetimes of cities can encompass centuries.

Although some of the new innovative approaches to energy use are discussed in this book, it should be realized that an extensive treatment is outside the scope of this chapter. Likewise, while it is recognized that the *practical implementation* of efficient energy use technologies in cities carries with it concomitant innovations in the political, social, cultural, and economic spheres, an extensive discussion of these issues is beyond the scope of this book. An effort has been made, however, to acknowledge the interrelationships between the technological options for efficient energy use and the complex, changing mix of nontechnological concerns which characterize all human communities.

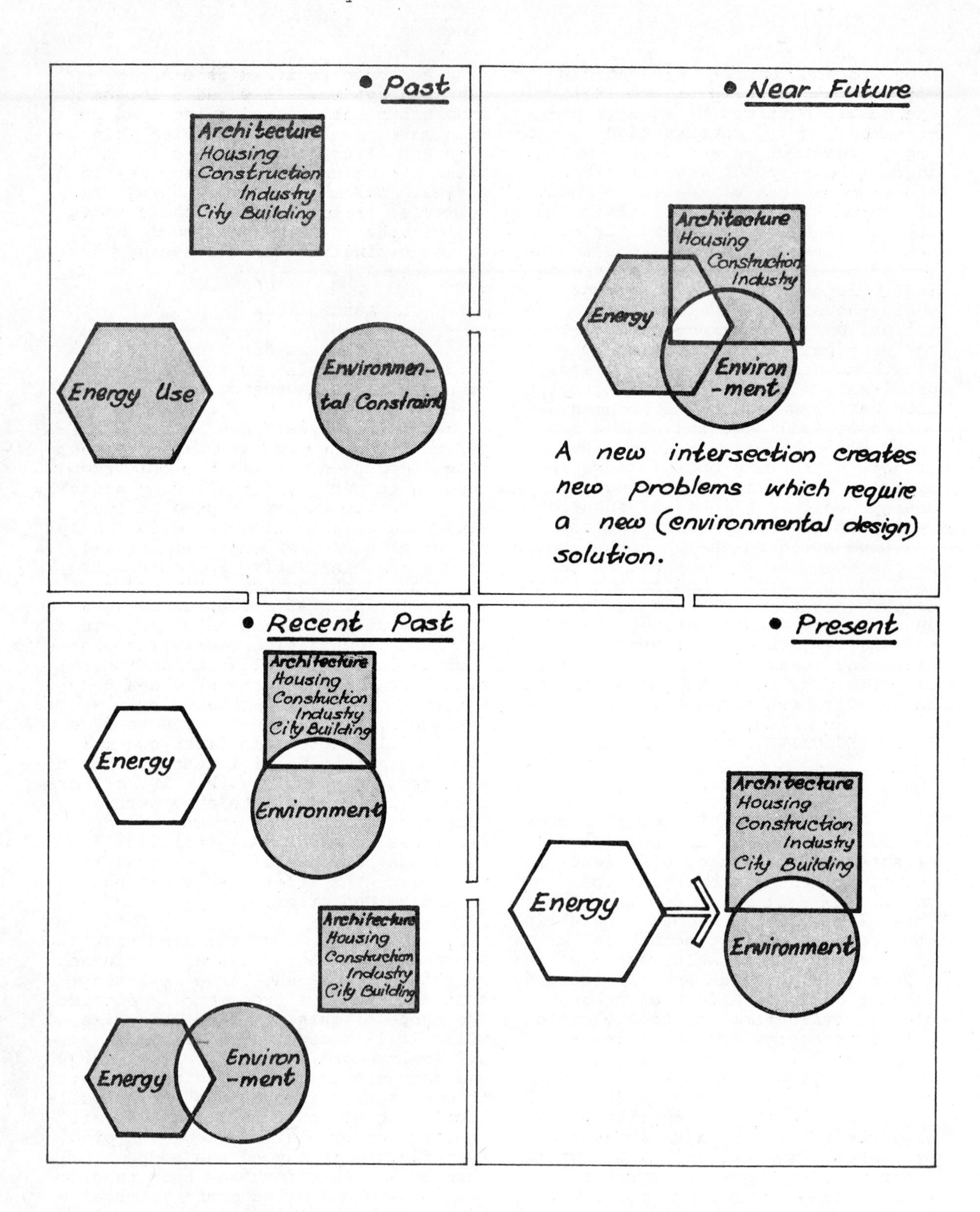

EVOLUTION OF ENERGY USE AND CITY DESIGN FIGURE 13.1

The wide variety of actors on the urban decision-making stage suggests that this chapter be directed to a broad spread of professional and public readers. Hopefully, it will be of interest to:

- technologists (physical scientists, researchers, and engineers);
- architects, urban designers, and urban planners;
- policy-makers, legislators, and administrators; and,
- utility company executives and utility regulators.

13.3 EFFICIENT ENERGY USE IN CITIES--A HISTORICAL PERSPECTIVE

Although there are many apparent similarities among cities across the world, individual cities are quite different. They have been classified by various schemes, ranging from political organization ("imperial" cities, colonial settlements); cultural norms ("sacred" cities, communes); functional needs (common defense, market towns, points for necessary modal changes in transportation or transhipment of trade); historical (primitive, classical, medieval, neoclassical, industrial, post-industrial); and finally to physical patterns (organic, radial, grid, linear).

In addition, as economic, political, cultural, functional, or historical imperatives change, both in an absolute sense and in relation to each other, cities are reconfigured accordingly. Some cities outlived their usableness and simply disappeared; other new cities have been raised upon the dormant foundations of predecessors wherever locational values have endured; and still other cities have new arrangements overlaid upon or expanded from what currently exists, as various mixes of those forces ebb and flow over time.

In every case, the particular set of *forces which shape cities result in and are reflected by the city's physical configuration, form, or pattern.* This construct is taken as being the most useful for the discussion which follows, especially since infrastructure services and therefore energy use can also be best understood and represented in terms of their physical organization and that of the cities they serve.

A word concerning the condensed discussion which follows: obviously the history of electricity use in cities is brief, not yet being a century old. Still, in human settlements dating back at least one hundred centuries (the Holocene Period), early man confronted energy needs which were resolved by trial-and-error or logical approaches. Some of these early lessons are reflected in the design of modern cities; but others have either been forgotten or cast aside in favor of new approaches. For the sake of completeness--and to spare repeating history--these historical developments are considered worthy of a brief review.

Even before the advent of permanent settlements, nomadic hunting and food-gathering tribes of the Paleolithic Era would return seasonally to a common habitat or series of campsites.[16] Not only did the need *to select a supportive local environment* characterize man's earliest gathering places, it continued to be a requirement wherever the existence of nomadic tribes persisted, even to almost our own times--e.g., the annual ceremonial and trade spring meeting ground of the Cheyenne and other buffalo-hunting Great Plains Indians on the Green River in Wyoming.[17] Attempts have been made to describe those temporary encampments in terms of their physical layout as a response to climatic forces. Much has been made of the location of temporary campsites in areas sheltered from wind and shaded from the sun and both in relation to upwind areas around a central fire, and of the location of openings into individual shelters in terms of the fire.[18]

Another basic principle is the concept of *minimal disturbance of the natural environment.*[19] Throughout the history of city-building, man first acceded to and then increasingly violated this precept--at considerable, even fatal cost to the cities he erected. Villages, towns, and cities are a human imposition upon the natural systems of a region.

This supposition causes the local natural systems to be disturbed. Generally, two expenditures are involved, first in initially disturbing the natural system by new developments, and second in attempting to correct this disturbance and to provide stability at a new equilibrium point. Disturbance and necessary correction are strongly correlated, and the greater the disturbance the greater the correction required. This leads to the concept that natural systems in equilibrium if minimally disturbed, require minimal correction to restore equilibrium.[20]

For example, the replacement of natural terrain and vegetation by roofs and pavement drastically changes the natural drainage systems of a region by increasing relative runoff and reducing critical storm duration times; these effects are cumulative. Expensive modifications to the drainage systems are then required, and natural water courses can no longer remain in their original states without remedial action. Both the initial impact and subsequent corrective measures require the expenditure of energy. Thus, minimal disturbance should lead to minimal energy requirements.

The dawn of the Neolithic Era saw the beginnings of agriculture, the domestication of animals and, thereby, the establishment of permanent primitive villages on all continents. These too depended upon the existence of a hospitable local environment, including sufficient arable land and dependable supplies of fresh water. While the interrelationship of specific structures in such villages often reflected microclimatic conditions, they were equally the result of tradition, culture, social arrangements, or religious beliefs. "Such settlements always occupied a selected environment in which the most advantageous location was chosen to bring nature under the control of man."[21] However, it might be more appropriate to refer to *the human ability to adapt to the natural environment.*

Within the United States, ruins of various primitive villages remain that exhibit *specific responsiveness to natural energy flows.* Many examples exist, such as the Longhouse at Mesa Verde in southern Colorado and the Acoma Pueblo in New Mexico.[22] Studies and climatological research indicate that the successful primitive man built communities in ways highly adaptive to seasonally recurring natural forces at the site (see Case Study 13-1).

In a desert climate where the extremes of temperature are great from day to night, the ability of a wall to store heat is useful. For the arrangement in Acoma Pueblo, in which the wall receives a high percentage of the winter energy, thick masonry helps to maintain an internally steady state by virtue of the time lag required for heat transfer. Once a steady state is achieved, the critical factor becomes the transmission coefficient of the material, which describes the rate of energy transfer.

Since the summer sun is more direct upon the horizontal surfaces and the winter sun is more direct upon the south-facing walls of the tiered section, it would be most reasonable if the vertical walls receiving winter sun had a high transmission coefficient and a high heat-storage capacity. Conversely, the horizontal surfaces receiving their maximum energy in the summer should exhibit a low heat storage capacity. A comparison of the materials used by the Indians reveals this to be the case.

Finally, if the efficiency of the form in transmitting energy from outside to inside is studied for winter and summer, the winter profile shows a 50 percent increase over the summer.[23]

Millenia later, we appear to be on the brink of realizing what primitive man must have intuitively understood as a precondition for survival--namely, that *human control of natural forces is limited as is the availability of natural resources.*

The two examples cited above suggest two principles of energy/resource conserving urban design. First, *site selection is a necessary passive design step to improve energy use efficiency.* Second, diurnal temperature swings can be averaged by buildings which have a large thermal capacity. Also, the change in angle of winter and solar sun can be exploited for partial solar space conditioning. Both of these concepts are illustrative of the next principle, namely that: *Energy use efficiency can be improved by correlating dwelling design with seasonally recurring natural phenomena.* This concept has obvious implications for electricity use in air conditioning and space heating.

While Athens had a population of between 100,000 and 150,000 during the fifth and fourth centuries B.C., few towns of the Hellenic Period had more than 10,000 souls. Hippodamus theorized that this was an appropriate size and Plato later concluded that city size should range between 5,000 and 10,000 colonists from the mother state to establish distant colonies.[24] The availability of food and water established feasible sizes of urban populations. Dependency upon the primitive hand plow, horse-cart, wells, springs, and occasionally a gravity water supply created significant limits in these terms.

Another principle of energy/resource conserving urban planning emerged: *Limiting the size of urban settlements according to available resources.* This principle was often applied in the past; it may once again become a significant planning determinant in the future design of cities and their infrastructures.

Ancient cities managed to remain more or less in balance with the local resource base which both sustained them and received their wastes. However, the dawn of the industrial revolution brought a great influx of people into most major cities. The resultant overburdening of natural systems brought about highly unsanitary conditions and serious epidemics. It was at this time that the first well-developed manmade forms of urban infrastructure began to take shape, beginning with the London Water Works and Distribution System. By the time the industrial revolution was in full flower, crude gas, electrical, and mass transit systems had taken their place along with the already well-established water, sewer, and cobblestone road systems in most larger cities. Through the ensuing years, technological progress and continued expansion of capacity enabled those systems to more or less keep pace with the phenomenal growth of urban development.

The extensive rebuilding of cities following World War II was mainly along pre-war lines using evolutionary rather than revolutionary technology. In the United States "urban renewal" of the older city centers became fashionable. This led to some changes and improvements in city design but rarely to the development and implementation of technologically innovative service systems.

Urban renewal has since fallen into severe disrepute, and almost all projects originally funded, at least in part, by government support or guarantees have been completed or redirected. What has followed has been a succession of different approaches including those known as selective renewal, rehabilitation, new satellite towns, occasional stand-alone new cities, model cities, and the most recent city-building innovation--paired towns and the new-town-intown. However, the amount of construction actually put into place under each approach has been minimal.[25,26,27]

Many estimates have been made of the means by which the future population of the United States will be housed. For the year 2000, the expectation is that around 300 million persons will be living in the US. Various estimates of new cities (the total amount depending upon the size of the community chosen) indicate that on the order of one hundred to several hundred new towns of at least 100,000 population and on the order of ten new cities of at least one million population must be constructed between now and the end of the century. Yet, these prognostications are not being realized because, at the present time, virtually all federal support for new community programs has been stopped and most existing projects are in serious financial trouble.[28,29,30,31]

The consequences of these changes and lack of a comprehensive land-use policy are such that the US might appear to be foredoomed to a straight-line continued projection of already well-established patterns of urban sprawl, with attendant problems and inherent inefficient use of energy. This leads to still another fundamental principle: *Energy use efficiency is related to population density* (see Case Study 13-2).

In terms of efficient use of electrical and other energy forms, whether in existing cities or in new towns, land use poses issues of particular concern.

There are many reasons behind the pattern of development which has emerged in the United States and other industrialized countries. Particularly in the US, the availability of cheap and undeveloped land has been a major factor. Other factors include the role of the FHA and guaranteed loan programs, the rapid growth of private ownership of automobiles, a preference for single-family dwellings, and low property tax assessments.

One has only to look at the plight of large cities surrounded by suburbs as well as the highly developed central area such as New York, to realize that historical patterns will not be successful models for the future and that new approaches are required. The phenomenon of urban sprawl may, in fact, be changing due to the pressure of energy costs and availability, as well as other forces such as increased land values, a growing awareness and concern for agricultural land adjacent to the cities, and the cost and difficulty of providing services to support further development.

As opposed to the reality of urban sprawl, the density concept--centralizing or concentrating people and the services they require--appears to offer greater efficiency in some aspects (see Case Study 13-2). However, as the Case Study points out, these advantages are not gained without additional energy, economic, social, and cultural costs. The main advantages of higher densities are:

- more efficient use of fuels, energy, and other resources;
- greater ease of accessibility to a variety of local activities and scenic resources;
- more intensive use of these activities and resources;
- allowance for conservation of lower-intensity use areas; and,
- reduced transportation requirements.

Three alternatives can be identified immediately:

- disperse population by generating growth in sparsely populated areas;
- foster the growth of existing small cities and towns in non-metropolitan areas; and,
- build new cities outside large metropolitan areas.

These are complex choices and no single answer will suffice. It should be clear, though, that the resolution of the problem of accommodating population growth and the related issues of renewing existing cities require the best thinking of government, industry, business, universities, and other sectors of society. Energy will certainly be one of the major constraints in this problem.[32,33,34]

In summary, when reviewing certain historical aspects of cities, one is impressed by the orderly and evolutionary nature of energy use which persisted up to fairly recent times. Even in ancient periods, energy played a significant role in the design of cities. Electrical energy, while a new form, is subject to many of the same principles which were followed historically. These are summarized in Table 13.1.

13.4 TECHNOLOGICAL OPTIONS FOR EFFICIENT ENERGY USE IN CITIES

A. Potential for Energy/Resource Conservation at the Municipal and Urban Scale

In the near-term, since energy use in the commercial, residential,

- Selection of a supportive local environment;
- Minimal disturbance of the natural environment;
- Adaptation to the natural environment;
- Human control of natural forces is limited as is the availability of natural resources;
- Appropriate site selection and response to its inherent microclimate and topographical chara-teristics are necessary passive design steps in improving energy use efficiency;
- Energy use efficiency can be improved by corre-lating dwelling design with seasonally recurring natural phenomena (as well as, to a lesser degree, the design of other larger and more complex building types);
- Limitation of the size of urban settlements ac-cording to available resources;
- Energy use efficiency is related to population density; and,
- Forces which shape cities result in and are re-flected by the city's physical configuration, form, or pattern

EFFICIENT ENERGY USE-
A HISTORICAL PERSPECTIVE

TABLE 13.1

industrial, and transportation sectors can be seen as taking place in or near urban centers, cities and towns, the latter become significant targets for efficient use of energy. These sectors have been dealt with individually in the preceding chapters and will not be discussed here except in terms of energy use implications stemming from the ways in which they may be located and grouped together.

In fact, it is the means by which such groupings of human activity are serviced that is the focus of this chapter. These services, or "infrastructure," which allow the city to function, include: electricity, natural gas, and fuel-oil utility service; intra-city transportation; waste removal and disposal; snow removal and storm drainage; water; schools; libraries; health care facilities; police and fire protection; telephone; radio; television; and public administrative services. Electrical energy is critical to most of these services.

Urban and municipal systems making up the infrastructure are of interest not only because they are in themselves significant users of energy, but because their size, load diversity, and systemic characteristics suggest new and highly promising options for energy efficiency and materials/resource recycling and reuse which is simply not possible at the scale of individual elements within the urban fabric.

When considered in combination with components of the built environment they serve, opportunities for energy efficiency tend to multiply as the scale increases from the single-family detached house to the urban and regional scale, as shown in Table 13.2. The options listed are primarily technological in nature. Each of the major options is discussed in the following pages. Equally significant, policy-oriented innovations are also possible. These include: urban design and planning methodologies; preferential interest rates; time-oriented utility rate structures; and use tax and tax forgiveness incentives. However, these are generally outside the scope of this chapter and are not discussed further.

Difficulties in implementing energy efficiencies increase commensurately, if not geometrically, especially in the short-term. These problems stem from the very economic, social, cultural, and institutional characteristics which create diversity, activity, delight, and opportunity in the city. They give rise to potential barriers to implementation of fuel conserving and efficient energy use measures. These barriers are primarily non-technical in nature. In addition, efficient energy use is but one of many conflicting goals which city planners will seek. Implicit in what follows in this chapter is the understanding that these constraints condition the application of any purely technological solution to efficient energy use.

B. Utility Systems

A complete energy system consists of all components necessary to convert and transfer a basic energy resource from its natural state to a point of end use. The end use is useful work and the by-products include various residues and waste products.

Much of this takes place beyond the confines of a single urban region. However, excluding distant hydro-electric projects, remote-sited facilities such as nuclear power plants, and large-scale system inter-ties, it can be said that production of usable energy, transmission, and use occur in areas roughly corresponding to the end-use region. This can involve both a number of adjacent municipalities *and* neighboring utility companies, both public and private, with adjacent or interlocking service territories. To improve the efficiency of a large-scale energy system, innovative savings are being introduced within its component elements (production, use) by improving the linkages between them (transmission).

Urban and municipal service systems are of interest not only because they are significant users of energy, but because their size and nature make possible savings in addition to those achievable within the individual energy use sectors. At this scale, however, distinctions between production, transporting, transmission, and use become increasingly blurred. In fact, it is through readjustment of these elements and their interrelationships, especially

System	Techniques for Improved Energy/Resource Management[a]
Element/Appliance (Furnace, refrigerator, heat pump, water heater, solar collector, lights)	(1) Improved efficiency of individual elements
Living Unit (Single-family detached through small-scale low-rise multi-family apartment buildings)	(1+2) Some integration of elements into efficient subsystems (exhaust refrigerator outside in summer, inside in winter; water-cool air conditioning and preheat domestic hot water); built form configured to appropriate site selection; and passive micro-climatic/topographical design approaches.
Clusters (Larger multi-family and planned unit developments, high-rise apartment buildings and required associated communals, industrial, and institutional support facilities)	(1+2+3) More efficient heating/cooling systems, e.g., central heat pump installations, energy storage capability, technologies requiring large scale applications to be economic--solar/electric generation.
"Village" Size Communities (Tapiola, other European new-towns)	(1+2+3+4) Community systems for heat recovery and energy storage, combustion of refuse where environmentally feasible, application of electronics communications and electric vehicles.
Regional Design/Development (TVA, cooperative planning at the Council of Governments--COGS--level of administration)	(1+2+3+4+5) Integrated materials and energy management systems, mobility systems, electronic communications as alternatives to physical transportation.

(a) For each level of size and complexity of an energy system, from an individual appliance or element to a large region, such as the developing megapolitan urban areas noted earlier, there are increasing possibilities for energy management and conservation and increasing difficulties in achieving such goals.

OPPORTUNITIES FOR EFFICIENT ENERGY USE IN BUILT ENVIRONMENTS **TABLE 13.2**

in conjunction with residues produced at various steps along the way, that new combinations and interrelations offer the potential for significant energy savings overall.

The technologies for many of the applications listed below are not new. In fact, some of the previously used technologies were once well established. With costs of various inputs changing at different rates, alternate technologies became more attractive. One major change may have been due to the decreasing relative cost of fuel. Now that fuel costs have risen more rapidly than many other goods and services, such previously used technologies may again be attractive. Some would claim that the principal difficulties impeding their re-implementation today are legal, political or economic relations.[35] Others would claim that changes in locational structure, environmental restrictions, and scarcities of the most readily adaptable primary fuels are more important factors.

One such technology is combined power cycles which could increase significantly the efficiency in the conversion of the primary fuel. The technological aspects of this are discussed in Chapter 2. Steam electric generating stations convert fuel efficiently into electricity in terms of the physical laws of the process. Modern generating stations achieve efficiencies of around 40 percent. However, in the process the residual heat (60 percent of the input fuel) is rejected. Most of the heat is discharged by the use of cooling water at relatively low temperatures above ambient, 5-10°C (10-20°F). If this discharged heat can be effectively used in another process (the basic idea of combined power cycle--see, for example, Case Study 2-7 in Chapter 15), then the overall efficiency in the use of the primary fuel can be improved.

The effective use of this discharged heat is the heart of the effective capturing of the added overall system efficiency. One problem is that the current discharge temperatures are not very warm. If, to make effective use of the heat, the temperature of the discharge is raised, the generating efficiency will be more and more isolated from other activities. The low heat content of the discharge does not allow efficient long-distance transport. A third problem is that if the need for electricity and heat are not well matched, the overall gain in efficiency will be small. Finally, as noted above, since the electric power producer is infrequently the same entity as the consumer with use for the excess heat, there are legal, political, and economic problems in trying to insure that both parties receive a fair share of the benefits and liabilities from the combined system.

Most individual consumers are not large enough to participate individually in the development of a combined power cycle. This means that if such options are to be developed, they must be approached from a systematic means.

It could be possible, for example, that industries with need for heat could co-locate with a new electric power plant. Or, in a large-scale new development, an energy corridor could provide for the efficient distribution and possibility of use of an electric power plant's discharged heat. Such district heating systems were once more prevalent in large cities than they are today. In part, this was due to the declining costs of delivery of alternate fuels, particularly natural gas and electricity. In part, this was due to the smaller and relatively less efficient generating stations located in the centers of metropolitan areas.

Opportunities for an effective combined power cycle system also exist with an electric utility's distributed peaking units. Conversely, users with large heat requirements may have opportunities to generate electricity while raising steam for their other requirements. However, because of one or more of the constraints listed above, few such systems have been implemented in the recent past. The current interest in using energy more efficiently is causing many firms to evaluate such opportunities again to determine if they are now economic in the light of the current cost picture.

To the extent that such combined cycle systems improve overall system efficiency, they reduce the overall requirement for fuel. Moreover, they simultaneously reduce the amount of waste heat discharged in the total system. In the case of generating stations, the productive use of discharged heat reduces the current problem of waste heat disposal.

While this option does not exist for the *individual user*, in the *systemic approach* for municipal energy use it is possible. This is particularly true for municipally owned utilities which have the option of meeting the city's other, nearby, lower quality energy needs (such as hot water for snow removal, and low temperature heat for adjacent buildings). Environmental restrictions and economics will probably limit the number of successful locations.

As a more elaborate approach, an *energy corridor* could be used to transport energy in several forms from the power plant to the city and to return the by-products. Fuel, heat, steam, or electricity for various uses would be bled off at appropriate temperatures and in appropriate forms along the way in both directions, with each user paying for the needed energy quality. *District heating* is a related technique which allows many buildings to be served from a single nearby steam plant, in lieu of heating systems in each building. This approach, for certain system sizes, may be efficient for heating, and could reduce the waste heat disposal problem for the power plant.

Another variation involves the use of *fuel cells* (see Chapter 12) which would provide (when improved units are developed) conversion of fuel directly into electricity. In principle, this conversion process could permit an aggregated approach to meeting energy needs with a single fuel, thus resulting in greater efficiency if the waste heat were utilized at the point of end use.

C. The Recovery of Energy from Solid Wastes

With the tripling of the price of Number 6 fuel oil in the 1974-75 period (and increases in other energy prices), there is renewed interest in 4.4×10^5 GJ (4.18×10^{11} Btu) which the US incinerates or buries each year in the form of 8.6×10^8 mt (8.5×10^8 tons) of dry municipal solid wastes. [36] The problem of getting rid of solid waste from major cities had long been evident. Two years ago, San Francisco conducted some widely published negotiations with rural and sparsely populated northern California and Nevada counties for sites to receive its "garbage trains." It had also become clear that the conventional sanitary landfill process was increasingly inadequate for ever-growing volumes of municipal solid wastes. Close-in areas suited for landfill are being depleted at a rapid rate. In the US, 32 kha (80 k acres/yr) are currently being filled to a depth of 2 m (6 ft) with solid waste.[37]

As shown in Figure 13.2, urban areas produce airborne, liquid, and solid waste by-products in amounts beyond the absorption capabilities of natural systems (unlike the primitive settlements discussed earlier). Environmental concerns of recent years over clean air and clean water have caused a redoubling of efforts to minimize discharge and to make handling of each type of waste more effective and efficient. These steps in themselves require energy. Solid waste management is of particular concern, due to recent technologicial developments and favorable economic shifts relating both to energy used in processing and, more recently, to primary and secondary forms of energy which can be extracted from wastes.

All solid waste management processes (including the most traditional--symbolized by the weekly curbside pickup) can be characterized by four successive elements: generation, collection, treatment, and disposal. There have been recent innovations in the first two stages. For example, household garbage disposers to separate and dispose organic wastes and, more recently, sealed in-house incinerators for waste disposal and trash-compactors to reduce the collectable volume have been claimed to reduce energy use. Similar claims have been made for vacuum collection systems which automatically gather solid wastes from individual buildings and deposit them in central pickup stations for building groups served by the system. However, it is the treatment of solid wastes and the disposal or partial reuse of the residue that are of special interest.

Traditionally, municipal solid wastes have been trucked directly to municipally owned landfills or private operations under city contract and buried without further treatment.

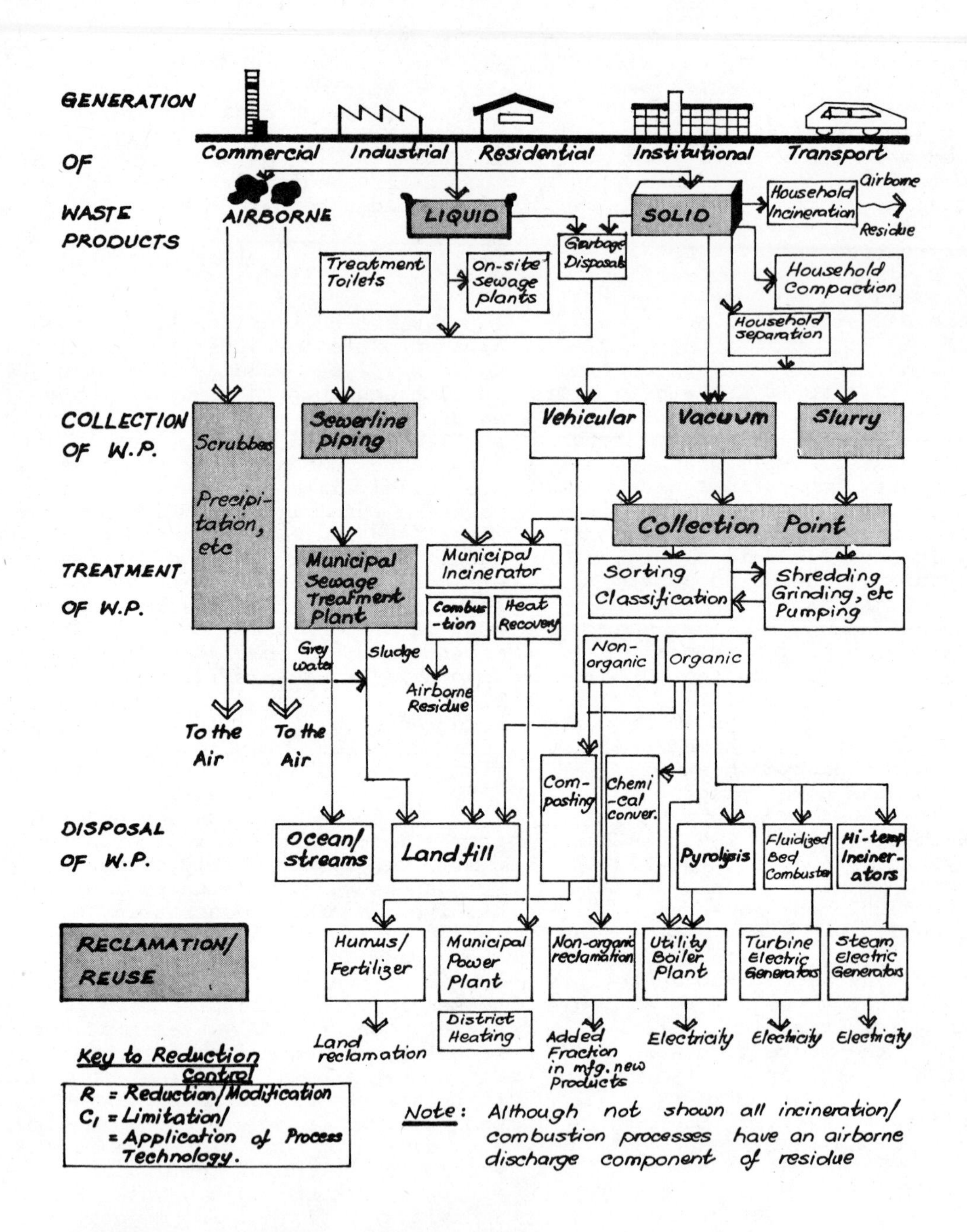

URBAN WASTE PRODUCTS GENERATION-DISPOSAL / REUSE CYCLES

FIGURE 13.2

Due to a growing scarcity of acceptable landfill sites, communities have built municipal incinerators to burn the wastes first. This process reduces the waste to an inert residue that is approximately 5 percent of the original volume, with ultimate disposal of the remainder still by landfill. However, the burning of municipal wastes is expensive both in financial costs (8.80 $/mt to 27.50 $/mt) and in terms of the levels of air pollution generated. Significant effort is being expended to develop incinerators that achieve a high volume reduction with minimum economic and environmental impact.[38] Unfortunately, current air pollution devices are costly to install and reduce the functional and energy efficiencies of the conventional incinerators on which they are used. In comparison with the costs of burning solid wastes, burying them costs about 3.30 $/mt ($3.00/ton)--based upon a 900 mt/day (1000 ton/day) sanitary landfill operation.[39] As of 1972, burning accounted for only 10 percent of solid waste disposal.[40]

A broad variety of technical improvements to incinerator operation (dual chambers, moving grates, mechanical stokers, after burners, rotary kilns, suspension burners, high temperature burning, and fluidized beds) as well as certain fundamentally new processes have been combined to convert the conventional municipal incinerator into a new type of treatment unit. In addition, similar innovations have been developed in the shredding and separation of solid wastes into organic and inorganic fractions (air classification, cyclone, magnetic, and optical separators). Separation is done either before or, less typically, after these new forms of treatment. The result has been increased viability of reclaiming inorganic fractions (metals and glass) and the direct use of shredded wastes before treatment--as partial power plant fuel or, after treatment, the use of oil or gas by-products to fuel the treatment process and to provide steam for power generation and building heating or cooling. Also, oil, gas, or fuel "pellets" can be produced to fuel more distant power plants. In most of these new systems, the final amount of residue is relatively minimal and can be reused rather than requiring landfill.

The various systems are generally classified by the basic, innovative treatment element involved. The many possible combinations of shredding and classification options prior to treatment, the extent of treatment, the nature of the residue output, and the use to which the output is put, all combine to create a broad variety of complete treatment-disposal systems. Therefore, actual design of such systems is based upon:

- the make-up and volume of solid wastes;
- transport distances involved;
- local costs of landfill, fuel oil, labor, and construction;
- local power generation modes and location of power plants in relation to the treatment/reclamation facility;
- availability of local markets for reclaimed metals and glass and the going market rate for those materials; and,
- the feasibility of nearby district heating applications, if the latter is the approach to be taken.

While systems are usually classified by total capacity of raw refuse treatable per day, they are comparatively evaluated according to the following values, all reduced to per metric ton daily capacity of raw intake refuse:

- processing and firing facilities' capital costs;
- operating costs;
- financing costs;
- metals and/or glass recovery value; and,
- recovered fuel value (see for example, Table 13.3).

This in turn results in a frequently used final figure of merit--"net cost of disposal per ton" (see Case Study 13-3).

The simplest basic system and that closest to commercial use is to burn trash directly in the furnaces of generating plants as supplementary fuel, generally providing from 10 to

Annual Costs*		Total	Per Ton of Waste Input
Operating Costs		$1,748,000	$ 5.83
Fixed Costs		212,000	0.70
Capital Charges			
Amortized Investment	$ 480,000		
Fixed Investment	1,415,000		
Recoverable	37,000		
Total Capital Charges		1,932,000	6.44
Total Annual Cost of Operation		$3,892,000	$12.97
Value of Recovered Resources			
Electrical Energy	$1,200,000		
Gross Value of Recovered Resources		$1,200,000	$ 4.00
Net Annual Cost of Operation at 1,000 TPD		$2,692,000	$ 8.97

Effect of System Capacity on Operating Costs

	250 TPD	500 TPD	1,000 TPD	2,000 TPD
Total Annual Cost	$1,316,000	$2,263,000	$3,892,000	$6,700,000
Resource Value	300,000	600,000	1,200,000	2,400,000
Net Annual Cost	$1,016,000	$1,663,000	$2,692,000	$4,300,000
Net Cost Per Ton	$13.55	$11.08	$ 8.97	$ 7.17
Net Gain Over Conventional Incineration Per Input Ton	($3.18)	($2.10)	($1.29)	($0.52)

*300,000 ton/yr (TPY) raw waste input

Reference 39

ANNUAL OPERATING COSTS-INCINERATION WITH ELECTRIC GENERATION

TABLE 13.3

20 percent of the total heat generated. In this process, refuse is shredded, air-classified to remove metals and other inorganics, and then fed into the furnace along with conventional fuel. Cities find that this process can result in a net disposal cost of less than 5.50 \$/mt (5 \$/ton) as opposed to 7.70 to 27.50 \$/mt (7-25 \$/ton) for incineration.[41] The total volume of the refuse, however, is only minimally reduced and the local steam-electric utility plant must be within a very few miles of the waste processing facility, or transportation costs will quickly make the process uneconomic.

One means to improve transport economics is to compact the shredded and air-classified waste in an extruder to form cubettes or pellets. The pellets can also be stored temporarily without bacteriological decomposition.[42]

Attempts have been made to produce steam from refuse municipally for resale to local utilities. However, it has generally cost more than utilities are willing to pay and the steam produced was not of the superheated 530.8 °C/41.4 x 10^5 N/m^2 (1000 °F/600 psi) quality required. In addition, untreated refuse does not burn well in superheaters and at high furnace temperatures refuse tends to produce abnormal heat-transfer surface corrosion.[43]

The burning of refuse to make lower temperature steam for the direct heating and air conditioning of buildings has been more successful. A plant in Nashville, Tennessee, consumes 655 mt/day of refuse and produces steam at a rate of 10^5 kg/hr. In winter this is used directly to heat buildings, while during the summer it drives two water chillers to cool them. Utilizing distribution lines under the streets of downtown Nashville, the steam and chilled water are distributed to 38 buildings within the Nashville district heating project.[44] While the Environmental Protection Agency estimates that 48 cities in the US could utilize trash for district heating, it should be noted that most of the buildings in the Nashville project are government owned (see Case Study 13-4).

District heating systems are more prevalent in Sweden than in most other nations and are used for major building groups (e.g., college campuses). One such system exists in lower Manhattan (New York). The large initial capital cost and the severe difficulty of backfitting such systems, as well as the problems of multiple private ownership of adjacent commercial buildings and the staging of their construction over time, should not be minimized.

Another basic process is hydrogenation, developed by the US Bureau of Mines Pittsburgh Energy Research Center. Waste is loaded into a reactor (with 5 percent catalyst), then pressurized to as much as 250 atmospheres and heated to 371 °C (700 °F) for one hour. This converts about 85 percent of the dry waste into a heavy fuel oil with a sulphur content of less than 0.4 percent and a heat value of 35 MJ/kg (15 kBtu/lb) as compared to about 42 MJ/kg (18 kBtu/lb) for common Number 6 fuel oil. Because some of the oil is recycled to heat the reactor during subsequent operations, the net yield is about 200 ℓ/mt (1.25 bbl/ton) of waste. The basic drawback has been cost. With the current price of fuel oil at ∿15 \$/bbl, the economics appear more favorable, since the cost of producing the oil is approximately 7-8 \$/bbl, due largely to the cost of the reactor.[45]

Pyrolysis is a straight disposal system which takes unsorted solid waste directly from collection trucks, shreds it, then subjects the organic fraction to destructive distillation (heating in the absence of oxygen). This results in a chemical breakdown into three streams: oil, low-energy gas, and char, along with various amounts of other liquids, light oils, and tar, depending upon the make-up of the refuse, temperatures attained, and heating rate of the pyrolizing chamber. Pyrolysis should be less expensive than hydrogenation since it is done at atmospheric temperatures and capital-intensive reactor vessels are not required. The process is self-sustaining (through use of its own gaseous products as a heat source) and is capable of recovering more than 90 percent of the energy value of the raw refuse as useful products.[46] Because three "fuels" result, however, collection and storage are more difficult.[47]

Although many basic as well as operational problems remain to be worked out on these various energy/reclaimed-materials-from-wastes systems, current fuel and materials prices (and the cost of energy to produce some of the latter, such as aluminum and glass) appear to be combining to prove or improve their dollar and energy saving potential. One survey in March 1975 indicated that some 15 manufacturers had two facilities in conceptual development and nine pilot plants. They were negotiating for municipal contracts to build four others, were breaking ground on one, had three more approaching completion, and had put seven on line as of that date.[48]

D. Energy Efficient Urban Transportation

As indicated in Chapter 5, 25 percent of the US energy budget is for transportation. Another 15 percent is utilized to build and maintain transportation systems. No other sector of the US economy is more sensitive to energy limitations or has greater potential for improved energy efficiency.

Strategies which minimize transportation energy use while maintaining an acceptable level of service include both short- and long-term plans, some of which are being developed in anticipation of possible future shortages. The first is to design overall systems which minimize energy use. This can both decrease the transportation related load placed on society and, at the same time, decrease the relative impact upon such systems of any shortage which may occur. The second long-term strategy is to develop systems with as much diversity in energy forms as possible. In the US transportation is 95 percent dependent on petroleum for its energy supply. Electricity is a particularly attractive alternative because it has inherent diversity in terms of its primary fuel supply which can be petroleum, natural gas, coal, nuclear energy, the sun, or anything else that can generate electricity. Hence, electric vehicles have a much greater fuel versatility than gasoline or diesel vehicles.[49] In addition, their drive systems have an overall efficiency of 25 to 30 percent rather than the 15 to 20 percent of the internal combustion vehicle.

Another long-term strategy is to develop a system with the capacity for a rapid shift from one mode to another in the event of shortages of one energy form. For example, during the gasoline shortage of 1973-1974, many automobile drivers commuting to and from urban areas joined carpools, while others switched from cars to buses, while still others shifted to mass transit where that option was available. The capability for modal shifts should be designed into all future transportation systems.

Short-term strategies largely deal with existing systems and generally are ways to alter setting or use to achieve greater efficiency. To some extent, this means changes in business practices and life styles related to transportation use. For example, downtown areas (or any other high traffic areas) can ban *all* on-street parking, thereby doubling the carrying capacity of existing four-lane streets and decreasing traffic congestion approximately 25 percent. This strategy could be augmented by city financing of public multi-story garages on many of the on-grade parking lots and service station locations within at least the center city. The gas stations could be moved within the ground floor of such structures, and offices could occupy upper floors over the parking levels--to help defray costs of both construction and operation (or playgrounds and parks could be constructed on the top deck).[50] This form can be seen in buildings such as the John Hancock Building or Marina Tower in Chicago, as well as in other structures in Atlanta, New York, etc.

In designing a system the planner utilizes various population characteristics and technological features of the options open. Note that minimizing transportation energy cannot be a sole criterion for planning decisions. "Planning factors" range from "directional preferences" and "trip times" to vehicle and seat passengers per km. These must be considered in establishing what is an acceptable level of service. Once this has been done, it is reasonable to minimize the energy use needed to provide the minimum level.

The principal technological factors open to the system designers are:[51]

- select optimum vehicle type -- Table 13.4 shows vehicle forms, uses, and transportation options. (Table 13.5 defines the options.) See also Chapter 5.
- use lightweight vehicles -- Weight is a major factor in determining transportation energy. Weight reductions involve a trade-off in convenience, comfort, and safety.
- maximize vehicle loading -- This is the most immediate and effective technique for improving energy efficiency.

Finally, in keeping with an earlier part of this chapter, it should be emphasized that transportation energy cannot be considered outside the context of the urban setting. For example, greater savings could be achieved in a low population density area by compacting the city than by more efficient vehicles. Conceivably, higher land prices could have a greater effect than years of technological development.

E. Cities and Underground Urban Service Systems

Increased land costs in dense urban areas and increasing concerns over all forms of environmental blight, including visual, have led to a growing demand for the undergrounding of all service network systems. Poles and overhead power and telephone wires are considered by many to be a serious visual distraction in both urban and rural settings. For a number of years, utilities have had undergrounding of distribution lines as a condition for building.

Historically, many utilities have been buried under city streets, including sewers, gas lines, water systems, and storm drains. In recent years the intense competition for every square meter of high cost urban land has led to innovative proposals for underground communication links, vacuum-tube parcel delivery, vacuum-to-slurry solid waste transport, freight delivery tunnels, and more underground transit such as subways. In an effort to achieve a more efficient means for initial placement and later ease of repair and revision of this growing underground network, the common utility tunnel, or "utilador" has been proposed (see Case Study 13-5). Environmental concerns over despoilment of the rural environment have created similar pressures for the undergrounding of long distance, high voltage transmission lines, even as their number and length increase under dual forces of growth and increasing distances from source to use. The large line losses of these long distance lines have created a complementary impetus to push the development of superconducting and cryogenic technology which would facilitate and require undergrounding.

To these established forces for undergrounding, the energy problem has added new impetus for expansion of the concept to include new functions: houses, buildings, factories, and even entire cities.[52] Significant energy could be saved through this approach because the mass of the surrounding earth acts as a heat sink whose temperature varies only slightly from the yearly average. Underground, less heating is required in winter and less cooling in summer. In addition, subsurface construction presents less area (little or no walls and even roofs) to the direct radiation of the sun or to prevailing winds, both major contributors to heating and cooling loads. The long thermal lag (or heat retention) of the earth also allows extended periods of shutdown of HVAC equipment with imperceptible changes resulting within the conditioned space. [53]

As two brief case studies indicate (See Case Study 13-6 for an underground frozen food storage facility and Case Study 13-7 for an underground manufacturing plant), significant savings in operating and energy costs can be achieved. Reduction of heat losses to and heat gains from surroundings can provide most significant savings in energy. Yet, many architects and engineers claim that it is more economical and effective to use more wall and roof insulation than to build underground. As Case Study 13-7 shows, the heat flow rate in Minneapolis on a cold winter day for a wall with

	Size	Power	Operation
Automobile, Internal Combustion	Small (private)[c]	Fuel Combustion[f]	Demand Responsive Route[g]
Automobile, Electric	Small (private)	Energy Storage[e]	Demand Responsive Route
PRT, Personal/Private	Small (private)	Electric Pick-Up[d]	Express Route (no stops)
PRT, Group/Shared	Medium (public)[b]	Electric Pick-Up	Demand Responsive Route
Rail, Commuter	Large (public)[a]	Electric Pick-Up	Express Route (no stops)
	Large (public)	Fuel Combustion	Fixed Route with Stops
Rail, Trolley/LRV	Medium (public)	Electric Pick-Up	Fixed Route with Stops
Bus, Fixed Route	Medium (public)	Fuel Combustion	Fixed Route with Stops
Bus, Express	Large (public)	Fuel Combustion	Express Route (no stops)
Bus, Dual Mode	Medium (public)	See Table 13.5	See Table 13.5
Bus, Demand Responsive	Medium (public)	Fuel Combustion	Demand Responsive Route
Bus, Jitney	Medium (public)	Fuel Combustion	Fixed Route with Stops

Notes:

(a) 40-160 passengers
(b) 6-40 passengers
(c) 1-6 passengers
(d) Overhead, Trolley, Third Rail, In-Ground Pick-Up
(e) Battery, Inertial Flywheel
(f) Gasoline, Diesel
(g) One to many
Many to one
Many to many

Reference 49

TRANSPORTATION OPTIONS **TABLE 13.4**

- *Electronic Automobile*--This is a vehicle similar in structure and form to the common gasoline automobile except that it uses electric motors to propel the car, and storage batteries to drive the motors. In a second form, the electric car might operate in a "dual-mode" (see below), running on a guideway with electric pick-up for part of its trip. Conceivably it might be made compatible with PRT vehicles so that it could enter and leave PRT guideways as desired.

- *PRT (Personal Rapid Transit)*--This term has become very common in recent years. Unfortunately, it is used quite ambiguously in the literature. The term almost always implies a system which has fully automatic control (no operator). It usually implies a demand responsive operation, wherein vehicle routing depends on the trip demands of the rider. It sometimes, though not always, implies a small family-sized vehicle. Two sub-definitions of PRT are in the next two items.

- *Personal (Private) Rapid Transit*--We use this term to define a class of fixed guideway systems in which automated vehicles no larger than small automobiles carry people or goods nonstop between any pair of stations in a network of slim guideways. The vehicle is occupied by an individual or a private group travelling together. It seats about one to six persons.

- *Group (Shared) Rapid Transit*--Group Rapid Transit is demand responsive, automated, and guided, but its vehicles serve more people than PRT, in a shared mode. Each car may carry perhaps six to forty seats. Different riders usually enter and exit at different stations. The vehicle may follow a fixed route, or the route may be variable, depending on the travel demands of the mix of riders. This class has been called "people mover systems." However, the term "people mover" has been used in so many senses that it has lost almost all consistent meaning. This type of system is in operation at the Tampa Airport and the Seattle-Tacoma Airport, and is under construction or consideration at a number of other sites, many of which are airports.

- *Commuter Rail*--This category describes a great number of systems which have been in operation for years. It includes, for example, the Chicago and Northwestern, the Lindenwold Line, and the Southern Pacific service to San Francisco. Power can be electric or internal combustion (diesel locomotive). Train frequency increases during commuter hours. Average speed is 50-65 km/hr (30-40 mph).

- *Rail Rapid Transit*--This category is perhaps best exemplified by the new BART system in and around San Francisco. It has many characteristics in common with "commuter rail." The exceptions are that it will probably always be electric, speeds will be higher, service during off-peaks more frequent, and headways shorter.

- *Trolley (Light Rail Vehicle - LRV)*--LRV's have been around since the turn of the century but they survive now in only a few cities, including San Francisco, Philadelphia, and Boston. LRV's are experiencing a resurgence of interest. Expansion will take place in some existing system, and some new systems may be developed.

DEFINITION OF TRANSPORTATION OPTIONS **TABLE 13.5**

- *Bus-Express*--This one-stop or few-stop bus is one which carries people from one central location to another. It is used in a number of cities where blocks of transit customers can be identified and efficiently served.

- *Dual-Mode*--This is a very complex term embracing many proposed new modes during a given trip. For example, a bus might enter a guideway or track line and be moved along, possibly by an electric system, to an exit point where it would continue normal operation on the street as a bus. In another configuration, cars would operate as they do now until they reach a guideway. They would then drive onto a pallet which would carry the car along the guideway to its desired exit point.

- *Bus Demand Responsive*--This category again includes a number of modes of operation. The essential feature is that the bus responds on any trip to the particular demands of customers by going to their house or some other personalized spot to pick them up or drop them off. The bus may go to a number of houses to pick people up and then take them all to the same place (called many-to-one or "gather"). The bus may pick up a number of passengers and drop them at their homes (called one-to-many, or "scatter"). Finally, the bus may pick up people at a number of places and drop them off at a number of other places using some efficient routing pattern (called many-to-many).

- *Jitney*--This term is quite old, and is used in many senses. It often means a small bus, or van which travels a fixed route serving a fairly small number of passengers. There has been some tendency to use the term to represent the demand responsive vehicle described above.

TABLE 13.5 CONT'D

20 cm (8 in.) of insulation will be 5.5 times greater above ground and for a wall with 10 cm (4 in.) of insulation will be 8.4 times greater. The heat flow rate can be 10 to 22 times greater for a roof than underground.[54]

Much of the resistance to the concept of underground construction for human work and habitation is based upon subjective opinion and vague recollections of supposed emotional disturbances experienced in windowless defense plants during World War II and windowless schools and other buildings built during the days of civil defense concern in the 1950s.

Yet, in a detailed study of pupil achievement, anxiety, and mental health in the first completely underground elementary school (and fall-out shelter) in the United States, opinions about the school, psychological effects, and general health recorded in 1972 led its authors to conclude:

> It seems that after ten years of experience with children attending an underground and windowless elementary school, the professionals concerned with the health care of children in Artesia, New Mexico, the location of the Abo school, are generally convinced that not only is the school not detrimental to the physical and mental health of their students, but it is actually a benefit to some.
>
> Although not as supportive of the school/fall-out shelter facility as the parents of pupils who attended, the sample of the public clearly favored the school. Nine out of ten recommended that other schools be built like Abo, if such schools cost no more to build than other schools.[55]

One problem with underground construction is the lack of information. A review of the primary handbook on heating and cooling for architects and engineers, produced by the American Society of Heating, Refrigerating and Air-Conditioning Engineers (ASHRAE) (1972 version in a four volume series comprising 2,478 pages), indicates that only one-half page was devoted to basements, the nearest approach to subsurface space, and begins, "Unfortunately complete data on ground temperatures adjacent to buildings are not available."[56] There are various reports that do give some useful information. Interestingly enough, more information is now being developed in relation to underground storage for solar energy systems.

Significant benefits can be gained from buildings depressed below the earth's surface as well as those which use large earth berms built up around them on grade. For example, Case Study 13-8 shows how the "Ecology House" manages to be below grade and yet is arranged around a open-air, sunken atrium. Case Study 13-8 also indicates how these techniques of berming and partial undergrounding were used in the design of an office building.

There are a number of variations on these possibilities. For example, in a single-family residence, those rooms most needing windows and views, such as the living areas, could be placed at grade or above ground, with bedrooms, bathrooms, and storage areas being placed in a basement with considerable energy savings. It would be a most exciting and challenging architectural prospect to deal with groups of buildings or residential areas with perhaps one-half to one-third of the space above ground, and the remaining extending into various terraces in below grade areas which have openings into outdoor sunken terraces or gardens. This would also allow less than one-third of the building form to project above the ground and would create more outdoor space.

As has been shown previously, however, housing cannot be built alone but needs to be developed in a context of commerce, industry, and activity centers for cultural and institutional usage. The design possibilities and challenges for partial or complete undergrounding implied in this larger grouping of buildings and the minimization of impact upon the landscape begin to suggest radical changes in the physical form of all communities and even cities. This is not without precedent; it has been done in ancient cities in some parts of the world.

A number of alternatives would have to be investigated in any serious undergrounding effort. Obviously, smaller air-conditioning and heating units would be required to condition spaces partially or fully underground, as opposed to those exposed above grade. Not only would they take less energy to operate, being smaller, but they would require less energy to manufacture. Similar savings would be attributable to reduced amounts of insulation required. However, the extra energy required and cost for underground construction would have to be balanced against those potential savings. The possible need for sumps to protect against water intrusion and pumps to lift sewage up would also have to be taken into account. It is possible that technological improvements will significantly narrow the difference in cost between underground and surface construction. As Dr. Starr points out in the Prologue to this book:

> These concepts are leading to an examination of the rationale for our present life style which encourages individual dwellings, separation of work and industrial activities, and deurbanization of our high density population centers. The outcome could be an "energy constrained" society as discussed in Chapter 14 of this book.
>
> If one assumes that visual amenities, privacy, and avoidance of cross-pollution are the basic motivations for such separation, then an extreme solution, such as the underground cities described in Chapter 13 might provide an environmentally and energetically acceptable solution. The surface would be devoted to parks, food production, leisure facilities and natural scenery. The underground would provide a stable heat reservoir with a long time constant for a heat pump system and privacy would now be compatible with a concentration of energy-using functions. Transport requirements would be drastically reduced and made compatible with increased accessibility to leisure and recreational activities.

With the exception of some of the modern mechanical equipment mentioned by Dr. Starr, the city described in this passage greatly resembles Case Study 13-1--the Longhouse at Mesa Verde in southern Colorado. Dr. Starr goes on to point out:

> ...The question remains as to whether such underground cities require less total integrated energy for their construction and operation than that required by our present apparently inefficient surface cities.
>
> The suggestion of a compact underground city, with completely integrated and aggregated activities, represents a wide departure from presently accepted life style concepts. Nevertheless, this extreme scenario illustrates most of the concepts which are a consequence of optimizing energy and resource conservation. It deserves serious thought as the end point limit of the spectrum of options between the present energy-abundant life styles and the most energy-constrained system.

Underground cities may be the extreme in any energy-limited world of the future, or they could be the mean, if for no other reason than to release land for food production.

13.5 INFERENCES

Efficient use of energy in cities has been reviewed from both a historical and modern perspective. Due to the complexity of the problems related to urban energy use, as well as the complexity of cities and urban systems which supply, control, or use energy, it is impossible to provide a comprehensive treatment in a single chapter. The approach adopted was to delineate general principles relating efficient energy use and city design, and then to illustrate the application of these principles both in the text and with case studies.

For the near-term future, the greatest improvements in energy efficiency in cities can come about from changes in utility systems, solid waste disposal and recovery, urban transportation, site sensitive design, and undergrounding. These also represent extensions of the general principles delineated earlier in the chapter. Other innovations are no doubt possible, but have not been described either because it was believed that the technology is not fully developed or the time scale to implement them is too distant.

Success in the future depends on several factors. Too often innovations such as waste recycling or total energy systems consider only vertically integrated processes, with little reflection upon horizontal interactions and impacts upon related urban systems, such as the design of buildings to utilize waste heat or the location of waste energy plants in relation to point of use. Where possible, system interaction effects should be considered in the approach so that institutional barriers (building codes, labor union regulations) or other non-technological factors do not obviate innovative solutions. Narrowly construed approaches--whether by architects, engineers, economists, or city planners--must be replaced by interdisciplinary and systemic investigations, research, and response at all levels--from the laboratory to the construction site. Only in this way will solutions of sufficient power and responsiveness be obtained.

Due to the scale of energy use in cities and the potential for improvements in efficiency, it is imperative that investigations of improved efficiency begin immediately. Not only are present rates of energy use high, but due to their long lives, every building built and every community constructed with present technology and current inefficiencies make the job of modifying these patterns more difficult and prolonged.

* * *

REFERENCES

1. Dantzig, George B. and Saaty, Thomas L., *Compact City: A Plan for a Livable Urban Environment*, (San Francisco, California: W.H. Freeman & Co., 1973).

2. Halprin, Lawrence, *Cities*, (New York: Reinhold Publishing Co., 1964).

3. Dantzig and Saaty.

4. McAllister, Donald M., ed., *Environment: A New Focus for Land-Use Planning*, (Washington, D.C.: National Science Foundation, 1973).

5. Owen, Wilfred, *A Fable: How the Cities Solved Their Transportation Problems*, (Washington, D.C.: Urban America, 1967).

6. *American Almanac 1970*, "Statistical Abstract of U.S.," Table 5 (Series D), (New York: Grosset and Dunlap, 1970).

7. Ehrlich, Paul R. and Ehrlich, Anne H., *Population, Resources, Environment*, 2nd edition, (San Francisco, California: W.H. Freeman & Co., 1972).

8. McAllister.

9. Reilly, William K., ed., *The Use of Land: A Citizens' Policy Guide to Urban Growth*, Task Force Report sponsored by the Rockefeller Brothers Fund, (New York: Thomas Crowell Co., 1973).

10. Ewald, William R., Jr., ed., *Environment and Policy: The Next Fifty Years*, American Institute of Planners' Fiftieth Year Consultation, (Bloomington, Indiana: Indiana University Press, 1968).

11. Lynch, Kevin, "The Possible City," in ibid, pp. 137-157.

12. Gruen, Victor, "Comments on Lynch," in Ewald, pp. 157-161.

13. Caudill, William W.; Lawyer, F.D.; and Bullock, T., *A Bucket of Oil; The Humanistic Approach to Building Design for Energy Conservation*, (Chicago, Illinois: Cahners Books, April 1974).

14. McAllister.

15. Schoen, Richard, "Aspects of the Production of Housing--A Systems View," (Masters Thesis, Second Professional Degree [M'Arch II] Program, UCLA School of Architecture and Urban Planning, June 1971).

16. Mumford, Lewis, *The City in History--Its Origins, Its Transformations, and Its Prospects*, (New York: Harcourt, Brace, and World, 1961).

17. Fraser, Douglas, *Village Planning in the Primitive World*, (New York: George Braziller, Inc., 1968).

18. Banham, Reyner, *The Architecture of the Well-Tempered Environment*, (Chicago, Illinois: University of Chicago Press, 1969).

19. Brotchie, J.F., "Some Systems Concepts for Urban Planning," *RAPIJ - Royal Australian Planning Institute Journal*, 12(April 1974): 43-50.

20. Ibid.

21. Moholy-Nagy, Sibyl, *Matrix of Man--An Illustrated History of Urban Environment*, (New York: Praeger Publishers, 1968).

22. Knowles, Ralph L., *Energy and Form, An Ecological Approach to Urban Growth*, (Cambridge, Massachusetts: MIT Press, 1974).

23. Ibid.

24. Gallion, Arthur B. and Eisner, Simon, *The Urban Pattern; City Planning and Design*, (Princeton, New Jersey: D. Van Nostrand Company, Inc., 1950).

25. Picard, Jerome, *Dimensions of Metropolitanism*, (Washington, D.C.: Urban Land Institute, 1967); and Advisory Commission on Intergovernmental Relations, *Urban and Rural America: Policies for Future Growth*, (Washington, D.C.: US Government Printing Office, 1968).

26. Alonzo, William, "The Mirage of New Towns," *The Public Interest*, No. 19 (Spring 1970).

27. Canty, Donald, ed., *The New City*, Report of the National Committee on Urban Growth Policy, published for Urban American, Inc., (New York: Praeger Publishers, 1969).

28. Picard.

29. Advisory Commission on Intergovernmental Relations, *Urban and Rural America*.

30. Jackson, Samuel C., "Population Growth and New Communities, A New Venture," *HUD International* (18 May 1970).

31. Alonzo.

32. Perloff, Harvey S.; Berg, T.; Fontain, R.; Vetter, D.; and Weld, J., *Modernizing the Central City: New Towns Intown . . . and Beyond*, (Cambridge, Massachusetts: Ballinger Publishing Company, 1975).

33. Citizens' Advisory Committee on Environmental Quality, *Report to the President and to the Council on Environmental Quality*, (Washington, D.C.: 1974).

34. Miller, Brown; Pinney, N.J.; and Saslow, W.S., *Innovation in New Communities*, MIT Report No. 23 (Cambridge, Massachusetts: MIT Press, 1972).

35. Schoen, Richard; Hirshberg, Alan; and Weingart, Jerome, *New Energy Technologies for Buildings: Institutional Problems and Solutions*, A Report to the Ford Foundation Energy Policy Project, (Cambridge, Massachusetts: Ballinger Publishing Company, 1975).

36. Bryson, Frederick E., "Garbage Power: There's Heat in 'Them Thar Heaps'," *Machine Design*, 47(9 January 1975): 20-26.

37. Miller, Pinney, and Saslow.

38. Perloff, Berg, Fontain, Vetter, and Weld.

39. Reese, Terrence G. and Wadle, Richard C., *General Survey of Solid Waste Management*, NASA Technical Memorandum JSC-08696 NASA TM X-58133, (Houston, Texas: May 1974).

40. Miller, Pinney, and Saslow.

41. Reese and Wadle.

42. Midwest Research Institute, *Resource Recovery, The State of Technology*, prepared for the Council on Environmental Quality, (Washington, D.C.: US Government Printing Office, February 1973).

43. Reese and Wadle.

44. Ibid.

45. Bryson.

46. Reese and Wadle.

47. Bryson.

48. State of California Solid Waste Management Board, *Current Status of Resource Recovery Systems and Processes*, Bulletin No. 5, (Sacramento, California: Technical Information Series, February 1975).

49. Healy, Timothy J., *Energy Use of Public Transit Systems*, Report prepared for CALTRANS Division of Mass Transportation, (Santa Clara, California: University of Santa Clara, 1 August 1974).

50. Owen.

51. Healy.

52. Bligh, Thomas P. and Hamburger, Richard, "Conservation of Energy by Use of Underground Space," in *Legal, Economic, and Energy Considerations in the Use of Underground Space*, ed: National Academy of Sciences, (Washington, D.C.: 1974).

53. Ibid.

54. Ibid.

55. Lutz, Frank W.; Lynch, Patrick D.; and Lutz, Susan B., *Abo Revisited*, (Washington, D.C.: Defense Civil Preparedness Agency, 1972).

56. Bligh and Hamburger.

CHAPTER 14

AN ENERGY CONSTRAINED WORLD

C.B. Smith*

CHAPTER CONTENTS

KEY WORDS

Energy Accounting
Energy Management
Energy Sources
Government Policy
Industrial Use
International Energy Use
Projections

SUMMARY

The world's energy resources are either continuous or exhaustible. Early man depended on continuous solar energy (a fraction of more than 3×10^{15} GJ/yr [$\sim 3 \times 10^{15}$ MBtu/yr] incident on the earth and seas) to meet his needs. Industrialized man differs from his predecessors in the use of energy in that he depends to a much greater extent on fossil and nuclear fuels than on continuous sources of solar energy.

World energy use is estimated presently to be 3×10^{11} GJ/yr. The United States alone uses approximately 1×10^{11} GJ/yr, an increasing fraction of which is in the form of electricity. In the United States in 1975, electricity generation used 28 percent of the equivalent fuel input to the economy--surpassing industry as the largest single fuel user. Electricity generation is expected to approach 50 percent by the end of the century.

Current energy use in the United States exceeds 300 GJ/person-year (300 MBtu/person-year).* If the entire world population had the same requirement, the world's inexpensive fossil fuel reserves would either be near exhaustion or be very expensive in approximately 30 years. Probably more important, the global annual fuel bill would soon exceed the entire Gross National Product (GNP) of the United States. Thus, it is possible to speculate that the world of tomorrow could consist of many energy constrained societies. Possible constraints would include economics, self-sufficiency in natural resources, socio-cultural limits, and finally an upper limit of energy use which is biologically tolerable. As a lower limit, some minimum expenditure of energy is necessary for subsistence.

---

*Principal, Applied Nucleonics Company, Inc.

*For the reader accustomed to American/British units, the GJ is equal to 0.95 MBtu; for practical purposes the two units can be considered equivalent.

Improving efficiency is of fundamental significance in terms of resource utilization, cost, and the environment. In the United States a 15 percent increase in electricity use efficiency by the year 2000 could reduce the projected number of 1000 MWe generating plants by between 300 and 400. The savings in capital cost alone would be \$150 to \$200 billion; savings in fuels ($\sim 3 \times 10^{10}$ GJ/yr [$\sim 3 \times 10^{10}$ MBtu/yr] or \$6 to \$12 billion/yr) would be significant as well.

14.1 THE ENERGY CYCLE OF THE EARTH

The urgent need for efficient electricity use is a result of both electricity's growing popularity as an energy form and a growing awareness of the finite nature of energy resources. This chapter provides a global backdrop against which a particular energy use can be compared and assessed. It is intended to provide a perspective on energy use and to propose a rationale for evaluating national energy use. The concept of an energy constrained world is described. In such a world improved energy use efficiency is not a matter of morality, but of practicality. It provides an immediate technological tool for maintaining the best life quality consistent with available resources.

The world's energy resources are of two types: continuous (renewable) and exhaustible (nonrenewable). Solar energy is of the first category, while fossil and nuclear fuels fall into the second category. Figure 14.1 shows the earth's total energy flux, illustrating the input of solar radiation, the storage and recycling of energy bound up in earthly systems, and the radiation of degraded energy from the earth. It is important to remember that energy is not consumed; it is simply transformed from one form to another. In the process of being used, however, the quality of energy is continually downgraded; that is, its capability to perform useful work becomes more and more limited following each use. In its most degraded and least useful form, it is eventually radiated from earth into space.

Not all of incident solar energy is readily available since part of it is outside the useful frequency band and part is incident on water rather than on land (see Figure 14.2). In addition, exhaustible fuels are not distributed uniformly over the earth; many countries have either limited stocks or no national supplies of stored energy at all.

Table 14.1 shows the biosphere's estimated annual gross productivity (storage of solar energy). This amounts to about 4.2×10^{12} GJ/yr, or roughly 0.13 percent of incident solar energy (3.24×10^{15} GJ/yr, Figure 14.1). In contrast to the energy used by the biosphere, man's total energy use has not yet reached 10^{12} GJ/yr (see Figure 14.3). In fact, in Figure 14.3 we postulate a scenario in which human energy use tends to level out at $\sim 3 \times 10^{12}$ GJ/yr, or 0.1 percent of incident solar energy. By comparing Figure 14.3 with Figure 14.1 it is apparent that human throughput of energy is small compared to that resulting from natural processes.

As Figure 14.3 indicates, an increasing trend both in the United States and in the world is to use a larger and larger fraction of total energy output in the form of electricity. The figure also indicates what would happen if *all* energy use became electric, with an efficiency of 3 GJ of fuel per GJ of electricity, by the year 2050.* Although looking this far into the future is pure speculation, at the present time world electricity use is increasing faster than that of the US. By the year 2000 we still would expect the US to use 20 to 30 percent of all the electrical energy produced in the world. The use of other energy forms is also increasing although less rapidly than electricity. If industrialization continues, this demand is expected to grow. Figure 14.4 shows energy use and Gross National Product

*This assumes hydroelectricity's share decreases to about 2 percent of the total, and values the hydro and nuclear input at the same average heat rate as fossil fuels.

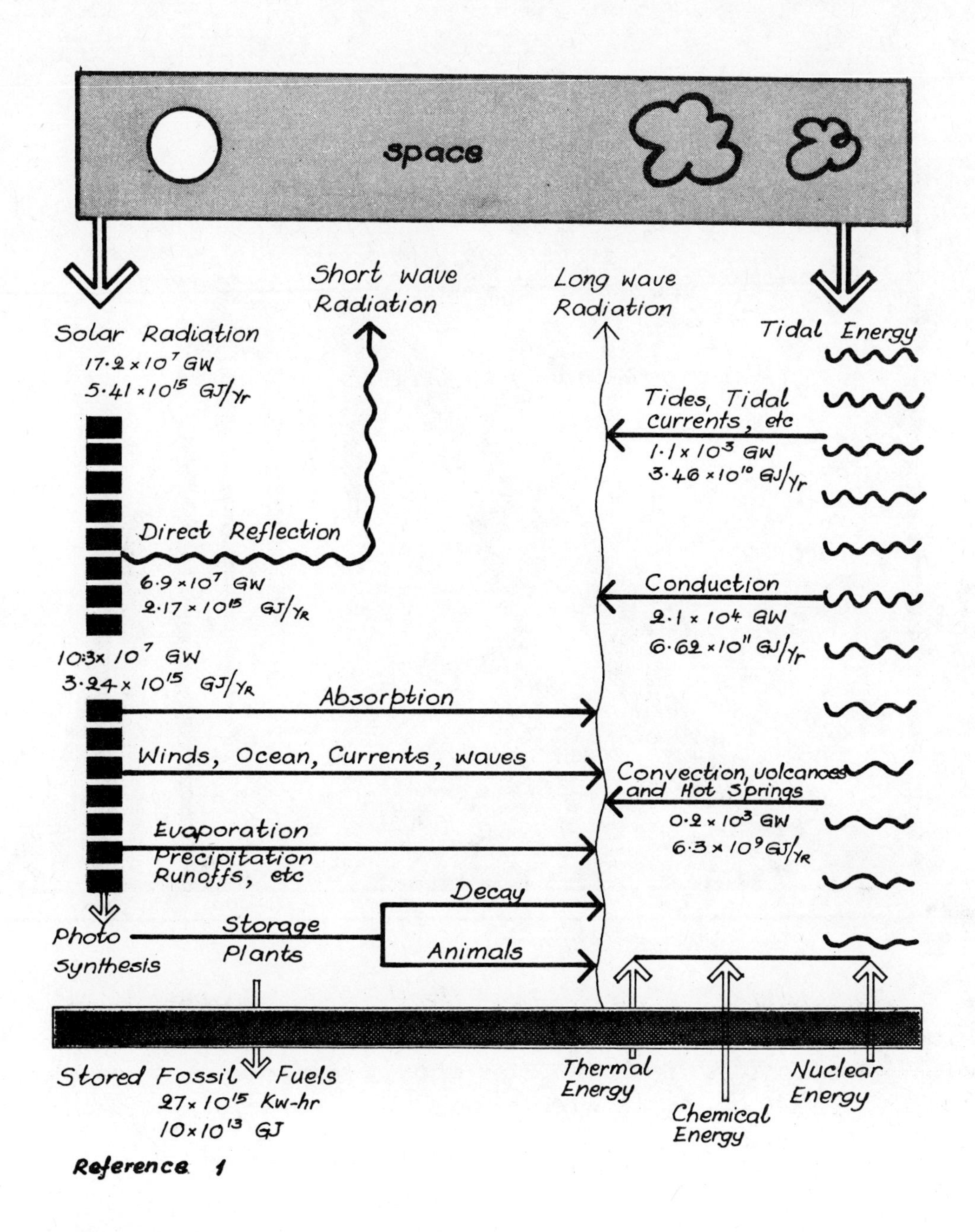
space
Short wave Radiation
Long wave Radiation
Solar Radiation
17·2 × 10⁷ GW
5·41 × 10¹⁵ GJ/Yr
Tidal Energy
Tides, Tidal currents, etc
1·1 × 10³ GW
3·46 × 10¹⁰ GJ/Yr
Direct Reflection
6·9 × 10⁷ GW
2·17 × 10¹⁵ GJ/Yr
Conduction
2·1 × 10⁴ GW
6·62 × 10¹¹ GJ/Yr
10·3 × 10⁷ GW
3·24 × 10¹⁵ GJ/Yr
Absorption
Winds, Ocean, Currents, waves
Convection, volcanoes and Hot Springs
0·2 × 10³ GW
6·3 × 10⁹ GJ/Yr
Evaporation
Precipitation
Runoffs, etc
Decay
Photo Synthesis
Storage Plants
Animals
Stored Fossil Fuels
27 × 10¹⁵ Kw-hr
10 × 10¹³ GJ
Thermal Energy
Chemical Energy
Nuclear Energy
Reference 1

ENERGY FLOW DIAGRAM FOR THE EARTH

FIGURE 14.1

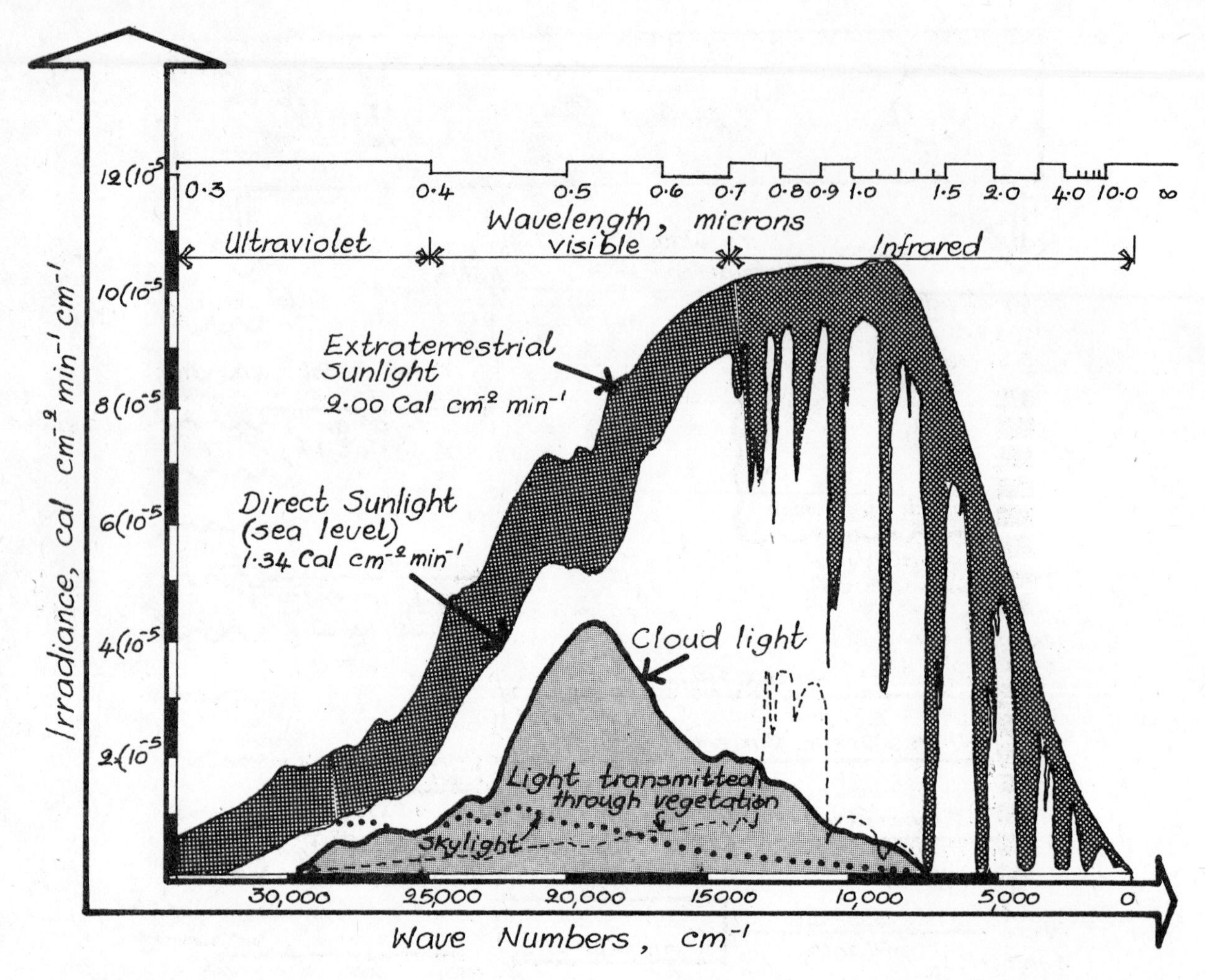

Spectral distribution of extraterrestrial solar radiation, of solar radiation at sea level for a clear day, of sunlight from a complete overcast, and of sunlight penetrating a stand of vegetation. Each curve represents the energy incident on a horizontal surface.

Reference 2

SPECTRAL DISTRIBUTION OF EXTRATERRESTRIAL SOLAR RADIATION

FIGURE 14.2

ECOSYSTEM	AREA (10^6 KM^2)	GROSS PRIMARY PRODUCTIVITY (KCAL/M^2/Yr	TOTAL GROSS PRODUCTION 10^{16} KCAL/YR	TOTAL GROSS PRODUCTION 10^{10} GJ/YR
Marine				
Open ocean	326.0	1,000	32.6	136.6
Coastal zones	34.0	2,000	6.8	28.5
Updwelling zones	0.4	6,000	0.2	0.8
Estuaries and reefs	2.0	20,000	4.0	16.8
Subtotal	362.4	-	43.6	182.7
Terrestrial				
Deserts and tundras	40.0	200	0.8	3.4
Grasslands and pastures	42.0	2,500	10.5	44.0
Dry forests	9.4	2,500	2.4	10.1
Boreal coniferous forests	10.0	3,000	3.0	12.6
Cultivated land with little or no energy subsidy	10.0	3,000	3.0	12.6
Moist temperate forests	4.9	8,000	3.9	16.3
Fuel subsidized (mechanized) agriculture	4.0	12,000	4.8	20.1
Wet tropical and sub-tropical (broadleaved evergreen) forests	14.7	20,000	29.0	121.5
Subtotal	135.0	-	57.4	240.6
Total for biosphere (not including ice caps) [round figures]	500.0	2,000	100.0	420.0

Reference 2

SOLAR ENERGY STORED ANNUALLY BY THE BIOSPHERE **TABLE 14.1**

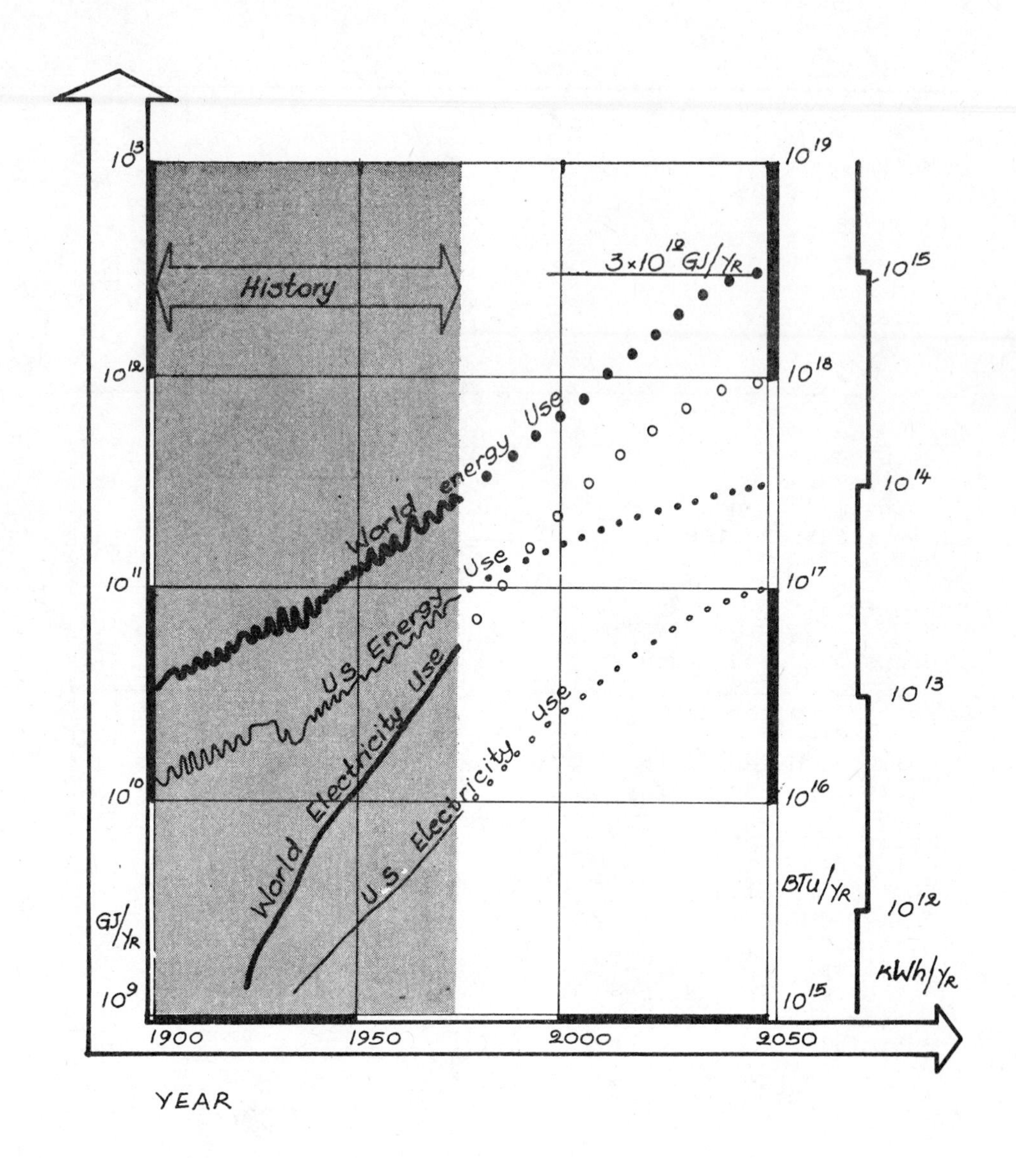

Reference 3

ANNUAL ENERGY OUTPUT FIGURE 14.3

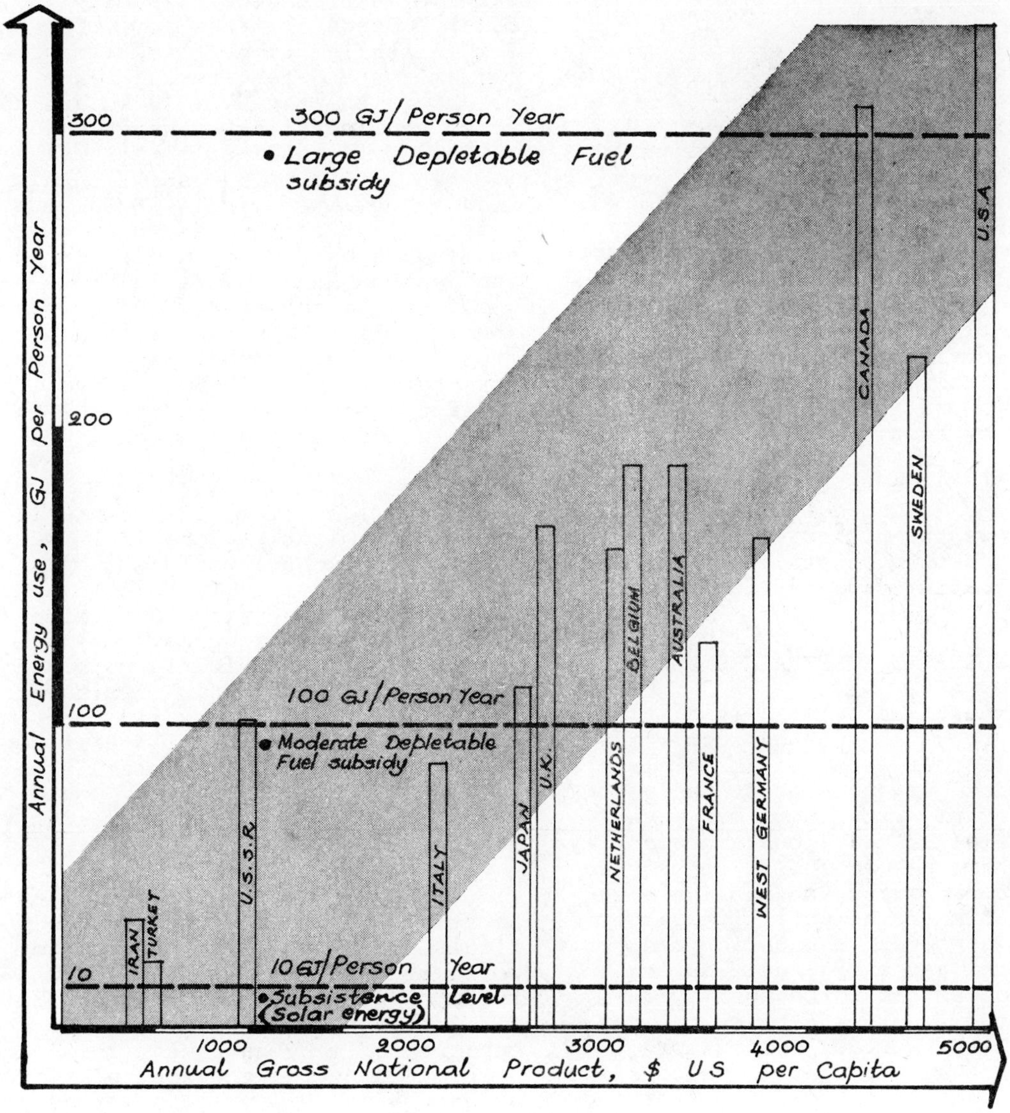

300 GJ/Person Year

• Large Depletable Fuel subsidy

100 GJ/Person Year

• Moderate Depletable Fuel subsidy

10 GJ/Person Year

• Subsistence Level (Solar energy)

Annual Energy use, GJ per Person Year

300
200
100
10

IRAN
TURKEY
U.S.S.R.
ITALY
JAPAN
U.K.
NETHERLANDS
BELGIUM
AUSTRALIA
FRANCE
WEST GERMANY
CANADA
SWEDEN
U.S.A.

1000 2000 3000 4000 5000

Annual Gross National Product, $ US per Capita

Reference 4

PER CAPITA ENERGY USE AND GNP

FIGURE 14.4

on a per capita basis for various nations.

14.2 ENERGY USE IN THE UNITED STATES

Figures 14.5 to 14.7 show US energy use on a total and per person basis. Figure 14.8 shows the estimated breakdown of energy use for 1968. For the eight-year period of 1961 to 1968, the annual growth rate in total energy use averaged 4.3 percent per year.[7]

Electricity in the US is generated by public and private utility systems operating generating stations powered by fossil fuels, hydropower, nuclear power, and geothermal power. The plants in operation today are summarized in Table 14.2. Figure 14.9 shows the historical and projected growth of installed electrical generating capacity in the US.

14.3 AN ENERGY CONSTRAINED WORLD

Industrialized man is distinguished from his predecessors and from all other forms of animal life by his intensive use of stored energy forms. Early man, in common with other animals, had a certain quantity of biochemical (metabolic) energy to spend on procuring and storing energy in the form of the nutritive value of food. To prevent exhaustion and death, the net energy gain during this process had to be positive. This pattern persists today in many parts of the world. For example, members of an African agropastoralist community in Angola exist by farming and cattle-raising. They "manage" solar energy through the use of plants and animals to meet their needs for food, shelter, and wealth. In this process, they use 20 to 30 GJ/person-year, derived almost exclusively from solar energy.[10]

In an industrialized country such as the US, total energy use has averaged more than 300 GJ/person-year. Per capita use ranges from one-third to one-half the national average to two or three times greater than the national average, or as high as 1000 GJ/person-year. On a per capita power basis, 300 GJ/person-year corresponds to 10 kW/person for total energy use, of which approximately 2.5 kWe is electrical capacity (1975 data). Since it takes an average of 2 kW installed to meet peak load requirements and to generate 1 kWe, 1.2 kW of the total of 10 kW are devoted to electricity generation.

Three limits are shown in Figure 14.4. The lower limit, 10 GJ/person-year, is the energy use which would be achievable using only the solar energy incident on land. This is based on a world population of 3×10^9 persons, and is obtained by taking the energy stored by the biosphere and correcting for that which would not be available to land-based herbivores.* The balance is $\sim 3 \times 10^{10}$ GJ/year, which gives 10 GJ/person-year.

The middle limit in Figure 14.4 shows what might be done if it were decided to use the world's supply of fossil fuels sparingly for the next 100 years while trying to develop alternative energy sources. Using a hypothetical stable world population of 4.5×10^9 persons, and assuming economically recoverable world resources of 45×10^{12} GJ of fossil fuels (coal, oil, gas), we find that global use of up to 100 GJ/person-year could be permitted for the next 100 years if someone would pay the bill.†

This "moderate" subsidy of exhaustible fuels would be, in fact, difficult to achieve in many parts of the world because fossil fuels are distributed unevenly and the economic cost for importing fuels would exceed the capacity of many nations. This can be seen by computing the cost of energy using the current price of crude oil, 3 $/GJ (∿15 $/bbl). This gives (3 $/GJ)($4.5 \times 10^9$ persons)(100 GJ/person-year) = $1350 billion/year, or an amount equal to the US Gross National Product in 1973. An energy bill of this magnitude is untenable.

---

*This is on the optimistic side since it neglects the use by nonhuman herbivores.

†Estimates for "eventually recoverable" fossil fuels range from $\sim 23 \times 10^{12}$ GJ to $\sim 230 \times 10^{12}$ GJ. (Reference 11)

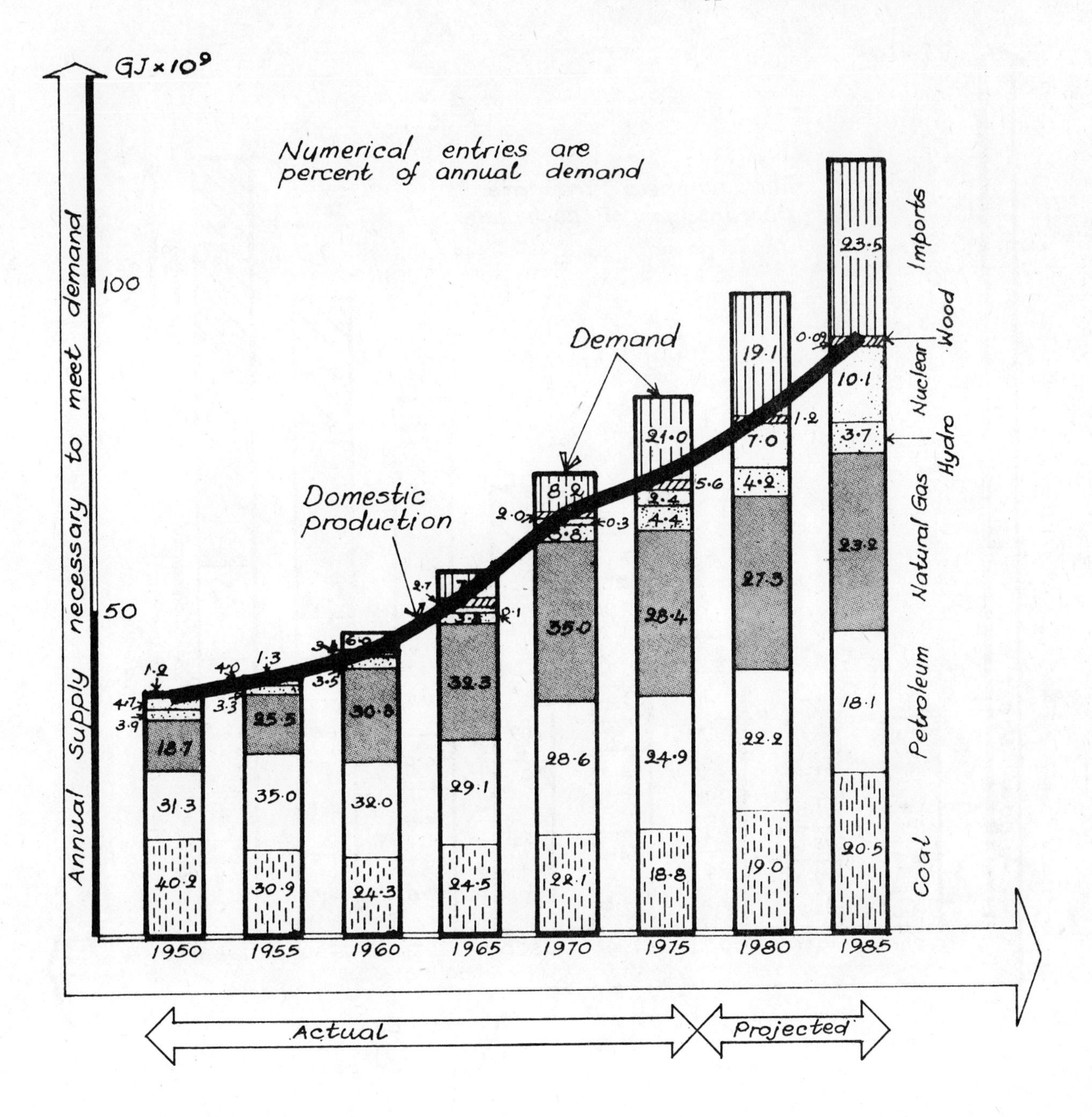

Reference 5

SOURCES OF U S ENERGY SUPPLY

FIGURE 14.5

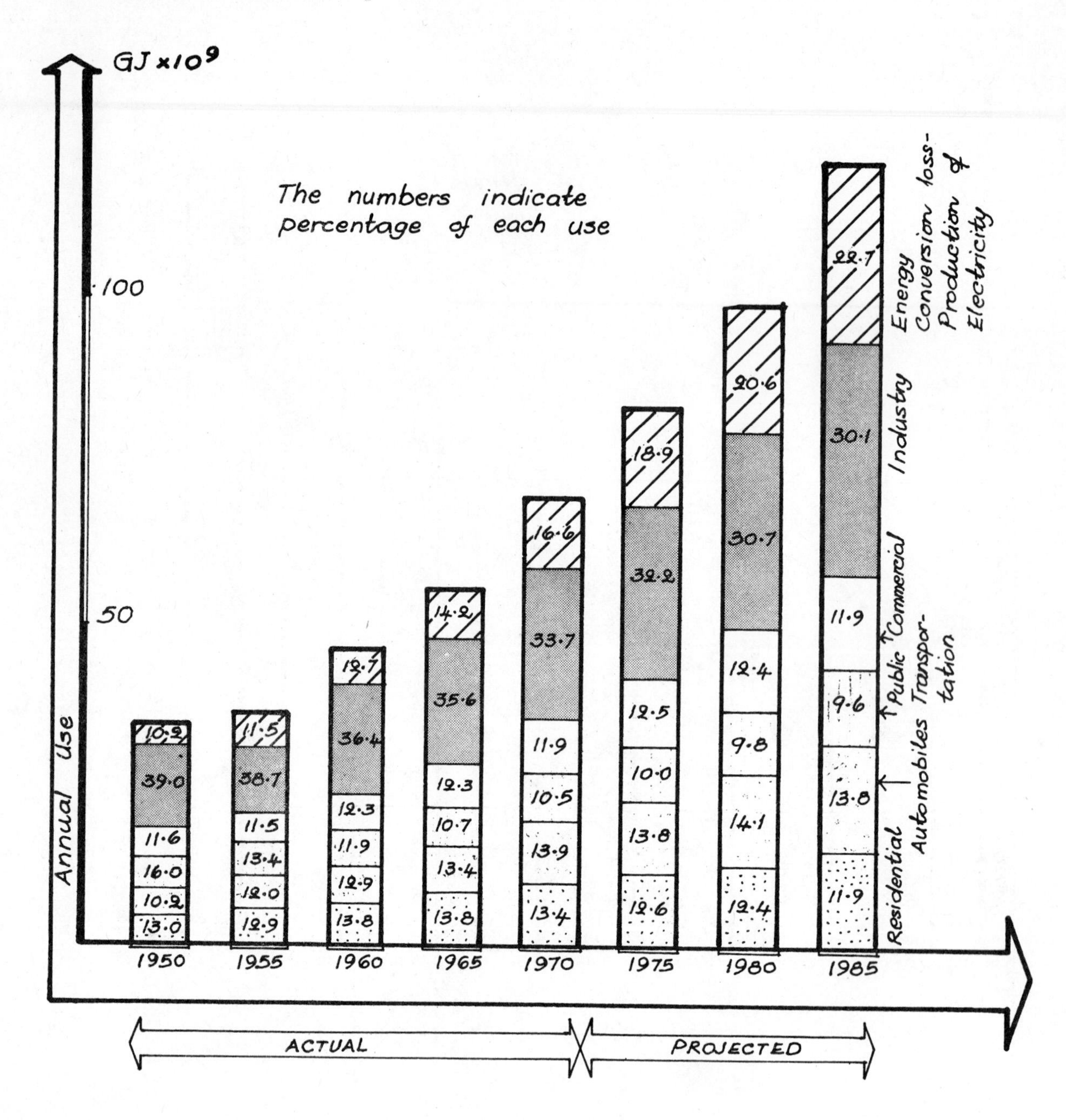

Reference 5

ANNUAL USE OF ENERGY IN THE US

FIGURE 14.6

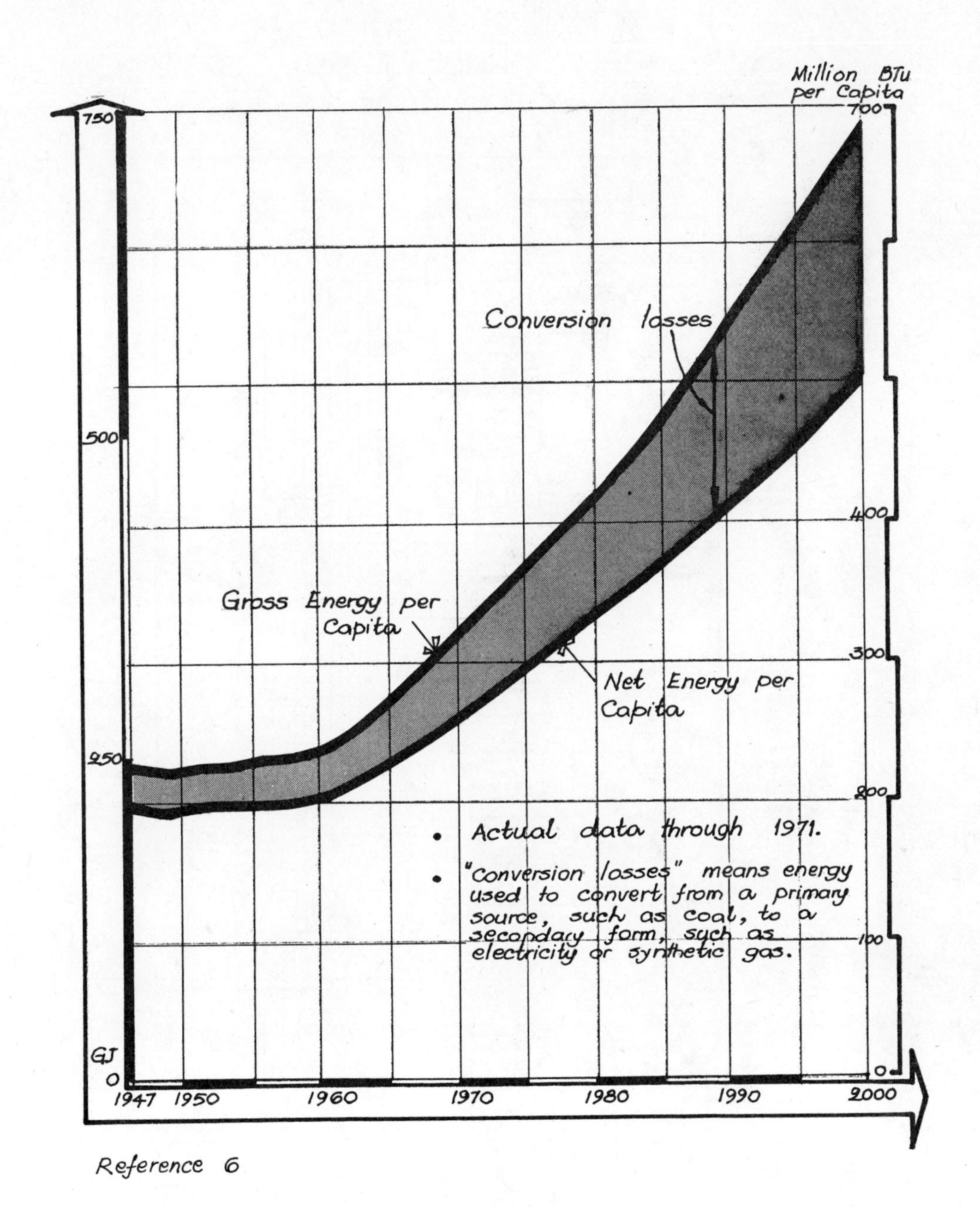

Reference 6

US NET AND GROSS ENERGY INPUTS PER CAPITA **FIGURE 14.7**

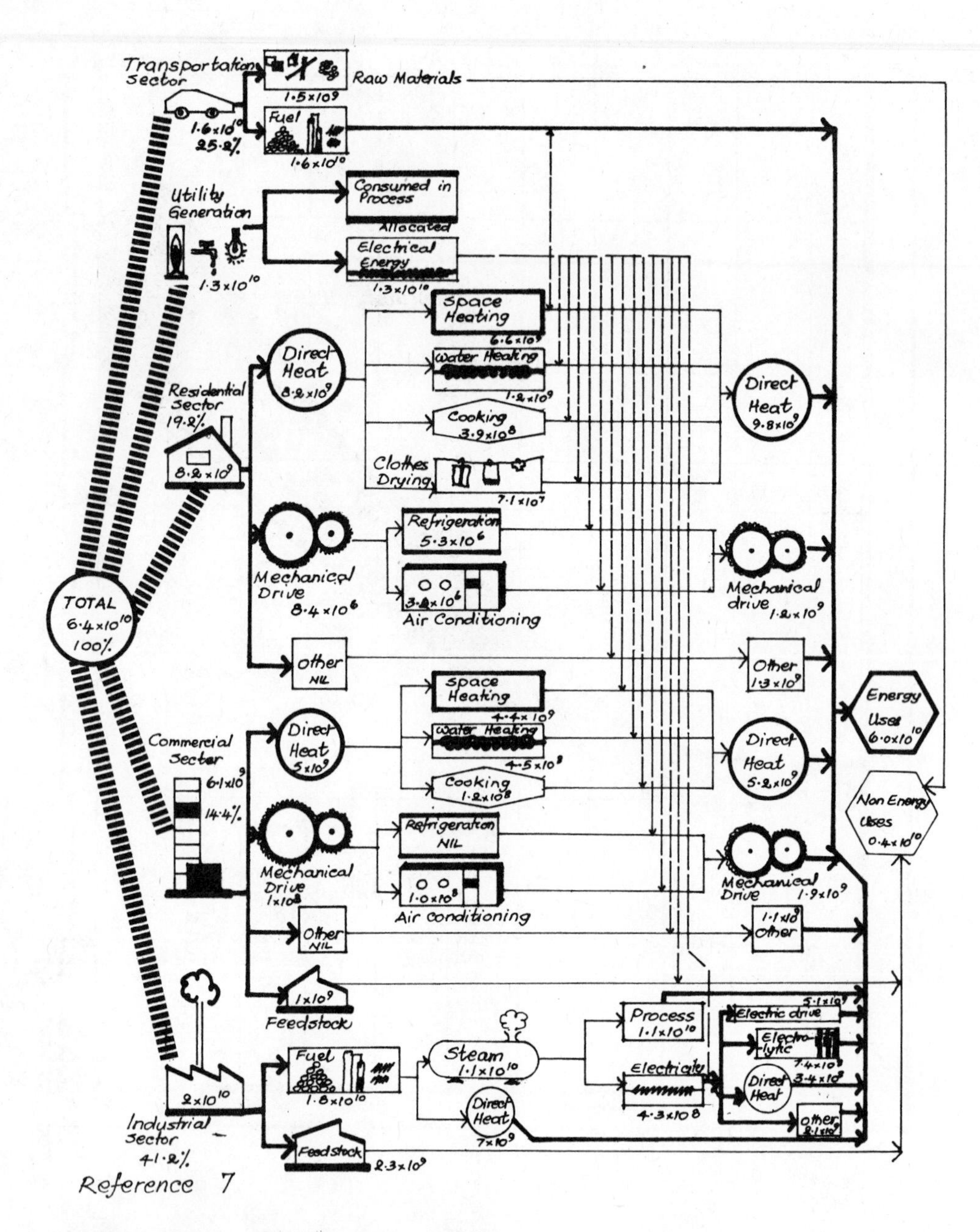

1968 US ENERGY USE ALL FUELS (in GJ)

FIGURE 14.8

Item	1950	1955	1960	1965	1970	1975
Production of Electricity (10^9 kWh)						
Industrial Plants(1)	60	82	88	102	108	85
Electric Utility (for Public Use):						
Privately owned(2)	267	421	579	809	1,183	1,486
Publicly owned(2)	62	126	175	246	348	429
Total Production (10^9 kW-hr)	389	629	842	1,157	1,639	2,000
Source of Energy (%)						
Coal	{47.1	55.1	53.6	54.5	46.2	44.7
Nuclear					1.4	8.9
Oil	10.3	6.8	6.1	6.1	11.9	15.1
Gas	13.5	17.4	21.0	21.0	24.3	15.7
Hydro	29.2	20.7	19.3	18.4	16.2	15.8
Total (%)	100.0	100.0	100.0	100.0	100.0	100.0
Installed Capacity (10^6 kW)						
Industrial Plants(1)	14	16	18	18	19	19
Electricity Utility (for Public Use):						
Privately owned(2)	55	87	128	178	263	398
Publically owned(2)	14	28	40	59	78	108
Total Installed Capacity (10^6 kW)	83	131	186	255	360	525

Notes: (1) Plants of 100 kilowatts and over, including stationary power plants of railroads.

(2) Noncentral stations included only in total prior to 1955; distributed to other publicly owned classes thereafter.

ELECTRIC ENERGY PRODUCTION AND INSTALLED GENERATING CAPACITY, 1950 - 1975

TABLE 14.2

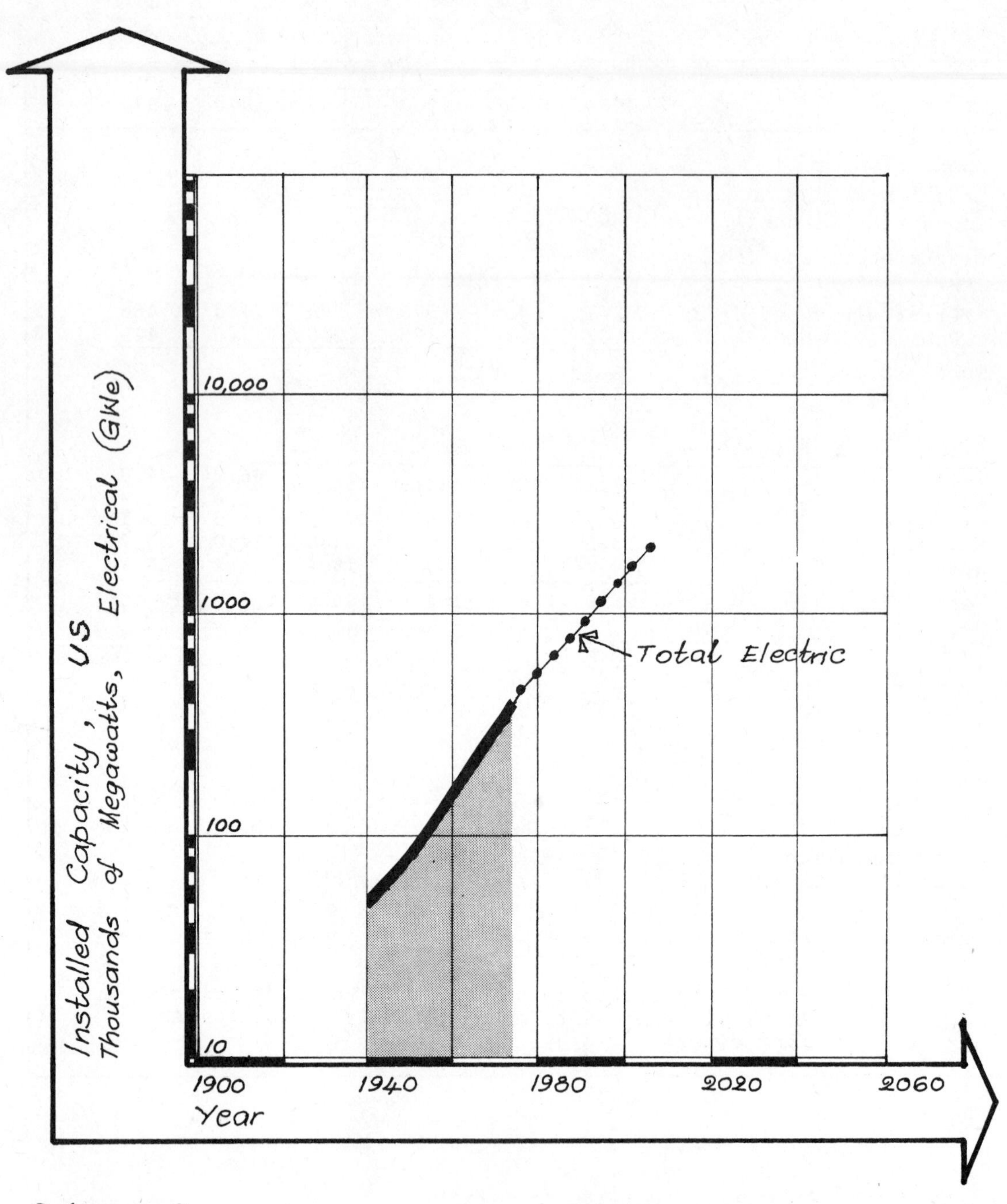

Reference 9

PROJECTED ELECTRICAL GENERATING CAPACITY-US

FIGURE 14.9

If the imported fuels were coal and uranium, rather than 15 $/barrel oil, the resulting cost might be lower. However, further price increases could worsen the situation.

Furthermore, energy use above 100 GJ/person-year, such as the 300 GJ/person-year upper limit shown in Figure 14.4 implies: 1) reducing the time span of dependence on low-cost fossil fuels from 100 years to 30; 2) anticipating even greater expenditures for energy; or 3) development of alternative low-cost energy sources. History would seem to indicate that the third alternative could not be accomplished in less than 30 to 50 years. Thus, the harsh economic fact is that energy use at the scale currently practiced by the US is unlikely to be possible for much of the world in the near-term.

There are other limiting factors besides economics, making it inevitable that the world of tomorrow will consist of many energy constrained societies. These constraints include:

- *a biological limit*, determined by the level of energy use which ceases to be biologically tolerable;
- *an economic limit*, which depends on national wealth, energy prices, technological know-how, and national resources;
- *a self-sufficiency limit*, which depends on the availability of domestic versus imported energy forms;
- *a socio-cultural limit*, which depends on the social tolerance of energy use and the social costs which must be paid in exchange for a given level of energy use; and,
- *a substistence limit*, which is based on the minimum energy input necessary to feed the population and provide for minimum amenities of life.

The *biological limit* will depend on one of several factors such as heat disposal, water availability, waste disposal, pollutants, or possible effects on weather. To illustrate the concept, suppose that the ultimate limit on energy use depends on biological tolerance to heat. Thus, global heat production which approached the natural heat balance would be viewed with concern. Below this value, effects less significant than those associated with diurnal and seasonal solar variations would be expected.

The average solar power density on the surface of the earth is 100 W/m^2.[11] Man-made power densities range from less than 0.1 W/m^2 (global average) to 1 W/m^2 (Federal Republic of Germany) to 17 W/m^2 (industrial park), up to 1000 W/m^2 or greater for an electric power producing area.[11]

Clearly, on a global basis, the world is far from the solar power density of 100 W/m^2. On a local basis, however, population density is such that the limit is being approached today and problems related to heat disposal may be predicted. For example, consider New York (Manhattan) where the population density is ∿30,000 persons/km^2. This gives:

$$(30{,}000 \text{ persons/km}^2)(10^4 \text{ W/person}) = 300 \text{ W/m}^2$$

which exceeds the natural solar surface power density.

On the basis of these calculations, one would predict energy problems for New York during the summer when the solar heat load is greatest. Note the "vicious circle" which could and has resulted: rising temperature causes greater use of air conditioning which leads to greater energy dissipation and hence further temperature increases. Thus, in areas of high population density, heat rejection will be a concern although some other factor could turn out to be the limiting one. The point is that energy use cannot increase indefinitely.

To understand the concept of an *economic limit*, consider current (1977) US energy prices:

Food	250-500	$/GJ
Electricity	5-30	$/GJ
Gasoline	4-8	$/GJ
Natural Gas	2-4	$/GJ
Crude Oil	2-4	$/GJ

These prices could double or triple

in the next several years to several decades due to scarcity, political confrontation, and increased difficulty in resource development and exploitation. Current prices in Europe for some of these energy forms are already roughly twice the US values, indicating the direction US prices will eventually move. The effect of energy prices alone on the US standard of living (through impact on exports, food prices, balance of payments, and loss of jobs) is potentially of great significance, neglecting other important considerations such as energy availability and national security.

It would be difficult to contemplate any nation paying more than 100 $/GJ for energy in any form other than food. This value will be used as a hypothetical upper limit on energy cost for the ensuing discussion. Note that this constraint means that electricity would be about ten times as expensive as it is now, costing about as much (per unit of energy delivered) as food products derived from cereal grains.

The *self-sufficiency limit* presumes that each nation will increasingly prefer to meet the majority of its energy needs with its share of incident solar energy plus domestic (rather than imported) fuels. An attempt would be made to minimize imports to essentials. This constraint would vary from country to country, depending upon available supplies.

In the United States, for example, domestic fuel resources are estimated as $20\text{-}40 \times 10^{12}$ GJ of fossil fuels (oil, gas, and coal) and $10\text{-}20 \times 10^{12}$ GJ of nuclear fuels (used in thermal and fast breeder reactors).[12] Assuming an ultimate population of 400 million persons and a 100-year strategy to conserve these resources while alternative energy sources (solar, fusion, or whatever) may be developed, we obtain for the 100-year period:

Only oil and gas:

$$\frac{1\text{-}2 \times 10^{12} \text{ GJ}}{400 \cdot 10^6 \text{ persons} \cdot 100 \text{ years}} =$$

25-50 GJ/person-year

Oil, gas, coal, and thermal reactors:

$$\frac{20\text{-}40 \times 10^{12} \text{ GJ}}{400 \cdot 10^6 \text{ persons} \cdot 100 \text{ years}} =$$

500-1000 GJ/person-year

If we permit extensive use of coal and fast breeder reactors, we have:

All fuels:

Oil, gas, coal, and thermal plus fast breeder reactors

$$\frac{30\text{-}60 \times 10^{12} \text{ GJ}}{400 \cdot 10^6 \text{ persons} \cdot 100 \text{ years}} =$$

750-1500 GJ/person-year

Thus, if the US continues to depend primarily on oil, gas, and coal, imported fuels will be necessary to avoid exhausting economically recoverable resources at an early date. If the US is to provide fuels to sustain current levels of energy use for a period of time sufficiently long enough to develop alternative energy sources, then extensive use must be made of domestic coal and fast breeder reactors, or massive fuel imports must be obtained. Note that consideration of solar, geothermal, wind, or other advanced power sources is academic since a time scale of at least 50 years would be required to build a significant number of power stations.* However, these sources may be useful in certain limited applications.

---

*This is amply demonstrated by the nation's nuclear energy program. The first nuclear reactor was built in 1941-42 and a massive development program has gone on since then. Yet by 1975, 35 years later, nuclear power accounted for less than 9 percent of electricity supply and only about 2 percent of total US energy use. Only by the year 2000--60 years after the first nuclear power plant--is it conceivable that nuclear power could provide 25 to 50 percent of the nation's total electricity needs.

The *socio-cultural limit* depends on the availability of land for power plant siting, alternative land uses, the esthetics of power plants, transmission lines, and right-of-ways, and on other social compromises associated with energy use. Clearly there are conflicting opinions over the priorities for open beaches versus inexpensive electricity, or for the use of domestic coal fired power plants and the higher level of pollutant emissions versus the use of imported low sulfur oil, and so on. Ultimately, the availability of land and water will become the limiting factor since increasing energy use must compete for land and water against an increasing population and a growing need to provide greater quantities of food. In the western United States, projected needs for power plant cooling water begin to approach a sizeable percentage of the annual run-off, while public pressure to move the plants inland and deny them the available ocean water creates an even greater demand for potable water resources, even though power generation and other water uses are not necessarily mutually exclusive.

There is considerable evidence of the upper limit on the social acceptance of energy systems. Clearly the population does not want a society where condominiums are sited on transmission line right-of-ways between every other tower, or where utility poles replace trees. In practice this constraint already has been reflected in the lengthening of the licensing process for new power plants, in public criticism of plant designs and sites, and in rising costs for sites.

Finally, there is the *subsistence limit*, or that energy use which must exist in any society to permit survival. This lower limit varies with climate and food supply but can be estimated as 2000 Kcal/person-day of food inputs plus 1000 Kcal/person-day of other energy inputs for shelter, clothing, and other needs. This gives:

$$(3000 \text{ Kcal/person-day})(4.19 \times 10^3 \text{ J/Kcal})(365 \text{ days/year})(10^{-9}) = 4.6 \text{ GJ/person-year.}$$

These concepts are shown schematically in Figure 14.10 for the hypothetical fiefdom of Grand Vinland which, for a number of years, enjoyed abundant supplies of energy but recently has encountered more difficult circumstances.

14.4 THE BENEFITS OF IMPROVED EFFICIENCY

Given the constraints discussed above, methods for increasing energy use efficiency are urgently needed. To establish priorities, note that about 28 percent of US energy use in 1975 was as electricity and this share is increasing. The greatest user of electricity in the US in 1975 was the industrial sector, which accounted for roughly 42 percent of the electricity used in the US and required approximately 12 percent of the total fuels used in the US.[13] The most significant industrial electricity use is electric drives, while electrolytic processes rank second in importance. Following industry, the next significant category is the residential sector in which about 10 percent of the nation's fuel use is in the form of electricity. Major uses within the residential sector are refrigeration, air conditioning, water heating, space heating, and clothes drying. Air conditioning and refrigeration are also important in the commercial sector, which uses about 5 percent of US fuel in the form of electricity.

A 15 percent overall improvement in electrical use efficiency would result in saving 250×10^9 kWh or the equivalent of nearly 60 1000 MWe power plants, based on 1975 US statistics (see Table 14.2). If historical growth rates continued and there were no efforts to conserve fuels or to improve energy use efficiency, the US installed generating capacity would be expected to exceed the equivalent of 2000 1000 MWe plants by the year 2000. Then, a 15 percent increase in use efficiency could reduce the projected number of 1000 MWe equivalent generating stations by between 300 and 400. The savings in capital cost alone would be 150 to 200×10^9 \$; savings in fuels would be of the same order of magnitude over the lifetime of these plants.

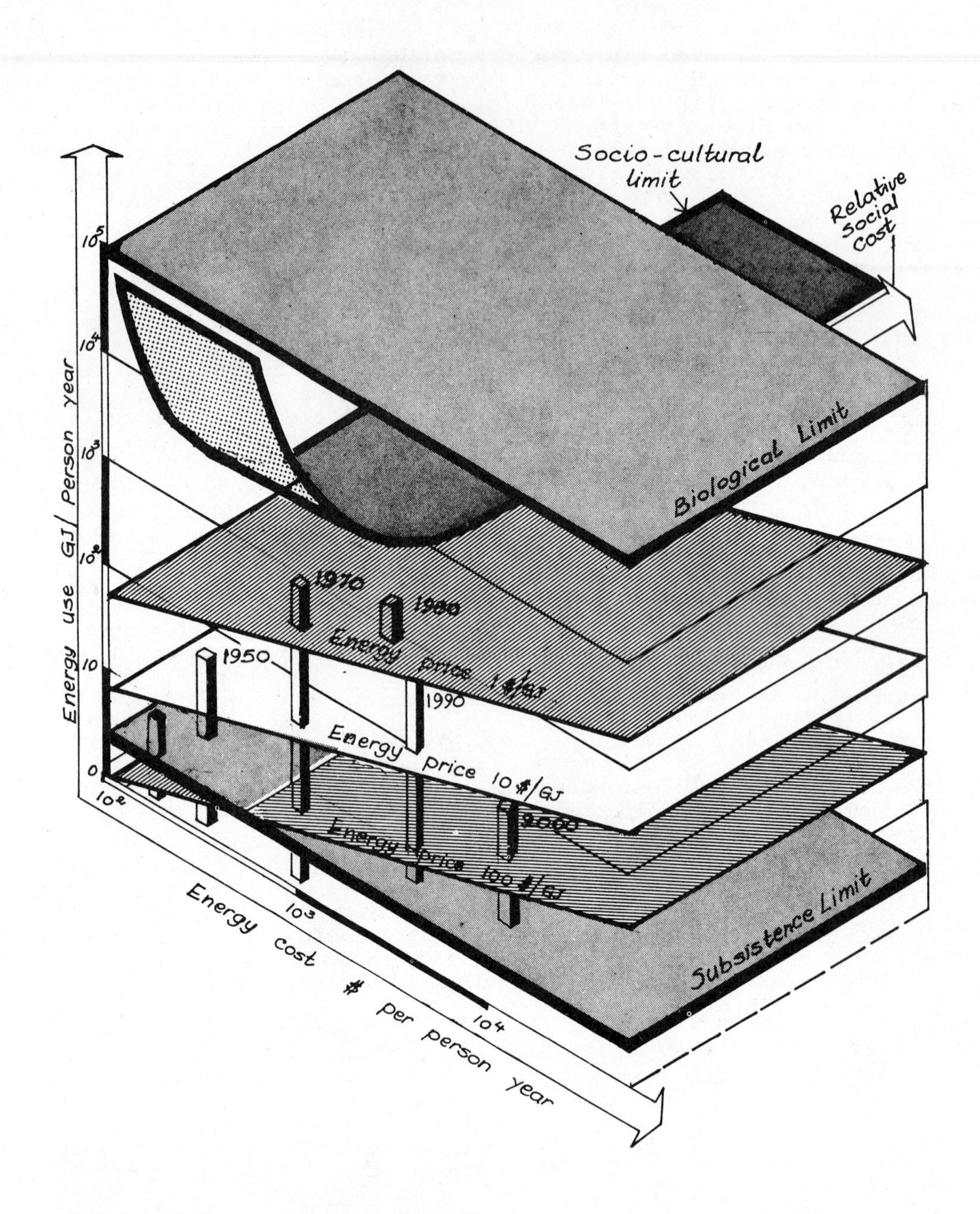

ENERGY CONSTRAINTS IN THE FIEFDOM OF GRAND VINLAND, 1900-2000

FIGURE 14.10

Based on the findings in this book (see Chapter 1), total energy and electricity savings as high as 25 to 30 percent (compared to historical extrapolations) appear to be technically and economically feasible. The incentive for improving electrical energy use efficiency is quite large when considered on a national basis.

* * *

REFERENCES

1. Hubbert, M. King, *Energy Resources*, Publication No. 1000-10, (Washington, D.C.: National Academy of Sciences - National Research Council, 1962).

2. Odum, Eugene P., *Fundamentals of Ecology*, 3rd edition, (Philadelphia, Pennsylvania: W.B. Saunders Co., 1971).

3. Based on Figure 3 in Starr, C. and Smith, C.B., *Energy and the World of 2000 AD*, (Los Angeles: University of California, Department of Engineering, September 1967).

4. US Atomic Energy Commission, *News Release*, 2(Week Ending 24 November 1971).

5. Citizens' Advisory Committee on Environmental Quality, *Citizen Action Guide to Energy Conservation*, (Washington, D.C.: n.d.).

6. Dupree, Walter G., Jr. and West, James A., *United States Energy Through the Year 2000*, (Washington, D.C.: US Department of the Interior, December 1972).

7. Stanford Research Institute, *Patterns of Energy Consumption in the United States*, Report prepared for the Office of Science and Technology, (Menlo Park, California: January 1972).

8. US Department of Commerce, Social and Economic Statistics Administration, Bureau of the Census *Statistical Abstract of the United States*, 1976, (Washington, D.C.: US Government Printing Office, 1977).

9. Szego, G.C., *The U.S. Energy Problem*, Vol. 1: *Summary;* Vol. 2: *Appendices--Part A;* Vol. 3: *Appendices--Part B*, (Warrenton, Virginia: Intertechnology Corporation, November 1971).

10. Carvalho, Eduardo Cruz de and Silva, Jorge Vieira da, "The Cunene Region: Ecological Analysis of an African Agropastoral System," in *Social Change in Angola*, ed: Franz-Wilhelm Heimer (Munich: Weltforum Ferlag, 1973), pp. 145-192.

11. International Institute for Applied Systems Analysis, ed., *Proceedings of IIASA Planning Conference on Energy Systems*, Report No. IIASA-PC-3, (Laxenburg, Austria: 17-20 July 1973).

12. Compiled from various sources. See, for example, Hubbert, passim; Idem, "Energy Resources," in *Resources and Man*, (San Francisco: Freeman, 1969), Chapter 8; Dupree and West, passim; Gillette, Robert, "Oil and Gas Resources: Academy Calls USGS Math 'Misleading'," *Science* 187(28 February 1975): 723-727; US Atomic Energy Commission, Division of Production and Materials Management, *Nuclear Fuel Resources and Requirements*, Report No. WASH-1243, (Washington, D.C.: US Government Printing Office, April 1973); and Idem, *Nuclear Fuel Supply*, Report No. WASH-1242, (Washington, D.C.: US Government Printing Office, May 1973).

13. US Department of Interior, *Energy Perspectives 2*, (Washington, D.C.: US Government Printing Office, 1976).

Chapter 15

CASE STUDIES

CHAPTER CONTENTS

KEY WORDS

Case Studies
Efficient Energy Use
Energy Conscious Design
Life Cycle Costing

SUMMARY

More than 50 case studies have been included in Chapter 15 to illustrate the methods described in this book. The case studies demonstrate the implementation, economic savings, and in many cases, environmental benefits of the suggested energy saving techniques. Examples include single family dwellings, apartments, commercial buildings, dairies, farms, and large industrial plants. Savings vary, ranging from a few percent to more than 50 percent, with paybacks ranging from immediate to three to five years.

15.1 INTRODUCTION

It is one thing to talk about saving energy, and another thing to do it. Yet, throughout the United States and other nations there is a growing recognition that energy is a valuable resource which must be used effectively. Efforts to improve efficiency and reduce waste have been stimulated not only by escalating energy costs, but also by concern over the future availability of electricity and fuel.

The authors of this book were asked to submit case studies describing actual applications of some of the important ideas discussed in their chapters. Examples were taken from published literature, industry sources, personal experience, or in some cases were provided by consultants or reviewers.

Obviously a case study is an example of a *particular* solution to a specific problem and may not be applicable directly to another situation. For example, heat recovery is a concept with a wide variety of potential applications. However, the specific heat recovery system utilized for one building may not be suited to another building having different construction, occupancy, climatic conditions, etc.

Yet, the plethora of energy uses which confront the engineer or architect is staggering; it would take ten books this size to begin to discuss in any detail the energy management techniques and their applications specific to one industrial classification, e.g., organic chemical production. The advantage of case studies is that they inform by example; what has been done in one application can be done in another, even if some modifications to the approach are required.

To keep these examples to a reasonable size, they have purposely been greatly condensed. Refer to the references in each case study for additional details. Case studies are grouped in sections according to the relevant chapter; the reader seeking case studies related to residences (Chapter 4) would look under Section 15.4 and so on. The chapter concludes with a list of case studies.

15.2 INDUSTRY (Chapter 2)

CASE STUDY 2-1: AVAILABLE WORK AND OXYGEN SEPARATION[1]

THE CONCEPT

Available work provides a method for evaluating and optimizing processes where the efficient use of energy is critical.

BACKGROUND

The production of liquid oxygen--by the compression and subsequent fractionation of air--is an energy-intensive process. The available work of oxygen and residual gases (primarily nitrogen) separated from 454 gram-mole (1 lb-mole) of air at one standard atmosphere and a temperature of 27°C (80°F) is 588 kJ (557 Btu). This is the theoretical minimum work required under the conditions specified above to produce the two gases from air.

In a practical system much more energy is required to produce oxygen. Figure 1 is a flow sheet of one such system. This plant produces 345 mt (380 ton) of oxygen per day with a recovery of 96 percent.

Inlet air is compressed by a blower to a pressure of 20.7 psi and forced through the heat exchangers and fractionating towers of the plant. In the main heat exchanger, the incoming air is cooled by the outgoing oxygen and nitrogen products. In the tower, the air is separated into oxygen and nitrogen fractions and then discharged. The nitrogen is also used to subcool tower reflux and is ultimately discharged at atmospheric pressure.

FINDINGS

An analysis of this process was made using the concept of available work. This was done by determining the work required at each major step in the process and by evaluating the entropy production of each element of the process. The main irreversible changes which contribute to the process entropy are the main exchanger, the nitrogen compressors, the nitrogen coolers and the fractionating tower.

To improve the process, it is necessary to determine the extent to which the entropy production is characteristic of the process and the extent to which it is a consequence of equipment inefficiencies which could be improved. If this is done, the idealized process shown in Figure 2 results. It is clear from a comparison of Figures 1 and 2 that the greatest potential for the process improvement is in the main heat exchanger.

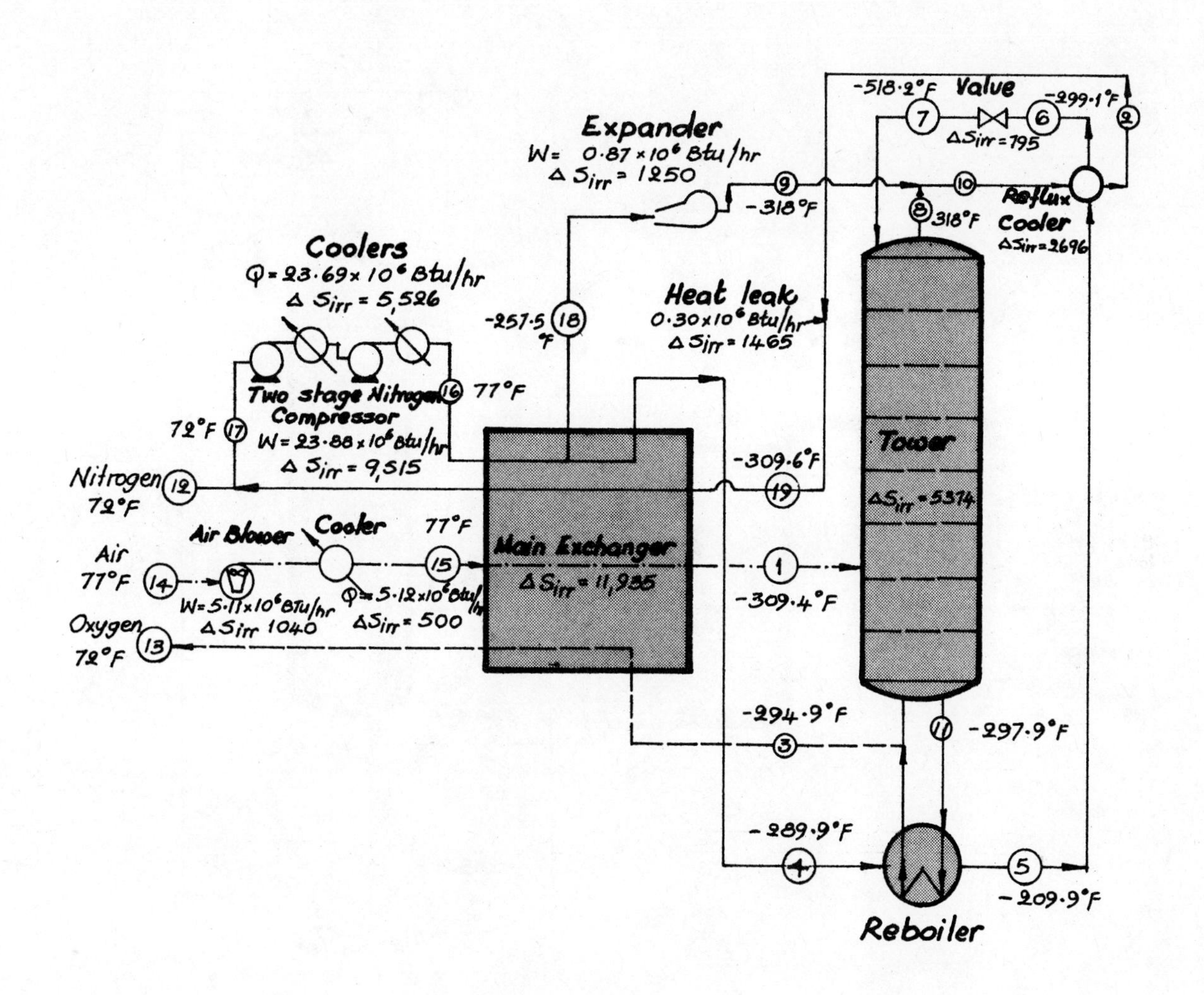

FIGURE 1

PRACTICAL OXYGEN SEPARATION PROCESS

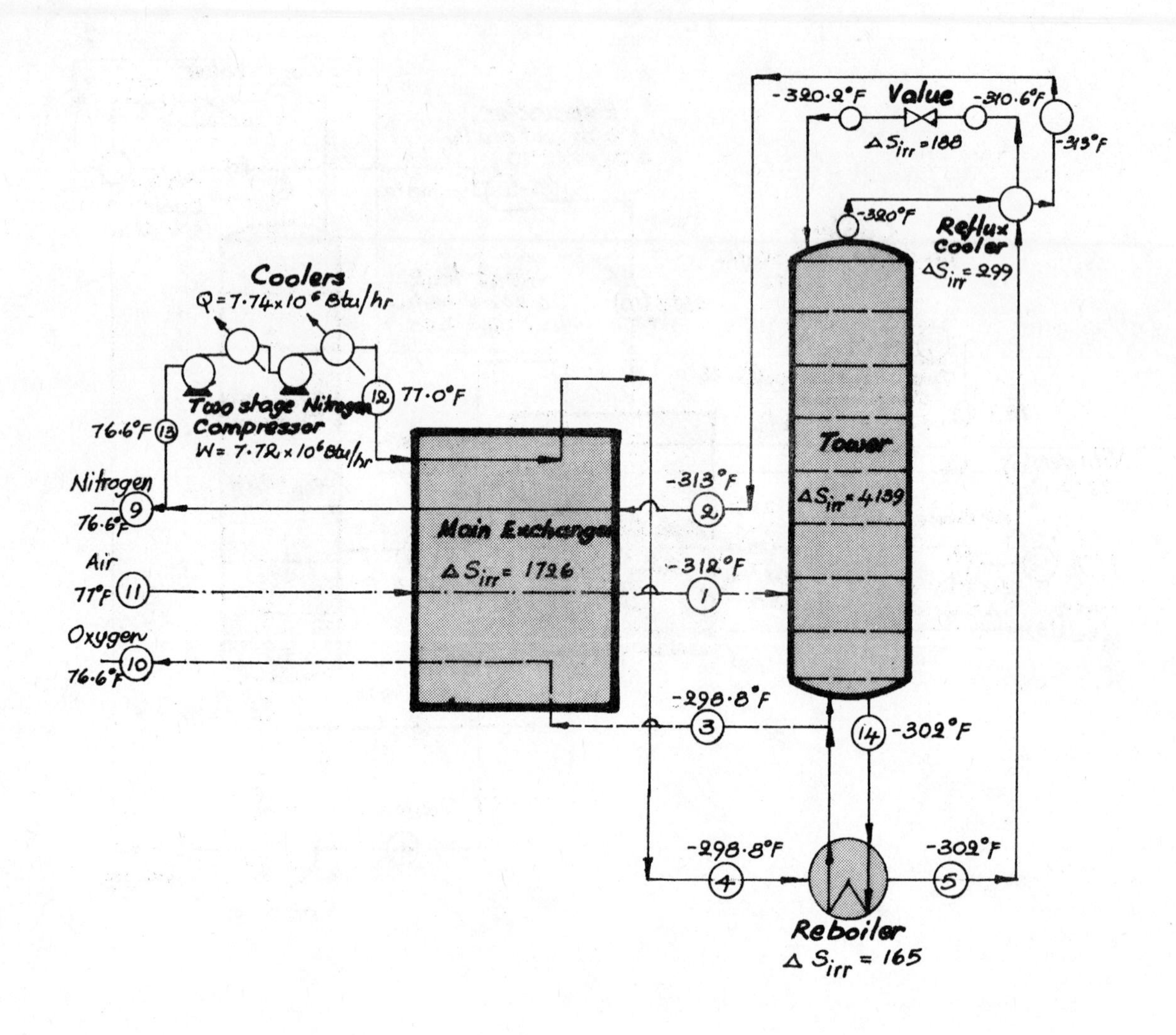

FIGURE 2

IDEAL OXYGEN SEPARATION PROCESS

If the cost versus performance characteristics of the heat exchanger are analyzed, the system performance can be optimized. Parameters of importance include operating cost, capital cost, pressure drop, temperatures, and heat exchange surface. Parametric studies showing the relationship between these variables and entropy production indicate both how efficiency can be improved and what the cost impact will be.

RECOMMENDATIONS

The use of the concept of available work provides a useful tool for evaluating energy-intensive processes. Although the theoretical efficiencies may not be significant in a practical sense, they provide a convenient basis for comparing alternative processes and equipment.

References

1. Gyftopoulos, Elias P.; Lazaridis, Lazaros J.; and Widmer, Thomas F., *Potential Fuel Effectiveness in Industry,* A Report to the Energy Policy Project of the Ford Foundation, (Cambridge, Massachusetts: Ballinger Publishing Company, 1974).

CASE STUDY 2-2: ENERGY AUDIT IN AN INTEGRATED STEEL PLANT[1]

THE CONCEPT

An *energy audit* is a valuable tool for identifying equipment or processes which are major energy users. A systematic approach is important for industrial plants which generally involve complex interrelated operations.

BACKGROUND

This case study is based on an integrated steel plant. Material and energy balances were made so that areas of intensive energy use could be identified. The results are shown in Figure 1 and Table 1.

FINDINGS

When considering energy use in iron and steel production, the blast furnace and the rolling operations should be given top priority, since they require the bulk of the energy. If efficiency is improved, a 5 to 25 percent energy reduction on most operations could be achieved by modifying existing equipment and improving operating procedures.

The Blast Furnace

Due to the extreme complexity of the blast furnace, it is difficult to propose energy saving methods which reduce specific energy use. However, it is generally true that substituting a richer ore in the furnace will lower energy requirements. Such was the case for the UK steel industry which was able to achieve a 37 percent decrease in specific energy requirements. The considerable reduction from 49 MJ/kg (42 MBtu/ton) of ingot steel in 1950 to 31 MJ/kg (27 MBtu/ton) of ingot steel in 1964 can be explained principally by the gradual switch from home to rich foreign iron ore.

Soaking Pits

The greatest potential for energy savings in the pit area is from improved timing and movement of the ingots. Soaking pit operators need to know well in advance when and where the ingots should be moved so as to limit the exposure of the ingots to ambient conditions, reduce the charging of cold ingots into the pits, and maintain proper arrangement of the ingots to achieve a good heat circulation. If programmed practices such as these are put into operation, energy savings of up to 10 percent could be achieved.

Reheating Furnaces

Reheating furnaces increase the malleability of the steel after it has cooled from previous rolling. For a typical furnace, energy savings can be realized through improved insulation coverage outside the furnace and improved temperature control within the furnace. Present practice is to completely reinsulate the furnace every 15 to 18 months. Toward the end of

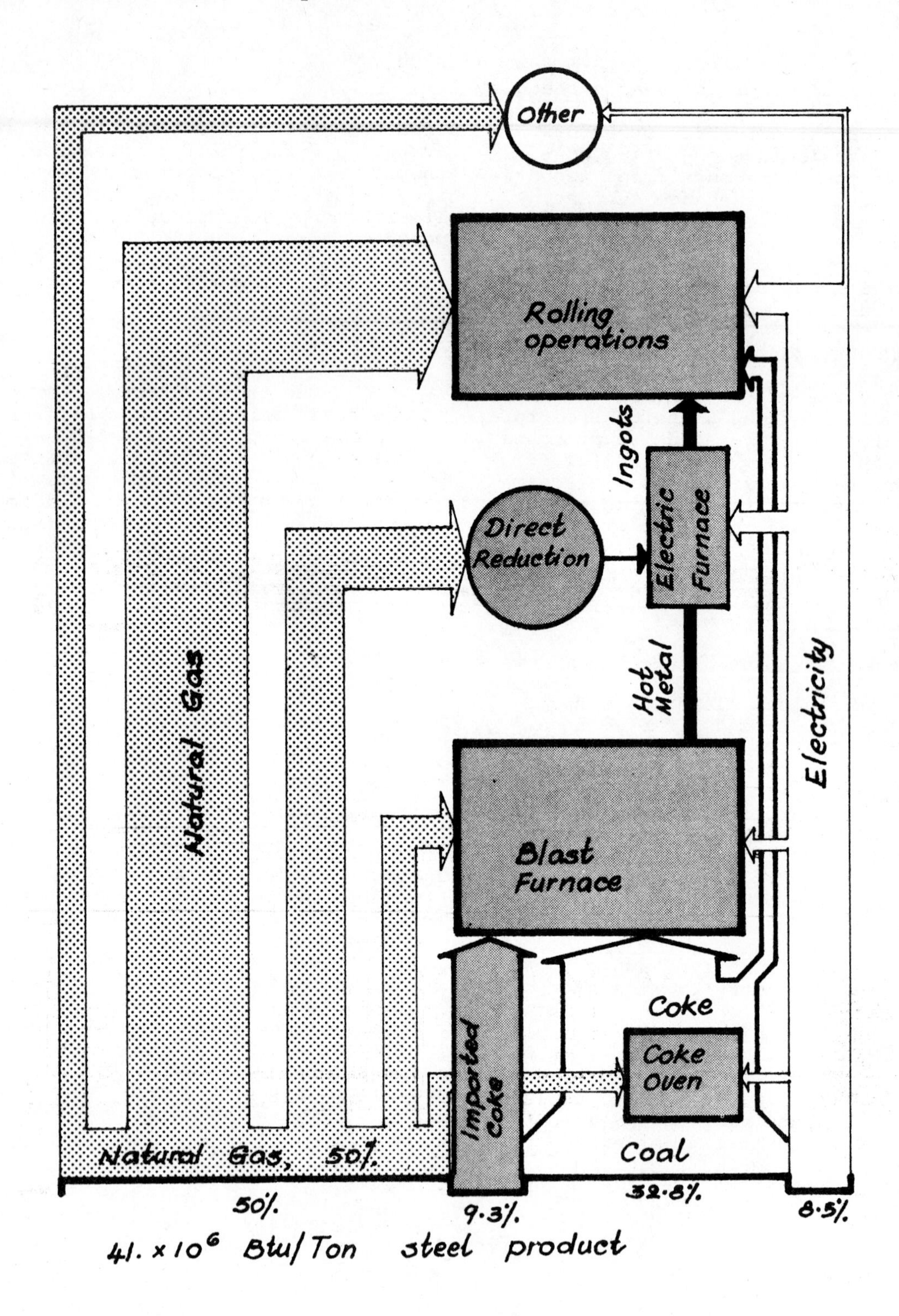

FIGURE 1

ENERGY DISTRIBUTION

Operation	Materials/ton of Product				Energy Used--MBtu/ton of Product					Net Energy Used*
	Consumed	lb	Produced	lb	Natural Gas	Elec-tricity	Coal	Coke	Coke Gas	MBtu/ton of Product
Coke Oven	Coal	1060	Coke	650	0.6	0.05	13.2	(8.70)	(1.0)	4.15
	Coke Gas	57	Coke Gas	147						
Blast Furnace, Including Stoves	Coke	926	Iron	1700	0.8	0.05		12.50		13.35
	Flux	140								
	Ore and Agglomerates	2000								
Direct Reduction	Ore	1300	Iron	720	6.0					6.00
Electric Furnaces	Iron	2600	Ingots	2600		1.73				1.73
	Scrap	1200								
	Flux	75								
	Oxygen	161								
Rolling	Steel Ingot	2600	Finished Product	2000						
Operations	Coke Gas	90								
• Soaking					2.5					2.50
• Reheating					9.0				1.0	10.00
• Electric Drives						1.30				1.30
Other	Water	27000		0	1.6	0.37				1.97
TOTALS					20.5	3.50	13.2	3.80	0.0	41.00

*Note: Multiply MBtu/ton by 1.162 to get MJ/kg.

TABLE 1

MATERIALS & ENERGY USAGE IN PROCESS

this period a substantial amount of insulation is lost, resulting in an overall average coverage of 50 percent. With maintenance every 6 months the overall average could be increased to about 80 percent coverage.

An optimal criterion for the reheat furnace is for the steel to attain the desired temperature just prior to exiting ("good control"). If the loading of the furnace is reduced or if the temperature profile inside the furnace is high enough, the steel will reach the desired temperature prematurely ("poor control"). As the furnace loading approaches 80 percent, the unfavorable effects of poor temperature control are minimized. Thus, if nothing more is carried out than operating the reheat furnace at 80 percent insulation coverage and 80 percent design capacity, energy savings of 10-20 percent are feasible.

Recuperators

Recuperators are continuous heat exchangers commonly used in the steel industry to preheat combustion air. In the past few years these units have been installed on new combustion equipment. As a result, energy savings of 15-20 percent have been achieved. Generally older furnaces do not utilize recuperators because of high initial cost; however, based on current and projected fuel costs, economic calculations indicate the feasibility of installing recuperators.

If these measures were applied to a typical steel plant, it is estimated that savings in specific energy (MJ/kg) of 11 percent could be achieved.

RECOMMENDATIONS

Energy-intensive industrial processes should be reviewed to establish efficiency goals and identify areas where significant improvements are possible.

References

1. Prengle, H. William Jr.; Crump, Joseph R.; Fang, C.S.; Grupa, M.; Henley, D.; and Wooley, T.; *Potential for Energy Conservation in Industrial Operations in Texas*, Report No. S/D-10, (Houston, Texas: University of Houston, Department of Chemical Engineering, Cullen College of Engineering, November 1974).

CASE STUDY 2-3: A COMPUTER MODEL OF ENERGY USE IN A STEEL PLANT[1]

THE CONCEPT

To develop a computerized model of energy flow in a large steel plant.

BACKGROUND

There is actually little flexibility for large energy users in conversions between oil, gas, and electricity since a shortage of one affects the availability of the others. For the individual firm, the complexity of establishing management policies in the face of rising prices, declining energy availability, and rapidly changing government regulations is almost overwhelming. The possibility that energy costs might double is "indeed sobering to the steel industry." To cope with these problems, one steel company decided to develop a computerized model of a steel plant which could handle a large number of variables at one time and could provide answers in a short period of time.

Figure 1 is a schematic diagram of the energy flow at the steel plant, which includes 8 blast furnaces, 6 coke batteries, 4 steel making shops (open hearth, electric, and basic oxygen furnaces), billet and slab casting, 40 different rolling mills, and 3 electrical power generating plants.

FINDINGS

The plant was modeled by developing linear equations relating energy inputs, mass flows, and product output. Operational constraints (fuel types, fuel substitution, or unit capacities) were reflected in the equations. The equations were developed from operational records, historical data, theoretical relationships, and regression analysis. These

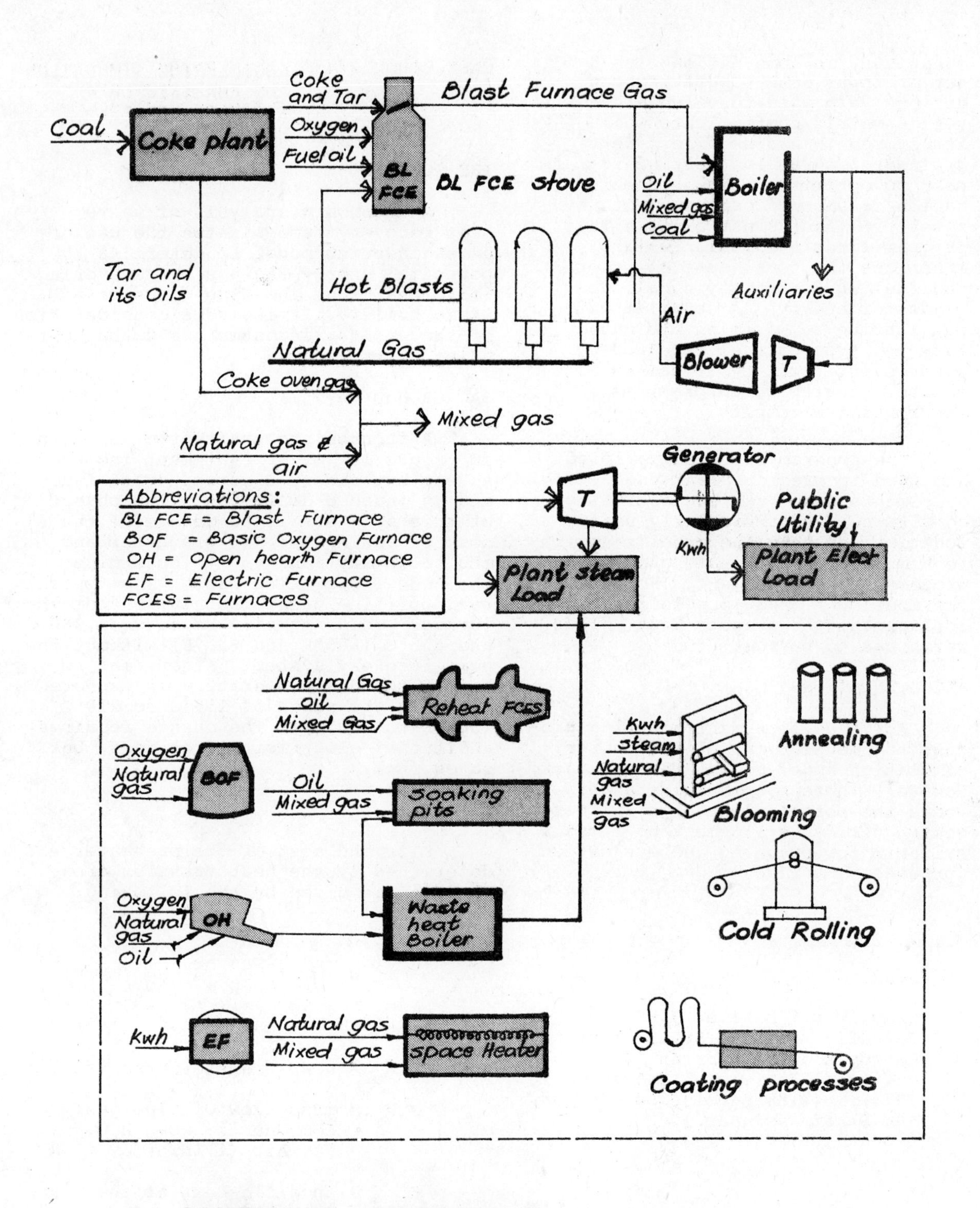

SCHEMATIC DIAGRAM OF ENERGY FLOW AT THE INDIANA HARBOR WORKS

FIGURE 1

equations, in combination with a set of input data (operating conditions) are used to calculate a set of matrix coefficients which are stored in a disk file. Next, a linear programming routine is used to optimize a linear combination of energy costs subject to a set of linear constraints. This program finds a solution which minimizes the total cost of power and fuel while satisfying all demands for energy. A detailed output listing summarizing the quantity and type of fuel required by every plant facility is passed to another program which prepares a report of the results.

The program has been verified and used to predict plant energy requirements under varying operating conditions. A particularly useful function has been the model's ability to simulate and evaluate changes in processes and equipment. For example, the model was used to select the best replacement for a waste heat boiler which was to be shut down.

RECOMMENDATIONS

For a complex manufacturing process the development of a computer simulation model is not an easy task. However, there are many instances where the potential savings in energy costs justify development of a model suitable for planning and evaluation purposes.

References

1. Gray, W.R.; Fekete, J.D.; and Tarkoff, M.I., "A Steel Plant Energy Model," Reprint from *Iron and Steel Engineer* 51(November 1974). (With permission of *Iron and Steel Engineer.)*

CASE STUDY 2-4: ENGINEERING SIMULATION AND ANALYSIS OF A STEAM BOILER PREHEATER[1]

THE CONCEPT

An economic analysis of waste heat recovery demonstrates the use of an engineering model to determine the potential energy savings by recovering waste heat from the flue gases of a steam boiler. It includes consideration of the capital investment and the operating costs.

BACKGROUND

A furnace preheat system is shown in Figure 1. Major equipment includes a heat exchanger, a forced draft fan, and an induced draft fan. The object of this analysis is to determine the cost of adding the air preheater and the necessary fans and to determine the annual operating cost. The typical range of flue gas temperatures from the furnace would normally be between 360°C and 455°C (600°F and 850°F) without the heat recovery system. After heat recovery the gas temperature is expected to be in the range of 150°C to 210°C (300°F to 400°F). The change requires additional electricity (for fans) but saves fuel.

FINDINGS

Size and cost of the preheater are determined by the heat transfer area. The area is given by the following:

$$A_o = \sum_{i=1}^{4} W_i \int_{T_{g2}}^{T_{g1}} C_{p,i} dT / U_o \text{ (LMTD)}$$

where A_o = heat transfer area (ft^2)

W_i = mass flow of flue gas components; CO_2, H_2O, N_2, Air (1 lb/hr)

$C_{p,i}$ = specific heat of the ith gas component (Btu/lb°F)

U_o = overall heat transfer coefficient

LMTD = logarithmic mean temperature difference.

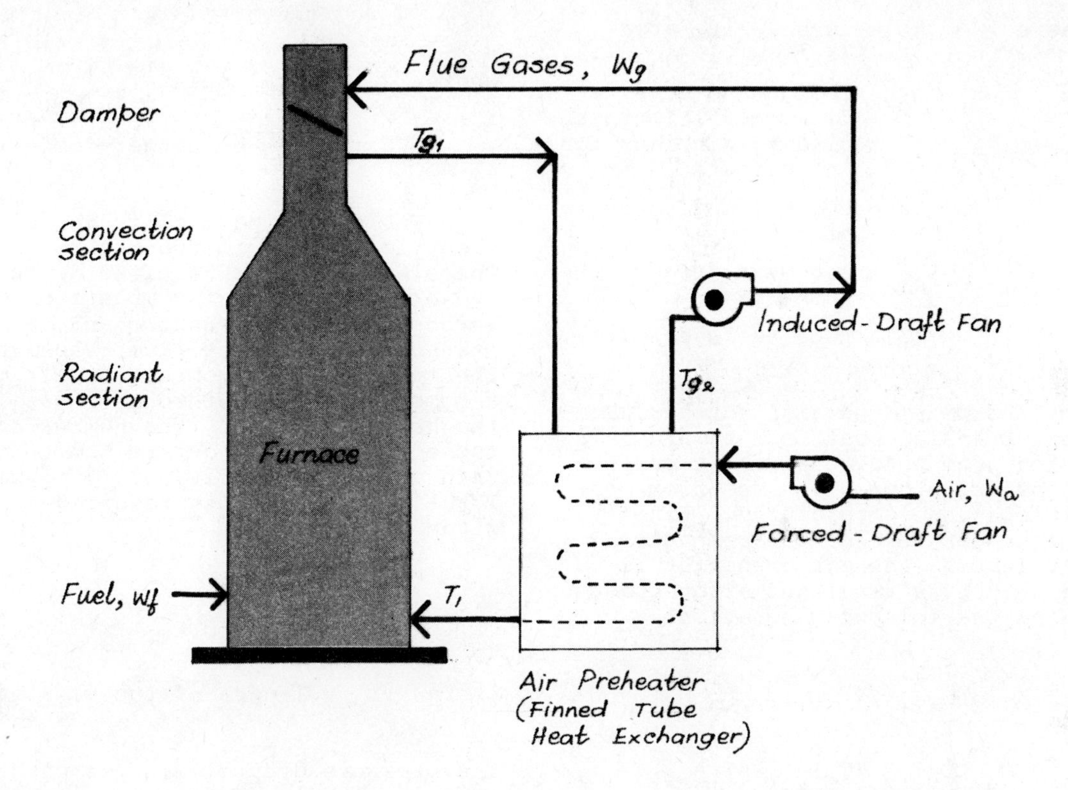

AIR PREHEAT SYSTEM FIGURE 1

The overall heat transfer coefficient is given by:

$$\frac{1}{U_o} = \frac{A_o}{h_i A_i} + \frac{bA_o}{kA_{1m}} + \frac{1}{h_o}$$

where A_o = tube inside area (ft^2)

h_i, h_o = inside and outside film coefficients including dirt and fouling effects

k = thermal conductivity of the tube

A_{1m} = log mean area of the tube (ft^2)

b = tube thickness (ft).

The inside and outside convective heat transfer coefficients are calculated from nondimensional correlations which depend on the flow conditions and temperatures. Finally, the temperature of the combustion air leaving the air preheater is calculated by a trial and error procedure using the following equation:

$$W_a \int_{T_2}^{T_1} (C_p)_{air}\, dT = \sum_{i=1}^{4} W_i \int_{T_{g2}}^{T_{g1}} C_{p,i}\, dT$$

where W_a = mass flow rate of the combustion air, lb/hr.

Given the stated flue gas temperatures and flows and the rate of combustion air intake, the necessary heat transfer area A_o and the heat recovered can be computed. Given A_o, the cost of the heat exchanger can be computed.

The total initial capital investment of the air preheater system includes the finned tube heat exchanger, fans, motors, instrumentation, and installation labor. This is found to be 2.994 times the heat exchanger cost:

$$C_T = 2.994\, C_E$$

where C_T = total cost

C_E = heat exchanger cost.

The operating cost is the cost of electricity used to run the motors. The electric power required depends on the pressure drop of combustion air through the air preheater and the pressure drop of the flue gases through the furnace. The combustion air makes a number of passes through tubes in the heat exchanger. The number of tubes per pass was determined to maintain a linear velocity of 75 ft/sec. Total number of tubes required is given by:

$$N_t = \frac{A_o}{\pi D_o L_t}$$

where L_t = length of tubes.

The pressure drop of air per pass is calculated using Fanning's equation:

$$(-\Delta P)_p = \frac{\rho_a f L_t u_b^2}{2 g_c D_i}$$

where $f = 0.046\, R_e^{-0.2}$.

From this the electric power required for the forced draft fan can be computed:

$$E_f = \frac{(\Delta P)_p\, N_p\, W_a\, (24\text{ hrs})\, (0.7455)\, (d)}{\rho_a \eta_p\, (3600)(550)} = \text{kWh/yr}$$

where N_p = number of passes

d = days of preparation

η_p = pump efficiency = 0.70.

Similarly, the induced draft fan power requirements can be calculated and the annual operating cost of both systems can then be computed utilizing the cost of electricity. Parametric studies can now be made using different materials of construction and design flow conditions to determine initial investment and operating costs.

Results are shown in Table 1. The payout time of the capital investment in each case is less than three years, indicating a favorable rate of return.

RECOMMENDATIONS

Engineering simulation studies are useful as a tool for evaluating efficiency options. Important to such studies is inclusion of the capital and operating costs in the model.

References

1. Prengle, H. William Jr.; Crump, Joseph R.; Fang, C.S. Grupa, M.; Henley, D.; and Wooley, T., *Potential for Energy Conservation in Industrial Operations in Texas*, Report No. S/D-10, (Houston, Texas: University of Houston, Department of Chemical Engineering, Cullen College of Engineering, November 1974).

CASE STUDY 2-5: ENERGY MANAGEMENT IN AN INDUSTRIAL PLANT SAVES 46 PERCENT[1]

THE CONCEPT

Energy management efforts, including an energy audit, lighting system evaluation, HVAC system review and modification, heat recovery, and use of outside air, permitted saving 46 percent of electricity and 20 percent of natural gas use compared to a base year.

BACKGROUND

The plant occupies 10,500 m^2 (113,000 ft^2), employs 350 persons, and manufactures medical instruments including blood analyzers and centrifuges. Facilities include a 600 m^2 (6,500 ft^2) manufacturing and assembly area; a 325 m^2 (3,500 ft^2) machine shop; a 6,000 m^2 (65,000 ft^2) warehouse; a 465 m^2 (5,000 ft^2) research and development laboratory, and 9,300 m^2 (100,000 ft^2) of parking space. During the course of the study, the plant was operating on a single shift. It is located in New Jersey.

The energy management program used in this plant followed the outline described in Chapter 2. Major steps undertaken included:

- review of historical energy use;
- energy audit of all process equipment, lighting, and building services;
- comparison of estimated loads based on energy audit with historical records;
- identification of energy management opportunities in lighting, HVAC, etc.;
- implementation of energy management techniques; and,
- personnel training, follow-up, etc.

FINDINGS

Lighting was found to use 482 2-lamp, 277 volt, 8 foot industrial-type fixtures in the warehouse, machine shop, and assembly areas. The office and research laboratory areas used 1,123 4-lamp, 277 volt, 4 foot fluorescent fixtures. By checking illumination levels and using IES-recommended values (refer to Chapter 9), it was found that illumination was excessive in some areas. For example, the average value in the warehouse was 1,076 lux (100 fc). This load was reduced by 50 percent by disconnecting alternate fixtures (ballasts and lamps) in each row. This brought the new level of illumination to 500 lux (∿50 fc), well within IES guidelines.

Similar reductions were found to be possible in much of the office space, where illumination levels ranged from 1,076 - 1,500 lux (100 - 140 fc). Also, efforts were made to employ task-oriented lighting by selecting certain lamps for removal or by providing supplementary lighting where required. Further

Case	Heat Transfer Area, ft^2	Heat Recovered $/yr($0.80/ MBtu)	Cost of Preheater and Fans, $	Annual Operating Cost, $/yr	Payout Time, Year
1	75,330	147,686	166,374	5,226	1.17
2	31,224	88,013	81,918	2,523	0.96
3	55,537	117,331	130,192	4,146	1.15
4	53,426	89,107	126,194	3,774	1.49
5	22,261	53,124	62,396	1,817	1.22
6	39,529	70,848	99,033	3,082	1.46
7	43,764	89,107	107,485	3,175	1.25
8	32,295	70,848	84,171	2,367	1.23
9	18,164	53,124	52,978	1,547	1.03
10	11,634	14,769	37,019	907	2.67
11	4,862	8,801	18,347	385	2.18
12	8,621	11,733	29,086	706	2.64

SUMMARY **TABLE 1**

modifications were made in corridors, in the cafeteria, and in the parking lot. The parking lot lamps (80 - 175 W mercury vapor lamps on 3 m [10 ft] standards) were replaced by two 1,500 watt tungsten-halogen floodlights installed 60 m apart at a height of 6 m on the building. Besides greatly reducing the load (by a factor of ∿5), the new lamps had an automatic photocell/time clock controller rather than a manual control as was used with the old system.

Heating and cooling energy was saved by developing improved control methods for the existing equipment. There are seven hot/cool air handling units serving seven zones of the plant. These are supplied from a central boiler/hot water system. There are also seven cooling compressors, located on the plant roof.

Time clocks were installed to shut the system down in the afternoon, just before the end of the shift, and to start them 30 minutes before work begins in the morning. New thermostats were installed, permitting nighttime setback.

The system previously used an air-modulating arrangement to mix heated and cooled air to provide the appropriate temperature for each zone. Now, the compressors are shut down at night during the summer and turned off during the winter. The boiler is turned off during the summer. The make-up air fan was disconnected; make-up air is now drawn from the shipping area (where large loading doors are normally open), reducing fan pumping power and recovering some heat. During the winter, outside air is now used in the air modulating system.

Finally, an employee education plan was developed. All employees were informed of the need to save energy and were encouraged to assist by submitting ideas and ensuring that lights and equipment were turned off when not in use. Posters, a car-pool program, an energy saving club, and award certificates to employees for energy management efforts in their homes were also developed.

RECOMMENDATIONS

Every industrial operation is different and has its own unique features, needs, and problems. The Energy Management Coordinator needs to develop a plan appropriate to his operation, and then see that all employees are aware of the program, understand its goals, and know what they can do to contribute to it.

If these steps are followed, low cost changes and minor modifications to existing equipment can be made to yield substantial savings, with no sacrifice in product quality or production levels.

References

1. Eckert, William H., "How an Industrial Plant Saves Energy," *Electrical Construction and Maintenance*, (June 1974): pp. 78-81.

Acknowledgments

This Case Study was provided through the courtesy of Mr. Orin F. Zimmerman, General Manager, Conservation and Energy Management Department, Portland General Electric Company.

CASE STUDY 2-6: DELIBERATE PLANNING FOR EFFICIENT ENERGY USE IN PETROLEUM REFINERY OPERATIONS REDUCES ENERGY USE BY 16 PERCENT AND SAVES MONEY[1]

THE CONCEPT

Refinery costs are significantly reduced when energy management is a basic criterion.

BACKGROUND

A typical overall refinery heat balance (Figure 1) includes input of methane and nitrogen from the air in addition to crude oil for an output

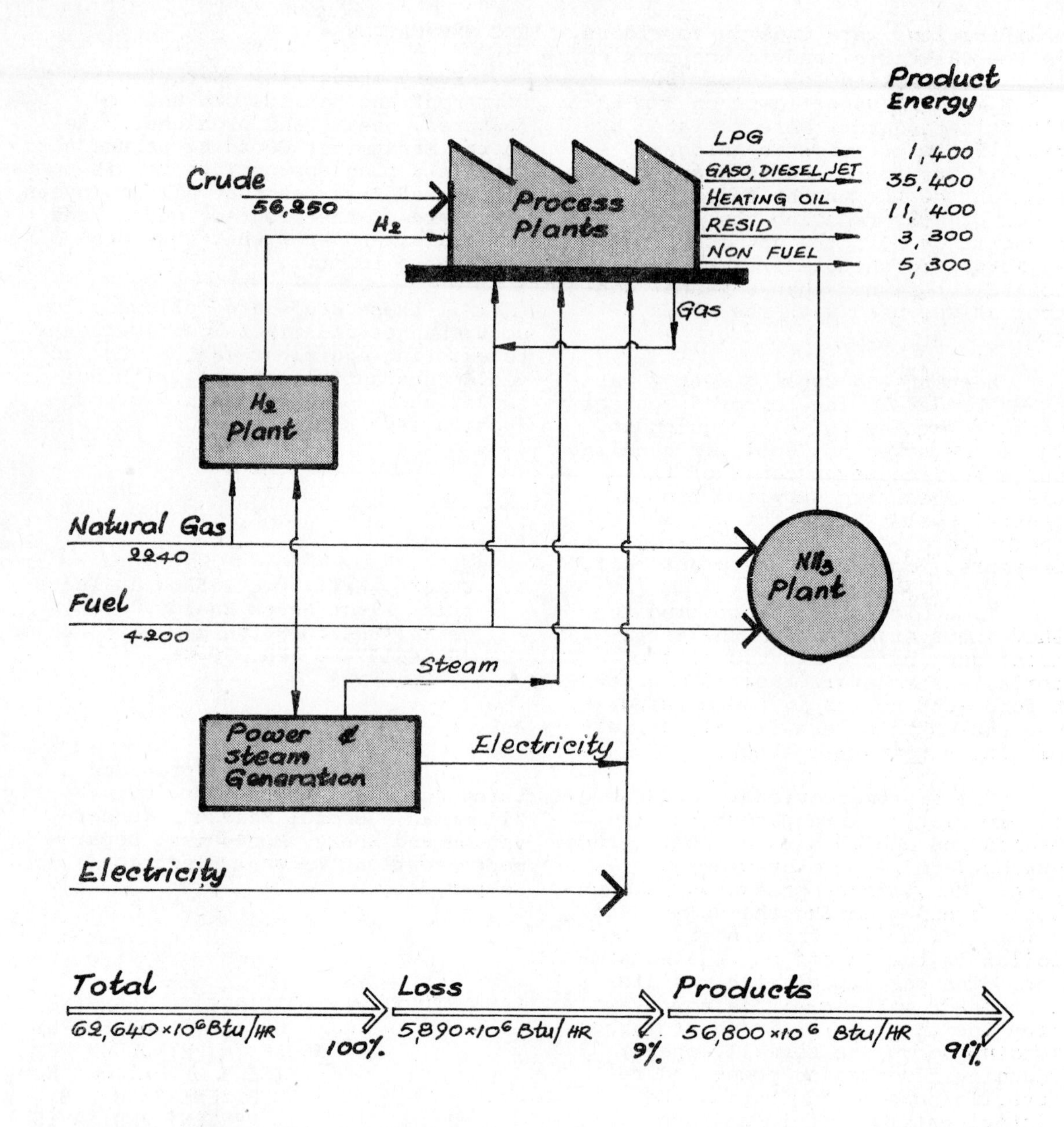

OVERALL ENERGY BALANCE, PASCAGOULA REFINERY

FIGURE 1

of 0.237 Mbpd compared to a 0.225 Mbpd input. However, the energy content of the end products is less than the energy input, resulting in a 91 percent conversion efficiency.

There are several alternatives for energy supply to a refinery. These include purchased power, locally produced power, use of recovered heat, and combinations of these. If all waste streams are used, efficiency is higher. Greater efficiency requires more extensive investment in heat recovery equipment, with correspondingly greater initial capital investment.

FINDINGS

Two crude units with identical capacity were built in the refinery. In the second unit, special emphasis was placed on efficient energy use. This was achieved largely by improvement in four areas: (1) crude preheat; (2) vacuum operations; (3) pump drives; and, (4) steam stripping.

Design of the second crude unit included optimization of heat exchanger performance and overall heat utilization. Improvements in fouling control and optimization of shell-side design resulted in a 10 percent increase in recovered heat. Increasing the operating pressure from 1.4 to 4.0 psia improved operating efficiency and lowered operating costs. Efficient double-ended low-pressure steam condensing turbine drives were used for primary feed, booster, and spare pumps. This was both more efficient and more economical than electrical drives with high pressure steam turbine backup drives. Steam stripping was found to improve recovery of jet fuel and was more efficient than using a fired reboiler.

Table 1 compares the efficiency and operating costs associated with units 1 and 2. The more efficient unit (2) has 16 percent less energy use and 18 percent lower annual operating costs.

RECOMMENDATIONS

If consideration is given to efficient energy use in the *design* phase of a project, it is often possible to save energy and reduce operating costs with little or no capital costs. Even if capital costs are greater, they are often justified by the resulting savings in energy cost.

References

1. Hayden, J.E. and Levers, W.H., "Design Plants to Save Energy," *Hydrocarbon Processing* (July 1973): 72-75.

CASE STUDY 2-7: EFFICIENT CYCLES FOR INDUSTRIAL POWER AND STEAM GENERATION SAVE 14-31 PERCENT[1]

THE CONCEPT

As fuel and energy costs increase, new approaches are needed to reduce operating costs.

BACKGROUND

Power plant cycles in a large US chemical company have changed over time as more efficient equipment has become available. Existing equipment types can be classified into three cases:

- Case 1 - Pre-World War II
- Case 2 - Post-World War II
- Case 3 - Mid-1960s to present

Each of these cases will be described to show how improvements have been obtained. The process requirements in each case are:

Electric Power	Steam
120 MWe	1.14×10^6 N/m^2 (165 psi) 163 mt/hr (360 klb/hr)
	2.41×10^5 N/m^2 (35 psi) 78 mt/hr (172 klb/hr)

Figure 1 shows schematic diagrams for each case.

Utility	Consumption, 10^6 Btu/hr Unit 1	Consumption, 10^6 Btu/hr Unit 2	Operating Costs, thousand $/yr Unit 1	Operating Costs, thousand $/yr Unit 2	Savings 10^3 $/yr	Savings %
Electricity	13	8	226	143	83	37
Cooling Water	2	1	80	42	38	5
Fuel Gas	597	477	2,710	2,160	550	20
Boiler Feed Water	21	12	121	71	50	41
Steam						
600 psi	118	61	581	300	281	48
150 psi	(116)	31	(477)	128	(605)	(136)
40 psi	0	(52)	0	(164)	164	inf.
Condensate	(5)	(8)	(27)	(43)	16	59
Total	630	530	3,214	2,637	577	18

RESULTS OF DESIGN FOR ENERGY CONSERVATION

TABLE 1

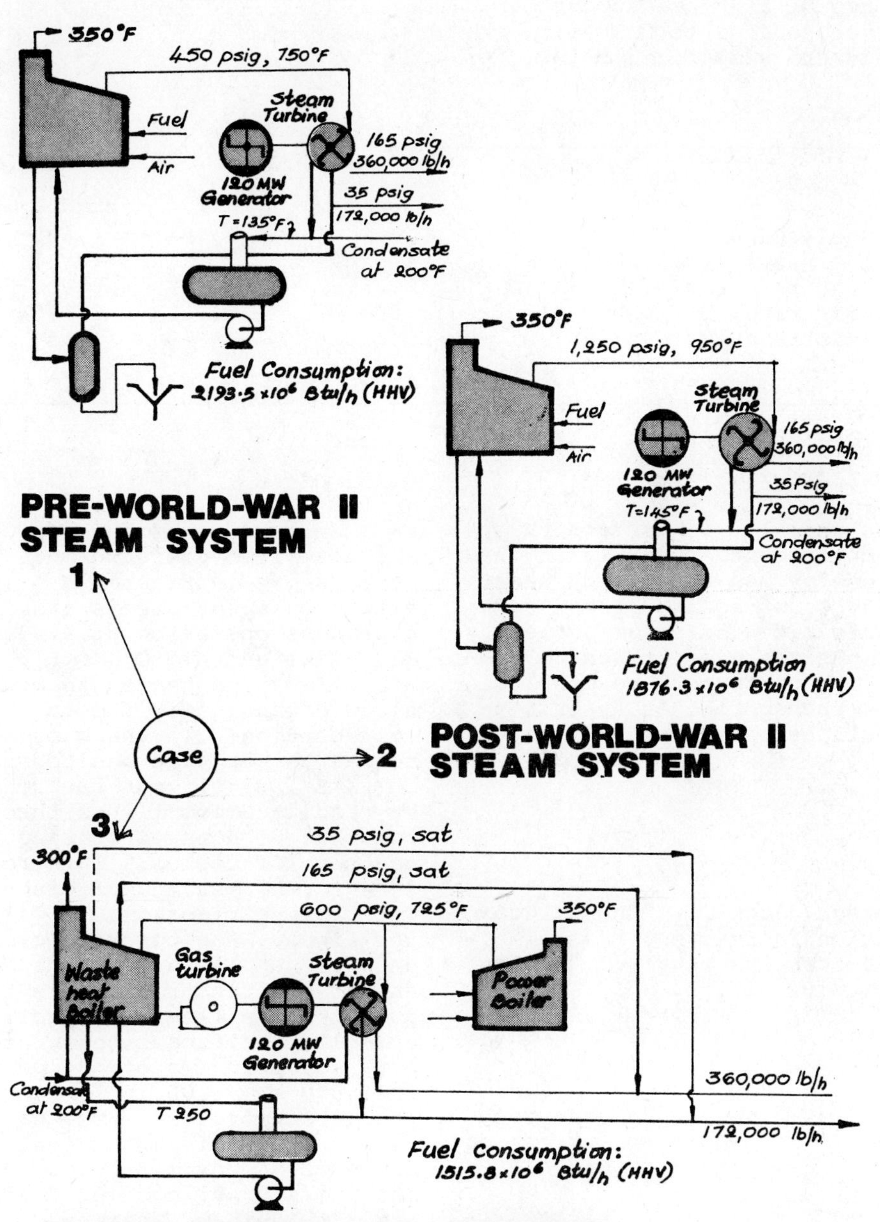
350°F
450 psig, 750°F
Fuel
Air
Steam Turbine
120 MW Generator
165 psig
360,000 lb/h
35 psig
172,000 lb/h
T = 135°F
Condensate at 200°F
Fuel Consumption: 2193.5 × 10⁶ Btu/h (HHV)
PRE-WORLD-WAR II STEAM SYSTEM
1
350°F
1,250 psig, 950°F
Fuel
Air
Steam Turbine
120 MW Generator
165 psig
360,000 lb/h
35 Psig
172,000 lb/h
T=145°F
Condensate at 200°F
Fuel Consumption 1876.3 × 10⁶ Btu/h (HHV)
Case
2
POST-WORLD-WAR II STEAM SYSTEM
3
35 psig, sat
300°F
165 psig, sat
600 psig, 725°F
350°F
Waste heat Boiler
Gas turbine
Steam Turbine
Power Boiler
120 MW Generator
Condensate at 200°F
T 250
360,000 lb/h
172,000 lb/h
Fuel consumption: 1515.8 × 10⁶ Btu/h (HHV)
STEAM GENERATING SYSTEMS FROM MID 1960s TO PRESENT

FIGURE 1

FINDINGS

The main improvement has resulted from the use of high pressure steam, gas turbines, and waste heat boilers. The savings are summarized below:

	Case 1	Case 2	Case 3
Fuel, GJ/hr	2,316	1,981	1,600
Fuel savings, %	base case	14	31
Heat rate, J/Je	3.75	2.99	2.10
Heat rate, Btu/kWh	12,800	10,200	7,175
Heat rate improvement, %	base case	20	44

RECOMMENDATIONS

These examples illustrate the potential improvements possible in industrial energy use efficiency when improved types of equipment and combined electricity and steam cycles are used. As the need arises to expand capacity or to replace obsolete equipment, the approaches described above should be considered.

References

1. Robertson, Jack C., "Energy Conservation in Existing Plants," *Chemical Engineering* (21 January 1974): 104-111.

CASE STUDY 2-8: POWER RECOVERY FROM HIGH PRESSURE GAS OR LIQUID STREAMS[1]

THE CONCEPT

Energy can sometimes be recovered from process streams.

BACKGROUND

High pressure process streams in chemical plants and other industrial operations sometimes have energy which can be recovered. Figure 1 shows two such systems, one based on spent-air power recovery and the other based on hydraulic turbine recovery systems.

FINDINGS

An example of power recovery from a pressurized gas stream is the situation shown in Figure 1A. In this situation spent-air is available from a process operation at 4.83 x 10^5 N/m^2 (70 psig) and 4.5°C (40°F). This stream must be heated in order to allow expansion in the turbine without condensation. In the case shown the reheat temperature was considered to be 93°C (200°F), but the reheat could be readily accomplished through heat exchange with any process stream that needs to be cooled. This results in a further saving of either cooling water or refrigeration. As shown in the figure, the power recovered is used to supply part of the energy for an air compressor which is tandemly driven by an electric motor. In this example, the expansion turbine recovered approximately 20 percent of the 5970 kW (8,000 hp) required for the air compressor. The capital cost is $300,000, giving capital costs of approximately 60,000 $/yr. The utility savings (at 0.03 $/kWh) are 296,000 $/yr, giving a net savings of 236,000 $/yr. Stated another way, the simple payout period for this investment is about 15 months. An example of a hydraulic turbine installation is shown in Figure 1B.[2] The turbine works in tandem with the electric motor which is driving the pump for the process. The motor-load is reduced by power developed in the turbine. A clutch is placed between the turbine and the motor which allows the charge-pump to be started independently of the

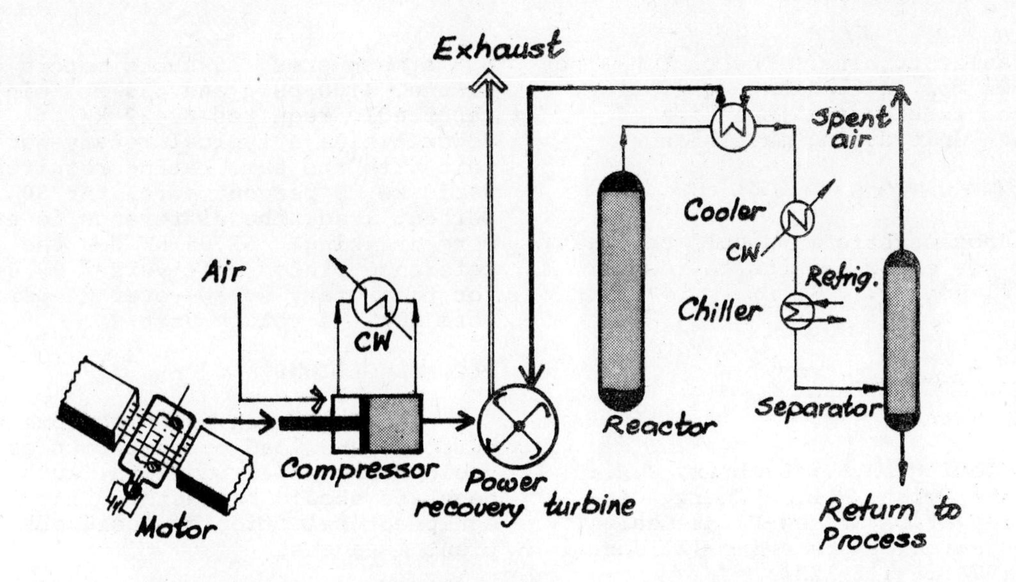

TYPICAL SPENT-AIR POWER-RECOVERY SYSTEM FIGURE 1A

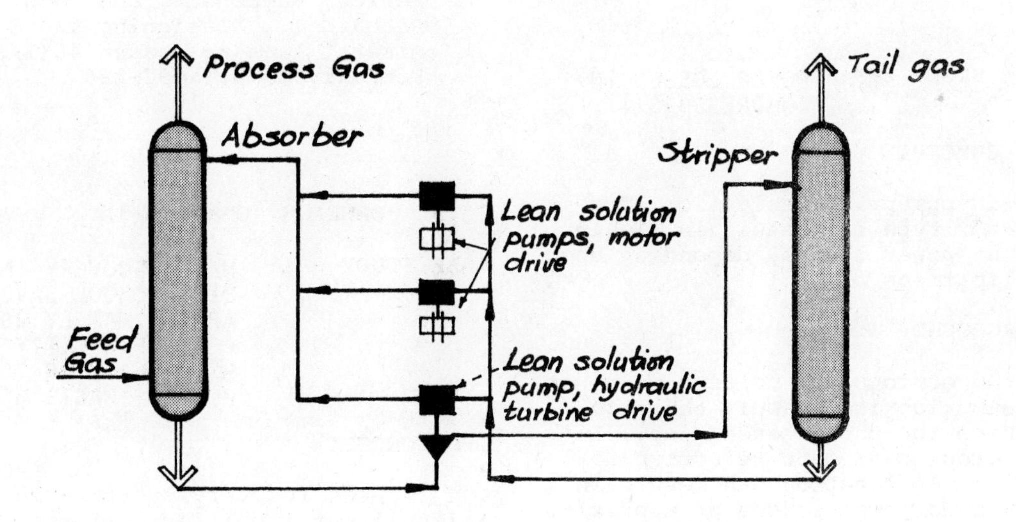

HYDRAULIC TURBINE POWER-RECOVERY SYSTEM FIGURE 1B

hydraulic turbine. The clutch also allows for independent activation of the turbine and isolation of either unit during maintenance.

RECOMMENDATIONS

Process streams should be evaluated for opportunities to recover useful energy.

References

1. Fleming, J.B.; Lambrix, J.R.; and Smith, M.R., "Energy Conservation in New-Plant Design," *Chemical Engineering* (21 January 1974): 112-122.

2. Braun, S.S., "Power Recovery Cuts Energy Costs," *Hydrocarbon Processing* (May 1973): 81-85.

CASE STUDY 2-9: POWER LOSSES IN COMPRESSORS[1]

THE CONCEPT

Inappropriate selection of compressor type can result in significant power losses, depending upon applications.

BACKGROUND

An obvious way to save on power consumption is to apply the right machine for the intended job. Injudicious equipment selection for a given application can result in increased power losses of approximately 10 percent or more.

FINDINGS

To evaluate performance differentials between compressor types with the same rating, a rotary compressor was compared to a piston design. Whereas a rotary compressor design is considered an "energy miser" in an air-conditioning application, a piston compressor may be better than rotary-screw types for plant air applications. For example, at 100 percent load one particular reciprocating compressor rated at 6.8 atmospheres, 15 cubic meters per minute, (100 psig and 535 ft^3/min) reportedly requires 82.22 kW of power, while a typical rotary-screw unit with the same rating requires 89.90 kW, 9 percent more. At 80 percent load, the difference is even more striking: 67.05 kW for the reference piston unit versus 90.04 kW for the rotary screw--over 34 percent more for the rotary unit.

RECOMMENDATIONS

Careful matching of equipment, purpose, and load characteristics can yield substantial benefits; such benefits should be factored into equipment selection by designers and plant managers.

References

1. Herzog, Raymond E. and Dann, Richard T., "Designing the Energy Miser," *Machine Design* 46(21 February 1974): 97-106.

15.3 COMMERCE (Chapter 3)

CASE STUDY 3-1: HEAT RECOVERY IN A HIGH SCHOOL SAVES APPROXIMATELY 29 PERCENT OF UTILITY BILL BY RECLAIMING 70 PERCENT OF WASTE HEAT.[1]

THE CONCEPT

A heat recovery system using a centrifugal water chiller heat pump, double bundle condensers and evaporators, hot water storage, and a deep well heat sink was designed for a new high school in Wisconsin.

BACKGROUND

This approach was selected in view of several unique design requirements:

- the school had no windows;
- it was the first compact, two-story school in the state;
- it was the first air-conditioned school in the state; and,

- it was one of the few schools to be built without a conventional fossil fuel heating plant.

These design requirements were chosen in an effort to minimize capital and operating costs while providing a more effective heating system that would avoid the need for cooling interior zones of the school during winter.

The well pump system was planned to provide a source of heat when the temperature dropped below -3°C (27°F). However, experience has shown that this system is only needed when the temperature is less than -15°C (5°F). Likewise, the auxiliary electric heaters which were provided have been used frequently.

FINDINGS

The performance of the system has been excellent. Over a nine year operational period annual costs averaged 32,500 $/year for electricity usage of about 2 x 10^6 kWh/yr. This included lighting, miscellaneous power, heating, cooling, hot water, the well pump, and auxiliary heaters. Total annual operating costs average about 0.21 $\$/ft^2$. It is estimated that heating and cooling required about 10,000 $ per nine month school year. This corresponded to a saving of 4,000 $ (29 percent) compared to 14,000 $ per nine month school year for a conventional system.

Maintenance costs have also been low, averaging about 3000 $/year, some of which is attributed to the well and back-up systems and could be avoided in future designs.

Experience has shown that several improvements to the system could be made. The fact that the back-up systems had unneeded capacity has been mentioned previously. The centrifugal water chiller also has some drawbacks, notably a tendency to surge at low loads and the need to purge noncondensables from the system. Modifications were made to correct these difficulties, but this increased both capital and operating costs. (Note that the choice of equipment was limited by what was available at the date of installation: 1963.)

In the newer systems installed in Wisconsin schools, supplemental booster heaters have been placed in the hot water storage tanks. Variable pitch fans and dampers (rather than the double duct fan system) have been used, resulting in a 40 percent reduction in fan power. Low static pressure, constant volume regulators on mixing boxes were used to allow a pressure drop reduction from 19 to 9 mm of water (0.75 to 0.35 inches of water). Fewer control valves were used in the newer designs. Finally, introduction of the screw or rotary-type compressor eliminated several of the problems associated with the centrifugal chiller. The rotary compressor has a lower first cost; operating above atmospheric pressure, it can expel noncondensables without purge; being a positive displacement machine, it eliminates the surge problem.

RECOMMENDATIONS

Designers should consider heat recovery from HVAC systems in both new designs and retrofit projects. As fuel and electricity prices escalate, potential savings will be even greater than indicated in this example.

References

1. Ratai, Walter R., "Heat Reclaim Case Study: Kimberly High School, After 9 Years," *ASHRAE Journal* (February 1973): 40-42.

CASE STUDY 3-2: INTERNAL HEAT GAINS FROM LIGHTING[1]

THE CONCEPT

High illumination levels in buildings reduce heating requirements but are an added air-conditioning load.

BACKGROUND

Illumination levels in office buildings have been prescribed according to Illuminating Engineering Society standards. Design conservatism, or uniform illumination throughout the buildings, can sometimes lead to excessive illumination levels.

Though this is still a subject of current research, a reduction in lighting levels by as much as 40 percent may be possible. Besides reducing electricity required for lighting, the heat load on air conditioning would be reduced, although supplemental heat could be needed.

FINDINGS

Architects, engineers, and building operators have found that alternative lighting levels are often feasible. It is possible to calculate the electrical energy savings due to reduced lighting and the energy savings of reduced air conditioning. The air-conditioning savings are equal to the total energy saved by the lights divided by the COP of the air conditioner (e.g., 2.25 on the average).

As an example consider a building consisting of:

- general offices;
- drafting rooms;
- accounting rooms;
- elevators, corridors;
- rest rooms; and,
- closets and storage areas.

General Electric type lamps and the *GETP606 General Lighting Design* are used for coefficients of utilization (COU) (includes luminaire characteristics and mounting heights, room size, and ceiling, wall and floor reflectances) and lamp maintenance factors (MF). The following relations allow calculation of the total initial lamp lumens (TILL) necessary to create the desired illumination level in each space (equation [1] below), the number of lamps (given the lumens per lamp) and the power requirements in W/ft^2 (given the power for lamp and lamp ballast listed by the manufacturer). In addition, the assumption is made that this building is in a warm climate where cooling is the major energy use. The heat release of the lamps is equivalent to the power density computed in equation (3). These calculations are carried out for uniform levels and task-oriented lighting (TOL) and summarized in Table 1. In addition, the annual cost savings on air conditioning is determined by equation (4).

$$\text{TILL} = \frac{\text{Footcandles x Floor Area (ft}^2\text{)}}{\text{Coefficient of Utilization x Maintenance Factor}} \tag{1}$$

$$\text{\# of Lamps} = \frac{\text{TILL}}{\text{Lumens per Lamp}} \tag{2}$$

$$\text{W/ft}^2 = \frac{\text{\# of Lamps x Watts per Lamp plus Ballast}}{\text{Floor area in ft}^2} \tag{3}$$

$$\frac{\text{(Uniform kW-TOL kW) x 100 hr equivalent full-load cooling}}{\text{COP}} = \$2{,}625 \tag{4}$$

Space Use	Floor Area (ft^2)	TOL LEVEL fc[a]	W/ft^2	Total kW	UNIFORM LEVEL fc	W/ft^2	Total kW
General Offices	70,000	50	1.9	133	100	3.8	266
Drafting Rooms	2,000	50	1.9	4[b]	200	6.1	12
Accounting	8,000	50	1.9	15[b]	150	5.3	42
Elevators, Corridors	14,000	10	0.5	7	20	1.1	15
Restrooms	4,000	15	0.7	3	30	1.5	6
Closets, Storage	2,000	5	0.1	0.2	5	0.1	0.2
Total Area	100,000						
Day Totals (kW)				162			341
7 A.M.--6 P.M. (11 hours)				4[b]			
Night Totals (kW)				25			51
6 P.M.--7 A.M. (13 hours)							
(@ 15% of Day-time)							
Total Cost (@ 3¢/kWh) per day				$64/day			$132/day
Total annual cost (@ 255 working days 110 days of night operation only)				$17,393			$35,848

Notes: (a) fc = Footcandles of illumination.
(b) kW = 20% (4 + 15) kW = added power for task-oriented lighting (TOL)

SUMMARY OF LIGHTING CALCULATIONS **TABLE 1**

RECOMMENDATIONS

This calculation shows a total annual operating cost saving of approximately 60 percent which can be achieved by delamping and a certain amount of task-oriented illumination. Building operators should measure current lighting levels and determine the minimal acceptable lighting levels practical in their building. The cost difference, as calculated in the foregoing example, should indicate the advisability of proceeding with a revised lighting program.

References

1. Taussig, Robert T., ed., *Energy Conservation in High Density Urban Areas: A Study of the Energy Crisis in Metropolitan New York City*, (New York: Columbia University, School of Engineering and Applied Science, 20 September 1974) pp. 39-42.

CASE STUDY 3-3: AN ENERGY AUDIT OF A 15 STORY CITY ADMINISTRATION BUILDING[1]

THE CONCEPT

Energy audits are an effective tool for implementing energy management programs, discovering energy saving possibilities, establishing energy management goals, and measuring performance.

BACKGROUND

Since 1973 a series of operational changes was made in the San Diego Community Concourse, a group of five municipal buildings (City Administration Building, City Operations Building, Theater, Exhibition Hall, and Parking Structure) served by a central plant. By delamping, informing employees through educational programs, increasing use of outside air in the HVAC system, and "fine-tuning" the HVAC system by manual controls, electricity use in the Community Concourse for 1975 was reduced by 37 percent compared to the 1973 base year (Figure 1).

Early in 1976 it was decided to perform an energy audit of the City Administration Building to determine if additional savings were possible and to establish how they might be obtained.

The first step in the audit was to review historical energy use. Calculations were made to estimate the significance of energy use for HVAC, lighting, computers and other equipment. Following the initial analysis the audit was performed. About 20 man days were spent in the audit and in other investigations in the building. Portable instrumentation was brought in and installed to measure the heating and cooling inputs to the City Administration Building from the central plant. Data obtained during the audit were processed using a computer program and then compared with historical records. Other measurements included the installation of separate electrical meters for each of the Community Concourse buildings and current readings taken in lighting and power panels during the audit.

FINDINGS

Table 1 indicates energy management options during the audit. These are in addition to the housekeeping and operational changes already achieved. The savings potential for the City Administration Building alone amounted to an estimated 1.9×10^6 kWh/yr, or approximately 50 percent of the 1976 electricity use of 3.7×10^6 kWh/yr. To achieve these savings, a capital investment estimated to be $23,000 would be required. This would save at least 70,000 $/year in electricity alone and would be paid back in four months. In addition to the above savings, further savings that would affect the central plant energy bill were identified. These involved modifications to the building envelope and to the HVAC system. Savings were approximately: 300,000 kWh/yr, about 4000 MBtu/yr, for a total cost savings of about 25,000 $/yr at a capital cost estimated to be $30,000. These changes had a payback of 15 months.

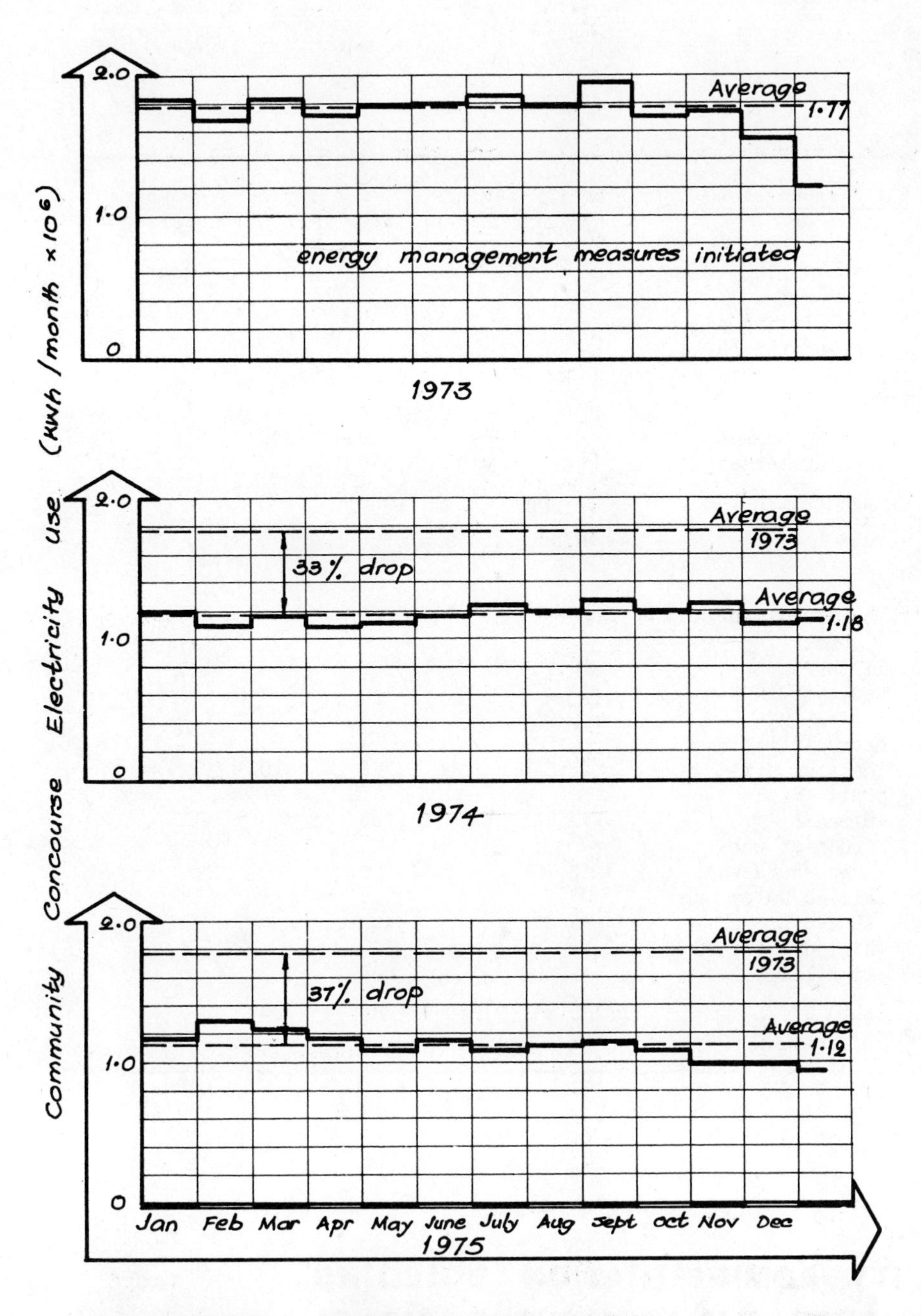

HISTORICAL ENERGY USE **FIGURE 1**
SAN DIEGO COMMUNITY CONCOURSE

Building Envelope	Energy Savings kWh/yr	MBtu/yr	$/yr	Imple-mentation Costs $	Payback Period Months
Reduce glazing and volume of the building and add insulation*					
Heating season		99	450	20,000	
Cooling season		910	5,650		
Change design temp 2°F*	40,000		1,200	0	
Subtotal	40,000	1,009	7,300	20,000	33
HVAC					
Fan Operation Cutback (to 12 hr/day)	290,000		11,600	0	
Decrease supply air	615,000		23,700	5,000	
Decrease minimum outside air*	41,750		1,250	0	
Modify mixed air plenum*	142,500	2,940	10,700	10,000	
Condenser water pump shutdown*	68,750		2,750	0	
Chilled water pump shutdown*	35,000		1,400	0	
Chilled water pump impeller trim*	45,000		1,800	100	
Subtotal	1,238,000	2,940	53,200	15,100	4

CITY ADMINISTRATION BUILDING ENERGY MANAGEMENT OPTIONS **TABLE**

	Energy Savings kWh/yr	MBtu/yr	$/yr	Implementation Costs $	Payback Period Months
Computer Facility					
Climate control		933	6,283	3,000	
Lighting modifications	82,600		2,217	1,000	
Subtotal	82,600	933	8,500	4,000	6
Electric Power Systems					
Delamping and relamping	340,000		10,200	1,910	
Disconnect ballasts	130,000		3,900	1,250	
Light timeclocks	120,000		3,600	3,000	
Modify custodial procedures	250,000		7,500	0	
Replace lenses	100,000		3,000	7,500	
Subtotal	940,000		28,200	13,660	6
City Administration Building Totals	1,927,600	933	72,000	22,660	4
*Central Plant Total (asterisked items)	373,000	3,950	25,200	30,100	15
GRAND TOTAL	2,300,600	4,883	97,200	52,760	7

CITY ADMINISTRATION BUILDING ENERGY MANAGEMENT OPTIONS

TABLE 1 (CONT.)

RECOMMENDATIONS

Even though substantial savings had been achieved in this modern office building, an energy audit revealed the potential for additional savings. Energy audits are an important tool in implementing energy mangement programs in commercial buildings.

References

1. Applied Nucleonics Company, Inc., *Efficient Energy Use at the San Diego City Administration Building: A Summary Report,* Report No. 1168-2, (Santa Monica, California: April 1977).

CASE STUDY 3-4: ENERGY USE IN A SMALL ENGINEERING OFFICE

THE CONCEPT

During the fuel shortage of December 1973, commercial buildings in Los Angeles were required to reduce electricity use by 20 percent. This case study describes energy use in a small engineering office and the measures used to cope with the electricity cutback.

BACKGROUND

The building described in this case study is a two-story, flat roof, uninsulated office building. It has brick walls, wood floors, partitions, and roof, and some interior plaster walls. The floor space is 279 m^2 (3000 ft^2), of which 186 m^2 (200 ft^2) upstairs and downstairs is office space and 93 m^2 (100 ft^2) (downstairs) is shop and laboratory space. Air conditioning (upstairs only) is electric and heating is by three gas heaters. From 1973 to the end of 1974, the occupancy of the building increased from ten to a maximum of 20 persons. The average full-time equivalent number of employees was 8.2 (1973) and 12.2 (1974). All of the employees were rarely in the office at one time due to field projects. Figure 1 shows energy usage.

The inefficient design of this building is reflected in the following typical 1973 winter-summer operating patterns. Due to lack of insulation, the temperature of the building on a winter morning would be in the range of 10 to 16°C (50 to 60°F). A gas heater was used to bring the temperature up to 19 to 20°C (66 to 68°F), which typically required one to two hours. At this time, in the upstairs offices, the heat input from the high output overhead fluorescent lamps began to have an effect and the temperature of the offices became uncomfortable. The heater was turned off by a thermostat control and the air-conditioning unit removed the heat to return the temperature to 19 to 20°C.

During this time the outside temperature was 5 to 10°C cooler than the temperature of the office. However, due to the lack of windows (except for some small vents), no use could be made of the cooler outside air. The air-conditioning unit was a closed system which recirculated the air in the building. During the summer, the air conditioner had to be operated to remove the combined solar and lighting heat load, even on cool days.

FINDINGS

Two measures were employed in this building to reduce electricity use. The first was to install a transfer fan in the upstairs area. This fan was used in place of the air conditioner. It transferred warm air from the upstairs to the downstairs shop area. Outside air was drawn in upstairs through vents whenever the outside air was cooler than the inside air. In addition, the thermostat settings were adjusted to 20°C (68°F) for heating and 24°C (75°F) for cooling.

The major energy saving method was to reduce lighting. Measurements indicated that illumination levels on desk-top working surfaces produced by two overhead rows of high output fluorescent lamps were in the range of 1100-1500 lux (100-150 fc). This was found to be more than needed for office work and 50 percent of the lamps were removed, bringing illumination levels down to the range of 550-1100 lux (50-100 fc).

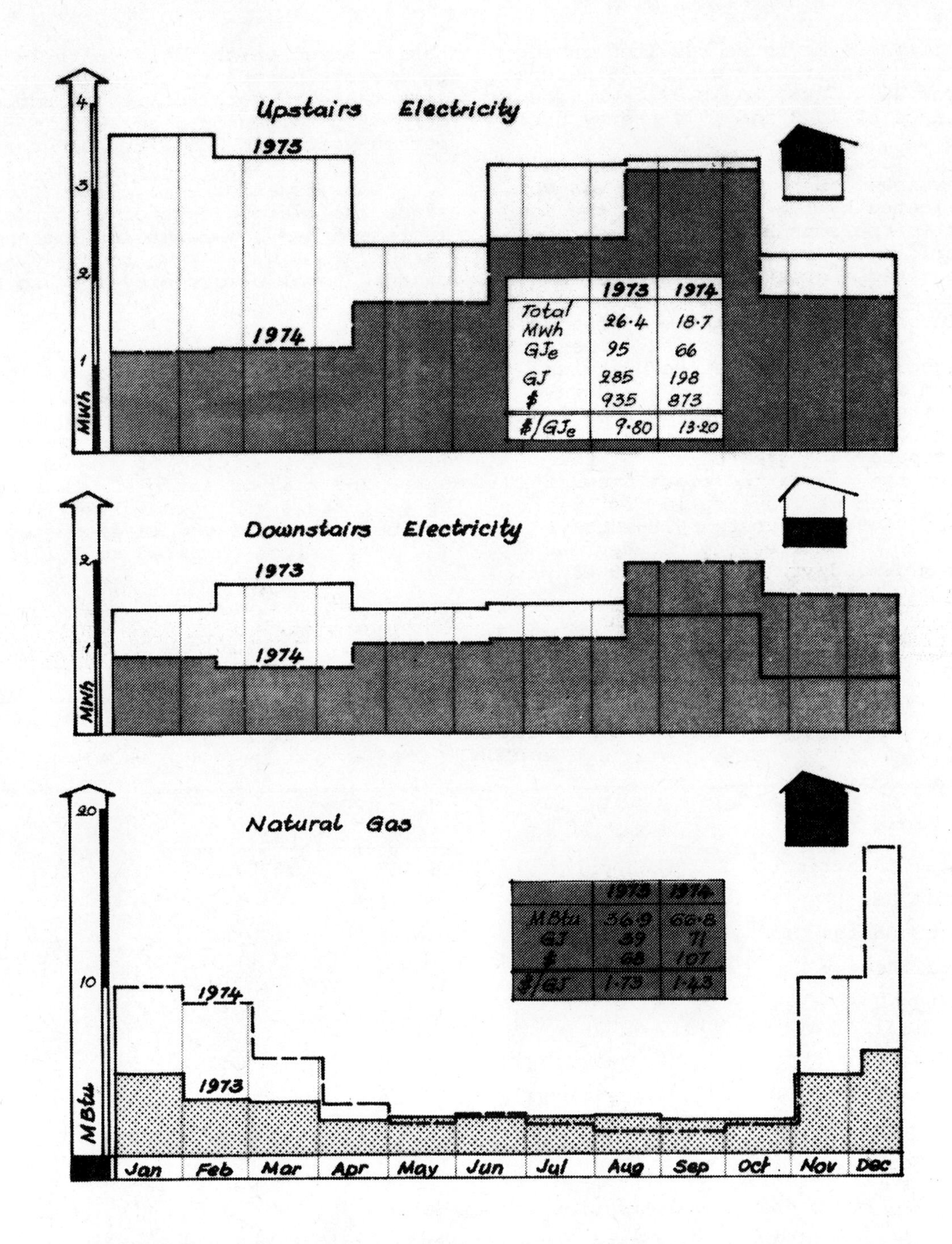

FIGURE 1

BUILDING ENERGY USE

These measures resulted in reducing electricity use by 28 percent, from 26,407 kWh to 18,772 kWh (see graphs of 1973 and 1974 energy use).

Total energy use was also determined. Here the picture was complicated by the increase in the number of persons being served by the building. By 1974 the larger staff required a greater use of the downstairs area, which was colder in the winter for lack of the very features--solar heat on the roof and excessive lighting--which made the upstairs warm in the summer. In the downstairs area, the lighting heat load was negligible in the winter. The occupancy and greater use of the downstairs area therefore necessitated increased gas consumption for heating, from 36.9 MBtu/yr to 66.8 MBtu/yr (1974). Total energy use decreased from 324 GJ/yr (1973) to 269 GJ/yr (1974).

More revealing is the energy usage and cost on a per person-year basis, based on the average full-time equivalent occupancy of the building (see table). For 1973, these numbers were 39.5 GJ/person-year (37.6 MBtu/person-year) and 125 $/person-year.

This is all the more significant since the average cost of electricity increased by 31 percent for the same period, from 0.035 $kWh to 0.046 $kWh, causing total electricity cost to be only $62 less.

The savings resulting from these measures can be estimated as follows. Assume 1973 electricity use (26,407 kWh, 285 GJ) remained the same in 1974 and include actual 1974 gas usage of 71 GJ, giving a total of 356 GJ, or 356 - 269 = 87 GJ savings. This saving is equivalent to approximately two tons of fuel oil and saved an estimated $350 (at 1974 prices compared to 1973).

ENERGY USE DATA FOR 279 m^2 (300 ft^2)
ENGINEERING OFFICE

		1973	1974
Total Electricity	(kWh/yr)	26,407	18,772 (1)
Total Gas	(MBtu/yr)	36.9	66.8 (2)
Total Energy Use (3)	(MBtu/yr)	308	256
Total Energy Use (3)	(GJ/yr)	324	269
Energy Use	($MJ/m^2 \cdot yr$)	1,161	964
Performance Factors (4)	(GJ/person·yr)	39.5	23.8
Total Cost	($/person·yr)	125	81

Notes

(1) Decrease due to conservation measures.

(2) Increase due to increased occupancy and possibly reduced lighting.

(3) Total energy used. Includes both electricity and gas.

(4) The number of persons is average full-time-equivalent employees for year.

RECOMMENDATIONS

The building in this case history was a small one; clearly the potential for similar or greater savings in large buildings is significant. This example points out that electricity reductions in office buildings of 20-30 percent without impairing performance are possible. However, the exact savings in a particular case depends on the type of building and the energy uses involved. Energy management techniques should be applied to use energy effectively and reduce costs. An average use index provides a measure of the performance of the building in providing occupant services efficiently.

CASE STUDY 3-5: SOLAR RETROFIT TO 16-STORY APARTMENT BUILDING PROVIDES HOT WATER AND SAVES ELECTRICITY[1,2]

THE CONCEPT

A solar water heating system was added to an existing 16-story all-electric apartment building. The system was expected to reduce hot water electricity usage by as much as 60 percent.

BACKGROUND

The building, located in Brookline, Massachusetts, consists of 230 apartments for the elderly. The solar heating system was designed to supply domestic hot water to the entire building and to supplement the existing hallway space heating system. The total cost of the installation was $96,000, with an estimated 8.5 year payback based on escalated electricity savings. The system became operational in 1976.

FINDINGS

The system uses 88 flat plate collectors (170 m^2, 1,830 ft^2) mounted on the roof of the building. The collectors are tilted at an angle of approximately 42° above the horizon (latitude angle) to provide a relatively consistent hot water supply throughout the year.

A closed heat transfer loop is used to couple the solar collectors to the heat storage tanks (see Figure 1). The system uses an existing 2000-gallon electric domestic hot water tank in addition to three new 350-gallon storage tanks. Also, there is a circulating pump and a shell and tube heat exchanger. A non-toxic anti-freeze under pressure is the heat transfer fluid.

The heat exchanger is connected to the three 350-gallon preheat tanks, located in the basement and connected in series. Energy is transferred to the three storage tanks when circulating pumps 1 and 2 are operational. The circulators are activated when a differential thermostat, which senses and compares the collector plate temperature and the storage tank temperature, provides a signal.

The solar heated water then enters the electric water heater. If the incoming water temperature exceeds 60°C (140°F), electric heating is not used. Automatic mixing with supply water is used to regulate the temperature. Below 60°C, supplemental electric heating is used.

The roof-top heat exchanger is also connected (through a third circulating pump and a water-to-air heat exchanger) to the existing fan-coil unit used for hallway heating. If both the solar unit and the fan-coil unit are operating, solar heated water is used to preheat the cold air entering the fan-coil unit.

The solar unit was financed by the Massachusetts Housing Finance Agency, which financed the original building. Under the terms of the mortgage (40 years at 8.75 percent interest) the annual cost of the system is 8,310 $/year. First year savings for the system are estimated to be 8,331 $/year, net of maintenance and operating costs. Using an 8 percent annual escalation for electricity price (including inflation), the saving grows to 36,000 $/year at the end of a 20-year projected equipment life. Cumulative savings pay back the original investment in 8.5 years.

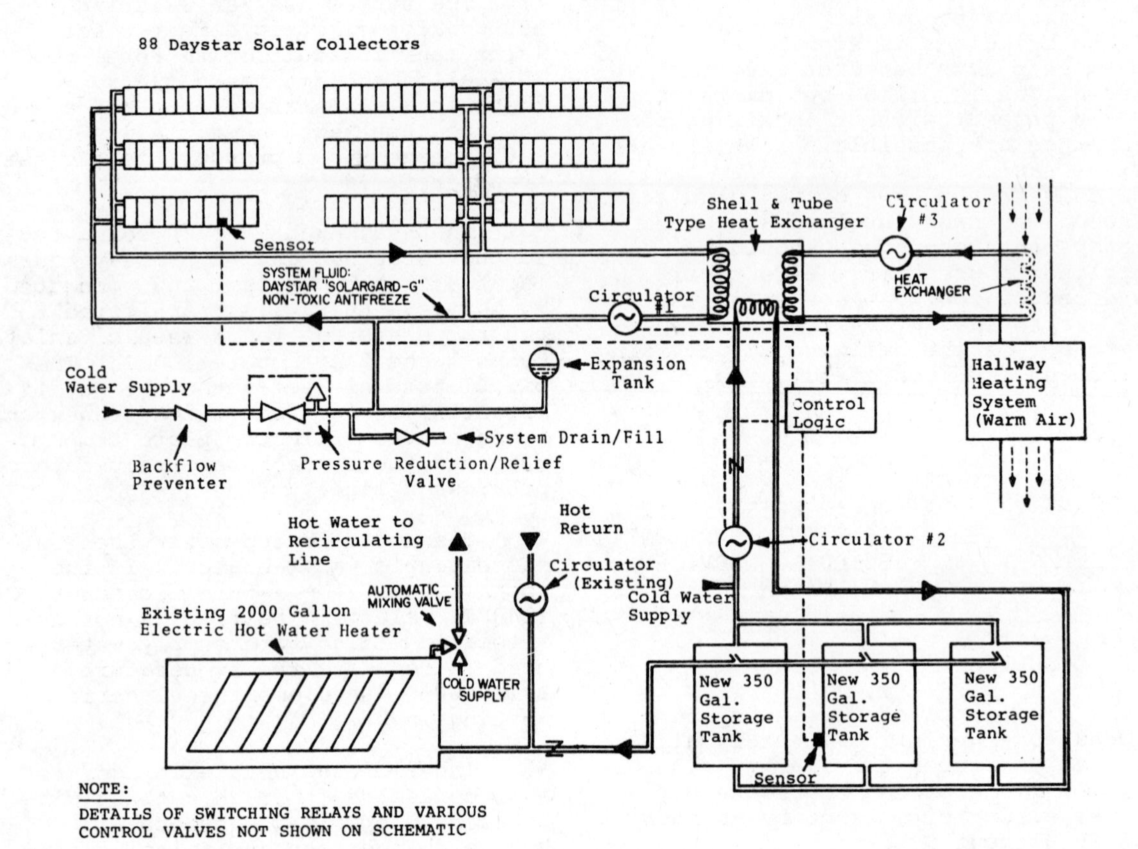

SOLAR SYSTEM SCHEMATIC FIGURE 1

RECOMMENDATIONS

Alternative energy forms such as solar should be considered for low temperature heating purposes where the high price of electricity or conventional fuels, or limited availability, makes the alternative form economically viable. Extensive research and development of solar equipment is bringing about rapid technological changes in this field. Today there are literally hundreds of demonstration projects scattered throughout the world. As results from these projects become available, new opportunities for energy management using alternative energy forms will become available.

References

1. Anonymous, "Brookline High Rise Apartments Put Solar System into Operation," *Solar Engineering*, (November 1976): 14.

2. Anonymous, "One Solar System Heating Water for 230 Apartments," *Boston Sunday Globe*, (17 October 1976).

Acknowledgments

This case study was provided through the courtesy of Mr. Orin F. Zimmerman, General Manager, Conservation and Energy Management Department, Portland General Electric Company. In addition, we are indebted to Dr. Richard Cummings, Executive Vice President of Daystar Corporation, Burlington, Massachusetts, for providing technical and economic details on the project. Daystar Corporation designed the system and fabricated the solar collectors.

15.4 RESIDENCES (Chapter 4)

CASE STUDY 4-1: SELECTION OF RESIDENTIAL AIR-CONDITIONING UNITS ON A LIFE CYCLE COST BASIS

THE CONCEPT

Life cycle costing would reduce the incentive to purchase inefficient room air conditioners having lower first cost.

BACKGROUND

Room air-conditioner units retail for \$125-\$250 each (1973-74) and have energy efficiency ratios (EER) ranging from 1.5-3.0 J/J (5-10 Btu/hr per watt electrical). Despite these variations, most units are purchased on initial cost considerations only, ignoring significant differences in energy costs over the lifetime of the equipment.

FINDINGS

Table 1 shown on the next page illustrates the procedure for sizing a room air conditioner. Using this method, the proper unit capacity can be selected. In the comparison that follows, we assume a cooling capacity requirement of 1465 W (5000 Btu/hr), use of 10 hours per day, 6 months per year, that the compressor is on 70 percent of the time, and that electricity costs \$0.05 per kWh.

A high efficiency 5500 Btu/hr unit has an EER of 8.8, needs 625 W input, draws 5.5 A, and typically costs \$200. The annual energy use for this unit is approximately:

$$(625)(10 \text{ hrs})(180 \text{ days})(0.70)\left(\frac{5000}{5500}\right)(3600)$$

$$= 2.58 \text{ GJe } (716 \text{ kWh/yr})$$

This corresponds to an annul operating cost of \$36.*

---

*See Chapters 4 and 8 for additional data on air conditioners and typical EER values.

COOLING-LOAD ESTIMATE FORM FOR ROOM AIR-CONDITIONERS (1)

See the instructions for each step in the list starting at the bottom of the page.

HEAT GAIN FROM	QUANTITY	FACTORS				QUANTITY X FACTOR
1. WINDOWS: Heat gain from sun.		No. shades*	Inside shades*	Outside awnings*	(Area x Factor)	Make only one entry: largest figure from column to left
Facing northeast	______ sq. ft.	60	25	20	______	
Facing east	______ sq. ft.	80	40	25	______	
Facing southeast	______ sq. ft.	75	30	20	______	
Facing south	______ sq. ft.	75	35	20	______	
Facing southwest	______ sq. ft.	110	45	30	______	
Facing west	______ sq. ft.	150	65	45	______	
Facing northwest	______ sq. ft.	120	50	35	______	
Facing north	______ sq. ft.	0	0	0	______	
2. WINDOWS: Heat gain by conduction (total of all windows)						
Single glass	______ sq. ft.			14		______
Double glass or glass block	______ sq. ft.			7		______
3. WALLS: (Based on linear feet)		Light construction		Heavy construction		
a. Outside walls						
North exposure	______ ft.	30		20		______
Other than north exposure	______ ft.	60		30		______
b. Inside walls (between conditioned and unconditioned spaces only)	______ ft.		30			
4. CEILING: (Use one only.)						Enter one figure only
a. Uninsulated with no space above	______ sq. ft.			19		
b. Insulation 1 inch or more, no space above	______ sq. ft.			8		
c. Uninsulated with attic space above	______ sq. ft.			12		
d. Insulated with attic space above	______ sq. ft.			5		
e. Occupied space above	______ sq. ft.			3		______
5. FLOOR: (Disregard if floor is on ground or over basement.)	______ sq. ft.			3		______
6. NUMBER OF PEOPLE:	______			600		______
7. ELECTRICAL EQUIPMENT:	______ watts			3		______
8. DOORS AND ARCHES OPEN TO UNCOOLED SPACE:	______ ft.			300		______
9. SUBTOTAL:	xxxxxxx			xxxxxxx		______

These factors are for single glass only. For glass block, multiply the above factors by 0.5; for double glass or storm windows, multiply above factor by 0.8.

10. TOTAL COOLING LOAD: Btu per hour to be used for selection of room air-conditioner(s) ______ (Item 9) X ______ (Factor from map) = ______

INSTRUCTIONS

The number of each of the following paragraphs refers to the correspondingly numbered item on the form:

1. Multiply square feet of window area for each exposure by applicable factor. Window area is the area of wall opening in which window is installed. Include the window in which the air-conditioner is to be installed. For windows with inside shades or blinds, use factor for "inside shades." For windows with outside awnings (with or without shades or blinds), use factor for "outside awnings." Only one number should be entered in the right hand column for item 1, and this number should be the largest "area x factor" figure.

2. Multiply total square feet of all windows in room by the applicable factor taken from the column at right.

3a. Multiply total length (linear feet) of all walls exposed to the outside by the applicable factor. Doors should be considered as part of the wall. Walls shaded by adjacent structures should be considered as having "north exposure." Do not consider trees and shrubbery as shading. An uninsulated frame wall or a masonry wall eight inches thick or less is considered "light construction." An insulated frame wall or a masonry wall more than eight inches thick is considered "heavy construction."

3b. Multiply by the factor given the total length (linear feet) of all inside walls that separate the space to be cooled from adjacent uncooled spaces.

4. Multiply total square feet of ceiling area by the factor given for the type of construction most nearly describing that of your own ceiling. Use one line only.

5. Multiply total square feet of floor area by the factor given. Disregard this item if the floor is directly on the ground or over a basement.

6. Multiply the number of people who will normally occupy the cooled space by the factor given. Use minimum of two people.

7. Determine the total number of watts for lights and electrical equipment in the cooled area (except the air-conditioner itself) that will be in use when the conditioner is operating. Lights are marked with their wattage demand. Allow 200 watts for a black-and-white TV set. 350 watts for a color-TV set, 300 watts for a radio-phono system. 50 watts for a table radio using tubes. Other appliances may give wattage on their nameplates; if not, multiply the nameplate amperage by the voltage for a rough estimate. Multiply the total wattage by the factor given.

8. Multiply by the factor given the total width (linear feet) of any doors or arches continually open to an uncooled space. Where such a door or arch is more than five feet wide, the rooms so connected should be considered as a single large room, and the room air-conditioner unit, or units, should be selected according to a calculation made on both rooms.

9. Total the loads estimated for items 1 through 8.

10. Multiply the figure obtained in item 9 (subtotal) by the proper correction factor, selected from the map at right for your locality. The result is the required design cooling load estimated in Btu per hour. For best results select a room air-conditioner unit or units having rated cooling capacity as close as possible to the estimated load. In general, a greatly oversized unit would operate intermittently and be much less satisfactory than a slightly undersized unit, which would operate more nearly continuously.

NIGHT COOLING. The preceding provides Btu/hr. figures for daytime cooling. Where an air-conditioner is to be used only for night cooling, the following revisions should be made. Disregard item 1. In item 3, the factor for all walls is 30. In item 4, the factors are 5, 3, 7, 4, and 3 respectively for a, b, c, d, and e. The factor for item 8 is 200. All other factors remain the same.

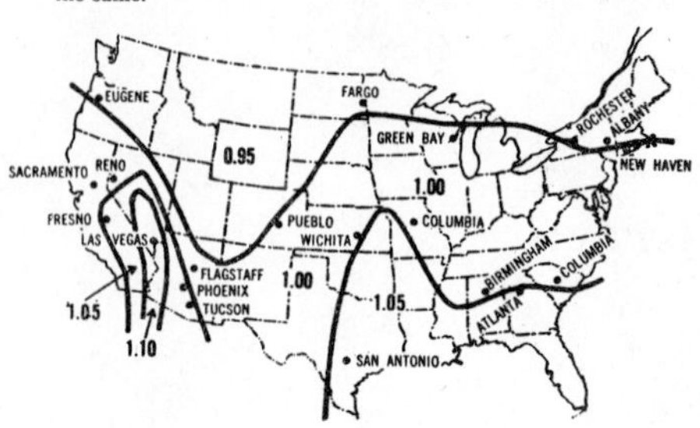

Now consider a typical less efficient unit, rated at 5000 Btu/hr, with an EER of 5.5, requiring 910 watts input power, and costing $150. The annual energy use for this unit is:

$$(910\ W)(10\ hrs)(180\ days) \times (0.70)(\frac{5000}{5000})(3600) = 4.1\ GJe\ (1146\ kWh/yr)$$

This corresponds to an annual operating cost of $57.

The more efficient unit has a housing guarantee of 10 years and other parts guaranteed for 1 and 4 years. Thus an operating life of 10 years is reasonable. The economic decision is one of determining whether it is justified to spend $50 now to recover $57-36=$21/yr for 10 years. Neglecting interest, the more efficient unit will pay for itself in two years, five months.

Comparison of the two units on a life cycle cost basis is shown in Table 2. The results indicate that the inefficient unit would use about 58 percent more energy over its lifetime and would cost about 28 percent more over its lifetime than the more efficient unit.

Another approach is to determine the possible savings between two units and use this information to determine how much greater the first cost could be and still be competitive (see Table 3).

These numbers indicate that a purchaser who demands an 8 percent return on his money could afford to spend $67.10 more for an air-conditioner unit which would be expected to save 200 kWh/yr over its 10-year life. In the example given above, the saving is more than 400 kWh/yr, meaning that the greater cost ($50) of the more efficient unit is economically attractive.

RECOMMENDATIONS

Consumers should be provided data to evaluate life cycle costs of air conditioners and other household appliances. Appliance manufacturers should be given legislative or tax incentives to improve device efficiency.

References

1. Anonymous, *Consumer Reports* 38(July 1973): 443-451. Table reproduced through courtesy of publisher.

CASE STUDY 4-2: RESIDENTIAL HEAT PUMPS[1,2]

THE CONCEPT

For many years the heat pump has been recognized as an efficient heating system. It will normally produce two or more units of energy in the form of heat for every unit of electrical energy required in its operation. The Tennessee Valley Authority (TVA) has a program for informing the public, training installation personnel, certifying competent dealers, and inspecting heat pumps after installation. The TVA program has become a model for other utilities.

BACKGROUND

As of 1976, it was estimated that there were approximately 75,000 electric heat pump installations in the TVA service area. Since the heat pump heats a home for approximately 50 percent less, when compared to electric resistance heating, this corresponds to saving more than 500 x 10^6 kWh of electricity every year (equivalent to 260,000 tons of coal saved each year).

Typical efficiency curves for a 2.5 ton heat pump (1 ton = 12,000 Btu/hr = 3.5 kW) are shown in Figure 1. The figure indicates that the COP (coefficient of performance) decreases as the outside air, which is the heat source, decreases in temperature.

The COP refers to the refrigeration unit operating alone. The auxiliary heat and the defrost cycle also use energy throughout the heating season. Another measure of efficiency, the seasonal performance factor (SPF), includes the additional energy used by this equipment to give a more accurate

	Cost, $	
	Efficient Unit	Inefficient Unit
Initial Cost	200	150
10 Year Operational Cost	360	570
Life Cycle Cost	560	720
	Energy, GJ	
Energy to Manufacture (∿100 MJ/kg)	2.7	2.2
10 Year Operational Energy	77.4	124.0
Life Cycle Energy	80.1	126.2

LIFE CYCLE COST COMPARISON OF TWO AIR CONDITIONERS **TABLE 2**

Electricity Savings in kWh/yr	200	400	600	800
Value of Electricity saved in $/yr (@ 0.05 $/kWh)	10	20	30	40
Acceptable difference in $ (life = 10 years interest @ 8%)	67.10	134.20	201.30	268.40

DIFFERENCE IN ANNUAL COST BETWEEN EFFICIENT AND INEFFICIENT ROOM AIR CONDITIONING UNITS **TABLE 3**

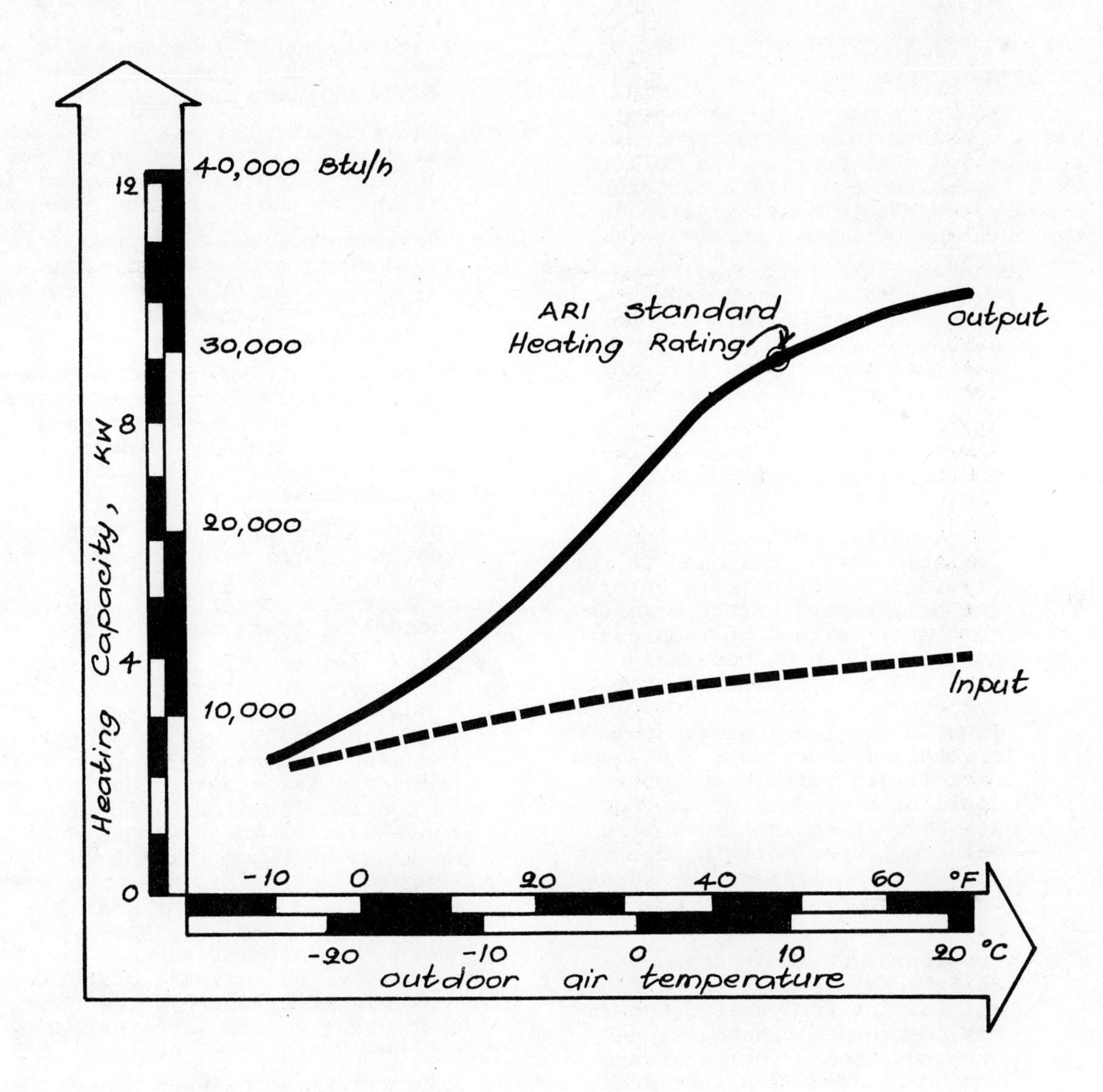

TYPICAL 2 1/2 TON HEAT PUMP PERFORMANCE FIGURE 1

estimate of seasonal energy use.

FINDINGS

The Tennessee Valley Authority has a consumer information program for residential heat pumps. The following is a list of some of their recommendations which should be considered in the purchase of a home unit.

General: The following items are recommended heat pump features. The absence of certain of these features, however, does not indicate that a heat pump will not give reliable and good service.

-All Temperature Compressor: Should operate at outdoor temperatures from -18 to 40°C (0 to 105°F)

-Suction-line Accumulator or other reliable device designed to keep liquid refrigerant from entering the compressor: Since compressors are designed to pump gas, not liquid, it is necessary to have a dry compressor operation through its complete range. Thus, a suction-line accumulator or other device traps and holds back liquid refrigerant which could otherwise enter the compressor. Keeping liquid refrigerant from entering the compressor also lessens the stress on the compressor and reduces wear on its parts.

-Crankcase Heater or some other reliable means of keeping liquid refrigerant from collecting in the compressor crankcase: A crankcase heater helps assure that compressor bearings are properly lubricated by removing liquid refrigerant which can dilute the oil necessary for proper lubrication.

-Liquid-line Filter/Dryer: Helps keep system clean and free of acid and moisture.

The following items are external to the heat pump itself, but they should be given important consideration in order to obtain a good total heat pump system.

- Duct System: Properly designed and adequately sized ducts are of great importance. A heat pump's performance is hampered and some experts assert mechanical problems are hastened by an inadequate duct system, particularly one with undersized air ducts. Ducts must be large enough to provide a minimum of 400 CFM per ton.
- Indoor Thermostat: Should have two stages of heating, one stage for cooling. This is a convenient location for the emergency heat switch.
- Supplemental Heat: Automatic resistance-type heaters in unit or duct. Operate when necessary to maintain the temperature level you've selected.
- Heat Switch: For manual activation of the resistance-type heaters in case the compressor becomes inoperative. Usually referred to as emergency heat switch. It is recommended.
- Outdoor Thermostat(s): Used to prevent the supplemental heaters from switching on before needed, helping to insure a comfortable, efficient, economical operation.
- Size: Let your power distributor assist you in determining capacity. Since an electric heat pump both heats and cools, how do you calculate the size needed? The normal procedure is to determine the amount of cooling required. Then resistance heaters are added to supplement the heat pump at low outside temperatures. This way, the system will produce the right amount of cooling and heating. Once you've found the watts or tonnage required, don't make the mistake of thinking that bigger is better. Get the exact size needed. The same applies to the supplemental heaters.
- Manufacturer: Compare brands, price, and *value*. Look for certification seals such as UL (Underwriters' Laboratories) and ARI (Air-Conditioning and Refrigeration Institute).
- Efficiency Ratings: The energy efficiency ratio, or EER, is an indication of the heat pump's cooling efficiency. The EER is calculated by dividing the cooling capacity in Btu/hr by the power input in watts, and is expressed in Btu/hr per watt.

  The coefficient of performance, or COP, is an indication of the heat pump's heating efficiency. The COP is calculated by dividing the heating capacity in watts by the power input in watts; thus COP is the watts output divided by watts input. The higher the EER and COP, the greater the cooling and heating efficiency of the unit.

RECOMMENDATIONS

Heat pumps offer important benefits to homeowners, both in terms of energy savings and economics. Select reliable, properly sized units for best performance.

References

1. Tennessee Valley Authority, Power Marketing Division, *A Heat Pump Program for the TVA Area*, (Chattanooga, Tennessee: August 1975).

2. Tennessee Valley Authority, Division of Power Utilization, "Guide for Buying an Electric Heat Pump," Brochure #B/7752/MET.

Acknowledgments

This case study is based on information provided through the courtesy of Mr. J.W. Ward, Chief, Electrical Demonstration Branch, TVA, Chattanooga, Tennessee.

CASE STUDY 4-3: ENERGY MANAGEMENT APPLIED TO RESIDENCES

THE CONCEPT

This case study shows how to save energy and money by the use of energy management practices in eight residential energy using areas. Measurements were conducted in a residence under typical operating conditions. Energy management options and efficient operational strategies are described and energy and dollar savings reported.

BACKGROUND

The study was conducted in a residence located in Southern California. The family (four persons) lives in a single story, wood frame house. The major electric appliances include lighting, a range with a microwave oven, dishwasher, refrigerator, clothes washer and dryer, and miscellaneous appliances. Space and water heating are natural gas. Calculations of savings and payback on investment are based on 1976 energy prices (5 ¢/kWh and 21 ¢/therm).

An energy audit was made on all energy using appliances and systems. Complete utility records were examined and meter readings were taken using the household gas meter, watt-hour meter and a portable appliance watt-hour meter. A temperature recorder was used to sense refrigerator and water heater operations. Light meter readings were taken to determine lighting levels.

FINDINGS

Energy Savings From Home Insulation

The residence is 150 m^2 (1600 ft^2) and has two gas-fired wall furnaces. One is rated at 13.2 kW (45 kBtu/hr) and the other at 10.3 kW (35 kBtu/hr). In 1973, 8.9 cm (3.5 in.) (R-11) of foil-backed fiberglas batting was added to the attic of the original house and to the dining room walls which were ripped apart during remodeling. (Two bedrooms which had been added to the original house already were insulated.) The family installed the insulation themselves, and the total cost was $150.

The added insulation saved 18 GJ (17 MBtu) in 1974, which amounted to $35 per year at 2.10 $/MBtu. This resulted in a four year payback. The following year the family turned off the gas pilot lights from June to October, recalibrated the thermostat and set it at 20°C (68°F) during the day and 15.6°C (60°F) at night. The wall furnace in the rear of the house was turned on only when someone was using that area. That year 23 GJ (22 MBtu) were saved which amounted to $46 per year. If these additional energy management strategies are taken into account, then the payback is reduced to 3.2 years at present prices.

It may be concluded that the savings due to insulation are significant even in a mild climate. The payback in several California homes which were studied ranged from three to six years. If the heating were electric, the weather more extreme, or air conditioning were used, the payback period would be greatly reduced.

The price of natural gas is expected to rise due to shortages and price deregulation. Escalation of natural gas prices can probably be expected to double fuel prices, cutting the payback in half. When this occurs, there will be a greater incentive to reduce thermal losses in the home.

The Efficiency of Water Heater Insulation

This section reports on the actual energy and dollar savings resulting from water heater insulation. The gas water heater holds 0.114 m^3 (30 gal) and is rated at 13.2 kW (45,000 Btu/hr). It is located outside the house (west side) in a galvanized steel enclosure. The measurements were made in May, in generally mild (typically 15-21°C;

60-70°F) weather. A seven day temperature recorder was placed on top of the water heater. Following the measurements the heater was insulated with 8.9 cm (3-1/3 in.) of R-11 foil backed, fiberglas insulation, held in place with duct tape.

The week before adding insulation, gas usage for water heating was measured at 0.95 GJ (0.9 MBtu). The week after insulating the tank, usage was down to 0.74 GJ (0.7 MBtu) --a saving of 0.21 GJ per week. The number of "turn-ons" (as evidenced by spikes in temperature records) went from 50 down to 37 after insulating. This is a 26 percent reduction.

The savings from insulation are estimated to be about 22 percent averaged over the year--about 11 GJ/yr. The cost impact of this would be to save approximately $21 per year. The cost of the insulation and tape was $10. Thus the payout period is 10 ÷ 21 $/yr = 0.48 years, or about six months.

Relamping With Fluorescent Lights

The most frequently used lights were replaced by fluorescent lamps over a two year period (1973-1975). To summarize, 1.11 kW of incandescent lighting was replaced by 0.515 kW of fluorescent lighting--a difference of 0.595 kW. In rooms with two or more fixtures, separate switches were installed.

The energy audit method was used to determine lighting use before and after adding fluorescent lamps. This was compared with historical records for additional verification.

Lighting levels for both fluorescent and incandescent lamps were measured. In most cases the fluorescent fixtures had a diffuser covering the tubes. The incandescent light was measured using a bare bulb.

The light meter measurements indicated substantially equivalent light levels in most rooms. In some areas illumination levels were reduced intentionally. In others, lower wattage (compared to incandescent) fluorescent lamps provided more illumination. Lighting levels (measured 75 cm [30 in.] above the floor) ranged from 110 lux (10 fc) in the halls to 320 lux (30 fc) in the study.

Light meter tests showed the greater efficiency of fluorescent lamps as compared with incandescent lamps. In some cases fluorescent fixtures gave off twice the amount of light for the equivalent wattage or gave the same amount of light as an incandescent bulb with twice the wattage.

Relamping with fluorescent lamps saved about 60 kWh per month. An additional 60 kWh was saved by the use of separate switches, delamping, turning off lights in unoccupied rooms, and reducing use of outdoor lights.

Savings of 120 kWh per month or 1,440 kWh per year translates into a saving of $72 a year. The cost was $220 for the fluorescent lamps, new switches, miscellaneous hardware and labor. The payback on this investment was roughly three years.

Energy Use in a "Frost Free" Refrigerator

Electricity use was measured and energy management opportunities in a refrigerator under actual operating conditions were investigated.

The 0.4 m^3 (19 ft^3) refrigerator is located on the west side of the residence. It is near a window and, therefore, receives some direct sunlight in the late afternoon. The measurement was made in May in generally mild (15.6-23.9°C [60-75°F]) weather. The ambient temperature in the kitchen varied from 19°C (66°F) to 22°C (72°F). The temperature inside the refrigerator averaged 4.4°C (40°F) and the freezer temperature averaged -13.9°C (7°F) when the compressor was off.

From 7:30 p.m. Sunday, May 2, to 7:30 p.m., Sunday, May 9, the refrigerator used 28.2 kWh or 4 kWh per day. This agreed with the data published in Reference 1. The total household electricity use averages 321 kWh per month, or approximately 10 kWh daily. The refrigerator accounts for 37 percent of the electricity used.

In the household where the experiments were made, this appliance used more electrical energy than any other appliance or system. At 120 kWh per month with a charge of 5¢/kWh, it costs $72 per year to operate.

Experiments show that opening the door has a moderate effect and should be minimized. The addition of warm or hot food should be avoided. However, even though the data show increased energy use from adding food and opening the door, the total daily energy use is little affected by these actions.

The most efficient strategy for saving energy with the refrigerator comes not so much from efficient operation but from making a wise purchase. The refrigerator should suit the family's needs but should not be larger than necessary. The larger the refrigerator and the more work-saving features it has, the more electricity it needs to operate.

In California, new efficiency standards have been proposed for refrigerators. Appliance manufacturers are developing more efficient, better insulated refrigerators that may save half the energy. These refrigerators may cost from 75 to 100 dollars more but the difference will be paid back in two to three years in reduced energy bills.

Energy Efficient Cooking

The energy use of typical cooking appliances was measured in order to determine more energy efficient cooking practices.

Measurements were obtained in a home during the process of preparing foods for a family of four. The appliances used in the study were: an electric range consisting of a smooth "Corning" cooktop (1.50 kW/burner), a large continuous cleaning lower oven (5.2 kW), and an eye-level microwave oven (1.3 kW); a toaster oven (1.0 kW); an electric fry pan (1.15 kW); a 5 liter crock pot (0.15 kW); and a 2 slice, pop-up toaster (1.0 kW).

The results of this study have been tabulated in Table 1. The microwave oven was found to be the most energy efficient and, as an extra bonus, cooked food more quickly. The toaster oven was efficient for some uses. Using the crock pot was another energy saving way to cook. It took longer but used much less energy than a conventional oven. The electric fry pan was another excellent energy saver. It used less energy than all other appliances tested for cooking chicken and stew.

The large conventional oven proved to use the most electricity. It took 0.5-0.7 kWh to heat the cavity to the desired temperature and then it cycled on and off every one to three minutes. In order to make proper use of a big oven, large amounts can be cooked and then saved for another day's meal. Or, a complete oven meal can be planned where the main dish, vegetable, potatoes and dessert are all cooked at the same time. Tabulated results show that the microwave oven used less (1.2 kWh) even though the items have to be cooked one at a time. For most families who do not have this option the complete oven meal is the best method (2.5 kWh). The worst case would be to bake the dessert earlier in the day and then use one oven for the main dish, one for the potatoes and the stove top for the vegetable (5.2 kWh)(see Table 2).

Better energy management in cooking can save money. For this family, estimated savings were about 50 kWh per month which translates into $30 per year. This was accomplished by purchasing a range with an eye-level microwave oven in place of an eye-level conventional oven. This range cost $150 more but saves about $30 per year and will be paid back in five years. A typical cost for a portable microwave oven is $300. This ten year payback is long but with energy prices continuing to grow it will be reduced. Considering the savings of time and human energy, this longer payback period can be justified.

Additional savings come from using the large oven only when cooking several dishes, keeping pots covered and using small portable appliances where appropriate.

Food item	Appliance	kWh	Min	Temp°F	Comments
4 baked potatoes	Large oven	2.3	60	400	No preheat
	Microwave	0.3	16		
	Toaster oven	0.5	75	425	
Chicken (cut up)	Large oven	2.0	65	350	
	Microwave oven	0.55	25		
	Fry pan	0.5	60	300-350	Two chickens were used
	Crock pot	0.7	6.5 hr.	lo & hi	in crockpot experiment
Meat loaf	Large oven	2.0	60	350	
	Microwave oven	0.22	10		
	Toaster oven	0.35	60	350	
Brownies(9 X 13"pan) 23 X 33cm	Large oven	1.6	30	350	Microwave quality not
	Microwave oven	0.22	10		as good as conv. oven
Corn bread	Large oven	1.5	20	350	
	Microwave oven	0.15	7		Not browned well in MW
Frozen peas(petite)	Stove top	0.16	6	hi	Brought water to boil &
	Microwave oven	0.11	5		turned off burner. Peas continued to cook.
Beef stew	Large oven	2.7	5 hrs.	275	
	Crock pot	1.15	12 hrs.	10	Beef is browned first
	Electric fry pan	0.7	4 hrs.	300	in fry pan method.
Toast (2 slices French bread)	Pop up toaster	0.035			
	Toaster oven	0.060			
Toast (2 slices regular bread)	Pop up toaster	0.025			It takes over 17 pieces
	Large oven broiler	0.2			of toast in broiler to toast more efficiently than 2 slice toaster.
Water (2 cups cold start)	Stove (lid off)	0.176	7	hi	
	Stove (lid on)	0.151	6	hi	
	Microwave	0.121	5.5		

COMPARISON OF ENERGY USE IN COOKING APPLIANCES **TABLE 1**

Meal	Cooked together in a large oven	Cooked separately in microwave oven	Cooked in separate oven, stove top & toaster oven
2.3 kg (5 lb) ham (canned)	2.5 kWh	0.55 kWh	2.5 kWh (large oven)
0.23 kg (0.5 lb) frozen petite peas		0.11 kWh	0.16 kWh (stove top)
4 medium yams		0.35 kWh	0.5 kWh (toaster oven)
23 X 23 cm (9 X 9") pineapple upside down cake		0.22 kWh	2.0 kWh (large oven)
	2.5 kWh	1.23 kWh	5.16 kWh

ENERGY USED IN COOKING A COMPLETE MEAL THREE DIFFERENT WAYS

TABLE 2

Energy Management Strategies in Clothes Drying

Clothes drying represents 20 percent of the electrical energy used in the residence.

The dryer power was 6.0 kW. It used approximately 0.3 kWh to heat up before cycling off for the first time. Typical performance was as follows:

- 5.75 pounds of "permanent press" fabric (sheets, pajamas, pillow cases, blouses) took 25 minutes to dry at the "wash and wear" setting and used 1.7 kWh. (Wet weight 8.5 lbs - dry weight 5.75 lbs = 2.5 lbs water removed.)
- A 6.75 pound mixed load (T-shirts, blouses, socks, trousers, handkerchiefs) took 40 minutes to dry at the "wash and wear" setting and used 2.6 kWh. (Wet weight 10 lbs - dry weight 6.75 lbs = 3.25 lbs water removed.)
- 6.6 pounds of thick towels took one hour to dry at the hot setting and used 3.4 kWh. (Wet weight 12.3 lbs - dry weight 6.6 lbs = 5.7 lbs water removed.)

Based on a theoretical calculation for the vaporization of water, it should take 0.73 kWh to remove one kilogram of water. Averaging the above three experiments it took 1.52 kWh to remove one kg of water. Therefore, 52 percent of the energy input was dissipated as waste heat.

Obviously, it takes a significant amount of energy to heat up the dryer (0.3 kWh). Part of this heat can be saved each time by doing one load immediately after another. Note that most dryers have a cool-down cycle. In order to save the heat for the next load the dryer needs to be shut off before the cool down begins.

Sort the laundry so that quick drying items such as "permanent press" shirts and sheets are together. The "permanent press" load used almost one kWh less than the mixed load and 50 percent less than the load of towels.

The heavier the clothes the greater the amount of water they hold and the more energy it takes to remove it. Therefore, be sure that the washing machine goes through the complete spin cycle before clothes are placed in the dryer. A complete spin cycle in the washing machine measured less than 0.1 kWh.

These strategies do help but only save a small percent of the total energy used in residences. There is cause for concern when one takes into account all of the waste heat that is vented outside. With space heating the number one energy user in residences, it is important for engineers to design some practical ways for recovering dryer waste heat for household heating purposes. Alternatively, dryer heat recovery (about 50 percent of the heat is lost) may be feasible as a method of increasing dryer efficiency.

Electric Dishwashing

Dishwasher energy use was monitored during the wash and dry cycles in order to discover opportunities for more efficient use of energy.

The dishwasher in this study was a 1973 model rated at 1.15 kW. It does not contain a water heating element. The full wash cycle uses 51 liters (13.5 gals) per load and requires 0.125 kWh (25 min at 5 W/min). The pre-wash cycle uses 21 liters (5.5 gals) per load and requires 0.035 kWh. During the dry cycle, 0.45 kWh are required (25 min at 18 W/min).

By eliminating the dry cycle the consumer can save 162 kWh per year or \$8. By eliminating the pre-wash an additional 12 kWh can be saved, but more importantly, approximately 6800 liters (1800 gals) of hot water per year can be saved.

$$\frac{(1800)(8.3\ \text{lbs/gal})(1\ \text{Btu/lb/°F})(80°\Delta t)}{0.8\ \text{efficiency}}$$

$$= 1.57\ \text{GJ}\ (1.49\ \text{MBtu})$$

Eliminating the pre-wash saved approximately 1.6 GJ (1.5 MBtu) natural gas or 440 kWh of electricity per year for water heating. In dollars this amounts to \$3 saved on natural gas or \$22 saved if water is heated electrically.

Energy Use in Clothes Washers

A 1965 washing machine with a 68 ℓ (18 gal) tub capacity was used in this study. A watt hour meter was read during a typical eight minute load. Hot water energy use was determined by reading the gas meter before and after a hot water wash and warm water rinse and by calculations based on the following equation:

$$\frac{(\#\ \text{gal})(8.3\ \text{lbs/gal})(1\ \text{Btu/lb/°F})(\Delta t)}{0.8\ \text{efficiency}}$$

where Δt is 140°-60° or 80°F

Electricity use for washing was determined to be 155 watt hours per load. During the eight minute wash and two minute deep rinse, 72 watt hours were measured. Drain and spin time lasted a total of 13 minutes and measured 82 watt hours. Each additional minute of washing time is 7.2 watt hours. This summarizes all of the energy used during a cold water wash.

A hot water wash uses 68 ℓ (18 gal) of hot water in the wash cycle plus 34 ℓ (9 gal) of cold water in the rinse cycle. This is equivalent to natural gas usage of 23 MJ or 22 kBtu per load. If the water heater were electric, 6.5 kWh per load would be needed--more than a 14 ft^3 refrigerator uses in a day.

If three loads per week were changed from hot water to cold water, the resulting yearly savings would be 3.7 GJ (3.5 MBtu) or $7 per year. With an electric water heater, 1,024 kWh or $50 would be saved.

Converting all washing to cold water (assuming an average of five loads per week) would save 6.1 GJ (5.8 MBtu) or $12 per year. With an electric water heater, savings would be 1,700 kWh or $85 per year.

RECOMMENDATIONS

The family of four described in this study reduced energy use significantly by using the following energy management practices. First, operational strategies were developed to use appliances more efficiently. This required no capital investment. Secondly, on-site modifications were made, such as delamping and adding insulation. This required a modest capital investment. Finally, by the purchase of new energy efficient equipment, such as a microwave oven and fluorescent lamps, additional energy savings were realized. This required a larger capital investment but resulted in additional energy savings.

Household electricity use, as compared with use before the 1973 oil embargo, was reduced to 56 percent and natural gas use was reduced to 60 percent. Some of these results were reported previously in a 1975 study (Reference 2). Additional savings have been realized since the earlier study due to the addition of water heater insulation and implementation of new operational strategies (see Figure 1).

The total estimated savings are $200 per year. Seven hundred dollars were invested. This yields a 3.2 year payback. The results are summarized in Table 3.

References

1. Association of Home Appliance Manufacturers, *1976 Directory of Certified Refrigerators and Freezers*, Edition 1, (Chicago, Illinois: January 1976).

2. Smith, Nancy J. and Smith, C.B., "Direct and Indirect Use by Metropolitan California Families," in *Q=E³ The Future is Now, Greater Los Angeles Area Energy Symposium*, Los Angeles Council of Engineers and Scientists, Proceedings Series, Volume 1, pp. 11-27, A.D. Emerson, ed., (North Hollywood, California: Western Periodicals Company, 1975).

3. Smith, Nancy J., "Energy Management and Appliance Efficiency in Residences," *Energy Use Management, Proceedings of the International Conference*, (New York: Pergamon Press, 1977).

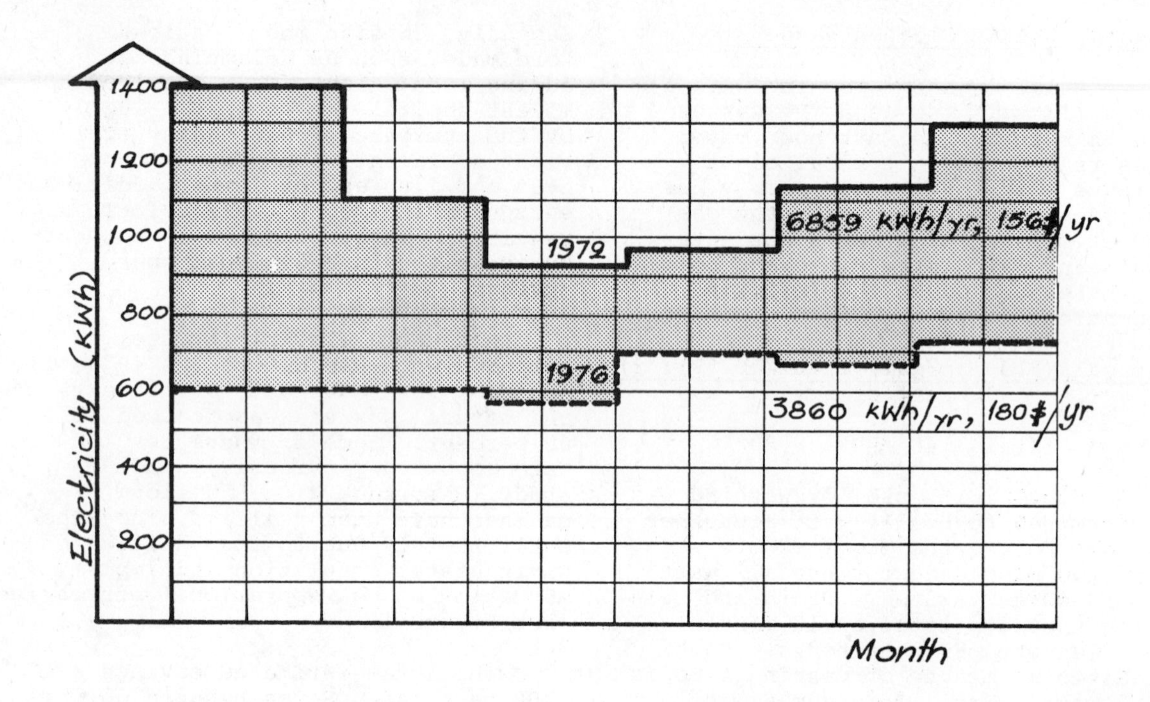

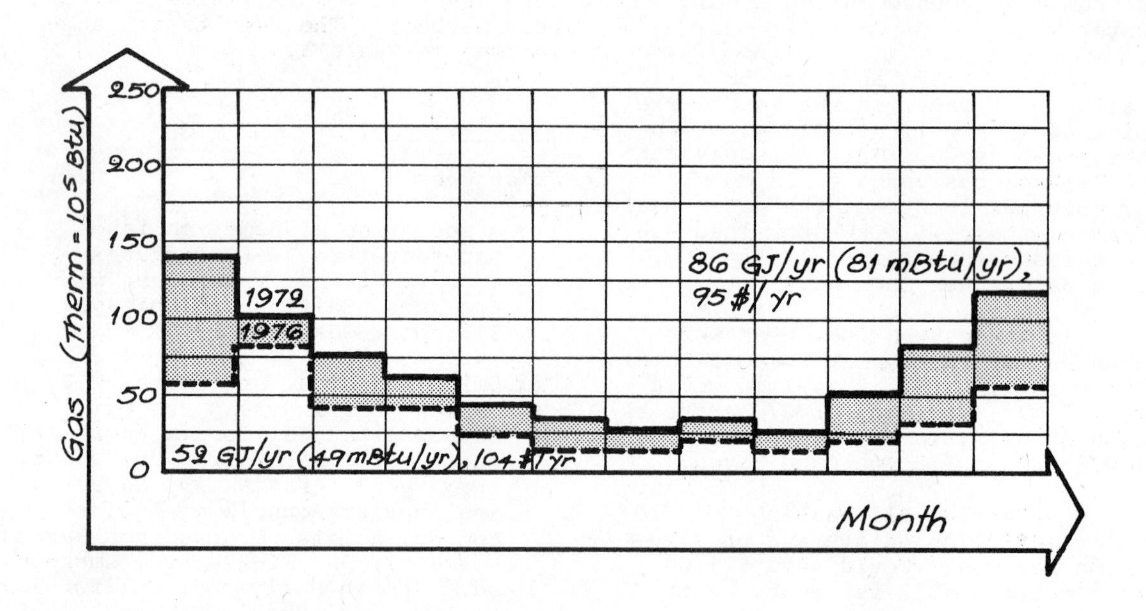

HOUSEHOLD UTILITIES FIGURE 1

ENERGY MANAGEMENT PRACTICE	CAPITAL COST $	YEARLY SAVINGS energy	$	PAYBACK YRS.
1. Install R-11 insulation in attic and lower thermostat to 68°F.	150	220 therms	46	3.3
2. Insulate water heater with R-11 insulation.	10	100 therms	21	0.5
NATURAL GAS SAVINGS	160	320 therms	67	2.3
3. Change 1/2 of lighting load to fluorescent, delamp, install separate switches.	220	1,440 kWh	72	3.0
4. Purchase microwave oven and use extensively.	300	600 kWh	30	10
5. Disconnect electric heaters in bathrooms.	20*	500 kWh	25	0.8
6. Eliminate dry cycle on dishwasher.	-	160 kWh	8	-
7. Miscellaneous: unplug TV cable, use appliances more efficiently, turn off unnecessary lights.	-	300 kWh	15	-
ELECTRIC SAVINGS	540	3,000 kWh	150	3.6
TOTAL INVESTMENT, SAVINGS AND PAYBACK	700		217	3.2

*labor cost
prices based on 5¢/kWh and 21¢/therm (2.10 $/MBtu). Note that 1 therm = 10^5 Btu.

RESIDENTIAL ENERGY MANAGEMENT OPPORTUNITIES

TABLE 3

CASE STUDY 4-4: THE "FAMILY CIRCLE" ENERGY HOUSE--A TOTAL ENERGY DESIGN

THE CONCEPT

To design a low-cost ($30,000) house which makes use of currently available, proven, energy-saving techniques.*

BACKGROUND

The working plans for an energy-saving house are available from the *Family Circle* magazine or the American Plywood Association.† This specially sponsored private architect/engineer design is estimated to save 50 percent of the electricity bill, plus additional savings in fuels and water.

Only present-day technology is used, but all advantages are pursued. The floor-plan and cut-away views shown in Figures 1 and 2 suggest how the heat re-use system, natural lighting, and central services are exploited.[1]

In summary, features include:

- Size: Wood frame, with 4-inch fiberglas insulation, caulked joints.
- Windows/Doors: Double frames, wood frames, weatherstripped, shaded overhangs.
- Climate Control: Minimum inside partitions; exterior wind screens and vestibules; zoned for temperature control.
- Plumbing: Located in central core; local hot water with flow control; basically one line supply.
- Appliances: Energy economical models; cool top range; microwave oven; tumbler washer.

*Price does not include cost of land.

†House plans are available from STJ, Inc., Family Circle Plans, Dept. 949, Box 450, Teaneck, New Jersey 07666. Cost is US $10.60.

FINDINGS

The solution adopted in this case was to consult with architects, engineers, designers, and manufacturers and to compile all currently available techniques for reducing energy use. The basic shape of the house is essential to its efficient use of energy, minimizing winter heating requirements and eliminating the need for mechanical cooling equipment. For summer cooling, the house functions like a chimney. Overhangs optimize winter solar heating and provide summer shading. Windows and frames reduce heat losses and weatherstripping eliminates air leaks. Appliances such as the microwave oven, cook-top, dishwasher, and clothes washer were selected to provide reduced energy use.

Interest in the demonstration house has been high, as evidenced by a report that within four weeks of publishing a report on the house, more than 3,000 copies of the house plans were sold.[2]

Annual water savings for a family of four are estimated at 72,000 ℓ (19,000 gal) per year. Total electricity savings are estimated as 1687 kWh/yr due to reduced hot water waste; additional savings result from use of low wattage lamps, fluorescent lamps, efficient heating, efficient appliances, and the microwave oven.

RECOMMENDATIONS

This example illustrates how the ideas described in Chapter 4 can be applied to home modification or new home design and can lead to significant improvements in energy efficiency. Those ideas should be considered for use by contractors or homeowners planning home improvements or new construction.

References

1. Anderson, Robert L. and Bishop, Carolyn, "Family Circle's 'Energy House' Slashes 50% Off Utility Bills," *Family Circle* (August 1974): 92.

2. "Energy House: 3000 Opt for Plan That Saves Fuel," *Professional Builder* (October 1974): 91.

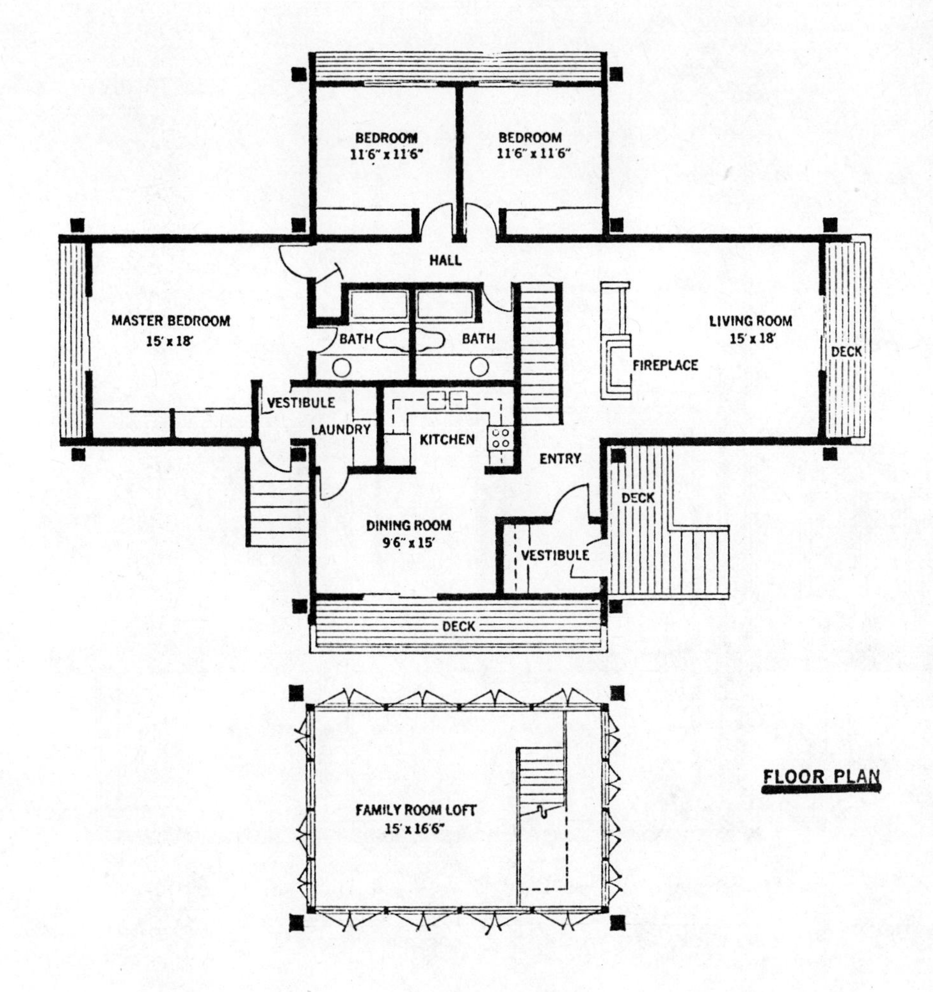

FAMILY CIRCLE'S 'ENERGY HOUSE'

FIGURE 1

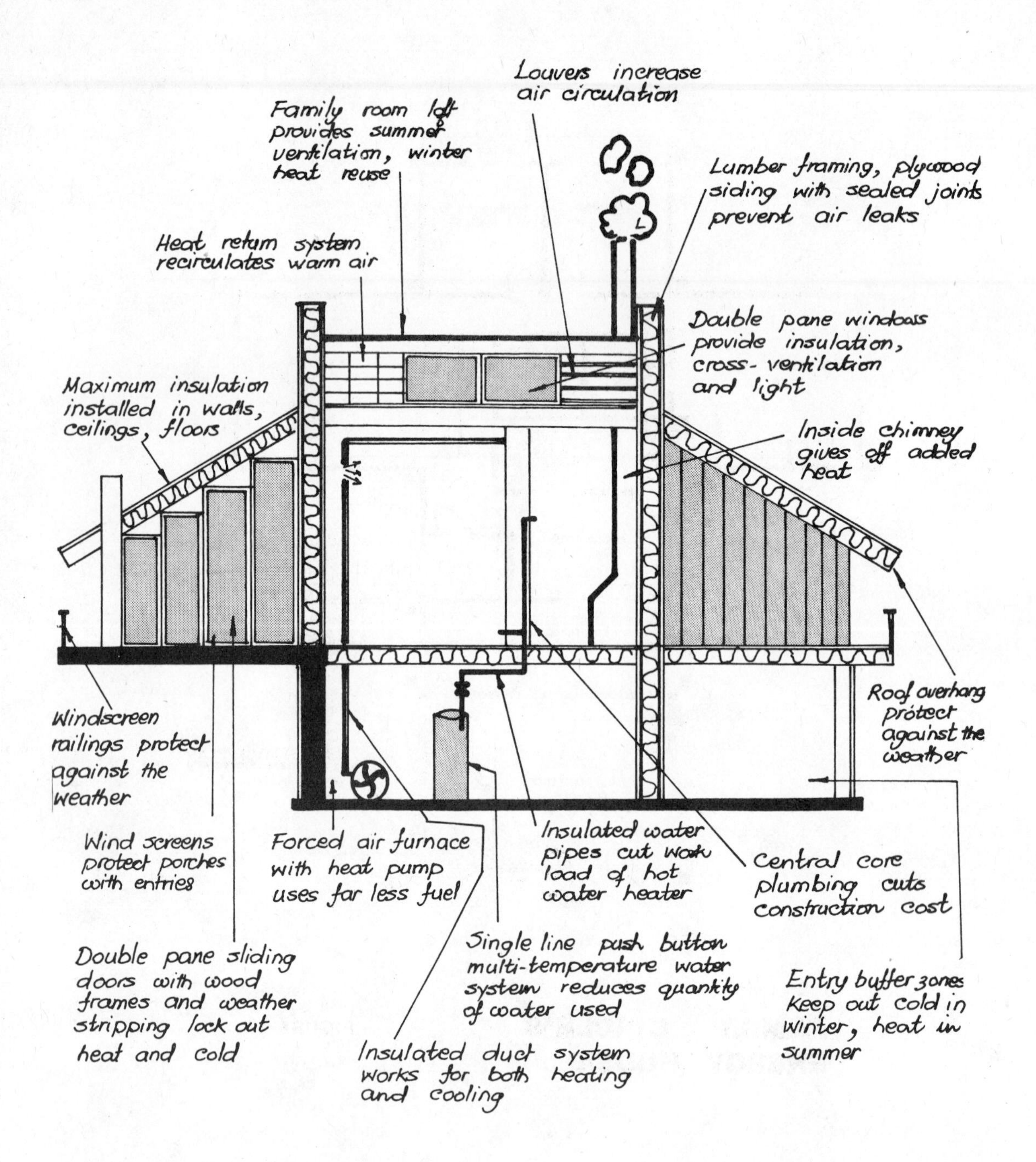
Louvers increase air circulation
Family room loft provides summer ventilation, winter heat reuse
Lumber framing, plywood siding with sealed joints prevent air leaks
Heat return system recirculates warm air
Double pane windows provide insulation, cross-ventilation and light
Maximum insulation installed in walls, ceilings, floors
Inside chimney gives off added heat
Windscreen railings protect against the weather
Roof overhang protect against the weather
Wind screens protect porches with entries
Forced air furnace with heat pump uses far less fuel
Insulated water pipes cut work load of hot water heater
Central core plumbing cuts construction cost
Double pane sliding doors with wood frames and weather stripping lock out heat and cold
Single line push button multi-temperature water system reduces quantity of water used
Entry buffer zones keep out cold in winter, heat in summer
Insulated duct system works for both heating and cooling

EFFICIENT FEATURES OF ENERGY HOUSE FIGURE 2

CASE STUDY 4-5: RESIDENTIAL HEAT RECOVERY USING A HEAT PUMP

THE CONCEPT

Waste heat in residences accounts for a significant fraction of household energy use inefficiency.

BACKGROUND

Heat recovery has been practiced in industry for many years, but only recently has it been considered for use in commercial and residential applications.

One approach adopted by several industry-utility groups has been that of developing experimental "demonstration" homes to try out new technological approaches.

For example, Westinghouse Electric Corporation and Pennsylvania Power and Light Company (PP&L)--among others--are planning and building energy-saving model residences designed to reduce electricity use by 30 to 50 percent. These homes will be open to the public, and will also serve as demonstration homes for additional new approaches for improving energy use efficiency.[1]

FINDINGS

The Westinghouse home, called *Electra III*, has these features:

- Total architectural design and siting of the structure to take maximum advantage of sunlight, shadows, and natural breeze ventilation.
- Use of fluorescent and mercury-vapor lamps.
- Use of waste heat and solar heat for both tap and swimming pool water.

The home, to be built in Florida, includes almost 267 m^2 of indoor living space, plus more than 184 m^2 of screened patio and pool area--really an "extension" of the indoor space.

The cost of the home's energy-saving components and subsystems will be evaluated and compared to a buyer's initial investment in a conventional house. The potential benefit of the home to overall US energy use will also be appraised.

A final site has not yet been selected for the PP&L house, but its approximate location will be in the Lehigh Valley area of Pennsylvania. In addition to the fact that *Electra III's* location is semi-tropical, and PP&L's home is in an area subject to severe winters, there are other notable differences. For example, the latter residence will be designed around two heat pumps for HVAC and energy conservation (see sketch), supplemented by thermal collectors. The utility believes that one third of the energy requirements will be furnished by the heat pumps, one third by the energy-reclamation systems, and one third by the solar collectors. Figure 1 shows a line diagram of the primary equipment, components, and elements to be contained in the PP&L model. The utility company hopes that, conservatively speaking, one third of the energy normally used in an all-electric home will be saved--and hopes are for a saving as high as 50 percent.

The house itself will be of conventional size; three bedrooms, 1-1/2 baths, dining room, living room, and kitchen; it will be a two-story type, with an area of about 147 m^2. It will be equipped with standard major appliances, including clothes washer and dryer, dishwasher, and electric range--plus fire-and-smoke-detection and burglar-alarm systems. Insulation will consist of 2.5-cm-thick styrofoam sheathing in walls; and urethane insulation will be foamed around the framing areas to minimize heat-cold infiltration.

Among the unusual features of the home will be:

- a fence-like solar thermal collector to furnish auxiliary heating requirements;
- a heat reclamation system to utilize waste heat from the refrigerator and clothes dryer;

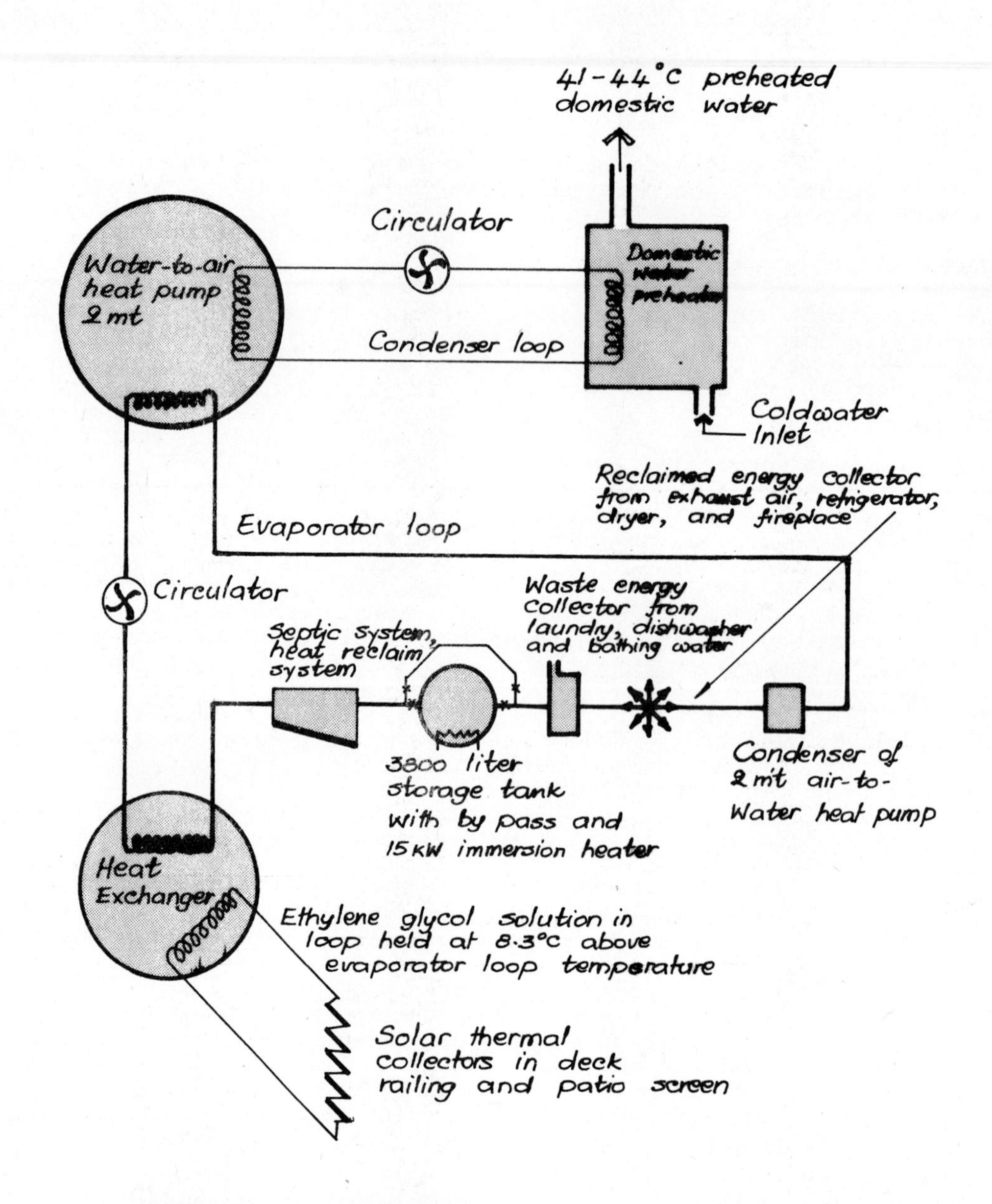
41-44°C preheated domestic water
Circulator
Water-to-air heat pump 2 mt
Domestic water preheater
Condenser loop
Coldwater Inlet
Reclaimed energy collector from exhaust air, refrigerator, dryer, and fireplace
Evaporator loop
Circulator
Waste energy collector from laundry, dishwasher and bathing water
Septic system, heat reclaim system
3800 liter storage tank with by pass and 15 KW immersion heater
Condenser of 2 mt air-to-Water heat pump
Heat Exchanger
Ethylene glycol solution in loop held at 8.3°C above evaporator loop temperature
Solar thermal collectors in deck railing and patio screen

ENERGY SYSTEM BLOCK DIAGRAM

FIGURE 1

- a "heat pipe" that will collect additional energy from waste hot water (draining from tub and shower); and,
- reclamation of heat from the septic (sewage) system.

The fundamental element of PP&L's energy-saving system is a 2 mt water-to-air heat pump that supplies heat to a domestic water preheater; it is also the heart of the HVAC system. This primary heat pump transfers energy from a complex water loop that draws heat energy from six sources:

1. solar-thermal collectors;
2. the sewage septic system;
3. a 3785-liter storage tank (containing a 15 kW immersion coil heater);
4. a waste-water energy collector;
5. the condenser of a 2 mt heat pump; and,
6. a reclaimed energy collector.

In normal function, the heat pump transfers energy from the water loop whose temperature is raised by the solar-thermal collectors and the other five subsystems. If the heat pump proves inadequate to meet requirements, heat to make up the difference can be obtained by the auxiliary conventional systems.

About 18 m^2 of thermal-collector solar modules will be required. As shown in the figure, an ethylene glycol solution will be transferred through the loop to the heat exchanger (at about an 80 percent efficiency of absorption of solar energy). The overall efficiency of heat input to heat output through the heat exchanger is expected to be about 50 percent.

A unique feature of the PP&L system will be its potential capability of reclaiming heat from the septic system's bacterial action. To do this, coils will externally encircle the main holding tank to capture heat given off by the bacterial action. If the temperatures produced are sufficiently high, the heat energy will be controlled by a valve system and used as a supplemental input.

Other installations include an electric motor that will open and close insulated draperies automatically in winter and summer. This will be accomplished by thermostatic control to retain heat in winter and at night, and to exclude heat during the summer. A central fireplace will be interconnected to the house's heating unit. A coil in the chimney will reclaim flue heat and recirculate it through the HVAC system. Fluorescent lighting will largely replace incandescent lamps, and thermopane fenestration will be installed closer to the corners of rooms to permit greater ease of furniture placement and rearrangement.

RECOMMENDATIONS

Builders and homeowners should consider the feasibility of new techniques for heat recovery in residences.

References

1. Friedlander, Gordon D., "Energy-Conserving 'Model Homes'," *IEEE Spectrum* (November 1973): 41-42.

15.5 TRANSPORTATION, COMMUNICATION AND COMPUTERS (Chapter 5)

CASE STUDY 5-1: DEVELOPMENT OF A PROTOTYPE ELECTRIC VEHICLE

THE CONCEPT

The Copper Electric Town Car, a small two-passenger prototype automobile suitable for use in urban and metropolitan areas, was designed and built in 1974 as a prototype project of the Copper Development Association, Inc. (CDA). It was introduced in May 1975. This case study reports on design improvements that have been made on the Copper Electric Town Car through mid-1977 and describes its present performance.[1,2,3]

BACKGROUND

Performance--Performance has now been improved to the point where the Copper Electric Town Car can handle a high percentage of metropolitan driving trips, based on the information on US city driving presented by H.J. Schwartz of NASA at the Fourth International Electric Vehicle Symposium.[4] Schwartz's analysis of automobile usage was

> . . . a car in the United States with a practical daily range of 82 miles (132 km) can meet the needs of the owner on 95% of the days of the year, or at all times other than on his long vacation trips. Increasing the range of the vehicle beyond this point will not make it more useful to the owner because it will still not provide intercity transportation.

Speed Control--Many speed control systems are possible for an electric vehicle; eleven were evaluated for the Copper Town Car (Table 1). Number 11 was chosen. The system chosen is basically a field control system.

The basic reason for selecting field control is that it allows acceleration at constant current at speeds above full-field base speed and drawing energy from the battery at constant current is most economical. Also, motor efficiency at cruising speed is higher because full battery voltage applied to the motor terminals reduces motor current requirement and, thereby, i^2r losses.

Regenerative Braking--Regenerative braking is an added feature of the new system. The controller adds the negative regenerative current to a fixed positive signal and modulates the sum to the driver's demand signal. Maximum regeneration is achieved simply by removing the foot from the accelerator pedal. This is set to approximate a normal, gentle brake stop. No attempt is made to make the occasional high-deceleration stops completely regenerative. Regeneration current is limited to 160 amps, giving only 0.92 - 1.8 m/sec (3 - 6 fps) deceleration, depending upon speed.

The brake pedal has no connection with the regenerative system. Normal friction braking is used to bring the car to a stop from speeds below 21 mph. With regenerative braking down to 21 mph, 78 percent of the vehicle's kinetic energy on the 45 mph SAE Schedule "D" test procedure is recovered. The value of 34 km/hr (21 mph) is set by speed ratio limits and could, no doubt, be decreased with more attention to commutation.

When regenerative braking was incorporated, tests were run to determine its effect on both motor and battery temperature. Motor temperature was measured with a thermocouple at the interpole winding and battery temperature with a thermocouple immersed in the battery electrolyte. Results for motor temperature are shown in Figure 1. There was no detectable difference in battery electrolyte temperature with or without regenerative braking, probably because of the low internal resistance of the batteries used.

Batteries--Four different batteries have been evaluated. Three were lead-acid and one nickel-zinc. The nickel-zinc batteries were tested briefly according to SAE Recommended Procedure J227A, Schedule D and in 64 km/hr (40 mph) cruise tests. At a steady 64 km/hr (40 mph) they gave a range of 232 km (144 miles). In the J227A, Schedule D test, the high internal resistance of the nickel-zinc batteries resulted in such high current draw to achieve the 72 km/hr (45 mph) top speed that the motor burned out at 80 km (49.7 miles). Also, acceleration was slow due to the high internal resistance.

Two strings of the originally installed golf-cart type batteries were tested. These have been removed and replaced with a special 21-plate design. After seasoning with 25 cycles and allowing for test weight differences and the effect of other changes, these batteries have improved the range on the SAE J227A, Schedule D 72 km/hr (45 mph) test by 23 percent.

The configuration of these 21-plate batteries is the same as the golf-cart type. A major feature is their previously mentioned low internal resistance. Cycle life of those batteries is

No.	Motor Type	Speed Reduction	Speed Control
1	Series	3-Speed, Automatic	SCR Chopper
2	Permanent Magnet	Infinitely Variable Belt	Voltage Switching
3	Permanent Magnet	1-Speed, Chain	Starting Resistors & Voltage Switching
4	Series	3-Speed, Chain	Starting Resistors & Voltage Switching
5	Shunt	Infinitely Variable Hydrostatic	Hydrostatic
6	Permanent Magnet	5-Speed, Manual	Resistors
7	Series	4-Speed, Manual	Resistors
8	Separately Excited	1-Speed, Belt	Manual Field Control & Voltage Switching
9	Separately Excited	1-Speed, Chain	Automatic Field Control & Voltage Switching
10	Series	4-Speed, Automatic	SCR Chopper & Centrifugal Clutch
11	Separately Excited	1-Speed, Chain	Transistor Chopper & Voltage Switching

ELECTRIC DRIVE SYSTEMS EVALUATED IN THE COPPER ELECTRIC TOWN CAR TABLE 1

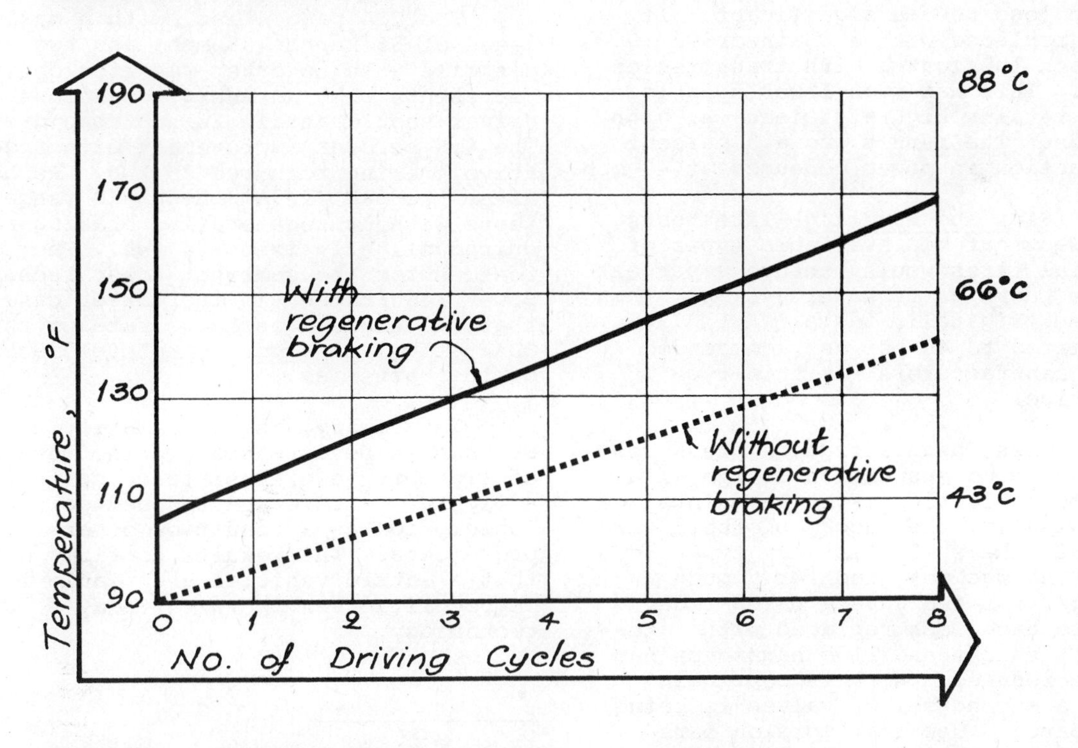

MOTOR TEMPERATURE RISE FIGURE 1

not yet known. Another of their virtues is a weight reduction totaling 42 kg (92 lb) for the 18 batteries used in the Town Car. Flame and acid-vapor-blocking caps have reduced electrolyte loss and vehicle structural corrosion. If, in service, the 90 percent deep-cycle life of these batteries is 500 cycles, as estimated by the manufacturer, they will make the Copper Electric Town Car a highly practical electric vehicle.

Overall Drive System--The electric motor-gear reduction axle package has also been developed further. The separately excited motor was dynamometer tested to "tune" it to maximum efficiency. Still further improvement is possible, for example, by optimizing the air gaps.

The original reduction between motor and final drive gears was a rubber belt drive. This eliminated the need for lubrication and was highly efficient at high power levels. However, at low power levels, its spin loss became significant. It was replaced with a chain drive splash lubricated with transmission oil. This has much lower spin loss and retains high efficiency at high loads. The result was a 5 percent reduction in power consumed at 40 mph.

Ride and Handling--Tire tests showed that two available types of radial tires would consume 4 percent less power at 64 km/hr (40 mph) than those originally installed. All were inflated to 40 psi as recommended by the manufacturers for this type of service.

These high inflation pressures demand good suspension design if a comfortable ride is to be maintained. Originally, the shock absorbers were an off-the-shelf air-lift type. Valving was designed for a much larger car and gave a harsh ride. These have been replaced with take-apart shocks and ride harshness has been reduced. Work is continuing and a second set of valves is being prepared. New rear torsion bars have also been installed to improve ride.

FINDINGS

The primary goal of the development program has been to improve the performance and extend the range to the point where the Copper Electric Town Car will be a practical vehicle for use today. With the improvements discussed above, the power consumption of the car at various speeds and driving cycles meets the original design parameters. Its present performance is compared with its original performance test results in Table 2, which also shows the range improvement attributable to regenerative braking. Vehicle weight for the 1977 tests was 1464 kg (3227 lb). Batteries weighed 490 kg (1080 lb), 33.5 percent of the total weight.

Improvement in the 0 to 48 km/hr (0 - 30 mph) acceleration time is the result of control changes and lower internal resistance in the batteries. The improvement in range at a steady 64 km/hr (40 mph) is mainly due to lower battery resistance.

The CDA range test, with a maximum speed of 56 km/hr (35 mph) and two stops per mile with no other restrictions, represents what an energy-conscious driver should attain in suburban driving. The 6.6 percent improvement with regenerative braking required 26 kWh. Recharge after the SAE J227A Schedule D range tests without regenerative braking required slightly less--24 kWh. The reasons for the somewhat lower recharge power requirement in the latter case is that the slower discharge rate of the CDA test gives a more complete discharge of the batteries.

In summary, these dramatic increases in performance are the result of lighter, lower-resistance batteries, a lighter vehicle and reduced power consumption as a result of other improvements. The results confirm that electric vehicles are viable for metropolitan service with today's technology.

References

1. Copper Development Assoc., Inc., "Application Data Sheet: The Copper Electric Town Car--Recent Developments," (New York, 1977).

	1975	1977 (Without Regenerative Breaking)		1977 (With Regenerative Breaking)	
	Performance	Performance	Improvement over 1975	Performance	Improvement over 1975
Acceleration: 0-30 mph	11.8 sec	8.8 sec	25%	8.8 sec	25%
Range: 40 mph Cruise	80.0 mi	103.0 mi	29%	103.0 mi	29%
Range: CDA 2 Stop/Mile, 35 mph	-	68.7 mi	-	73.3 mi	-
Range: SAE J227A "D", 45 mph	34.0 mi	47.9 mi	41%	54.5 mi	60%

COPPER ELECTRIC TOWN CAR RANGE AND ACCELERATION **TABLE 2**

2. Armstrong, D. and Pocobello, M., "A Quiet, Efficient Controller for Electric Vehicles," SAE Paper No. 750470 (Dearborn, Michigan: Triad Services, Inc.).

3. Pocobello, M. and Armstrong, D., "The Copper Electric Town Car," SAE Paper No. 760071 (Dearborn, Michigan: Triad Services, Inc.)

4. Schwartz, H.J., "The Computer Simulation of Automobile Use Patterns for Defining Battery Requirements for Electric Cars," NASA-Lewis Research Center, Cleveland, Ohio. Presented at the Fourth International Electric Vehicle Symposium, Dusseldorf, August 31-September 2, 1976.

CASE STUDY 5-2: CONSIDERATIONS AFFECTING ELECTRIC TRAIN EFFICIENCY[1]

THE CONCEPT

Various considerations affect the electrical power requirements and efficiency of an electric train system. Examples of important system design parameters are included here to illustrate the effect on energy use efficiency.

BACKGROUND

Forces which can accelerate a train are the tractive effort developed by the motors and the components of the weight along the track on the downgrades. The train's motion is retarded by the various frictional forces such as that caused by braking, and the component of the weight along the track on upgrades. All the various frictional forces except the braking resistance, such as track friction, journal friction, air friction, etc., which oppose the motion of the train on a straight track are usually considered together and referred to as the "track resistance." The extra friction due to track curvature is usually considered as an equivalent upgrade.

Train Resistance for Electric Train.--Let

N = number of cars (including electric locomotive, if any);

W = total weight of train, in tons;

w = average weight of train, in tons (W/N);

r = train resistance in lbf per ton;

v = speed in mph;

a = cross-section of car in ft^2;

A and B are constants in the formula; and,

K is taken as 0.0030 throughout.

$$r = A + Bv + \frac{Ka(0.09 + 0.1\ N)v^2}{Nw}.$$

The constant A depends chiefly upon the average total weight of car and load. Thus, for:

w = 15	20	25-30	35	40-45	50	70
A = 2.0	5.5	5.0	4.5	4.0	3.5	3.0

The constant B depends primarily upon the nature of the track and roadbed and, to some extent, upon the weight and type of the car. The following values are typical:

Passenger cars on excellent track....... 0.06-0.11

Passenger cars on ordinary track 0.10-0.15

Freight cars on ordinary track 0.05-0.06

The heavier the car the higher the value of this coefficient.

Grades and Curvatures--An actual upgrade of G% produces a retarding force of 20xG lbf per ton; and a downgrade of G% produces an accelerating force of 20xG lbf per ton. A curve always gives rise to a retarding force, which ranges from 0.05 to 1 lbf per ton per degree of curvature. Using the higher figure, each degree of curvature may be taken as equivalent to an upgrade of 0.05 percent. Note that for angles of curvature up to 12 degrees, the angle in degrees may be taken as equal to 5730 ÷ R, where R is the radius of curvature in feet.

Average Acceleration Rates

Service	Miles per hour per second (mph/s)
Steam locomotive, freight service	0.1 to 0.2
Steam locomotive, passenger service	0.2 to 0.5
Electric locomotive, passenger service	0.3 to 0.6
Electric motor cars, interurban service	0.8 to 1.3
Electric motor cars, city service	1.5 to 2.0
Electric motor cars, rapid transit service	1.5 to 2.0
Highest practical rate	2.0 to 2.5

Acceleration Constant--The tractive effort required to give to 1 ton (2000 lbm) a linear acceleration of 1 mph/s is 91.2 lbf. To accelerate a train of W tons requires a tractive effort of 91.2 aW lbf to produce a linear acceleration of a mph/s; but, to account for the accompanying angular acceleration of the rotating parts, an additional force is required.

The acceleration constant is raised by the flywheel effect by about 5 percent (i.e., Wr/W = 0.05) for heavy cars and locomotives, and between 5 and 10 percent for light, low-speed cars, 8 percent being an average figure. However, C is usually taken as 100, corresponding to an increase in effective weight of about 10 percent. A given linear acceleration of a mph/s then requires an accelerating force of 100 a lbf per ton.

Tractive Effort and Adhesion Coefficient--Let

F = tractive effort, in lbf/ton, exerted by motors;

G = percentage actual grade (+ for upgrade);

g = degrees of curvature;

r = train resistance, in lbf/ton; and,

a = acceleration in mph/s (- for retardation).

Then the tractive effort required per ton of total train weight is

$$F = 100a + r + 20G + g \text{ lbf/ton.}$$

The adhesion or "tractive" coefficient is the quotient (expressed usually as percent) of the tractive effort in pounds which will slip the drivers, divided by the weight in pounds on the drivers. The values in the accompanying table are typical. The maximum possible tractive effort is the product of the adhesion coefficient (as a decimal fraction) by the weight (in pounds) on the drivers.

Adhesion Coefficients

Condition of Track	Without Sand	With Sand
Most favorable condition	35	40
Clean, dry rail	28	30
Thoroughly wet rail	18	24
Greasy moist rail	15	25
Sleet-covered rail	15	20
Dry snow-covered rail	11	15

The Weight of Locomotive--The weight of locomotive required to accelerate a train weighing W tons at the rate of a miles per hour per second up a grade of G% on a g degree curve against a frictional resistance of r lbf per ton, when q% of the weight is on the drivers and the coefficient of adhesion is p%, is given by the following formula:

$$\text{Weight of locomotive} = \frac{5\ W}{pq}\ (100\ a + r + 20G + g).$$

Example--What weight of locomotive is required to accelerate a 400-ton train at the rate of 0.05 mile per second up a 0.1 percent grade against a fractional resistance of 8 lbf per ton, when 80 percent of the weight is on the drivers and the coefficient of adhesion is 20 percent?

$$\text{Weight of locomotive} = \frac{5 \times 400}{20 \times 80}\ (50 + 8 + 2) =$$

75 tons.

Maximum Overall Efficiency of Motors and Gears at Rated Voltage

Horsepower 1-hour rating	Kind of Motor	Maximum Efficiency Percent
30-100	dc geared	83-88
100-250	dc geared	88-89
250-500	dc gearless	91-83
50-200	ac series geared	70-80*
200-500	3-phase induction geared	85-89

*Including step-down transformers.

FINDINGS

Power Required at Given Speed--Let

r = train resistance in lbf/ton of total weight of train;

G = percentage grade;

g = degree of curvature;

a = degree of acceleration in mph/s;

v = speed in mph; and,

W = total weight of train in tons.

Then the power required *at the rims of the drivers* is 1.99 v (r + 20xG = g + 100a) horsepower, total.

The power input, p_i to the car or locomotive is equal to the power at the rims of the drivers divided by the overall efficiency E of the controller, motors and gears, that is,

$$p_i = \frac{1.99\ Wv(r + 20XG + g + 100a)}{1000XE} \text{kW}$$

Approximate Method of Calculating Energy Use--The following method is based upon simple kinetic principles and, if certain characteristics of the run are known, gives the actual energy output at the wheel rims. This fact makes the method useful, not only for rough calculations, but also for checking calculations made by the more accurate step-by-step method.

When the method is applied for checking purposes, the column of the table below, headed "Actual energy output," should be used and the input calculated from the known efficiencies. When applied to rough calculations, the column headed "Approximate electrical energy input," should be used. In the latter case the maximum speed and length of run with power on are not known, but it is possible to assume certain values, based upon experience, which will give a rough approximation of the energy required. The total energy in watt hours per ton-mile will be the sum of the amounts required for acceleration and overcoming frictional train resistance, grades, and curves. Let

V = average running speed in miles per hour;

V_m = maximum speed in miles per hour;

L = length of run in miles;

L_p = distance traveled, with power on, in miles;

$n = 1/L$ = number of stops per mile including one terminus;

r = average train resistance, in pounds per ton (say that corresponding to a speed from 10 to 20 percent greater than the average speed);

G = average equivalent grade, in percent;

g = average curvature in degrees;

$K = \frac{V_m}{V}$ = ratio of maximum to average speed; see accompanying table of values of K; and,

$Q = \frac{L}{L_p}$ = ratio of length of run to distance traveled with power on.

(See table of values of K and Q.)

Output at Wheel Rim and Input to Cars in Watt-hours per Ton-mile

Energy for	Actual energy output at wheel rims of cars	Approximate electric energy input to cars
Acceleration	$\frac{V_m^2}{36.2\ L}$	$\frac{K^2nV^2}{25}$
Frictional train resistance	$\frac{1.99\ r\ L_p}{L}$	$\frac{2.9\ r}{Q}$
Grades	$\frac{39.8\ GL_p}{L}$	$\frac{57\ G}{Q}$
Curves	$\frac{1.99\ gL_p}{L}$	$\frac{2.9\ g}{Q}$
Total	Sum	Sum

Note: 25 = 36.2 E, 57.8 = 39.8 ÷ E, and 2.9 = 1.99 ÷ E, where E is the efficiency, taken as 0.7. The formula for energy due to curves assumes each degree of curvature to be equivalent to a train resistance of 1 lbf per ton, which is probably high.

Values of K and Q

Stops per mile, n	K: Locomotive passenger trains	K: Single car multiple unit trains and freight trains	Q: All trains
0	1.0	1.0	1.0
0.1	1.2	1.1	1.1
0.2	1.35	1.2	1.25
0.3	1.5	1.25	1.4
0.4	1.6	1.3	1.5
0.5	1.7	1.35	1.7
0.6	1.75	1.4	1.8
0.7	1.8	1.45	1.9
0.8	1.85	1.45	2.0
0.9	1.9	1.5	2.1
1.0	1.95	1.5	2.15
1.2	1.95	1.6	2.2
1.4	1.95	1.6	2.3
1.6	1.95	1.6	2.4
1.8	1.95	1.65	2.5
2.0	1.95	1.7	2.6
2.5	1.95	1.75	2.7
3.0	2.0	1.8	2.8
3.5	2.0	1.85	2.9
4.0	2.0	1.9	2.9
4.5	2.0	1.95	2.95
5.0	2.0	2.0	3.0

More exact methods involve a knowledge of the motor characteristics, gears, brakes, and control.

Power Required for Car Heating and Lighting--In addition to the energy required for propelling the cars, a very appreciable amount is also required for heating them in the winter and a small amount for lighting at night. (see the table below.) In making up a load diagram this energy should be included.

Lighting of Cars

Length of car, feet	Average kW for lighting*
14-20	0.25
20-28	0.35
28-34	0.55
34-40	0.70

*During the hours lights are on, using tungsten lamps.

Street cars should be heated to a temperature of 55 to 60°F which is found to be comfortable for passengers in normal clothing. Suburban and interurban cars which have longer runs should be heated to between 65 and 70°F in order that the passengers may be comfortable without additional warm clothing.

Three systems of heating have been used on railways: the steam boiler, hot water furnace, and the electric heater. Steam is used on locomotive trains, hot water for some interurban cars making long runs, and electric heaters for urban cars and most interurban cars. In locomotive trains the steam boiler is oil-fired and built into the locomotive or carried on a tender behind the locomotive.

Electric Heating--The electric system of heating, although more expensive to operate than the others, finds the greatest favor for urban and interurban service and has the advantage of good heat distribution, cleanliness, ease of regulation, low fire hazard, and no attendance.

Electric heaters are all equally efficient with regard to the amount of heat developed but many differ in respect to durability and maintenance cost. They are usually made of resistance wire wound on porcelain forms, but sometimes the wire is embedded in insulating material. The latter type has greater heat storage capacity than the former.

The power required for car heating is as follows:

Power for Car Heating

Length of car, feet	Average Kilowatts for Heating: Average conditions	Severe conditions
14-20	3.5	4.0
20-28	4.5	5.5
28-34	5.5	7.5
34-40	7.5	10.5

RECOMMENDATIONS

When electric train systems are being considered for urban and interurban transportation means, the total electrical energy requirements for operation should be accounted for in selecting various alternatives. Slight differences in electrical energy requirements and utilization efficiencies of electric train components and equipment can have a significant economic effect over the lifetime of the operating system.

References

1. Merriman, Thaddeus (editor-in-chief), *American Civil Engineer's Handbook*, 5th edition (New York: John Wiley and Sons, 1941).

CASE STUDY 5-3: VIDEO-TELEPHONE VS PERSONAL TRAVEL COMPARATIVE ANALYSIS[1]

THE CONCEPT

As fuel costs continue to increase, communication alternatives to personal travel must be considered. The video-telephone offers one such viable alternative for the future.

BACKGROUND

The background information for the video-telephone concept of communication is contained in Chapter 5. In this case study an analysis of the costs based solely upon the relative monetary costs of travel and telecommunications to the individual consumer is made first. Then a comparison is made considering the energy use of the alternatives.

FINDINGS

A. Cost Comparison of Travel and Long Distance Video-Telephone

A sample calculation comparing the cost of air travel and the video-telephone is given below. The following assumptions are made for the calculations:

- an eight-hour work day;
- most of the travel occurs during the work day rather than on late night flights;
- no allowance is made for the time spent traveling to restaurants and overnight lodging;
- value of time spent in the act of travel is $10/hour;
- the cost of scheduling a group, video-telephone call balances the cost of scheduling travel;
- no surcharges for video-telephone conference calls;
- no cost is allocated for subscription to video-telephone service, only the charge for long distance service is considered;
- long distance video-telephone rates are 10 times audio-telephone rates. Day rate, station-to-station audio-telephone rates are used as a basis; and,
- January 1972 rates for both travel and audio-telephone will be used to approximate future rates. This neglects an expected rise in the cost of jet fuel, and other changes in relative costs.

Example:

Round trip between New York, N.Y., and Chicago, Illinois, one night stay.

Itemization: (based on published data for cities involved) [2]

Outward Bound: 8 A.M. departure from home or office

	Time (minutes)	Cost (dollars)
Local ground travel (origin)	42	3.00
Departure terminal allowance	30	-
Air travel	144	59.00
Arrival terminal allowance	18	-
Local ground travel (distance)	48	2.75
Totals (one way)	282 (4.7 hours)	64.75

Contact Hours--Time zone adjustment of one hour leaves (8 - 4.7 + 1 =) 4.3 hours of time available for working contact on day of departure. Overnight lodging and meals cost $18 and $10, respectively. Anticipation of time zone reversion during the day means (4.7 + 1 =) 5.7 hours must be allowed for return travel: this leaves (8 - 5.7 =) 2.3 hours for working contact before departure from central Chicago. Total working contact time available, 6.6 hours.

Return Travel--Inverse of outward bound travel:

Time, 4.7 hours; cost, $64.75

Opportunity Cost for Time--Total time spent in the act of traveling is 9.4 hours valued at $10 per hour. The opportunity cost for time is $94.

Total Cost of Trip:

Direct cash outlay	$162.50
Opportunity cost	94.00
Total	$256.50

Video-Telephone--Audio-telephone daytime, station-to-station rate: 2 minutes for \$0.95; each additional minute, \$0.30. Federal excise tax (10 percent) and state and local tax (3.75 percent) additional. Total cost for video-telephone use projected at \$3.40/minute (10 times audio cost). The number of video-telephone minutes obtainable at a cost equal to the cost of travel is therefore (\$256.50/3.40 = 75 minutes).

B. Energy Usage Comparison of Travel and Long Distance Video-Telephone Service

The following is a comparison of the energy requirements of travel and communication by video-telephone.

Transcontinental Air Travel--Data for Boeing 747 operation by American Airlines, Inc. in the first quarter of 1972:[3]

Off-on speed 805 km/hr (497 mph)
14,140 liters of fuel per total (off-on hour)
Seating capacity per plane, 307
Passenger load factor 50.1 percent

(See Table 1 on next page for typical performance data for other aircraft.) The distance by air between New York, N.Y. and Los Angeles, California, is 4008 km (2,474 mi). For a round trip the volume of fuel (kerosene) per passenger is obtained from the calculation:

$$\frac{4{,}008 \times 2 \times 14{,}140}{805 \times 0.50 \times 307} = 917 \text{ liters (241 gal)}$$

Since the energy content of kerosene is 38 GJ/m^3, the energy requirement per passenger for a round trip is 35 GJ.

Transcontinental Video-Telephone--Power consumption of the on-premises video-telephone apparatus is reported to be the following:[5]

terminal device,

4 watts in standby condition
40 watts in service

service unit,

21 watts in standby condition
35 watts in service.

Thus, in service, each video-telephone uses 75 watts and for a two-party conversation 150 watts are required. The equalizers (repeaters) currently in use for local video-telephone service use about 3 watts for each direction of video transmission. Within the local area, a signal might pass through three equalizers. Thus, local transmission of a video-telephone conversation would use about 18 watts. Microwave radio links, such as Bell's TD-2 system, would be suitable for long distance transmission. Radio relay stations typically occur at 40 km (25 mi) intervals and each station uses about 20 kW. Transcontinental telephone transmission does not follow the shortest path but rather zig-zags between cities. Consequently, the actual transmission distance between New York and Los Angeles will be assumed to be 6400 km (4,000 mi). When bandwidth reduction techniques now demonstrated in the laboratory are put into service, the TD-2 system could accommodate 72 two-way video signals--each at 1.5 megabits per second (Mb/s). Thus, the transcontinental power use per call can be determined from the sum of the power requirements of the station set, local transmission, and long distance. The latter is given by:

$$\frac{20 \text{ kW} \times 6400 \text{ km}}{40 \text{ km} \times 72 \text{ channels}} = 44 \text{ kW}.$$

This gives, per hour of operation, 44 kWh = 0.158 GJe.

When corrected for the efficiency of electricity generation in a central power plant (using 33 percent), the energy requirement is about 0.475 GJ/hr. Therefore, 8 hours of video-telephone conversation would require about 3.8 GJ of energy. This is about one-ninth

Parameter	Boeing Jet 747	McDonnell Douglas DC-10	Boeing Jet 707	Boeing Jet 727	Boeing Jet 737
Length (m)	70.5	55.4	46.6	42.0	28.7
Maximum range (km)	9,700	10,190	9,260	3,660	2,120
Maximum altitude (m)	13,740	12,200	12,800	12,800	10,500
Seating capacity	361	252	148	103	96
When Cruising:					
Speed (km/hr)	940	965	900	930	900
Fuel consumption (ℓ/h)	14,700	10,100	6,800	4,125	3,200
Fuel consumption (ℓ/km)	15.6	10.5	7.55	4.45	3.55
Passenger-km (full)/ℓ	23.1	24.0	19.6	23.2	27.0

ℓ = liter

Reference 4

TYPICAL PERFORMANCE DATA, MODERN AIRCRAFT

TABLE 1

(11 percent) of the 35 GJ calculated above for air transportation.

Local Automobile--Automobiles differ greatly in their fuel consumption. However, a value of 6-12 km/ℓ (15-30 mpg) is representative.

Local Video-Telephone--For local service involving about three repeaters, the video-telephone power requirement is about 170 watts and the energy input to the central electricity generating station is therefore about 1.8 MJ/hr. The energy content of one liter of gasoline is 38 MJ and is sufficient to provide about 21 hours of video-telephone, while a gallon of fuel would provide nearly 80 hours.

Table 2 summarizes the results of the energy use analysis.

RECOMMENDATIONS

The video-telephone and other forms of telecommunication should be considered in future planning as alternatives to long distance personal travel. With the inevitable increases in fuel costs, non-personal travel/communication alternatives will have increased economic advantages. Electric utilities should plan for this increased utilization of electric energy for expanded telecommunications purposes in the future.

References

1. Dickson, Edward M., (in association with Raymond Bowers), *The Video Telephone: Impact of a New Era in Telecommunications*, (New York: Praeger Publishers, 1974).

2. R.H. Donnelley Corporation, *Official Airline Guide*, North American edition, (1 December 1971).

3. "747 and DC10 Operating and Cost Data--First Quarter 1972," *Aviation Week and Space Technology*, (31 July 1972): 40-41.

4. *Logbook*, Lufthansa Airlines Inflight Magazine, (1973).

5. "The Picturephone System," *Bell System Journal*, Special Issue (February 1971): 219+

CASE STUDY 5-4: EFFICIENCY OF MASS TRANSIT[1]

THE CONCEPT

Various forms of mass transit can improve on the efficiency of the automobile in personal transportation within urban areas.

BACKGROUND

Recently interest has been revived in mass transit systems to decrease urban congestion and pollution. The Bay Area Rapid Transit System (BART) is the first major new rapid transit system in the US in decades. Systems are now being installed in other cities including Washington, D.C., Atlanta, Caracas, and São Paulo.

FINDINGS

The efficiencies of existing mass transit systems are shown in Table 1. Energy use by BART is from preliminary studies during partial operation; therefore, a range of probable values is given. This shows the energy used in "traction" only, which does not include energy used at the station, in maintenance, or in construction. The estimated total energy use by BART over 50 years of operation is shown in Table 2.

It is of interest to compare energy demands of BART with other systems, specifically buses and automobiles. The following is from Reference [1].

> A decision must be made at the outset about what components of energy cost will be compared. To compare only propulsion (or traction) energy is somewhat limited and perhaps misleading. However, the inclusion of factors such as operation, maintenance, and construction costs makes the analysis more complex and speculative. For example, although construction energy for BART can be estimated, how does one estimate energy used to build the freeways, roads, parking lots, garages, and

Contact Time	Energy Requirements		Ratio
	Boeing 747	Video-Telephone	Boeing 747 / Video-Telephone
Hours	GJ	GJ	
8	35	3.8	9.2
16	35	7.6	4.6
24	35	11.4	3.1

ENERGY USE FOR VIDEOTELEPHONE VERSUS AIR TRAVEL **TABLE 2**

System	kWh / car mile	Full Load kWh (a) / passenger mile	Part Load kWh (b) / passenger mile	Seats/car	Weight/car (metric tons)
Philadelphia	5.9	0.105	0.420	56	22
Chicago	4.5	0.090	0.360	51	19
New York	5.4	0.115	0.460	47	36
Cleveland	3.6	0.067	0.268	54	25
Toronto	5.2	0.084	0.336	83	26
BART					
Probable Lower Bound	3.2	0.045	0.180	72	26
Probable Upper Bound	5.5	0.076	0.304	72	26

(a) Full load is for all seats filled, no standees.

(b) Part load uses a rough national average of 25% of seats filled for all trips.

EFFICIENCY OF ELECTRIC MASS TRANSIT SYSTEMS (VEHICLE PROPULSION ONLY)

TABLE 1

ENERGY USE	10^6GJ	10^6MBtu	Percent
Construction	116	110	44
Propulsion	106	100	40
Operation and Maintenance	42	40	16
TOTAL	264	250	100

TOTAL BART ENERGY REQUIREMENTS **TABLE 2**

VEHICLE	PROPULSION ENERGY	
	MJ/ passenger-km	Btu/ passenger-mile
BART (a)		
PROBABLE LOWER BOUND	1.19	1800
PROBABLE UPPER BOUND	2.37	3600
BUS (a)	1.45	2200
AUTOMOBILE	5.34	8100

(a)Assumes a load factor of 25%

COMPARATIVE PROPULSION EFFICIENCIES OF TRANSIT SYSTEMS **TABLE 3**

driveways required by buses and automobiles? Two comparisons are made, and a series of assumptions is made in each case. First, propulsion energies only are compared. Then total energies are compared.

For BART traction energy, we use the probable upper and lower bounds from Table 1. We assume a load factor of 25 percent or 18 passengers per vehicle and an energy conversion rate of 2.93 J/Je (10,000 Btu/kWh) (corresponding to a power plant/distribution system efficiency of 34 percent). For buses we assume 2 km/liter (5 miles/gal), a gasoline conversion rate of 3.7×10^7 J/liter (136,000 Btu/gal), a 50-seat vehicle, and a load factor of 25 percent or 12.5 passengers/vehicle. For the automobile we assume 5 km/liter (12 miles/gal), 3.7×10^7 liter (136,000 Btu/gal), and 1.4 passengers per vehicle. The resulting propulsion energy intensities are given in Table 3 on a passenger-mile (kilometer) basis.

A number of facts should be kept in mind. First, the loading factors are assumed somewhat arbitrarily, although they tend to approximate national averages. Different loading factors could drastically change these comparisons. A fully loaded automobile (car pool) could be about as efficient as BART or a bus that is 25 percent loaded. However, a fully loaded bus or BART with standees could be as much as 10 to 15 times more energy efficient than the average automobile. The great sensitivity of energy comparisons to loading should be kept in mind at all times in transportation energy studies.

A second important fact is that, even though we have referred propulsion energy to power plant input energy, it is not obvious that the resulting comparison is completely valid. The energy source for the power plant could, in general, be coal, oil, natural gas, uranium, or hydroelectric power, each with different processing energy requirements. Gasoline must also be processed in a refinery and is about 85 percent efficient. Hence, comparing fuel energies at the input to the power plant with energy into the automobile has some real limitations.

Finally, we consider the question of comparative total energy use. In the case of BART and automobiles, propulsion energy tends to be about 50 percent of total energy. Although these data are probably fairly accurate for automobiles, BART must reach completion and experience some months of normal operation before comparative total energy use can be tested for BART. A similar analysis has not been made for buses. If buses have about the same 50-50 energy split, and if the data for BART and automobiles are reasonably accurate, then we can obtain an estimate of total energy comparison by simply doubling the propulsion energy figures given in Table 3. This is fine for buses and automobiles but is ambiguous in the case of BART since we have two bounds. The traction estimates used previously assume that auxiliary energy is accounted for as traction energy and that regenerative braking is partially successful. On the basis of these assumptions, we assume a traction energy of 10.7 MJ/km (4.7 kWh/car-mile) for purposes of calculating total BART energy. Again, the reader must be cautioned that these results are based on some major assumptions: The extensive use of light, compact automobiles could reduce automobile energy use by more than two times; the allotment of energy costs to roads and garages is quite arbitrary; and the impact energy for BART and for other vehicles has not been considered quantitatively.

RECOMMENDATIONS

The operating efficiency of rapid transit systems, such as BART or large bus networks, is from two to four times greater than that of the private automobile. The use of this type of transit should be promoted both by the development of new transit systems and by extensive use of existing systems. Furthermore, electric transit systems expand fuel supply options by permitting substitution of hydro, coal, or nuclear fuels for gasoline and diesel.

References

1. Healy, T.J., and Dick, D.T., "Total Energy Requirements of the Bay Area Rapid Transit System," *Transportation Research Record* #552, (1975): 40-56.

CASE STUDY 5-5: ON-LINE COMPUTER CONTROL OF THERMAL PROCESSES[1]

THE CONCEPT

A series of thermal processing steps in a steel plant required large blocks of energy. Methods were sought for reducing energy needed per product unit processed.

BACKGROUND

The use of computer controls in the operation of large thermal process plants offers a potentially powerful means for reducing fuel consumption and improving production efficiency. The method was applied to a large European steel plant.

FINDINGS

In the steel plant studied, an on-line computer control system was used to execute a carefully devised program of operation for steel reheating.

The functions monitored by the on-line system included furnace idling temperatures, the charge temperature in the furnace and in passage from furnace-to-furnace, and the speed of charge passage. One of the virtues of using the computer control system was that once a planned schedule of operation was programmed, it could be met. This was a key element in increasing the plant's productivity.

Operational experience with the on-line system has clearly shown to the plant management that the computer system investment was justified. The system resulted in a 25 percent reduction in fuel consumption per ton of production and was accompanied by a 12 percent increase in the rate of production.

RECOMMENDATIONS

Large thermal processing facilities with furnaces operating at high temperatures should be carefully reviewed by facility management. Very attractive rates of return may be available through increased thermal and production efficiencies achieved by using on-line computer control of processing.

References

1. Berg, Charles A., "Conservation in Industry," *Science* 184(19 April 1974): 264-270.

CASE STUDY 5-6: INCREASING COMBUSTION EFFICIENCY IN INDUSTRIAL FURNACES BY ON-LINE CONTROL

THE CONCEPT

Combustion equipment is sensitive to changes in atmospheric conditions, resulting in either incomplete fuel combustion or fuel waste through heating of excess air. Significant fuel cost savings are possible at advantageous rates of return on investment by on-line control of the combustion process.

BACKGROUND[1]

Within the constraints of air pollution control, regulation of combustion air is important for the efficient operation of high temperature industrial furnaces. Excess air quenches the flame temperature and reduces the efficiency of heat transfer to the furnace; insufficient air results in incomplete combustion of the fuel. Combustion equipment, particularly oil-fired equipment, can be thrown out of adjustment by rapid changes in atmospheric conditions and by other phenomena (such as progressive fouling). Industrial experts have studied the significance of proper burner adjustment and maintenance, and many have concluded that diligent application of exacting adjustment procedures could save 5 to 10 percent of the fuel consumed. Some field measurements have shown fuel savings of as much as 30 percent.

On-line computer controls are extremely useful for regulation of combustion equipment. Wide application would appear to offer a powerful means for increasing industrial fuel efficiency. The fundamental question is whether or not it pays. The following study investigates the economics of computer controls.

FINDINGS

To examine the possible economic benefits of on-line computer-directed combustion control, two steel treatment furnaces operating in parallel were investigated. Each of the two furnaces had an assumed throughput of 145 mt/day (160 tons/day). The operating parameters used in the study were as follows:

- average fuel savings of 7 percent will be achieved by on-line minicomputer control;
- furnace currently requires 1000 Btu/lb of steel treated; and,
- fuel oil energy content is 138,500 Btu/gallon when burned.

For a throughput of 160 tons/day of steel, approximately 3.2×10^8 Btu/day is currently required. With computer control, 7 percent will be saved, or 2.24×10^7 Btu/day. This corresponds to an annual saving of 7.84×10^9 Btu or 1347.8 bbl of fuel oil. At $10/bbl, the annual saving in fuel cost is $13,478.

As shown in Table 1, the annual benefit of the hypothesized $48,400 investment in a minicomputer directed combustion control system is 12 percent per year.

The payback period, at $10/bbl of oil, is 4.9 years. For a fuel cost of $15/bbl, the payback period would be 2.1 years. It should be noted that 32K memory minicomputers are currently on the market for $10,000 or less. It also should be noted that for nearly the same costs, the control system would be capable of operating treatment furnaces with higher throughput than assumed herein

$$(2 \text{ furnaces} \times 160 \text{ tons/furnace-day} = 320 \frac{\text{tons}}{\text{day}}).$$

Larger throughputs and energy costs above $12/bbl of oil could result in payback periods of less than two years.

RECOMMENDATIONS

Attractive rates of return on investment are possible through automated on-line combustion control of high temperature industrial furnace operations. Application of automated control methods should receive careful attention by plant managers.

References

1. Much of the background material is taken from Charles A. Berg, "Conservation in Industry," *Science* 184(19 April 1974): 264-270.

15.6 AGRICULTURE (Chapter 6)

CASE STUDY 6-1: HEAT RECOVERY IN A DAIRY

THE CONCEPT

The common practice in milk rooms is to immediately remove the heat from the milk after each milking by means of mechanical refrigeration. The heat removed is then dissipated to the atmosphere using an air-cooled condenser. By using a water-cooled condenser, heat can be recovered for water heating or space heating.

BACKGROUND

Typically 35-45 kg (80-100 lb) of milk are cooled per kWh of electricity expended. Studies carried out on dairy farms in North Carolina, Kentucky, and Tennessee indicated that water at a temperature of 60°C (140°F) can be produced in the process of cooling milk to 3.3°C (38°F).

Capital and Installation Costs	
Capital Costs, Minicomputer, Sensors and Mechanical Controls	$35,000
Installation Costs[a]	13,400
	$48,400
Annual Operating Costs	
Maintenance, 10%[b] of Capital & Installation Costs (C & I)	4,840
Taxes and Insurance, 4% of C & I	1,936
Interest, 9% of C & I	4,320
Depreciation[c] in one year (48,400/8)	6,050
Total Annual Costs	$17,146
Annual Economic Benefit of Fuel Use Reduction	
Annual Fuel Use Reduction[d] (2 x 13,478)	26,956
Annual Costs	17,146
Annual Benefit	$ 9,810
Annual Rate of Return (uniform series, present worth)	12%
Payback Period	4.9 years

Notes:

(a) Includes system engineering costs of $9,000; system functional checkout tests and debug costs of $3,000; and operational monitoring costs of $1,400.

(b) Reflects computer maintenance contract support.

(c) Eight year average life.

(d) Fuel savings of 7 percent; based on nominal furnace requirement of 1000 Btu/lb of steel, furnace throughput of 160 tons/day, 350 days/year, fuel oil energy content of 138,500 Btu/gal (5.8×10^6 Btu per 42 gal. barrel), $10/barrel.

ON-LINE COMPUTER SYSTEM COST — **TABLE 1**

Hot water is used in dairies for the purposes of warming drinking water for the cows, for "prepping," for heating milking rooms and other areas, or for preheating hot water used for cleanup purposes. (Normally water used for tank washing and line washdown must be 71°C [160°F].) Research also indicated that the hot water demands occur at a time when milk is being chilled. One of the design problems is to balance the heat recovery system capacity with the hot water demand. Frequently more electricity is used for water heating than is used for cooling milk.

In this case study the bulk milk processing arrangement of a dairy farm was reviewed. The owners previously had been using a 3.7 kW (5 hp) air-cooled condensing unit to chill a 3785 liter (1000 gal) bulk storage milk tank to 3.3°C (38°F). The heat removed from the milk was being rejected through the vapor compressor to the atmosphere. Figure 1 shows the processing system prior to the energy management conversion.

At the time of an upgrading of the current milk-handling system, the owner inquired as to the possibility of heat recovery. The milk system was being upgraded to replace the existing bulk storage tank with one having a capacity of 5678 liters (1500 gal).

FINDINGS

An investigation revealed that a water-cooled condensing unit could produce 60°C (140°F) water in the process of chilling milk.* The customer's water heating requirements were then reviewed, and it was determined that a need existed for at least 71°C (160°F) water to be used in the tank wash and line washdown. The prepping water could be 38°C (100°F). Based on the estimated need for high temperature water, it was found that a water-cooled condenser unit would not be necessary to do all of the milk chilling since it would generate more 60°C (140°F) hot water than needed for the processing. Therefore, to avoid the expense of excess water-cooled condenser capacity, the customer elected to install one 3 kW (4 hp) water-cooled condensing unit and a conventional 3 kW (4 hp) air-cooled condensing unit.

The existing 5.5 kW water heater was maintained to increase the 60°C (140°F) water temperature to the 71°C (160°F) necessary for the tank and line washdown. The elimination of the requirement for 77°C (170°F) water resulted in the elimination of the 3.5 kW water heater unit. Cow prepping, which required as a minimum 38°C (100°F) water, was accomplished by mixing the 60°C (140°F) water coming off the water-cooled condenser with normal cold water make up to attain the required temperature. As a backup measure in the event of an equipment failure, a pipe connection was made directly to the 5.5 kW water heater (see Figure 2).

RECOMMENDATIONS

As a result of these modifications to the system, electricity used for water heating was reduced from the previous level of 53 kWh per day to a new level of 8 kWh, for a net 45 kWh per day savings. In addition, the electrical demand for water heating was reduced from 9 kW to 5.5 kW. This reduced demand was offset by the increase in condensing unit power due to expansion of the system. However, without heat recovery, there would have been 9 kW demand for water heating in addition to 6 kW (8 hp) of air-cooled condensing equipment. Heat recovery should be considered for dairy operations.

Acknowledgments

Information for this case study was provided through the courtesy of Mr. Jim Ward, Chief, Electrical Demonstration Branch, Tennessee Valley Authority, and Mr. Ernest C. Dowless, Manager, Commercial Power Department, Duke Power Company.

*Several types are commercially available; this installation used a "Mueller Fre-Heater.

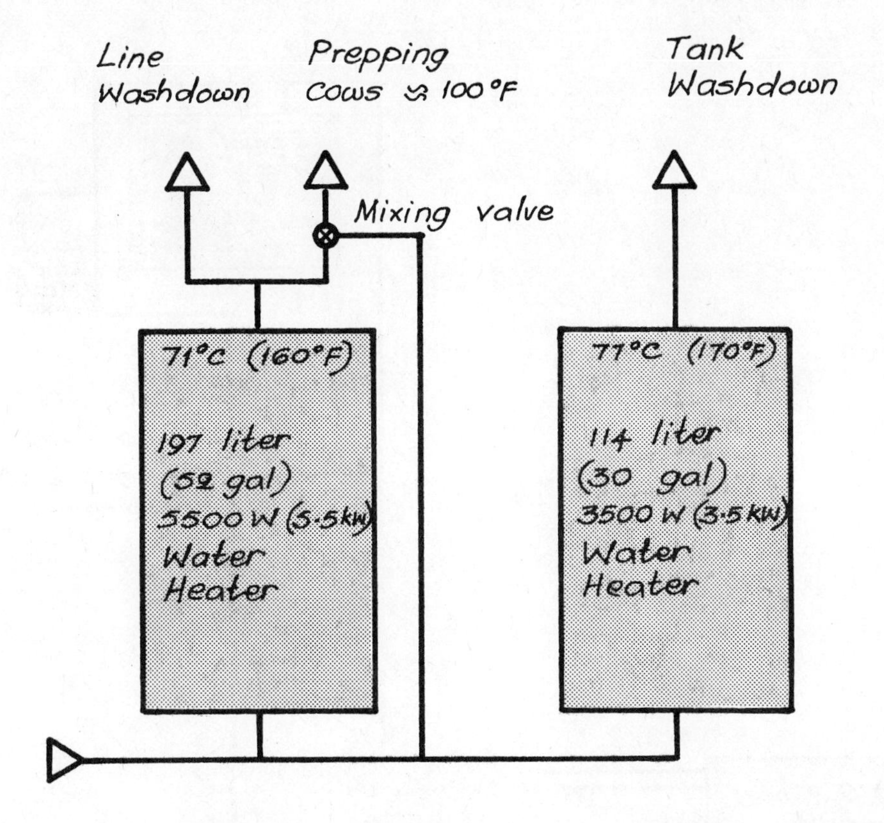

A. Water Heating and Piping Diagram

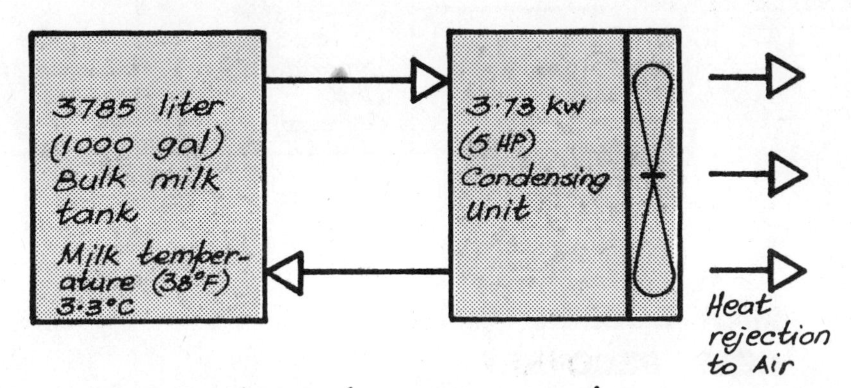

B. Conventional Milk Cooler

ORIGINAL MILK COOLING SYSTEM **FIGURE 1**

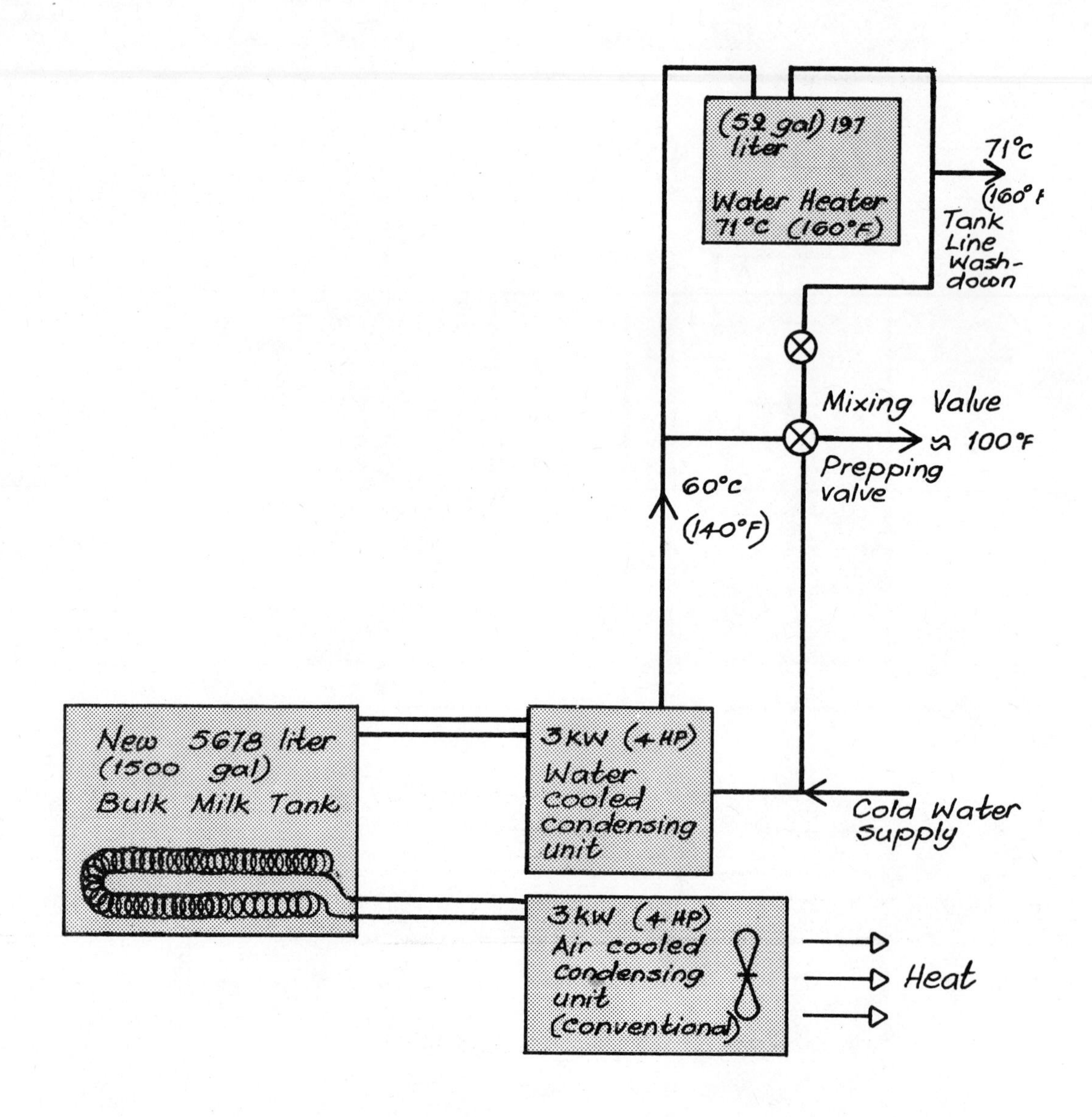

MILK COOLING WITH HEAT RECOVERY FIGURE 2

CASE STUDY 6-2: IRRIGATION SYSTEM SCHEDULING AS A LOAD MANAGEMENT TOOL[1]

THE CONCEPT

In the State of Nebraska, the electrical energy available for irrigation is severely limited during the summer due to the peak summer power demand. A series of studies and field tests was carried out with a number of irrigation systems. The irrigation schedule was arranged to coincide with hours of lower electrical demand. It was found that the peak electrical demand was reduced while the crop yields were maintained. Off-peak scheduling permitted significant savings in peak demand, electricity costs, and water usage.

BACKGROUND

Due, in part, to the anticipated reduction of available supplies of diesel and propane, many existing pumping systems are converting to electricity in order to insure a dependable power supply. However, the increase in demand and additional hookups being requested are resulting in a waiting period of several months. New irrigation loads also create a capacity requirement for the utility. Careful scheduling of irrigation loads can help ease these problems.

Center-pivot irrigation machines typically use multiple motors (up to 10) in the range of 0.37-1.1 kW (0.5-1.5 hp) to operate the sprinkler pipeline. Motors cycle on and off, with the motors most distant from the pivot being on for the longest times.

FINDINGS

Scheduling center-pivot irrigation systems for interruptible service, coupled with publicity on savings possible using off-peak electrical power, were effective in significantly reducing peak demand, electricity, and water use.

An average reduction of 39 hours of operation per center-pivot system was obtained for 20 center-pivot systems that were scheduled when compared with 47 center-pivot systems that were not scheduled.

Another interesting finding shows low correlation between electrical demand and evapo-transpiration and a low correlation between electrical demand and temperature, indicating that neither short-term evapo-transpiration nor temperature alone should be used to predict power demand.

It has also been shown that center-pivot irrigation machines have relatively low power factors (0.3-0.5) under typical field operating conditions.[2] This suggests the need to carefully select motor size and type to be appropriate for the application.

RECOMMENDATIONS

Irrigation is an energy use of growing importance in agriculture. More effective methods for pumping and irrigation are likely to have attractive rates of return in the future and should be considered. Sometimes this may only involve different use patterns for existing equipment.

References

1. Stetson, L.E.; Watts, D.G.; Corey, F.C.; and Nelson, I.D., "Irrigation System Management for Reducing Peak Electrical Demands," *Trans. of the ASAE*, 18(March-April 1975): 303-306, 311.

2. Steson, L.E. and Nelson, S.O., "Power Factors and Electrical Demands of Center-Pivot Irrigation Machines," *Trans. of the ASAE*, 18(July-August 1975): 673-676.

CASE STUDY 6-3: ENERGY SAVING PRACTICES IN CATTLE FEEDLOTS[1]

THE CONCEPT

Fattening cattle in feedlots requires relatively large quantities of energy to produce the desired weight increases. What then are the energy saving practices which can be employed to reduce the energy input to the feedlot?

BACKGROUND

Four cattle feedlots in the western US were studied in order to determine the direct energy inputs to each feedlot and the resulting energy output of the feedlot in the form of increased weight of the cattle. This study was performed by Mr. E.W. Greninger, a consultant for ANCO, who spent many weeks at the various feedlot sites. With the cooperation of the management, Mr. Greninger was able to review in detail the books for the year 1975 for one of the lots and was able to compile an accurate and detailed listing of all commodities purchased during the year, estimates of all transportation costs, and a detailed log of all the cattle passing through the lot during the year. As a result, tabulations of energy inputs to and outputs from the feedlot were constructed. From these tabulations, it is possible to determine the direct energy inputs to the feedlot and the resulting energy output in the form of increased weight of the cattle.

In addition, from the tabulation, it is possible to determine the overall efficiencies of producing cattle in the feedlot.

FINDINGS

Table 1 contains some of the results of the feedlot energy balance. The feedlots ranged from a relatively small operation (1,000 to 4,000 cattle) to a relatively large operation containing 15 to 25,000 cattle. The feed which is used to fatten the cattle was carefully tabulated for lot Number 2. Seventy-nine percent of of the feed was either alfalfa hay or barley, both of which could be used to feed other domestic animals, ranging from dairy cows to sheep or rabbits. In addition, the barley, which comprised 62 percent of the total feed to the cattle, could be consumed by humans. The rest of the food was relatively unimportant and in many cases was considered as waste. This material ranges from animal fat to beet pulp pellets and any other vegetable, fruit, or fibrous material which was available at low cost in the area.

The efficiencies listed in Table 1 assume that all feed supplied to the cattle is of some importance in human consumption whereas actually only about 62 percent of that food could be directly consumed by humans. Therefore, the tabulated efficiencies are relatively low if one considers only the barley which is consumed by the cattle. However, all the indirect energy sources, such as the energy required to produce capital equipment (machinery, transport vehicles, buildings, conveyers, and other systems), have not been included as energy inputs. In addition, this table does not account for the energy required to maintain the labor force, nor to deliver all the direct energy (electricity, propane, methane, gasoline, diesel, oil, grease, and hydraulic fluid). If one were to add all the indirect energy costs to the direct energy costs, the efficiency of producing cattle in a feedlot would be very small, possibly only 50 percent of the values in the table.

RECOMMENDATIONS

This case study points out that the efficiency of fattening beef in a feedlot is quite low compared to range fed cattle. The feedlot manager should investigate possibilities of reducing the direct energy cost to the feedlot in order to save energy and increase profits. Since the efficiency is relatively low, it appears that a number of improvements are possible, ranging from more efficient use of night lighting to increased sales of waste by-products such as manure. The table also shows that transportation requires a sizable amount of energy in the form of direct energy, primarily fuel and oil for the cattle trucks and feedlot vehicles. Improved maintenance procedures, more frequent inspection, and, most importantly, full utilization of the cattle trucks hauling loads to and from the feedlot would greatly reduce the energy used for transportation.

Another way to reduce energy inputs is to take cattle directly from range land for slaughter. This procedure would not only save energy but would result in a leaner, tougher beef product possibly healthier for the consumer (although also possibly meeting consumer resistance).

	Lot #1	Lot #2	Lot #3	Lot #4
1. Approximate number of cattle in yard at one time (thousands)	1-4	4-8	8-15	15-25
2. Average weight in (lb)	633	660	650	670
3. Average weight out (lb)	822	826	973	958
4. Total cattle on feed during 1973	6,073	21,739	30,631	71,114
5. Total feed consumed (lb)	14,935,150	29,831,400	79,001,000	172,232,760
6. Total weight gained (lb)	1,432,420	3,735,265	9,458,243	20,109,960
7. 43 percent of item (6)--edible meat gain (lb)	615,940	1,606,160	4,067,065	8,647,283
8. Energy in edible meat (GJ)*	4,125	10,757	27,240	57,916
9. Energy contained in feed (GJ)	125,160	217,320	663,840	1,333,650
Direct Energy Use (GJ)				
10. Electricity	270	1,020	1,860	4,770
11. Propane	None	4,700	1,020	6,080
12. Methane	None	None	7,560	24,080
13. Gasoline, 14. Diesel, 15. Oil, 16. Grease, 17. Hydraulic Fluid	2,370	3,540	3,630	2,170

*Assuming 100 kcal/oz

ENERGY BALANCE FOR FOUR FEEDLOTS (1973) **TABLE 1**

	Lot #1	Lot #2	Lot #3	Lot #4
18. Total Direct Energy (GJ)	2,640	9,260	14,070	37,100
19. Transportation of feed and cattle* (GJ) (fuel and oil only)	5,585	12,194	31,234	103,581
20. Efficiency of producing edible meat considering cattle feed only as input (percent)	3.30	4.95	4.10	4.34
21. Efficiency of producing edible meat considering all direct energy inputs (feed, transportation, fuel, electricity, propane, etc.)	3.09	4.51	3.84	3.93
22. Feed costs per pound gained (includes feed lot profits) (¢)	46.3	36.3	41.7	N.A.
23. Feed cost per head per day (¢)	95.4	88.5	105.3	N.A.

*Measured in Lot #2 and scaled with respect to relative number of cattle and feed input for Lots #1, #3, and #4.

ENERGY BALANCE FOR FOUR FEEDLOTS (1973)

TABLE 1 (Cont.)

References

1. Personal communication, E.W. Greninger, Santa Monica, California, June 1975.

15.7 BUILDING ENVELOPES AND SITES (Chapter 7)

CASE STUDY 7-1: SUPER INSULATED HOUSES [1]

THE CONCEPT

Proven insulating techniques combined with the efficiency of the heat pump can significantly reduce energy use for heating and cooling houses.

BACKGROUND

The Tennessee Valley Authority has developed the concept of "Super Saver" electric homes for its customers. The general specifications for building such a house are listed in Table 1. To determine which insulation meets the standards described, refer to Table 2.

The R-19 wall insulation can be achieved by using 2 in. by 6 in. wall studs with R-19 friction fit batts installed between them, or by using another construction method that will give an equivalent insulating value. For example, an alternate method would be to use 2 in. by 4 in. studs, insulating board in place of standard sheathing, and R-13 friction fit batt between the studs.

In selecting the heat pump, careful consideration should be given to the unit's energy efficiency ratio of cooling output and the coefficient of performance of heating output. The higher the EER and COP, the greater the efficiency.

FINDINGS

In the first heating season for most of the Super Saver homes, preliminary data indicate that annual energy requirements for heating and cooling a Super Saver home will be reduced by 60-70 percent over a conventionally constructed house.

Metered data from two homes in northeastern Tennessee indicate that 2,880 kWh and 2,550 kWh respectively were used by the heat pumps, including supplementary heat, from September 30, 1976, through January 3, 1977. These homes contain approximately 1,800 square feet of heated space. During this interval 2,274 degree days were recorded in the area. Based on a "conventionally" constructed home of similar size and heated with an electric furnace, these homes would have used approximately 11,230 kWh for space heating during this interval. Therefore, the 2,880 kWh used in one of the residences represents about 26 percent (74 percent reduction) of "normal" (with electric furnace) and the 2,550 in the other residence represents about 23 percent (77 percent reduction).

Many of the vocational technical schools throughout the Valley are constructing Super Saver homes for energy educational purposes and to help in providing impetus for energy efficient housing.

Those people living in Super Saver homes with whom we have spoken are pleased with the comfort and quietness of their homes, as well as the operating cost for heating and cooling.

RECOMMENDATIONS

The techniques described in this case study can be applied to new home construction, additions, or to renovation projects. The basic concept is to use existing methods and materials to build a thermally "tight" residence. Heating and cooling energy is reduced by minimizing both conduction and infiltration/exfiltration losses. Caution: With a thermally tight structure, special attention must be given to humidity problems, ventilation and make-up air, and odor control.

References

1. "Super Saver Electric Home Program - Update," Tennessee Valley Authority News Release, (February 10, 1977).

SUPER SAVER ELECTRIC HOME

Specification Sheet

All building construction must meet or exceed HUD minimum property standards:

1. INSULATION:

   1.1 Minimum required R-values— Ceiling R-30; Floors R-19; Walls R-19

   1.2 Concrete slab floor perimeters— 1½" urethane (R-10).

   1.3 Non-conditioned space ductwork— 3" external or 1½" internal liner or equivalent combination.

   1.4 Conditioned space ductwork— ½" duct liner.

   1.5 Water heater— should have 2" external layer.

   1.6 Hot water pipes in non-conditioned space— ½" insulation.

2. VAPOR BARRIER:

   2.1 Walls, ceilings, and floors— positive vapor barrier covering entire surface and having transmission rate not exceeding one perm.

   2.2 Polyethylene— lapped 6" at all joints.

   2.3 Foil-backed drywall may be used.

   2.4 Concrete slab— must rest on complete vapor barrier.

   2.5 Crawl space— covered with polyethylene vapor barrier of at least 6-mil thickness.

   2.6 Crawl space— ventilated to HUD standards.

3. WINDOWS AND DOORS:

   3.1 Not to exceed total of 10 percent of floor area. Windows and glass doors double-glazed or storm sashed.

   3.2 Exterior doors— core-filled with rigid insulation, or wood with storm door.

   3.3 Outside windows and doors— must be weatherstripped and caulked.

4. HEATING AND AIR CONDITIONING:

   4.1 Provided by an electric heat pump.

5. VENTILATING:

   5.1 Exhaust fans— in kitchen and all bathrooms and vented to outside, (may be omitted if dehumidifier and electrostatic air cleaner are used).

6. CONSIDER:

   6.1 Fluorescent lighting; 6.2 insulated draperies; 6.3 a power attic ventilator; 6.4 energy efficient appliances; 6.5 house shape and orientation; 6.6 fireplace— if used, should have its own duct for combustion air and should be be equipped with a heat exchanger or equivalent device.

SUPER SAVER HOME **TABLE 1**

	BATTS OR BLANKETS		**LOOSE FILL			
	glass fiber	rock wool	glass fiber	rock wool	cellulosic fiber	
R-11	3½"-4"	3"	5"	4"	3"	R-11
R-19	6"-6½"	5¼"	8"-9"	6"-7"	5"	R-19
R-22	6½"	6"	10"	7"-8"	6"	R-22
R-30	9½"-10½"*	9"*	13"-14"	10"-11"	8"	R-30
R-38	12"-13"*	10½"*	17"-18"	13"-14"	10"-11"	R-38

*two batts or blankets may be required.
**(must be poured or blown to manufacturer's specification for correct density).

RECOMMENDED INSULATION **TABLE 2**

Acknowledgment

This case study is based on materials provided through the courtesy of Mr. Jim Ward, Chief, Electrical Demonstration Branch, Tennessee Valley Authority.

CASE STUDY 7-2: THE USE OF BUILDING ENVELOPE COMPONENTS FOR SOLAR HEATING

THE CONCEPT

Integration of a solar heating system into the building envelope can provide a passive system for the temperature control of small buildings, saving electricity and fuel.

BACKGROUND

The use of solar heating is not limited to flat plate or concentrating collectors. The building itself can be a heat collecting and storage unit. This concept has been used to design structures that combine living space requirements and solar collector concepts.

FINDINGS

Included in this case study are two types of houses which use passive solar systems to supply most of the heating requirements. They are both extracted from Reference 1.

The Odeillo Solar House

Dr. Felix Trombe has for many years been the Director of Solar Energy Research for the Centre Nationale de la Recherche Scientifique (CNRS) of France.

The climate in the area near his laboratory at Montluis in the Pyranees is moderate in summer and the nearby town of Font Remeu is regarded by Parisiens as a summer resort because the temperature and humidity are both relatively low by French standards. The town becomes a ski resort in winter because of the heavy snowfalls which the region experiences. Winter temperatures in the valley below Font Remeu do not fall as low as they do in the much higher alpine terrain and the hours of sunshine experienced per year are the highest in France.

Because of the favorable climatic conditions in the area around Odeillo, where his solar furnace is located, Dr. Trombe began experiments in "climatization solaire." His concept, implemented by his collaborating architect, M. Jacques Michel, uses a massive south-facing concrete wall, single-glazed, as the solar energy absorber in winter (see Figure 1). The space between the glazing and the concrete becomes an air passage through which air from the floor of the single family residence can rise as it becomes heated by contact with the dark concrete wall. There are entry ports at the bottom and top of the wall, with dampers to control the air flow, and the remainder of the structure is well insulated so that the heat requirements are minimal. The warmed air reenters the room through ports at the top of the wall and natural convection, unaided by a fan, takes care of the air movement.

During the summer, the overhang of the roof shades the south wall completely, and there are vents opening outward which enable the heat generated by the wall to establish a desirable natural ventilation by bringing in fresh air from the north side of the house.

Dr. Trombe uses the massive south wall as the heat storage means, and the warmth from the inner surface of the concrete provides most of the heat which is needed at night during the winter. The time lag established by the heavy concrete wall means that the maximum heat input comes six to eight hours after the maximum irradiation, which takes place at noon.

The US modification of the Trombe-Michel concept involves the use of additional thermal storage in the form of a rock bed located beneath the floor. By the use of blowers (which admittedly complicate the system but also make it much more versatile) the heat from the south wall can be stored in a rock bed located beneath the house and thus the storage capability of the system can be considerably increased.

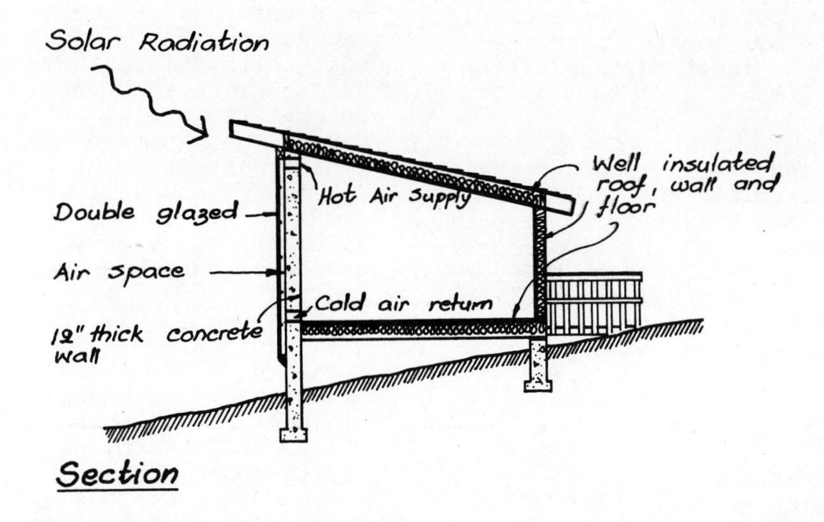

Section

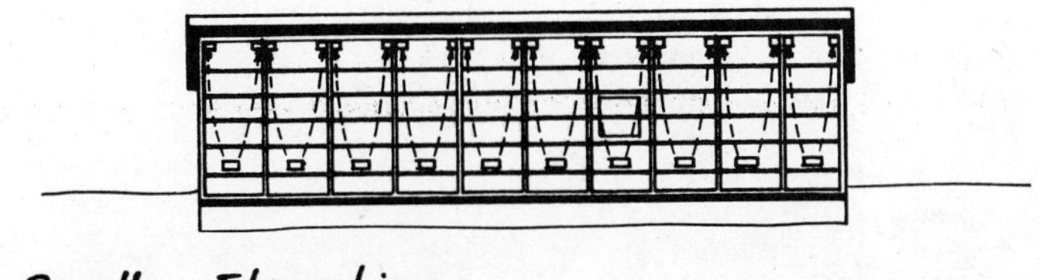

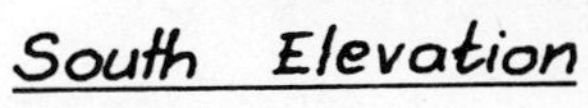

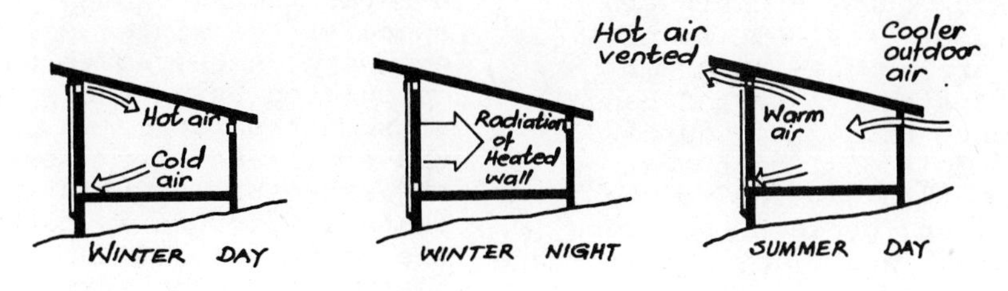

System Schematics

ODEILLO SOLAR HOUSE **FIGURE 1**

Auxiliary heat in the Trombe-Michel houses at Odeillo, which are the residences for the staff of the large solar furnace, comes from electric convectors located near the floor, at the inner surface of the concrete walls. (This heat is seldom needed.)

The mass of the concrete walls serves to insulate the south wall of the Odeillo houses quite effectively. The dampers at the floor-level entrances to the air space between the wall and the glazing are closed at night and the glazing prevents the loss of heat by radiation from the outer surface of the wall.

The David Wright Residence

The design of the David Wright residence is extremely simple and open, as shown in Figure 2. The sun dictated the design of the house from the very beginning. The solar aspects are so well integrated that the house actually becomes the solar collector and the heat storage system. The south wall of the house is 384 sq ft of "thermopane" glass. The solar heat gain on a cold winter day is 385 MJ/day (365,000 Btu/day). Total heat loss for a typical 24 hour period is 298 MJ (282,000 Btu).

The walls are made of 35.6 cm (14 in.) thick adobe and beneath the brick floor is 61 cm (24 in.) of adobe. Several 55 gallon drums filled with water are buried beneath an adobe banco along the south wall to provide additional heat storage. Around the entire outside of the adobe walls, and beneath the adobe floor, is 5 cm (2 in.) of polyurethane insulation. This insulation minimizes the flow of heat from the walls and floor to the cold outdoor air and the ground. Thus the heat is stored until the temperature inside the house drops and then the stored heat is radiated and convected into the space. The fabric of the building is capable of storing enough heat to keep the home comfortable for three or four sunless days.

Houses utilizing this principle of direct solar irradiation were first conceived, built and tested in the late 1930s. It was found that although it was possible to collect a large amount of heat during the day (sometimes to the point where it became exceedingly warm), the heat losses at night through the glass often exceeded the gains. Many houses were built with this principle, and sold as "solar houses," but the problems of what to do with the excess heat during the day, and how to keep from losing heat at night were never fully solved. Perhaps it was because the architects of these buildings looked at solar energy only as an auxiliary source of heat but not the prime source.

David Wright has solved the problem of heat storage by using the mass of the building itself. He has also minimized heat losses through exterior walls and floor by externally insulating the house well. Finally, to reduce the heat loss through the vast expanse of double glass, Mr. and Mrs. Wright have developed an ingenious system of vertical folding insulating shutters made of canvas and 2 in. polyurethane. These shutters can be easily raised and lowered with a simple small homemade device. The whole house is a completely passive system. The only moving parts are the shutters which control the heat gains and losses.

The depth to the back wall of the house was calculated by determining how far the sun would penetrate the space at the winter solstice. An overhang on the south wall was also calculated using the solar altitude at summer solstice in order to keep the summer sun from penetrating into the structure. Other energy conserving features include an earth berm along the north wall of the house to reduce heat losses even further, an air lock entry to reduce heat loss through air exchangers, and operable windows placed to maximize cross-ventilation for summer cooling.

The same factors of heat storage in the winter can be used for cooling in the summer. The massive adobe walls and floor can be cooled down at night by natural ventilation and they will absorb heat generated during warm summer days to provide the house with a comfortable condition.

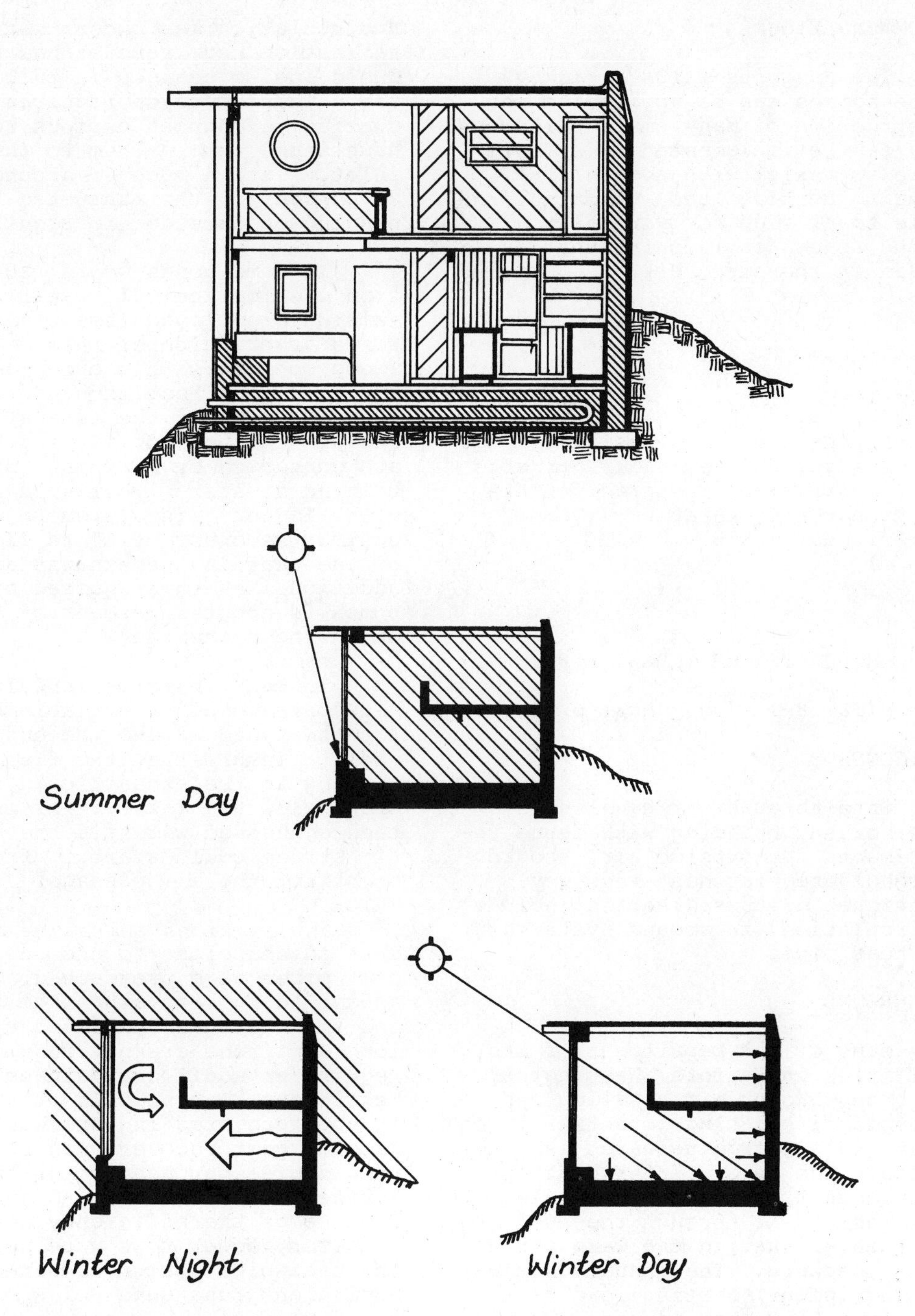

THE DAVID WRIGHT RESIDENCE

FIGURE 2

RECOMMENDATION

The concepts illustrated by these houses can be applied to the construction of many small buildings. They can be implemented on a small scale to assist with some of the building heating load, or on a large scale to provide for virtually all of the space conditioning that is needed in the structure.

References

1. Arizona State University College of Architecture, *Solar Oriented Architecture*, prepared for AIA Research Corporation, (Tempe, Arizona: 1975).

15.8 HVAC SYSTEMS (Chapter 8)

CASE STUDY 8-1: HVAC HEAT RECOVERY[1]

THE CONCEPT

Once-through air handling systems, or any building with large requirements for outside air, should be considered for heat recovery. Techniques discussed herein include recirculation, runaround systems, and heat pipes.

BACKGROUND

Many of the Department of Energy facilities in Hanford, Washington, use once-through air handling systems to avoid circulating potentially contaminated air from research laboratories. The major winter heating load comes from heating the inlet air, passing it through the building and then exhausting the warm air to the atmosphere. The annual outside air (equipment) heating load is roughly 0.56 GJ per m^3/sec (0.3 MBtu/cfm). All of the buildings are operated 24 hours a day.

FINDINGS

The different HVAC configurations already existing resulted in the conceptual design of three different types of heat recovery systems. Recirculation was chosen where contamination requirements had been reduced due to changes in building zone use. This approach requires insulated ducting to connect various roof exhausts and return them to the air inlet. It can recover around 80 to 90 percent of the exhausted heat depending on outside air requirements.

In some zones where recirculation was impractical, runaround systems were employed (see Chapter 8 for a description of this method). These operate with one counterflow air to liquid heat exchange coil of 10 to 12 rows in the exhaust duct and one in the supply inlet. They are connected by a piping loop containing a water glycol solution. The economics of this system showed an optimum recovery of 60 to 72 percent of the heat in the exhaust air. The additional energy required by fans and pumps is about 2 percent of the energy which the system saves.

In many installations it is necessary to add a variable bypass for the fluid around the supply coil to insure that the fluid returning to the exhaust coil does not drop below 0°C (32°F). If it does then condensed water in the exhaust air stream could start to freeze and constrict the air passage.

In locations where the exhaust duct passed close to the supply duct, heat pipes were prescribed. These appear to be less expensive to install and operate than runaround systems if long duct runs are not required. These require an additional exhaust port before the heat pipes so that heat is not recovered in warm weather. The rate of heat recovery can be adjusted to a degree but heat pipes cannot be "turned off" as runaround systems can. Because of the efficiency of heat transfer, fewer rows were needed in the heat pipe design compared to the equivalent runaround coils.

These various techniques will save from 0.32 to 0.48 GJ per m^3/sec (0.18 to 0.27 MBtu/cfm) of exhaust air per year. For the entire project the total energy saving is 2.1 x 10^5 GJ/yr (2 x 10^{11} Btu/yr). The average payback for the sixteen buildings included is an estimated 4.3 years.

RECOMMENDATIONS

Heat recovery should be considered for hospitals, kitchens, and other facilities which continuously exhaust heated air to the atmosphere.

References

1. Applied Nucleonics Company, Inc., "Heat Recovery from the Ventilation System of Nuclear Facilities," Report No. 1178-12, (Santa Monica, California: September 1977).

CASE STUDY 8-2: SPLIT FLOW MODIFICATION OF DUAL DUCT AND MULTIZONE SYSTEMS [1]

THE CONCEPT

The energy efficiency of double duct and multizone HVAC systems can be substantially improved with various modifications, one of which involves the separation of return and outside air flows through the air handler. Examples are presented in which 50 percent of heating energy use has been eliminated, and a static simulation offers a conceptual understanding of how these reductions were possible.

BACKGROUND

Over the years, multizone and dual duct systems have been used extensively in medium and large buildings across the country. This popularity results from their allowing both precise temperature control and centralized mechanical equipment, the latter offering low initial cost, the ability to provide simultaneous heating and cooling for different zones (i.e., interior versus perimeter spaces), ease of maintenance, reduced pump energy requirements, and sometimes reduced fan energy consumption. However, concern over the cost and availability of energy resources has brought to light one disadvantage. Whenever air is cooler outside than inside, these systems must mix heated and cooled air together to maintain comfortable temperatures. This is a serious problem, since much energy is used even when it seems that neither heating or cooling should be required.

A split flow strategy was developed in the course of research at the Claremont Colleges in California, where virtually all buildings are fitted with economizer systems. The absence of buildings with extensive interior zoning offered little opportunity to apply variable air volume modifications. The primary need was for a system that would transfer heat from the sunny sides of buildings to the cooler, shady sides in order to reduce energy use for simultaneous heating and cooling.

The theoretical model that follows will consider three system variations. Following the model, an actual installation will be presented, with energy savings, initial costs, and comfort considerations. A static simulation has been used to calculate heating and cooling curves as a function of outside dry bulb temperature for a theoretical building modeled after a five-story library.

The basic system under consideration is the dual duct or multizone system illustrated in Figure 1. It is equipped with a hot plenum reset controller that resets the hot deck temperatures upward on a set schedule as the outdoor dry bulb temperature falls when heating is required, constant cold plenum temperature, and constant 10 percent outside air for makeup. Its heating and cooling energy use appears in Figure 2. The notable point is the large amount of energy put in for heat that is then pumped out again with air-conditioning equipment. This can be eliminated without changing the building's skin characteristics.

Variable air volume modifications deal with this situation by eliminating or limiting the mixing of heated and cooled air at zone mixing boxes. This reduces the area under both curves in Figure 2. While heated and cooled air must still be provided with variable air volume, buildings with large interior zones can show substantial

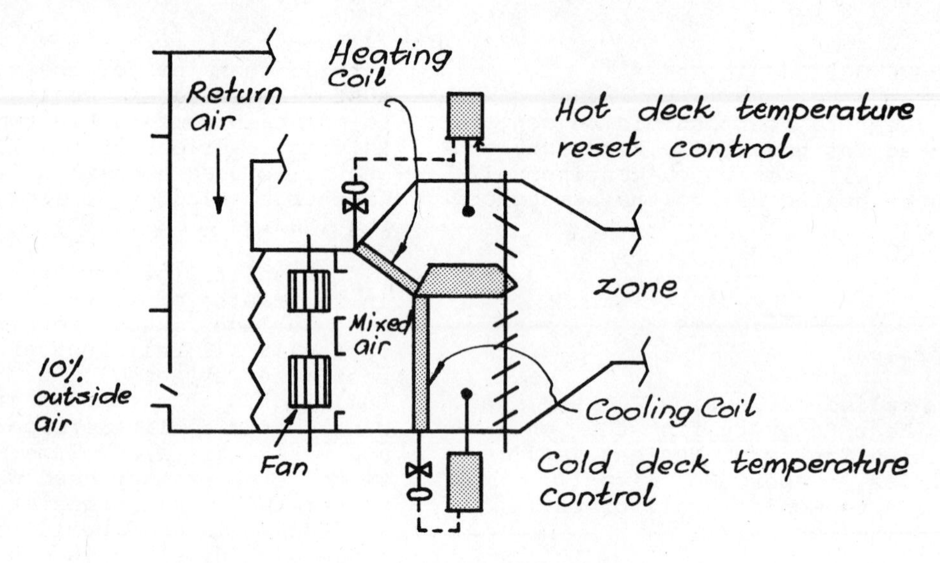

(A) A dual duct or multizone system set to operate with a fixed amount of outdoor air.

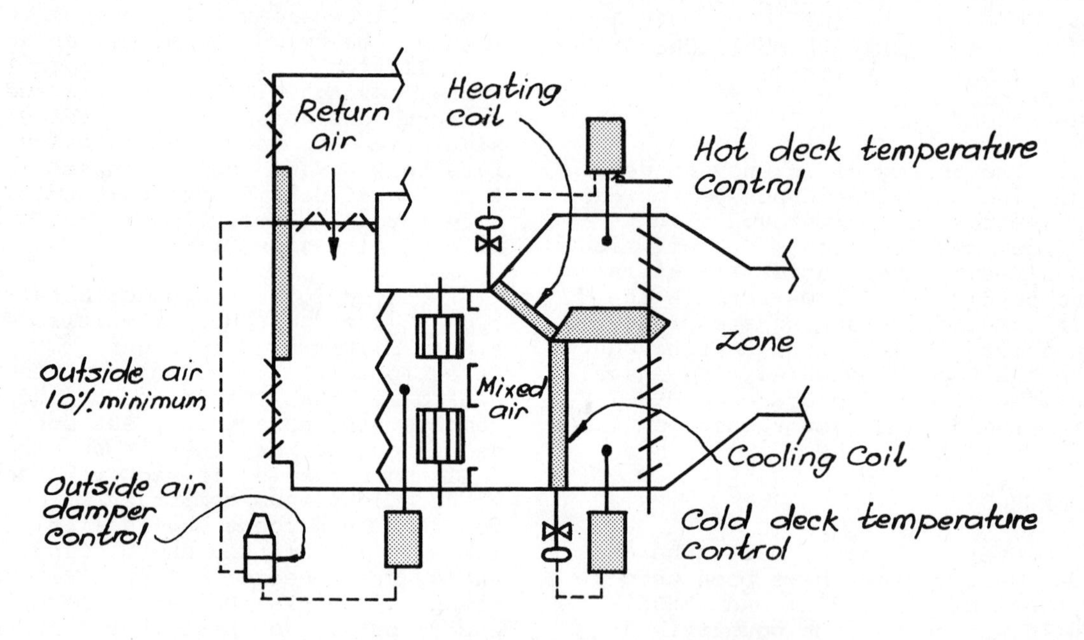

(B) A dual duct or multizone system with an air side economizer system permitting variable quantities of outside air to be used for cooling when outside dry bulb temperature is below the return air temperature.

DUAL DUCT AND MULTIZONE MODIFICATIONS **FIGURE 1**

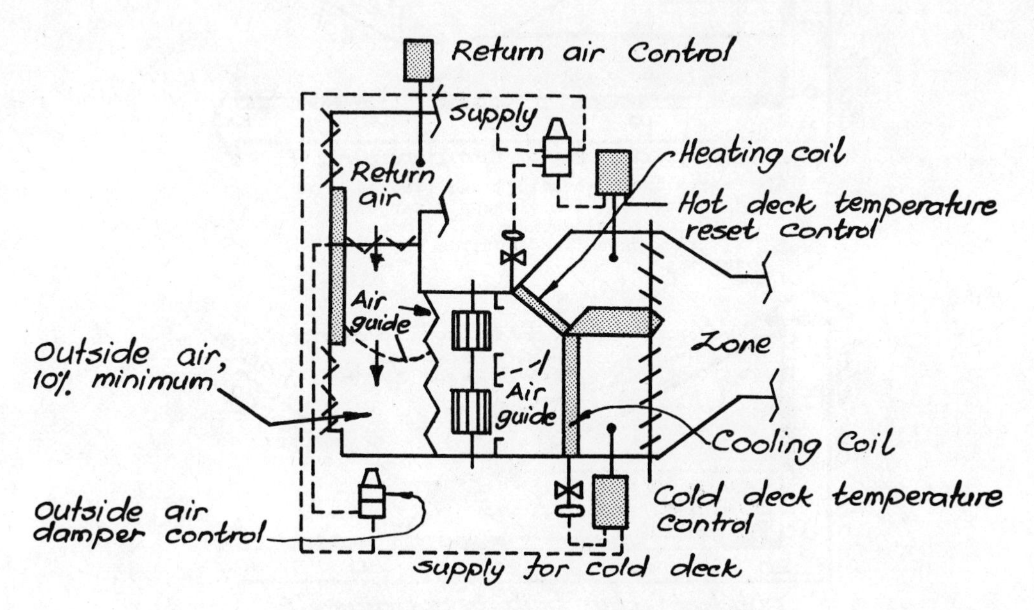

(C) A dual duct or multizone system with the split flow modification. Air guides direct return air to the hot deck and virtually all the outside air and some return air to the cold deck.

FIGURE 1 CON'T.

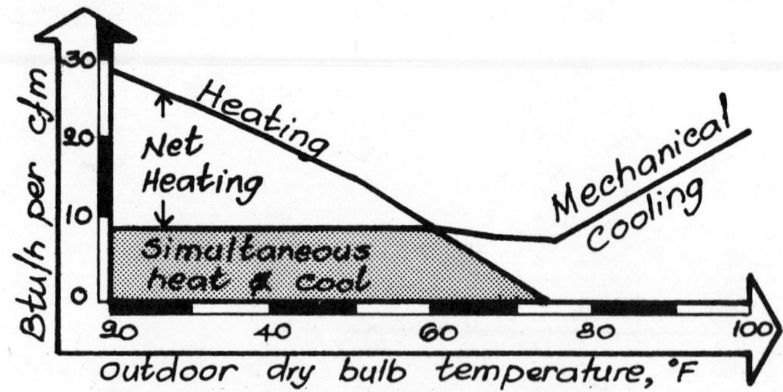

(A) Energy added (heating) and removed (cooling) per cfm of supply air in a typical dual duct or multizone system with a fixed outside air setting.

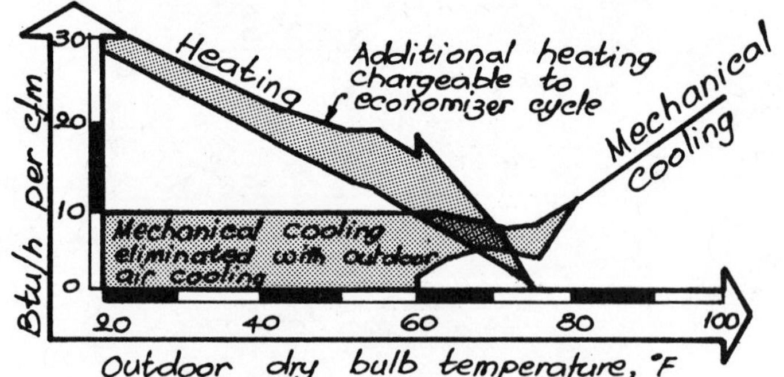

(B) Energy added (heating) and removed (cooling) per cfm of supply air in a typical dual duct system with an economizer cycle. Mechanical cooling requirements are greatly reduced, but space heating energy increases.

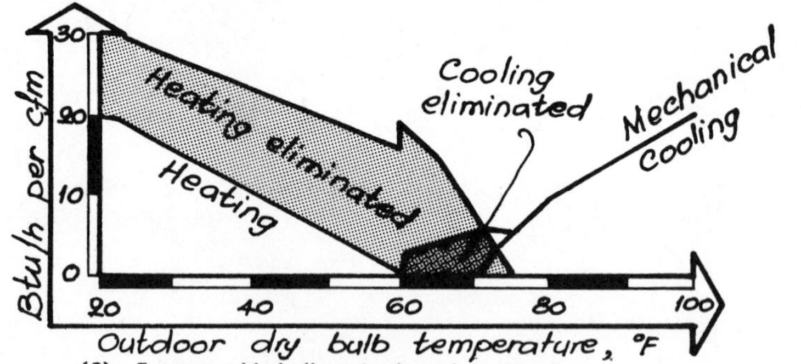

(C) Energy added (heating) and removed (cooling) per cfm of supply air in a dual duct system equipped with split flow modification. Values are the lowest for the three system variations discussed. Note that the heating energy curve is essentially equal to the net heating energy curve in Figure (A)

ENERGY SAVINGS FROM SPLIT FLOW MODIFICATION **FIGURE 2**

savings of energy.

The air side economizer added to the basic dual duct or multizone system appears in Figure 1. In this case, mixed air is maintained at 60°F by modulation of the outside air and return air dampers. Above 60°F outside temperature, the cold plenum is maintained at 55°F by mechanical refrigeration. As in the first system, the hot plenum temperature is reset inversely as the outside air temperature changes. This system's energy use is given in Figure 2. Mechanical cooling has been eliminated when outside air temperatures fall below 60°F, saving not only cooling energy, but also many hours of operation of the auxiliary cooling equipment. It should be noted, however, that heating energy has increased because of the effect the additional outside air has on lowering the mixed air temperature. Heating energy has historically been far less expensive than cooling energy, so this modification has been easily justified in the past.

The split flow modification is illustrated in Figure 1. Air guides discourage the mixing of return and outside air, directing them respectively to the hot and cold plenums. A return air controller keeps the average building temperature constant by controlling the hot plenum, the cold plenum, and the economizer. The theoretical results of this change can be observed in Figure 2. While there is a reduction in cooling energy and equipment operating time, the most notable reduction in this example is the heating energy use during moderate weather. At 48°F, the average winter temperature in Claremont, a theoretical reduction of 70 percent is possible.

FINDINGS

In Claremont research, this reduction was not fully realized, but results were significant. Fifteen air handlers delivering about 225,000 cfm were modified to correspond to the split flow pattern shown in Figure 1. Because winter temperatures will vary from year to year, energy use per degree day is shown in Table 1, for two different building complexes. The normalized reduction achieved was roughly 50 percent.

To illustrate typical results, Figure 3 shows energy savings in an all-electric building due solely to a split flow modification. In this building the total monthly electricity use was cut by an average of 40 percent.

Comfort conditions have improved in all areas. Overheating and overcooling problems have been all but eliminated. The only notable disadvantage is that an individual usually cannot obtain a temperature more than two degrees greater or less than the average building temperature.

The cost of installing this split flow modification is probably the most impressive aspect of the program. While variable air volume refitting for Project Libra would cost tens of thousands of dollars, material and labor to modify the seven air handlers came to roughly $900. Thatcher Building (all electric) is now saving approximately $1000 per month after a modification costing less than $200, giving a payback period of one week.

RECOMMENDATIONS

Buildings can be converted quickly and, as a rule, without the occupants realizing that a change is being made. Large amounts of energy can be saved. The dual duct or multizone system can again become an energy efficient system because of its ability to recover heat from warm zones for heating and to use outside air for cooling.

References

1. Lee, Richard H., "Energy Analysis of Double Ducts and Multizone Systems," in *Energy Use Management--Proceedings of the International Conference*, edited by Rocco A. Fazzolare and Craig B. Smith, (New York: Pergamon Press, Inc., 1977), Vol. I, pp. 259-265.

Natural gas usage versus degree-days for Project Libra during November, December, and January for the past three years.

	74-75	75-76	76-77
Therms	35,700	30,200	14,200
Degree-days	962	860	847
Therms per degree-day	37.1	35.1	16.8

Natural gas usage versus degree-days for Bauer Center between August and January for the last two years.

	74-75	76-77
Therms	16,600	7,900
Degree-days	980	894
Therms per degree day	16.9	8.8

GAS SAVINGS FOR TWO SPLIT FLOW PROJECTS **TABLE 1**

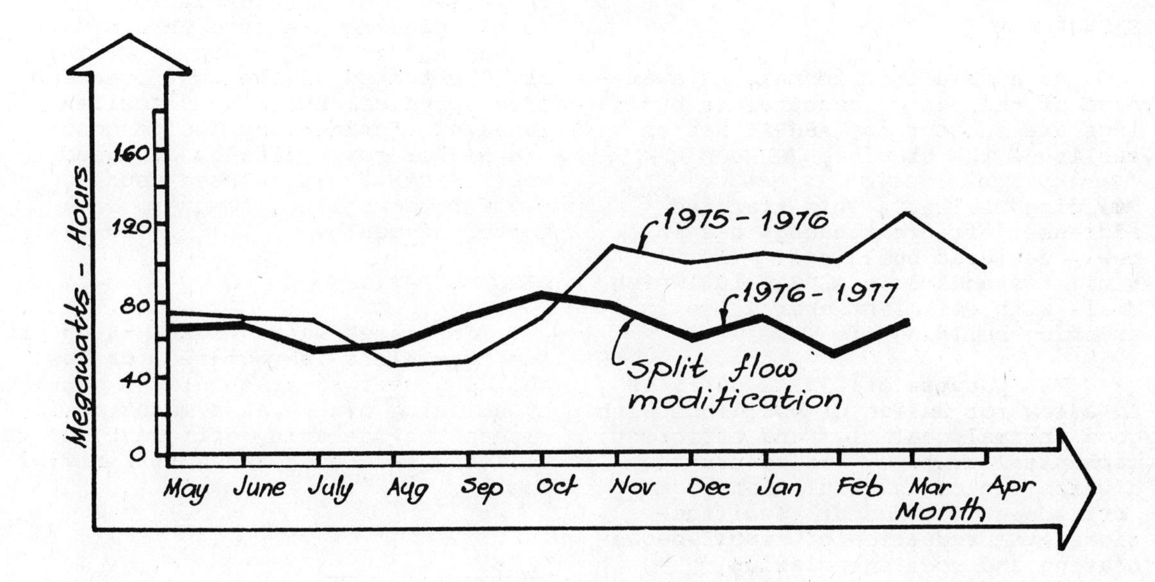

*An all electric building (Thatcher)

TOTAL ELECTRICAL USAGE PER MONTH*

FIGURE 3

CASE STUDY 8-3: THE IMPACT OF ASHRAE STANDARD 90-75[1,2,3]

THE CONCEPT

The American Society of Heating, Refrigerating, and Air Conditioning Engineers (ASHRAE) has established standards for energy efficient new building design. These standards not only reduce energy costs, but also reduce construction costs in most cases.

BACKGROUND

As a result of increasing awareness of the energy constraints buildings are subject to, ASHRAE has established the standard ASHRAE 90-75, "Energy Conservation in New Building Design." This standard addresses efficient energy use in newly designed buildings. (A similar standard is ASHRAE 100 which deals with efficient energy use in existing buildings.)

The purpose of ASHRAE 90-75 is to allow for design of buildings with good thermal qualities and efficient mechanical systems. It covers such diverse topics as insulation, comfort zones, thermostat specifications, the reduction of simultaneous heating and cooling, glazing, lighting and mechanical equipment efficiency. It also contains a section dealing with non-depleting forms of energy such as solar and wind power.

To determine the effect ASHRAE 90-75 could have on building energy use, the A.D. Little Company performed a special analysis, *Impact Assessment of ASHRAE Standard 90-75*. This study evaluated the results of changes dictated by the standard for three prototypical buildings (office, retail store, and school) located in various regions. In addition to computer simulation of the building energy performance, changes in engineering and construction costs were evaluated.

FINDINGS

Under a strict interpretation of ASHRAE 90-75, annual energy use is significantly reduced. The unweighted average reduction in annual energy use for each building compared to typical (pre-90-75) designs is shown in Table 1. The monetary savings which can be expected in utility costs were also calculated and are detailed in Table 2.

The initial construction costs of the modified buildings turned out to be less than the costs for a conventional building. The construction cost changes are shown by building and region in Table 3. In general, the increased costs of insulation and double glazing are more than offset by savings in HVAC equipment. This is illustrated by the cost breakdown from a typical building described in Table 4. Engineering design costs are higher for buildings designed to meet ASHRAE 90-75. These usually pay for themselves, however, in a matter of months.

RECOMMENDATIONS

In states where ASHRAE 90-75 or its equivalent is required for new building design, it should be considered by building owners as a means to increase the operating efficiency of the building and to decrease its initial costs.

References

1. American Society of Heating, Refrigerating and Air Conditioning Engineers, *Energy Conservation in New Building Design*, ASHRAE Standard No. 90-75, (New York: 1975).

2. Arthur D. Little, Inc., *Impact Assessment of ASHRAE Standard 90-75, "Energy Conservation in a New Building Design,"* (Cambridge, Massachusetts: 1976).

3. "ASHRAE Standard 90-75 Seen to Affect Design Engineer, Building Project Industry," *Professional Engineer* 46(February 1976): 35-37.

	Northeast	North Central	South	West	Unweighted Average
Office Building	61.5	61.2	58.7	56.9	59.7
Retail Store	41.6	42.5	37.9	38.5	40.1
School Building	45.6	44.4	51.5	51.1	48.1

REDUCTION IN ANNUAL ENERGY USE (%) **TABLE 1**

	Northeast	North Central	South	West	Unweighted Average
Office Building	0.72	0.35	0.29	0.24	0.40
Retail Store	1.05	0.67	0.58	0.41	0.68
School Building	0.30	0.14	0.14	0.12	0.18

SAVINGS IN ANNUAL ENERGY COST-ASHRAE 90-75 vs CONVENTIONAL DESIGN ($\$/ft^2$) **TABLE 2**

	Northeast	North Central	South	West	Unweighted Average
Office Building	-0.35	-0.29	-0.94	-0.93	-0.63
Retail Store	-0.11	+0.04	-0.32	-0.33	-0.18
School Building	-0.56	-0.39	-0.46	-0.33	-0.44

CHANGE IN UNIT CONSTRUCTION COST-ASHRAE 90-75 vs CONVENTIONAL DESIGN ($\$/ft^2$) **TABLE 3**

	Change in Cost ($ per sq ft Floor Area)
Exterior Walls	+0.33
Exterior Glass	+0.24
Roof	+0.15
HVAC Equipment	-0.39
HVAC Distribution	-0.48
HVAC Controls	+0.12
Lighting	-0.04
Electrical Distribution	-0.28
Domestic Water Heating	+0.02
Hot Water Distribution	+0.04
Net Change	-0.29

CHANGES IN UNIT COST FOR AN OFFICE BUILDING **TABLE 4**

CASE STUDY 8-4: ENERGY SAVINGS IN FAN OPERATIONS [1]

THE CONCEPT

Energy use in fans, especially in high pressure systems, can be a significant portion of building energy use. This case study describes the energy saving technique of slowing fans and a comparison of theoretical to actual energy reductions.

BACKGROUND

Ventilation rates in many buildings are being reduced due to standards and code changes which now require less ventilation, decreased cooling loads from reduced light levels, a wider accepted comfort range, and conservatism in past designs. The technique chosen to reduce the air flow will have a large effect on the actual energy savings, as discussed in Chapter 8.

In a college building complex the fans were slowed down by pulley changes in order to realize energy savings. The pulleys on the fan were increased by 11 to 45 percent to derate the fan, since the air flow is proportional to fan rpm. If the motor pulley is decreased in size, excessive squeaking at startup and accelerated wear may result.

FINDINGS

The results of these modifications are shown in Table 1. Table 1 shows that the reductions in energy use follow the cubic law prediction within several percent. The fan motors exhibit efficient unloading even in these low load regions.

RECOMMENDATIONS

Fan pulley changes represent an inexpensive technique for saving energy in buildings where reduced ventilation rates are permissible.

References

1. Personal communication, Richard H. Lee, Energy Conservation Specialist, Claremont Colleges, October 1977.

CASE STUDY 8-5: IMPROVING EFFICIENCY OF PACKAGE RECIPROCATING CHILLERS [1]

THE CONCEPT

Simple changes can be made to some reciprocating chillers to increase their energy efficiency during partial load operation.

BACKGROUND

This case study is based on experiments conducted at the campus of the Claremont Colleges. Modifications were made to several chillers in buildings throughout the campus. The result was energy reduction in chiller systems of between 30 and 40 percent.

FINDINGS

Two major areas of inefficiency characteristic of 4, 6 or 8 cylinder reciprocating chillers equipped with four staged temperature controllers are:

- low load operation inefficiencies, and
- the use of 25°C (85°F) cooling tower water when lower temperatures are available.

The low load inefficiencies result partially because chillers typically are oversized. It was found that no buildings in this study ever required full load operation, and few buildings required 3/4 load operation. In fact, 1/2 load operation proved more than adequate for most buildings, even during design weather conditions. This means that during more moderate weather, the chillers spend a great deal of time in the 1/4 load condition, cycling off when the cooling load is satisfied. Running chillers at higher loads for shorter periods of time offers increased efficiency of system operation as is illustrated in the following example:

Fan	Fan Speed Before (rpm)	Fan Speed After (rpm)	Speed Reduction (%)	Theoretical Savings (Reduction)[3] (%)	Power Draw Before (kW)	Power Draw After (kW)	Actual Savings kW Reduction (%)
P AC-1	1040	830	80	51	6.3	3.2	51
P RE-1	620	470	76	44	1.3	0.6	46
P AC-2	670	500	75	42	13.0	5.5	42
P RE-2	580	520	90	73	3.9	2.8	72
S AC-4	760	620	82	55	11.0	6.1	56
S AC-5	680	470	69	33	9.0	3.0	33
G AC-01	850	600	70	34	10.7	3.6	34
G RE-01	520	390	75	42	3.0	1.3	43

EFFECT OF FAN SPEED REDUCTION

TABLE 1

An 88 kW reciprocating chiller typically runs at 1/4 load continuously. In addition, a 10 kW chilled water pump, a 15 kW condenser water pump, and a 3 kW cooling tower fan must operate simultaneously. Alternatively, the same cooling would be provided if the chiller ran at 1/2 load for 30 minutes per hour. The chilled water pump runs continuously in both cases. Note also that the chiller operates less efficiently at low loads.

Item	1/4 Load (for 1 hr)	1/2 Load (for 1/2 hr)
Chilled Water Pump	10 kW (1 hr) = 10 kWh	10 kW (1 hr) = 10.0 kWh
Condenser Water Pump	15 kW (1 hr) = 15 kWh	15 kW (1/2 hr) = 7.5 kWh
Cooling Tower Fan	3 kW (1 hr) = 3 kWh	3 kW (1/2 hr) = 1.5 kWh
Chiller	34 kW (1 hr) = 34 kWh	52 kW (1/2 hr) = 26.0 kWh
Total	62 kWh	45.0 kWh

Therefore, operating the system at higher loads saves 17 kWh every hour (27 percent).

Thus, operating the chiller only at higher loads saves energy for two reasons. The chiller performs more cooling per kWh input at high loads and the peripheral pumps must be operated for shorter periods of time. Since modulating valves thermostatically control the temperature of the cold coils, in these chilled water systems the temperature of air delivered to the building can remain constant during the cycling.

Cooling tower water temperature also has a direct effect upon the energy used by the chiller. While the lowest available temperatures provide the most efficient operation, expansion valve sizing typically requires that a minimum of 30°C (85°F) tower water be provided at full load to prevent suction pressure depression. At 3/4 loads, 18°C (65°F) water is adequate for proper operation. This 12°C (20°F) drop represents a 40 percent decrease in theoretical energy use for compression. It is thus advantageous to utilize this extra cooling whenever the wet bulb temperature allows.

By rewiring the four stage temperature controller to eliminate 1/4 load and full load operation and by resetting the tower water controller to maintain 18°C (65°F), the advantage of both increases in efficiency can be obtained while still satisfying building requirements. Overall building electrical energy savings resulting from this study have ranged between 10 and 30 percent.

RECOMMENDATIONS

Energy efficiency in existing chillers can be increased dramatically by inexpensive modifications. In new chiller design, the inefficiencies discussed here should be avoided.

References

1. Lee, Richard H., "Energy Use Reductions in Air Conditioning Compressor Operation," in *Energy Use Management--Proceedings of the International Conference*, edited by Rocco A. Fazzolare and Craig B. Smith, (New York: Pergamon Press, Inc., 1977), Vol. 1. pp. 265-270.

CASE STUDY 8-6: PROTOTYPE COMMERCIAL HVAC SYSTEMS[1,2]

THE CONCEPT

Layered, floor by floor, in an office building of the General Services Administration (GSA) are several examples of energy efficient strategies and equipment, including a variety of HVAC systems.

The Manchester, New Hampshire building is a demonstration project planned for use by the building industry with the goal of more efficient commercial buildings based on the lessons learned in the project.

BACKGROUND

The building is seven stories high with a two-level basement parking garage located on a site of about 3700 m^2 (40,000 ft^2) as seen in Figure 1. The approximate gross floor area is 15,800 m^2 (170,000 ft^2) including approximately 6500 m^2 (70,000 ft^2) of underground garage. The building is occupied by some 400 people and the construction cost is estimated to be $9.7 million.

There are a number of unique features incorporated into the building. All core elements (stairs, elevators, toilets, mechanical rooms, etc.) are located adjacent to the north exterior wall thus creating one large vertical zone through the building that does not require close environmental control. Different environmental systems are provided on various floors to permit direct comparison of performance and efficiency. All systems included are expected to be energy efficient.

Various arrangements of unitary closed loop water-to-air heat pumps are utilized for HVAC on floors 1 through 3. Rejected heat from all heat pumps operating in the cooling mode during occupied periods is piped to a large insulated hot water tank for use at night, when heating is required. HVAC for floors 4 through 7 is provided by two central chillers and by four modular boilers with various types of distribution systems on the different floors. One central chiller is driven by a gas engine generator. The second chiller is of the absorption type, which operates from the waste heat from the engine generator. Condenser water from both chillers is piped to a heat storage tank (heat recovery) and to a cooling tower (excess heat rejection). Stored heat is used to the maximum extent possible.

Waste heat is used for space heating at night and for domestic hot water. Chillers have been deliberately undersized in relation to peak load. They are operated at night when the electrical load is light and production of chilled water is more efficient. Stored chilled water produced at night is used the next day during peak load periods. Various low-wattage lighting systems are provided on the different floors to permit direct comparison of performance and efficiency. Boilers and pumps are of the modular type to permit operation of only the minimum number required to gain high efficiency.

Figure 2 is a schematic of the central mechanical plant which serves the top four floors of the building. Approximately 460 m^2 (5000 ft^2) of flat plate solar collectors are mounted on the roof of the building and integrated with the mechanical plant servicing these floors. Table 1 summarizes the types of systems which will serve the various floors of the building.

FINDINGS

A computer system that runs the building systems is feeding data back from 900 sensors located throughout the building. It will be used to determine the optimal orchestration of the variables such as outside air use, illumination, solar collectors and the thermal storage tanks.

For the first three years of its operation (1977-1979) the Center for Building Technology will monitor the energy efficiency of the building systems. A special staff will analyze the operating characteristics and apply the information from the Manchester building to improve nationwide building efficiency guidelines.

RECOMMENDATIONS

Building designers should review the results of the Manchester project as they are obtained to evaluate cost effectiveness of each concept and to determine their applicability to other new building designs.

References

1. Hill, James E. and Kusuda, Tamami, "Manchester's New Federal Building! An Energy Conservation Project," *ASHRAE Journal* (August 1975): 47-54.

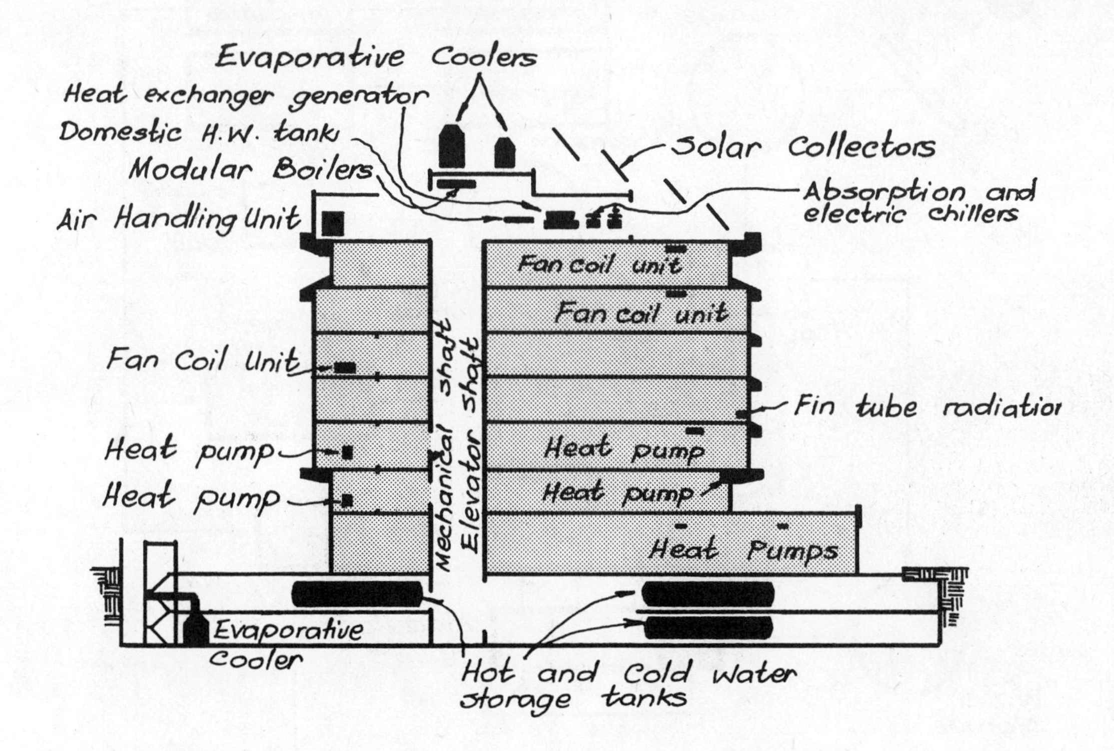

HVAC COMPONENTS OF THE GSA BUILDING (MANCHESTER, NH)

FIGURE 1

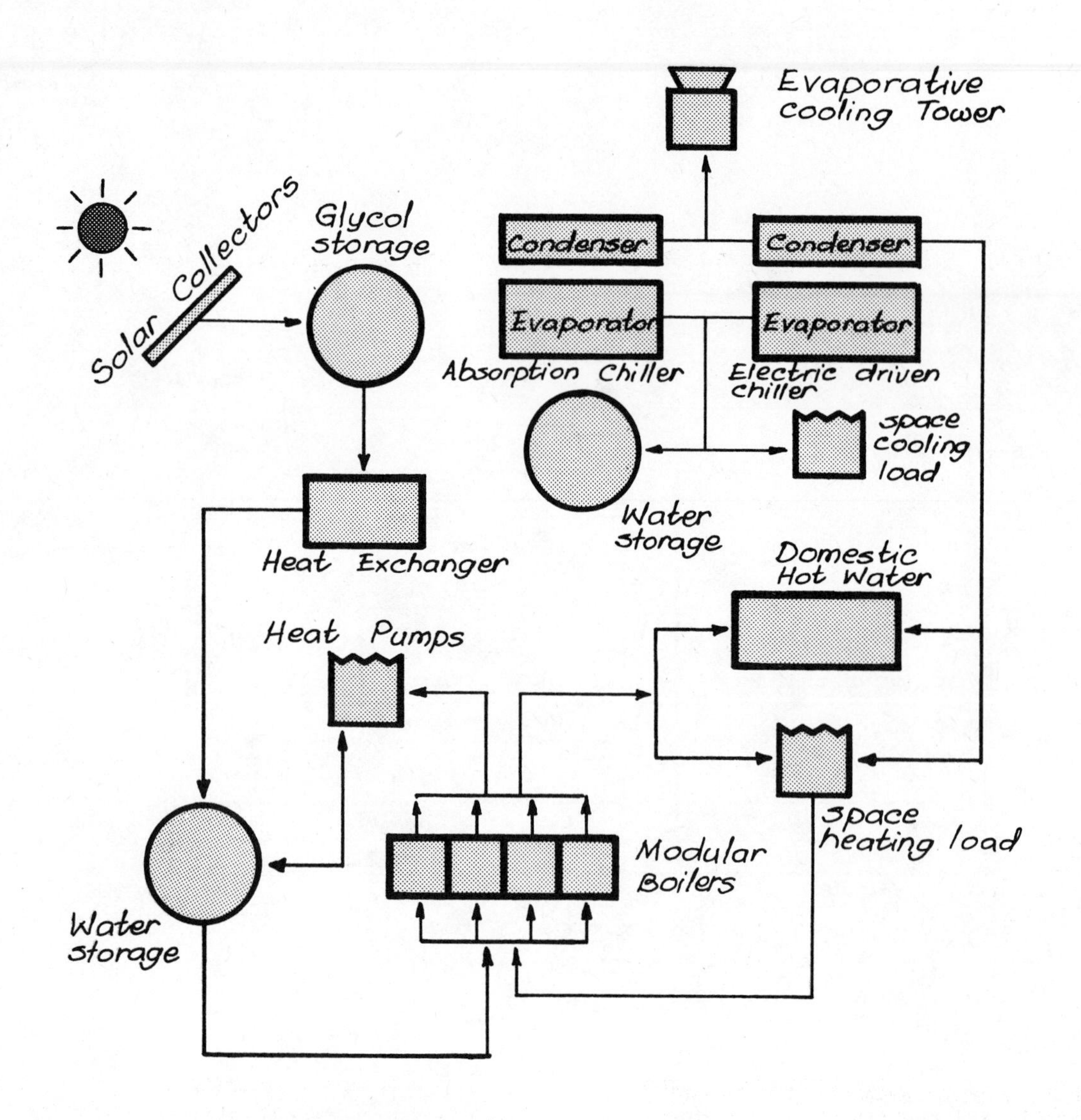

MECHANICAL PLANT SERVING TOP FOUR FLOORS

FIGURE 2

Floor	Core	Perimeter	Air Supply System
1	2 heat pumps closed water loop	6 heat pumps closed water loop	Variable Air Volume (VAV)
2	2 heat pumps closed water loop	34 heat pumps closed water loop	VAV
3	4 heat pumps closed water loop	6 heat pumps closed water loop	VAV
4	Central air handling system	fin-tube radiator	VAV
5	Engine-generator driven chiller and absorption	double duct system	Double duct
6	Chiller using engine-generator waste heat	3 fan coil units 4 pipe	VAV
7	None	3 fan coil units 4 pipe	VAV

HVAC SYSTEM TYPES TABLE 1

2. "Energy-Conserving Office Building--May Be Prototype for Thousands," *Center for Building Technology News*, (May 1977): 3.

CASE STUDY 8-7: AUTOMATIC CLEANING OF CONDENSER TUBES IN REFRIGERATION EQUIPMENT

THE CONCEPT

The condensers of water-cooled mechanical refrigeration equipment are subject to fouling which results in reduced thermal efficiency. To maintain equipment design capacity, the condenser tubes must be cleaned at regular intervals.

BACKGROUND

The period of time between cleaning the tubes is a function of local water quality and the type of water treatment. The operating parameter generally used as an indicator of the necessity to clean condenser tubes is the condensing pressure of the refrigerant at full load. As the tubes foul, the condensing pressure rises. The higher condensing pressure results in higher power demand by the refrigerant compressor. In terms of electrical demand, a condenser with clean, bright tubes will require ∿1 kW of power per ton of refrigeration. When the tubes have reached the maximum recommended level, the power demand would typically be 1.33 kW per ton. The average power requirement during the interval between cleaning of tubes would be 1.8 kW per ton.

FINDINGS

Patented devices have been commercially available to provide continuous cleaning of condenser tubes for a number of years. Although specifically designed for use in industrial process systems, they are directly applicable to centrifugal water chillers of the types used in air-conditioning systems.

The Lockheed Missile and Space Company at Sunnyvale, California began installing cleaning systems on chillers several years ago. Prior to installing the necessary equipment, they scheduled tube cleaning of chillers at twelve-month intervals. Inspection of condensers in which the cleaning systems have been installed for one year disclosed that the tubes were being maintained in a clean, bright condition by the automatic cleaning equipment.

Consider a typical centrifugal chiller of 600 tons capacity. The chiller operates 2600 hours at an average load of 300 tons. Reduced energy use would be 109.2 MWh per year.

The installed cost of the automatic cleaning system is currently (1975) approximately $8500. The reduction in annual operating costs would consist of electrical demand savings of $395 per year, electrical energy savings of $2,185 per year (at 0.02 $/kWh) and elimination of the annual cleaning procedure at a saving of $2500 per year for a total of $5080 per year. This yields a payback of the investment in approximately 20 months. If electricity costs increase to 0.03 $/kWh, the payback period would be more like 16 months.

RECOMMENDATIONS

Automatic cleaning devices for refrigeration equipment should be investigated for energy saving potential. The use of such equipment to augment heat transfer and improve energy use efficiency can lead to energy savings and operating economies.

Acknowledgments

This case study was provided by Facility Systems Engineering Company, Los Angeles, California through the courtesy of E.F. Slattery.

CASE STUDY 8-8: CONVERSION OF CONSTANT VOLUME AIR DISTRIBUTION SYSTEMS TO VARIABLE VOLUME

THE CONCEPT

Constant volume air systems of the double duct or multizone types are extremely wasteful of energy in that they mix mechanically refrigerated cold air with heated air for temperature control in space conditioning.

BACKGROUND

Double duct systems are energy intensive since they require heating and cooling of all air supplied and the long branch lines which are maintained in a hot or cold condition continuously dissipate energy. Conversion of double duct systems with zone mixing dampers to variable volume control is sometimes a useful approach to reduce energy usage. A study conducted on a series of buildings using double duct systems resulted in the following findings.

FINDINGS

The interior zones of certain air distribution systems can be converted to a variable air volume system by deactivating the hot duct damper and allowing only the cold deck damper to remain in operation. The modification would result in a reduction of electrical energy usage of 10 to 15 percent and a reduction of heating energy of as much as 35 percent depending upon the layout and usage of the building. The costs to convert the system could be paid back in energy savings in a period of one to three years, also depending upon the characteristics of the building and fuel and electricity prices.

RECOMMENDATIONS

The interior zones of the air distribution system should be converted to variable air volume by deactivating the hot duct damper and allowing only the cold deck damper to remain in operation. The exterior zones of the system would remain in service as double duct distribution.

Acknowledgments

This case study was provided by Facility Systems Engineering Company, Los Angeles, California, through the courtesy of E.F. Slattery.

CASE STUDY 8-9: RECONDITIONING OF AIR ECONOMIZER SYSTEMS

THE CONCEPT

A survey of the air-conditioning systems at a major electronics firm led to the discovery that 45 of the 90 systems had been originally designed to include the necessary equipment and controls to provide air economizer systems. None of these systems, however, were in operating condition at the time of the survey.

BACKGROUND

An air economizer system is one in which the internal heat load of the building is cooled using outside air when the ambient temperature is low enough. Control systems are available to allow use of 100 percent outside air at any time the enthalpy of the outside air is less than that of the return air. This effects a reduction of the mechanical refrigeration load and saves energy. Such control systems are termed enthalpic controllers.

FINDINGS

In the plant where this study was made, the majority of the air-conditioning systems had been installed during the period 1962 to 1966. A combination of a low level of preventive maintenance and a corrosive salty environment led to a failure of control devices and freezing of damper shafts.

Surveys of similar facilities on the West Coast of the US indicate that maintenance of control items is generally poor. Subsequent analyses indicated that not only do uncalibrated or inoperative control devices waste energy, but they are costly to the owner of the system as well.

Comparative analyses were conducted of typical systems with and without air economizer systems. The analyses were performed using a computer program having the ability to determine annual energy use. The data input included weather data on an hour-by-hour basis using ten year averages for the plant locality.

The results of the comparative analyses indicated that the use of an economizer system with enthalpic control results in a ∿10 percent reduction in annual energy use. In the plant studied, the cost of repairing or replacing the inoperative control components required to return the air economizer equipment to operating condition would be recovered by the reduced cost of energy in approximately 16 months based on 1975 costs.

RECOMMENDATIONS

Air economizer systems should be considered for facilities where new air-conditioning systems are planned or where modifications are intended. Existing systems with air economizers should be inspected to insure that the systems are operating properly.

Acknowledgments

This case study was provided by Facility Systems Engineering Company, Los Angeles, California, through the courtesy of E.F. Slattery.

15.9 LIGHT ENERGY (Chapter 9)

CASE STUDY 9-1: USE APPROPRIATE OUTSIDE LIGHTING LEVELS[1]

THE CONCEPT

A study was made on the outside lighting of a 34 ha (83 acre) shopping center to check the lighting during the hours after sundown and before sunrise.

BACKGROUND

The shopping center includes a total of 34 ha (83 acres) of which approximately 27 ha are used for outside customer parking and material and equipment storage. The outside area has a total of 2,250 lights ranging in type from flood lights to 10.6 m (35 ft) pole lights. The existing lights were the subject of a thorough study over a considerable period of time which considered such factors as the distribution of lights, which lights were wired on time clocks, which on photocells, and which were wired singly, in pairs, or in clusters.

FINDINGS

As a result of this study, only 1,401 of the total of 2,250 lights are normally used. Of these 1,401 lights, 1,230 operate from dusk until 10:00 P.M., one half-hour after the shops close. Thirty-eight lights operate until 11:00 P.M. and 133 operate all night. This plan provides adequate but not uniform lighting for nighttime operation and security. There have been no complaints from customers or merchants nor has there been an increase in accident frequency or vandalism.

Method for Calculating Lighting Requirements

The amount of light received from any arrangement of standard lighting fixtures and known wattage can be calculated as follows:

$$F = (N \times L \times D)/A$$

where F = footcandles of illumination,

N = number of lighting units of same type and wattage (refer to *IES Lighting Handbook--1972)*,

L = Output of lamp, lumens,

D = depreciation factor for installed lights--use 0.7,

A = lighted area, ft^2.

$$\text{Power Usage (kWh)} = N \times W \times \text{Usage (hrs)}/(1000\ \text{W/kW})$$

where W = watts.

$$\text{Power cost (\$)} = \text{Usage (kWh)} \times \text{unit cost from local rate schedule (\$/kWh)}.$$

Recommended illumination for parking lots is 1 to 2 fc. (Reference Figure 9-80, *IES Lighting Handbook--1972)*.

RECOMMENDATIONS

- Analyze outdoor lighting to

determine whether use of some lights can be eliminated, or rescheduled, without going below adequate, safe lighting levels.

- Consult the power company or a lighting consultant for additional information and assistance. Also, refer to Chapter 9 for more suggestions.

References

1. Gatts, Robert R.; Massey, Robert G.; and Robertson, John C., *Energy Conservation Program Guide for Industry and Commerce (EPIC)*, NBS Handbook 115, (Washington, D.C.: US Department of Commerce, National Bureau of Standards, September 1974) pp. 3-62.

CASE STUDY 9-2: COMPARATIVE ECONOMICS OF LIGHTING SYSTEMS[1]

THE CONCEPT

The major influence in the cost of a lighting system is the efficacy (lumens per watt) of the type of lamp which is used; i.e., cost decreases as efficacy increases.

BACKGROUND

The Illuminating Engineering Society has developed a procedure for determining lighting system costs. This procedure includes consideration of the principal factors affecting cost (see Table 1). This case study describes the results of comparative studies made for various types of lighting systems. In preparing these analyses, there were several important assumptions:

- fluorescent lamps are operated at six hours per start;
- burned-out lamps are "spot" replaced (at a labor cost of $3 per lamp) and fixtures are cleaned at that time;
- the lighted space is representative of a large industrial space (Room Cavity Ratio equals 1);
- the ceiling, wall, and floor reflectances are 50 percent, 30 percent, and 10 percent, respectively, "medium" dirt accumulation;
- luminaire and lamp costs are typical "net" prices;
- wiring costs are $150/manhour;
- owning cost is 10 percent per year carrying charges (equal to 15 percent of capital cost: initial cost less lamps);
- lighting system is operated on a two-shift basis (4,000 hours per year); and,
- energy cost is 3¢/kWh.

Results are shown in Tables 2 and 3.

FINDINGS

Direct comparison of the energy usage of various systems is given on Line "N" of Table 2. This number is on a "per luminaire basis." This can be converted to the energy use for the same level of illumination by dividing by the corresponding figures in Line "D". The result will be an arbitrary number which, if put on a relative basis, would show the relative energy required. For example, in Column 1, $57 would be divided by 109.8 fc, in column 2, $55.20 would be divided by 67.9 fc, and so on.

The preceding analysis compares other systems to high pressure sodium (HP-Na) (Column 1) because it has the lowest cost (Lines E, J, P, and R); however, any system can be used as the "base."

Although this summary concerns 400 watt high intensity discharge lamps, the same format may be applied to any combination of lamp and luminaire. Table 3--a "summary of a summary"--shows the relative data only for several other systems (the "basic data"--lines A-E and lines J, P, Q, and R).

Table 3 further indicates the effect of lamp efficacy on lighting system cost; the 1000 watt high pressure sodium system shows a substantially more favorable total annual cost for equal footcandles (lines P and R) than the 400 watt high pressure sodium "base." (A note of caution: the wider spacings which would be used for the larger lamp may not be suitable for low mounting heights.)

Whereas the above analyses pertain to *new* installations, similar economies may be achieved by lighting *renovation* in existing plants. For example, in old systems which consist

	Lighting Method #1	Lighting Method #2
Installation Data		
Type of installation (office, industrial, etc.) . .		
Luminaires per row		
Number of rows		
Total luminaires		
Lamps per luminaire		
Lamp type		
Lumens per lamp		
Watts per luminaire (including accessories) . . .		
Hours per start		
Burning hours per year		
Group relamping interval or rated life		
Light loss factor		
Coefficient of utilization		
Footcandles maintained		
Capital Expenses		
Net cost per luminaire		
Installation labor and wiring cost per luminaire		
Cost per luminaire (luminaire plus labor and wiring)		
Total cost of luminaires		
Assumed years of luminaire life		
Total cost per year of life		
Interest on investment (per year)		
Taxes (per year)		
Insurance (per year)		
Total capital expense per year . . .		
Operating and Maintenance Expense		
Energy expense		
Total watts		
Average cost per kWh		
Total energy cost per year*		
Lamp renewal expense		
Net cost per lamp		
Labor cost each individual relamp		
Labor cost each group relamp		
Per cent lamps that fail before group relamp . .		
Renewal cost per lamp socket per year		
Total number of lamps		
Total lamp renewal expense per year		
Cleaning expense		
Number of washings per year		
Man-hours for each (est.).		
Man-hours for washing		
Number of dustings per year		
Man-hours per dusting each		
Man-hours for dustings		
Total man-hours		
Expense per man-hour		
Total cleaning expense per year		
Repair expenses		
Repairs (based on experience, repairman's time, etc.)		
Estimated total repair expense per year		
Total operating and maintenance expense per year		
Recapitulation		
Total capital expense per year		
Total operating and maintenance expense per year		
Total lighting expense per year . .		

*Total energy cost per year =

$$\frac{\text{Total watts x burning hours per year x cost per kWh}}{1000}$$

Reference 2

LIGHTING COST COMPARISON TABLE 1

ITEM (Symbols in equations refer to "Item" descriptions)	1 (Base) HP Na (HPS Ballast)	2 Metal Halide (Metal Halide Ballast)	3 Metal Halide (CWA Mercury Ballast)	4 Metal Halide (Metal Halide Ballast)	5 Metal Halide (Metal Halide Ballast)	6 DELUXE MERCURY VAPOR	7 INCANDESCENT (1000 Watt)
	Lighting System Description (Industrial Luminaires: Shielded, Open, 5-15% Uplight)						
A Rated initial lamp-*lumens per luminaire*	50,000	34,000	34,000	34,000	34,000	22,500	23,740
B Rated lamp life (*hours*)	20,000	15,000	15,000	10,000	10,000	24,000	1,000
C Input *watts* per luminaire (including ballast losses)	475	460	455	460	460	445	1,000
D Average *foot candles* on work surface for equal number of luminaires	109.8	67.9	52.2	73.4	72.5	41.3	68.4
E *Relative number of luminaires* for equal footcandles	1.000	1.617	2.102	1.495	1.514	2.659	1.604
F Net *cost* of luminaire, installation and wiring system	$251.00	$184.00	$172.50	$184.00	$184.00	$167.00	$190.00
G Net initial lamp *cost* per luminaire	$ 30.20	$ 15.52	$ 15.52	$ 15.52	$ 17.43	$ 7.49	$ 2.35
H Total initial *cost* per luminaire (F+G)	$281.20	$199.52	$188.02	$199.52	$201.43	$174.49	$192.35
I Annual owning *cost* per luminaire (Fx15%)	$ 37.65	$ 27.60	$ 25.88	$ 27.60	$ 27.60	$ 25.05	$ 23.50
J *Relative* initial *cost* for equal footcandles (HXE↑H of Col. 1, BASE)	1.000	1.148	1.406	1.061	1.085	1.650	1.097
K *Number of lamps* spot replaced per year (4000 hrs ↑ B)	0.200	0.267	0.267	0.400	0.400	0.167	4.000
L Lamp replacement *cost* per year (KXG)	$ 6.05	$ 4.14	$ 4.14	$ 6.21	$ 6.97	$ 1.25	$ 9.40
M Labor *cost* for lamp for replacement and cleaning (KX$3 per lamp)	$ 0.60	$ 0.80	$ 0.80	$ 1.20	$ 1.20	$ 0.50	$ 12.00
N Annual energy *cost* (C x 4000x$.03per KWH ↑ 1000)	$ 57.00	$ 55.20	$ 54.60	$ 55.20	$ 55.20	$ 53.40	$120.00
O *Total* annual operating *cost* per luminaire (L+M+N)	$ 63.64	$ 60.14	$ 59.54	$ 62.61	$ 63.37	$ 55.15	$141.40
P *Relative* annual operating *cost* for equal footcandles (OxK↑O of Col. 1,BASE)	1.000	1.528	1.967	1.471	1.508	2.304	3.565
Q *Total* annual operating *cost* per luminaire (I+C)	$101.29	$ 87.74	$ 85.41	$ 90.21	$ 90.97	$ 80.20	$169.90
R *Relative* total annual cost for equal foot candles (QxE ↑ Q of Col. 1, BASE)	1.000	1.401	1.773	1.332	1.360	2.105	2.691
Number of luminaires required for 100 footcandles in area of 10,000 ft^2 (10,000x100 ↑D)	40	65	85	61	61	108	65

Notes: Incandescent column included only as a basis of reference. "Footcandles" refers to *maintained* value of illumination. Note that all "relative" values relate to Column No. 1.

Reference 1

COST OF LIGHTING ANALYSIS (SUMMARY) — HIGH INTENSITY DISCHARGE LAMPS (400 WATTS) **TABLE 2**

ITEM (Symbols in equations refer to "Item" designations)	8 (BASE)* High Pressure Sodium (400 watt)	9 High Pressure Sodium (1000 watt)	10 Metal Halide (1000 watt)	11 Fluorescent (1500 ma.) (2-lamp)	12 Fluorescent (800 ma.) (2-lamp)	13 Fluorescent (430 ma.) (2-lamp)	14 Fluorescent (40 watt) (4-lamp)
A Rated initial lamp-*lumens per luminaire*	50000	140000	100000	32000	18400	12600	13000
B Rated lamp life *(hours)*	20000	15000	10000	13500	15000	14000	15000
C Input *watts* per luminaire (including ballast losses)	475	1110	1080	455	250	175	194
D Average *foot-candles* on work surface for equal number of luminaires	109.8	326.1	216.0	57.9	38.8	28.3	29.4
E *Relative number of luminaires* for equal footcandles	1.000	0.337	0.508	1.898	2.827	3.884	3.733
J *Relative* initial *cost* for equal foot-candles (HXE + H of Col.1 BASE)	1.000	0.559	0.636	1.166	1.241	1.400	1.495
P *Relative* annual operating *cost* for equal footcandles (O xK + O of Col. 1, BASE)	1.000	0.806	1.115	1.750	1.451	1.445	1.609
Q *Total* annual owning and operating *cost* per luminaire (I + O)	$101.29	$217.03	$192.18	$83.43	$50.59	$38.37	$43/73
R *Relative* total annual cost for equal footcandles (Q x E + Q of Col. 1, BASE)	1.000	0.721	0.964	1.563	1.412	1.471	1.615
Number of luminaires required for 100 foot - candles in area of 10,000 sq.ft. (10,000 x 100 + D	40	14	21	77	114	157	151

*Same as Column 1 in Table 1.
Fluorescent lamps (Columns 11-14) are "Cool-White".

Reference 1

COST OF LIGHTING ANALYSIS (CONDENSATION) – OTHER LIGHTING SYSTEMS **TABLE 3**

of combinations of 400 watt mercury vapor and 750 watt incandescent lamps, the latter may be replaced by 400 watt high pressure sodium. The result is a much higher level of illumination with considerable savings in energy usage.

RECOMMENDATIONS

Today's requirement for effective energy management necessitates investigating the potentials for obtaining adequate illumination at the lowest feasible annual cost. Where lighting loads are an important cost item, the cost of alternative lighting systems should be evaluated.

References

1. This information was contributed by Mr. G.D. Rowe and Mr. R.T. Dorsey, General Electric Company, Cleveland, Ohio, 28 August 1974.

2. Kaufman, J.E., ed., *IES Lighting Handbook*, 5th edition, (New York: Illuminating Engineering Society, 1972).

CASE STUDY 9-3: EFFICIENT INDUSTRIAL LIGHTING WITH HIGH PRESSURE SODIUM LAMPS

THE CONCEPT

The design objective was to provide 540 lux (50 fc) of illumination in a manufacturing plant of 4,300 m^2 (46,000 ft^2) with the most energy efficient design.

BACKGROUND

The rising cost of electricity in Burbank, California has made manufacturers aware of the need for energy efficient designs. Thus, when Menasco Manufacturing Incorporated, a supplier of aircraft components, decided to construct new manufacturing facilities, consideration was given to techniques for using fuels and electricity as efficiently as possible. This included studies of alternative lighting systems (high pressure sodium vs fluorescent) and installation of infrared gas fired heaters with electric igniters rather than pilot lights.

FINDINGS

Figure 1 shows the two lighting configurations considered for this installation. The facility was designed for a 4000 A, 480/277 V service. The calculated load was 1989 A due to large machine tools, plating equipment, and lighting.

As the figure indicates, the two configurations studied involved: (1) 0.95 A,* high output fluorescent luminaires; (2) 1.44 A,* high pressure sodium lamp luminaires. These alternates are compared in the table. Alternative #1, using fluorescent lamps, had a lower initial cost. The equipment was less expensive and the installation was cheaper since the wiring is placed in raceways in the fixtures. Due to the long runs in some locations, #10 AWG wire was necessary to meet voltage drop limitations. In general, #12 wire was adequate and was used for most of the installation.

With alternative #2, the initial cost was twice as much due to more expensive fixtures and greater installation cost. (More conduit was needed and receptacles were provided to facilitate lamp maintenance.) This was partially offset by reduced labor costs to install fewer luminaires.

One question which arises with high pressure sodium lamps is the possible effect of light color on worker performance. Upon first entering the building, a difference is noted, but within a few minutes the sensation disappears and one is oblivious to any difference.

The return on the investment was studied in two ways. First, neglecting maintenance costs, the added capital cost was found to be paid out in 1.6 to 7.9 years, depending on the cost of electricity and the number of

*Includes ballast losses.

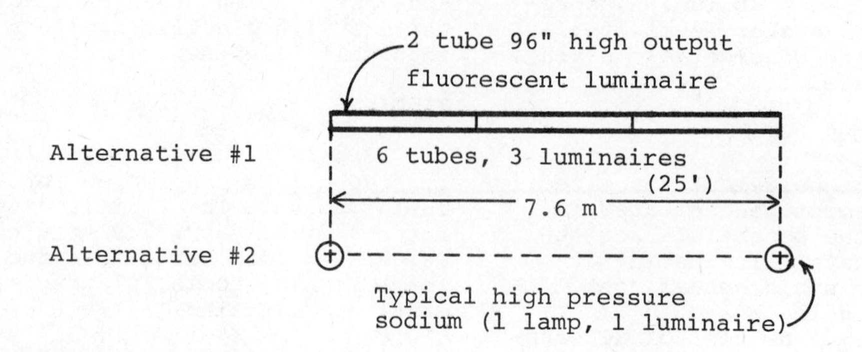

Comparison of Lighting Alternatives

Alternative #1 High Output Fluorescent	Alternative #2 High Pressure Sodium
269 Luminaires	90 Luminaires
35 Luminaires/row	12 Luminaires/row
110 We per tube	400 We per lamp
9050 Initial lumens	50,000 Initial lumens
18,000 hr average life (a)	20,000 hr average life (b)
71 kW load	40 kW load
480/277 Volt	480/277 Volt
148 A Current at full load	83 A Current at full load
#12, #10 AWG	#12 AWG
24 $ Luminaire cost (c)	153 $ Luminaire cost (c)
1.60 $ Lamp cost (c)	30 $ Lamp cost (c)
15,550 $ System cost (d)	30,240 $ System cost (d)

Notes: (a) 12 hours operation per start
(b) 10 hours operation per start
(c) Wholesale cost, 1975
(d) Exclusive of circuit breaker panel (existing)

ALTERNATE LIGHTING CONFIGURATIONS **FIGURE 1**

shifts worked. This is calculated from the following equation:

$$\text{Payout period (years)} = \frac{\$30{,}240-15{,}550 \text{ cost difference}}{(31 \text{ kW})(2000 \text{ shift hours/yr})(\# \text{ of shifts})(\$/\text{kWh})}$$

These results were obtained:

Payout Period, Years			
	Electricity Price $/kWh		
Number of Shifts	0.03	0.04	0.05
1	7.9	5.8	4.7
2	3.9	2.9	2.4
3	2.6	1.9	1.6

Based on expected electricity prices and projected plant usage, a payout period of two to three years was expected. With consideration of reduced maintenance costs (longer lamp life and one-sixth as many lamps to clean and replace, but more expensive lamps), the payout period would be reduced still further.

Another approach is to make a present worth cost comparison. Assuming two-shift operation and electricity at 0.04 $/kWh, the annual savings are:

$$(30 \text{ kW})(4000)(0.04) = 4960 \ \$/\text{yr}$$

Since the initial investment was $14,690, the capital recovery factor is 4960 ÷ 14,690 = 0.34. Interpolation in interest tables gives these rates of return: facility life = 5 years, return ∿20 percent; facility life = 20 years, return ∿34 percent. Thus, this is an attractive investment from a management point of view. This design resulted in a lighting load of about 10 We/m^2 (0.9 We/ft^2), which is an efficient design for an industrial facility.

The analysis described above has been simplified for clarity. It does not include consideration of the costs of replacement lamps, labor costs related to lamp cleaning and replacement, taxes, escalation, inflation, etc. For consideration of these factors, refer to Chapter 9, Case Study 9-2, and Appendix E.

In the initial design it was recognized that fluorescent lamps possessed certain potential advantages besides color, e.g., less reflected glare, reduced strobe effect, and instant restart. Experience has shown that none of these proved to be a problem except for the delayed restart of the high pressure sodium lamps following a loss of power. To overcome this limitation, trickle-charged, battery powered lamps were installed to provide emergency illumination in the event of a power failure.

RECOMMENDATIONS

This case study indicates that up to 50 percent of industrial electricity use for lighting could be eliminated by more efficient designs using currently available components and technology. For many types of lamps and luminaires, more efficient lamps are now available that can be readily installed in existing fixtures. Examples are more efficient fluorescent tubes that require less electricity; more efficient high pressure sodium lamps that work on existing mercury lamp-type ballasts, and so on.

For new facilities installation of more efficient equipment and lamps is usually a good investment. The alternative should be evaluated on a life cycle cost basis rather than initial cost.

Acknowledgments

This case study was provided through the courtesy of T.O Smith, A.R. Truger, and Ms. O. Swope, of Smith Electric Company, Inc., Glendale, California.

CASE STUDY 9-4: RELAMPING OPPORTUNITIES

THE CONCEPT

The objective of this study was to reduce energy use and still maintain lighting levels in existing installations without capital expenditures.

BACKGROUND

In existing facilities (commercial buildings, industrial facilities, parking lots, streets, garages), it is expensive to replace inefficient luminaires. To avoid the initial cost of new equipment, more efficient lamps can be substituted in some cases. This approach is inexpensive, has an immediate payback, and does not require highly skilled personnel.

FINDINGS

Most lamp manufacturers now have, or are developing, higher efficiency lamps which can be used in existing installations.

Examples:

- A high pressure sodium lamp replacement for mercury street lamps.

  Lamps are available which operate off the original mercury lamp ballast and which can be screwed into the same socket. (Typical trade name: GTE Sylvania "Unalux".) One type will work on either lag-type autotransformer ballasts or 240 V reactors. The sodium lamps have efficacies in the range of 100 lm/We and can produce twice the light output (when compared to mercury lamps) with less electricity use (360 We vs 400 We for mercury).*

- 150 We high-pressure sodium lamp.

  This is a smaller high-pressure sodium lamp which can replace 175 We and 250 We mercury vapor lamps. It produces approximately twice the light output with about 13 percent less electricity use. Although the same fixture can be used, ballasts must be changed. (Typical trade name: Westinghouse Ceramalux(TM).)

- More efficient lamps for residential and commercial applications.

  These include low-wattage (50, 75, 100 We) mercury vapor lamps and more efficient fluorescent lamps. The low-wattage mercury vapor lamps have approximately twice the light output and ten times the life of incandescent lamps of the same wattage. Although they fit ordinary medium screw bases, they must be operated with ballasts. Failure to use a proper ballast will cause burnout or failure to operate. (Typical trade name: Westinghouse Low Wattage Mercury Vapor Lamps.)

  Another type is the new lower wattage fluorescent lamp. These lamps reduce wattage by 10 to 20 percent with lamps that fit into existing fixtures. No ballast change is required. Light output as well as energy use is reduced. Moreover, the advantage of this approach, compared to lamp removal, is that uniform light distribution can be maintained. (Typical trade name: General Electric Company "Watt-Miser".) No special ballasts are needed.

RECOMMENDATIONS

Lighting efficiency in many existing installations can be improved by 10 to 50 percent simply by installing more efficient lamps in existing fixtures. This should be considered for application in residential, commercial, and industrial facilities.

CASE STUDY 9-5: MAXIMUM USE OF NATURAL LIGHT FOR AN INDUSTRIAL FACILITY

THE CONCEPT

Maximizing use of natural lighting resulted in reduction of energy use for

*Actually 106 lm/We (sodium) vs 58 lm/We (mercury).

lighting (the major single electricity load in this plant).

BACKGROUND

A new large equipment fabrication shop approximately 30 m wide by 80 m long was being planned for a manufacturing plant which uses 300-400 MWeh per month. This shop was to be used for the assembly of large steel pressure vessels and air filtration equipment. Since lighting was a major energy user in the plant, and since the new shop had a high ceiling--12 m (40 ft)--to permit movement of an overhead crane, a study was made of ways to reduce the energy requirement for lighting. It was found that by maximizing use of natural light, the electricity needed could be reduced by 38 percent.

FINDINGS

Three methods were used to minimize artificial lighting needs in this facility:

- Natural light was utilized by installing large fiberglas skylights in the roof. Two rows of these ran the length of the building.
- Natural light was also admitted by providing a 7.3 m (24 ft) high opening the entire length of the south wall of the shop. (Either this or doors were needed in any event to permit entrance of heavy equipment.)
- Electric lighting was divided into two circuits controlled by a photocell. When the natural light levels were adequate, the electric lamps were automatically switched off.

The shop was designed for illumination of 2150 lux (20 fc). On dark or overcast days or during the night shifts, this was provided by 50 overhead metal halide lamps requiring 426 We (lamp plus ballast). The photocells which control the system are installed outside the building on the east and west ends and are positioned so their field of view is down toward the ground. They have polarized filters which can be mechanically adjusted so that the trigger level corresponds to an illumination level within the building of 2150 lux.

In an actual operation it has been found that it is possible to operate the shop with no electric lighting about 38 percent of the time (6 hr out of average work day of 16 hr). Full load for the system is:

50 lamps x 426 We x 250 days/year x

16 hr/day = 852 x 10^2 kWhe/yr

852 x 10^2 kWhe/yr x 0.03 $/kWhe =

2,556 $/yr

This calculation is based on the present cost of electricity, which is 0.03 $/kWh including fuel adjustment charges, average demand charges, and local taxes. Thus the savings in electricity cost for this one building are about 1000 $/yr.

RECOMMENDATIONS

In many plants an increased use of natural light is possible, either by appropriate design for new facilities or by modification of existing facilities. The potential for use of natural light should be reviewed since significant cost savings may be possible.

Acknowledgments

This case study was provided through the courtesy of Mr. Clyde Booth, Farr Company, El Segundo, California.

15.10 PROCESS HEAT AND HEAT RECOVERY (Chapter 10)

CASE STUDY 10-1: IMPROVING FURNACE INSULATION

THE CONCEPT

New ceramic fiber products offer improvements in insulating high temperature electric industrial furnaces compared to insulated fire brick.

BACKGROUND

In recent years various types of ceramic fiber insulation have entered the insulator/refractory market. With the advantage of payback periods of one to three years, they are becoming increasingly popular in electric and fuel fired processes.

The major advantage of ceramic fibers is low heat conductivity, roughly two-thirds the conductivity of insulated fire brick (IFB). Typical examples of thermal conductivity of various forms of ceramic fibers are shown in Figure 1.

Another advantage is in ease of installation. A number of schemes have been developed to install the insulation which reduce down time by up to 50 percent. This can greatly increase the productivity of a furnace.

In addition, ceramic fibers have a much lower density and heat capacity than brick refractories. The heat capacity is roughly 75 percent less than that of IFB and 95 percent less than dense fire brick. This allows the ovens and furnaces to heat more quickly, reducing energy use and increasing productivity. Here ceramic fibers lend themselves to electric furnaces which themselves conserve energy by starting up and shutting down readily.

Some ceramic fiber materials are also machineable, allowing more elaborate forms in bricks. This can cut down on energy loss between stacked bricks by using a tongue in groove type fit. Ceramic fibers are also virtually immune to thermal shock, which has been known to collapse fire brick after a number of heating and cooling cycles.

Ceramic fiber insulation is not suited for furnaces subject to high velocities, particulates, corrosive atmosphere or ferrous metals. It can sometimes be employed in these situations if isolated between a primary wall of fire brick and the furnace wall. This provides greater overall thermal resistance while the fire brick takes the wear.

FINDINGS

In the furnaces where ceramic fiber can replace fire brick, it will provide greater insulation for the same amount of space. Alternatively if the same amount of heat resistance is required, using ceramic fiber will increase the usable volume of the furnace. Some types of ceramic fibers can also be applied over the fire brick to supply additional insulation with a quick retrofit.

At the Kingsport Foundry and Manufacturing Corporation in Tennessee, ceramic fiber insulation was installed in a large car-bottom annealing furnace. Due to maintenance problems with fire brick, ceramic fiber was installed, taking about a week. Installation of new fire brick would have taken at least three weeks. As a result of the increased insulation and shorter startups, the amount of energy required per operation of the furnace was reduced by 68 percent.

Another example is shown in Table 1, which lists comparisons between different linings in a ceramic kiln application. In addition to heat loss, heat storage, and surface temperature, the table presents energy usage for a firing (1315°C [2400°F] peak temperature) on a square foot of lining basis and for a 27 cu ft kiln with 200 lbs of ware. These are calculated comparisons and will vary with firing practice. The materials in each example are listed starting from the hot face.

RECOMMENDATIONS

Improved insulation should be considered for electric ovens and furnaces. In addition to reviewing

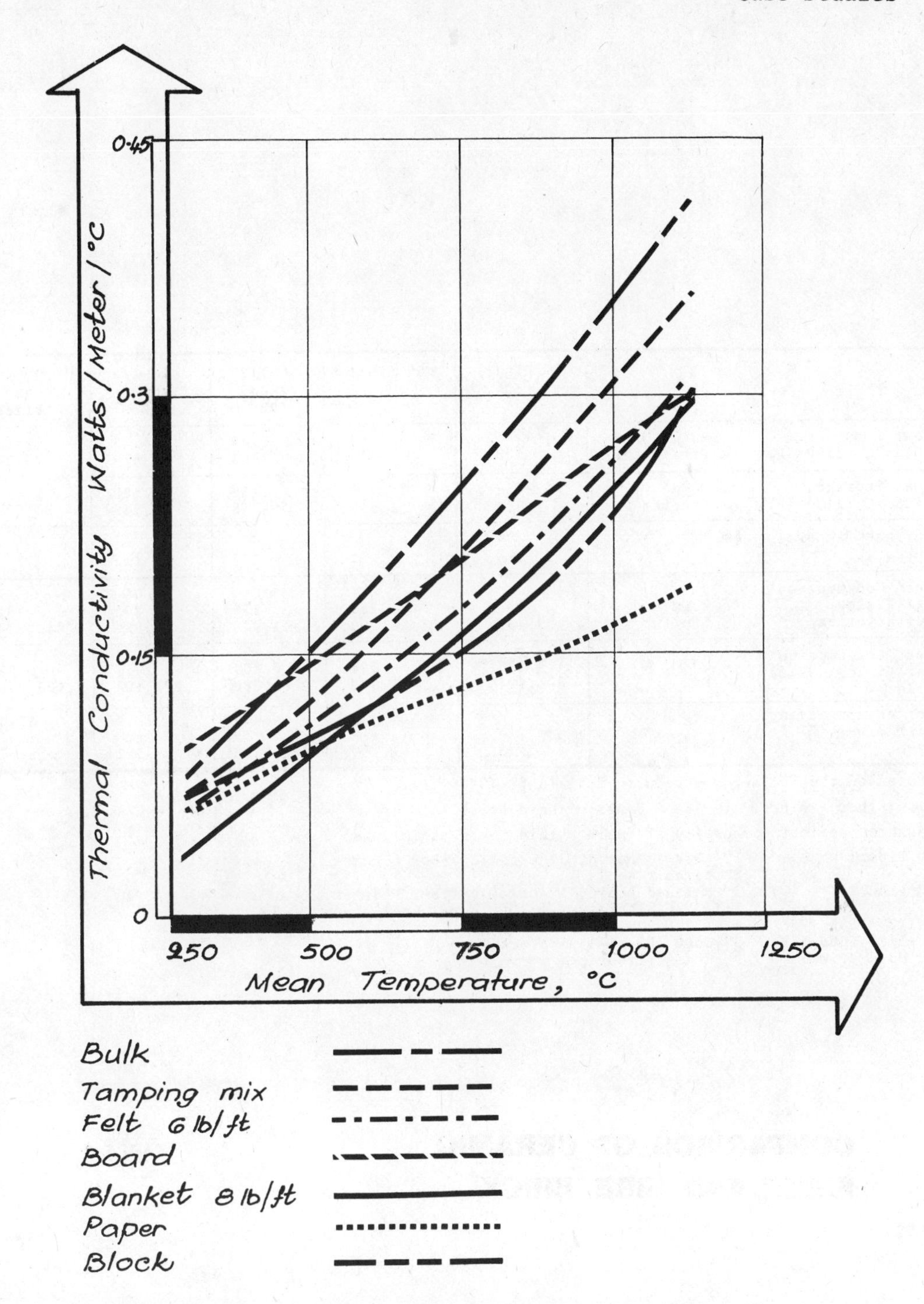

THERMAL CONDUCTIVITY OF CERAMIC FIBER FORMS

FIGURE 1

		5" Fiber	6" Fiber	6-1/2" 2600 IFB	10-1/2" 2600 IFB	9" Hard Brick
Heat Loss, 1315 °C (2400 °F)	MJ/m^2 hr Btu/ft^2 hr	4.89 430	3.57 314	6.40 563	4.25 374	21.2 1,864
Heat Storage, 1315 °C (2400 °F)	MJ/m^2 Btu/ft^2	18.8 1,650	22.3 1,958	110 9,718	189 16,580	375 33,000
Surface Temperature, °C		125	94	147	113	268
Heat Consumption* per Firing, Refractories	MJ/m^2 Btu/ft^2	39.3 3,461	32.9 2,900	136 11,970	205 18,076	481 42,320
Heat Consumption* per Firing, 27 ft^3 Kiln	GJ Btu	0.318 301,894	0.286 271,600	0.804 761,380	1.15 1,091,104	2.53 2,400,280
Percentage Savings with Fiber, 27 ft^3 Kiln		---	---	60.4 to 64.3%	72.4 to 75.1%	87.4 to 88.7%

*Based on the following schedules to 1315 °C (2400 °F):

Fiber - 6 hrs; IFB - 8 hrs; hard brick - 10 hrs.

Assumes perfect combustion of fuel. Linings are as follows:

6" Fiber: 1" ceramic board, 3" ceramic fiber blanket, 2" ceramic wool,
5" Fiber: 1" ceramic board, 1" ceramic fiber blanket, 1-1/2" 6 blanket, 1-1/2" mineral wool block
6-1/2" 2600 IFB: 4-1/2" 2600 IFB, 2" loose fill
10-1/2" 2600 IFB 9" 2600 IFB, 1-1/2" block
9" Hard Brick: 9" dense firebrick

COMPARISON OF CERAMIC FIBER AND FIRE BRICK

TABLE 1

insulation periodically from a maintenance viewpoint, it should be reexamined whenever electricity utility rates change. Higher electricity prices may make improved insulation cost effective.

CASE STUDY 10-2: HEAT RECUPERATION AND FURNACE EFFICIENCY IN INDUSTRIAL OPERATIONS[1]

THE CONCEPTS

The efficiency of industrial furnaces varies from approximately 30 percent to 60 percent. Typically, 50 percent or more of the energy used in such furnaces goes up the stack.

BACKGROUND

A very large part of the heat (electricity or fuel) used in high temperature furnaces is lost, either up the stack or by radiation, conduction, or convection from high temperature equipment. These losses can be reduced by use of more efficient equipment, better insulation, or by heat recovery techniques.

FINDINGS

Direct heating operations in industry, such as heat treating, smelting, and glass melting, account for approximately 11 percent of the total fuel consumption in the United States. It appears possible that as much as 30 percent of the fuel in certain direct heating operations can be saved through the use of recuperators, as discussed below.

There are an estimated 900,000 radiant tubes in heat treating furnaces in the United States, and few are equipped with heat recuperators. Industrial estimates indicate that each recuperator, at a 1973 cost of $1000-$1500 per unit, can save energy equivalent to approximately 0.5 bbl of oil per day. Thus, the total potential (equivalent) fuel savings of all radiant tubes in US operations today are approximately 450,000 bbl per day.

For many furnaces, it would be possible to use heat recovery equipment, such as heat recuperators, to recapture some of the heat normally lost in venting the oven or furnaces. One useful approach for radiant tube furnaces would be the use of the high temperature products of combustion for preheating air for the combustion process. For example, for a process at 2500 °F, use of the exhaust gases to preheat combustion air to 1000 °F could reduce total furnace fuel consumption by 30 to 40 percent. Air (or product) preheating is also valid for electric ovens or furnaces.

Assuming the upper recuperator cost of $1500 per unit, installation costs of $600 per unit, savings of 0.5 bbl of oil per day per unit, and approximately $10 per bbl of oil, one may evaluate the economic impact of recuperators as shown in Table 1. This example indicates an annual rate of return of almost 50 percent, with recovery of costs in 2.1 years. Projecting several years hence to the possibility of substantially higher fuel costs, the rate of return would be significantly increased. For example, at $12/bbl, the payback period is 1.6 years; at $15/bbl, the payback period is 0.7 years (8 months).

RECOMMENDATIONS

Process engineers should review existing process operations to evaluate the possibility of significant gains in fuel usage through use of such currently available technology as recuperators.

References

1. Much of the material in this case study is based on: Charles A. Berg, "Conservation in Industry," *Science* 184(19 April 1974): 264-270.

Capital and Installation Costs	
Capital Costs (Multiple Units)	$1,500
Installation Costs[a]	600
	$2,100
Annual Operating Costs[b]	
Maintenance, 10% of Capital & Installation Costs (C & I)	210
Taxes and Insurance, 4% of C & I	84
Interest, 9% of C & I	189
Depreciation in one year (2100/8)	263
Total Annual Costs	$ 746
Annual Economic Benefit of Fuel Use Reduction	
Annual Fuel Cost Reduction[c]	1,750
Annual Costs	746
Annual Benefit	$1,004
Annual Return on Investment	48%

(a) Calculated at 25 percent of new equipment cost for dismantling and reassembly and 15 percent of new equipment cost for installation.

(b) Eight year life, no salvage value

(c) One-half barrel of oil per day, 350 days operation per year, $10/barrel for No. 2 fuel oil.

COST CALCULATIONS FOR INSTALLATION OF RECUPERATORS

TABLE 1

CASE STUDY 10-3: PROCESS WASTE HEAT RECOVERY FOR BUILDING HEATING

THE CONCEPT

Process heat vented through stacks or other exhausts results in significant energy loss.

BACKGROUND

In many industrial operations, 30 percent or more of the heat generated in the furnace escapes through the stack. Typical industrial furnaces without recuperators have stack losses of: 56 percent, steel and alloy annealing; 41 percent, aluminum ingot heating, 34 percent, glass annealing, 58 percent, copper heating.[1] Various techniques are available for recovery of a significant fraction of rejected heat; if used for building heating, building electricity costs can be reduced 10 percent or more.

FINDINGS

A variety of proven methods for heat recovery from exhaust stacks has been developed for application to heating and ventilating systems.[2] These include heat wheels, runaround systems, and heat pipes.

A. Heat Wheels

Heat wheels consist of a motor driven wheel frame packed with such heat absorbing materials as aluminum or stainless steel mesh, or corrugated asbestos-type materials. The wheel is designed for installation in an immediately adjacent duct in a ventilation system, with the outdoor and exhaust air kept separate. As the wheel rotates, any given section passes first through one duct and then the other, transferring heat from the exhaust gases to the intake air. Cross contamination is normally low enough for most applications; returning a portion of the make-up air to the exhaust stream (after intake has passed through the wheel) provides additional purging, as required. Wheels may be designed to transfer only sensible heat (due to temperature gradients) or be total transfer devices in that they handle both sensible and latent heat.

B. Runaround Systems

A runaround system circumvents the necessity of close proximity of exhaust and inlet ducts. A system has two connected heat exchangers (one in each duct) driven by a motor-driven pump. The system circulates an antifreeze solution through the heat exchangers, cooling the exhaust while heating the inlet air. The system can be as efficient as the heat wheel if the heat exchangers are properly sized. Gains in exchanger efficiency through increased transfer surfaces must be balanced against increased fan size to compensate for pressure drops.

C. Heat Pipes

Heat pipes are passive devices which can transfer heat at efficiencies exceeding 90 percent. The pipe consists of a sealed hollow container (such as a finned copper tube) containing a capillary wick secured to the interior wall and a working fluid (e.g., refrigerant, liquid metal, deionized water). One end is placed in the hot exhaust stream while the other sees a cooler environment. Heat vaporizes the working fluid, which then condenses at the cooler end releasing the stored heat; the working fluid is then returned to the hot end by the cylindrical interior wick.

In one such application, heat pipes installed in a furnace flue transferred heat to a fresh air duct where the fresh air was blown over the pipes. The savings in building fuel costs resulting from this recovery operation are reported to reach 15 percent.

RECOMMENDATIONS

Facility designers and plant managers should evaluate their process heat systems in view of the savings possible by currently available recovery techniques. Significant savings are available through proven techniques such as those discussed above.

References

1. Berg, Charles A., "Conservation in Industry," *Science* 184(19 April 1974): 264-270.

2. Greiner, P.C., "Designing Sophisticated HVAC Systems for Optimum Energy Use," *ASHRAE Journal* (February 1973): 27-31.

CASE STUDY 10-4: HEAT RECOVERY FROM A RESTAURANT SAVES KILOWATT HOURS

THE CONCEPT

The use of a heat pipe heat recovery unit will reduce electricity used for space heating in a restaurant.

BACKGROUND

Certain types of facilities exhaust warm air nearly continuously during the winter months. (Hospitals and restaurants are examples.) Part of this heat can be recovered economically, reducing the electrical energy needed for space heating.

This case study concerns a restaurant in Toledo, Ohio. A packaged heat pipe recovery unit was installed on the kitchen air supply duct (Figure 1). The system was used to exhaust 142 m^3/min (5000 cfm). The air supply system was rated at 133 m^3/min (4700 cfm). A two stage (30 kW + 30 kW) electric duct heater is used to heat the air supplied to the kitchen.

FINDINGS

The heat pipe recovery unit cost $3000 to supply and install. This was a cost of 21 $ per m^3/min of capacity (in 1973). The unit was capable of recovering 55 to 60 percent of the exhaust heat.

Under winter (January) operating conditions, performance was found to be:

- outdoor temperature: -3.9°C (25°F)
- supply air temperature exiting heat recovery unit: 13.9°C (57°F)
- exhaust air temperature entering heat recovery unit: 26.7°C (80°F)
- exhaust air temperature exiting heat recovery unit: 11.7°C (53°F)

Therefore, the recovered heat added to the total cold supply air is:

$$(133\ m^3/min)(0.072 \times 10^6)(17.8°C) =$$

$$171.5\ MJ/hr$$

For a month this is equivalent to:

$$\frac{(171.5\ MJ/hr)(744\ hr/mo)}{3.6\ MJ/kWh} = 35,400\ kWh/mo.$$

The savings at 0.05 $/kWh would be:

$$(35,400\ kWh/mo)(0.05\ \$/kWh) = \$1770.$$

This was an extreme condition; under more typical winter conditions the temperature rise across the heat recovery coil was 12.8°C (23 °F), giving savings of

$$(133\ m^3/min)(0.072 \times 10^6)(12.8°C) =$$

$$122.6\ MJ/hr \cong 25,000\ kWh/month$$

RECOMMENDATIONS

Heat recovery units should be considered for buildings which exhaust warm air in cold climates. Payback periods of less than one year can be anticipated for many installations.

Acknowledgments

This case study was provided by Toledo Edison Company, Toledo, Ohio, through the courtesy of W.S. Shay.

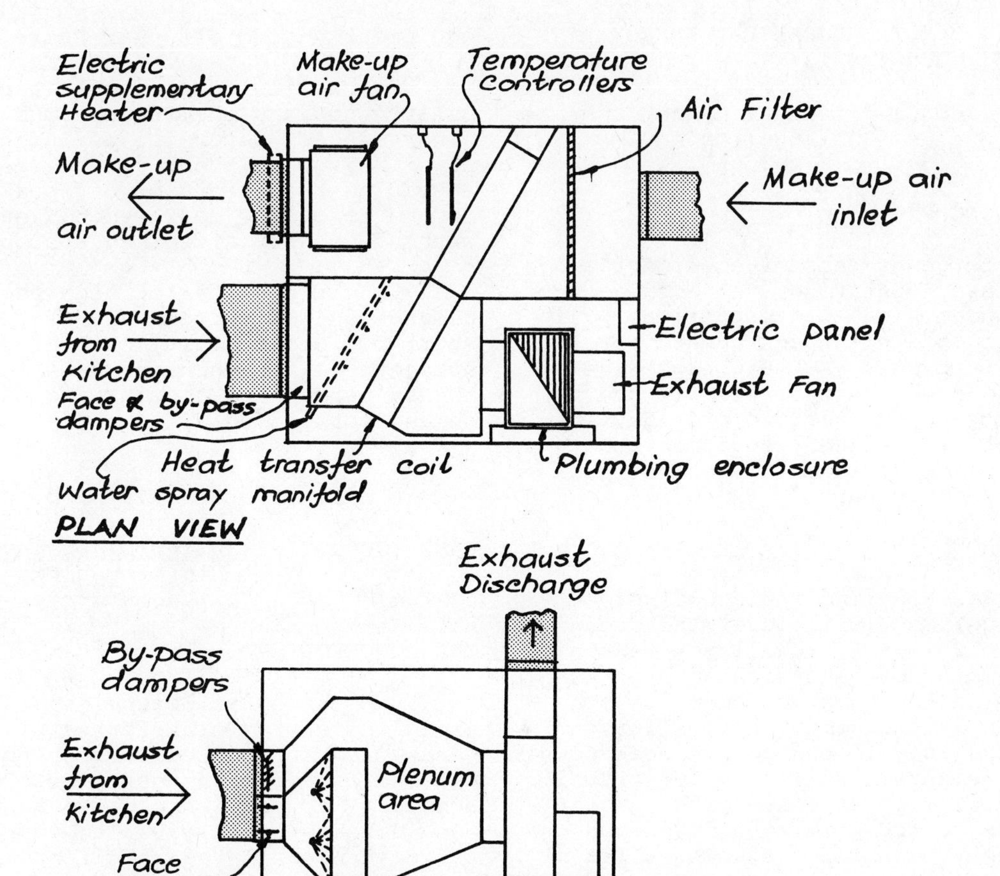

FRONT SECTION

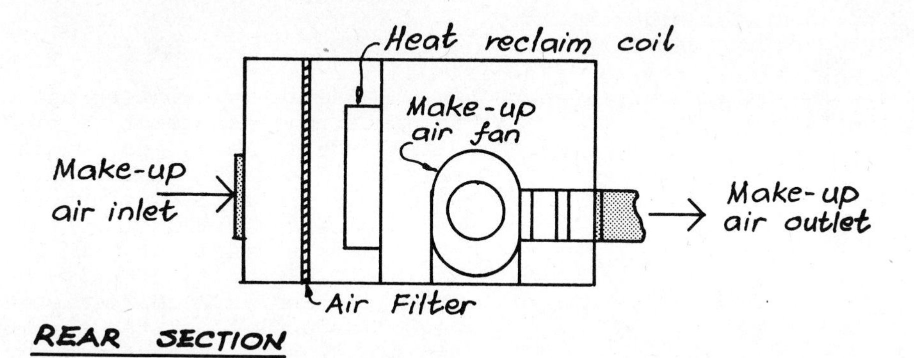

HEAT PIPE HEAT RECOVERY UNIT **FIGURE 1**

15.11 ELECTROMECHANICAL ENERGY (Chapter 11)

CASE STUDY 11-1: PEAK DEMAND REDUCTION FOR ELECTRIC MOTORS

THE CONCEPT

Large numbers of equipment and machinery operating at the same time and under high load conditions will result in a high peak demand for utility power. A "surcharge" (demand charge) must be paid to the utility by the user to provide, in effect, the standby equipment that must be maintained to meet the *peak* demand for power.

BACKGROUND

Rescheduling the use of electrical equipment will lower the demand peak. This action will not reduce the amount of electrical energy used, assuming the same equipment is continued in operation, but will even out the load to reduce the peak power requirements. While the individual industrial power consumer's contribution to the utility-wide demand peak may be small, reducing a particular user's peak demand can be financially rewarding.

FINDINGS

A small city utility uses a 600 kW (800 hp) pump for eight hours each day. By operating the pump only at night, an off-peak reduction of $1.40/kW-month in the demand charge results in the following annual savings:

Power demand = 600 kW

Charge cost = 1.40 $/kWh-month x 600 kW x 12 mo/yr

Savings ∿ $10,000 per year.

RECOMMENDATIONS

1. A plot of demand vs time is helpful in evaluating the possibilities for savings. If one is not available, the local power company will usually cooperate in preparing such a plot. If the plot shows some high cyclical peaks, usually some savings are possible by altering equipment use or possibly scheduling the use of equipment during off-peak hours.

2. Review equipment in the facility and rank them by maximum power usage. Determine if the higher power-using equipment can be rescheduled to periods of off-peak demand to reduce peak demand levels without greatly affecting the operation.

3. If a number of high power demand equipment items are used for short periods of time, maintain an accurate log on machine usage to adjust scheduling to minimize peak demand requirements.

CASE STUDY 11-2: MECHANICAL ENERGY TRANSFER EFFICIENCY IMPROVEMENTS

THE CONCEPT

Increasing efficiencies all along the energy transfer path, from generation through conversion to end use, will reduce the net energy expended to accomplish a given task. Much discussion is given to improving efficiencies related to direct electrical energy use. Improving mechanical energy transfer efficiency after conversion of electrical energy to mechanical energy (by means of an electric motor) is also an important consideration.

BACKGROUND

Valuable information can be obtained from the automotive field since efficiency of mechanical energy transmission from the engine to the drive wheels is a significant aspect of automobile performance. This is especially significant with electric vehicles. Examples of using new materials and advanced design techniques as applied to the automotive field can be applied to industrial machine design to improve the efficiency of the mechanical energy transfer systems.

FINDINGS

The following are three examples of new innovative designs in automotive transmissions that were developed as a result of the energy crisis. These examples were abstracted from Reference 1.

1. Traction Drives

For lightweight automobile models, traction drives may prove to be practical automatic transmissions. Presently, a traction-drive control system for an automobile is being tested at Tracor, Inc. The new drive system, unlike conventional automatics, is continuously variable and designed to hold engine speed at the optimum point for either economy or throttle response. The result is a power train that provides a combination of better fuel economy and improved acceleration.

2. Minicomputer Controlled Clutch

A bus transmission developed by Ford and Ferranti in Britain has most of the outward operating characteristics of an automatic, but is based on a fuel-saving standard clutch and gearbox. The conventional friction clutch is controlled by a minicomputer that monitors accelerator position, engine speed, road speed, and gear-selector setting. The driver still selects gears manually (through electro-hydraulic servos), but all declutching is handled automatically. The control system can hold engine speed so closely to gearbox speed during shifts that synchromesh cones have been eliminated from the gearbox.

3. Efficient Transmissions

The first five-speed transmission ever built in the United States is being manufactured by Borg-Warner Corp. The transmission was developed in response to the expanding market for efficient small cars. Top gear is an overdrive integrated into the transmission rather than being bolted on behind it as were previous United States overdrives. The shape of the housing--the gearset slides into one end rather than being dropped in from the top--permits a rugged, structurally efficient case.

RECOMMENDATIONS

Applications from all fields employing new materials and design concepts for mechanical energy transfer systems should be studied to obtain ideas for improving the efficiency of systems using the motive power supplied by electric motors. New products and equipment should be developed to incorporate mechanical energy saving concepts into machine designs so that total energy requirements are reduced.

References

1. "Mechanical Drives," *Machine Design* 1975 Reference Issue, 47(19 June 1975).

CASE STUDY 11-3: SEAL LOSSES IN ROTATING EQUIPMENT [1]

THE CONCEPT

Seal friction is a source of significant energy losses in many equipment designs.

BACKGROUND

Frictional losses in seals in large pieces of rotating equipment can result in power losses of 10 kW or more. Careful control of seal deflections or the selection of non-conforming face seals can substantially improve energy performance.

FINDINGS

To examine the potential benefits of alternative seal selection in the equipment design process, a 219 mm (8-5/8 in.) diameter face seal was studied. Operating at 1200 rpm (20 Hz) and approximately 90 atmospheres (1300 psig) pressure, the seal was found to require 15 kW (20 hp). By use of a load support mechanism in the seal face, and by carefully controlling seal deflection, the loss could be reduced to approximately 2.2 kW (3 hp).

However, by elimination of contact friction in the seal, even greater gains appear to be possible. For example, one large US manufacturer of seals asserts that through the use of a nonconforming face seal (which can handle pressures of 500 psig at speeds of 150 m/sec [500 ft/sec]), power dissipation can be reduced to less than 0.11 kW (0.15 hp).

RECOMMENDATIONS

Design engineers should examine the possibility of introducing design modifications which have a potentially high impact on equipment efficiencies to existing or planned product lines. As shown above, small details may produce significant gains in performance.

References

1. Herzog, Raymond E. and Dann, Richard T., "Designing the Energy Miser," *Machine Design* 46(21 February 1974): 97-106.

CASE STUDY 11-4: EFFECTIVE LUBRICATION PROGRAMS SAVE ELECTRIC ENERGY

THE CONCEPT

Every mechanical motion involves friction; friction causes heat, which constitutes net energy losses. Effective lubrication can develop measurable energy savings. Therefore, lubrication deserves special engineering consideration in energy efficient designs.

BACKGROUND

Energy measurements of electric power input and the recording of temperatures in bearings and gear drives of mechanical systems made in the course of comparisons between conventional and high performance lubricants have demonstrated appreciable energy savings, while using the high performance lubricants containing lubricating solids in suspension.

FINDINGS

(a) A wire drawing machine, operated in Pennsylvania, has five motors geared to five wire-drawing heads. The product load on the machine is stable, because the product output is constant for the same wire product. The plant electrical personnel made power input measurements for periods before and after a change of lubrication program. The engineering comparison indicates a power input reduction and energy saving of approximately 9 percent. The machine operates 4000 hours per year.

(b) A well-known electronic parts manufacturer in New England converted selected units of his production machinery to a different lubricant. The plant had the necessary in-plant instrumentation to monitor precisely the temperature results of the change of lubricants. Briefly: four bearing temperatures were measured. With the same outside ambient temperatures, all bearing temperatures dropped after the change of lubricant. For example, bearing "D" temperature dropped from (165°F to 130°F) after changing lubricant. The reduction of temperature clearly indicates energy saving. Electrical power comparisons for this mechanical system were not recorded. Note, however, that such a temperature rise in motor windings would cause a 6 percent increase in power usage due to electrical resistance increase alone.

RECOMMENDATIONS

Plant engineers should evaluate in-plant mechanical systems for lubrication performance, labor costs, mechanical failure incidence, and unscheduled production equipment down time for repair. When lubrication performance is questionable, and maintenance change advisable, the comparison of power input on a before-after change of lubricant may develop important energy savings, as well as fewer equipment breakdowns, and lower maintenance labor and repair parts costs. The before and after measurement of the power used by a mechanical system is always advisable. The kilowatt-hour meter is one method of measurement of power input which gives fully compatible comparisons.

Acknowledgments

This case study was provided through the courtesy of Mr. Al Mark, Imperial Oil & Grease Company, Los Angeles, California, 90024.

CASE STUDY 11-5: EFFICIENT MOTOR DESIGN AND RETROFIT

THE CONCEPT

While large electric motor designs currently provide efficiencies in the 90 - 95 percent range, smaller motors (less than 10 kW) have lower efficiencies. Recently several new designs have been developed to increase the efficiencies of these smaller motors.

BACKGROUND

A number of electric motor manufacturers are developing more efficient small motors. Several examples will be described to illustrate the possibilities and future trends.

Basically three options exist:

- retrofit an existing motor,
- replace an existing motor, or
- purchase a new motor.

For the first option, the Wanlass "controlled torque" motor concept can be used to retrofit certain motor types. For the second option, Gould, Westinghouse, Baldor Electric Company, Lincoln Electric Company, and other motor manufacturers have introduced more efficient motor designs. Some of these are also of potential interest for use in new installations, either in a plant or in a piece of equipment.

For new motor manufacture, rare earth permanent magnets make possible improved designs (smaller, lighter motors, less rotor inertia, greater torque rating, and better heat dissipation).

FINDINGS

The "controlled torque" motor design offered by the Cravens Wanlass Corporation depends on capacitors and modified windings.[1,2,3] The Wanlass approach is of interest because many motors potentially can be retrofitted. It is estimated that efficiency improvements of up to 30 percent and higher power factors are possible. Tests performed by the Southern California Edison Company have confirmed these claims.[4]

The Wanless Controlled torque motor has at least two windings per phase for either single-phase or polyphase motors. One of these windings, referred to as the main winding, is in series with a capacitor. The second winding is referred to as the control winding. This design achieves greater efficiency over the full operating range of the motor. Additional advantages claimed are:

- adjustable start up torque;
- lower starting current;
- less increase of current at reduced speeds;
- less overheating on over voltage;
- motor overload burnout avoided by controlled stall;
- better starting characteristics and speed control;
- reduced heating and increased life; and,
- little or no increase in manufacturing costs.

By adjusting the winding capacitance, a motor can be made to perform at maximum efficiency for any given load.

Not all motors can be easily retrofitted to the "controlled torque" design at this time. Those that can not include 3-phase motors and permanent split-phase capacitor motors. Hermetically sealed motors may present difficulties in conversion. Those that can be converted include single-phase motors, capacitor-start motors and split-phase motors. The Controlled torque concept can also be used in the manufacture of new motors.

A typical cost to convert a motor to the controlled torque design is quoted as $65.[3] It is difficult to establish exact payback periods but as a rough rule of thumb, payback in two years is claimed for the following cases:

1/8-1/6 hp motors run at least 16 hours/day

1/4-1/3 hp motors run at least 8 hours/day

1/2 and larger hp motors run at least 4 hours/day

While a number of different manufacturers are currently offering more efficient small motors in the range up to 100 kW, only one type will be described here due to space limitations. Refer to Reference 5 for an overview of several types.

These motors are all four-pole, three-phase units rated from 0.75 kW to 18.7 kW (1-25 hp). Compared to industry averages, they have improved efficiency and power factor as shown below.[6]

Capacity	Industry Average		New Motor	
	Efficiency	Power Factor	Efficiency	Power Factor
0.75 kW	76%	0.71	81.5%	0.84
18.7 kW	89%	0.83	91.0%	0.865

More efficient design was the result of reducing core losses and conductor resistance. Stator and rotor cores were lengthened, copper or aluminum was added to the stator and rotor conductors, and air gap and slot configurations were optimized. In addition to saving energy, these motors run cooler (longer life), are quieter, and have less power factor sensitivity to line voltage variation.

They are more expensive than less efficient motors. Under typical operating conditions, however, (3.7 kW [5 hp], 4000 hours/year of operation, 0.03 $/kWh), the price premium is paid back in less than 1.5 years.

An interesting development in dc motors is the increasing use of rare earth magnets. Due to recent process improvements, the cost of these magnets has declined so they are now feasible for dc motors.[7] Rare earth magnets combine the advantages of high coercivity (resistance to demagnetization) and good magnetic induction. Using better magnets, it is possible to manufacture smaller, more efficient motors with the same performance capabilities.

RECOMMENDATIONS

When replacing or purchasing new electric motors, consider new, more efficient types on a life cycle cost basis.

References

1. "New Electric Motor May Be Huge Energy Saver," *Southern California Industrial News*, Vol. XXIX, No. 35, 9 May 1977.

2. "New Electric Motor Idea Hailed as Energy Saver," *Los Angeles Times*, 26 April 1977.

3. Material provided by Cravens Wanlass Corporation, Tustin, California.

4. Personal communication with Southern California Edison Company, 1977.

5. Minerbrook, Scott and Edwards, Paul L., "Motor Market Still Slow for Efficient Units," *Energy User News*, Vol. 2, No. 10, 14 March 1977.

6. "Energy-Saving Motor Ready for Work," *Machine Design* 48(8 July 1976): 8.

7. Rashidi, Abdul S., "Better Motors with Rare-Earth Magnets," *Machine Design* 48(8 July 1976): 70-73.

15.12 ELECTROLYTIC AND ELECTRONIC PROCESSES (Chapter 12)

CASE STUDY 12-1: ELECTRICITY STORAGE AS AN ENERGY MANAGEMENT TOOL[1]

THE CONCEPT

Energy storage (using off-peak power) is a possible technique for reducing dependence on inefficient peaking units which require scarce or imported fuels. It may also have economic benefits for large electricity users.

BACKGROUND

A number of methods of storing energy have been studied. They include pumped hydro, compressed air, flywheels, steam, hot oil, batteries, chemical cells, and superconducting magnets. The only method which is economic for large-scale use at present is pumped hydro. However, recent research results indicate that off-peak storage of electricity in batteries may soon be commercially feasible.

FINDINGS

Table 1 summarizes the status of advanced battery research and development. Within the next two to three years (by 1980), it is anticipated that an advanced zinc-chlorine battery rated at 1 MWe and capable of storing up to 5 MWh may be available for testing. The next step would be the demonstration of a full-scale (e.g., 20 MWe and 100 MWh) battery plant on a utility system before 1985.

RECOMMENDATIONS

Advanced battery concepts may make on-site storage of electricity economically feasible within the next decade. Commercial and industrial facilities with high peak demands might consider the economic benefits of electrolytic energy storage.

References

1. Birk, J.R. and Pepper, J.W., "Energy Storage on Electric Utility Systems," *Energy Use Management--Proceedings of the International Conference*, edited by Rocco A. Fazzolare and Craig B. Smith, (New York: Pergamon Press, Inc., 1977), Vol. II, pp. 265-275.

CASE STUDY 12-2: EFFICIENCY IMPROVEMENTS POSSIBLE WITH FUEL CELLS[1]

THE CONCEPT

Fuel cell power plants are an emerging option for efficient energy use. Key features include high efficiency independent of size and constant generation efficiency over a wide load range (20 - 80 percent).

BACKGROUND

The development of fuel cell generators has been underway since the early 1960s. Early units were in the 10 - 40 kW capacity range. In the early 1970s the demonstration of a 1 MWe pilot plant was initiated and has since been achieved.

Current efforts are directed at development of a 27 MWe unit, called the FCG-1 generator. Now under construction, it will consist of six 4.5 MWe modules. The estimated installed cost is 250 $/kW (1975 $). It will have a life of 20 years and a heat rate of 2.7 J/Je (9300 Btu/kWh). Commercial introduction is planned for 1980. Fuel is expected to be naptha or clean coal fuels.

FINDINGS

Fuel cell generating plants can save energy in several ways. In addition to the inherent greater efficiency possible (compared to conventional units), fuel cell generators can be dispersed to provide on-site generation

Battery Type	Round trip Efficiency (%)	Capital Costs $/kW	$/kWh	Development Status
Lead-acid	60-75	60-100	70-80	State-of-the-art
Advanced aqueous	60-75	60-100	30-40	Small prototypes
High-temperature	70-80	60-100	30-50	Laboratory cells
Redox	60-70	100-120	15-35	Conceptual and Laboratory Studies

CHARACTERISTICS OF ELECTROLYTIC ENERGY STORAGE SYSTEMS **TABLE 1**

and therefore reduce transmission and distribution losses. (Consideration would have to be given to fuel transportation, which would offset this advantage.) By reducing generating capacity requirements--particularly spinning reserve--system efficiency could be improved through the elimination of the use of older, less efficient equipment.

The FCG-1 unit has been designed to permit recovery of waste heat. Both hot water and steam can be produced during normal operation. The heat exchanger equipment is required in any event and does not impact on plant cost or performance. When waste heat recovery is included, overall plant efficiency is estimated to be 73 percent.

RECOMMENDATIONS

Fuel cell generation plants should be considered for on-site generation where appropriate fuels are available and there is a need for electricity and process heat.

References

1. Gillis, E.A. and Rogers, L.J., "Fuel Cell Systems for Dispersed Generation of Electric Power," *Energy Use Management--Proceedings of the International Conference*, edited by Rocco A. Fazzolare and Craig B. Smith, (New York: Pergamon Press, Inc., 1977), Vol. II, pp. 283-288.

15.13 ENERGY MANAGEMENT IN CITIES (Chapter 13)

CASE STUDY 13-1: THE "CLIFF PALACE," PRE-COLUMBIAN CLIFF-DWELLING AS AN EXAMPLE OF DESIGNING FOR EFFICIENT ENERGY USE, MESA VERDE NATIONAL PARK, COLORADO, USA

THE CONCEPT

Primitive man, lacking other energy sources, adapted to the environment to achieve satisfactory housing, heating, cooling, and protection.[1,2]

BACKGROUND

The remains of permanent, Pre-Columbian habitation within the protective overhang of large caves geologically cut into massive sandstone cliffs (and therefore known as "cliff-dwellings") dot the southwestern area of the United States. Recent research has shown that at least one such community, the "Longhouse" or "Cliff Palace"(see Figure 1) located in a cave 150 m (500 ft) across, 40 m (130 ft) deep, and 60 m (200 ft) high at its opening on the face of the cliff, enjoyed a beneficent south to southwest orientation.[1] During a 400-year period of continuous occupancy (beginning about 1000 A.D.), cave inhabitants were able to farm the flat table land on the mesa above while being safe from increasingly frequent raids by roving marauders in their almost inaccessible cave, midway between the top of the mesa and the valley floor far below.

The orientation and cross-section of the cave protected its internal structures from the direct summer sun, which was high in the sky during the hottest part of the day and cut-off from the interior of the cave by its looming brow overhead (see Figures 2 and 3). During early mornings and late afternoons of those summer days, the sun would warm the buildings, while in the winter the sun was low enough in the sky to warm the dwellings throughout most of the day.

Because of construction with sandstone cut from the rubble within the cave, the heavy masonry walls of the building within possessed a large thermal capacity (not generally found in contemporary residences of lighter construction). Thus, the Longhouse dwellings responded relatively slowly to swings in temperatures, further modulating the effects of temperature extremes already modified by the cave's orientation and configuration. The heavy walls acted as solar collectors, gradually absorbing the heat and energy of the hot southwest sun during the day, instead of allowing it to pass through the walls to heat up the building interiors. By the cool of the evening, the walls

Cliff Palace. Mesa Verde National Park, Colorado.

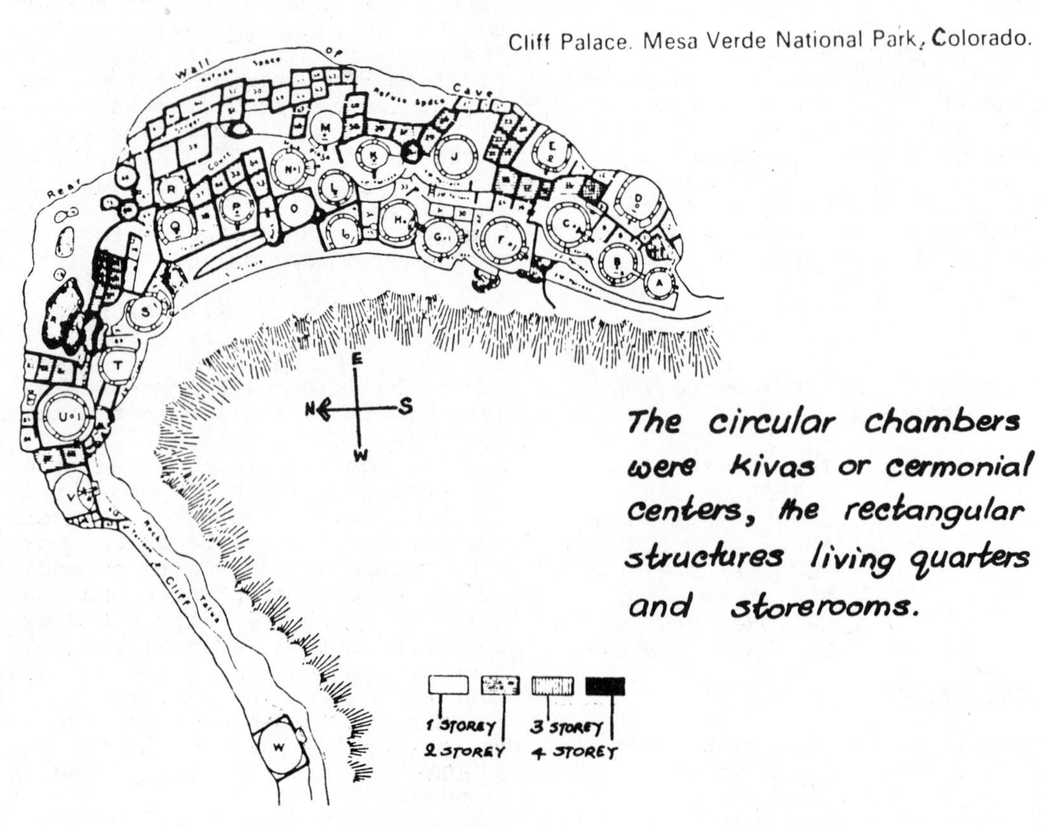

CLIFF PALACE

FIGURE 1

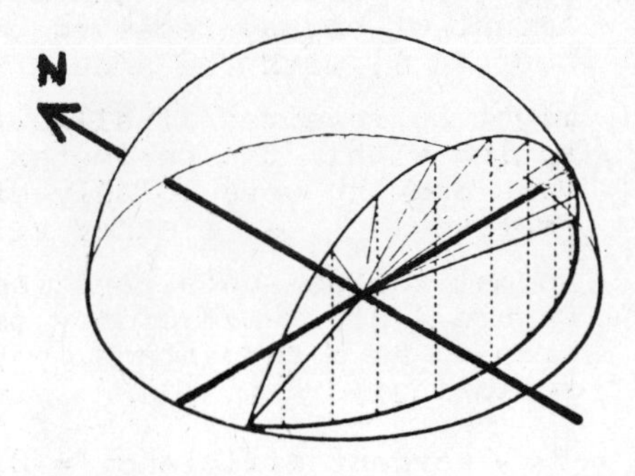

Winter, 30°+ Noon Altitude

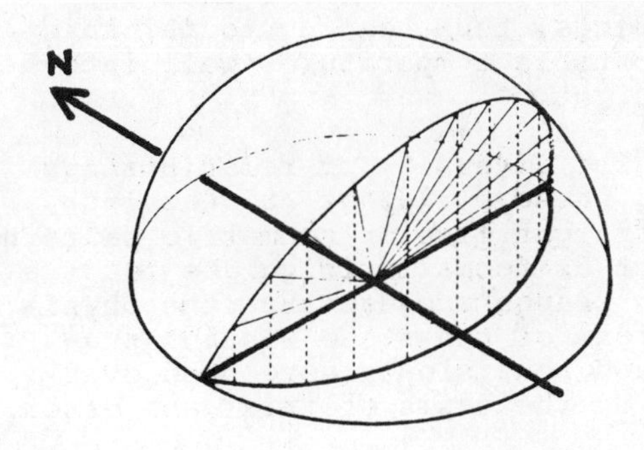

Equinox

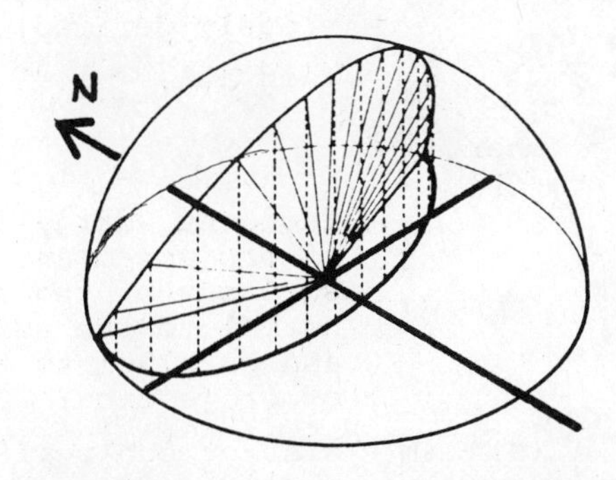

Summer, 78° Noon Altitude

SEASONAL DIFFERENCE ON INCIDENT ENERGY

FIGURE 2

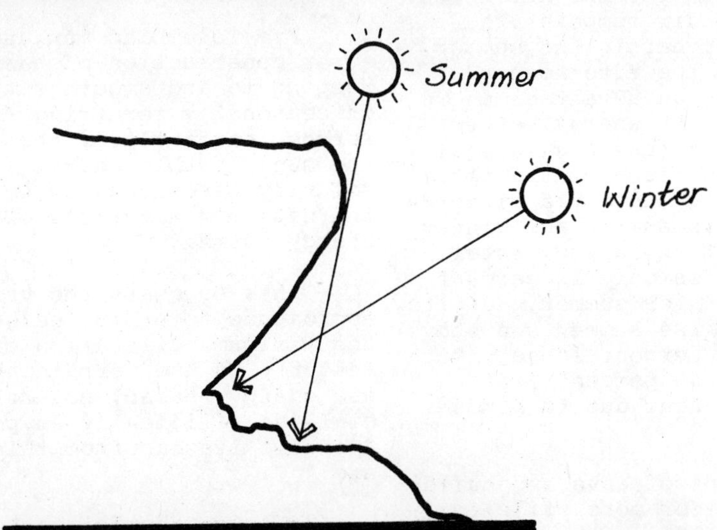

EFFECT OF SUN ELEVATION ON WINTER-SUMMER HEATING

FIGURE 3

began to reradiate the energy they had collected and stored into the buildings, thus helping to maintain comfortable temperatures well into the night.

The diversity of relationships among these buildings and the lack of straight line or geometric patterns to act as formalistic guides for systemic change suggest that the physical form of Longhouse was not preplanned, but slowly developed over time on the basis of trial and error.

FINDINGS

Longhouse was a self-organized development which responded favorably to seasonal and diurnal variations in the sun's impact. However, the very physical dimensions of the cave which further enhanced its climatic response also created finite and invariant limits upon its expansibility.

The building and cave arrangement was "56 percent more efficient as an energy collector in the winter than it was in summer."[3]

The following quotation explains the efficiency of this approach:

> A graph of comparative energy-profiles for summer and winter indicates two significant facts. First, incident energy is higher on a summer morning than on a winter morning while just the reverse is true for the afternoon. Since ambient air temperature tends to be higher in the summer, especially in the afternoon, there would be an advantage in reducing incident energy between noon and sunset (see Figure 4[a]). Second, a comparison of the total amount of energy received directly from the sun in summer and winter indicates that the energy total in the winter is only 12 percent less than that for summer, despite the fact that the summer sun stays in the sky 30 percent longer, with an average of 50 percent less reduction per hour due to atmosphere.
>
> The arrangement of cave and buildings provides for more efficient energy collection in the winter when such efficiency is of great advantage (see Figure 4[b]). The degree of efficiency can be plotted by comparing the actual amount of energy received per hour (E_i), with the amount that might be received if all surfaces acting within the perimeter of the cave opening were totally effective (E_m), as if they all lay normal to the sun's rays (see Figure 4[c]). Efficiency is expressed as a percentage by the following:
>
> $$\text{Percent efficiency} = E_i/E_m$$
>
> where
>
> E_i = incident energy in units of projected area (previously derived), and
>
> $E_m = (\sin a)A$;
>
> where
>
> E_m = incident energy in units of projected area if all surfaces of the form were normal to the sun's rays and parallel to the plane of projection;
>
> $(\sin a)$ = sine of sun's altitude angle to correct for atmosphere; and,
>
> A = maximum projected area of all exposed surfaces of the form.

RECOMMENDATIONS

Pre-Columbian man used trial-and-error construction of habitation to respond to and smooth out the impact of seasonally recurring climatological forces, particularly those created by the sun's path. Heating, cooling, and security were achieved by utilizing the site and naturally occurring energy forms.

This suggests the validity of approaches to urban design which stress the maximum utilization of natural features of the terrain and site to use energy--human, natural, or artificial--as efficiently as possible. Specific lessons from this example are:

- large thermal capacity of building gives advantageous averaging of diurnal temperature

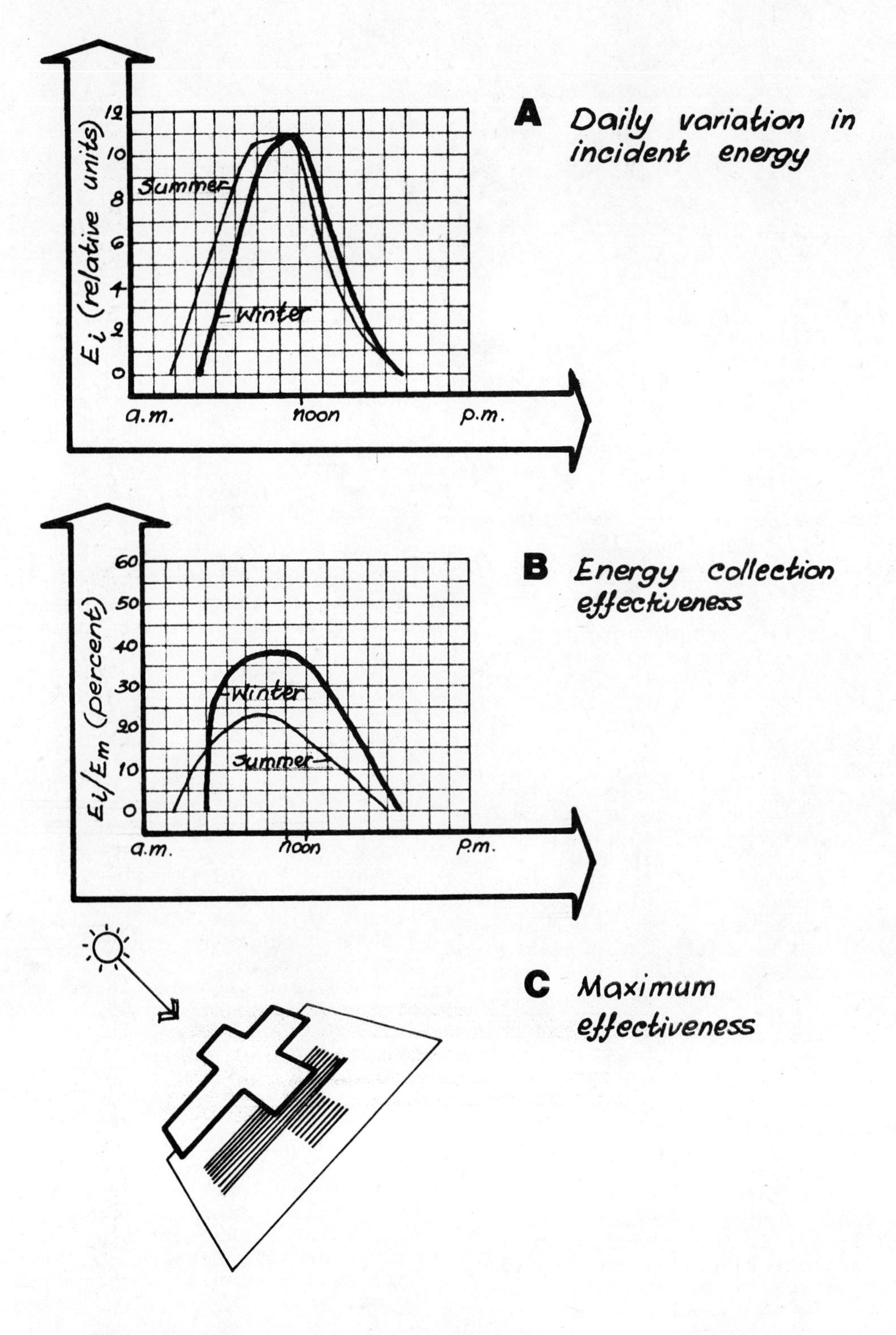

MEASURES OF EFFICIENCY FIGURE 4

swings; and,

- the change in angle at summer and winter can be used to some advantage for partial solar conditioning.

References

1. Knowles, Ralph L., *Energy and Form, An Ecological Approach to Urban Growth*, (Cambridge, Massachusetts: MIT Press, 1974).
2. Fraser, Douglas, *Village Planning in the Primitive World*, (New York: George Braziller, Inc., 1968).
3. Knowles.

CASE STUDY 13-2: ENERGETIC, ECONOMIC, ENVIRONMENTAL AND OTHER COSTS OF URBAN SPRAWL

THE CONCEPT

A wide variety of housing proposals containing a mix of various housing types--clustered, single family, townhouses, walk-up apartments, and even high-rise apartments--are being presented to local officials for approval of construction in various communities. Increased concerns relating to this new form of community development have, until recently, been limited mostly to potential economic impacts. Recently, these economic uncertainties have been joined by environmental and other concerns such as air, water, noise, pollution, erosion and the impact upon vegetation, wildlife, as well as resulting visual effects. The availability of water and energy supply and demand have become especially worrisome. Finally, there are questions as to how the development will affect the lives of people who live in it as well as those who live near it.

BACKGROUND

A pioneering study in this field attempts to summarize what is known about different costs as they apply to different neighborhood types and to different community development patterns, as well as indicating whether those costs are incurred publicly or privately.[1] Not all costs associated with residential development have been included, but the major ones have. The study is an analysis of prototype development and does *not* examine actual developments. However, empirical data derived by others is utilized in this report. The approach was to assume typical site conditions and an absence of any existing infrastructure such as power supply, roads, sewers, and so on at the site. Standard unit cost figures were used to estimate the cost of building alternative types of development.

The various costs were first estimated for different *neighborhood* types, each neighborhood being composed of 1,000 dwelling units of only one of the housing types listed across the top line of Table 1. Since many environmental and economic costs cannot be clearly identified at such a small scale, neighborhoods were aggregated into different communities, each of which contained 10,000 dwelling units or a total population of 33,000 people for each community. Six community types were analyzed, each containing a mixture of the various neighborhood housing types but differing in average density and in the amount of what the study calls "community planning." The term here is used to mean increased clustering and a general compactness of development. While the different neighborhood types require different amounts of land for 1,000 units, all six communities were limited to a mix of neighborhood types on a total of 6,000 acres. Population and other variables within the neighborhoods differed slightly as well. Table 1 indicates the specific land use, housing, and population characteristics of both different neighborhood types and different community development patterns.

The executive summary for the study (from which this case study is drawn) examined three of the six community types analyzed in the full effort. Referring to Table 1 they are: Community II--a "combination

NEIGHBORHOOD AND COMMUNITY COST ANALYSIS

	Neighborhood Housing Types					
	A Single-Family Conventional	B Single-Family Clustered	C Townhouses Clustered	D Walk-Up Apartments	E High-Rise Apartments	F Housing Mix (20% Each A-E)
Dwelling Units	1,000	1,000	1,000	1,000	1,000	1,000
Average Floor Area Per Unit (square foot)	1,600	1,600	1,200	1,000	900	1,260
Total Population	3,520	3,520	3,330	3,330	2,825	3,300
Persons per Unit	3.5	3.5	3.3	3.3	2.8	3.3
School Children	1,300	1,300	1,100	1,100	300	1,100
Total Acreage	500	400	300	200	100	300
Residential	330	200	100	66	33	145
Open Space/Recreation	45	90	90	73	32	66
Schools	29	29	26	26	15	26
Churches	5	5	5	5	5	5
Streets and Roads	75	60	45	30	15	45
Vacant	16	16	34	0	0	13
Residential Density						
Units per Gross Acre	2	2.5	3.3	5	10	3.3
Units per Net Residential Acre	3	5.0	10.0	15	30	6.9

	Community Development Patterns					
	I Planned Mix	II Combination Mix (50% PUD, 50% Sprawl)	III Sprawl Mix	IV Low Density Planned	V Low Density Sprawl	VI High Density Planned
Dwelling Units	10,000	10,000	10,000	10,000	10,000	10,000
Housing Types[(1)]	20% - Type A 20% - Type B 20% - Type C 20% - Type E	Same as I.	Same as I.	75% - Type B 25% - Type A	75% - Type A 25% - Type B	10% - Type B 20% - Type C 30% - Type D 40% - Type E
Total Population	33,000	33,000	33,000	33,000	33,000	30,000
School Children	11,000	11,000	11,000	11,000	11,000	11,000
Total Acreage	6,000	6,000	6,000	6,000	6,000	6,000
Residential	1,450	1,450	1,450	2,333	3,000	733
Open Space/Recreation	660	530	400	660	400	660
Schools	260	260	260	260	260	260
Other Public Facilities	140	140	140	140	140	140
Streets and Roads	530	530	530	720	790	380
Vacant, Improved[(2)]	152	213	278	206	459	109
Vacant, Semi-Improved[(3)]	456	922	1,390	617	951	326
Vacant, Unimproved	2,352	1,955	1,522	1,064	0	3,392

Notes: (1) Type A - single-family, conventional; Type B - single-family, clustered; Type C - townhouses, clustered; Type D - walk-up apartments; Type E - high-rise apartments.

(2) Includes all roads and utilities.

(3) Includes only arterial roads and trunk utility lines.

NEIGHBORHOOD AND COMMUNITY CHARACTERISTICS

TABLE 1

mix"; Community V--"low-density sprawl"; and, Community VI--"high-density planned." In order to establish mix and physical layout, the three communities were characterized as follows:

- *Low-density sprawl*--The entire community is made up of single family homes, 75 percent sited in a traditional grid pattern, with the rest clustered. Neighborhoods were assumed to have been sited in a "leap frog" pattern with little contiguity. This represents the typical pattern of much current suburban residential development.
- *Combination mix*--This community consists of a housing mix of 20 percent of each of the five types of dwellings, half located in planned-unit development and the other half in traditional subdivisions.
- *High-density plan*--In this community, housing is composed of 40 percent high-rise apartments, 30 percent walk-up apartments, 20 percent townhouses, and 10 percent clustered single family homes. All the dwelling units are clustered together into continuous neighborhoods much in the pattern of a high-density "new community."

FINDINGS

Land Use

Although all the communities cover the same area, over 50 percent of the land in the high-density planned community (Community VI in Table 1) remains completely undeveloped whereas all of the land is at least partially developed in the low-density sprawl community (Community V in Table 1). Figure 1 shows that although four times as much land is used for residential purposes in the low-density sprawl community compared to the high-density planned community, only two-thirds as much is actually dedicated to open space. However, one can consider back yards as a form of open space, although private rather than public in nature. If this land is included the low-density sprawl community has twice the public and *private* land devoted to open space as the high-density community. (Note that over half the high-density community has not been developed and therefore is not considered as *dedicated* open space.) The amount of land used for schools and other public buildings is the same in all communities, but the high-density community uses about half as much land for transportation compared to the low-density community.

Energy and Water Use

Energy use is determined primarily by residential heating and air-conditioning environments and by automobile use. Heating and air-conditioning requirements are related primarily to the type of dwelling unit. Denser developments have lower demands than single family units. For one thing, they are generally smaller in area (m^2) and have more common walls than is true of the single family detached home (which presents a great deal of exterior wall and roof area to the outside environment).

Transportation demands are affected both by the degree of clustering and community planning and by density. "Planning" alone can save nearly 14 percent of total energy required. "Planning" combined with increased density can save up to 44 percent. Water consumed in cooking, drinking, and so forth is not affected either by planning or by density. This is a reflection of cultural patterns of use. However, water for lawns is affected by both. Figure 2 indicates the variations in requirements for these two important resources.

Economic Costs

In terms of total investment costs, the high-density planned community requires 21 percent less investment than the combination mix community, and 44 percent less than the low-density sprawl community. Much of these savings result from differences in development density. Savings of about 3 percent of total development cost result from "better planning" whereas those from increased density amount to 41 percent. (Note that "good

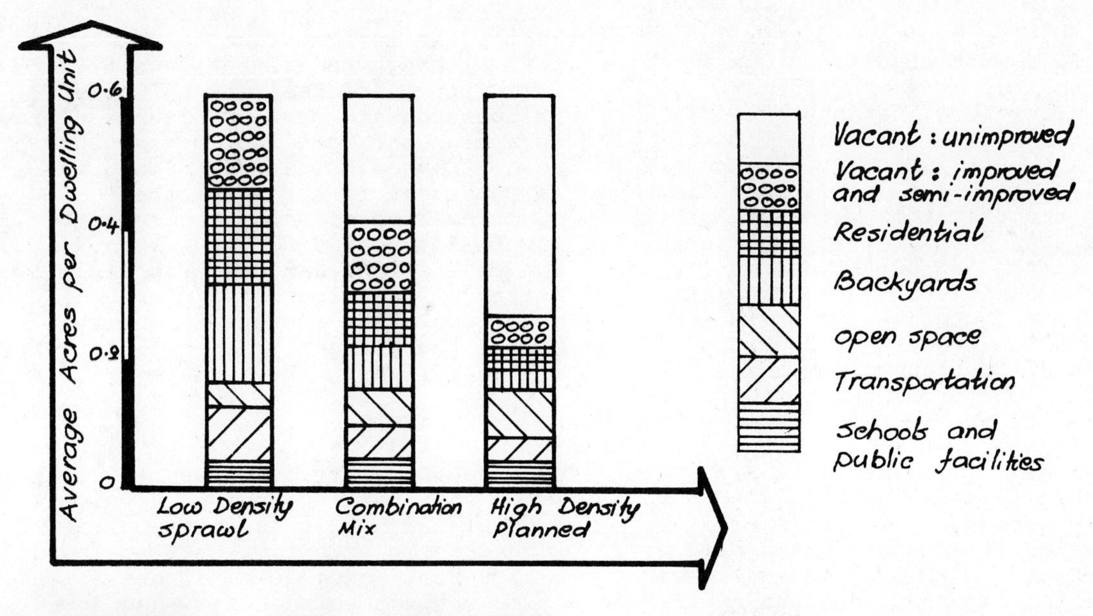

COMMUNITY COST ANALYSIS LAND USE FIGURE 1

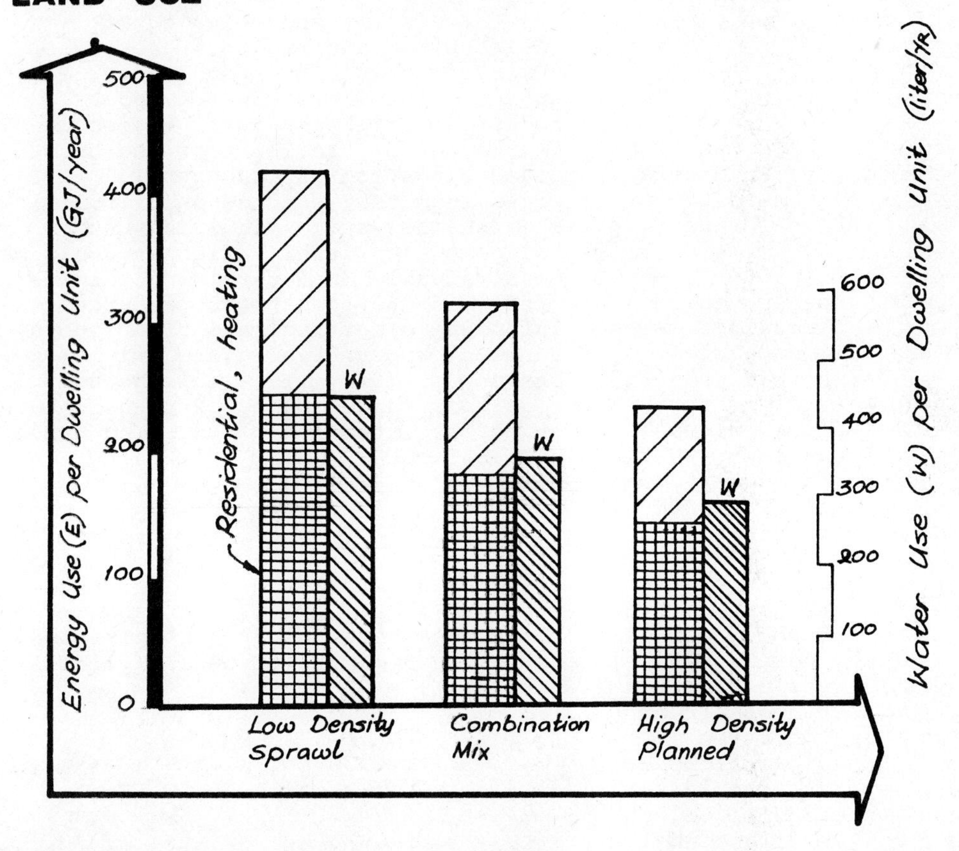

COMMUNITY COST ANALYSIS ANNUAL ENERGY AND WATER USE FIGURE 2

planning" includes much more than just "clustering" and may well result in more significant cost savings than shown in this study.) The largest cost savings are in construction of residential dwellings although important savings are attributable to reduced costs for roads and utilities which are about 55 percent lower in the high-density than in the low-density community. Figures 3 and 4 summarize investment and operating costs for the three communities. Not only does the high-density planned community cost less to construct, but a lower proportion of the cost is likely to be borne by government. Similarly higher density communities are again somewhat less expensive in terms of total operating and maintenance costs including those costs paid by government. Implicit in the lower capital and operating costs of the high-density community is a reduced energy commitment for direct and indirect uses such as building materials, street maintenance, municipal lighting, etc.

Environmental Cost

Air pollution has two major sources--automobiles and residential heating. Higher density developments require less energy for heating. Higher density, better planned communities generate about 40 percent less air pollution than the lower-density sprawl community. Although careful and more compact planning has a slight effect upon the amount of pollution resulting from residential heating, it can reduce the amount from automobiles by 20 to 30 percent. The amounts of air pollution generated by different communities are shown in Figure 5. Figure 6 indicates a similar pattern of water pollution generated by different development patterns. The type of development has no effect on the amount of sanitary sewage generated because this is a function only of population. It does affect the important problems of storm water pollution and sediment; the less paved area there is, the less storm water runoff there will be. Clearly, there are also direct energy savings--for air filtration and water clean-up--reflected in the lower rates of pollution.

For both air and water pollution, it is important to note that although the higher density community generates less pollution, it does so in a smaller area resulting in a higher rate of pollution generated per area developed. Planning is important in terms of other environmental effects: noise problems, preserving wildlife and vegetation, and creating visually attractive developments. For a given developed area, increased density allows the planner greater flexibility in accomplishing these goals.

RECOMMENDATIONS

To summarize, the results of the study show that "planning" to some extent, but higher densities to a much greater extent, result in lower energy use, lower economic costs, lower environmental costs, and lower natural resource consumption, for a given number of dwelling units.

Simulation and modeling of entire urban and regional areas grew during the late 1960s and early 1970s to the point where the tremendous amount of variables involved became unmanageable. In light of the undecided situation of urban growth, it would seem that studies and simulations at a smaller scale requiring better management of fewer variables would be more valuable in determining the most efficient use of natural resources, including electrical and other energy forms. Properly integrated with broader studies, they would be useful to either large-scale new town building or to the extension of existing urban communities.

References

1. Real Estate Research Corporation, *Urban Sprawl*, (Chicago, Illinois: 1971).

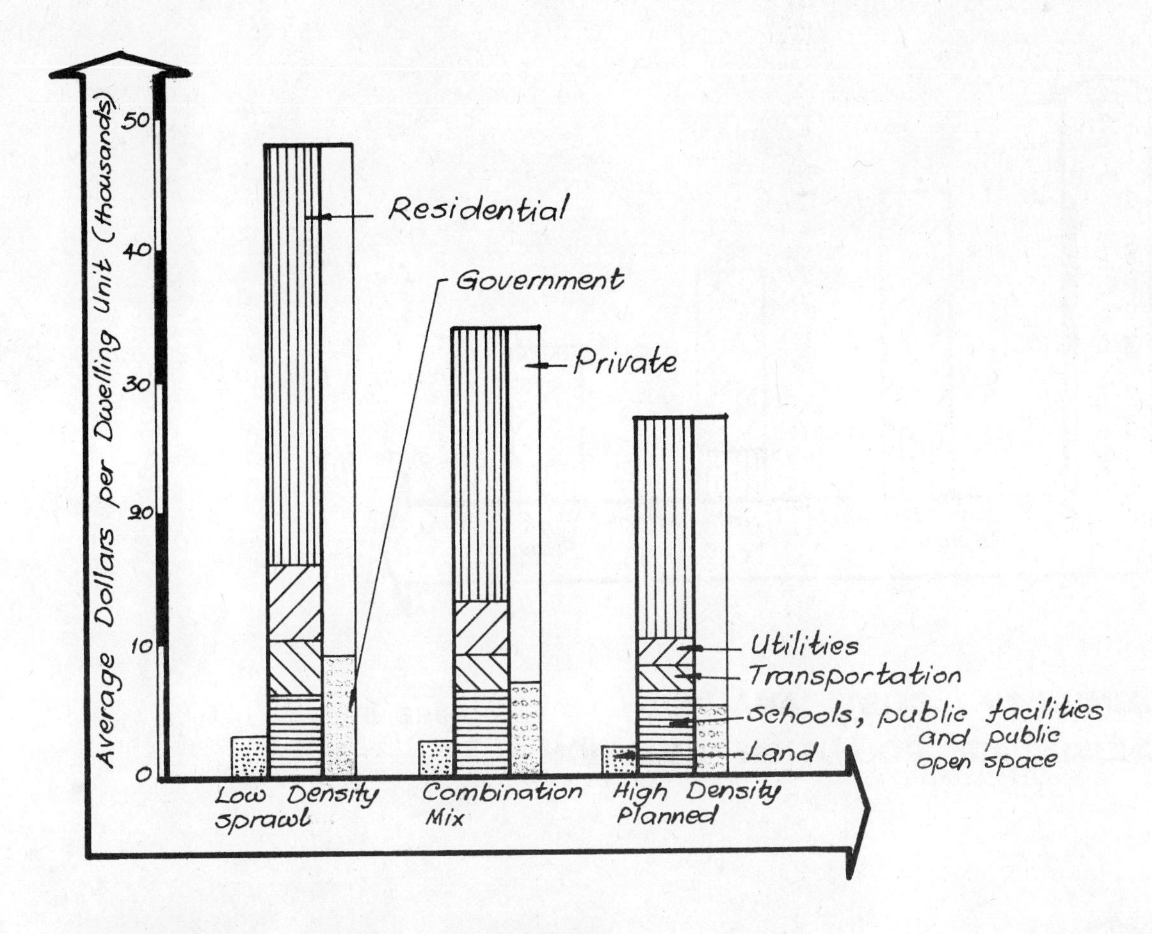

COMMUNITY COST ANALYSIS CAPITAL COSTS FIGURE 3

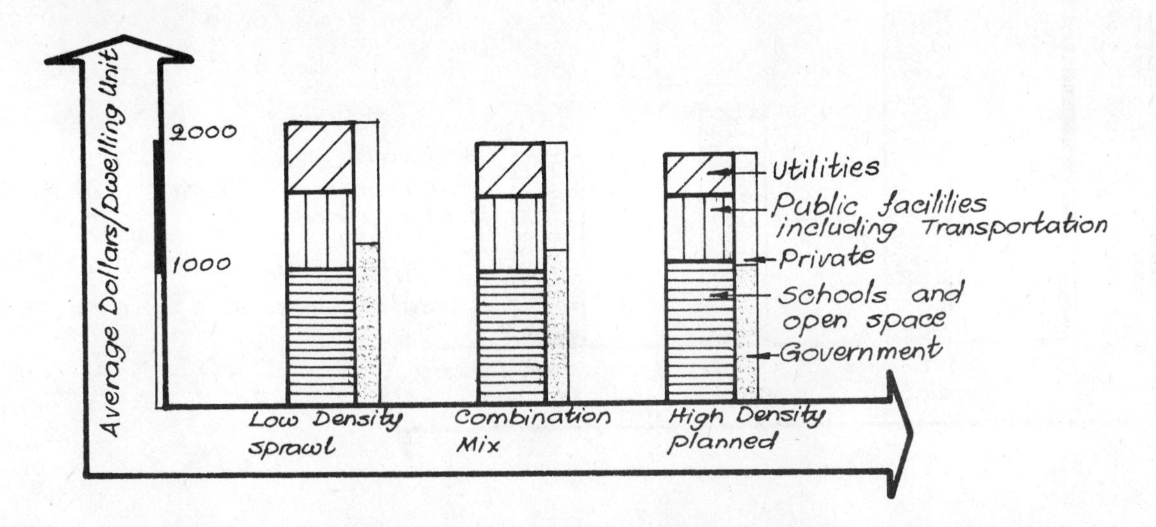

COMMUNITY COST ANALYSIS ANNUAL OPERATING & MAINTENANCE COSTS FIGURE 4

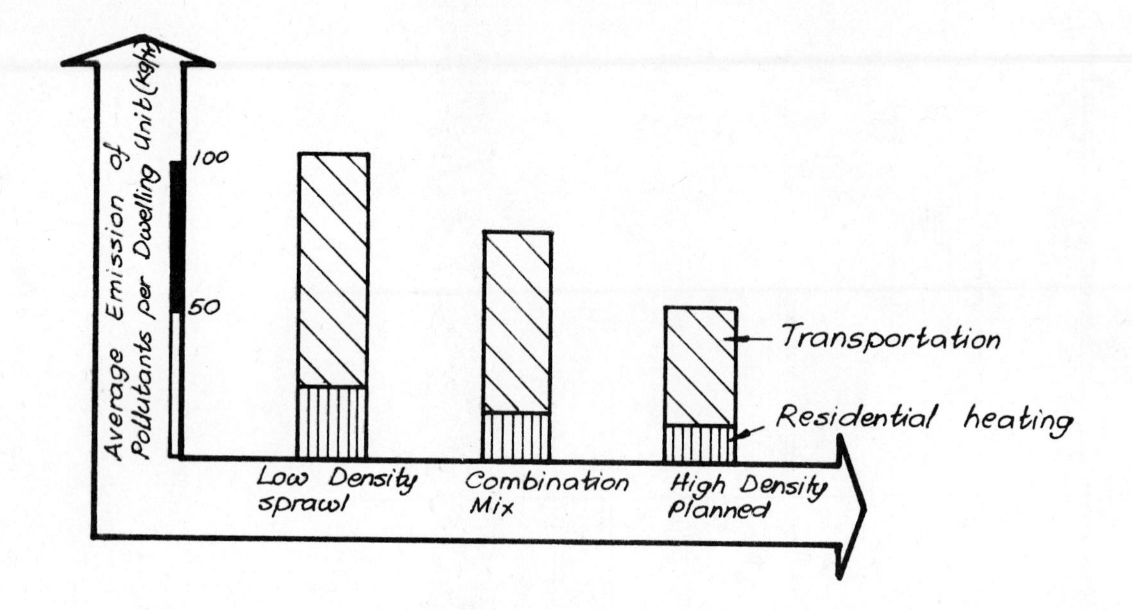

COMMUNITY COST ANALYSIS FIGURE 5
ANNUAL AIR POLLUTION EMISSION

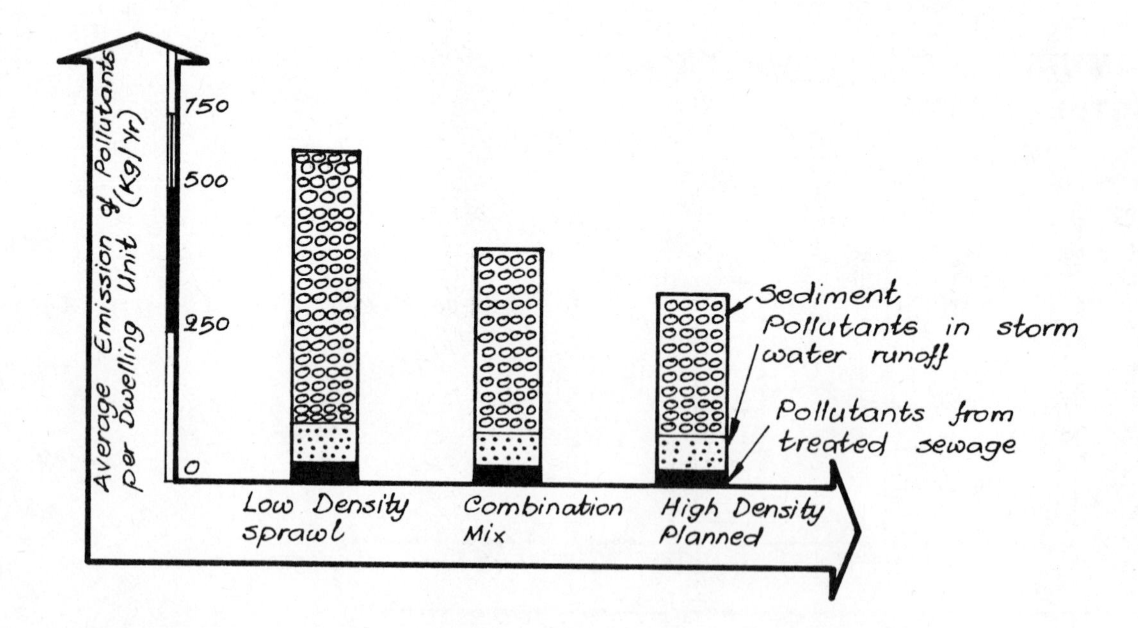

COMMUNITY COST ANALYSIS FIGURE 6
ANNUAL WATER POLLUTION GENERATION

CASE STUDY 13-3: SEPARATION OF MUNICIPAL SOLID WASTES INTO RECOVERABLE METALS AND SUPPLEMENTARY POWER PLANT BOILER FUEL [1]

THE CONCEPT

Solid municipal waste is shredded and subsequently classified or separated into heavy and light fractions. The heavy fraction is reclassified for recoverable metals while the light fraction refuse is stored and then transported to an electric power plant. There it is fed into the boiler and provides between 10 and 20 percent of boiler fuel requirements.

BACKGROUND

This waste-processing facility, recently constructed adjacent to one of two City of St. Louis incinerators, was funded jointly by the city and the US Environmental Protection Agency as a demonstration project in cooperation with the Union Electric Company (see Figure 1).

The main technical parameters for the plant were:

- *Process Products:* The output per 100 mt waste input is:

solid waste fuel	80 mt @ 10-14 MJ/kg (4.5-6.0 kBtu/lb)
ferrous metal	7 mt

- *Process By-Products:* Thirteen percent fly-ash/solid residue after incineration. Water effluents from boiler bottom-ash sluicing process.
- *Developmental Status:* The facilities were operated at full capacity 6 hrs/day to produce enough fuel for a two shift/day Union Electric burn. Now the project has been terminated and the plant shut down.
- *Process Description:* Refuse is unloaded onto a conveyor equipped with a belt scale and proceeds to the hammermill feeder which supplies refuse in a uniform manner. The hammermill reduces the size of the refuse to 4 cm or smaller. The refuse falls through grates at the bottom of the hammermill and proceeds to the surge bin via a conveyor. Refuse is transported from the surge bin by vibrating conveyor to the infeed air lock of the air density separator. From here the refuse goes directly into the air-separation zone where light and heavy fractions are separated (light = 84-86 percent, heavy = 14-16 percent by weight). After density separation the heavy fraction undergoes magnetic separation for removal of ferrous metals.

  The refuse then passes through a nuggetizer and magnetic separation again. The marketable metal would be transported by truck to local markets. The light fraction proceeds to the cyclone separator to separate the light refuse from the conveying air. This material goes into the light fraction storage bin where it would be compacted and transported to the Meremec Power Plant 18 miles away.

  The light refuse would be deposited in a receiving bin at the Meremec Power Plant and pneumatically fed into the surge bins which regulate the amount of refuse pneumatically fed into the tangentially fired boilers. The refuse fuel made up about 15 percent of the fuel required.
- *Cost Information:*

Capital Investment	$4,300,000*
Maintenance and Operation/Year	$1,900,000 - $2,000,000
Credits	
Fuel, at $1.36/million Btus	$3,900,000
Metals (ferrous)	$ 340,000
Total Credits	$4,240,000

  The value of the solid waste fuel increases proportionately to the increased value of

---

*Initial working capital excluded.

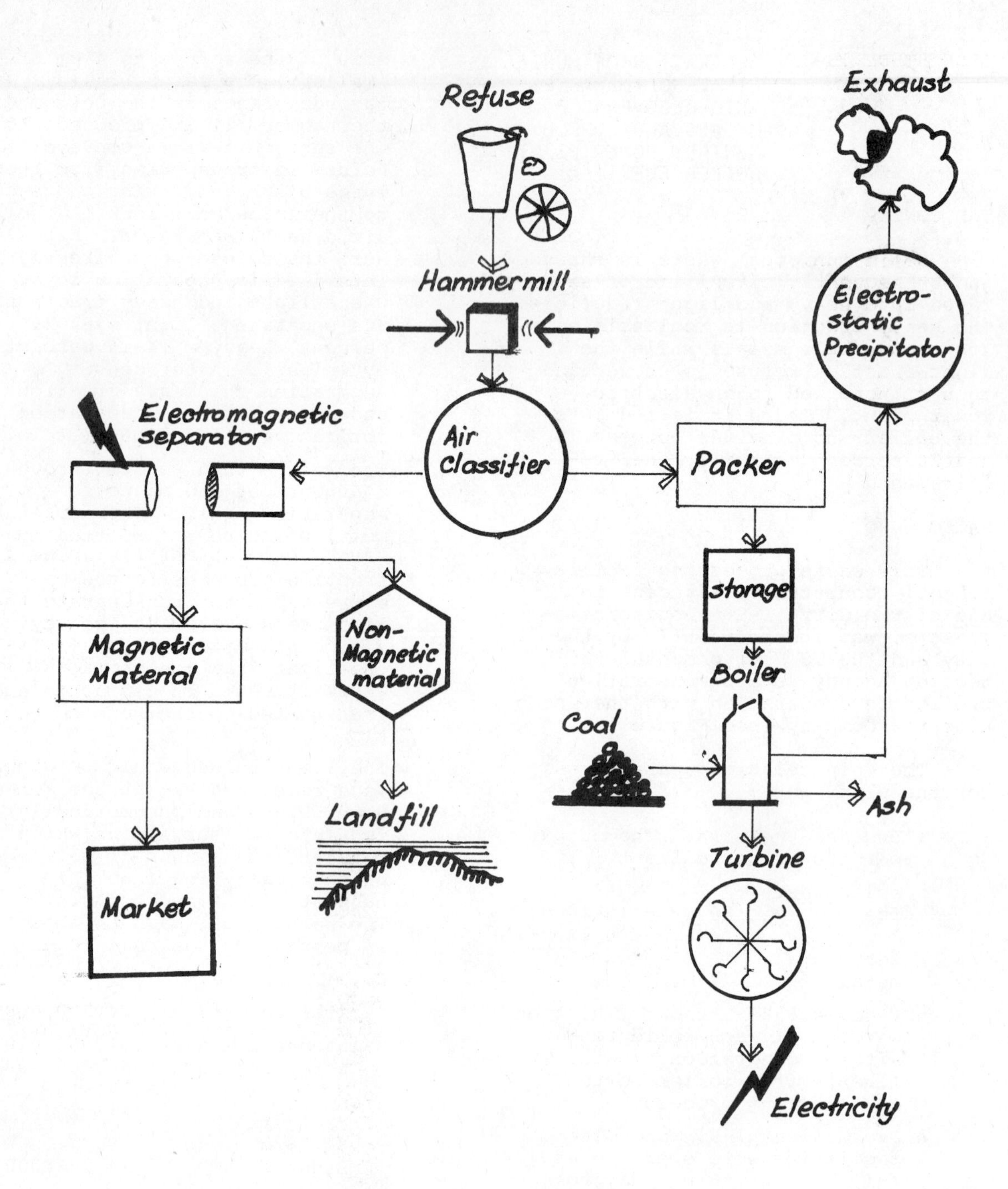

ST. LOUIS / UNION ELECTRIC WASTE PROCESSING FACILITY

FIGURE 1

coal. As the metal recovery technology improves for more efficient separation, the metals credit will increase accordingly.

FINDINGS

Air pollution tests indicated that some increased particulate emission occurs, but this can be reduced by fine tuning the electrostatic precipitator. No significant change was found in gaseous pollutants. The increased particulate levels were thought to be due to increased flow rates to the precipitators resulting from the higher moisture content of the refuse-based supplemental fuel. EPA plans to make further emissions tests. Combustion efficiency was found to range between 60-95 percent. Fuel mixing patterns may account for most of this variation. Surprisingly, no correlation could be found between refuse moisture content and the degree of burnout of the refuse. The distance between the refuse treatment facility and the Meremec Power Plant (18 miles) is the first disadvantage of this system, but the project was such that closer proximity was not feasible.

The St. Louis project had problems with downtime due to mechanical and structural inadequacies. Also significant is the fact that the utility attempted to go ahead with an 8000 ton per day plant at its 2400 MW Labadie Power Plant. However, in April 1977, four months before start-up, the utility announced plans to abandon the project due to political and economic problems.

RECOMMENDATIONS

Municipal refuse may represent an important utility fuel, thus saving energy two ways. However, there are many technical, economic, and political obstacles to be overcome.

References

1. Data taken and extrapolated from International Research and Technology Corporation, *Problems and Opportunities in Management of Combustible Solid Wastes,* (Washington, D.C.: October 1973).

CASE STUDY 13-4: INCINERATION OF MUNICIPAL REFUSE WITH THERMAL ENERGY RECOVERY AS A BY-PRODUCT

THE CONCEPT

Municipal solid refuse is gathered, stored, and fed to combined incinerator-boilers in order both to reduce landfill requirements and to produce steam for heating of buildings and chilled water for cooling of buildings within the immediate district.

BACKGROUND

The Nashville (Tennessee) Thermal Transfer Corporation contracted with I.C. Thomasson and Associates (ICTA) for the design of this combined operation (see Figure 1).

The main technical parameters for the plant are:

- *Process Products:* Steam (500,000 lbs/hr distributed at 150 lbs pressure and 600°F), chilled water (41°F from two 15,000 ton chillers).
- *Process By-Products:* Incineration residue (5 percent of original volume). Recovery of ferrous metals is under study.
- *Developmental Status:* This $16,500,000 project was financed by revenue bonds and a $650,000 Ford Foundation grant. I.C. Thomasson reported on December 3, 1974, that the plant was operational and burning solid waste 24 hours per day.
- *Process Description:* Solid waste is continuously fed by an agitating grate into the incinerator boilers operating at 1800°F and at a slight negative pressure. The 1800°F temperature removes odiferous gases and the negative pressure eliminates odors from escaping into the general area. Ash from the incineration process drops through grates into an ash hopper and is sprayed with

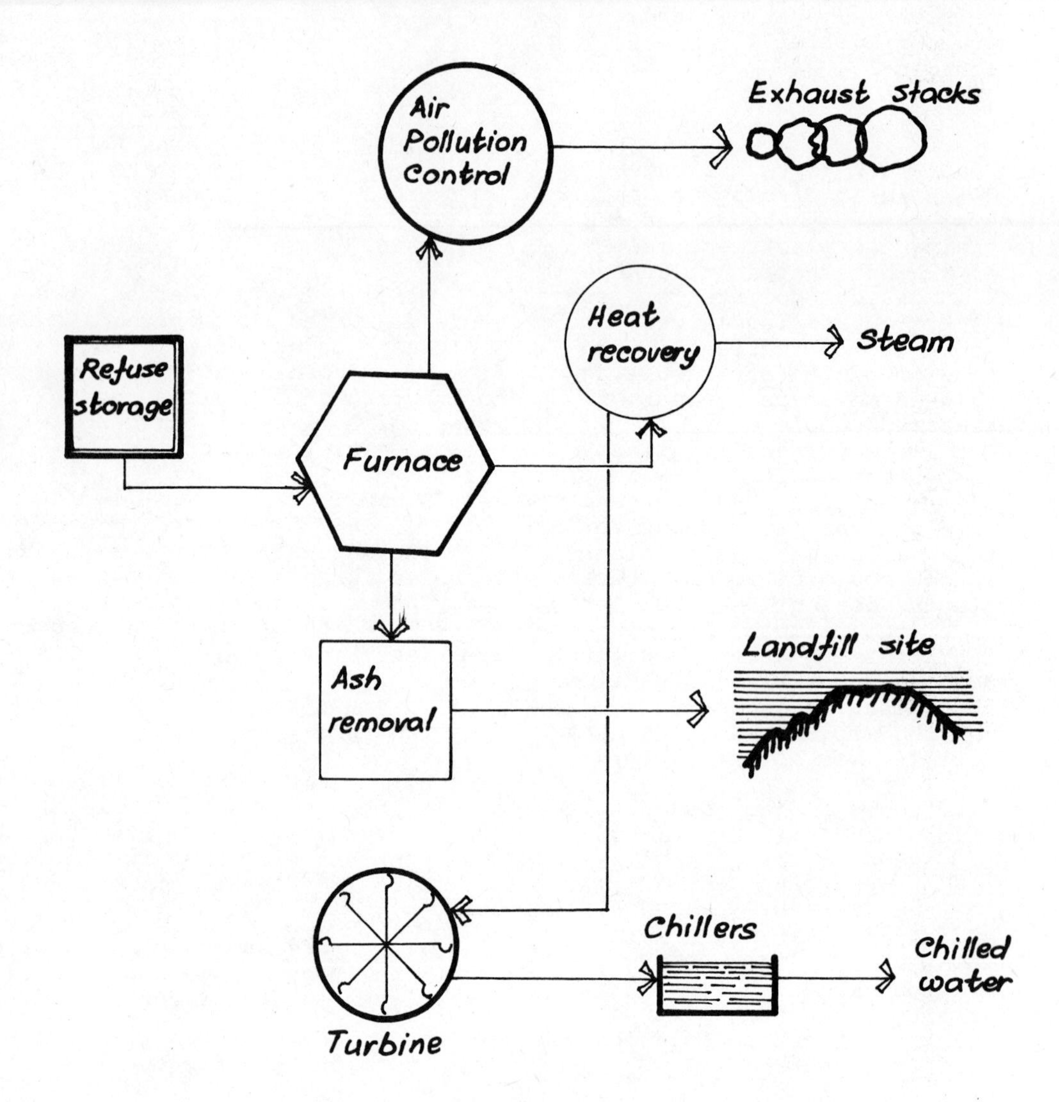

REFUSE INCINERATION SCHEMATIC

FIGURE 1

water to both cool the ashes and eliminate dust problems. The steam produced from the incinerator-boilers is piped to non-condensing turbines. The exhaust steam from the turbines is used to drive two condensing-steam-turbine-driven Carrier Corporation chillers (14,000 ton capacity). The resulting chilled water and steam are then directed to the distribution system to serve outlying areas in Nashville central heating and cooling systems.

The flue gas is subjected to an economizer bank where the temperature is dropped to 500°F. After heat reduction, the air goes through a dry "cyclone" separator to remove large particulate matter and then through "wet scrubbers" to remove remaining particulate matter and soluble gas. The gas exits at 140°F.

- *Cost Information:*

Land Requirements	3 acres for steam generation only and 5 acres for steam and chilled water production combined.
Capital Costs	$15,000-16,000/ daily ton capacity (southeastern US).
Staff Requirements	Plant requires 25-30 employees to operate 1000 ton/day facility.

FINDINGS

An immediate market for steam and chilled water with appropriate distribution system potential must be available for economic feasibility. Gaseous emissions seem to be reduced by the wet scrubber, but the control of submicron particulates may present some problems with the current design. The large water requirement (1,500,000 gal/day) could represent a problem in applicability to regions with water availability problems.

RECOMMENDATIONS

Where suitable economic conditions exist, incineration of municipal refuse with thermal energy recovery as a by-product should be considered in addition to landfill operations.

CASE STUDY 13-5: COMMON CORRIDORS FOR MUNICIPAL UTILITY SERVICE LINES*

THE CONCEPT

Although a majority of municipal service networks of water piping, telephone lines, sewers, gas lines, etc. are already located underground, their grouping into a single tunnel, rather than in separate trenches, is a well-known concept with a variety of claimed advantages as well as some potentially serious drawbacks. One of the potential advantages is improved energy efficiency, both direct (by reducing losses) and indirect (less energy and energy intensive materials to construct the system).

BACKGROUND

Conventional utility installations are disruptive and costly, requiring separate digging operations (providing all are underground) for water, gas, electricity, sanitary sewer, storm sewer and telephone and inconvenient interruptions to street use for installation of laterals or repair (see Figure 1).

Furthermore, while most power transmission lines in the United States (especially long-distance voltage lines) are above ground, there are increasing

*Grouped utility corridors are sometimes known as "utiladors." That name has been trademarked by RIC-WIL Inc. and they are now variously called utility tunnels (below grade) and utility corridors (above grade). In this case study they are called utility tubes to indicate the generic category of below-grade and above-grade horizontal and vertical ground utilities. This case study has been extracted from Reference [1].

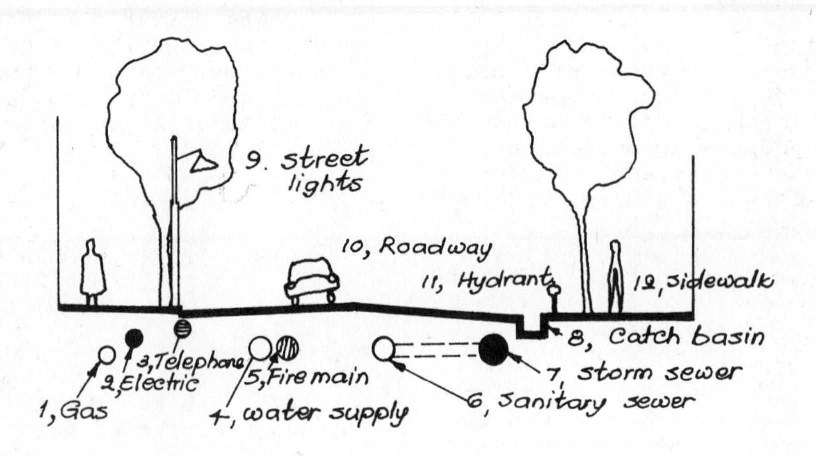

CONVENTIONAL UTILITY INSTALLATIONS FIGURE 1

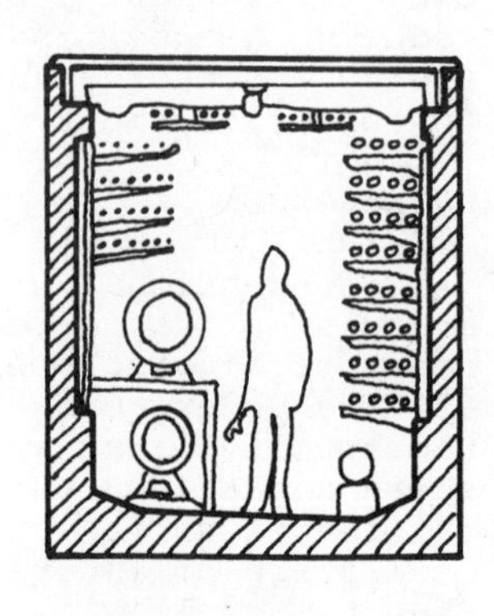

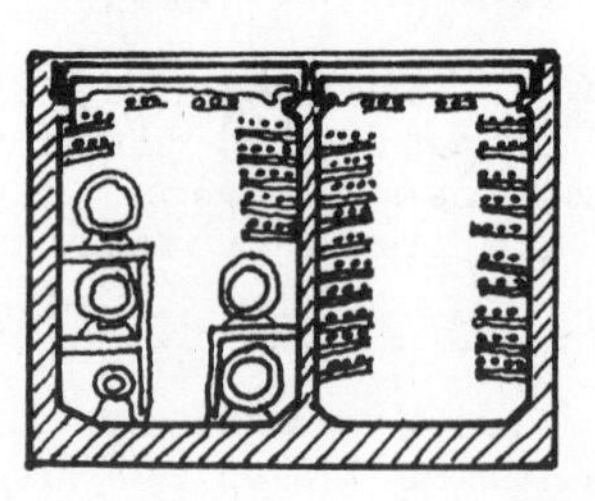

TYPICAL UTILITY TUBES FIGURE 2

environmental and other pressures (land availability and cost, for example) to put them underground, particularly in urban areas. The extent to which undergrounding will take place depends in large measure upon the success of technological research aimed at reducing costs and energy losses in underground facilities. [2,3]

Figure 2 shows several conceptual designs for utility tubes. These are not unlike existing systems which have been used in New York, London, and Moscow.

The American Public Works Association, in Project #68-2, has recently concluded an 18-month investigation on utility tubes.[1] One major conclusion of the study was that utility tubes can provide a channel for electric power, telephone, telegraph, gas, water, sanitary and combined sewers, storm sewers, police and fire alarm lines, steam lines, street lighting and traffic signal lines. They can also accept newly emerging services such as coaxial cables, waveguides, heating and cooling lines, secondary quality water systems, vacuum and slurry solid waste collection lines, snow removal, mail delivery tubes, and package delivery tubes. Further, utility tubes are adaptable to presently unidentified systems which might be developed in the future, with diminished likelihood of interruption of other services or street surface activity. If combined with a transit system they could be considered to be a total delivery system including delivery of cargo and persons in a multiple-use right-of-way (for example, in combination with a "guideway"). With transportation and utilities both strongly influencing urban growth form, their combination makes for an even stronger form determinant.

Where service networks serve highly loaded activity centers or dwelling areas, there are possibilities for integrating them. For example, where an elevated guideway is used, it could provide the structural support for an above-grade utility tube with new service tubes for an automated vacuum tube trash collection and pneumatic tube delivery (see Figure 3). An integrated system of this type could be located above or below ground.

As an example of a future innovation which would be compatible with the utility tube concept, research is being conducted to develop new transmission channels capable of handling large blocks of electrical power with fewer losses. Typical are: cryogenic resistive conductors, compressed-gas-insulated cables, and, superconductors. (Figure 4 shows several possible configurations.) Presently, underground high power transmission is accomplished with oil-paper insulated cables, which cost 5 to 25 times as much as equivalent overhead lines.

FINDINGS

Utility tubes have advantages and disadvantages, which can be summarized as:

Advantages

- potential for energy savings in construction, operation, and maintenance;
- joint use of right-of-way;
- energy and cost saved in street cuts for installing laterals or new systems;
- noise pollution from street cuts avoided;
- trip time shortened by avoidance of traffic interference;
- man hours saved and overall efficiency achieved by ease of inspection and maintenance of utilities and aggregation (instead of fractionalization) of administration and maintenance responsibility;
- downtime saved from outages and repairs;
- cost savings in reduction of utilities' and services' length by means of unifying them in one channel rather than two or more (especially in the case of those utility lines customarily running along both sides of the street); and,

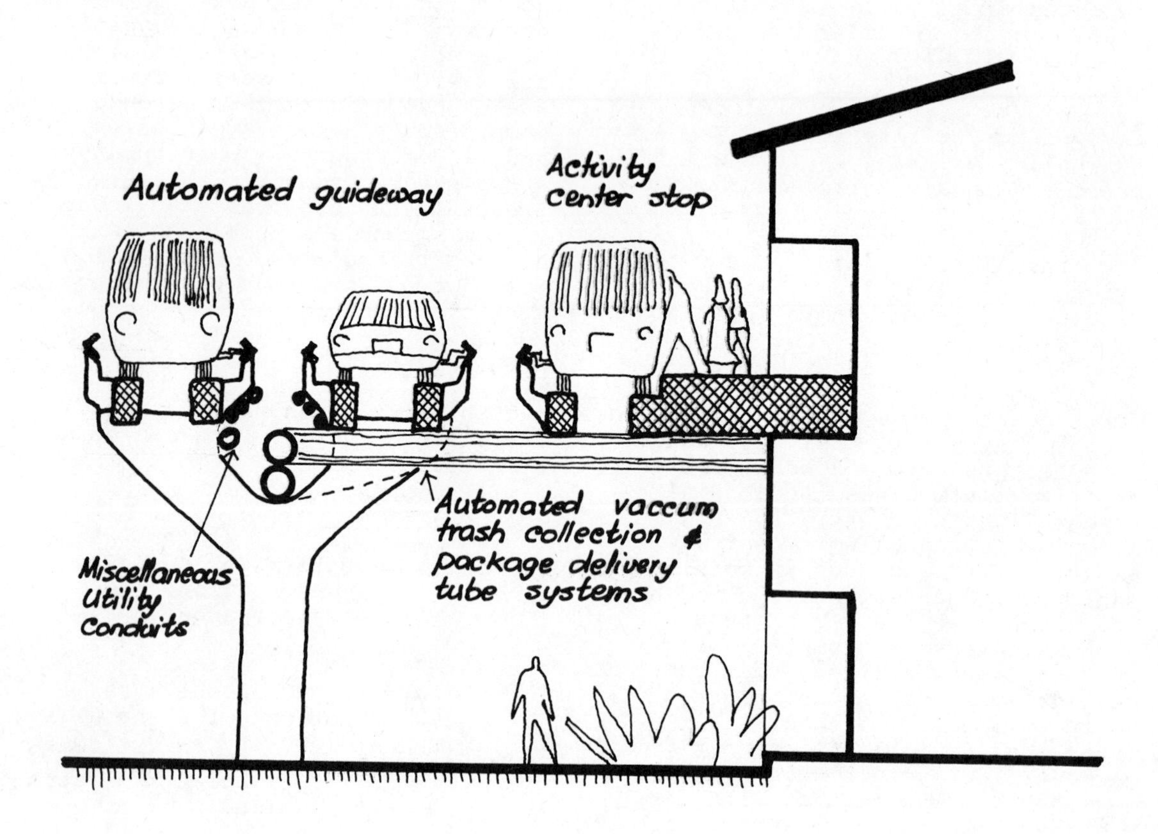

GUIDEWAY-UTILITY TUBE INTEGRATION

FIGURE 3

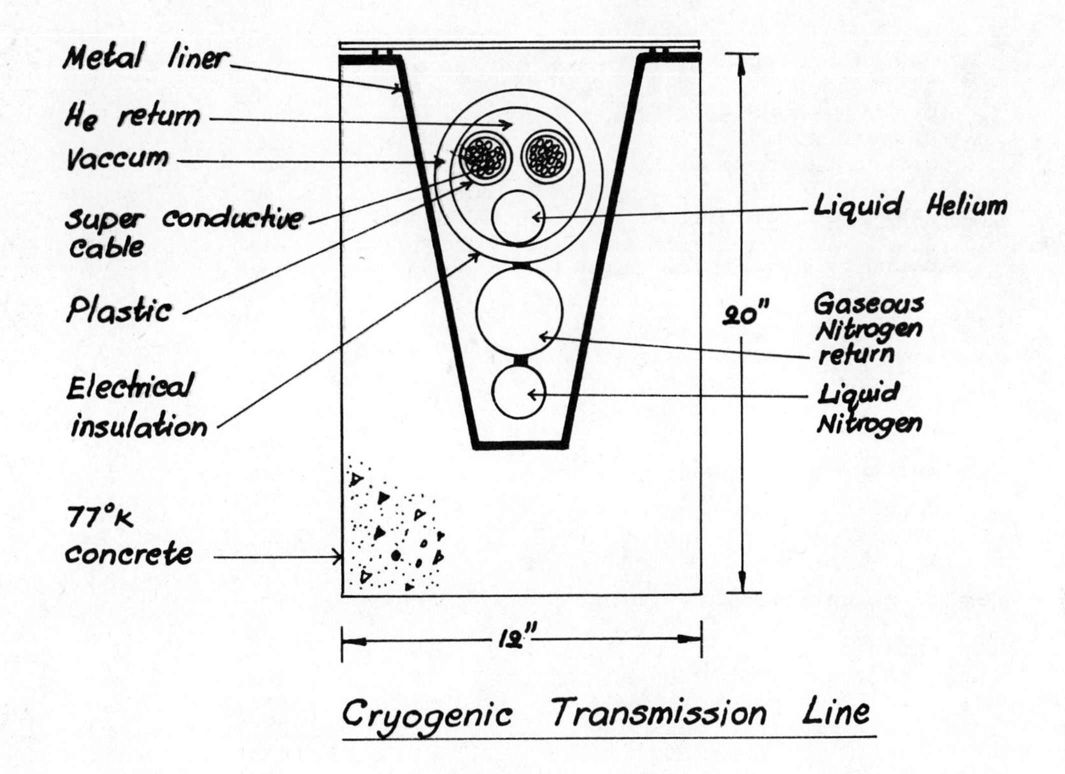

Cryogenic Transmission Line

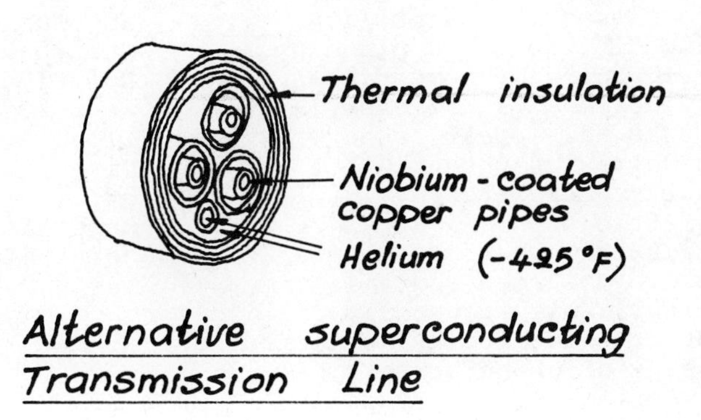

Alternative superconducting Transmission Line

TWO CONCEPTS FOR MORE EFFICIENT UNDERGROUND CONDUCTORS

FIGURE 4

- potential for incorporating new technological advances at reduced capital costs, e.g., without tearing up streets every time a new system becomes available.

Disadvantages

- ruptures and explosion of gas lines;
- leakage of water or sewer lines and possible resultant flooding, contamination, and short-circuiting of electrical lines;
- leakage of cryogenic lines;
- conflicts resulting from overlapping maintenance and responsibility; and,
- large capital investment.

With the use of modern materials and techniques these problems can be minimized. In their present level of development, utility tube networks are not necessarily compatible with expressway networks; but the potential and benefits of compatibility can be enhanced by combined planning. They are more generally compatible with primary collection and distribution networks and appear to have the highest degree of potentiality in high-density urban areas. In secondary distribution and local street systems they are fully compatible, as every street contains most utilities. However, utility tubes cannot be economically justified for all classes of urban streets; and in secondary street systems, easily accessible small conduits, instead of utility tubes, would offer major long-term economics.

Conceptually, a combined transit and utility tube system could be a visible service spine for public and private buildings; and where laterals are required, they need not be excavated for, but rather could be made visible. It would be possible for some buildings to "plug" on to this utility/movement service spine and for the spine to pass through some public buildings.

RECOMMENDATIONS

Municipal utility service systems both require and furnish energy. In high density areas aggregation of service systems in utility tubes has potential for improving energy use efficiency, reducing costs, and making better use of available land. Such systems may be more important in the future if there is a widespread use of superconducting or other types of underground power distribution systems.

References

1. Miller, B., Pinney, N.J., and Saslow, W.S., *Innovation in New Communities*, MIT Report #23, (Cambridge, Massachusetts: MIT Press, 1972).

2. Herbert G. Poertner, Director of Research, APWA, "Evaluation of the Feasibility of Utiladors in Urban Areas," paper delivered at the conference on "Joint Utilization of Right-of-Way for Utilities and Municipal Services in Urban, Suburban and Rural Environments," July 1969.

3. Lloyd A. Dove, Assistant Executive Director, APWA, "Feasibility of Utility Tunnels in Urban Streets," paper delivered to AASHO/APWA joint meeting, November 1970.

CASE STUDY 13-6: UNDERGROUNDING OF SPECIAL-PURPOSE PHYSICAL FACILITIES REQUIRING MODULATION OF TEMPERATURE EXTREMES--KANSAS CITY COLD STORAGE

THE CONCEPT

Certain kinds of physical facilities requiring relatively constant temperatures the year around can be put underground, resulting in more efficient energy use and lower operating costs.

BACKGROUND

The insulating capability of earth results in subsurface temperatures which vary only slightly from the yearly average above ground in a given area. At depths greater than ∿1 m, not only are the daily variations greatly damped, but seasonal changes are also reduced (see Figure 1). Less heating in winter and less cooling in the summer are required for underground spaces, with particular advantage gained through avoidance of the direct radiant energy of the sun in the summer.

In addition, since the surrounding mass of earth acts as a heat sink, standby refrigerating equipment normally required for special-purpose facilities of this type would not be needed. Cooling plants can be shut down for days for repair or even unintentionally (due to power or mechanical failures), without adverse effects on frozen goods stored. In the Kansas City Cold Storage Facility, the temperature rises typically 0.5°C (1°F) per day after plant shutdown. Similar above-ground facilities rise 0.5°C (1°F) *per hour*, making standby equipment essential. Comparative experience of Spacecenter, Inc. in the operation of their underground Kansas City facility and their above-ground operation in St. Paul, Minnesota is:

FINDINGS

Operating costs for special-purpose underground facilities (including savings due to reduced energy usage) are typically one-tenth of those for above-ground facilities. Capital outlay is also reduced considerably.

RECOMMENDATIONS

Where large-scale refrigeration or cold storage capability is required, undergrounding, leading to more efficient use of energy, should be considered.

References

1. Bligh, T.P., and Hamburger, R. "Conservation of Energy by Use of Underground Space," in *Legal, Economic, and Energy Considerations in the Use of Underground Space*, edited by: National Academy of Sciences, (Washington, D.C.: 1974).

2. Kenagy, G.J., Smith, C.B. *Depth and Activity Measurements of a Heteromyid Rodent Community Using Radioisotopes*, UCLA-ENG-7075, (Los Angeles: University of California: August 1970).

Cost Comparison of Above- and Below-Ground Storage and Refrigeration (dollars per foot^2)[1]

Kind of Storage	Installation Costs[a,b]		Operating Costs	
	Above Ground	Underground	Above Ground	Underground
Dry Storage	10	2.50	0.03	0.003
Refrigeration	30	8.10	0.12	0.010

[a]Excluding cost of land or underground space.

[b]No standby equipment required underground.

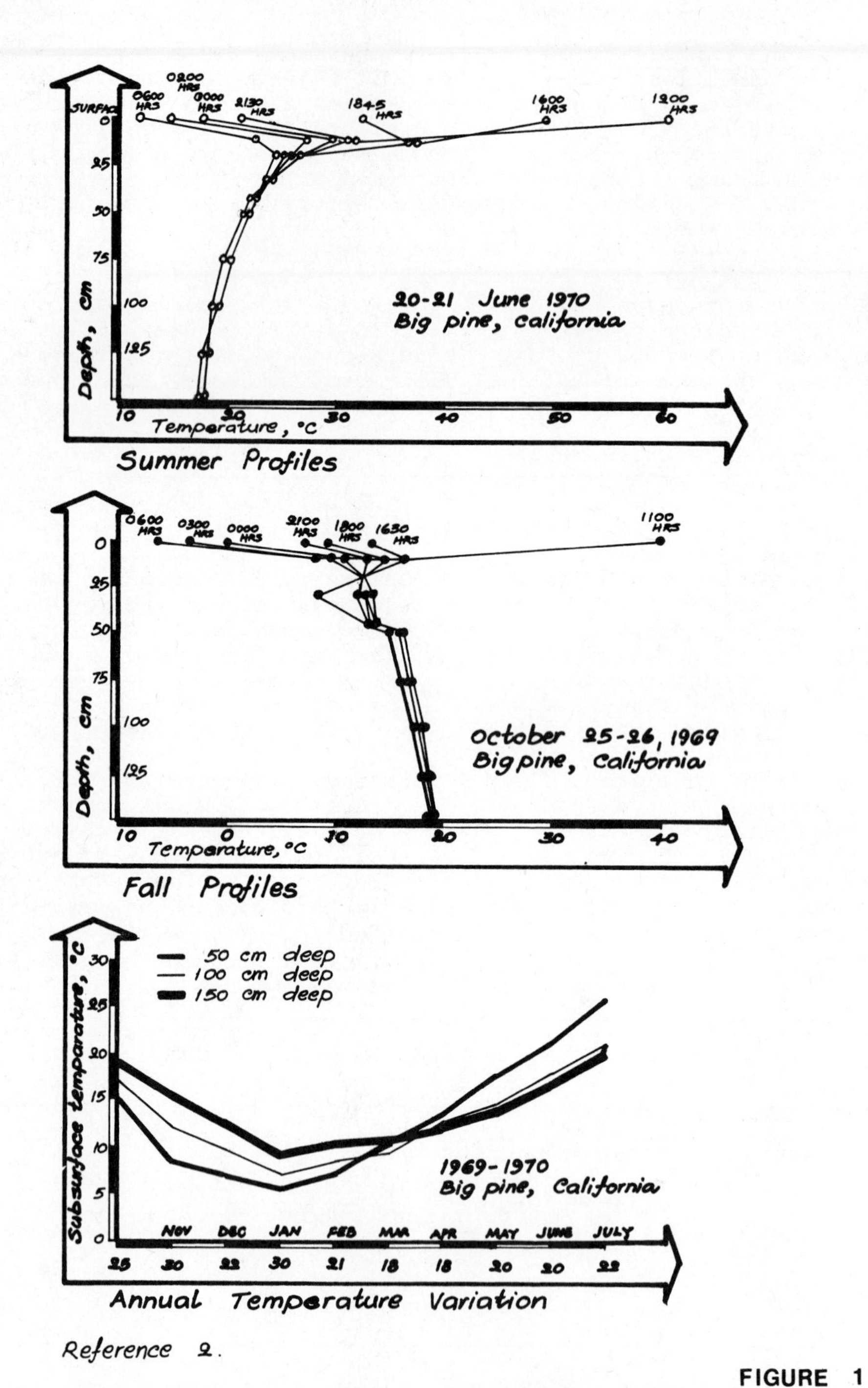

Reference 2.

FIGURE 1

SUB SURFACE SOIL TEMPERATURES

CASE STUDY 13-7: UNDERGROUNDING OF SPECIAL-PURPOSE PHYSICAL FACILITIES REQUIRING A VIBRATION-FREE ENVIRONMENT--BRUNSON INSTRUMENT COMPANY, KANSAS CITY, MISSOURI

THE CONCEPT

Apparently, most existing underground, non-residential facilities were built in order to achieve some special objective. Subsequently, a variety of bonus features have been discovered, such as savings in operating costs brought about by more efficient energy use.[1]

BACKGROUND

The greatly reduced energy requirements for an underground manufacturing plant are illustrated in Table 1. At full capacity, the underground facility will employ 500 people and install more machines. "It has been estimated that with underground conditions, no more heating equipment will be needed, due to the added heat input from people and machines and only two-thirds more air conditioning plant will be required."

The heat-sink characteristics of the surrounding earth are of equal significance to its capabilities of modulating and integrating above-ground ambient air temperatures. Other than the energy content of domestic hot water tanks (and the relatively few solar-equipped facilities that have been built), energy storage in buildings is almost nonexistent in the US. Since load patterns within a given building type and in a particular area are relatively consistent, the peak-loads which result are critical in establishing a utility system's total generating capacity needs. To the extent that underground facilities *shift* their own peaking periods out of phase with conventional facilities, demand factors upon the local system(s) are little impacted by the addition of below-grade facilities.

FINDINGS

- *Maintenance*--Everything underground is protected from wear and tear of weather extremes--wind, moisture, heat, freezing; no roof or exterior walls to maintain.

- *Utility Savings*--Energy savings have been described above. There is an additional advantage--electrical utilities, pipes, sewers, and drains can be either hung from the ceiling or put in shallow ditches; there is no problem of freezing.

- *Insurance*--Fireproofing construction costs are less, and windstorm hazard is nonexistent, making excellent insurance rates available.

- *Strength*--Floor loads are almost unlimited. Heavy machinery does not require elaborate foundation support; e.g., in Kansas City the shale can be loaded to 200 tons per square foot.

- *Stability*--There is less vibration. Delicate machines and instruments need much less isolation, thus avoiding an expensive and difficult task.

- *Operating Savings*--Machines remain accurate for much longer without realignment, due to very stable temperature and humidity conditions.

- *Energy Savings*--Approximately 60 percent in heating (Btu/hr) saved from that which could be expended in a similar aboveground facility and 90 percent of refrigeration dehumdification tonnage saved (see Table 1).

While energy savings inherent in underground construction are widely acknowledged, it is often asserted that equal or better results can be achieved by simply increasing the amount of insulation in otherwise conventional structures. However, as shown in a simple example, underground structures are far superior to insulation from an efficient energy use viewpoint.

An equation for the heat flow rate is

Item Compared	Above Ground (Estimate)	Underground (Brunson Instrument)[a]
Heating units (Btu/hr	∿ 2,000,000	750,000
Refrigeration (tons) for dehumidification	∿ 500-700	57
Operating costs (dollars/year)	∿50,000-70,000	3,200[b]
Fire insurance (dollars/$1,000)	2.85	0.10

(a) Brunson Instrument Co., conditions: 140,000 ft^2; 125 employees; 77 ft below surface; 54°F initial rock temperature.
(b) This figure is particularly low since the air conditioning plant is operated only at night to bring temperature and humidity below that required. Because of the heat capacity of the rock, temperature and relative humidity of the air then slowly rise during the day. This technique reduces the electrical demand factor

ENERGY AND COST-ABOVE GROUND Vs UNDERGROUND

TABLE 1

$$q = UA(t_1 - t_2)$$

or

$$Q = q/A = U(t_1 - t_2),$$

where q = heat flow rate, Btu/hr
Q = heat flow rate per unit area, $Btu/hr \cdot ft^2$
U = thermal transmission coefficient, $Btu/hr \cdot ft^2 \cdot °F$
A = area through which heat is transferred, ft^2
t_1 = inside temperature, °F
t_2 = outside temperature, °F.

In a given region the temperature difference is determined by weather extremes for above-ground structures. Underground, however, as noted, the temperature remains almost constant at the yearly mean temperature. For example, the temperature 10 feet underground in the Minneapolis area varies from 47 to 51°F, whereas the daily temperature varies from -30 to 95°F. Table 2 lists typical thermal-transmission coefficients (U values).

Table 3 gives Q, the heat flow rate per unit area, above and below ground in Minneapolis, for the mean, maximum, and minimum daily temperatures in winter and summer. This shows, for example, that on a cold winter day the heat flow rate per unit will be 5.5 times greater above ground for a wall with eight inches of insulation (wall 3) and 8.4 times greater for a wall with four inches of insulation (wall 2), compared with an uninsulated wall underground, and Q can be 19-22 times greater through a roof than underground.

During summer a large amount of heat that must be removed flows into a building above ground, whereas heat flows out of an underground structure, lowering the cooling load. The ratio Q above/Q below is not given in summer because heat flow underground is out of a building, which is desirable since heat is produced by lights, cooking, machines, and people, whereas heat flow above ground is into a building, which is undesirable as it adds heat to the internal heat load. On a hot summer's day, for example, to maintain an above-ground building (of wall 2 construction) at the same temperature as a similar underground building, (4.0 x 2.5) $Btu/hr/ft^2$ of roof area would have to be removed by an air-conditioning plant, assuming the heat loss through the floor to be comparable to that in the underground building.

In conclusion, the calculation shows that improved insulation and above-ground construction do not compare favorably with sub-surface structures from the viewpoint of efficient energy use.

RECOMMENDATIONS

Additional effort should be instituted to obtain data concerning underground structures, thermodynamic data for soils, and economics of undergrounding. Based on this particular example, undergrounding offers a unique tool for using energy more efficiently in industrial facilities.

References

1. Bligh, Thomas P. and Hamburger, Richard, "Conservation of Energy by Use of Underground Space," in *Legal, Economic, and Energy Considerations in the Use of Underground Space*, ed: National Academy of Sciences, (Washington, D.C.: 1974).

CASE STUDY 13-8: BELOW GRADE CONSTRUCTION TO SAVE ENERGY (ECOLOGY HOUSE, CAPE COD, MASSACHUSETTS, AND ARCHITECTURAL OFFICES, CHERRY HILL, NEW JERSEY)

THE CONCEPT

Underground architecture, potentially involving almost all building types including office buildings, offers immediate, practical improvements in efficient energy use. These include:

Material	U Btu/hr·ft^2·°F
Roof, asphalt plus 1-in. timber	0.45-0.53
Windows, double glazed, 70 percent glass	0.45-0.55
Wall 1, no insulation	0.30-0.45
Wall 2, 4-in. insulation	0.20
Wall 3, 8-in. insulation	0.13
Basement, in contact with soil, no insulation	0.10

TYPICAL 'U' VALUES **TABLE 2**

	Above Ground[a]				Below Ground[b]
	Roof	Wall 1	Wall 2	Wall 3	(t_2 = 50°F)
Winter (January) mean,[b,c] t_1 = 75° F		t_2 = 10°F (t_1 - t_2) = 65°F			(t_1 - t_2) = 25°F
Q, Btu/hr/ft^2	29-35	19-29	13.0	8.5	2.5
Ratio Q above/Q below	12-14	8-12	5.2	3.4	
Winter (January) minimum,[d] t_1 = 75°F		t_1 = -30°F (t_1 - t_2) = 105°F			(t_1 - t_2) = 25°F
Q, Btu/hr/ft^2	47-56	32-47	21.0	13.7	2.5
Ratio Q above/Q below	19-22	13-19	8.4	5.5	
Summer (July) mean,[e] t_1 = 75°F		t_2 = 80°F (t_1 - t_2) = -10 °F			(t_1 - t_2) = 25°F
Q, Btu/hr/ft^2	-4.5 to -5.3	-3.0 to -4.5	-2.0	-1.3	2.5
Ratio[f]					
Summer (July) maximum,[e] t_1 = 75°F		t_2 = 95°F (t_1 - t_2) = -20°F			(t_1 - t_2) = 25°F
Q, Btu/hr/ft^2	-9.0 to -11.6	-6.0 to -9.0	-4.0	-2.6	2.5
Ratio[f]					

[a]Negative sign indicates heat gained.

[b]An inside temperature of t_1 = 75°F and an underground temperature of t_2 = 50°F were used throughout.

[c]In the winter or heating cycle, the mean temperature for the full 24-hr period averaged over the month was used since buildings must be heated continuously; here t_2 = 10°F.

[d]A minimum winter temperature of t_2 = -30°F and a maximum summer temperature of t_2 = 95°F were used as an example of the maximum heat flow rate conditions. The heating and cooling plant size must be sufficient for these extremes.

[e]During summer the mean temperature during the day was used since buildings need cooling only when the outside temperature exceeds 75°F; here t_2 = 85°F.

[f]A ratio Q above/Q below is not listed for summer since above-ground heat flows into a building, while underground heat flows out of it (see text).

HEAT FLOW RATE Q ABOVE AND BELOW GROUND

TABLE 3

- beneficial use of the earth as a constant-temperature heat sink, resulting in lower heating and minimal cooling requirements in most areas of the world;
- very little outside maintenance (no snow removal or lawn sprinkling);
- isolation from ambient noise and vibration;
- savings on construction costs--simpler, cheaper building materials, no exterior painting or maintenance, better use of natural rainfall, including avoidance of elaborate storm drains and handling of runoff waters; and,
- reduction or elimination of outside lighting needs.

In this case study two such structures are described; the first is a private residence, the second, an architect's office.

BACKGROUND

To evaluate the problems of building underground, first consider a shelter as a means of resisting natural forces. *Structure* resists the natural forces of gravity, (vertical loads) and earthquakes or winds (horizontal loads). *Environmental controls* mediate against natural forces of heat, cold precipitation, and humidity. *Enclosure,* whether non-loadbearing infill between structure or structure itself, contributes to mediation of both kinds of forces. *Cost* is obviously an important criterion. Typical light frame construction easily and inexpensively resists most of these forces to an acceptable level, while contributing relatively little to the structural load due to its own dead weight.

Building underground requires resistance to additional forces. These forces include the horizontal load created by the weight of the surrounding earth itself, which is heightened by attendant hydrostatic pressure. The need to stem leakage from water under pressure is constant. Thus, heavy masonry or poured-in-place concrete walls must be utilized. They must be designed as retaining walls or be integrated into a box-like web of cross-walls, exterior walls, floor, and roof which together can resist these imposed loads, where such walls bear directly and entirely upon surrounding earth.

The roofs of such structures would be even less like typical light constructed housing. Here a cave into a hillside is not being spoken of, but an excavation from above. In order to maximize both benefits of earth insulation and of *replacing* earth, top soil, and plant life, at least one meter of solid earth must be replaced over the roof structure. Only concrete or some combination of steel structure and concrete-fill membrane slabs can take the combined forces of the hydrostatic and soil dead loads. The special waterproofing procedures normally used for below-grade, retaining-wall construction have to be even more carefully and completely applied to such roofs. There are additional costs of excavation and disposal of excavated earth as well. It is doubtful that the savings which accrue in the form of no exterior wall painting, no conventional roofing materials, and fewer windows and doors, could make up for the additional costs noted above. Additional less obvious but nonetheless real costs would stem from the fact that this is unconventional housing construction and would, if pursued on a large scale, require skilled trades and construction materials not widely found or used in housing today.

There would be institutional resistances not related to the above issues which would have to be taken into account in large-scale production of underground housing. The skilled trades and materials producers and suppliers involved with conventional housing could create strong resistances. Architects who do not understand it may fear that a kind of non-architecture is being created. Users themselves may perceive all underground construction to be the kind of environment they connect with subway stations, tunnels, and basements. Despite these concerns, actual experience with underground dwellings has been positive, particularly where the house surrounds an open, underground atrium.

Certain practical problems will have to be resolved as well. Other than for very small houses, all rooms in a facility canot be placed immediately adjacent to open atrium courtyards. Excessive use of multiple atriums will or could over-extend circulation and require extensive construction. Insofar as all below grade space must be excavated and, at its eventual perimeter, be surrounded by expensive retaining walls, *all* below grade areas, indoor and out, will add substantially to total costs. While building into hillsides can reduce the cost of roof structures, it is difficult to say just how much geologically suitable, preferably south-facing, hillside land exists for development. In either kind of construction, the ability to gain even lighting distribution and cross-ventilation will be both critical and difficult. Considerable thought will have to be given to carefully designed use of light wells and ventilation shafts and perhaps the use of such devices as interior reflecting plastic light conduits to conduct light from the exterior to interiors in ways that both recreate and simulate the qualities of natural light.

Finally, it is not quite correct to suggest that underground construction will leave the existing natural attribute of the site undisturbed. Since on flat or slightly sloping ground, complete excavation will be required, the land must be cleared just as if an above-grade house was being built. While some of the natural growth may be saved and replaced, mature trees and the like could not be saved. To minimize water pressure on below grade walls, rock-fills and drain tile to conduct subsurface water around the structure will usually be required and will have to be carefully designed, depending upon existing water-table and soil conditions.

The fact is that underground construction requiring the excavation of large amounts of earth may have as much or more of an impact upon near- and sub-surface ecology as does a conventional house upon the surface characteristics of the land.

FINDINGS

The Ecology House (see Figures 1 and 2) is an embryonic attempt to take advantage of the energy-conserving environment-preserving aspects of underground house construction, while minimizing as much as possible at least some of the potentially negative issues just discussed. It was designed by architect John E. Barnard, Jr. in order to create a house that "would have privacy, be dust- and pollen-free, and be easy and inexpensive to build, heat, and maintain." He also thought it should be comfortable, attractive, and "marketable." He decided on an underground house because he thought it was one solution to such problems as conservation of land in dense population areas, of energy, and of wood products. [1,2]

For a test of his ideas, architect Barnard designed the smallest unit that he thought would prove the usefulness of the house. In 1200 ft^2 of enclosed living space he has provided a bedroom, bath, kitchen, and living/dining room. All but the bathroom have floor-to-ceiling sliding glass panels opening out onto the 300 ft^2 atrium. The house is entered by descending stairs into the atrium/garden/sun-bathing area. The latter could conceivably be covered with an air-supported plastic bubble during winter, keeping out rain and snow and making the space usable the entire year.

Walls are cast-in-place concrete, utilizing typical foundation wall forming and reinforcing techniques. They were covered with rigid styrofoam boards and the latter were hot-mopped with pitch prior to backfilling with earth. The floor slab is 3-1/2 inches of concrete poured directly on earth. The roof is constructed of standard precast 8-inch thick cored concrete planks. Rigid styrofoam planks are utilized above the precast concrete elements. They are waterproofed with three layers of asbestos felt mopped-in with hot pitch. (This is equivalent to a standard, high-grade built-up roofing method, without the normal final mineral cap and gravel sheet required to resist ultra-violet radiation and abrasion due to walking for repairs, etc.) A 12-18 inch earth cover overall

Precast concrete planks form roof of house

UNDERGROUND ECOLOGY HOUSE FIGURE 1

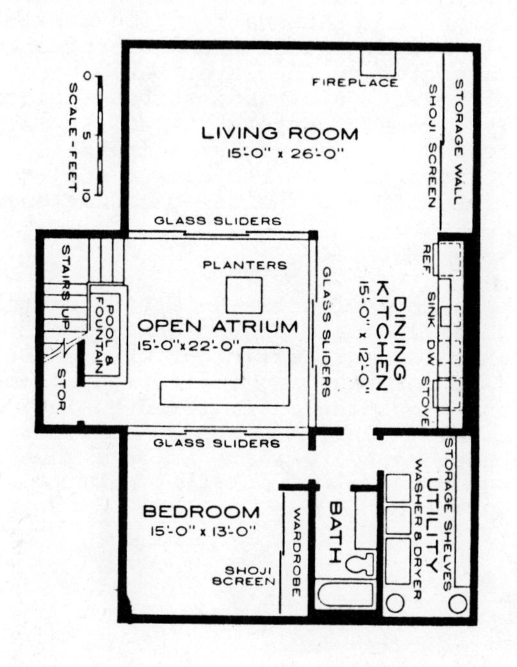

Walls are poured concrete. Styrofoam insulation, water-proofing, and earth (12 to 18 inches deep) cover all.

ECOLOGY HOUSE FLOOR PLAN FIGURE 2

brings the "roof" to the level of surrounding finish grade of the site.

Air is mechanically induced and exhausted from the house which is equipped with an electronic air filtering device and a dehumidifier which maintains a constant 50 percent humidity all year. The house is also air conditioned. Sewage goes into a sump, is pumped into a septic tank, and then goes to a leaching field.

A fire-rating bureau has given the Ecology House a fireproof dwelling rating which is one-half that of a typical brick residence.

Architect Malcolm B. Wells, long a believer in and practitioner of minimal disturbance or "gentle" architecture, has attempted to put into practice his beliefs by constructing his own office building below grade, in Cherry Hill, New Jersey. Unlike the preceding example of an underground residence, this building is not buried. Instead, it is built into one side of a larger excavation, onto which the enclosed spaces look (and draw light and air from) on the three remaining sides. In this case, other than for protection from prevailing winds, wind-driven snow and rain, etc. above-grade, the building appears to benefit less from direct contact with and enclosure by the surrounding earth. However, close examination of the building's details (Figures 3, 4, and 5) reveal a number of interesting points. First, the effect of the solar heat load is mitigated by 1 m (3 ft) of earth over the roof. This protection is extended over extremely broad and generous roof overhangs, which cantilever beyond the building glassline as much as 2.3 m (7 ft). Thus, as would be the case in *any* building using such expansive overhangs, he is able to open up the structure and yet protect it from the direct radiant energy of the sun during all seasons. This feature is, of course, assisted by the early cutoff of the sun offered by opposite sides of the retained sunken-court onto which the building opens. But he has also included full-height insulated shutters to be closed after dark (and on cold, winter days) to reduce night and winter heat losses.

The second point exemplified by this building is that efficient energy use, as important as it is, cannot be the *only* environmental concern to which this type of architectural design is to respond. Thus, along with concerns for immediate reduction of fossil fuel use, Wells' building also responds to concerns for preservation of natural plantings, undisturbed absorption of rainwater by the water table, and preservation of undisturbed (relatively) views of nature. Also, it is intended to eventually deal with or incorporate waste recycling, soil enrichment, and the use of solar energy.

RECOMMENDATIONS

Both of these underground structures indicate the potential, not only for more efficient energy use, but also for reducing environmental impact and providing an improved quality of life. Architects should consider undergrounding where economically viable and when client interest permits this approach.

References

1. "Saving by Going Underground," *AIA Journal* (February 1974).

2. Smay, V. Elaine, "Underground Living in This Ecology House Saves Energy, Cuts Building Costs, Preserves the Environment," *Popular Science* (June 1974).

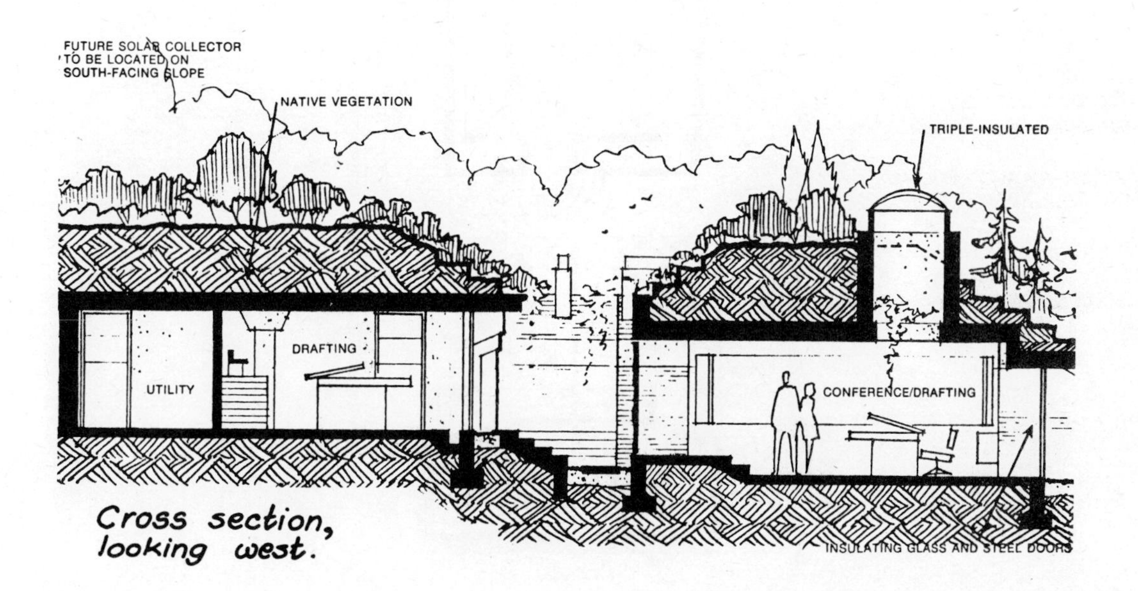

UNDERGROUND OFFICE, SECTION FIGURE 3

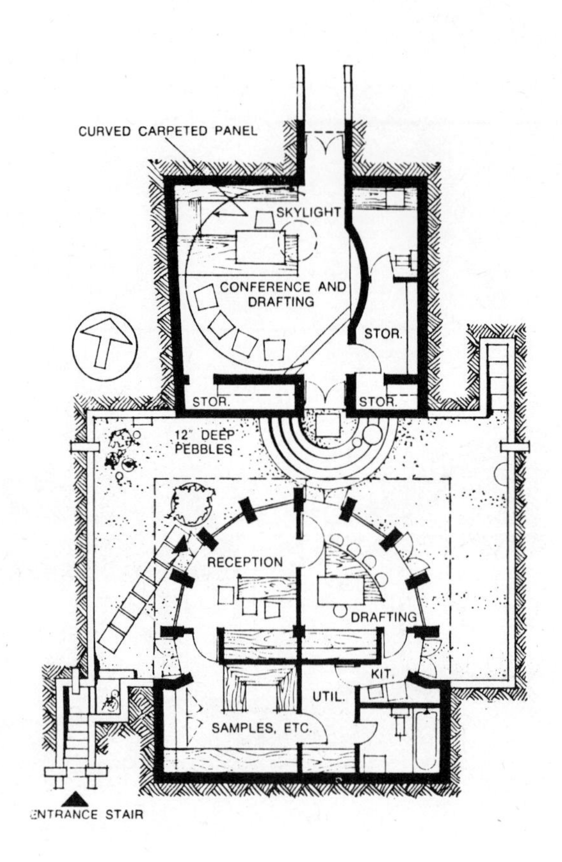

PLAN VIEW, UNDERGROUND OFFICE FIGURE 4

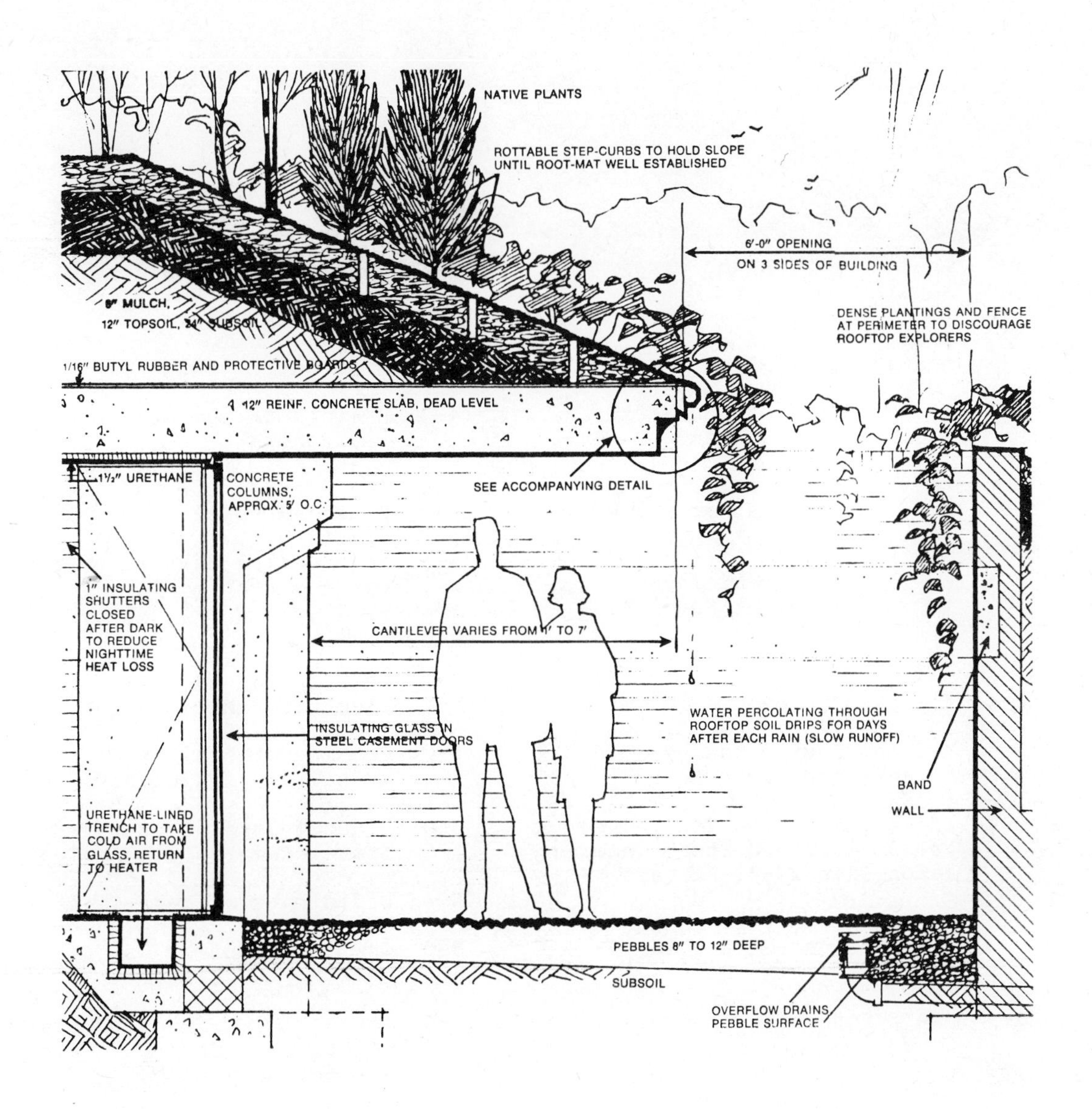

DETAILS, UNDERGROUND OFFICE FIGURE 5

LIST OF CASE STUDIES

8-2 Split Flow Modification of Dual Duct and Multizone Systems

8-3 The Impact of ASHRAE Standard 90-75

8-4 Energy Savings in Fan Operations

8-5 Improving Efficiency of Package Reciprocating Chillers

8-6 Prototype Commercial HVAC Systems

8-7 Automatic Cleaning of Condenser Tubes in Refrigeration Equipment

8-8 Conversion of Constant Volume Air Distribution Systems to Variable Volume

8-9 Reconditioning of Air Economizer Systems

Chapter 9

9-1 Use Appropriate Outside Lighting Levels

9-2 Comparative Economics of Lighting Systems

9-3 Efficient Industrial Lighting With High Pressure Sodium Lamps

9-4 Relamping Opportunities

9-5 Maximum Use of Natural Light for an Industrial Facility

Chapter 10

10-1 Improving Furnace Insulation

10-2 Heat Recuperation and Furnace Efficiency in Industrial Operations

10-3 Process Waste Heat Recovery for Building Heating

10-4 Heat Recovery from a Restaurant Saves Kilowatt Hours

Chapter 11

11-1 Peak Demand Reduction for Electric Motors

11-2 Mechanical Energy Transfer Efficiency Improvements

11-3 Seal Losses in Rotating Equipment

11-4 Effective Lubrication Programs Save Electric Energy

11-5 Efficient Motor Design and Retrofit

Chapter 12

12-1 Electricity Storage as an Energy Management Tool

12-2 Efficiency Improvements Possible with Fuel Cells

Chapter 13

13-1 The "Cliff Palace" Pre-Columbian Cliff Dwelling as an Example of Designing for Efficient Energy Use

13-2 The Energetic, Economic, Environmental and Other Costs of Urban Sprawl

13-3 Separation of Municipal Solid Wastes into Recoverable Metals and Supplementary Power Plant Boiler Fuel

13-4 Incineration of Municipal Refuse with Thermal Energy Recovery as a By-Product

13-5 Common Corridors for Municipal Utility Service Lines

13-6 Undergrounding of Special-Purpose Physical Facilities Requiring Modulation of Temperature Extremes

13-7 Undergrounding of Special-Purpose Physical Facilities Requiring a Vibration-Free Environment

13-8 Below Grade Construction to Save Energy

APPENDICES

ENERGY CONSERVATION POLICY REQUIRES A DELICATE BALANCE BETWEEN WHAT IS REALLY DESIRED AND WHAT IS REALLY POSSIBLE.

--René Malès, 1977

APPENDIX A

CONVERSION FACTORS

BY C.B. SMITH*

CONTENTS

A.1 INTRODUCTION

This book has been prepared using the International System of Units (SI), which has been or is being adopted by all nations of the world.[1,2,3] In SI practice, the approved units for energy and power are the joule and the watt:

Energy, heat, work: Joule (J)= 1 meter · newton

Power: Watt (W) = 1 joule/ second

*NOTE: The megajoule is preferred usage rather than the kilowatt-hour

When SI units are used for calculations, they are frequently preceded by a prefix which is a multiple of ten. Then the same basic unit can be used to measure a very large or very small quantity. Thus, a millimeter is slightly more than one-sixteenth of an inch, while a kilometer is slightly more than one-half a mile. The prefixes, symbols, and multipliers are:

Recommended			(Alternate Prefix--Not Recommended)
tera	(T)	10^{12}	(million million or trillion)
giga	(G)	10^{9}	(billion)
mega	(M)	10^{6}	(million)
kilo	(k)	10^{3}	(thousand)
hecto	(h)	10^{2}	(hundred)
deca	(da)	10^{1}	(ten)
deci	(d)	10^{-1}	(one-tenth)
centi	(c)	10^{-2}	(one-hundredth)
milli	(m)	10^{-3}	(one-thousandth)
micro	(μ)	10^{-6}	(one-millionth)
nano	(n)	10^{-9}	(one-billionth)
pico	(p)	10^{-12}	(one-trillionth)

The alternate prefixes are not recommended because of ambiguity in their meanings as used in different countries.

For readers who are not familiar with the International System (SI) of units, it has several important advantages: (1) it is universal, with a single unit for each quantity (e.g., m for length); (2) it uses decimal arithmetic, facilitating changes, calculations, and conversions; and (3) it is coherent, meaning that when two units are multiplied or divided the product or quotient has the units of the resultant quantity (e.g., m and ha are coherent while ft and acre are not).

The common units of the SI system are the kilogram (mass), the meter (length), the second (time), the newton (force), the watt (power), and the joule (energy). To understand what these units signify, some approximate equivalents are given:

Kilogram (kg)	equals about 2 pounds (mass)
Meter (m)	equals about 3 feet or 1 yd

*Principal, Applied Nucleonics Company, Inc.

Second (s)	equals one second
Newton (N)	equals about one-fourth pound (force)
Watt (W)	equals the power required to operate a table radio or an electric clock
Joule (J)	equals the energy used by a table radio or electric clock while operating for one second or the energy required to lift a one-quarter pound weight three feet (1 N-m)

A.2 CONVERSION FACTORS

Conversion factors are shown in three tables. Table A-1 has conversion factors for the energy and power units commonly encountered. Table A-2 shows equivalencies for two convenient units, the gigajoule (GJ) and kilowatt (kW). Table A-3 lists miscellaneous factors. Finally, Table A-4 lists other useful SI conversion factors.

A.3 ELECTRICAL ENERGY CONVERSION AND OTHER CONFUSING MATTERS

Perhaps the greatest advantage of SI units is to permit a systematic and consistent treatment of energy use. It is not uncommon to read reports dealing with energy use in which:

- diverse units such as kcal, kWh, and Btu are intermingled;
- conversions to one or another unit take place, resulting in confusion e.g., expressing heat [Btu's] in kWh and then combining with electricity in kWh); and,
- the efficiency of converting fuel to electricity is ignored.

The subject of energy units was discussed in an amusing letter by Barrow who pointed out that the practice of using different types of energy units often leads to unnecessary confusion while consistent use of the appropriate units gives insight into the relationship involved.[4]

For example, Barrow points out that heat rates for thermal power plants are expressed as so many Btu per kilowatt-hour. If we assume a typical value of 10,200 Btu per kilowatt-hour, which is 1.08×10^7 joules per 3.6×10^6 joules electric, we finally discover that it is 3 joules per joule electric. Then it is clear that the heat rate is the reciprocal of the thermal efficiency, which is 3^{-1} or 33.3 percent.

We have not entirely resolved these difficulties in this book, but we have adopted two conventions which we hope minimize the problem and provide consistent treatment:

- All electrical energy use is measured in joules electric (Je) or kilowatt-hours (kWh) and is so identified in the text. Electrical power is denoted as watts electric (We).
- Where total energy use is calculated or electricity is compared with direct fuel use, energy is expressed in joules (J) and the conversion of fuel to electricity is arbitrarily assigned a value of 33 percent. Thus, Je is multiplied by three to get J. These conversions are footnoted for clarity.

Relative heat content (Joules or Btu's) of an energy source can be assessed differently depending on its capability to do a specific job. This is a difficulty when trying to establish some form of common denominator for comparing different fuels--particularly electricity, which is processed fuel--to the raw input fuels.

For national energy comparisons the equivalent fuel input to generate one kWh, or 10.8 MJ (10,200 Btu), has been used. This leads to the 3J per Je rule described above, and is generally a useful measure when the total energy input into the economy is of concern. However, even here it is not strictly accurate; for example, should

ENERGY, HEAT, WORK

Multiply	By	To obtain
Btu (mean)	1.056×10^3	Joule
Calorie	4.190	Joule
Kcal	4.190×10^3	Joule
Ft·lbf	1.356	Joule
N·m	1.000	Joule
Kgf·m (or Kilopond·m)	9.807	Joule
Erg (1 dyne.cm)	1.000×10^{-7}	Joule
Electron volt (ev)	1.602×10^{-19}	Joule
Special case:		
kilowatt·hour electric	3.6×10^6	Joule electric
Joule electric	3.0*	Joule

POWER

Multiply	By	To obtain
Btu/s	1.056×10^3	Watt
Btu/min	17.60	Watt
Btu/hr	0.2933	Watt
Calorie/s	4.184	Watt
Ft·lbf/s	1.356	Watt
Ft·lbf/min	2.260×10^{-2}	Watt
Horsepower (550 ft.lbf/s)	7.46×10^2	Watt
Kgf·m/s	9.807	Watt
Erg/s	1.10^{-7}	Watt

*Typical value. Actual value varies widely. This value has been used throughout this book.

CONVERSION FACTORS FOR ENERGY AND POWER

TABLE A-1

ENERGY (Suggested unit: *The Gigajoule* = 10^9 J)

Multiply	By	To obtain
MBtu ("million" Btu)	1.056	Gigajoule
Gcal ("million"Kcal)	4.190	Gigajoule
MWeh (thousand kWeh)	3.600	Gigajoule electric

The following are approximate values for comparative studies (the exact value is given in parentheses):

1 GJe is roughly equal to 300 kWh (277.8)

1 GJ is roughly (within ± 35%) equal to any of the following:

$$\frac{100\ kWh}{(92.6)} \text{ or } \frac{30\ kg\ of\ coal}{(32.7)} \text{ or } \frac{30\ \ell\ of\ oil}{(25.7)} \text{ or } \frac{30\ m^3\ of\ gas}{(28.3} \text{ or }$$

$$\frac{10\ mgm\ of\ \mu\text{-}235}{(13)}$$

In American/British units,

1 GJ is roughly equal to any of the following:

$$\frac{100\ kWh}{(92.6)} \text{ or } \frac{100\ lbm\ of\ coal}{(72)} \text{ or } \frac{10\ gal\ of\ oil}{(6.8)} \text{ or } \frac{1\ kft^3\ of\ gas}{(0.92)} \text{ or }$$

$$\frac{1/5\ grain^*\ of\ U\text{-}235}{(exact)}$$

*7000 grain = 1 lbm

POWER (Suggested unit: *The Kilowatt* = 10^3 W)

Multiply	By	To obtain
Btu/s	1.056	Kilowatt
Kcal/s	4.184	Kilowatt
Kft·lbf/s	1.356	Kilowatt
hp	0.746	Kilowatt

1 kW is roughly equal to any of the following:

$$\frac{1\ Btu/s}{(1.056)} \text{ or } \frac{1\ Kft\cdot lbf/s}{(1.356)} \text{ or } \frac{1\ hp}{(0.746)}$$

ENERGY AND POWER EQUIVALENCIES

TABLE A-2

- 1 therm = 10^5 Btu = 0.1056 GJ
- 1 ton of refrigeration = 12 kBtu/hr ∿ 3.5 kW cooling
  (to produce 1 ton requires ∿ 1.0 kW of air conditioning)
- 1 degree day = # of days x 65°F - mean temperature per day
- 1 ton hard coal equivalent = 26 x 10^6 Btu ≃ 27 GJ
- 1 barrel (42 gal) of crude oil = 5.8 x 10^6 Btu ≃ 6 GJ
- One thousand cubic feet of natural gas = 1 x 10^6 Btu ≃ 1 GJ
- 1 lbm of steam ≃ 1000 Btu ≃ 1 MJ

MISCELLANEOUS ENERGY CONVERSION FACTORS

TABLE A-3

Multiply	By	To obtain
Length:		
Inch	0.0254	meter
Foot	0.3048	meter
Yard	0.9144	meter
Mile	1.609×10^3	meter
Area:		
in^2	6.452	cm^2
ft^2	0.0929	m^2
acre	0.4047	ha
m^2	1.0×10^{-4}	ha
ha	$1.0 \times 10^{+4}$	m^2
Volume:		
ft^3	0.02832	m^3
gal (US liquid)	3.785	ℓ
gal (US liquid)	3.785×10^{-3}	m^3
gal (UK liquid)	4.546	ℓ
gal (UK liquid	4.546×10^{-3}	m^3
barrel (42 US gal)	0.159	m^3
liter	1.0×10^{-3}	m^3
m^3	1.0×10^3	ℓ
Mass and density:		
ounce (avoirdupois)	28.35	gm
pound (avoirdupois)	0.4536	kg
ton (short 2000 lb)	0.9072	mt
ton (long 2240 lb)	1.016	mt
kg	1.0×10^{-3}	mt
mt	1.0×10^3	kg
pound/ft^3	16.02	kg/m^3
pound/in^3	27.68	g/cm^3
g/cm^3	1.000	mt/m^3
Pressure:		
Psi	6.895×10^3	N/m^2
N/m^2	1.000	Pa*
Energy Values:		
Btu/lbm	2.323×10^{-3}	MJ/kg
Btu/ton	1.162×10^{-6}	MJ/kg
Kcal/kg or cal/g	4.19×10^{-3}	MJ/kg
Btu/gal (US)	2.79×10^{-4}	GJ/m^3
Btu/ft^3	3.72×10^{-5}	GJ/m^3
kWeh/ton	3.96×10^{-3}	MJe/kg

*Pa = pascal

OTHER SI CONVERSION FACTORS

TABLE A-4

hydroelectric capacity be treated in this fashion?* It is also questionable if uranium, which has no fuel use except the production of process heat or electricity, is best viewed from this perspective. International comparisons of fuel use are particularly complicated by the amount of hydroelectric energy used by each country. To some extent the price of a fuel will reflect its scarcity and utility, except that fuel pricing has not been determined entirely in a free market.

For energy comparisons at a user's premise, the use of the input fuel requirement at the beginning of the entire fuel process is of uncertain value. Although this comparison gives the consumer a signal as to which energy form requires the least overall Joules or Btu's, it gives no indication of the relative scarcity or preciousness of fuel, and could be said to be analogous to a comparison of diamonds and steel for use as drill bits on the basis of least weight. Although price may reflect preciousness in this case, it is not a proper yardstick for assessing the scarcity of energy. Another alternative is to merely consider the metered energy input into the consumer's premise, but this fails to recognize the higher cost associated with the higher value of electricity.

We have tried to overcome this dilemma by showing both energy input figures, when appropriate, and adding energy cost comparisons when trade-offs between energy forms are the question. This is not a fully adequate solution but the best that can be done given the energy market imperfections and the fact that all Joules (or Btu's) are not created equal.

---

*One argument is that this is "value" in the sense that hydro can replace scarce or imported fossil fuels.

* * *

REFERENCES

1. American Society for Testing and Materials (ASTM), *Standard Metric Practice Guide E380-72*, (Philadelphia, Pennsylvania: 1972).

   Also see American National Standards Institute (ANSI), *American National Standard Z210.1*, (New York).

2. *ISO International Standard 1000* (available from ANSI).

3. US Department of Commerce, National Bureau of Standards (NBS), *The International System of Units (SI)*, (Washington, D.C.).

4. Barrow, R.B., "Letters," *Science* 179(23 March 1973).

APPENDIX B

ENGINEERING DATA

BY C.B. SMITH*

CONTENTS

B.1 INTRODUCTION

The purpose of this appendix is to present engineering data useful for energy management, energy accounting, life cycle costing, and evaluation of indirect energy use.

These values should be regarded as "typical" since there is a wide variation in the energy content of fuels, materials, and processes, depending on the specific application. However, it is believed that these numbers are representative and are useful for comparative studies and order-of-magnitude estimates.

B.2 ENERGY CONTENT OF FUELS

Energy content of fuels is shown in Tables B-1, B-2, and B-3. Table B-1 is an abbreviated table listing energy content of major fuels. Table B-2 shows ranges for a wide variety of fuels compiled from several sources. Table B-3 shows equivalencies between coal and other fuels.

B.3 ENERGY CONTENT OF MANUFACTURED MATERIALS

These tables show typical energy contents of various types of manufactured products including:

- Table B-4 Energy in Mining Operations
- Table B-5 Energy in Chemical Processing
- Table B-6 Energy in Raw Material Production
- Table B-7 Energy in Manufactured Products
- Table B-8 References for Tables B-4 to B-7

B.4 ENERGY INVESTMENT IN POWER PLANTS

Comparative studies have been made of nuclear and coal-fired power plants to estimate the energy which must be invested to build the plant. These studies indicate that for all three plants, the energy input is (rounded numbers);

	Energy as % of Energy Produced During 30-Year Plant Life
Energy to Build Coal or Nuclear Plant	2%
Energy Used By Fuel Cycle	5%
Total	7%

From the numbers, it takes the first two years operation to "repay" the energy invested in the plant. The next

*Principal, Applied Nucleonics Company, Inc.

28 years represent a net energy gain. Table B-9 shows the construction energy calculation for one of the nuclear plants analyzed.

B.5 ENERGY CONTENT OF STOCKPILED MATERIALS

The energy required to mine, manufacture, and machine metals represents a major part of primary metals electricity use. The energy content of materials is an important concept not generally recognized. For this reason, the wastage of primary metals (scrap, corrosion, etc.) results in a significant energy loss. This can be seen in Table B-10 which shows the Strategic Materials stockpile maintained by the US. As indicated in the table, the "energy equivalence" of these stockpiled materials is on the order of 0.3 percent of the US 1973 energy use--or 46×10^6 barrels of oil. Thus, it is an interesting point to note that materials stockpiles represent an energy reserve as well.

	MJ/kg
Hydrogen*	120
Gasoline	48
Natural gas*	47
Kerosene	46
Crude oil, fuel oil	44
Manufactured gas*	36
Coal, charcoal	30
Alcohol	27
Lignite, coke	25
Peat	15
Wood	13

*At standard temperature and pressure

TYPICAL SPECIFIC ENERGY CONTENT OF FUELS **TABLE B-1**

	Btu per gal, lb, ft^3 or ton	MJ/kg	GJ/m^3 or MJ/ℓ
80% Efficiency			
Propane	91,500 Btu/gal 21,500 Btu/lb	50	26
Butane	94,670 Btu/gal 19,520 Btu/lb	45	26
LPG	90,000-105,000 Btu/gal (average 21,500 Btu/lb)	50	25-29
Natural Gas	960-1550 Btu/ft^3 (average 1000 Btu/ft^3)	--	0.036-0.058
Manufactured Gas	460-650 Btu/ft^3 (average 565 Btu/ft^3)	--	0.017-0.024
Methane (from organic digester	500-700 Btu/ft^3	--	0.019-0.026
70-80% Efficiency			
Oil No. 1	136,000 Btu/gal 19,800 Btu/lb	46	38
Oil No. 2	138,500 Btu/gal 19,400 Btu/lb	45	39
Diesel	130,300 Btu/gal 18,400 Btu/lb	43	36
Gasoline	127,600 Btu/gal 20,750 Btu/lb	48	36
Kerosene	135,000 Btu/gal 19,810 Btu/lb	46	38
65-80% Efficiency			
Anthracite Coal	12,000-13,000 Btu/lb 24-26 million Btu/ton (average 100 lb/ft^3)	28-30	45-48
55-65% Efficiency			
Bituminous Coal (soft)	10,000-15,000 Btu/lb 20-30 million Btu/ton (average 85 lb/ft^3)	23-35	32-47
50-60% Efficiency			
Wood (12% moisture)	8,000-10,000 Btu/lb	19-23	8-10
Wood (on an air dry basis 12% moisture)	8050 Btu/lb (average 28 lb/ft^3)	19	8.4

Type	lb/cord	lb/ft^3	Btu/cord*	Btu/ft^3
Douglas Fir	2480	31	20,000,000	250,000
Aspen	2080	26	15,800,000	210,000
Cottonwood or poplar	2080	26	16,800,000	210,000
Elm	2880	36	23,200,000	290,000
Grand Fir	2160	27	17,400,000	217,000
Western Hemlock	2320	29	18,800,000	235,000
Western Larch	2880	36	23,200,000	290,000
Black Locust	3840	48	31,000,000	387,500
Lodgepole Pine	2320	29	18,800,000	235,000
Ponderosa Pine	2240	28	18,000,000	225,000
White Pine	2160	27	17,400,000	217,000
Western Red Cedar	1840	23	14,800,000	185,000
Engleman Spruce	1840	23	14,800,000	185,000

*Note: 1 cord - 128 ft^3 of stacked wood 8 ft x 4 ft x 4 ft; or 80 ft^3 of solid wood.

ENERGY CONTENT OF FUELS TABLE B-2

The average heat content of bituminous coal is 26,200,000 Btu/ton Based on this number, the following comparisons can be made:

1 Ton of Bituminous Coal Equals:		Btu Per Unit:
Pennsylvania Anthracite	1.031 tons	25,400,000 per ton
Crude Petroleum	4.517 barrels	5,800,000 per barrel
Natural Gas Liquids	6.532 barrels	4,011,000 per barrel
Natural Gas	25,372 ft^3 (a)	1,075 per ft^3 (a)
Natural Gas	25,388 ft^3 (b)	1,032 per ft^3 (b)
Coke	1.056 tons (c)	24,800,000 per ton (c)
Coke Breeze	1.248 tons (d)	21,000,000 per ton (d)
Blast Furnace Gas	262,000 ft^3	100 per ft^3
Coke Oven Gas	47,636 ft^3	550 per ft^3
Manufactured Gas	48,519 ft^3 (e)	540 per ft^3 (e)
Mixed Gas	28,887 ft^3 (e)	907 per ft^3 (e)
Tar and Pitch	3.899 barrels	6,720,000 per barrel
Kerosene	4.621 barrels	5,670,000 per barrel
Gasoline (motor fuel)	5.021 barrels	5,218,080 per barrel
Gasoline (aviation)	5.190 barrels	5,048,400 per barrel
Distillate Fuel Oil	4.498 barrels (f)	5,825,400 per barrel (f)
Residual Fuel Oil	4.167 barrels (g)	6,287,400 per barrel (g)
Acid Sludge	5.822 barrels	4,500,000 per barrel
Refinery (still) Gas	17,467 ft^3	1,500 per ft^3
Petroleum Coke	0.870 ton	30,120,000 per ton
Wood	1.250 cords	20,960,000 per cord

Notes:
(a) At wellhead - containing liquids.
(b) As delivered to consumer - most liquids removed.
(c) Assuming 4 percent moisture as consumed.
(d) Assuming 12 percent moisture as consumed.
(e) Sold by Utilities. Mixed gas includes a portion of natural gas, and this conversion factor applied to 1974 only.
(f) Weighted average of grades 1 to 4.
(g) Weighted average of grades 5 and 6.

Reference: Based on "Bituminous Coal Facts," National Coal Association, Washington D.C. (1972)

COAL EQUIVALENT AMOUNTS OF FUELS **TABLE B-3**

(includes mining, beneficiating, smelting and other)

Sector	MBtu/ton	MJ/kg
Aluminum: (Largely estimated)[1]		
Bauxite (Mining and processing)	2.39	2.78
Alumina	19.1	22.2
Aluminum	102	119
Blast furnace[2]	18.4	21.4
Cement	6.0	6.97
Clays	3.1	3.60
Coal (bituminous and lignite)	0.17	0.197
Coke	3.65	4.24
Copper:		
Ore and concentrate	27.2	31.6
Smelters	34.2	39.7
Ferroalloys	43.7	50.8
Gypsum	2.6	3.02
Iron ore	1.1	1.28
Lime	7.5	8.72
Phosphate rock	0.65	0.76
Salt:		
Solar, rock, and brine	0.048	0.056
Vacuum pan	4.2	4.88
Sand and gravel	0.055	0.064
Stone (crushed and broken)	0.048	0.056
Sulfur (French)	6.63	7.70
Zinc:		
Ore and concentrate	68.0	79.0
Smelters	680	790

(1) Includes estimates of fuels used by the aluminum industry to generate electricity as well as purchased electricity. Electricity (10% of total energy) does not include fuel conversion efficiency.

(2) Final data expected to be slightly lower when verified to eliminate duplicate reporting.

Reference (a)

ENERGY IN MINING OPERATIONS (1973)

TABLE B-4

Reference	Material Description	MBtu/ton#	MJ/kg#
	Petroleum Refinery		
(b)	Vacuum Distillation	0.4	0.5
(b)	Crude Distillation	0.7	0.8
(b)	Class A Refinery §	1.0	1.2
(b)	Hydro-cracking	1.3	1.5
(b)	Class B Refinery	1.5	1.7
(b)	Fluid Catalytic Cracking	1.7	2.0
(b)	Catalytic Reforming	2.2	2.6
(b)	Class C Refinery	3.0	3.5
(b)	Class D Refinery	4.5	5.2
(b)	Class E Refinery	5.0	5.8
(e)	Petroleum Refining (Overall)	4.4	5.1
	Chemicals		
(b)	Industrial Gases	1.2	1.4
(b)	Oxygen, Nitrogen	2.0	2.3
(b)	Acetylene	2.3	2.7
(b)	Formaldehyde	3.3	3.8
(b)	Normal Paraffins	3.5	4.1
(b)	Isopropanol	4.0	4.6
(b)	Benzene, Zylene (from Toluene)	4.4	5.1
(b)	Methanol	5.0	5.8
(b)	Wood Chemicals	6.5	7.6
(b)	Polyethylene	7.6	8.8
(b)	Styrene	8.5	9.9
(b)	Ammonia	9.0	10.5
(b)	Ethanolamines	11.0	12.8
(b)	Acetone	14.4	16.7
(b)	Ethylene	20.0	23.2
(b)	Agricultural Chemicals	20.0	23.2
(b)	Chlorine	21.0	24.4
(b)	Petrochemicals (Average for 1st Derivative)	27.0	31.4
	Inorganic Chemicals		
(k)	Phosphorus	200	233
(k)	Soda Ash	10.4	12.0
(k)	Chlorine	23.4	27.2
(k)	Ammonia	16.4	19.0
(k)	Nitric Acid	2.38	2.77
(k)	Ammonium Nitrate (prilled)	6.20	7.21
	Organic Chemicals		
(k)	Acetylene	59.6	69.3
(k)	Vinyl Chloride	43.2	50.2
(k)	Methanol	22.0	25.6
(k)	Benzene/xylanes	22.6	26.3
(k)	Para-xylene	48.1	55.9
(k)	Ethylene	14.7	17.1
(k)	Styrene	59.0	68.6
(k)	Butadiene	38.5	44.8
	Polyvinyl End Products		
(k)	Polyvinyl Chloride	59.9	69.6
(k)	Polyester Fibre (incl. DMT)	140.7	163.5
(k)	Polythene	37.9	44.0
(k)	Synthetic Rubber (SBR)	60.9	70.8
	Water Supply		
(n)	Potable Water, Typical Municipal System	1.7×10^{-3}	2×10^{-3}
(k)	Process water, Chemical Plant	7.2×10^{-3}	8×10^{-3}
(n)	Sewage Treatment, Typical Municipal Sewage Plant	0.9×10^{-2}	1×10^{-2}
(n)	Sewage Treatment, Home Plant (285 ℓ/day-75 gal/day)	1.5	1.7

† Note: In Tables B-5 through B-7, electricity is included at 10.8×10^6 GJ/kWhe, i.e., with a conversion efficiency of 33% taken into account. If this *has not* been done, or if it is not known, the data are marked with an asterisk.

§ API Refinery Classification: A to E, increasing in complexity of crude processing.

\# Includes energy used in providing basic raw materials, electric power (at 30% conversion efficiency), water, and all necessary production facilities.

ENERGY IN CHEMICAL PROCESSING **TABLE B-5**

Reference	Material Description	MBtu/ton	MJ/kg
	Metals (1)		
(b)	Steel, Primary	43.4*	50.4
(b)	Steel, Secondary	6.75*	7.8
(e)	Iron and Steel, Overall	26.5	30.7
(b)	Lead, Secondary	10.5*	12.2
(b)	Alumina	17.9*	20.8
(b)	Aluminum-Alumina (Bayer-Hall Process)	169*	196
(b)	Aluminum, Secondary	18.0*	20.9
(e)	Aluminum (Primary and Scrap)	155	180
(b)	Zinc, Primary (Electrolytic)	40.3*	46.8
(b)	Zinc, Primary (Horizontal Retort Furnace)	78.9*	91.7
(b)	Tin, Primary	90.0*	105
(b)	Magnesium, Primary	351*	408
(c)	Silver	87	101
(c)	Nickel	6	7
(c)	Manganese	46	53
(c)	Lead	44	51
(c)	Copper	75	87
(c)	Copper	25.8	30
	Cement		
(e)	Cement	7.9	9.2
	Metal (Ore or Main Source) (2)		
(d)	Magnesium (Sea Water)	310	360
(d)	Aluminum (Bauxite)	176-203	203
(d)	Aluminum (Clays)	225	236
(d)	Aluminum (Anorthosite)	246	286
(d)	Iron (High Grade Hematite)	10.9	12.6
(d)	Iron (Magnetic Taconites)	12.2	14.1
(d)	Iron (Iron Laterites)	17.7	20.5
(d)	Copper (1% Sulfide Ore)	46.2	53.6
(d)	Copper (0.3% Sulfide Ore)	84.5	98.0
(d)	Copper (98% Cu Scrap Recycle)	2.01	2.3
(d)	Copper (Impure Cu Scrap Recycle)	5.32	6.2
(d)	Titanium (High Grade Rutile)	431	500
(d)	Titanium (Ilmenite-Bearing Mineral [sands, rocks])	512-536	594
(d)	Titanium (High Grade Ti Soils)	706	622
(d)	Titanium (Ti Scrap Recycle)	133	154

(1) See also reference (m) for an extensive discussion of energy use in basic metals industries.
(2) See also reference (g).

ENERGY IN RAW MATERIAL PRODUCTION **TABLE B-6**

Reference	Material Description	MBtu/ton	MJ/kg
	Paper Products		
(b)	Wood Siding, Insulation (mechanical pulping)	3.0*	3.5
(b)	Paper, Recycled	10.0*	11.6
(b)	Newsprint from Pulp	11.3*	13.1
(b)	Pulp	12.3*	14.3
(b)	Paper, Including Pulping Process	44.0*	51.1
(b)	Bleached Paper, Including Pulping Process	56.0*	65.1
(b)	Paper and Paperboard (Overall)	24.5	28.5
	Glass		
(f)	Finished Plate	24.6	28.6
(f)	Container	27.3	31.7
(i)	1 pt. bottle (returnable)	22.3	25.9
(i)	1 pt. bottle (throw away)	19.2	22.3
	Plastics		
(f)	Average	9.9	11.5
(f)	Packaging and Containers	11.9	13.8
	Foods		
(f)	Average (does not include food value)	5.5	6.4
	Metal Products		
(i)	Steel Beverage Cans	69	80
(f)	Aluminum Cans and Packaging	283	329
(f)	Steel Cans and Packaging	6	7
(j)	Automobiles	19	22
(f)	Industrial machinery (average)	116	135
(f)	Agricultural Implements	68	79

ENERGY IN MANUFACTURED PRODUCTS

TABLE B-7

(a) U.S. Department of the Interior, Bureau of Mines, *Mineral Industry Surveys*, "1973 Fuel and Electrical Energy Requirements of Selected Mineral Industries Activities," (Washington D.C.: 7 May 1975).

(b) Prengle, H. William Jr., Crump, Joseph R., Fang, C.S., Grupa, M., Henley, D., and Wooley, T., *Potential for Energy Conservation in Industrial Operations in Texas*, Report No. S/D-10, (Houston, Texas: University of Houston, Department of Chemical Engineering, Cullen College of Engineering, November 1974).

(c) Rombough, Charles T. and Koen, Billy V., "Total Energy Investment in Nuclear Power Plants," *Nuclear Technology*, 26(May 1975), pp. 5-11.

(d) Seidel, Marquis R., Plotkin, Steven E., and Reck, Robert O., *Energy Conservation Strategies*, Report No. EPA-R5-73-021, (Washington D.C.: U.S. Environmental Protection Agency, Office of Research and Monitoring, Implementation Research Division, July 1973).

(e) Gyftopoulos, Elias P., Lazaridis, Lazaros J. and Widmer, Thomas F., *Potential Fuel Effectiveness in Industry*, A Report to the Energy Policy Project of the Ford Foundation, (Cambridge, Massachusetts: Ballinger Publishing Company, 1974).

(f) Makhijani, A.B. and Lichtenberg, A.J., *An Assessment of Energy and Materials Utilization in the U.S.A*, Memorandum No. ERL-M310(Revised), (Berkeley, California: University of California, College of Engineering, Electronics Research Laboratory, 22 September 1971).

(g) Bravard, J.C. and Portal, Charles (Revised by P.H. Wadia and J.T. Day), "Energy Expenditures Associated with the Production and Recycle of Metals," Preliminary Report No. ORNL-MIT-132, (Cambridge, Massachusetts: Massachusetts Institute of Technology, School of Chemical Engineering Practice, Oak Ridge Station, 26 May 1971).

(h) Tihansky, Dennis P., *Patterns of Energy Demand in Steelmaking*, Working Note No. WN-7437-NSF, (Santa Monica, California: The Rand Corporation, May 1971).

(i) Hannon, Bruce M., "Bottles-Cans-Energy," *Environment*, 14(March 1972), pp. 11-21.

(j) McGowan, Jon G. and Kirchhoff, Robert H., "How Much Energy is Needed to Product an Automobile?" *Automotive Engineering*, (July 1972), pp. 39-40.

(k) Smith, Harold, "The Cumulative Energy Requirements of Some Final Products of the Chemical Industry," *Proceedings, VII World Power Conference*, (Moscow, U.S.S.R., 1968).

(l) Drexel University, Franklin Institute Research Laboratories and Decision Sciences Corporation, *Proceedings of the Effective Energy Utilization Symposium*, (Philadelphia, Pennsylvania: 8-9 June 1972).

(m) Institute of Gas Technology, editor, *Efficient Use of Fuels in the Metallurgical Industries*, Symposium Papers, (Chicago, Illinois: 9-13 December 1974).

(n) Personal communication, C.B. Smith and various water authorities, June 1975.

REFERENCES FOR TABLES B-4-B-7 TABLE B-8

Industrial Classification	Cost (10^3 $)	Energy/Cost Ratio (10^4 Btu/$)	Total Energy (10^{10} Btu)
Stone and clay mining/quarry	2,526.5	9.4221	23.805
New construction, residential	150.0	6.0227	0.903
New construction, highways	242.0	9.8507	2.384
New construction, other	8,297.5	7.1266	59.133
Maintenance/repair construction	295.7	6.7117	1.985
Floor coverings	6.0	7.9340	0.048
Wood structure	5,617.3	4.7062	26.436
Wood products	169.0	5.2875	0.894
Metal office furniture	50.0	7.7441	0.387
Metal partitions and fixtures	8.5	8.3291	0.071
Furniture and fixtures	116.9	6.6302	0.775
Miscellaneous chemical products	922.0	28.5800	26.351
Paints and allied products	570.9	14.1320	8.068
Petroleum refining	25.0	19.6660	0.492
Glass containers	20.0	16.3060	0.326
Concrete block and brick	4,020.2	11.8910	47.804
Gaskets and insulation	1,386.5	7.9301	10.995
Nonmetallic mineral products	50.0	6.5012	0.325
Primary metal products	3,954.8	12.3330	48.775
Metal cans	815.0	13.6980	11.164
Plumbing fittings and brass	485.5	7.8266	3.800
Heating equipment (nonelectric)	3,212.5	7.3506	23.614
Fabricated structural steel	18,371.0	12.3390	226.680
Metal doors, sash, and trim	217.0	10.9910	2.385
Fabricated plate work	2,218.1	11.5620	25.645
Sheet metal work	492.0	11.4950	5.656
Architectural metal work	39.6	14.1830	0.562
Miscellaneous fabricated wire products	497.5	14.4650	7.196
Pipe, valves, and fittings	27,847.4	7.3742	205.352
Fabricated metal products	5,929.8	9.19930	54.550
Steam engines and turbines	47,307.0	8.4232	398.476
Internal combustion engine	940.0	6.0312	5.669
Construction machinery	6,780.0	7.3089	49.554
Elevators and moving stairways	75.0	5.9308	0.445
Conveyors and conveying equipment	120.6	6.4384	0.776
Hoists, cranes, and monorails	926.7	7.5412	6.988
Special dies and tools	57.1	5.3254	0.304
Special industry machinery	776.5	6.5080	5.054
Pumps and compressors	7,804.9	5.8254	45.467
Blowers and fans	3,043.3	6.3324	19.271
Power transmission equipment	926.7	6.6378	6.151
General industrial machinery	3,729.0	6.2497	23.305
Computers	1,684.7	2.7460	4.626
Refrigeration machinery	75.0	6.4015	0.480
Electric measuring instruments	40.0	3.8293	0.153
Transformers	653.7	8.1050	5.298
Switchgear and switchboards	1,507.6	4.8784	7.355
Motors and generators	884.5	6.5378	5.783
Industrial controls	4,713.0	3.8656	18.219
Electrical industrial apparatus	975.0	7.0497	6.873
Lighting fixtures	1,089.0	7.6642	8.346
Wiring devices	4,486.0	7.4282	33.323
Telephone and telegraph	85.0	4.2821	0.364
Storage batteries	30.0	7.5884	0.228
Electrical equipment	14.0	6.1728	0.086
Railroads and railroad cars	1,165.0	11.0610	12.886
Engineering and scientific instruments	2,480.0	4.1106	10.194
Miscellaneous manufacturers	10.0	6.7900	0.068
Water and sanitary services	1,365.0	11.6660	15.924
Miscellaneous business services	2,700.0	3.2067	8.658
Miscellaneous professional services	23,750.0	2.6554	63.066
Educational services	500.0	4.9807	2.490
Other nonenergy expenses	2,235.0	-	-
Totals	211,483.6		1,592.400

Reference C

ENERGY INVESTMENT IN POWER PLANT CONSTRUCTION

TABLE B-9

	Energy 10^6 GJ
Aluminum	96
Zinc	33
Magnesium	8.3
Titanium sponge	15
Lead	14
Ferromanganese	8.8
Copper	11.7
Ferrochromium	10.8
Beryllium-copper	9.3
Bauxite, Jamaica type	9.3
Manganese ore, metal grade	3.9
Tin	7.4
Bauxite, Surinam type	5.6
Ferrochrome	5.8
Silicon carbide	5.6
Tungsten, ores and concentrate	3.8
Chromite, metal grade	3.6
Fused alumina	3.3
Fluorspar, acid grade	2.7
Bauxite	2.5
Cobalt	2.2
Antimony	2.2
Lithium hydroxide	2.0
Mercury	1.9
Chrom-ferro-silicon	1.4
Chromite	1.3
Manganese ore - Metal (nonstockpile grade)	1.1
Berryllium metal	1.3
Totals	274

Reference: Metals Week, February 11, 1974

ENERGY EQUIVALENCE OF STOCKPILED MATERIALS

TABLE B-10

APPENDIX C

ENERGY EFFICIENCY--THE CONCEPT OF AVAILABLE WORK

The Role of the Second Law of Thermodynamics in Assessing the Efficiency of Energy Utilization

--American Physical Society[1]

CONTENTS

C.1 FIRST-LAW EFFICIENCY: THE NEED FOR A MORE GENERAL MEASURE OF EFFICIENCY

Consider the following examples which illustrate typical measures of energy utilization efficiency;

(1) A certain household furnace is described as being 60 percent efficient; this means that the ratio of the heat usefully delivered within the house to the heat of combustion* of the fuel burned is 0.6. This measure suggests that a 100 percent efficient furnace would be "perfect," which is incorrect. One could do better in various ways--for instance, by having the fuel power an engine to drive a heat pump (possibly with the intermediary of electricity) to provide more heat to the house than the fuel's heat of combustion.

(2) A certain air conditioner has a coefficient of performance (COP) of 2. This means that the ratio of heat extracted to the input electric work is 2. To assess this number as a figure of merit, one needs to compare it with a maximum COP. The maximum COP depends on temperature (actually on the inside-outside temperature ratio) and might be considerably greater than 2.

(3) A modern coal-fired power plant has an efficiency of 40 percent. This means that the ratio of its output electric energy to the input heat of combustion of coal is 0.40. As is well known, this efficiency is constrained by the Second Law of Thermodynamics and has an upper limit that is less than 1.

(4) A large motor is 90 percent efficient; the ratio of its output mechanical work to its input electric work is 0.9. Its ideal maximum efficiency is 1.†

*Heat of combustion = maximum heat provided to environment in constant-pressure combustion = decrease of enthalpy of fuel reacting ideally at constant temperature and pressure in standard air.

†In the context of concern for fuel conservation, it is often preferable to think of the motor as an extension of the thermal power plant that provides the electricity. The motor's "input" is then power-plant fuel and its efficiency is about 0.3, not 0.9.

The figure of merit applied in each of these examples--and in many more--is basically the same. We may call it the *first-law efficiency*. It is

η = energy transfer (of desired kind) achieved by a device or system/ energy input to the device or system (1)

(first-law efficiency)

When the maximum value of this ratio is greater than 1, it is usually called a coefficient of performance. When $\eta_{max}<1$, it is usually called an efficiency. Table C-1 shows numerators and denominators that define η for various classes of devices and also gives η_{max} and standard nomenclature. (Although η itself is defined without reference to the Second Law of Thermodynamics, its ideal maximum value--for all but the work-in, work-out devices--is limited by the Second Law. The temperatures that specify η_{max} are, of course, absolute temperatures.)

As a general figure of merit, the first-law efficiency has several drawbacks: (a) its maximum value depends on the system and on temperatures and may be greater than, less than, or equal to 1; (b) it does not adequately emphasize the central role of the *Second Law* in governing the possible efficiency of energy utilization; (c) it cannot readily be generalized to complex systems in which the desired output is some combination of work and heat; (d) it cannot be used realistically to compare one type of system with another in regard to the most energy efficient approach to achieve a given work output.

C.2 SECOND-LAW EFFICIENCY: THE CONCEPT OF AVAILABLE WORK

Various other special and general figures of merit have been proposed.* The one that is recommended for general adoption is a quantity that can be called the *second-law efficiency*.† It is a measure of performance relative to the optimal performance permitted by both the First and Second Laws of Thermodynamics. It is applicable equally to a simple device with a single output or a complex system with multiple outputs.

Second-law efficiency will first be defined for a device or system whose output is the useful transfer of work or heat (not both)--for instance a motor, refrigerator, heat pump, or power plant:

ε = Heat or work usefully transferred by a device or system/maximum possible heat or work usefully transferable for the same function using the same energy input as the given device or system (2)

(second-law efficiency for single-output system)

As is obvious from this definition, the maximum value of ε is 1 in all cases. The numerator in the defining ratio is the same as that in the first-law efficiency ratio (eq. 1). The new denominator represents a simple but powerful change because it brings the laws of thermodynamics directly into the definition of efficiency. The "maximum" in the denominator means the theoretical maximum permitted by the First and Second Laws. Note that this is a *task* maximum, not a *device* maximum. To maximize the heat delivered to a house by fuel, for instance, a furnace should be replaced by an ideal fuel cell and an ideal heat pump.

The second-law efficiency provides immediate insight into the quality of performance of any device relative to what it could ideally be.§ It shows how much room there is for improvement in principle. It measures the "waste" of fuel (some of it inevitable in practice, of course). For any specified task requiring heat or work, maximizing ε is equivalent to minimizing fuel consumption. In case no fuel consumption is involved--such as with hydroelectric plants, wind mills, geothermal sites or

*Recently, for example, D.O. Lee and W.H. McCulloch introduced the concept of "utility," defined as the ratio (work plus heat usefully transferred by a system)/(available useful work consumed by the system). Utility can have a maximum value greater than 1.[2]

†This quantity has previously been termed "effectiveness."[3]

§We acknowledge C.A. Berg's suggestions and contributions to this concept.[4,5] Also influential was a report by Keenan, Gyftopoulos and Hatsopoulos.[6]

TYPE OF DEVICE OR SYSTEM[1]	NUMERATOR IN RATIO DEFINING η	DENOMINATOR IN RATIO DEFINING η	η_{max}	STANDARD NOMENCLATURE
Motor (W/W)	Mechanical work output	Electric work input	1	Efficiency
Heat pump, electric (Q/W)	Heat Q_2 added to warm reservoir at T_2	Electric work input	$\frac{1}{1-(T_0/T_2)} > 1$	Coefficient of performance (COP)
Air conditioner or refrigerator, electric (Q/W)	Heat Q_3 removed from cool reservoir at T_3	Electric work input	$\frac{1}{(T_0/T_3)-1}$ (not restricted in value)	(COP)
Engine, power plant (W/Q)	Mechanical or electric work output	Heat Q_1 from hot reservoir at T_1	$1 - \frac{T_0}{T_1} < 1$	Efficiency (thermal efficiency)
Furnace[2] (Q/Q)	Heat Q_2 added to warm reservoir at T_2	Heat Q_1 from hot reservoir at T_1	$\frac{1-(T_0/T_1)}{1-(T_0/T_2)} > 1$	Efficiency or (COP)
Absorption refrigerator[3] (Q/Q)	Heat Q_3 removed from cool reservoir at T_3	Heat Q_1 from hot reservoir at T_1	$\frac{1-(T_0/T_1)}{(T_0/T_3)-1}$ (not restricted in value)	(COP)

NOTES:

(1) The symbols W and Q refer to work and heat, respectively.
(2) "Furnace" means any heat-powered device for heating.
(3) "Absorption refrigerator" means any heat-powered device for cooling.

FIRST-LAW EFFICIENCY **TABLE C-1**

solar collectors--maximizing ε minimizes capital investment in power-producing units. Inevitably, therefore, the maximization of ε becomes a matter of *policy* consideration. It is a technical goal to be placed alongside economic, environmental, and conservation goals.

Under certain circumstances, the first-law and second-law efficiencies can differ dramatically. For example, a furnace providing hot air at 43°C (110°F) to a house when the outside air temperature is 0°C (32°F) has a second-law efficiency ε = 0.082 if its first-law efficiency is η = 0.60.* For a closed, single-cycle power plant, on the other hand, the first-law and second-law efficiencies are nearly the same.

In order to understand and calculate second-law efficiencies, an important concept is B, *available work.*† Its definition in terms of thermodynamic quantities is given in Section C.4. In words, the definition is:

AVAILABLE WORK

B = maximum work that can be provided by a system (or by fuel) as it proceeds (by any path) to a specified final state in thermodynamic equilibrium with the atmosphere; interaction with the atmosphere is permitted but work done on the atmosphere is not counted.§

This quantity specifies only harnessable work, not work done on the atmosphere. The reference is to *work* (rather than heat) because work is the highest "quality" form of energy--equivalent to heat at an infinite temperature. Work is the best overall measure of capacity for doing *any* task. For a raised weight (to take an elementary example), the available work is simply mgh, the work that can be done by lowering the weight (ground level must then be specified as part of the "ambient" conditions). For heat Q_1 extracted from a hot reservoir at temperature T_1, the available work is

$$B = Q_1 (1 - [T_0/T_1])$$

For fuel with heat of combustion $|\Delta H|$, B turns out to be roughly equal to $|\Delta H|$. (The difference between a fuel's heat of combustion and its available work is discussed in Section C.5.) For a chemical cell, the available work is:

$$B \simeq -\Delta G_0$$

the change of Gibbs free energy in the reaction carried out at the ambient temperature and pressure.#

In terms of the available work concept, the definition of second-law efficiency can be restated much more simply and more generally.

Efficiency defined in terms of available work:

$$\varepsilon = B_{min}/B_{actual}\text{¶} \qquad (3)$$

(second-law efficiency in general)

This can be applied to anything from an electric motor to a nation's total energy economy. It states that the efficiency∫ is equal to the ratio of

---

*See Example 2, Section C.3, for the calculation.

†Also called "availability" in some texts and "exergy" in European and Soviet literature.

§In some cases, some base reservoir other than the atmosphere might be appropriate.

#The equation $B = \Delta G_0$ is not quite precise because of a small additional contribution to B from a term that measures the available work resulting from the diffusion into the atmosphere of the products of the chemical reaction. This is discussed in Section D.

¶For a device where output is heat or work (not both), it is easy to show the equivalence of Equations (2) and (3).

∫When no confusion is likely to result, "efficiency" or "efficiency ε" is used to mean "second-law efficiency."

the least available work that could have done the job to the actual available work that used to do the job.

Earlier in this book, the point was made that energy is *never* lost or *consumed* and is always *conserved*. It can be, however, inefficiently used or transferred. In contrast, available work (unlike energy) *is* consumed. The consumption of available work is related to total entropy change (see Section C.4).

Table C-2 displays available work provided by three kinds of sources and minimum available work needed by three kinds of end uses. To find the second-law efficiencies ε for single source-single output devices, it is only necessary to divide one of the end-use entries in Table C-2 by one of the source entries. This gives the set of nine efficiency formulas shown in Table C-3. For comparison, first-law efficiencies (or coefficients of performance) are also shown in this table.

To help emphasize the central role of temperature in efficiency considerations, Figures C-1 and C-2 show the minimum available work required to move a unit of heat from one temperature to another. Figure C-1 is explained in its caption. Note that over a wide summer-to-winter range, $B_{min}/Q \lesssim 0.15$, a very small value attributable to the small fractional changes of absolute temperature that are involved. The *actual* expenditure of available work in a typical furnace is $B_{actual}/Q \simeq 1.7$, more than ten times the minimum required. Note that the vertical scale in Figure C-1 is *not* efficiency. To find the second-law efficiency of any actual furnace, heat pump, or air conditioner, it would be necessary to divide the value B_{min}/Q shown in the figure by the quantity B_{actual}/Q for the device in question. The two different curves in Figure C-1 are calculated for different temperatures at which heat is transferred to or from a room. As the temperature at which the heat is transferred departs further from room temperature (higher or lower), the required expenditure of available work increases.

Figure C-2 shows the minimum available work required to provide heat at high temperature. In the range of temperature relevant to process steam production, $B_{min}/Q \simeq 1/3$. In current practice, $B_{actual}/Q > 1$. As the "quality" of the heat increases (T_2), the available work required to provide it increases. Thus, the inefficiency of current practice is more pronounced at lower end-use temperatures since the difference in the available work from conventional heating sources and that required by the end use increases as the temperature of the end use decreases. This difference is lost available work that comprises the second-law inefficiency.

C.3 SIMPLE EXAMPLES OF SECOND-LAW EFFICIENCIES

1. Power Generation

If the available work in fuel is approximated as being equal to the heat of combustion, then $\varepsilon \simeq \eta$ (square 2 in Table C-3); this is in the range of 0.3 to 0.4 for most operating power plants. Usually η is said to have a maximum value (the Carnot efficiency) that is less than 1. For a flame temperature of 2,000°C, the Carnot efficiency is 0.87; for a maximum steam temperature of 550°C, it is 0.64. What the second-law efficiency helps to emphasize is that even the Carnot efficiency is not an upper limit. The ideal fuel cell is a realization of a device that does not suffer the irreversibility of combustion and therefore has the higher upper limit $\varepsilon_{max} = 1$.

2. Oil- or Gas-Fired Furnace

It is in the provision of relatively low-temperature heat--in space heating and cooling and also in industrial process steam--that the most wasteful consumption of available work occurs (as suggested by Figures C-1 and C-2). Consider the same furnace cited earlier as an example: its first-law efficiency is $\eta = 0.6$; it provides air at 110°F ($T_2 = 316°K$); the outside air is at 32°F ($T_0 = 273°K$). From square 5 in Table C-3, we calculate $\varepsilon = 0.082$. If we postulate heat transfer at room temperature and set $T_2 = 70°F = 294°K$, the calculated efficiency is even less, $\varepsilon = 0.043$.

	WORK W_{in}	FUEL WITH HEAT OF COMBUSTION $\|\Delta H\|$	HEAT Q_1 FROM HOT RESERVOIR AT T_1
SOURCES	(e.g., water power, wind power, raised weight) [or electricity if the wall socket is treated as the source] $B = W_{in}$	(e.g., coal, oil, gas) $B \simeq \|\Delta H\|$ * (usually to within 10%)	(e.g., geothermal source, solar collector source) [also fission reactors and fossil-fuel plants if alternatives to thermal operation are excluded] $B = Q_1\left(1-\frac{T_0}{T_1}\right)$
	WORK W_{out}	HEAT Q_2 ADDED TO WARM RESERVOIR AT T_2	HEAT Q_3 EXTRACTED FROM COOL RESERVOIR AT T_3
END USES	(e.g., electric motor, diesel engine, all forms of transportation) $B_{min} = W_{out}$	(e.g., furnace, heat pump, oven) $B_{min} = Q_2\left(1-\frac{T_0}{T_2}\right)$	(e.g., refrigerator, air conditioner) $B_{min} = Q_3\left(\frac{T_0}{T_3}-1\right)$

*An exact consideration is required for each fuel. See Table C- 4

AVAILABLE WORK PROVIDED BY SOURCES AND NEEDED BY END USE

TABLE C-2

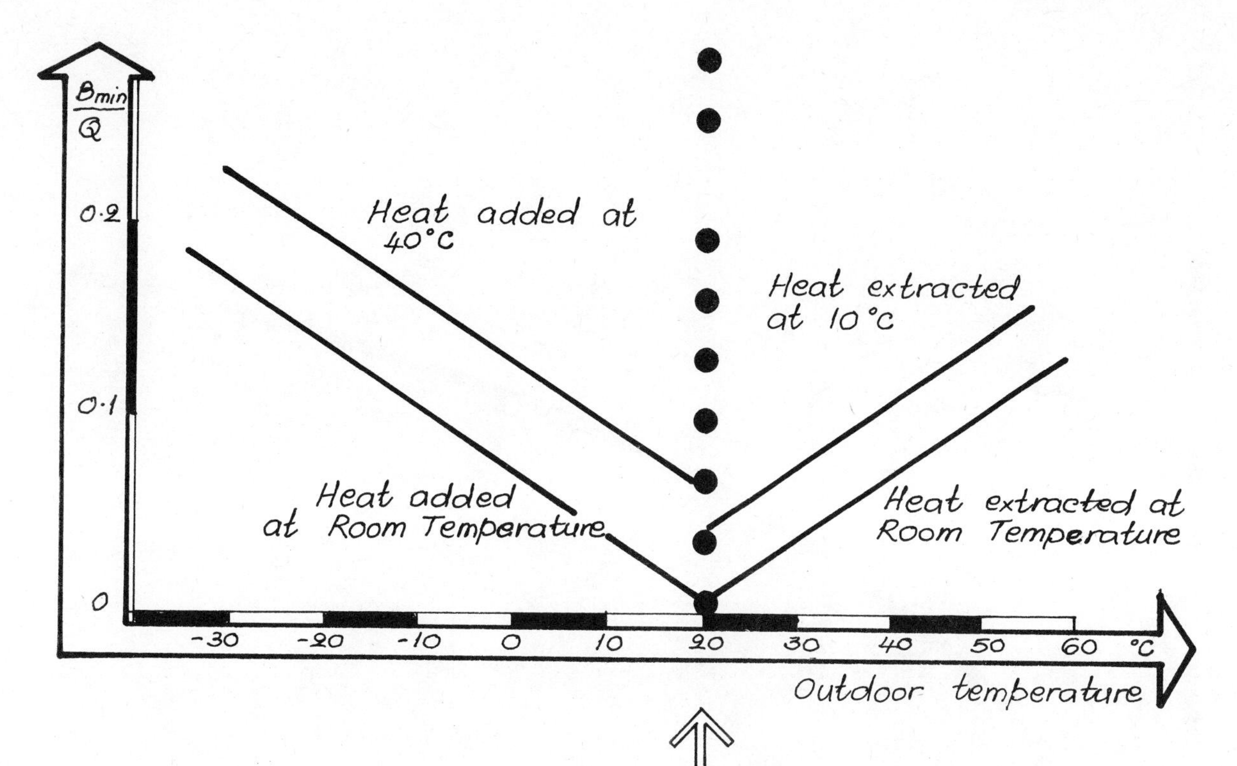

NOTES

- The vertical scale gives B_{min}/Q, the least available work required to add 1 unit of heat to a room in the winter (to the left of the dots) or to extract 1 unit of heat in the summer (to the right of the dots), as a function of the outdoor temperature. The vertical scale is not to be read as an efficiency; see the text.

- The upper curve refers to heat addition and extraction at typical practical temperatures.

- The lower curve shows the even smaller expenditure of available work that would be required if heat could be added and extracted at room temperature.

MINIMUM AVAILABLE WORK FOR SPACE CONDITIONING

FIGURE C–1

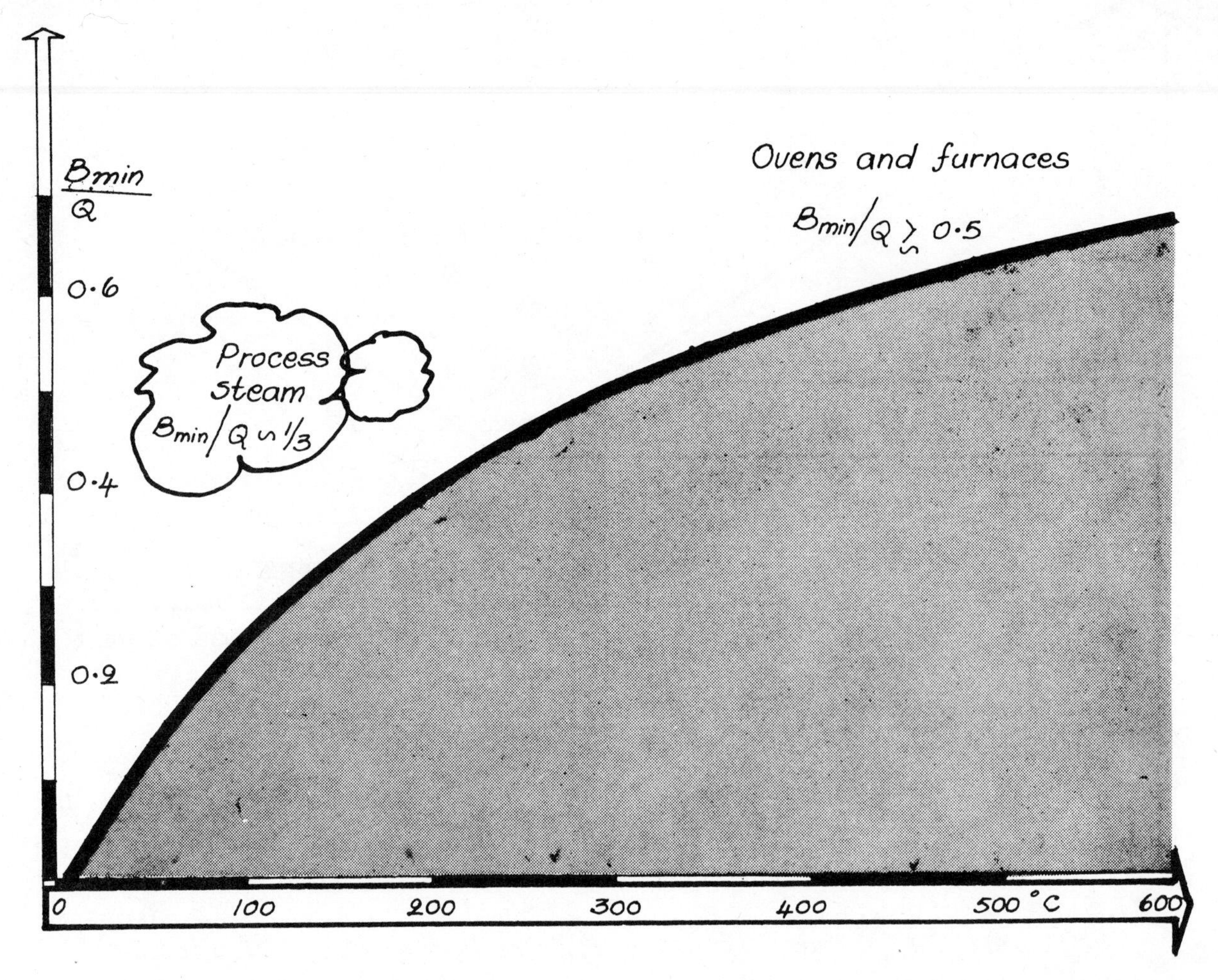

- The vertical scale gives B_{min}/Q, the least available work required to provide 1 unit of heat at the temperature shown on the horizontal scale. The vertical scale is not to be read as an efficiency.
- Ambient temperature is $T_o = 13°C$
- Note : $6{\cdot}9 \times 10^5$ N/m^2 = 100 psig

FIGURE C-2

MINIMUM AVAILABLE WORK FOR INDUSTRIAL HEAT

SOURCE / END USE	WORK W_{in}	FUEL: HEAT OF COMBUSTION $\lvert\Delta H\rvert$ AVAILABLE WORK B	HEAT Q_1 FROM HOT RESERVOIR AT T_1
Work W_{out}	1. $\eta = W_{out}/W_{in}$ $\varepsilon = \eta$ (e.g., electric motor)	2. $\eta = W_{out}/\lvert\Delta H\rvert$ $\varepsilon = \frac{W_{out}}{B}$ $(\simeq\eta)$ (e.g., power plant)	3. $\eta = W_{out}/Q_1$ $\varepsilon = \frac{\eta}{1-(t_0/T_1)}$ (e.g., geothermal plant)
Heat Q_2 added to warm reservoir at T_2	4. $\eta(COP) = Q_2/W_{in}$ $\varepsilon = \eta\left(1-\frac{T_0}{T_2}\right)$ (e.g., electrically driven heat pump)	5. $\eta(COP) = Q_2/\lvert\Delta H\rvert$ $\varepsilon = \frac{Q_2}{B}\left(1-\frac{T_0}{T_2}\right)$ (e.g., engine-driven heat pump)	6. $\eta(COP) = Q_2/Q_1$ $\varepsilon = \eta\frac{1-(T_0/T_2)}{1-(T_0/T_1)}$ (e.g., furnace)
Heat Q_3 extracted from cool reservoir at T_3	7. $\eta(COP) = Q_3/W_{in}$ $\varepsilon = \eta\left(\frac{T_0}{T_3} - 1\right)$ (e.g., electric refrigerator)	8. $\eta(COP) = Q_3/\lvert\Delta H\rvert$ $\varepsilon = \frac{Q_3}{B}\left(\frac{T_0}{T_3} - 1\right)$ (e.g., gas-powered air conditioner)	9. $\eta(COP) = Q_3/Q_1$ $\varepsilon = \eta\frac{(T_0/T_3) - 1}{1 - (T_0/T_1)}$ (e.g., absorption refrigerator)

FIRST-LAW & SECOND-LAW EFFICIENCIES FOR SINGLE SOURCE-SINGLE OUTPUT DEVICES

TABLE C-3

3. Heat Pumps

A heat pump is a device that transfers heat "uphill" from a cooler to a warmer place. Refrigerators and air conditioners use heat pumps. More often, however, the name "heat pump" is used when the goal is heating, not cooling. Consider a heat pump whose input energy is work W (electricity or mechanical drive) and whose output energy is heat Q_2 supplied at temperature T_2 when the cooler reservoir (e.g., the atmosphere) is at temperature T_0. Its coefficient of performance is $\eta(\text{COP}) = Q_2/W$ and its second-law efficiency is $\varepsilon = Q_2/W(1 - [T_0/T_2])$ (square 4in. Table C-3). Typical existing heat pumps have efficiencies ε of about 0.3 and COP's that depend on temperature:

$$\eta(\text{COP}) = 1 - \frac{\varepsilon}{(T_0/T_2)}$$

A heat pump is most effective over small temperature steps. *Example:* Let $\varepsilon = 0.3$ and $T_2/T_0 = 1.1$, a ten percent "boost" in temperature. Then $\eta(\text{COP})=3.3$. This heat pump would lose its "magnification" (i.e., $\eta \leq 1$) if the step up in absolute temperature exceeded 43 percent (if $T_2/T_0 \geq 1.43$).

C.4 THE THERMODYNAMICS OF AVAILABLE WORK

Consider a system characterized by energy E, entropy S and volume V (see Figure C-3). Its energy may also include, in addition to its internal energy, gravitational potential energy, kinetic energy of bulk motion, etc. (Temperatures, pressure and other intensive variables may vary from one part of the system to another.) Now let the system proceed via chemical reaction or other changes until it comes into thermodynamic equilibrium with the atmosphere. (By "atmosphere" we mean an appropriate large reservoir comprising the environment of the system; usually, in fact, it will be the earth's atmosphere.) The atmosphere has temperature T_0, pressure P_0, and a specified composition. The system, as it changes, may exchange heat and work with the atmosphere and it may also do work on other systems. The latter is called "useful work." Upon reaching equilibrium with the atmosphere, the system has energy E_f, entropy S_f and volume V_f. The maximum useful work that can be transferred by the system is called its *available work*. The available work is determined by the condition that the total entropy of the system plus the environment does not change. The maximum useful work is

$$B = (E-E_f) + P_o\ (V-V_f) - [T_o(S-S_f)] \quad (4)$$

(available work without diffusion)

The second and third terms on the right represent energy "bootlegged" from the atmosphere. If the system's volume decreases ($V-V_f > 0$), the atmosphere does work on the system which can be passed on as useful work (the second term is then positive). The diffusion term (not included) arises as the products in the final state diffuse into the atmosphere; it is relatively small and may be neglected in practical calculations.[10]

In certain applications (charging a battery, for example, or pumping water to a higher reservoir), it is useful to think of transferring available work from one system to another. Second-law efficiencies are then multiplicative. Let the efficiency of storage be $\varepsilon_1 = B_{stored}/B_{source}$ and the efficiency of secondary use be $\varepsilon_2 = W/B_{stored}$. The overall efficiency, $\varepsilon = W/B_{source}$, is then

$$\varepsilon = \varepsilon_1 \varepsilon_2 \quad (5)$$

This is a rather obvious point. It is made here only to emphasize that in an energy storage device, it is the stored available work, not the stored internal energy, that is the relevant quantity. Efficiencies of storage and reuse should be measured in terms of available work.

Most end uses, it should be noted, strongly increase entropy and do not store available work. Imagine a vehicle powered by a hypothetically perfect fuel cell and electric motor. This supercar operates with an efficiency $\varepsilon = 1$ and $\Delta S = 0$; that is, no entropy increase is associated with getting the fuel's energy to the drive wheels of the car. Cruising along the freeway,

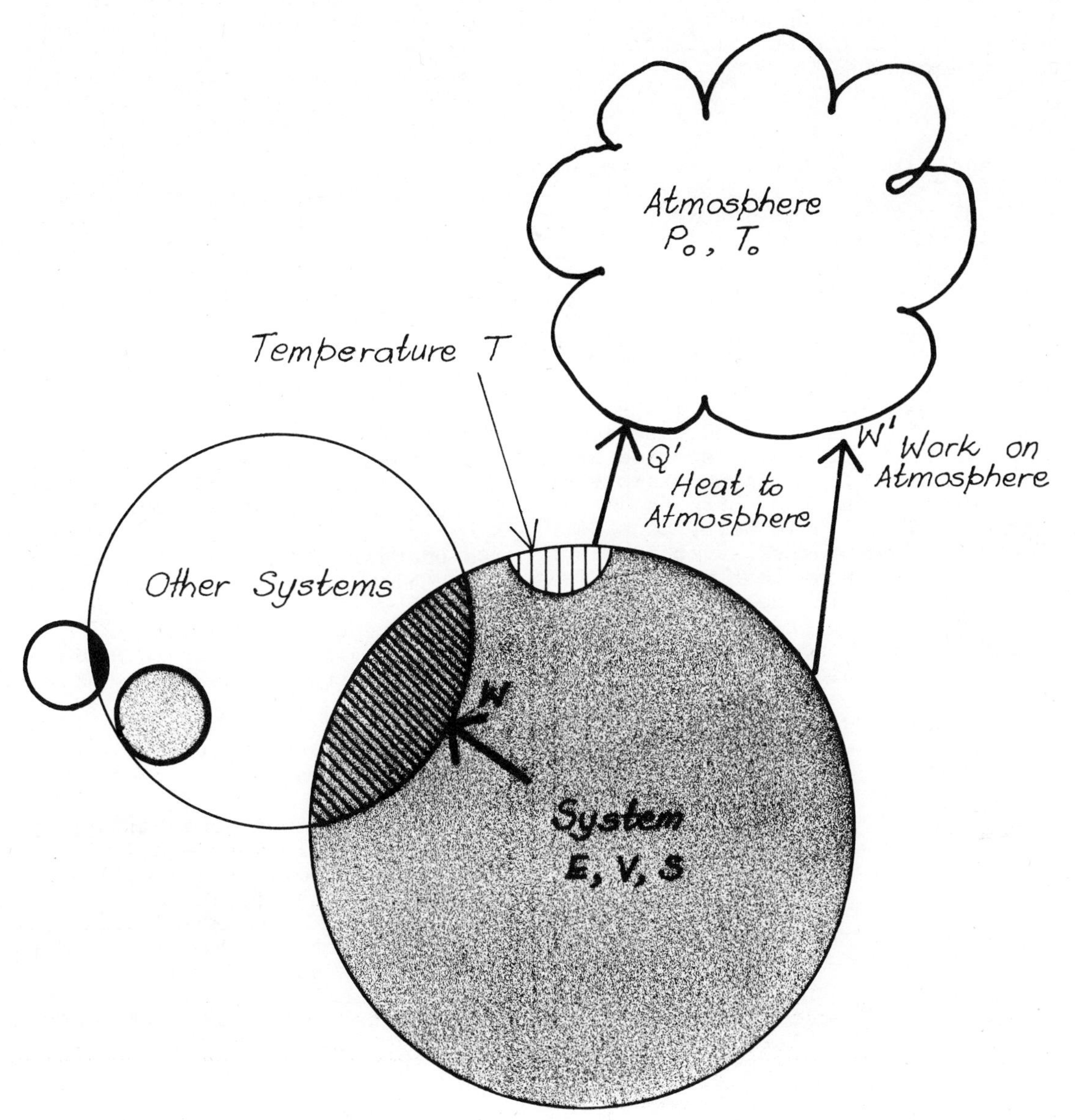

The available work of the system is the maximum useful work it can transfer to other systems as it comes to thermodynamic equilibrium with the atmosphere. For reversible change, the temperature T at which heat leaves the system must be equal to the atmospheric temperature T_o. This can be achieved, if necessary, by including one or more ideal heat engines in the system. In the absence of irreversibilities, $W = B$. Otherwise, $W < B$.

FIGURE C-3

INTERACTION OF SYSTEM WITH THE ATMOSPHERE AND OTHER SYSTEMS

however, the supercar is dissipating all of this hard-won energy to the environment so the entropy change of the universe is

$$\Delta S = B/T_0 \quad (6)$$

This again is a rather obvious point; but it is made as a reminder that available work (unlike energy) is a quantity that is consumed. Minimizing its consumption minimizes total entropy change.

An Example: Compressed-Air Energy Storage

Compressed air in large underground caverns is under serious consideration for energy storage at power plants. As a simple example to illustrate the link between entropy and available work, consider air that is compressed adiabatically and then loses some heat while it is in storage awaiting decompression. Initially the air is at pressure P_0, temperature T_0, and volume V_1. It is compressed adiabatically to P_2, T_2, and V_2. It then loses heat Q and its pressure and temperature fall to P_3, T_3 while its volume remains constant ($V_3 = V_2$). Set $P_2/P_0 = r$, the maximum compression ratio.

The fractional loss of available work during storage is

$$\frac{|\Delta B|}{B} = \frac{(T_2/T_0) - (T_3/T_0) - \ln(T_2/T_3)}{(T_2/T_0) - 1} \quad (7)$$

Graphs of this quantity as a function of compression ratio r are shown in Figure C-4 for complete cooling during storage ($T_3 = T_0$) and for partial cooling. (Note that $T_2/T_0 = r(\gamma-1)/\gamma$.) For complete cooling, the losses are large--more than 20 percent for a compression ratio of 5 and about 30 percent for a compression ratio of 10. If the stored gas loses only 25 percent of its sensible heat (the lowest curve in the figure), its loss of available work is 9 percent at r = 5 and 11 percent at r = 10. For compression ratios greater than r = 10, the loss of available work increases rather slowly with increasing r. This same loss in available work would also occur if the air is cooled after compression by a heat exchanger in which heat is dissipated to the atmosphere (or to river water). If the extracted heat were applied to some other use such as space heating (where the space heating would normally be done by a combustion source), then the overall available work is more effectively used. In addition to the losses shown here, there are, of course, pumping losses.

For comparison with these numbers, one may note that the *total* loss of available work in the storage of water in elevated reservoirs (electricity to electricity) is about 30 percent.

C.5 AVAILABLE WORK AND ENERGY UTILIZATION EFFICIENCY

Keenan and others have suggested that available work be used as a basis for assessing the efficiency of energy utilization at all levels (6,8). This appears to be a useful approach. As was emphasized earlier, the concept of second-law efficiency is an especially attractive way to bring the available work concept (and the Second Law of Thermodynamics) to bear in a unified way on efforts to use energy more efficiently.

1. Available Work and Power Generation

For power generation in particular, available work would clearly be a practical tool for evaluation. Useful work is exactly the goal of power generation. We recommend that available work, not heat of combustion, be adopted generally as the reference energy for computing the efficiency of power generating units. (Then the efficiency would indeed be the second-law efficiency.) It is sometimes said, for instance, that the "maximum" efficiency of a hydrogen-oxygen fuel cell is 0.83 (see, for example, Reference 9). The efficiency is not limited in any real physical sense to a value of 0.83. It is an apparent limitation imposed by adopting an inappropriate unit of measure, heat of combustion (enthalpy change), rather than available work.*

*This does not mean a definition of efficiency as $\varepsilon = W_{out}/|\Delta H|$ is in any sense wrong; it is a matter of arbitrary definition. However, it would appear that this time-honored way of defining efficiency is not the best way and should be changed.

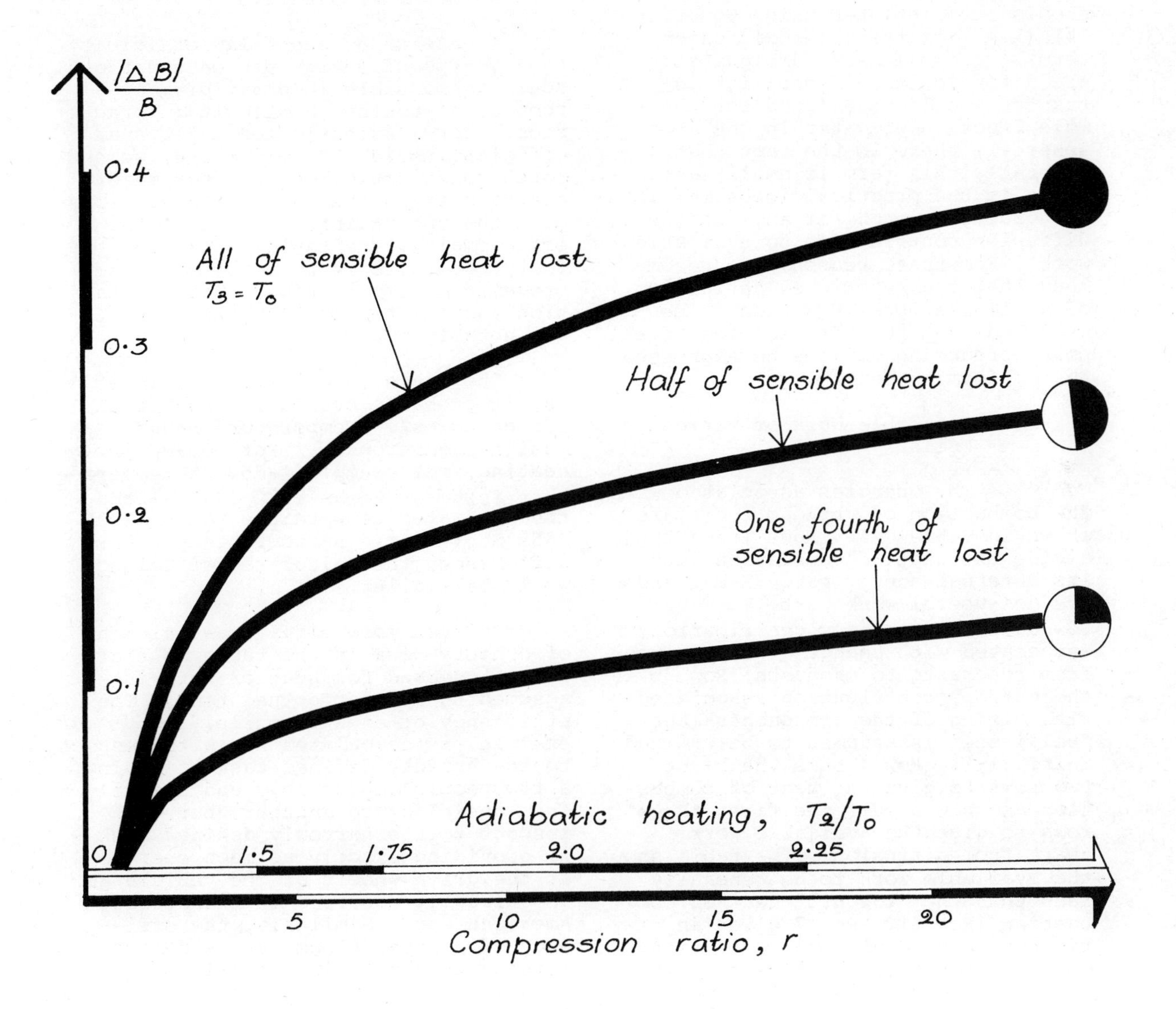

FRACTIONAL LOSS OF AVAILABLE WORK IN COMPRESSED AIR (storage at constant volume)

FIGURE C-4

However, for practical reasons, we suggest that the available work should be calculated using Equation (4) (i.e., omitting the diffusion term). There is some ambiguity in the diffusion term (what, for instance, should be adopted for the mole fraction of water in the atmosphere?); there is the fact that generally this term is small; and there is the practical impossibility of harnessing much, if any, of the diffusion contribution to available work. For these reasons, we recommend that the restricted definition of available work (Equation 4) be used and that the efficiencies of all power-producing devices be expressed in the form $\varepsilon = W_{out}/B$.

2. Available Work of Various Fuels

The energies associated with the combustion of various fuels are shown in Table C-4. Rows 1 and 2 divide the heat of combustion into its internal-energy part, $E-E_f$, and its non-useful work part, $P_0(V-V_f)$. Row 3a is the entropy contribution associated with change of composition from reactants to products; Row 3b is the entropy contribution associated with mixing of the components (the fuel itself is assumed to be unmixed initially). Row 4 sums the first two rows to give the heat of combustion and Row 5 sums the first three rows to give the available work apart from diffusion. The table shows the available work to be generally less than the "low Btu" heat of combustion (Row 6). Row 7 gives an estimate of the additional diffusion contribution to available work; it amounts to 3 to 6 percent of the "basic" available work. Finally, Row 5 shows the available work in other units.

Overlooking the diffusion terms, one sees that available work for hydrocarbon fuels is within a few percent of the heats of combustion. For carbon, B and $|\Delta H|$ are nearly identical. In Table C-5, heats of combustion and available work for several fuels are stated and compared for the alternative assumptions of condensed water products and high heating values. Hydrogen is an interesting special case; its available work is 18 percent less than its heat of combustion.

C.6 SECOND-LAW EFFICIENCY: A NATIONAL PERSPECTIVE

To assign a second-law efficiency to any type of energy use on a national scale is possible in principle, although, of course, difficult in practice. Berg has estimated a few such efficiencies.[5] In Table C-6, we set forth rough estimates of ε for major energy uses in the United States. Despite the uncertainty in some of the efficiencies, the table emphasizes the fact that there is room for great improvement in most areas. Transportation and low-temperature use of heat are notably inefficient.

The table is conservative (that is, it probably overestimates actual efficiencies). Temperature ranges are assigned generously. For hot-water heating, for example, ambient temperature is taken to be 55°F (386°K) and the hot-water temperature to be 180°F (355°K).* For a narrower temperature difference, the calculated efficiency would be smaller.

An even more significant element of conservatism in the table is that current demand for heat or work is assumed to reflect actual need. The efficiency of an automobile, for example, is calculated not with respect to the broadly defined task of moving a few people comfortably and speedily from one place to another, but with respect to the narrowly defined task of providing exactly as much energy at the drive wheels of the vehicle as is currently provided in the average American car. Similarly, the efficiency of process steam use is defined without regard to industrial goals (e.g., drying paper); it merely assumes a given demand at a given temperature.

In the case of industrial direct heat, we have used data provided by Berg for typical operating temperature (500-1,000°C) and typical first-law efficiencies ($\eta \simeq 0.4$).[11] The value $\varepsilon \simeq 0.3$ in the table does *not* mean that the ultimate saving could be more--if, for example, specific radiation were used to heat the working substances, or if alternative lower-temperature processes were developed.

---

*The appropriate ratio of gas and electric heaters in use is taken into account.

[Calculations are for combustion in air with gaseous H_2O product ("low heating value"). Energies in the upper part of the table are in kcal per mole of fuel. Available work is given in other units in the lower part of the table.]

	Energy Terms	Hydrogen H_2	Carbon C (to CO_2)	Carbon Monoxide	Methane CH_4	Ethane C_2H_6	Propane C_3H_8	Ethylene C_2H_4	Liquid Octane C_8H_{18}
(1)	$E - E_f$	57.5	94.1	67.3	191.8	341.5	489.1	316.2	1216
(2)	$P_o(V - V_f)$	+0.3	0	+0.3	0	-0.3	-0.6	0	-3
(3)	$-T_o(S - S_f)$ (a)	-3.2	+0.2	-6.2	-0.4	+3.3	+7.2	-2.1	+35
	(b)	+0.4	0	+0.4	+2.0	+3.3	+4.5	+2.5	+11
(4)	Heat of combustion, $-\Delta H$ = (1) + (2)	57.8	94.1	67.6	191.8	341.2	488.5	316.2	1213
(5)	Available work without diffusion, B = (1) + (2) + (3)	55.0	94.3	61.8	193.4	347.8	500.2	316.6	1259
(6)	Percentage change from $\lvert\Delta\rvert$ to B	-4.8%	+0.2%	-8.6%	+0.8%	+1.9%	+2.4%	+0.1%	+3.8%
(7)	Additional diffusion contribution B_d	1.9 (3.5%)	3.9 (4%)	4.0 (6%)	6.5 (3%)	11.3 (3%)	16.2 (3%)	9.8 (3%)	41 (3%)
(5)*	Available work in other units[a] kcal/gm	27.3	7.86	2.21	12.06	11.58	11.35	11.29	11.03
	MJ/kg	114	32.9	9.23	50.5	48.4	47.5	47.3	46.1
	kBtu/lb	49.1	14.1	3.97	21.7	20.8	20.4	20.3	19.9
	kBtu/lb-mole	99.0	170	111	348	626	900	570	2266
	eV/amu	1.18	0.341	0.096	0.523	0.502	0.492	0.490	0.478

*Available work in other units does not include the diffusion contribution B_d.

ENERGIES ASSOCIATED WITH COMBUSTION OF VARIOUS FUELS

TABLE C-4

(Energies are in kcal per mole of fuel. The columns correspond to Rows 4, 5, and 6 in Table C-4.)

Fuel	Heat of Combustion $-\Delta H$ Kcal/mole	Available Work B* Kcal/mole	Percentage Change from $\|\Delta H\|$ to B %
Hydrogen, H_2	68.2	55.9	-18
Methane, CH_4	213	194	-8.6
Ethane, C_2H_6	373	349	-6.4
Propane, C_3H_8	531	502	-5.5
Ethylene, C_2H_4	337	317	-5.9
Liquid Octane, C_8H_{18}	1307	1261	-3.5

* This is the "basic" available work, not including a diffusion contribution.

AVAILABLE WORK & HEAT OF COMBUSTION COMPARED FOR AIR COMBUSTION WITH WATER PRODUCT IN LIQUID FORM

TABLE C-5

Use	Percent of US Fuel Consumption (1968)	Estimated Overall Second-Law Efficiency ε
Space heating	18	0.06
Water heating	4	0.03
Air conditioning	2.5	0.05
Refrigeration	2	0.04
Industrial process steam	17	∿0.25
Industrial direct heat	11	∿0.3
Industrial electric drive	8	0.3
Auto transportation	13	0.1
Truck transportation	5	0.1

NATIONAL PERSPECTIVE ON EFFICIENCY OF ENERGY USE

TABLE C-6

Other examples of the use of available work to evaluate industrial processes can be found in the literature.[12,13,14] A review of the use of available work in the US and in the USSR has also been presented.[15]

There are, of course, technical as well as economic obstacles, some of them formidable, that prevent the achievement of second-law efficiencies close to 1. In economic terms, there may be no *reason* to press for efficiencies beyond a certain level. Nevertheless, as a guideline, it is helpful to think in terms of what nature permits.

* * *

REFERENCES

1. This Appendix is abstracted, with some changes and deletions, from Chapter 2 of Carnahan, Walter; Ford, Kenneth W.; Prosperetti, Andrea; Rochlin, Gene I.; Rosenfeld, Authur H.; Ross, Marc H.; Rothberg, Joseph E.; Seidel, George M.; and Socolow, Robert H., *Efficient Use of Energy: A Physics Perspective*, A Report of the Summer Study on Technical Aspects of Efficient Energy Utilization under the Sponsorship of the American Physical Society, (New York: January 1975).

2. Lee, P.O. and McCulloch, W.H., "A New Parameter for Evaluating Energy Systems," *8th Intersociety Energy Conversion Engineering Conference Proceedings*, (New York: American Institute of Aeronautics and Astronautics, 1973).

3. Keenan, J.H., *Thermodynamics*, (New York: John Wiley & Sons, Inc., 1948).

4. US Federal Power Commission, Office of the Chief Engineer, *A Technical Basis for Energy Conservation*, Staff Report, (Washington, D.C.: April 1974). Also see article by same title in *Mechanical Engineering*, (May 1974).

5. Berg, C.A., "A Technical Basis for Energy Conservation," *Technology Review*, (February 1974): 14.

6. Keenan, J.H.; Gyftopoulos, E.P.; and Hatsopoulos, G.H., "The Fuel Shortage and Thermodynamics," in *Proceedings of The MIT Energy Conference*, (Cambridge, Massachusetts: MIT Press, (1973).

7. For a brief review of the history of this concept's development, see Gyftopoulos, Elias P.; Lazaridis, Lazaros J.; and Widmer, Thomas F., *Potential Fuel Effectiveness in Industry*, A Report to the Energy Policy Project of the Ford Foundation, (Cambridge, Massachusetts: Ballinger Publishing Company, 1974), p. 11.

   Two other references of interest are Bosnjakovic, F., *Technical Thermodynamics*, (New York: Holt, Rinehart, and Winston, 1965); and Tolman, R.C. and Fine, P.C., "On the Irreversible Production of Entropy," *Reviews of Modern Physics*, 10(1948).

8. Private communication with Charles A. Berg, Princeton, New Jersey, July 1974.

9. Vielstich, W., *Fuel Cells*, (New York: Wiley-Interscience, 1970), pp. 19-20.

10. For a discussion of the computation of the diffusion term and some typical results, see Carnahan, et al, pp. 43-46.

11. Berg, Charles A., "Conservation in Industry," *Science*, 184(19 April 1974): 264.

12. El-Sayed, Y.M. and Aplenc, A.J., "Application of the Thermoeconomic Approach to the Analysis and Optimization of a Vapor-Compression Desalting System," ASME *Journal of Heat Transfer*, (May 1964).

13. Bejan, A. and Smith, J.L. Jr., "Thermodynamic Optimization of Mechanical Supports for Cryogenic Apparatus," *Cryogenics*, (March 1974).

14. Petela, R., "Exergy of Heat Radiation," ASME *Journal of Heat Transfer*, (May 1964).

15. Ahern, John E., "The Exergy Method of Energy Systems Analysis," Paper presented at the American Society of Mechanical Engineers Technical Division Conference, "Exergy...The Efficient Use of Energy," Los Angeles, California, 17 May 1975.

APPENDIX D

AN OVERVIEW OF GOVERNMENT REGULATIONS AND POLICIES AFFECTING EFFICIENT ELECTRICITY USE

Laurence H. Martin*

CONTENTS

D.1 BACKGROUND/STATUS OF ENERGY POLICY IN THE US

Government involvement in the nation's energy system has developed gradually over many years. This evolution has occurred through legislative action resulting in a multitude of acts, codes and ordinances, through administrative actions such as Presidential Executive Orders and Proclamations and judicially through creative interpretations of existing law. Current laws and bodies implementing these laws have their roots in wide and divergent areas of governmental responsibility, which may be summarized as follows:

- Public Land Management from which has come the Mineral Leasing Act of 1920 and such agencies as the Bureau of Land Management;
- Water Policy from which has come the Rivers and Harbors Act of 1899, Reclamation Act of 1902, and Federal Water Power Act and such agencies as the Bureau of Reclamation; Federal Power Commission; and the Bonneville, Alaska, Southeastern and Southwestern Power Administrations;
- Economic Regulation and Redevelopment from which has come the Utility Holding Company Act, Hepburn Act of 1906, and the National Gas Act and such agencies as the Rural Electrification Administration which greatly expanded the duties of the Federal Power Commission;
- Tax Policies such as the depletion allowance once applicable to taxes on income from oil and gas wells;
- National Security from which has come the Defense Act of 1916, Atomic Energy Act of 1946, Trade Agreement Extension Act of 1955 now amended to the Trade Expansion Act of 1962,

*Laurence H. Martin is a lawyer specializing in energy law. He has been retained by Applied Nucleonics Company to advise on the legal aspects of efficient energy use.

and such agencies as the Tennessee Valley Authority and Atomic Energy Commission;

- Health, Safety, and Environmental Management from which has come the Federal Coal Mine Health and Safety Act of 1969, National Environmental Policy Act of 1970, and; such agencies as Environmental Protection Agency and Council on Environmental Quality; and,

- Research and Development from which has come the National Science Foundation Act of 1950 and much of the recent legislation and such agencies as the Bureau of Mines, Office of Coal Research, and recently the Energy Research and Development Administration. [1]

Regrettably, the evolutionary process leading to regulations and regulatory bodies affecting the energy system was not the result of any conscious "energy policy" or objective, but rather was a byproduct of governmental action designed to achieve other and mainly unrelated objectives.[2] This long term, uncoordinated, piecemeal development of energy-related rules and regulations is the unstable foundation on which the US commenced to build a true national energy policy. Much of the recent legislation proposed or passed has been devoted, in substantial part, to undoing the historical structure and building an organized and coordinated basis on which to proceed in the future.[3]

In 1971, the US government's involvement in the energy field rapidly accelerated. On June 4, 1971, and for the first time publicly, the President of the United States acknowledged that the US had an energy problem and suggested a program to help alleviate it. This initial program consisted of the following goals:

- to facilitate research and development for "clean" energy;

- to make available the energy resources on federal lands;

- to assure a timely supply of nuclear fuels;

- to use energy more wisely, with particular reference to a new Federal Housing Administration standard requiring additional insulation in new federally-insured homes;

- to develop and publish additional information on how consumers can use energy more efficiently and to pursue other areas of energy conservation;

- to balance environmental and energy needs including a system of long-range open planning of electric power plant sites and transmission line routes with approval by state or regional agencies before construction; and,

- to organize federal efforts more effectively, particularly by creating a single structure and uniting all important energy resource development programs.[4]

The concepts advanced by the President in 1971 remain virtually unchanged today and until recently, little has been done to solve the fundamental problem of disorganization and duplication of action.[5] By the summer of 1973, there were approximately 46 agencies of the Federal Government that administered programs or effected policies that had specific impacts on the energy systems.[6] There were an additional 18 agencies that had jurisdiction over programs that were not intended to be energy-oriented but which nevertheless had an impact on the energy system. These agencies were proliferated among nine executive departments and 15 independent agencies in the Executive Office of the President. At the state and local levels, matters were no different.

In an effort to eliminate the chaos at the federal level, largely motivated by the sudden and serious fuel shortages in mid-1973, the administration created an Energy Policy Office (EPO) to coordinate the energy related activities of various agencies having an impact on fuel production and consumption. EPO also had jurisdiction over the voluntary oil allocation program and later the voluntary allocation program for middle distillates. In the fall of 1973, in response to a need created by the Administration's decision to impose mandatory allocations on more petroleum products, the Interior Department formed an Office of Petroleum Allocation (OPA) to take charge of the program.[7]

In December 1973, a Federal Energy Office (FEO) was created to take over the responsibilities of both EPO and OPA which were phased out as separate entities. FEO also assumed new powers (mandated by the Emergency Petroleum Allocation Act) vested in the Executive Branch for the mandatory allocation of crude oil and all refined petroleum products and the petroleum price control authority, formerly wielded by the Cost of Living Council. FEO, however, was considered to be an embryonic agency that would remain in operation only until statutory authority was provided for a more encompassing Federal Energy Administration (FEA). On June 27, 1974, FEO was actually abolished and FEA became the Federal Government's line agency for petroleum allocation and pricing, energy data collection and analysis, fuel conservation (shared with the Department of Commerce), and energy independency planning.[8]

On October 11, 1974, the President signed a bill (effective 120 days from that date) that abolished the Atomic Energy Commission and created an Energy Research and Development Administration (ERDA) to handle the vast majority of federal research and development projects and a Nuclear Regulatory Commission (NRC) to regulate the nuclear power industry. The FEA and ERDA were sister organizations, both technically reporting directly to the President, but theoretically under the policy control of the Energy Resources Council (ERC) established October 11, 1974. ERC, which was established to insure communication and coordination among federal agencies in energy matters and to set and implement national energy policy, is charged with overall control of the national energy effort.[9]

On the federal level, therefore, the development of a foundation for US energy policy began with the creation of the FEA, ERDA, and the Energy Resources Council. In practice, however, these agencies have not yet caused any meaningful impact on the problems discussed above and, at present, Congress has extended the FEA life only to December 31, 1977. [10]

On March 1, 1977, the President of the US sent to Congress proposed legislation to reorganize the Federal Government's energy agencies and programs. This bill anticipated the creation of a new Cabinet-level Department of Energy by combining the functions of the three major federal energy agencies along with energy-related functions of six other executive and independent regulatory agencies. This new Department would provide the organizational base and the programmatic authorities needed to develop and implement overall federal energy policies. Among the major responsibilities of the Department of Energy (DOE) will be:

- conservation,
- regulation,
- research and development,
- resource development and production, and,
- data management.

In submitting the proposed legislation, the President made the following pertinent comments regarding the state of US energy policy:

- nowhere is the need for reorganization and consolidation greater than in energy policy;
- all but two of the Executive Branch's Cabinet departments now have some responsibility for energy policy, but no agency, anywhere in the Federal Government, has the

broad authority needed to deal with our energy problems in a comprehensive way; and,

- the legislation will bring immediate order to this fragmented system.

The changes called for by the proposed legislation may be summarized as follows:

- abolishes the Federal Energy Administration (FEA), Energy Research and Development Administration (ERDA), and the Federal Power Commission (FPC);

- combines conservation programs which are now split between FEA and ERDA and creates an Assistant Secretary for Conservation, who will be personally responsible for seeing that the conservation program is carried out;

- places under one roof the powers to regulate fuels and fuel distribution systems, powers which are now shared by the FEA and FPC along with the Securities and Exchange Commission and the Interstate Commerce Commission;

- transfers to the new department several energy related authorities and programs, such as the:

  - building thermal efficiency standards from Housing and Urban Development; and

  - voluntary industrial compliance program from Commerce;

- provides for consultation between the Energy Department and the Department of Transportation on auto fuel efficiency standards;

- establishes a role for the Energy Secretary in the REA loan program at Agriculture;

- abolishes the Energy Resource Council (ERC);

- transfers certain parts of the Interior Department--those concerning fuel data collection and analysis and coal mine research and development--to the new department; and,

- will bring together all energy data gathering and analysis capabilities currently done by more than 20 executive departments and agencies who collectively operate more than 250 energy data programs (100 of these are FEA, ERDA, FPC, and Interior Department programs).[11]

The new US Department of Energy was approved by Congress and began operations on October 1, 1977. It has assumed responsibility for programs formerly carried out by ERDA, FEA, FPC, the Department of Interior, and the NRC.

D.2 REVIEW OF KEY AGENCIES AND CONGRESSIONAL COMMITTEES AT THE FEDERAL LEVEL--US

At present, most federal and many state and local agencies have already devised and implemented energy efficiency plans and have additional proposals in the formative stages. [12,13] However, these programs, whether in existence or in the development stage, are generally restricted to the agency's or governmental entity's specific area of responsibility.[14] For example, the Federal Housing Administration (FHA), General Services Administration (GSA), and the Veterans Administration (VA) just to name a few, are quite active in the area of improving the efficiency of new and existing commercial and residential structures insofar as energy use is concerned. As another example, the Federal Power Commission (FPC), Tennessee Valley Authority (TVA), and other agencies are active in the area of rate design to achieve more efficient energy use.[15] Therefore, it is important to determine which agencies have responsibility over a specific area of interest since regulations might progress faster and in a more comprehensive fashion in one area than in another. The great majority of federal agencies that are now directly involved in, or at least have a substantial impact on, the energy system are listed in reference [16].

It does appear, however, that certain federal agencies and congressional committees at the federal level of government have played a dominant and perhaps controlling role in the attempt to structure a true national energy policy.[17] It is important to be aware of the makeup of these agencies and committees and their basic functions since they furnish a wealth of diverse up-to-date information. These agencies and Congressional committees are described below.[18]

Federal Agencies

- Department of Energy (DOE)

  This new agency assumes responsibilities formerly exercised by ERDA, FEA, FPC, etc. At the time this Appendix was being revised, the organization of the DOE had not been finalized. It will include, however, groups with responsibility for fuel allocation and pricing regulations, energy data collection and analysis, energy research and development, energy conservation, and nuclear regulation and safety.

- Department of Commerce (DOC)*

  DOC involvement in the field of energy is geared mainly towards conservation and efficiency by industry and business. Under the auspices of the Department, the Office of Energy Programs has an energy planning division that is responsible for identifying and stimulating industry's voluntary implementation of energy conservation practices. A second division of the office, the energy resources division, handles oil policy committee responsibilities of the Department as well as a variety of other functions pertaining to energy policy information. Specialists provide data projections and analysis of the implications, foreign and domestic, on energy policy for the US economy as a whole and commercial enterprises in particular.

- Environmental Protection Agency (EPA)

  Since energy-producing facilities and activities normally produce air emissions, water effluents, and other production residuals, nearly all phases of the power generation are subject to regulations under the pollution control programs administered by the EPA. In its air pollution control program, EPA establishes national ambient air quality standards and the states enforce limits on emission to achieve these standards. If a state fails to set such limits or if the state limits are considered inadequate, the EPA is authorized to set the required emission limits for the state. In addition, EPA establishes uniform national air pollution control standards for new plants and factories, emission limitations for hazardous air pollutants, and motor vehicle standards. EPA administers a similar national program on water pollution control.

- National Science Foundation (NSF)

  The Office of Energy Research and Development Policy is part of the National Science Foundation. The office, established August 10, 1973, is charged with the following functions within the National Science Foundation:

  - providing analysis of specific issues and selective programs related to energy research and development including energy supply technology, energy demand and conservation,

*Note: Some of these functions will be incorporated within the new Department of Energy.

and energy-related areas on environmental and sociological research;

- developing a general systems framework for the evaluation of energy research and development programs and developing appropriate criteria for assessing the merits of individual technological approaches;

- identifying and recommending to the Executive Office of the President critical needs in energy research and development;

- identifying and evaluating significant research findings that could affect energy programs or policies;

- providing an individual assessment of environmental health and safety standards and identifying additional research to improve standards;

- maintaining an awareness of current plans and viewpoints of industry and associations in energy matters; and,

- determining ways in which universities and other research organizations can make the most effective contributions to better energy development.

- Treasury Department

An Office of Natural Resources and Energy has been established within the Treasury Department to monitor energy resources, supply and demand as well as the implications of the United States policies with respect to selected natural resources. The work of this office includes contact with other government agencies concerned with resources and energy on one hand and on the other with the various parts of the Treasury concerned with tax policy, trade policy, economic policy, and international and monetary affairs. Recommendations from the office will be made directly to the deputy secretary regarding the most appropriate policies to produce a favorable US supply-demand balance for energy and other selected resources including coal, oil, natural gas, copper, and other minerals of importance to the balance-of-payments. With respect to tax policies, the Office will advise on such matters as depletion allowances, accelerated amortization, investment credits, and production incentives. Tariffs, quotas, financial and investment policies, price controls, loans and guarantees as they relate to energy and natural resources will also be a part of the on-going studies and analysis of this new office.

Congressional Committees of Significance[19]

Senate Committees

- Energy and Natural Resources--Jurisdiction includes all proposals relating to energy policy, regulation, conservation and research, including those affecting solar power, non-military nuclear power, the Alaskan Naval Petroleum Reserve, oil and gas production and distribution, mineral extraction from the oceans, energy-related aspects of deepwater ports, hydroelectric power generation, coal production and utilization, mining lands and mineral conservation, and mining education and research. The Committee also has jurisdiction over bills relating to public lands, natural forests, and the territorial possessions of the United States.

- <u>Interior and Insular Affairs</u>--The jurisdiction of this committee includes lands (generally), mineral resources, petroleum and radium conservation.

- <u>Armed Services</u>--Jurisdiction in energy areas includes conservation, development and use of naval petroleum and oil shale reserves, and strategic and critical materials necessary for common defense.

- <u>Finance</u>--Jurisdiction generally includes revenue measures.

- <u>Commerce, Science and Transportation</u>--Jurisdiction includes interstate and foreign commerce generally and specifically covers the regulation of oil and gas pipelines, navigational aspects of deepwater ports, science and engineering research and development, nonmilitary aeronautical sciences, and the transportation aspects of matters relating to the Outer Continental Shelf.

<u>House Committees</u>

- <u>Interstate and Foreign Commerce</u>--Jurisdiction includes inland waterways, oil compacts, petroleum and natural gas, except on public land, and regulation of transmission of power (except installation of connections between government water projects).

- <u>Science and Technology</u>--Jurisdiction covers energy research and development including activities of the Energy Research and Development Administration (now DOE).

- <u>Interior and Insular Affairs</u>--Jurisdiction includes mineral lands (generally), mineral resources, petroleum and radium conservation.

- <u>Government Operations</u>--Jurisdiction includes conservation, development, and use of the naval petroleum reserve and oil shale reserve.

- <u>Small Businesses</u>--Jurisdiction includes matters dealing with small businesses.

- <u>Outer Continental Shelf</u>--Jurisdiction covers all matters concerning the Outer Continental Shelf, including oil and gas leasing.

<u>Alphabetical Listing of Agencies Affecting Energy Use</u>

Agriculture Department

Alaska Power Administration

Bonneville Power Administration

Commerce Department and particularly the National Bureau of Standards (DOC and NBS)

Defense Department

Department of Energy (DOE)

Department of Housing and Urban Development (HUD)

Energy Research and Development Administration (ERDA)*

Energy Resources Council (ERC)*

Environmental Protection Agency (EPA)

Executive Office of the President (EOP)

Federal Energy Administration (FEA)*

Federal Housing Administration (FHA)

Federal Maritime Commission (FMC)

Federal Power Commission (FPC)*

Federal Trade Commission (FTC)

General Accounting Office (GAO)

---

*Now the Department of Energy (DOE)

General Services Administration (GSA)

Interior Department--Geological Survey

Interior Department--Bureau of Land Management

Interior Department--Bureau of Reclamation

Interstate Commerce Commission (ICC)

Justice Department

Labor Department

National Aeronautics and Space Administration (NASA)

National Science Foundation (NSF)

Nuclear Regulatory Commission (NRC)*

Oil Policy Committee (OPC)

Securities and Exchange Commission (SEC)

Small Business Administration (SBA)

Solar Energy Coordination and Management Project

Southeastern Power Administration

Southwestern Power Administration

State Department

Tennessee Valley Authority (TVA)

Transportation Department (DOT)

Treasury Department

Water Resources Council

D.3 ANALYSIS OF PERTINENT REGULATIONS AND GOVERNMENTAL POLICIES--US

A. Methods Used by Government to Cause Energy End Use Efficiency--An Overview

Historically, government (at all levels) has promoted or attempted to initiate a desired type of citizen behavior by use of the following methods:

- direct regulation;
- use of financial pressures designed to cause action or inaction;
- educational programs;
- use of power in the marketplace as a major customer;
- selective granting of exceptions to or exemptions from existing law or procedures; and,
- establishment of major research and development programs with well defined priorities.[20]

The activity by government to cause energy end use efficiency on the part of citizens closely follows the above-described pattern and, as a result, can be effectively analyzed in that context.

B. Direct Regulatory Action

Direct regulation of end users to cause more efficient use of energy is a method which has been slow in developing. However, the following distinct patterns of action can be identified:

- maintenance and further sophistication of "emergency regulations";
- sponsoring voluntary conservation programs involving the private sector which gradually become mandatory in character and directive in nature; and,
- a gradual process of regulating out of existence structures and products of all kinds which do not meet government-imposed standards.

The so-called "emergency" or "standby" measures and certain aspects of the current Federal Energy Management Program (FEMP) (applicable to Federal employees and facilities only) constitute clear examples of direct regulation of end user behavior, i.e., take out light bulbs, turn down

*Now the Department of Energy (DOE)

or up thermostat, etc. On the other hand, controlling the end user's buying habits by forcing efficient products, structures, etc., upon them through a program designed to gradually eliminate all inefficient products or structures is no less of a direct regulation of behavior even though it is more subtle in application. Examples follow.

1. Emergency Measures and Federal Energy Management Program

The Federal Government and most, if not all, state and local governmental entities have "emergency" type laws which contain stringent and detailed regulations with reference to how, when, and in what manner end users of energy may or must use it. [21] These laws may now be dormant, but they are nevertheless "on the books" and can be activated swiftly when events dictate the need to do so. Furthermore, studies are continuing in most states regarding how to develop better programs to meet emergency needs.[22]

In the event other more gradual fuel conservation programs and legislation fail to control the existing energy problem, it is quite likely these "emergency measures" will be activated and used on a long-term if not permanent basis.[23] An example of such strict directive regulation is the Los Angeles City Emergency Energy Curtailment Plan. This plan is one of the broadest and most restrictive in the nation and, in fact, has been used as a model by the Federal Government, the Public Utilities Commission in California, and such states as Washington and New York in connection with the development of their own programs.[24]

Federal agencies and many state and local governmental agencies have programs designed to cause their employees to make more efficient use of energy. These programs are important to note since they too are an indicator of the course of future regulation of the general public. An example of this type of activity is the Federal Energy Management Program which involves sixteen of the most energy-intensive departments and agencies. These eleven cabinet departments and five large agencies make quarterly reports to the FEA, identifying energy savings in two major categories--(1) buildings and facilities, and (2) in the use of fuels in motor vehicles, aircraft, ships and equipment. Energy conservation strategies of the program fall into three general areas, to wit: building operations, transportation, and federal employee activities. Details of these areas are set forth below:

- Building Operations. This includes:
  - reducing illumination levels by delamping and practicing regular cleaning and replacement of lighting fixtures;
  - maintaining temperature settings during the heating season at a maximum of 65 to 68°F and improving maintenance of cooling equipment;
  - reducing the operating time of machines; and,
  - changing from nighttime to daytime cleaning.
- Transportation. This includes:
  - cutting overall travel budget and specifically reducing funds available for air travel, thus causing a shift to less energy-intensive modes (train or bus);
  - using the telephone or other telecommunications in lieu of travel whenever feasible;
  - encouraging official travel in off-peak periods by common carrier whenever possible;
  - reducing mileage of federal vehicles (now up to 15 percent);
  - buying and leasing smaller cars; replacing large sedans and limousines;
  - substituting bicycles for automotive vehicles for use at federal facilities;
  - requiring tune-ups on all

government vehicles every 12,000 miles (or once a year) for a maximum operating efficiency;

- establishing a maximum of 50 mph for motor vehicles (now 55 mph), and reducing cruising speeds for aircraft and ships;
- substituting simulators for many flight activities and excusing certain flight personnel from proficiency flying; and,
- reducing ship steaming time.

- Federal Employee Activities. This includes:

  - encouraging carpooling and use of public transit;
  - encouraging employees to ride bicycles to work and providing bicycle racks at federal facilities;
  - conducting employee awareness programs through printed and other in-house media to encourage energy conservation practices in the office, at home, and in their communities; and,
  - establishing employee suggestion systems and awards programs to induce greater involvement in energy conservation.[25]

2. Voluntary Program Leading to Direct Regulation

In addition to the above-described existing laws and governmental programs now providing some degree of direct regulation, there exist several "voluntary" programs sponsored by the Federal Government and some state governments which likely will lead to direct and very comprehensive regulations of end-user behavior. Examples of such programs together with an explanation of their purpose and function follow.

- Voluntary Industrial Energy Conservation Program [26]

  This program began in early 1973 and is co-sponsored by the Federal Energy Administration (FEA) and the Department of Commerce (DOC). The program consists of two parts:

  - an Energy Management Program which is designed to benefit industry as a whole and individual companies and plants within that industry; and,
  - a reporting system which furnishes data to the Federal Government to enable government policy makers to develop rational energy policies and programs. One of the major aspects of the program is the monitoring and periodic reporting of energy conservation progress.

- State Energy Conservation Program[27]

Recognizing the potentially important role of states in stimulating energy conservation, the Federal Energy Administration in early 1975 began developing a program to provide a framework for state/federal cooperation in the furtherance of national energy conservation goals. During these early stages, FEA worked closely with the National Governor's Conference and, in the spring of 1975, the National Governors' Conference meeting in New Orleans endorsed this plan for a state/federal energy conservation program. Initiation of the program was then announced simultaneously by Governor Salmon of Vermont, Chairman of the National Governor's Conference Committee on Natural Resources and Environmental Management, and Frank G. Zarb, Administrator of FEA. The voluntary state/federal program, as was then envisioned, relied heavily on the participation of the states in a data classification system to be used as a base for energy forecasts, on a state-by-state basis, to the year 1985. For the purposes of this program, a state is defined as any one of the 50 states, the District of Columbia, Puerto Rico, Guam, American Samoa, the Virgin Islands, and the Trust Territory of the Pacific. After the data system was in place, the states could establish energy conservation goals and develop specific programs to meet those goals. FEA would provide technical assistance to help the states in their efforts.

On December 22, 1975, then President Ford signed into law the Energy Policy and Conservaton Act (EPCA). Title III, Part C of the Act authorized FEA to establish guidelines for the development and implementation of state energy conservation plans. The program is in many respects similar to the voluntary program originally initiated by FEA. It calls for the establishment of energy conservation goals for the reduction by 5 percent of actual state energy consumption in 1980 from projected consumption. The program is voluntary; states are not penalized if they elect not to participate.

There are, however, some important differences between the original program and the program established by EPCA. First, Section 365(d) authorizes $50 million for each of fiscal years 1976, 1977, and 1978. States can only receive these funds if they commit themselves to the objectives of the program. Second, the provisions of EPCA require that every state plan include a minimum of five specific energy conservation actions, which are:

- mandatory lighting efficiency standards for non-federal public buildings;
- programs to promote the availability and use of carpools, vanpools, and public transportation;
- mandatory standards and policies relating to energy efficiency to govern the procurement practices of a state and its political subdivisions;
- mandatory thermal efficiency standards and insulation requirements for new and renovated non-federal buildings; and,
- traffic laws or regulations which to the maximum extent practicable and consistent with safety, permit motor vehicles to turn right at a red light after stopping.

The Act also lists the following additional actions that the states may (not mandatory), include in their plans for energy efficiency:

- restrictions governing the hours and conditions of operation of public buildings;
- restrictions of the use of decorative or non-essential lighting;
- transportation control;
- programs of public education to promote energy conservation; and,
- any other appropriate methods or programs to conserve and improve efficiency in the use of energy.

The FEA has recently proposed that three more mandatory energy conservation actions be added to the five now required, which are:

- inter-governmental coordination in the area of energy use efficiency;
- a public education and awareness program; and,

- an energy audit program for buildings and industrial plants.

- State Level Programs (California)[28]

  Both the California Public Utilities Commission (PUC) and the State's Energy Resources, Conservation and Development Commission (ERCD) have programs aimed at encouraging utilities to institute and maintain aggressive programs designed to cause energy use efficiency at the end user level. Although participation in the program is to a large degree voluntary in nature, reporting requirements regarding progress are mandatory and the degree of cooperation and results achieved receive heavy weight at agency hearings regarding applications for the siting of new power plant facilities or rate increases.

  The above-described voluntary programs and others like them will likely lead to comprehensive regulations requiring and directing that:

  - energy efficiency "report cards" be submitted to the FEA or like governmental agency on a regular basis;
  - a prescribed level of energy use efficiency be maintained; and,
  - a precise and detailed procedure be adhered to in maintaining the efficiency standards.

Already this evolutionary process can be seen in the industrial sector in the form of mandatory reporting requirements imposed by the Energy Policy and Conservation Act combined with stringent, albeit still voluntary, efficiency targets which are to be developed by the Federal Energy Administration. Furthermore, once the state's energy conservation programs are finalized (22 now under submission to the FEA for approval), they will cause much direct regulation of end users at the state level.

3. Regulation through the Gradual Elimination of Inefficient Products or Structures

Government action (all levels) in this area presently emphasizes the development of:

- efficiency-descriptive labels on all things using energy;
- efficiency-causing building codes for new and existing structures; and,
- regulations requiring motor vehicles to meet specified fuel consumption standards.[29]

At the moment, the Federal Government's energy efficiency labeling program is in a state of flux. Initially, the Environmental Protection Agency developed mileage labels for 1974 model automobiles and encouraged car makers to affix these labels to their products along with the price stickers. Later the Federal Energy Administration joined the EPA in this endeavor, and the two agencies administered the voluntary program through mid-March 1976. On March 21, 1976, however, the automobile labeling program became mandatory under the terms of the Energy Policy and Conservation Act, and the EPA recodified its voluntary regulations and made them applicable to all cars manufactured in or imported into the US (the voluntary program had had a 95 percent compliance rate, and that part of the industry that participated in the voluntary program was basically unaffected by the mandatory rules).[30]

As for the labels for other items, the Commerce Department currently administers a voluntary program for household appliances, a program that will be taken over and revamped by the Federal Energy Administration under another mandate contained in the Energy Policy and Conservation Act. FEA has been directed to establish test procedures, labeling rules and energy efficiency standards for the following products: refrigerators and refrigerator-freezers, freezers, dishwashers, clothes dryers, water heaters, room air conditioners, home heating equipment, not including furnaces, television sets, kitchen ranges and ovens, clothes washers, humidifiers and dehumidifers, central air conditioners, furnaces, and any other product that is determined to be likely to use more than 100 kW/hr (or Btu equivalent) a year.[31]

While the FEA goes about setting up its program, the Commerce Department's voluntary labeling regulations and specifications remain in effect. Commerce, prior to enactment of the energy policy law, had labels in effect for general guidance, air conditioners, refrigerators, refrigerator-freezers, and freezers, and had proposed guidelines for labeling water heaters and clothes dryers.

In addition to the Commerce Department's voluntary program, and prior to the new FEA labeling authority, the Federal Trade Commission proposed a set of mandatory labeling specifications for air conditioners. That proposal was later withdrawn by FTC.[32]

Eventually, FEA plans to have a comprehensive double-edged program which will include:

- the labeling of virtually everything manufactured which will be aimed to better inform the consumer; and,
- efficiency standards aimed at industry and requiring that products made comply with such standards.

In the case of automobiles, this approach has already been implemented. With reference to other products, a recent statement of current administration policy indicates that:

- the present appliance efficiency program will be strengthened;
- voluntary targets will be replaced by mandatory standards on certain home appliances, such as air conditioners, furnaces, water heaters, and refrigerators as soon as possible; and,
- the program under existing law to develop test procedures and to establish labeling requirements for appliances will be continued.[33]

Many states and municipalities are developing or already have voluntary labeling programs of their own. In some states, mandatory efficiency standards are either a matter of law or under serious consideration. For example, in California, minimum efficiency standards are now set by law for refrigerators, refrigerator-freezers, and air conditioners and soon the law will be expanded to cover water heaters, televisions, clothes dryers, washing machines, cooking appliances, dishwashers, plumbing fixtures, gas appliances and space heaters. Furthermore, California law will ultimately prohibit continuous burning pilot lights in favor of intermittent ignition devices.[34] It appears, however, that the federal programs will be expanded and will dominate the labeling and efficiency standards areas so that a consistency of action and legislation can be insured.[35]

Much work has been expended at the federal level to develop a basic national building code, but such a result has not yet been achieved. Generally, government at all levels has been studying both prescriptive and performance standards aimed at causing new residential and commercial structures to use energy more efficiently. The inclusion of such prescriptive or performance standards into design is being pursued through one of the three following approaches:

- by promoting their acceptance by professional organizations as standard operating procedures;
- by causing their inclusion in state and local building codes; or,
- by indirectly forcing their use through some conditioned mechanism in the institutional processes of the building industry.

In August of 1976, the Energy Conservation Production Act (ECPA) was passed. By this act the Department of Housing and Urban Development (HUD) has been charged with the responsibility of formulating national building standards by 1981. The standards developed by HUD must be performance rather than prescriptive in nature. In the meantime, the FEA is making every attempt to cause states to adopt building standards consistent with those propounded by the American Society of Heating, Refrigerating and Air Conditioning Engineers in their work, commonly known as ASHRAE-75. The ASHRAE standards are prescriptive rather than performance oriented. Aiding the FEA in their effort to cause acceptance by the states of ASHRAE-75 standards is a provision of the Energy Policy and Conservation Act (EPCA) to the effect that a state's participation in the State Energy Conservation program is conditioned on it adopting into law efficiency standards at least as restrictive as those contained in ASHRAE-75.[36]

The inclusion of prescriptive standards, such as the Department of Housing and Urban Development (HUD) "51 (B)" changes in the "Minimum Property Standards," in conditions for residential financing, is an example of the use of conditioned mechanisms in institutional processes.

At the state level, the development of codes designed to cause the building of energy efficient structures has progressed faster than at the federal level. As early as March of 1976, the following state progress has been achieved:

- Building Energy Regulatory Authority:
  - In statewide code--17 states;
  - In separate law--4 states;
- Regulations:
  - Promulgated--10 states;
  - Under consideration--14 states;
- Authority being considered:
  - By legislature--5 states;
  - By staff or study groups--3 states.[37]

Furthermore, twenty-two states have submitted programs to the FEA under the State Energy Programs.[38]

For the purposes of this discussion regarding direct regulation of end-user behavior, it is important to note the following typical provisions which exist or will exist in most of the new building codes:

- upon completion of the structure or a portion thereof, a card certifying conformance with the requirements of the regulations must be completed, signed by the builder, and approved by a representative of a responsible governmental agency;
- the card denoting approval of the structure (or part thereof) must be posted in a conspicuous location on or in the structure; and,
- until the above procedure has been complied with, the final inspection and ultimately, occupancy itself, cannot be commenced.[39]

The provisions set out above clearly demonstrate how building codes can and will effectively inhibit or prohibit action by end users.

In addition to the rapidly advancing growth of building codes in the "new" construction area, several retrofit programs are gaining momentum. For example, the Federal Energy Administration (FEA) and Department of Commerce (DOC) are aggressively pursuing the development of such a program for national use.[40] This area will most certainly lead to laws requiring that owners (and hence energy users) of structures make certain modifications to existing structures so as to meet prescribed energy efficiency standards as a condition of continued use and/or transfer of legal title.

C. Use of Financial Pressure

1. Overview--Over the years government has substantially influenced behavior through the use of direct or indirect financial pressure.

Sometimes the pressure has taken the form of a reward, i.e., tax rebate, and sometimes the form of a penalty, i.e., tax surcharge. In the energy field, numerous proposals exist but little law has evolved as yet to effectively cause energy use efficiency through financial pressure.[41]

Although the precise manner in which financial pressure will be exerted is still the subject of much debate, the basic approaches by which government will exert the pressure appear clear and may be summarized as follows:

- by use of a new utility rate (tariff) design that more accurately communicates the true value of energy to the end user;
- by use of the taxing system to encourage efficient and discourage inefficient action;
- by way of making loans or granting direct financial aid to persons and business to:
  - encourage voluntary efficiency causing expenditures; or
  - enable compliance with regulations requiring such expenditures to be made;
- by way of refusal or the threat of refusal of loans or aid; and,
- by way of termination or the threat of termination of existing lending, aid, or related funding programs.[42]

2. Overhaul of Tariff Structure--As distinguished from direct curtailment measures which at present are designed for emergency use only, this rate-design approach is one based on "cost analysis."[43] With the possible exception of those utilities under the jurisdiction of the Federal Power Commission (FPC), the majority of all proposals in this area contemplate that the actual tariff adjustment or restructuring be accomplished locally or regionally through appropriate legislation or utility rulings rather than by some "Federal Grand Design."[44] It has been determined that due to the unique problems existing in different areas of the country any national uniform-rate structure or procedure would be totally impractical and in fact unworkable.[45]

The first clear-cut expression of federal policy on this subject is found in Title II of the Energy Conservation and Production Act passed into law on August 14, 1976. The stated purpose of the Title II provisions is to require the Federal Energy Administration to:

- develop proposals for the improvement of electric utility rate design and transmit such proposals to Congress:
- fund electric utility rate demonstration projects;
- intervene or participate upon request in the proceedings of utility regulatory commissions; and,
- provide financial assistance to state offices of consumer services to facilitate presentation of consumer interests before such commissions.[46]

The Title II provisions go on to provide that:

- proposals shall be designed to encourage energy conservation, minimize the need for new electrical generating capacity, and minimize costs of electric energy to customers and shall include (but not be limited to) proposals which provide for the development and implementation of:
  - load management techniques which are cost effective;
  - rates which reflect marginal cost of service, or time of use of service, or both;
  - rate-making policies which discourage inefficient use of fuel and encourage economical purchases of fuel; and,

- rates (or other regulatory policies) which encourage electric utility system reliability and reliability of major items of electric utility equipment.

- the proposals prepared shall be transmitted to each House of Congress not later than 6 months after the date of enactment of the Act, for review and for such further action as the Congress may direct by law.[47]

A more recent expression of federal philosophy on this subject can be seen in the current administration's proposed energy policy, pertinent portions of which are set forth below.

- Conventional utility pricing policies discourage conservation. The smallest users commonly pay the highest per unit price due to practices such as declining block rates. Rates often do not reflect the costs imposed on society by the actions of utility consumers. The result is waste and inequity. The President therefore will submit legislation which contains the following provisions:

    - state public energy commissions must require their regulated electric utilities to phase out and eliminate promotional, declining, and other rates for electricity that do not reflect cost incidence;

    - to shift energy use from peak to non-peak periods, electric utilities would be required to offer daily off-peak rates to each customer who is willing to pay metering costs and to offer lower rates to customers willing to have their power interrupted at times of highest peak demand;

    - master metering for electricity would generally be prohibited in new structures;

    - state public utility commissions would require gas utilities to eliminate declining block rates and to implement such rules as FPC may prescribe with respect to master metering summer-winter rate differentials and interruptible rates; and,

    - by amendment to the Federal Power Act, the Federal Power Commission would be authorized to require interconnection and power pooling between utilities even if they are not presently under FPC jurisdiction, and to require "wheeling" (the transmission of power between two noncontiguous utilities across a third utility's system). [48]

3. Use of Taxing System--At present, there is no question that this device will be a major tool in the government's effort to cause consumer efficiency in the use of energy. The area of uncertainty is over how this tool will be used. The following proposals have been under consideration, for some time:[49]

- a windfall profits (income) tax on utilities designed to absorb extraordinary profits that may be generated through substantial modification of "base rate" computations;

- a tax credit or deduction for improvements to existing homes (retrofitting) to increase thermal efficiency;

- the investment tax credit for utilities increased so as to be comparable with that granted other businesses;

- a preferred stock dividend deduction, aimed primarily at the utilities, for the purpose of reducing the cost of capital and stimulating equity rather than debt financing;

- a tax credit or deduction for

expenditures in installing a new energy system, i.e., solar energy equipment, having a fuel consumption rate that complies with government established efficiency standards;

- a refundable tax credit for certain retrofit activities;
- an excise tax on use of electricity beyond a predetermined base; and,
- a real property tax exemption relative to the installation of a new energy system, or substantial retrofit activity. For example, the state of North Dakota already has such installation of machinery and equipment systems that utilize solar energy for heating and cooling in new and existing buildings. The North Dakota law allows the property owner to deduct annually (for five years following installation) from the assessed valuation of the property a sum equal to the lesser of:
    - the remainder of the assessed valuation of the property with the solar system included, *minus* the valuation of the property without the solar system; or,
    - $2,000.

Many of the above-listed tax proposals are likely to be combined in legislation so as to be more effective in achieving a desired goal. An excellent example of the use of a combination of various tax-related devices to accomplish an energy efficiency causing objective is the current administration (President Carter) energy plan submitted to Congress on April 20, 1977. Among the President's proposals were the tax-oriented devices described below.

- An exemption from federal and state public utility regulations would be available to industrial cogenerators.
- Tax increases of 7 to 11 cents on aviation fuel and an end to a two-cent-a-gallon rebate on gasoline used in motor-boats.
- Removal of a 10 percent excise tax paid by passengers who ride on intercity bus lines.
- A 10 percent tax credit for businesses that invest in energy conservation.
- Returning increased oil and gasoline taxes to consumers through a federal tax rebate, likely starting at $15 and rising to $25. The rebate is intended to compensate low income families hit hard by sharply higher energy costs.
- A standby gasoline tax increase, starting at five cents per gallon and rising to 50 cents after 10 years if gasoline consumption fails to decline.
- A tax rising to a maximum of $2,488 by 1986 on each gas-guzzling car, accompanied by a rebate of up to $493 for fuel-efficient cars.
- Tax credits of up to $410 for homeowners who insulate their houses or take other steps to make their residences more energy efficient.
- Tax credits of up to $2,000 for homeowners who install solar heating and cooling devices.
- Imposing a new tax on crude oil in addition to the hike in gasoline taxes. The crude oil tax could raise the price of gasoline another seven cents per gallon.[50]

4. Loans, Direct Financial Aid, or Discontinuance or Refusal Thereof

For some time, agencies of the Federal Government have been conducting lending and related programs designed to aid those suffering from energy shortages and those who desire to make expenditures designed to cause energy use efficiency. [51] An example of such a program is that sponsored by the Small Business Administration (SBA) to assist small firms adversely affected by a fuel shortage. The SBA loans can be used for working capital, to pay existing financial obligations, to refinance debts, or to convert operations to a different fuel source.[52]

Passage of the Energy Conservation and Production Act (Public Law 94-385) in 1976 greatly accelerated the use of this tool as a means to cause or encourage energy use efficiency. Fundamentally, ECPA provides for:

- FEA guarantee of a loan made by a conventional lending institution for:
  - retrofitting with energy efficient devices; or,
  - retrofitting with renewable resource devices, i.e., solar, geothermal or wind power generating sources; and,
- a Weatherization Assistance Program to be administered by the states, or if they fail to do so, local governmental entities, whereby such entities may apply to the FEA for a grant to be used to weatherize low-income housing.[53]

The current Administration suggests a more extensive use of this strategy as may be seen from the summary of the current proposals set forth below.[54]

- A federal grants program will assist public and non-profit schools and hospitals in installing conservation measures, funded at the rate of $300 million per year for three years.
- The President has requested that the national 55 mph speed limit be vigorously enforced by states and municipalities. The Secretary of Transportation may, if he finds it necessary, withhold highway trust fund revenues from states not enforcing the limit.
- The Secretary of Housing and Urban Development will advance by one year, from 1980 to 1981, the effective date of the mandatory standards required for new residential and commercial buildings by the Energy Conservation and Production Act, with funds to be made available to states to help them in this effort.
- By reducing gasoline consumption, state revenues from gasoline taxes would also be reduced. These funds are used by the states for repair and maintenance of highways. The Administration will develop a program which will reduce their hardships and, to insure adequate highway maintenance, will compensate them for this loss through sources such as the Highway Trust Fund.
- The Federal Government will remove the barriers to opening a secondary market for residential energy conservation loans through the Federal Home Loan Mortgage Corporation and the Federal National Mortgage Association. This action should help to ensure that capital is available to homeowners at reasonable interest rates for residential energy conservation through private lending institutions.
- Funding for the existing low-income residential conservation program (weatherization) will be increased to $130 million in fiscal year 1978; and $200 million in FY 1979 and in FY 1980 (budget).
- The Secretary of Labor will take all appropriate steps to ensure that recipients of funds under the Comprehensive Employment and Training Act (CETA) will supply labor for the residential conservation program. The CETA program's employment levels, as proposed by the Administration, would

meet the labor requirements of the program.

- The Secretary of Agriculture will vigorously implement a rural home weatherization program in cooperation with the nation's 1,000 Rural Electric Cooperatives, with loans provided through the Farmer's Home Administration.
- State public utility commissions will be required to direct utilities to offer their customers a residential energy conservation service performed by the utility and financed by loans repaid through monthly utility bills.

D. Educational Programs

Government at all levels is actively involved in the collection and dissemination of data in an effort to cause energy end-use efficiency. The private sector, particularly regulated utilities, has also been required to participate with governmental bodies.[55] Examples of government activities in this area are described below.

1. Big Three Program[56]

This program, started in late 1976, consolidated several existing FEA administered educational programs including:

- lighting and thermal operations for commercial buildings;
- vanpooling; and, to a degree,
- the voluntary industrial energy conservation program.

Basically, the program involves a series of seminars in major metropolitan areas of the US. The initial seminars are to involve chief executive officers of high-intensity, energy using firms of all types, where the general concept of conservation will be discussed from both an ideological and an economic viewpoint. There will be 100 seminars held throughout the country of this category. Following the chief executive seminar, there will be three other seminars which is the reason this program received the name "Big Three." These three additional seminars are to be directed to the following problem areas:

- transporation - 190 seminars will be held to discuss such things as vanpooling and other energy use efficiency causing techniques;
- general building maintenance and light retrofitting projects - 200 seminars will be held; and,
- industrial - 190 seminars will be held to discuss such things as equipment energy use efficiency.

There will be approximately five million dollars diverted to this program.

2. Small Business Conservation Program

Administered by the FEA, this program is designed to serve a similar service to the small business people that the "Big Three" program serves for the large energy users.[57]

3. Public Schools Energy Conservation Service Program (PSECS)

This program is administered by Educational Facilities Laboratories, Inc., in conjunction with the FEA. The Public Schools Energy Conservation Service (PSECS) is a product of eighteen months of research, development and testing, involving more than 200 school districts and 1500 elementary schools. By taking advantage of recent advances in engineering and computer technologies, PSECS is able to establish energy use guidelines for electricity and fuel for most elementary school plants. These guidelines take into consideration such factors as climate, pattern of use, and special physical characteristics of the facility. By comparing the actual use with the guideline, the plant's potential for both energy and dollar savings is determined. The Service has been designed to work interactively with school and district staff members, and provides modification information on two levels. On level one, the operating level, the PSECS Self-Audit Program provides suggestions for bringing the plant down to guideline levels with little or very modest expenditures. Experience to date indicates that savings of 25 to 30 percent can usually be achieved if the PSECS operating suggestions are followed. At level two, capital modifications requiring larger investments are analyzed, using simple life cycle cost procedures to obtain a selection of cost effective modifications specific to each school. This computer analysis is intended to indicate those modifications that have sufficient merit to justify the employment of design professionals to carry out a more detailed analysis.[58]

- All existing or proposed voluntary programs sponsored by the public and private utilities, Department of Commerce, or other agencies designed to increase the efficiency of energy use.
- FEA, GSA, and DOC guidelines dealing with improving the energy use efficiency of new and existing buildings.
- FEA sponsored "Conservation Award" which will be a highly publicized recognition of energy efficiency-causing efforts.
- Current FEA efforts to disseminate information regarding fuel conservation and use efficiency through advertisements on television and in newspapers throughout the US.
- FEA intervention in selected legal actions or related utility rate increase application hearings at the state level in order to advance its policies regarding utility tariff rates and related subjects.

The current administration is continuing to encourage the use of educationally oriented programs to cause energy use efficiency as can be seen from a review of the proposals for new action, set forth below.[59]

- The Federal Government will spend up to $100 million over the next three years to add solar hot water and space heating to suitable federal structures to help demonstrate the commercial potential of such measures.
- The utilities should be required to inform customers of all available residential conservation programs and how to obtain financing, materials and labor to perform residential conservation themselves. Other fuel suppliers would be encouraged to offer similar programs, with the help of their state energy offices.

E. Use of Power as a Major Customer

This is emerging as an effective method to expedite "market penetration" of energy efficiency-causing hardware. Already this approach has had significant impact through the Federal Energy Management Program and like endeavors. In the immediate future, predictable action in this area will include:

- the purchase of only that equipment which complies with the government's energy use efficiency standards;
- an express prohibition on the use of equipment that does not comply with energy use efficiency standards;
- the requirement that all structures occupied by the government meet government energy use efficiency standards;
- the requirement that only equipment or building and design techniques which comply with the government's energy use efficiency standards be used in connection with construction or like projects for or financed by the government; and,
- by way of a long-term commitment to buy or lease in large quantities, providing financial support of private industry in the development of products which use energy more efficiently.[60]

One of the more recent examples of how this approach may be used can be seen in the current Administration's energy plan advanced by the President on April 20, 1977 Details of the President's proposals are set forth below.

- The President will direct federal agencies to alter their auto purchasing practices so that new cars purchased by the Government will, on the average, exceed the average fuel economy standard under the EPCA by at least 2 mpg in 1980 and thereafter.
- The Secretary of Commerce will encourage state and local governments to include items that will contribute to energy conservation in their proposals under the Department's Local Public Works program.
- The President will direct all federal agencies to adopt procedures which aim at reducing energy use per square foot by 1985 by 20 percent from 1975 energy consumption levels for *existing* federal buildings and by 45 percent for *new* federal buildings. Investments which are not cost-effective would not be funded under the program. The Director of the Office of Management and Budget and the Administrator of the Federal Energy Administration will implement this program.
- Legislation will be proposed to initiate a federal vanpooling program. This program will demonstrate the energy conservation and pollution control potential of this form of commuter transportation by the largest employer in the nation. About 600 vans will be purchased by the Government and made available for use by the federal employees. All costs of the program will be repaid to the Federal Government by the riders.

F. Grant of Special Exceptions to or Exemptions from Existing Law or Governmental Policies

Often government has influenced behavior by excepting certain actions from the application of existing regulations. A classic example of this approach is in the tax field where deductions, exemptions, and similar devices abound. In this regard, the current Administration has made the following proposals in its energy plan unveiled on April 20, 1977:[61]

- removal of a 10 percent tax paid by passengers who ride on intercity bus lines; and,

- an exemption from federal and state public utility regulations would be available to industrial cogenerators.

In the area of pricing of energy, the use or proposed use of this approach is most evident. The current administration has stated recently that prices should generally reflect the true replacement cost of energy and that US citizens are only cheating themselves if they make energy artificially cheap.[62] The problem is that energy in the US is made artificially cheap, by various regulations, particularly when US pricing is compared with world market level energy costs.[63] Since deregulation of prices is not a viable solution politically in the US at this time, the current Administration is proposing to address the problem more conservatively by use of the process of selective exceptions, exemptions and related devices.[64] Examples of how the current Administration would like to use this technique to address the artificially low prices are set forth below.

- Regarding Oil Pricing[65]
  - Continue indefinitely the current price ceilings of $5.25 and $11.28 per barrel for previously discovered oil, subject only to escalation at the general rate of inflation.
  - Define newly discovered oil as oil from a well drilled more than 2-1/2 miles from an existing on-shore well as of April 20, 1977, or more than 1,000 feet deeper than any well within any 2-1/2 mile radius. New oil offshore will be limited to oil from lands leased after April 20, 1977.
  - Allow newly discovered oil to rise over a three-year period to the current world price (adjusted for the rate of inflation); thereafter, newly discovered oil would continue to be priced at the 1977 world price with adjustments for domestic increases in the general rate of inflation.
  - Incremental tertiary recovery from old fields and stripper oil would be free of price controls.
- Regarding Natural Gas Pricing[66]
  - Subject all new gas, sold anywhere in the United States, to a price limitation of the Btu equivalent of the average refiner acquisition cost (before tax) of all domestic crude oil. That price limitation would be approximately $1.75 per Mcf at the beginning of 1978; the interstate-intrastate distinction would disappear for new gas.
  - Define new natural gas using the same standards as are used to define newly discovered oil (2-1/2 miles, 1,000 feet, new leases).
  - Guarantee price certainty at current levels for currently flowing gas, with adjustments to reflect inflation.
  - Authorize the establishment of higher incentive pricing levels for specific categories of high cost gas.
  - Allow gas made available at the expiration of existing interstate contracts or by production from existing reservoirs in excess of contracted volumes to qualify for a price no higher than the current $1.42 Mcf ceiling adjusted for inflation; gas made available under the same circumstances from existing intrastate production would qualify for the same price as new gas; i.e., $1.75 per Mcf at the beginning of 1978.

- Allocate the cost of the more expensive new gas to industrial users, not to residential and commercial users.
- Extend federal jurisdiction to SNG facilities guaranteeing them a reasonable rate of return.

- Other Oil and Gas Measures[67]
  - Inclusion of North Slope Oil in the domestic composite price under the current provisions in EPCA would introduce a degree of unnecessary uncertainty into domestic crude oil pricing. The $5.25, $11.28 and new oil pricing tiers, adjusted for inflation, would be substituted for the composite average limitation. Alaskan Oil would be subject to an $11.28 wellhead ceiling price, but would be treated as foreign oil for purposes of the entitlements program. New Alaskan oil finds would be subject to the new oil wellhead price.
  - Because of the high risks and costs involved in shale oil development, shale oil will be entitled to receive the world price of oil in the United States.

Another area where this technique has been used is the patent field. In October 1973, a program was announced aimed at encouraging more rapid development and commercialization of energy-related inventions by according applications for patents on such inventions a "special" status that will permit them to be processed in much less time than is normally required. Any patent application for an invention which materially contributes to the discovery or development of energy resources, or the more efficient utilization and conservation of energy resources may, if the inventor wishes, be included in the program. Examples of inventions would be those relating to further developments in fossil fuels (natural gas, coal and petroleum), nuclear energy, solar energy, inventions relating to the reduction of energy consumption systems, industrial equipment, household appliances, etc. In the first year of the program, the Patent Office afforded special processing to over 100 applications.[68]

The likelihood of additional government action in this area is good, particularly in its effort to strike a balance between the developing fuel conservation laws and existing environmental protection rules and regulations.

G. Establishing Priorities for Research and Development

Although this area in part involves financial pressure in the form of large expenditures of money for predetermined research and development priorities, the expression of governmental emphasis alone is a potent "cause and effect" device. ERDA acknowledged the existence of this tool in its current plan (ED-76-1) where it states that the planning process is a useful mechanism because the Federal Government can use such an approach as one context for its own actions *and* as a way to promote consensus on the nation's approach to energy RD&D. Many states and utility companies have their own plans for research, development, and demonstration which are equally influential in directing the development and market penetration of both supply and conservation technologies. [69]

H. Summary of Strategy Used by Government to Cause Energy End Use Efficiency/Application in the Future

Up to the present time, the above-described regulatory and policymaking activities have been used primarily to *expedite market penetration* of end users, energy efficiency-causing devices or programs in the areas of:

- motor vehicles;
- major appliances, i.e., refrigerators, freezers, dryers,

air conditioners, etc.;

- building standards with particular emphasis in the area of insulation, lighting, and design;
- heat pumps and related devices;
- utility rate reform;
- voluntary energy use management in the commercial and industrial sectors; and,
- miscellaneous small hardware for use (voluntarily) by end users (particularly in the residential area) which expedite energy use efficiency, i.e., light bulbs, watt watchers, light dimmers and regulators, water flow regulators and restrictors, air deflectors, weather stripping terials, thermostats, air circulating devices (thermocyclers), filter clog control devices for air conditioner night set-back devices, and flow control devices.

Although vigorous governmental support for the above-described activities continues in the form of existing and proposed regulations and policy, a *significant increase in pressure* is emerging to expedite the development and implementation of *coordinated* energy conservation technologies at a level where end use efficiency can be effectively monitored and controlled and which have an acceptable impact on reducing the energy use growth rate.[70] This emphasis appears to be based in part on the factors described below.

- Causing significant voluntary participation by the mass population, particularly in the residential and small commercial areas, is *quite difficult* because the long-range success of such a program is dependent on the coexistence of the following factors:
  - people must be *convinced* there is a serious problem which affects them *personally*;
  - technology must provide tools for people to solve the problem;
  - the tools developed must have market penetration capability, i.e., people must be able to afford them;
  - when applied, the tools must have a visible and curative impact on the problem; and,
  - acceptance by people that the problem exists must continue for a period long enough that attempted solutions can achieve success.[71]
- Reduction in the growth rate of utility load demand, and if possible, overall energy use must be achieved to significantly control the need for new high cost generating facilities.[72]

With the passage of the Energy Policy and Conservation Act (EPCA) in December of 1975 (PL 94-163), the Federal Government placed the development of conservation technologies on a par with supply technologies. The EPCA has as its stated purpose to "reduce domestic energy consumption through the operation of specific voluntary and mandatory conservation programs." ERDA's current plan reflects the EPCA emphasis on the development of conservation technologies. Many states, including California, have passed energy legislation which places considerable emphasis on the research and development of conservation technology. California is significant to watch since its regulatory programs are advanced and accepted by the Federal Energy Administration (now DOE) as models for other states to follow in its State Conservation Program.[73]

Set forth below are anticipated future governmental activities which will have a significant influence on programs designed to develop or improve energy conservation technologies.

- The Federal Government will look to the states and utilities to develop and implement conservation technologies consistent with certain minimum goals.[74]

- The state utility regulating agencies will start, or continue to *instruct* the utilities in their jurisdiction to develop and implement vigorous energy efficiency-causing programs at the "end user" level which will require:

    - elaborate reporting procedures regarding their programs and the progress thereof;
    - stepped-up research and development programs;
    - mass media educational programs; and,
    - various market penetration programs for efficiency-causing systems and hardware, i.e., building insulation.[75]

- Permission to utilities for rate increases and new plant construction will be *conditioned* on the existence of acceptable energy efficiency-causing programs. In California, the Public Utilities Commission (PUC) and ERCDC clearly intend to pursue this type of approach.[76]
- Programs designed to *reduce* the utility load growth rate can be expected. For example, a study conducted by consultants to the California Assembly Committee on Resources, Land Use and Energy indicated that if all programs anticipated by the California Energy Resources Conservation and Development Act (Publ. Rec. Code 25,000 *et. seq.*) were implemented, there would be very little impact on the load growth over the next ten years.[77]
- Pressure is developing to cause at least limited decentralization of generating facilities. In California, legislation is now pending which would permit "wheeling" by major users (AB-4069-Warren) in order to meet growth needs. This concept has significant support from the large industrial users of electricity.[78]
- Consumer protection is receiving high priority. Solar equipment and home appliances are two areas of immediate emphasis.[79]
- Mandatory efficiency standards for all significant manufactured products, i.e., automobiles, home appliances, and buildings, will be developed.[80]
- Various utility rate reform plans are developing and, in some states, have been implemented.[81]
- Increase in government financial assistance or pressure in the lending and tax areas can be expected to expedite development of energy conservation technologies.[82]
- National minimum energy efficiency performance standards for all new buildings will be developed, combined with more emphasis on the thermal and lighting efficiency standards of existing structures.[83]
- Use of funds available for research will be used more effectively to implement policy designed to cause energy use efficiency.[84]
- Increased national emphasis will be placed on a combination of price/taxing strategies.[85]
- Use of monetary rewards for efficiency oriented consumers and products and like penalties for those that are wasteful will increase.[86]
- Numerous programs designed to increase utilization of coal, solar, and nuclear sources of energy will be pressed.[87]

D.4 NEED TO ELIMINATE OR MODIFY EXISTING LAWS AND GOVERNMENTAL POLICIES

One of the most important tasks for lawmakers in developing energy use efficiency programs will be to identify and cause the repeal or appropriate modification of existing laws and underlying policies which directly or indirectly prohibit or inhibit the expeditious implementation of proven technical solutions. Areas which need immediate attention include:

- environmental laws in general;
- procedures governing coal extraction and utilization;
- utility rate restructuring and procedures involved in computing rates;
- siting requirements and procedures for nuclear power plants;
- zoning, land use and related laws, with particular reference to their present application to the use of solar energy facilities or other new and exotic systems for creating energy;
- taxing laws with particular reference to the elimination of de-incentives;
- anti-trust laws with particular reference to the allowance of the pooling of information and technology and elimination of wasteful duplication of action; and,
- safety oriented laws, i.e., OSHA regulations, with particular reference to thermal and lighting standards.[88]

Studies have commenced in the indicated areas, but the development of meaningful conclusions will progress at a slow pace.[89] At this point in time, the great majority of action centers around the conflict between laws and policies designed to protect the environment and laws or proposed laws designed to solve the energy shortage problem.[90]

In 1970, the National Environmental Policy Act of 1969 (NEPA) was enacted to issue a mandate to all federal agencies to consider the environmental impact of their actions. NEPA requires federal agencies to prepare environmental statements on major federal action (including proposals for legislation significantly affecting the quality of the human environment). NEPA also established the Council on Environmental Quality (CEQ). Thereafter, the Clean Air Act was passed, designed to eliminate, among other things, air pollution and pollution causing devices. The Environmental Protection Agency (EPA) now coordinates the overall program in conjunction with the Federal Energy Administration (FEA). States and local municipalities followed suit by establishing comparable laws and agencies and the courts, through interpretation of laws, further expanded and strengthened regulations in this area.

Problems have now developed which center on:

- environment-protecting devices causing more inefficiency in energy use; and,
- environment-protecting regulations prohibiting or greatly inhibiting the use of a fuel, i.e., coal, or establishment of an energy creating installation, i.e., a nuclear power plant.

Steps to resolve this environmental protection/energy shortage conflict have already begun in the form of the Energy Supply and Environmental Coordination Act (ESECA) of 1974 and the Federal Non-Nuclear Energy Research and Development Act of 1974 (FNERDA). ESECA causes certain amendments to the Clean Air Act but expressly states its purpose to be to provide for a means to assist in meeting the essential needs in the US for fuels in a manner which is consistent to the fullest extent practical with existing national commitments to protect and improve the environment. FNERDA also purports to manifest the policy of Congress to develop, on an urgent basis, the technical capabilities to support the broadest range of energy policy options through the use of domestic resources by socially and environmentally acceptable means. Most certainly, the two above-described acts are indicative of how future laws, at all levels, will develop in the continuing effort to resolve the environmental protection/fuel shortage conflict.

In addition to the above-described legislation (ESECA and FNERDA), a methodology has been developed by the Federal Energy Administration (FEA) for assessing the environmental consequences of all national energy policies.

As touched on above, this FEA function is carried out pursuant to Section 102 (2) (c) of the National Environmental Policy Act of 1969 (NEPA) and Executive Order 11514 (35 FR 4247). The FEA approach is a multi-step process in which the relationship between the following elements are examined:

- proposed energy policy;
- economic implications on energy users and suppliers;
- impact on the production and consumption of the various fuels in the short- and long-term;
- impact on the rate of development of energy facility construction and application of pollution controls; and,
- resultant environmental impacts.

The environmental impact of three national energy policies has already been assessed by the FEA in the following statements:

- *Draft Environmental Impact Statement--Energy Independence Act of 1975 and Related Tax Proposals* (March 1975);
- *Draft Environmental Impact Statement--Mandatory Oil Import Program* (June 1975); and,
- *Draft Environmental Impact Statement--Electric Power Facility Construction Incentives Act of 1975* (July 1975).

The FEA is now assessing each of the state energy plans submitted to it in connection with the State Energy Conservation Program.[91]

There is mounting evidence indicating how policy makers intend to reconcile the conflict between laws designed to protect citizens and their environment and those which advance a worthy cause, i.e., energy supply and conservation technologies, but which might tend to threaten the safety of citizens or the quality of their environment. Fundamentally, the approach involves the requirement that in the process of propounding or implementing laws designed to protect citizens and their environment, the economic and employment impact must be carefully considered and given significant weight.[92] Pertinent examples on proposed laws addressing the conflict are set forth below.

- Toxic Substances Control Act [93]--Employment effects must be evaluated in implementing this Act, including reduction in employment or loss of employment from threatened plant closures. Any employee can cause an investigation to be made and public hearings to be held if faced with actual or threatened discharge or lay-off or otherwise faced with adverse or threatened adverse effects on employment.

- Resource Conservation and Recovery Act of 1976[94]--Under the provisions of this Act, the administrator (Environmental Protection Agency) must conduct continuing evaluations of potential loss or shifts of employment which may result from the administration or enforcement of the provisions of the Act and applicable implementation plans, including, where appropriate, investigating threatened plant closures or reductions in employment allegedly resulting from such administration or enforcement. Every employee who is discharged, or laid-off, or otherwise discriminated against by any person because of the alleged results of such administration or enforcement may request the administrator to conduct a full investigation of the matter including the holding of public hearings.

- HR 6161 (April 6, 1977) Proposed Amendment to the Clean Air Act[95]--This bill calls for an amendment to the Clean Air Act by adding a new section requiring that an "Economic Impact Statement" be made before publication of notice of proposed rulemaking with respect to any standard or regulation covered by the section. The required statement must contain an analysis of the following factors with respect to any standard or regulation:

- the costs of compliance with any such standard or regulation, including the extent to which the costs of compliance will vary depending on:
    - the effective date of the standard or regulation; and,
    - the development of less expensive, more efficient means or methods of compliance with the standard or regulation;
- the potential inflationary or recessionary effects of the standard or regulation;
- the availability of capital to procure the necessary means of compliance with the standard or regulation;
- the direct and indirect effects on employment of the standard or regulation;
- the effects on competition of the standard or regulation, particularly the effects on small business;
- the effects of the standard or regulation on consumer costs, including costs especially affecting economically vulnerable segments of the population;
- *the effects of the standard or regulation on energy use or availability;*
- the impact of the standard or regulation on productivity;
- the impact of the standard or regulation on the nation's balance of payments;
- the economic impact of postponing the standard or regulation or of not promulgating such standard or regulation;
- alternative methods to such standard or regulation for achieving equal or greater degree of emission reduction (or health or environmental protection) at lesser economic costs;
- comparative expenditures required to achieve incremental levels of reduction of emissions (or enhancement of health or environmental protection); and,
- any possible alternative for minimizing or eliminating part or all of any adverse economic impacts of such standard or regulation.

D.5 INTERNATIONAL ENERGY MANAGEMENT PROGRAMS AND POLICIES

A. General Approach

In evaluating US efforts to cause more efficient use of energy, it is helpful to view such activity from an international perspective. Internationally, US interaction with other countries of the world is coordinated by the Office of International Energy Affairs within the Federal Energy Administration. This office performs the following basic functions.

- identifies, analyzes, develops, proposes, and coordinates US international energy policies to assure appropriate interface between domestic and foreign energy entities as they relate to international industries, producer and consumer countries, their associations, and the relationships between and among these entities and the US Government;
- provides assessments of the international availability of all types of fuels as well as projections of the international environment in which the United States will seek to meet its future energy requirements;
- works worldwide to develop US positions in cooperation with

other major energy importers in the areas of demand restraint, emergency sharing, transportation of energy materials, and maritime and environmental questions; and,

- evaluates the adequacy of the following:
  - energy resources in physical terms;
  - stability of contractual arrangements for their acquisition;
  - the firms acquiring such resources for the United States; and,
  - the collateral logistics and refining systems.[96]

At the international level, efforts to resolve the many worldwide energy related problems appear to be developing the following definitive patterns of action:

- formation of energy management and related programs among countries that already have close economic ties, i.e., Common Market countries, within existing agencies;
- formation of new international agencies which involve high government level cooperation in developing and coordinating the implementation of energy policies; and,
- use of international agencies to cause the private sectors of the participating countries to join common energy-related programs.

A more detailed discussion of these areas follows.

B. Programs within Existing Structures

The program advanced by the energy commission of the European Communities (EC) is a good example of this approach. Pertinent aspects to the Commission's program submitted to the Council of Ministers are set forth below.

- Objectives for 1985 should be to:
  - keep energy usage 10 percent below pre-energy crisis forecasts;
  - increase electricity use by 10 percent to reach 35 percent of total use;
  - use nuclear energy for 50 percent of electricity production;
  - maintain internal solid fuel (coal, lignite, peat) production at its present level;
  - raise natural gas internal output and imports; and,
  - restrict oil consumption to specific uses such as auto fuel and as a raw material.
- To expedite efforts to achieve the above-described policy guidelines for each major energy source.
- By the end of the century, nuclear energy and gas should be the predominant sources of the Communities' energy. By the year 2000, nuclear energy could cover at least 50 percent of total energy requirements while natural or synthetic (oil- or coal-based) gas covers about one-third of total needs. Coal, therefore, would account for only one-quarter of EC needs by that time. The Commission does not expect non-conventional energy sources such as solar and geothermal energy to account for more than a small portion of EC energy sources.
- In the area of electricity, the following specific points were submitted:
  - Expansion of the use of electricity will depend largely on:
    - adequate financing and insured economic stability of nuclear power stations;

- more profitable and rational use of available power stations; and,
- adoption of appropriate price measures to ensure continued demand during non-peak periods.

- To prevent higher electricity production from increasing the demand for oil, oil burning stations would be restricted to using heavy residue oil from refineries, and would eventually be used only as medium and peak-load plants. Additionally, construction of oil-fired base load plants would be authorized only in exceptional cases. These measures would reduce oil-based electricity production from the current 30 percent to less than 20 percent by 1985. Fuel would be available for power plants only when supplies were interrupted or for economic, technical, or environmental reasons. Coal, on the other hand, would be given a larger share of the power station markets.
- The EC's total nuclear power station capacity will have to exceed 200 Gigawatts if nuclear energy is to supply 50 percent of energy needs by the end of the century. To reach that goal, the Commission will draw up proposals to: enable EC industry to build the needed stations, protect EC public health and environment, and guarantee adequate nuclear fuel supplies.
- The EC will act to encourage development of new and better breeder reactors. Work is already underway on the fast breeder reactor, which would increase nuclear fuel supplies and on the high temperature reactor, which could also be used as a source of process heat for industry.
- The creation of incentive devices would speed up investment and adoption of fiscal measures to prevent excess profiteering from low cost energy sources. The Community budget should be drawn on only if member states or EC action was essential for moderating or encouraging immediate energy related developments.[97]

C. Creation of New International Agencies for Government Interaction

1. Creation of New Agency--Background

The Organization for Economic Cooperation and Development (OECD) was set up under a Convention signed in Paris on December 14, 1960, which provides that the OECD shall promote policies designed:

- to achieve the highest sustainable economic growth and employment and a rising standard of living in member countries, while maintaining financial stability, and thus to contribute to the development of the world economy;
- to contribute to sound economic expansion in member as well as non-member countries in the process of economic development; and,
- to contribute to the expansion of world trade on a multilateral, non-discriminatory basis in accordance with international obligations.[98]

The members of OECD are Australia, Austria, Belgium, Canada, Denmark, Finland, France, the Federal Republic of Germany, Greece, Iceland, Ireland, Italy, Japan, Luxemborg, the Netherlands, New Zealand, Norway, Portugal, Spain, Sweden, Switzerland, Turkey, the United Kingdom, and the United States. The Socialist Federal Republic of Yugoslavia is associated in certain work of the OECD, particularly that of the Economic and Development Review Committee.[99]

The International Energy Agency (IEA) was established by decision of the OECD Council on November 15, 1974 as an autonomous body within the framework of the Organization. On November 18, 1974, the sixteen members of the organization then participating in the Agency entered into an agreement on the International Energy Program (IEP). The IEP is implemented through the IEA. The principal aims of the program are:

- development of a common level of emergency self-sufficiency in oil supplies;
- establishment of common demand restraint measures in an emergency;
- establishment and implementation of measures for the allocation of available oil in time of emergency;
- development of a system of information on the international oil market and a framework for consultation with international oil companies;
- development and implementation of a long-term cooperation program to reduce dependence on imported oil, including:
  - conservation of energy,
  - development of alternate sources of energy,
  - energy research and development, and,
  - supply of natural and enriched uranium;
- promotion of cooperative relations with oil producing countries, and with other oil consuming countries, particularly those of the developing world.[100]

The countries that participate in the IEA are: Austria, Belgium, Canada, Denmark, Germany, Greece, Ireland, Italy, Japan, Luxemborg, the Netherlands, New Zealand, Norway, Spain, Sweden, Switzerland, Turkey, the United Kingdom, and the United States.[101]

2. Establishment of International Guidelines for the Development of Energy Use Efficiency-Causing Policy and Regulation/Application

Since its creation, one of the major purposes of the IEA has been the promotion of strong energy use efficiency-causing programs. Within the IEA is a Conservation Sub-Group (CSG) in which all member countries are represented. The CSG leads the Agency's conservation efforts. In 1975, the CSG reviewed and evaluated each country's program and submitted a report to the Agency's Governing Board. After the 1975 review, it became apparent that a more uniform approach was needed in the members conservation programs, hence the CSG developed an "Indicative Test" of conservation measures and urged all members to consider adopting these measures. The suggested measures are set forth below.

- All energy priced at world market levels. This does not necessarily mean thermal equivalent pricing.
- Significant taxes on certain fuels to reinforce the effects of market prices where these prices are judged for national reasons to be inadequate signals (e.g., gasoline taxes).
- Changes in utility marketing practices and price structures to reward conservation by final consumers.
- Comprehensive public education programs with a conservation message, including programs specifically directed at schools.
- Specialized energy conservation education and/or training for such personnel as architects, engineers and building contractors, supervisors, and inspectors.
- Permanent full-time government conservation staff of adequate size.
- Consultants and other government staff spending a significan proportion of their time on energy conservation.
- Programs to increase use of waste heat from electrical generation and from industrial processing.
- Priority for government funding of energy efficient public transport (e.g., rail, bus) over funding for less energy efficient modes (e.g., air travel, highway construction).
- Programs to intensify government energy conservation R & D (excluding those encompassed within IEA R & D activities).

- Thermal and lighting efficiency in new commercial and public buildings and new residences through changes in building codes and standards.
- Incentives to increase retrofitting of existing residences and commercial buildings to improve thermal efficiency (e.g., loans or grants for insulation).
- An exemplary and effective effort to reduce all central government and local government energy use.
- Energy efficiency labeling for all major consumer appliances (e.g., water heaters, air conditioners, refrigerators, freezers, automobiles).
- Programs to improve the efficiency of heating/cooling devices and major appliances.
- Speed limits (of e.g., 90-110 km/hr) on all highways, including super-highways.
- Programs to increase automobile efficiency in countries where average new car efficiency is low (e.g., fuel economy standards, weight, horsepower or displacement taxes).
- Programs to increase load factors on transportation modes with excess capacity (e.g., car pools, public transit).
- Programs to stimulate energy efficiency in industrial production (e.g., target setting, loans for energy improvements, tax credits, rapid depreciation allowance, energy audits of individual companies, provision of information for small companies).
- Policies and programs to improve the efficiency of electrical generation such as peak load pricing, ripple load controls, thermal storage and other load management techniques.[102]

In 1976, the review process was repeated, this time based on the three criteria described below.

- To determine and measure energy use efficiency within a given country, the actual conservation results in each country over the past two years (1975/1976) were compared, including
    - changes in demand for energy relative to what would have been expected if 1968-1973 consumption trends had continued; and,
    - the relative change in the energy consumed/Gross Domestic Product (GDP) ratio.
- A comparison was made of specific energy efficiencies of major services or products within each energy use sector, including such items as:
    - energy used per passenger kilometer of automobile travel;
    - energy per ton of crude steel produced; and,
    - the capacity factor and conversion efficiency within a country.
- Each member's energy program was compared with the indicative list to assess the comprehensiveness and strength of the measures adopted to promote energy use efficiency.[103]

Table D-1 shows the overall conclusions drawn by CSG from its 1976 review based on the above-described criteria.[104]

Obviously, the program for each member country of the IEA could readily justify a lengthy detailed analysis, which is here inappropriate. However, a brief analysis of the programs is in order to better provide the reader with an overall view of the effort of government to cause energy use efficiency from an international perspective. To this end, consider the program summaries for Austria, Belgium, Canada, Denmark, Germany, Ireland, Italy, Japan, The Netherlands, New Zealand, Norway, Spain, Sweden, Switzerland, Turkey, United Kingdom, and the United States, set forth below.[105]

- Austria--The main element of the program is the pricing of

*See Code below table for significance of letter grades.

Member Country	Actual Conservation Results	Specific Efficiencies in Industry	Specific Efficiencies in Transportation	Status of Program
Austria	(A)	(A)	(B-)	adopted, but still important gaps
Belgium	(A)	(B)	(B-)	adopted and comprehensive
Canada	(B-)	(C)	(C)	adopted program--needs strengthening
Denmark	(A+)	(A+)	(A+)	adopted--excellent and comprehensive
Germany	(B)	(A)	(A)	adopted and fairly comprehensive
Ireland	(B)	(A)	(A)	adopted and fairly comprehensive
Italy	(A)	(A+)	(A+)	adopted and fairly comprehensive
Japan	(A)	(A)	(A)	adopted and fairly comprehensive
Netherlands	(A+)	(B-)	(A)	adopted--needs improvement in areas
New Zealand	(A)	(B-)	(B-)	adopted but has gaps--not comprehensive
Norway	(B)	(A+)	(A+)	adopted-many strong elements
Spain	(C)	(A)	(A)	adopted--needs improvement
Sweden	(C)	(B-)	(B-)	adopted and comprehensive
Switzerland	(C)	(--)	(C)	no overall plan as yet
Turkey	(B-)	(--)	(--)	adopted but needs substantial improvement
United Kingdom	(A)	(C)	(A+)	adopted but needs improvement in areas
United States	(B-)	(B-)	(B-)	adopted but needs improvement in areas such as pricing/taxes and buildings

Code: (A+) Above average (high)
(A) Above average
(B) Average
(B-) Below Average
(C) Below average (low)
(--) No information

SUMMARY OF CONSERVATION EFFORTS OF I E A MEMBER COUNTRIES

TABLE D-1

fuel at or above world market levels. Other aspects of the program include:

- a progressive tax on engine size of automobiles;
- interest-free loans for new and existing structures conditioned on compliance with mandatory efficiency standards; and,
- 100 percent loans at subsidized interest rate to companies making energy use efficiency-causing investment.

- Belgium--The main element of the program is energy priced at world market levels combined with substantial energy use taxes. Other aspects of the program include:
  - grants to industrial firms that demonstrate a 12 percent reduction in energy use per unit of output from the time assistance is requested;
  - energy use efficiency causing standards for new buildings; and,
  - 25 percent grant program for the retrofitting of existing residences.
- Canada--The program is still largely voluntary and maintains prices on oil and gas below world market levels. Other aspects of the program include:
  - regulations requiring a doubling of automobile efficiency by 1985;
  - a voluntary target setting and reporting system for industry has been established which includes:
    - accelerated capital cost allowance to encourage energy efficiency-causing changes; and,
    - efforts to encourage waste heat recovery;
  - a building code for new structures subject to acceptance by each province prior to implementation; and,
  - development of minimum efficiency standards for major household appliances, combined with energy efficiency labeling.
- Denmark--The program is based on a strong price/tax policy. Other aspects of the program include:
  - taxes that increase with engine size on both new and old automobiles;
  - monetary incentives to public transit combined with reduced funding for both highway and airport construction;
  - loans and grants to industries wishing to invest in energy efficiency causing equipment;
  - 25 percent grants to homeowners for retrofitting residences;
  - mandatory standards for new buildings have been further improved; and,
  - a well-developed district heating program (10 percent of homes).
- Germany--The basis for the program is energy pricing where world market levels apply except in coal which is held above international levels. Other aspects of the program include:
  - high taxes on gasoline and heating oil;
  - large research and development expenditures;
  - 7.5 percent tax allowance for industry and utility investments for dual use of power and other combined heat recovery techniques;
  - assistance program for small businesses;
  - progressive tax on cars to improve efficiency;
  - a new conservation law applicable to buildings and building appliances is in effect by virtue of which:

    - building codes will be upgraded for new buildings;
    - efficiency standards are to be set for appliances including air conditioning, furnaces and water heating;
    - mandatory annual maintenance is required for all building heating units including residences; and,
  - extensive use of peak load pricing and other load management techniques.
- Ireland--The program is premised on maintenance of pricing at world market levels, a strong educational campaign and an improved data operation. Other aspects of the program include:
  - fuel efficiency surveys for industry, one-third of which is paid for by the government;
  - a grant of up to 35 percent for industries that wish to make energy efficiency-causing improvements;
  - a new building insulation code has been mandated;
  - homeowners can obtain a grant of two-thirds of the cost, subject to a maximum of £400 for renovating their residences;
  - progressive taxes on automobiles according to engine size; and,
  - elimination of declining block rates combined with extensive use of peak load pricing by utilities.
- Italy--Major elements of this program are prices and taxes. Hydro-carbon fuels are priced at world market levels and all major refined petroleum products are taxed. Other aspects of the program include:
  - a high gasoline tax;
  - a prohibition on automobile use in some urban areas;
  - high annual license fees which increase sharply with engine displacements;
  - a new energy law passed in April, 1976, which:
    - regulates performance of existing heating systems by holding them to 20 °C;
    - establishes standards for new heating systems; and,
    - sets thermal insulation standards for new and renovated buildings;
  - a comprehensive mandatory reporting system for industry;
  - a major program of building pumped storage plants to deal with peak demand; and,
  - utility tariffs which dramatically increase with higher consumption in the domestic sector.
- Japan-- Pricing of energy at world market levels is the main area of emphasis in this program. Other aspects of the program include:
  - major use of rail transportation;
  - low "advisory" speed limit;
  - favorable tax treatment for the smaller and lighter automobiles;
  - automobile efficiency testing and labeling procedures are under development;
  - special bus lanes;
  - large funding for public transit;
  - ride sharing utilization of taxis;
  - major industrial program including:
    - a system of administrative guidance and

cooperation;

- accelerated depreciation (one-third first year write-off) for energy use efficiency-causing investments;
- loans for improving energy efficiency provided by the government bank at interest rates slightly below commercial rates;
- significant expenditures for research and development; and,

- increase in utility rates and a gradual elimination of declining block rates.

- The Netherlands--The program is still in the formative stages, at least in terms of specific proposals. Energy prices have been raised to world market levels except for natural gas where the price is moving upward more slowly. Other aspects of the program include:
    - progressive tax on car weight;
    - increased subsidies for public transit;
    - 33 percent grant for owners of residential and commercial buildings who want to improve the efficiency of their structures;
    - a plan to reinsulate all homes in ten years is under development;
    - mandatory building codes exist and have been greatly strengthened;
    - voluntary appliance labeling has been adopted; and,
    - peak load management techniques are in use.
- New Zealand--Until recently this program was retarded by the regulation of prices significantly below international market levels. At present, gasoline, manufactured gas and electricity are priced and/or taxed at high levels; however, coal and natural gas are still sold at below world market prices. Other aspects of the program include:
    - an elimination of declining block rates;
    - a graduated tax on engine size;
    - efficiency labeling for automobiles is under consideration;
    - establishment of an advisory board to work for more efficiency in the industrial sector; and,
    - interest-free loans to homeowners for retrofitting residences.
- Norway--The program relies heavily on pricing to cause energy use efficiency which is set and controlled by the government at or near world market levels. Other aspects of the program include:
    - a heavy tax on gasoline;
    - a 90 km/hr speed limit;
    - a 100 percent import tariff on automobiles;
    - low interest loans for energy efficiency-causing efforts in industry;
    - a strong national building code including strict insulation standards;
    - loan program for retrofitting existing buildings; and,
    - strict controls on growth of electricity use.
- Spain--Pricing at above world market levels combined with heavy emphasis on the reduction of imports provides the basis for the program. Other aspects of the program include:
    - a horsepower tax on automobiles;
    - a high parking tax in urban areas;

- the building of new subways in several cities;
- establishment of an institute to provide technical assistance to industry;
- a non-subsidized line of credit is available to companies to fund investments in new energy efficient technologies;
- a supply restriction which limits households to 80 percent of 1973 consumption;
- new building standards designed to cause energy use efficiency; and,
- a voluntary labeling program for appliances which is gradually becoming mandatory.

- Sweden--Pricing and taxes are providing the foundation for this program. Other aspects of the program include:
  - a 10 to 20 percent tax on electricity;
  - a goal to limit energy use growth to 2 percent per year through 1990 and zero thereafter;
  - a progressive tax on automobiles;
  - speed limits;
  - prior government approval of major energy intensive industrial facilities;
  - a new upgraded building code including both strict insulation standards and a requirement that new homes be individually metered;
  - grants and loans for improving efficiency in existing residences;
  - a study of appliance labeling and efficiency-causing regulations; and,
  - significant increase in funding of research and development projects.

- Switzerland--At present there is no federal government program. Energy use efficiency-causing policies and regulations do exist, however, and include:
  - high energy prices and taxes; and,
  - progressive tax on automobiles.

- Turkey--This program is based on pricing, with energy held to world market levels. Other aspects of the program include:
  - revised building insulation standards; and,
  - mandatory quality standards for new boilers and a training program for boiler operators.

- United Kingdom--Pricing energy at international levels combined with a strong and effective public information campaign provides the basis for this program. Other aspects of the program include:
  - government paid energy use audit schemes for industry;
  - a non-subsidized loan program to industry that makes energy use efficiency-causing investments;
  - strict mandatory thermal insulation requirements for new and altered or extended buildings;
  - renovation grants for existing homes conditioned on the achievement of a reasonable standard of roof insulation; and,
  - adoption of peak load pricing.

- United States--This program still emphasizes public education and is greatly inhibited by low taxes on all fuels and by regulations holding oil and gas prices below world market prices. Other aspects of the program include:
  - mandatory automobile efficiency standards to double efficiency by 1985;

- the mandatory labeling of automobiles regarding efficiency;
- increased public transit funding;
- voluntary efficiency target setting for the ten most energy intensive industries along with mandatory reporting schemes for the largest firms;
- a reasonably rapid elimination of declining block rates; and,
- sponsorship of innovative load management rate demonstrations.

3. International Strategy Used to Cause More Efficient Use of Energy--An Overview

Overall a number of conclusions were reached and observations made by the Conservation Sub-Group (CSG) after the 1976 review regarding patterns of governmental action in the international effort to cause more efficient use of energy. These conclusions and observations are set forth below.[106]

- Regarding the use of taxes and pricing:
  - While crude oil prices vary country by country, the prices of gasoline and gas oil lie in a similar price range for many IEA countries. Particularly in the case of gasoline, only three countries' prices are outside the range of 30 to 40 cents per litre.
  - In general, gasoline and motor diesel fuel are taxed at higher rates, while other fuels are rarely taxed except for a value added tax. Further, lower rates of value added tax are applied to those fuels than those applied to gasoline and motor diesel in some countries.
  - The price increases in fuel oil are much smaller than those in gasoline. In particular, it is noted that fuel oil prices in real terms were reduced in at least three countries during this period (1976).
  - In spite of fairly sharp price increases in nominal terms for both fuels, the real price increases are surprisingly modest even for gasoline in many countries. For many IEA countries, the price of energy in real terms decreased between 1974 and 1975. In particular, gasoline prices in real terms decreased in all countries but three. This is a result of the massive increase in the general price level.
- Regarding the transportation sector:
  - Automobile efficiency is a critical concern since autos by far are the dominant transport mode in both urban and rural use, and, thus, the dominant transport fuel user. High national gasoline prices and/or taxes have promoted the manufacture and purchase of relatively efficient autos notably in Western Europe, and low gasoline prices/taxes have led to large inefficient autos, notably in the United States and Canada. These two countries have established mandatory programs to upgrade the efficiency of their new cars, by standards and labeling. Other countries continue to rely on price effects to maintain already high efficiencies.
  - The auto is used for the majority of urban passenger travel in all countries reporting, even those with cheap and efficient urban mass transit, except for Japan. Load

factors for urban trips are uniformly low, all around 30 to 40 percent. Yet only the United States has anything approaching an organized, comprehensive program to promote carpooling or ride sharing. This is a major and surprising weakness of IEA conservation efforts, especially since effective ride sharing will not come about as a result of price effects alone; institutional organizational efforts are needed.

- Severe speed limits (e.g., 90 to 110 km/hr) are in force in roughly half of the IEA countries, and have resulted in energy savings variously estimated to be between 3 and 5 percent, as well as savings in death, injury and damage from accidents avoided that are in some cases quite remarkable. There seems, however, to be no discernible move among the other countries to adopt these lower limits.

- Urban public transit, both bus and rail, has steadily lost ridership in every IEA country except Japan since World War II. This appears largely to be the result of increased suburbanization and increased incidence of personal automobiles. As a result, transit is everywhere subsidized, both capital and operating costs, and many measures are suggested, both incentives (such as lowering fares) and disincentives (such as parking surcharges and auto free zones), to increase transit ridership. Yet the real problem of residential housing patterns and general urban planning does not seem to be addressed anywhere effectively.

- Trucks have encroached on rail freight markets in virtually every IEA country in the last 15 years, despite the fact they are on average less energy efficient per ton-kilometer of freight moved. The resulting decline in rail traffic has in general led to national rail subsidies, but few other policies are in place or being considered to promote rail freight.

- Regarding the industrial sector:

  - Energy intensive industries such as iron and steel, aluminum and other nonferrous metal fabrication, cement and pulp and paper account for a great part of national energy consumption in many countries although the number of firms is fairly small. Some countries, including the US, Canada, Japan, and the United Kingdom, pay special attention to energy conservation potential in these industries by introducing reporting/auditing schemes, and working out sectoral conservation targets.

  - Some countries pay particular attention to small and medium sized firms. These firms as a whole consume large amounts of energy although many of them do not belong to energy intensive industries. They are usually lacking in the knowledge of where energy waste takes place and how it can be reduced economically.

  - In several countries, voluntary or compulsory reports of energy conservation performance by industry are made. In Japan, the selection of a "heat manager" is compulsory.

  - Most countries provide some financial incentives to industry to invest in conservation, although in some countries these incentives are given only to investment for improvement of building insulation, heating and lighting.

- Regarding the domestic/residential sector:
    - Most IEA nations have some form of insulation building codes. However, the force and direction of the codes, the degree of enforcement, and the responsibility for administering the standards varies widely.
    - Most IEA members have initiated incentive programs to insulate existing buildings. The bulk of the programs work by a direct grant of from one-fourth to one-third of the total cost. A few programs accomplish the same results vis-à-vis tax reductions for insulation, etc. Most of the programs however are limited to a few years and/or have limits on the amount of financing available.
    - The most common form of member governments' internal conservation programs involve the use of circulars and directives to all departments. The circulars instruct the agencies to undertake conservation programs regarding lighting, temperature settings, insulation for public housing, etc. Where the effects of the programs are known, energy savings of from 9 percent to 24 percent have been achieved.
    - The only country to have actually initiated a program on energy efficiency labeling of appliances is the United States. However, most countries are reviewing such a program. In Canada, a program similar to that of the US will be introduced quite soon.
    - Most IEA members have yet to initiate a program in the area of appliance and heating efficiency. The US, under the Energy Policy and Conservation Act (EPCA), is developing efficiency improvement targets for all major appliances. Other nations have programs of public education on furnace servicing, etc. Germany has standards for industrial boilers, and Canada has a mandatory minimum furnace efficiency progam.
- Regarding the electric utility sector:
    - Most members have instituted some form of peak load management;
    - A small number of the members (approximately 5) are using ripple load controls;
    - The use or development of more sophisticated storage mechanisms is found in a majority of the member countries;
    - Techniques involving confined power-heat production exist or are the subject of research and development projects in a majority of the member countries;
    - About half the member countries have instituted or are in the process of developing other load management techniques such as:
        - development of inter-regional electrical connections;
        - superconduction for electricity transmission;
        - MHD power generation;
        - studies as to how transmission and distribution losses can be reduced;
        - audio-frequency power line carrier control;

- international electrical inter-connection;
- central dispatching control; and,
- grid lineage with neighboring countries with significant exchanges of power.

Certainly, the IEA has made significant progress in developing an international program to cause energy use efficiency which has, in turn, expedited the development and implementation of similar programs in the member countries. Expansion of the IEA membership and the Agency's involvement in energy affairs can be anticipated.

D. Use of International Agencies to Involve the Private Sector

Not only are governments in industrialized nations joining together to resolve energy related problems, but they also are using the newly created international agencies, i.e., IEA, to encourage and aid the private sector of participating countries to likewise cooperate and work towards common goals. By way of example, on March 28, 1975, final approval by the Attorney General of the US was given to the "Voluntary Agreement and Program Relating to the International Energy Program." This program was developed by the Federal Energy Administration and Department of State.[107]

Participation by the US in this voluntary program is premised on the realization by its government that it must be prepared to cooperate with other nations in the distribution of available supplies of fuel or sources of fuel on a rational and equitable basis in order to utilize them with maximum efficiency during any future supply interruption.[108]

The program evolved from a general policy of participating countries to reduce their dependence on foreign oil and to obtain the greatest quantity of supplies during an oil emergency. The US and certain other members of the Organization for Economic Cooperation and Development (OECD) first signed an agreement on November 18, 1974 on an International Energy Program (IEP) pursuant to which the International Energy Agency (IEA) was established as an autonomous institution within the OECD. It is a premise of the IEP Agreement that consultation and cooperation between oil companies and the IEA is essential to the effective functioning of the IEP, and thus to the solution of economic, strategic, and national security problems facing oil-importing nations. Accordingly, the President of the United States requested that this voluntary agreement and program be entered into by the US and that the participating oil companies undertake the actions contemplated thereby in order to further the objectives of the IEP and to implement the related policies and procedures of the IEA. The President's request was premised on the belief that such participation and action would be in the public interest and contributes to the national defense of the US.[109] Authorities, with respect to the IEP, are now well defined in the Energy Policy and Conservation Act (P.L. 94-163) passed December 22, 1975.

The basics of the IEP are set forth below.

- It is an entirely voluntary program, consistent with the purpose and scope of the Defense Production Act of 1950.
- It provides immunity from the anti-trust laws and the Federal Trade Commission Act with respect to acts or omissions by participants, and such of their affiliates as it may have been designated in accordance with its terms (Section 9) which are required to implement the objectives of the IEP.
- It contemplates that such acts by the participants will include:
    - membership in standing groups, working parties, advisory bodies or other bodies established at the request of the IEA;
    - consultations, plannings, and individual and joint

actions to implement the international allocation of petroleum pursuant to the IEP directives;

- the furnishing by participants of data and information, consultations, and planning in respect thereof; and,

- membership in ancillary industrial groups established by the US government. However, unlike those formed by the IEA, operations of these groups, if established by the State Department, would be governed by the provisions of the Federal Advisory Committee Act, 5 U.S.C. App. I (1973 Supp) and if established by the Federal Energy Administration, they would also be subject to the special provisions of Sections 17 of the Federal Energy Administration Act of 1974, 15 U.S.C.A. 776.

- The procedure for an oil company to become a member generally involves:

    - a request to join initiated by the Administrator (FEA) after approval by the Attorney General upon a finding that such participation is in the public interest and contributes to the national defense; or,

    - an oil company asking the Administrator to request that it become a participant which then triggers the above-described clearance procedures.

Notices of all requests and acceptances shall be published in the Federal Register.

- Any participant may withdraw from the Agreement upon at least 30 days notice to the Administrator (of the FEA) subject to fulfillment of the obligations incurred under this agreement prior to the date of such a notice except where emergency measures have been taken. When emergency measures have been undertaken, the effective date of the withdrawal may be postponed up to 60 days.

- Participation by the US in the program may be terminated at any time by the Administrator (FEA) after consultation with the Secretary of State upon notice by letter, telegram, or publication in the Federal Register. In no event, however, shall the program continue beyond June 30, 1985.[110]

The above-described program involving oil companies is significant because it provides a "blue print" for future participation of the private sector in attacking the energy problem on a world-wide front.

D.6 TESTING THE MERITS OF ENERGY MANAGEMENT STRATEGIES

Engineers, architects, scientists, urban planners, equipment and process designers, personnel at utilities, and other similar disciplines necessarily must play a vital role in the development and implementation programs designed to cause energy use efficiency. This participation will most certainly require the actual drafting of portions of legislation and comprehensive recommendations as to how and in what manner existing laws should be modified or repealed in order to enhance the success of new programs. Numerous existing private technical, design, and similar groups or societies have already recognized the need for their input and are actively participating in the development of policy and legislation. One example is the American Society of Heating, Refrigerating and Air Conditioning Engineers (ASHRAE) which for several years has played a significant role in the development of building codes designed to cause more efficient use of energy. Other such groups are the American Institute of Architects (AIA) and the International Conference of Building Officials. Groups or societies such as those mentioned above have been and will continue to be a creative and moving force in the formation and implementation of government policy and future legislation designed to cause more efficient use of energy.[111]

Furthermore, teamwork between the legislative bodies (many members of which are lawyers) and members of the technical and related professional fields is critical. Both groups must have at least a fundamental understanding of the basic principles of governing the other. Any proposed energy use efficiency-causing strategy must be tested against answers to such socio-political or legally oriented questions as:

- Where does the economic burden of pursuing the strategy lie?
- Who bears the burden of the extra "bother" which the strategy may cause?
- What alternative modes of behavior are people likely to adopt to accommodate themselves to the changed conditions which the strategy induces and can the energy consequences of those alternative modes of behavior be evaluated?
- Will the strategy require additional government or private sector manpower to implement?
- Is the strategy difficult to enforce?
- Are there any incentives built into the regulations to encourage compliance?
- How will it be financed?
- Has an appropriate agency been assigned responsibility for implementing and enforcing the strategy?
- Does the agency have a constituency which will enable it to resist attempts to change its mission?
- Are there inherent conflicts within the agency itself?
- Is there relevant expertise and power available to enforce actions?
- Are the administrative structures fair?
- Where, if at all, should judicial review of administrative action play a role?
- Does the law create an unnecessary bureaucracy?[112]

The answers to questions such as those set out above could determine the success or failure of any given energy efficiency-causing policy, regulation or technology.

Insofar as the citizenry at large is concerned, (particularly non-business "end users"), knowledge of and support for energy use efficiency programs and laws passed to implement them are essential. The present inability of the government to expedite well-organized and meaningful legislation programs is in large part attributable to a lack of public concern premised on the mistaken belief that the "energy crisis" has passed.[113] A continuation of this "head-in-the-sand" philosophy will only make the adjustment to energy use efficiency-causing rules and regulations more difficult.

* * *

REFERENCES

1. Commerce Clearing House (hereafter referred to as CCH), *Energy Management*, Chapters 2502-2510; and Senate Committee on Interior and Consular Affairs, *Federal Energy Organization*, Serial No. 93-6(92-41), (1973), as cited by ibid.

2. "Presidential Energy Message," (4 June 1971), reprinted in CCH, *Energy Management*, Chapters 403-414; and CCH, *Energy Management*, Chapter 2503.

3. "Energy Reorganization Act of 1974," reprinted in CCH, *Energy Management*, Chapters 10, 721-10 755; "Federal Non-nuclear Energy Research and Development Act of 1974," reprinted in CCH, *Energy Management*, Chapters 981-997; "Federal Energy Administration Act of 1974," reprinted in CCH, *Energy Management*, Chapters 10, 550-580; and "California Warren-Alquist State Energy Resources Conservation

and Development Act." Energy Policy and Conservation Act, reprinted in CCH, *Energy Management*, Chapters 10, 850-10, 968.

4. "Presidential Energy Message," (4 June 1971), reprinted in CCH, *Energy Management*, Chapters 403-414.

5. "Presidential State of the Union Message," (15 January 1975), summarized in the *White House State of the Union Fact Sheet*, reprinted in CCH, *Energy Management*, Chapters 692-698; US Federal Energy Administration, proposed *Energy Independence Act of 1975 and Related Tax Proposals*, Chapter 12; and, Presidential Energy Message Detailed Fact Sheet (April 2, 1977).

6. CCH, *Energy Management*, Chapter 2501.

7. CCH, *Energy Management*, Chapter 2511; and Presidential Executive Orders, Nos. 11712 and 11726, reprinted in CCH, *Energy Management*, Chapters 11,001 and 11,021.

8. CCH, *Energy Management*, Chapter 2511; and Presidential Executive Orders, Nos. 11743, 11748, 11775, and 11790, reprinted in CCH, *Energy Management*, Chapters 11,031, 11,041, 11,071, and 11,081.

9. CCH, *Energy Management*, Chapter 2511; Presidential Executive Orders, Nos. 11814, 11819, and 11834, reprinted in CCH, *Energy Management*, Chapters 11,091, 11,100, and 11,101; and the "Energy Reorganization Act of 1974."

10. Interview with Dr. Douglas C. Bauer, Associate Assistant Administrator, Utilities Programs, Federal Energy Administration Office of Conservation and Environment, Washington, D.C., 18 June 1975; CCH, *Energy Management*, Chapter 2501.

11. Presidential Message on Department of Energy (March 1, 1977), reprinted in CCH, *Energy Management*, Chapters 723-730.

12. Interview with Grant P. Thompson, Institute Fellow, Environmental Law Institute, Washington, D.C., 18 June 1975; Warren-Alquist State Energy Resources Conservation and Development Act (California, 1975); *Energy Conservation Project Report*, No. 2, October, 1975, published by the Environmental Law Institute, Washington, D.C.

13. Interview with Grant P. Thompson, Institute Fellow, Environmental Law Institute, Washington, D.C., 18 June 1975; Warren-Alquist State Energy Resources Conservation and Development Act (California, 1975); *Energy Conservation Project Report*, No. 2, October 1975, published by the Environmental Law Institute, Washington, D.C. Interviews with Dr. Douglas C. Bauer, Robert R. Jones, David Rosoff, Ted Farfaglis, and Dr. Melvin H. Chiogioji, Federal Energy Administration Office of Conservation and Environment, Washington, D.C., 18 July 1975; Phone conference with Robert R. Jones and David Rosoff, 30 June 1975; Interviews with Walter J. Cavagnaro, Chief Electrical Engineer, and Rufus G. Thayer, Jr., Esquire, Counsel, both of the California Public Utilities Commission, San Francisco, California, 7 June 1975; Interview with Sharon Sellars, Federal Energy Administration, Energy Conservation Division, Region IX, San Francisco, California, 27 April 1977.

14. Ibid.

15. Ibid.

16. CCH, *Energy Management*, Chapters 2533-2601.

17. Ibid; Interview with Grant P. Thompson, Institute Fellow, Environmental Law Institute, Washington, D.C., 18 June 1975;

Interviews with Dr. Douglas C. Bauer, Robert R. Jones, David Rosoff, Ted Farfaglia, and Dr. Melvin H. Chiogioji, Federal Energy Administration Office of Conservation and Environment, Washington, D.C., 18 July 1975; Phone conference with Robert R. Jones and David Rosoff, 30 June 1975; and Interviews with Walter J. Cavagnaro, Chief Electrical Engineer, and Rufus G. Thayer, Jr., Esquire, Counsel, both of the California Public Utilities Commission, San Francisco, California, 7 June 1975.

18. CCH, *Energy Management*, Chapters 2533-2535, 2540, 2541, 2552, and 2553.

19. CCH, *Energy Management*, Chapters 2904-2907, 2921-2927.

20. Interview with Grant P. Thompson, Institute Fellow, Environmental Law Institute, Washington, D.C., 18 June 1975; and a *National Plan for Energy Research and Development - Creating Energy Choices for the Future - 1976*, (ERDA-76-11).

21. Ibid; CCH, *Energy Management*, Chapters 9676-9689, 9690; Interviews with Dr. Douglas C. Bauer, Robert R. Jones, David Rosoff, Ted Farfaglia, and Dr. Melvin H. Chiogioji, Federal Energy Administration Office of Conservation and Environment, Washington, D.C., 18 July 1975; Phone conference with Robert R. Jones and David Rosoff, 30 June 1975; Phone conference with Steven Powers, Counsel for the Los Angeles Department of Water and Power, Los Angeles, California, 30 June 1975; Energy Policy and Conservation Act (P.L. 94-163); *Energy Conservation Project Report*, No. 2, October 1975, published by the Environmental Law Institute, Washington, D.C.

22. Ibid; and "California Warren-Alquist State Energy Resources Conservation and Development Act."

23. CCH, *Energy Management*, Chapters 9676-9689, 9690; Interview with Grant P. Thompson, Institute Fellow, Environmental Law Institute, Washington, D.C., 18 June 1975; Interviews with Dr. Douglas C. Bauer, Robert R. Jones, David Rosoff, Ted Farfaglia, and Dr. Melvin H. Chiogioji, Federal Energy Administration Office of Conservation and Environment, Washington, D.C., 18 July 1975; Phone conference with Steven Powers, Counsel for the Los Angeles Department of Water and Power, Los Angeles, California, 30 June 1975; Energy Policy and Conservation Act (P.L. 94-163); and *Energy Policy and Conservation Project*, No. 2, October 1975, published by the Environmental Law Institute, Washington, D.C.

24. "Staff Study of Impact of Energy Shortages on Los Angeles," prepared for the Permanent Subcommittee on Investigations of the Senate Committee on Government Operations, February 15, 1974, GPO Stock No. 5270-02211, as reported in CCH, *Energy Management*, Chapter 9758; and Chapter XIII, Los Angeles Municipal Code.

25. White House Fact Sheet, "Presidential Energy Message," (8 October 1974); reprinted in CCH, *Energy Management*, Chapters 655-679; Interviews with Dr. Douglas C. Bauer, Robert R. Jones, David Rosoff, Ted Farfaglia, and Dr. Melvin H. Chiogioji, Federal Energy Administration Office of Conservation and Environment, Washington, D.C., 18 July 1975; Phone conference with Robert R. Jones and David Rosoff, 30 June 1975; US Federal Energy Administration, Office of Conservation and Environment, *Federal Energy Management Program, First Annual Report, Fiscal Year 1974*, (Washington, D.C.: December 1974) and Interview with Sharon Sellars, Federal Energy Administration, Energy Conservation Division, Region IX, San Francisco, California, 3 May 1977.

26. US Federal Energy Administration, Office of Conservation and Environment and Office of Industrial Programs, *Fact Sheet*, "Voluntary Industrial Energy Conservation Program," (Washington, D.C.: June 1975); Interview with Fred King, Federal Energy Administration, Energy Conservation Division, Region IX, 2 June 1976 and Sharon Sellars of that agency on 3 May 1977; description of major programs, Office of Energy Conservation and Environment, Federal Energy Administration, Washington, D.C., November 1975; and U.S. Department of Commerce, Energy Conservation Programs, April 1975.

27. Federal Energy Administration, "State Energy Conservation Program Fact Sheet;" Energy Policy and Conservation Act (EPCA); Interview with Sharon Sellars, Federal Energy Administration, Energy Conservation Division, Region IX, San Francisco, California , 4 April 1977.

28. Interviews with Robert Watkins, Assistant Division Chief, and Bruce Rogers, Manager of Licensing and Siting Division, Energy Resources Conservation and Development Commission, on 2 July 1976; Interviews with Walter Cavagnaro, Rufus Thayer and George Amaroli of the California Public Utilities Commission on 25 June 1976 and 26 June 1976.

29. Interview with Grant P. Thompson, Institute Fellow, Environmental Law Institute , Washington, D.C., 18 June 1975; Interviews with Dr. Douglas C. Bauer, Robert R. Jones, David Rosoff, Ted Farfaglia, and Dr. Melvin H. Chiogioji, Federal Energy Administration Office of Conservation and Environment, Washington, D.C., 18 July 1975; Phone conference with Robert R. Jones and David Rosoff, 30 June 1975; Interviews with Walter J. Cavagnaro, Chief Electrical Engineer, and Rufus G. Thayer, Jr., Esquire, Counsel, both of the California Public Utilities Commission, San Francisco, California, 7 June 1975. *Energy Conservation in the International Energy Agency*, 1976 Review. Interviews with George Amaroli, California Public Utilities Commission, 6 June 1976. Interview with Warren Osborn, Energy Conservation Division, Federal Energy Administration, Region IX, 6 June 1976. Interview with Bert Gauger and Craig W. Hoellwarth, Conservation Division, California Energy Resources Conservation and Development Commission, 2 July 1976; Energy Policy and Conservation Act (42 U.S.C. 6201 et. seq.); the Motor Vehicle Information and Cost Savings Act, (15. U.S.C. 1901 et. seq.).

30. CCH, *Energy Management*, paragraphs 3901, 3902, 3905, 3907, 3909, 3911, 3913, 3914, 4201, and 4301; Interview, Sharon Sellars, Federal Energy Administration, Energy Conservation Division, Region IX, San Francisco, California, 27 April 1977; Energy Policy and Conservation Act (P.L. 94-163); and, *The White House Detailed Fact Sheet*, The President's Energy Program, April 20, 1977.

31. Ibid.

32. Ibid.

33. Interviews with Dr. Douglas C. Bauer, Robert R. Jones, David Rosoff, Ted Farfaglia, and Dr. Melvin H. Chiogioji, Federal Energy Administration Office of Conservation and Environment, Washington, D.C., 18 July 1975; Phone conference with Robert R. Jones and David Rosoff, 30 June 1975; Interview with Sharon Sellars, Federal Energy Administration, Energy Conservation Division, Region IX, San Francisco, California, 27 April 1977; Energy Policy and Conservation Act (P.L. 94-163); *The White House Detailed Fact Sheet*, The President's Energy Program, April 20, 1977.

34. Interview with Marshall F. Johnson, Conservation Division, Energy Resources Conservation and Development Commission, Sacramento, California, 2 July 1976; And phone conservation with Jon Leber, Conservation Division, Energy Resources Conservation and Development Commission, Sacramento, California, 4 May 1977.

35. *Energy Conservation Project Report*, No. 2, October 1975, published by the Environmental Law Institute, Washington, D.C.

36. Interview with Sharon Sellars, Federal Energy Administration, Energy Conservation Division, Region IX, San Francisco, California, 27 April 1977; Energy Conservation and Production Act (P.L. 94-385); Energy Policy and Conservation Act (P.L. 94-163).

37. Building Energy Authority and Regulations Survey, State Activity, Office of Building Standards, Code Services, National Bureau of Standards, March 1976.

38. Interview with Sharon Sellars, Federal Energy Administration, Energy Conservation Division, Region IX, San Francisco, California, 27 April 1977.

39. California's proposed "Energy Conservation Standards for New Nonresidential Buildings"; California Administrative Code, Title 25, Chapter 1, Subchapter 1, Article 5, Section 1094, "Energy Insulation Standards"; and City of Cerritos, California, Ordinance No. 475, "An Ordinance of the City of Cerritos Amending the Environmental Performance Standards of the Municipal Code by Incorporating a Section on Energy Conservation in Residential Dwelling Units."

40. Interview with Sharon Sellars, Federal Energy Administration, Energy Conservation Division, Region IX, San Francisco, California, 27 April 1977; Energy Policy and Conservation Act (P.L. 94-163), and Federal Energy Administration Fact Sheet, State Energy Conservation Program. (Undated)

41. Interview with Walter J. Cavagnaro, Chief Electrical Engineer, and Rufus G. Thayer, Jr., Esquire, Counsel, both of the California Public Utilities Commission, San Francisco, California, 7 June 1975; and CCH, *Energy Management*, Chapters 4503 et. seq.; *The White House Detailed Fact Sheet*, The President's Energy Program, April 20, 1977.

42. Ibid.

43. Interviews with Dr. Douglas C. Bauer, Robert R. Jones, David Rosoff, Ted Farfaglia, and Dr. Melvin H. Chiogioji, Federal Energy Administration Office of Conservation and Environment, Washington, D.C., 18 July 1975.

44. Ibid.

45. Ibid.

46. Energy Conservation Production Act (P.L. 94-385).

47. Ibid.

48. *The White House Detailed Fact Sheet*, The President's Energy Program, April 20, 1977.

49. CCH, *Energy Management*, Chapters 4503 et. seq.; Joint Economic Committee Staff Study, *The 1975 Budget: An Advanced Look*, 3 December 1973, reprinted in CCH, *Energy Management*, Chapter 9790; and CCH, *Energy Management* , Chapter 9657.

50. *The White House Detailed Fact Sheet*, The President's Energy Program, April 20, 1977.

51. Small Business Administration, *Fact Sheet*, No. 12, September 1974, reprinted in CCH, *Energy Management*, Chapter 9653.

52. Ibid.

53. Energy Conservation and Production Act (P.L. 94-385) as reprinted in CCH, *Energy Management*, Chapters 10, 450, et. seq.; and an interview with Sharon Sellars, Federal Energy Administration, Energy Conservation Division, Region IX, San Francisco, California, 27 April 1977.

54. *The White House Detailed Fact Sheet*, The President's Energy Program, April 20, 1977.

55. Interview with Robert Watkins, Conservation Division, Energy Resources Conservation and Development Commission, Sacramento, California, 2 July 1976; Interviews, Rufus Thayer, Esq., George Amaroli and Walter Cavagnaro, California Public Utilities Commission, San Francisco, California, 25 June 1976; and Interview with Sharon Sellars, Federal Energy Administration, Energy Conservation Division, Region IX, San Francisco, California, 27 April 1977.

56. Interview with Stacey Swor, Conservation Division, Federal Energy Administration, Region IX, San Francisco, California, 24 June 1976; Interview with Sharon Sellars, Federal Energy Administration, Energy Conservation Division, Region IX, San Francisco, California, 27 April 1977.

57. Interview with Sharon Sellars, Federal Energy Administration, Energy Conservation Division, Region IX, San Francisco, California, 27 April 1977.

58. Federal Energy Administration *Fact Sheet*, "Public Schools Energy Conservation Service," (undated).

59. *The White House Detailed Fact Sheet*, The President's Energy Program, April 20, 1977.

60. Interview with Grant P. Thompson, Institute Fellow, Environmental Law Institute, Washington, D.C., 18 June 1975; Interviews with Dr. Douglas C. Bauer, Robert R. Jones, David Rosoff, Ted Farfaglia, and Dr. Melvin H. Chiogioji, Federal Energy Administration, Office of Conservation and Environment, Washington, D.C., 18 July 1975; Interview with Sharon Sellars, Federal Energy Administration, Energy Conservation Division, Region IX, San Francisco, California, 27 April 1977.

61. *The White House Detailed Fact Sheet*, The President's Energy Program, April 20, 1977.

62. *The White House Detailed Fact Sheet*, The President's Energy Program, April 20, 1977.

63. *Energy Conservation in the International Energy Agency*, 1976 Review.

64. *The White House Detailed Fact Sheet*, The President's Energy Program, April 20, 1977.

65. Ibid.

66. Ibid.

67. Ibid.

68. Report by the US Department of Commerce published in 38 *Federal Registry* 29629 (26 October 1973); CCH, *Energy Management*, Chapter 9696; and Department of Commerce "Energy Conservation Programs" Fact Sheet, April 1975.

69. Interview with Michael De Angelis, California Energy Resources Conservation and Development Commission, 2 July 1976; Interview with Walter Cavagnaro, Rufus Thayer, Esq., and George Amaroli of the California Public Utilities Commission, 25 June 1976; Interview with Fred King, Conservation Division, Federal Energy Administration, Region IX, 24 June 1976;

Proceedings of an EPRI Workshop on Technologies for Conservation and Efficient Utilization of Electric Energy, prepared by Applied Nucleonics Co., Inc., Los Angeles, July 1976; *A National Plan for Energy Research, Development and Demonstration: Creating Energy Choices for the Future*, 1976, Vol. I, *The Plan* (ERDA, 76-1).

70. Interview with Emilio Varanini, Commissioner, California's State Energy Commission, Sacramento, California, 2 July 1976. *The White House Detailed Fact Sheet*, The President's Energy Program, April 20, 1977.

71. Ibid.

72. Ibid.

73. Ibid.; Interview with Stacey Swor, Conservation Division, Federal Energy Administration, Region IX, San Francisco, California, 3 May 1977.

74. Ibid.

75. Interview with Robert Watkins, Energy Resources Conservation and Development Commission, Sacramento, California, 2 July 1976; Interview, Walter Cavagnaro, Rufus Thayer, Esq., and George Amaroli of the California Public Utilities Commission, San Francisco, California, 25 June 1976.

76. Ibid.

77. Interview with Emilio Varanini, Commissioner, California's State Energy Commission, Sacramento, California, 2 July 1976. *The White House Detailed Fact Sheet*, The President's Energy Program, April 20, 1977.

78. Ibid.

79. Interview with Robert Watkins, Energy Resources Conservation and Development Commission, Sacramento, California, 2 July 1976; Interview, Walter Cavagnaro, Rufus Thayer, Esq., and George Amaroli of the California Public Utilities Comm., San Francisco, Calif., 25 June 1976.

80. Ibid.; Environmental Policy and Conservation Act (Public Law 94-163); CCH, *Energy Management* Chapters 3901-4306; and phone interview with Jon Leber, Energy Resources Conservation and Development Commission, 4 May 1977. *The White House Detailed Fact Sheet*, The President's Energy Program, April 20, 1977.

81. Interview with Robert Watkins, Energy Resources Conservation and Development Commission, Sacramento, California, 2 July 1976; Interview, Walter Cavagnaro, Rufus Thayer, Esq., and George Amaroli of the California Public Utilities Commission, San Francisco, California, 25 June 1976. *The White House Detailed Fact Sheet*, The President's Energy Program, April 20, 1977; *Proceedings of an EPRI Workshop* . . . , prepared by Applied Nucleonics Company, Inc., July 1976.

82. Ibid.

83. Interview with Sharon Sellars, Federal Energy Administration, Energy Conservation Division, Region IX, San Francisco, California, 27 April 1977. Energy Conservation and Production Act (P.L. 94-385); Energy Policy and Conservation Act (P.L. 94-163). *The White House Detailed Fact Sheet*, The President's Energy Program, April 20, 1977.

84. Interview with Robert Watkins, Energy Resources Conservation and Development Commission, Sacramento, California, 2 July 1976; Interview, Walter Cavagnaro, Rufus Thayer, Esq., and George Amaroli of the California Public Utilities Commission, San Francisco, California, 25 June 1976. *The White House Detailed Fact Sheet*, The President's Energy Program, April 20, 1977. *Proceedings of an EPRI Workshop* . . ., prepared by Applied Nucleonics Company, Inc., July 1976.

85. Ibid.

86. Ibid.

87. Ibid.; *Energy Conservation in the International Energy Agency*, 1976 Review.

88. Interview with Grant P. Thompson, Institute Fellow, Environmental Law Institute, Washington, D.C., 18 June 1975; Interviews with Dr. Douglas C. Bauer, Robert R. Jones, David Rosoff, Ted Farfaglia, and Dr. Melvin H. Chiogioji, Federal Energy Administration, Office of Conservation and Environment, Washington, D.C. 18 July 1975; Phone conference with Robert R. Jones and David Rosoff, 30 June 1975; Interviews with Walter J. Cavagnaro, Chief Electrical Engineer, and Rufus G. Thayer, Jr., Esq., Counsel, both of the California Public Utilities Commission, San Francisco, California, 7 June 1975; Interview with Sharon Sellars, Federal Energy Administration, Energy Conservation Division, Region IX, San Francisco, California, 27 April 1977; and interview with Tom Jones, Energy Coordinator, Environmental Protection Agency, Region IX, San Francisco, California, 29 April 1977.

89. Ibid.

90. Ibid.

91. Interview with Tom Jones, Energy Coordinator, Environmental Protection Agency, San Francisco, California, Region IX, 29 April 1977.

92. Ibid.; Interview with David R. Andrews, Esq., Regional Council, Environmental Protection Agency, Region IX, 29 April 1977; H.R. 6161 (Bill to amend Clean Air Act), dated April 6, 1977; Toxic Substance Control Act, October 11, 1976 (P.L. 94-469); Resource Conservation and Recovery Act, October 21, 1976 (P.L. 94-580).

93. P.L. 94-469, October 11, 1976, 15 U.S.C. 2601.

94. P.L. 94-580, October 21, 1976, 42 U.S.C. 6901, amends the Solid Waste Disposal Act (42 U.S.C. 3251).

95. US House of Representatives bill to amend the Clean Air Act and for other purposes, April 6, 1977 (pending).

96. United States Government Manual, 1976-1977, published by The Office of Federal Register, National Archives, General Services Administration.

97. European Community Information Service, "A New Energy Policy for the European Community," Background Note No. 13/1974, (15 July 1974), reprinted in CCH, *Energy Management*, Chapter 9697.

98. "Energy Crisis: Strategy for Cooperative Action," Address by Secretary of State Henry Kissinger in Chicago, Illinois, 14 November 1974, reprinted in CCH, *Energy Management*, Chapter 9644; and, *Energy Conservation in the International Energy Agency*, 1976 Review.

99. Ibid.

100. Ibid.

101. Ibid.

102. Ibid.

103. Ibid.

104. Ibid.

105. Ibid.

106. Ibid.

107. Ibid.; "Voluntary Agreement and Program Relating to the International Energy Program," reprinted in CCH, *Energy Management*, Chapter 10,822; Letter of the Administrator of the Federal Energy Administration to the Administrator of the General Services Administration, 6 March 1975, reprinted in CCH, *Energy Management*, Chapter 10, 822; and Letter of the Administrator of the General Services Administration to the Administrator of the Federal Energy Administration, 6 March 1975, reprinted

in CCH, *Energy Management*, Chapter 10,822; and Letter of the Administrator of the General Services Administration, 28 March 1975, reprinted in CCH, *Energy Management*, Chapter 10,822.

108. Ibid.

109. Ibid.

110. Ibid.

111. Interview with Grant P. Thompson, Institute Fellow, Environmental Law Institute, Washington, D.C., 18 June 1975. Interviews with Dr. Douglas C. Bauer, Robert R. Jones, David Rosoff, Ted Farfaglia, and Dr. Melvin H. Chiogioji, Federal Energy Administration Office of Conservation and Environment, Washington, D.C. 18 July 1975; and Interview with Fred King, Conservation Division, Federal Energy Administration, Region IX, San Francisco, California, 24 June 1976.

112. Interview with Grant P. Thompson, Institute Fellow, Environmental Law Institute, Washington, D.C., 18 June 1975; and Environmental Law Institute, "Energy Conservation Project," summary of a project being conducted under a grant from the National Science Foundation, (Washington, D.C.: n.d.).

113. Statement by Donald Craven, Assistant Administrator for Resource Development, Federal Energy Administration, before the Texas United Press International Editors Association, San Antonio, Texas, June 1975, as reported in CCH, *Energy Management*, "Weekly Bulletin," (18 June 1975).

APPENDIX E

ECONOMIC ANALYSIS FOR ENERGY MANAGEMENT DECISION MAKING

Steven L. Westfall*

CONTENTS

SUMMARY

Because the economic aspects of energy management are complex and constantly subject to change, there is a natural tendency to want to employ complex and sophisticated methods of economic analysis in support of energy decisions. Complexity and sophistication in analysis, however, can cause confusion and obscure the fundamental issues. What is needed is a clearer understanding of those issues and how they affect decisions, and this can best be done by employing simple, widely understood and widely practiced procedures for economic decision making. It is this simple and fundamental approach which is described in this appendix.

This treatment of the subject of economic decision making for energy management is based on four premises:

- decision making should be the process of choosing the best of several available alternatives;
- the results of the economic analysis that goes into decision making and the decisions themselves must be easy to communicate and easily understood by others;
- simplicity in analysis is better than complexity, because it promotes better understanding of the issues and facilitates communication of findings; and,
- most errors in decision making result from improper attention to assumptions and to incomplete development of all the available alternatives.

What the reader will learn from this section are the five basic principles for making good economic decisions, the fundamental methods of analysis for comparing alternatives, and how to communicate findings and decisions in succinct, easy-to-understand terms.

*Dr. S.L. Westfall is a consultant with broad industrial experience. He has been retained by Applied Nucleonics Company to advise on finance and financial planning for efficient energy use.

E.1 BIG DECISIONS AND LITTLE DECISIONS

Any practical approach to economic decision making needs to take into account that some decisions are major decisions and merit a great deal of effort and detail, while many decisions will have lesser impact and should be made quickly in a more routine manner. A big decision is one that has a major impact on the decision maker's organization. A small decision has a small impact. For example, the decision to change a firm's entire production process to economize on electricity usage is a major decision. If the plant manager has a 10 kW electric motor repaired rather than buying a new one, it is a small decision.

Two important differences between big decisions and small decisions are the amount of time that goes into making the decision, and the number of cost factors that are taken into consideration. Big decisions, as we are referring to them here, usually involve:

- effect on the future profitability of the firm;
- expectations for energy cost escalation;
- state and federal taxes and tax credits;
- amount of capital required and methods of finance;
- interest costs;
- time and its implications, including the time value of money, and the timing of cash flow;
- the effect of depreciation on taxes and cash flow;
- cost of retraining personnel;
- evaluation of risk; and,
- future capital and labor costs, and material costs.

Small decisions usually involve:

- estimates of short-term operating costs;
- one-time costs (e.g., the purchase price of equipment);
- economic life; and,
- a standard criterion for an acceptable return on investment.

E.2 THE ART OF DEVELOPING ALTERNATIVES

Economic decision making starts with the question, "what are the options?" The answer deserves considerable and careful consideration, because the quality of the options put forth will determine the quality of the final decision. For example, a manager who is considering a possible investment in a heat recovery system, and who limits his decision to that of approving or rejecting the proposal to purchase a particular piece of equipment is likely to make a bad decision. At the very least there is likely to be more than one piece of equipment or more than one technical approach to do the job. There is the choice of the level of recovery (10 percent recovery or 90 percent recovery), other energy saving projects that could be done instead of, or in conjunction with, this project. There is also the question of timing--when such a project should be undertaken.

A good start toward developing a full range of alternatives is to write down answers to the basic interrogatives: what?, when?, who?, how?, where?, how much?, and then look for the most promising combinations. To illustrate, consider the following list of questions that might be asked in connection with a decision concerning a project to reduce money spent on electric space heating.

Question:
What methods are available to reduce $$ spent on space heating?

Possible Answers:
1. HVAC heat recovery, System #1
2. HVAC heat recovery, System #2
3. Burn waste chemicals
4. Change equipment to a heat pump
5. Recover heat from a process and use for space heating

Question:
How much reduction should be targeted?

Possible Answers:
1. 10%
2. 50%
3. 90%

<u>Question</u>:
When should the project be undertaken?

<u>Possible Answers</u>:
1. Immediately
2. 6 - 9 months
3. In 2 years
4. Postpone indefinitely

<u>Question</u>:
Who should carry out the project?

<u>Possible Answers</u>:
1. Employee A
2. Department II
3. Outside Engineering Co. X
4. Outside Engineering Co. Y

<u>Question</u>:
Where should changes be made?

<u>Possible Answers</u>:
1. Entire company
2. Plant A only
3. Plant B only
4. Product line X only

<u>Question</u>:
How should the job be carried out?

<u>Possible Answers</u>:
1. Construct during routine maintenance period
2. Wait until next summer
3. Interrupt production for 3 weeks

The above list of possible answers describes, through all possible combinations, 2880 different alternatives, each with different costs and impacts. The art comes in applying judgment to eliminate those combinations that are infeasible or clearly impractical and to make those combinations that appear to have the highest potential benefit. The list gives a framework for quickly visualizing a broad range of alternatives and zeroing in on the ones that look promising.

In narrowing down the list of alternatives to the ones to choose from, many of the variables (possible answers) will become constraints not requiring decision, while others will be eliminated for reasons related to constraints that may already exist. The result might be a list of nine best alternatives described by the 3 x 3 decision matrix shown in Table E-1. At this point, the analyst should be confident that the nine possible decision combinations represent the best available alternatives within his organization's resources.

E.3 MAKING ASSUMPTIONS

Too often, when decisions are being made, the decision maker is not aware that there may be undisclosed assumptions, that some of his assumptions may not be valid, or that certain of his assumptions may have a critical impact on the results of his decision. Improper attention to assumptions is common because of the tendency to focus on the precise techniques and logic of analysis instead of on the imprecise guesses we call assumptions--on which the entire analysis is based!

A good start in defining assumptions is to make a list, and then keep it handy to add to as more assumptions become necessary as the analysis progresses. A list of assumptions might include:

- energy cost escalation;
- cost of the equipment or initial cost;
- useful economic life of equipment;
- salvage value;
- operating characteristics and operating costs of equipment (e.g., fuel consumption);
- level of operation (e.g., duty cycle, hours per year);
- rate of depreciation for tax purposes;
- tax rate;
- cost of capital; and,
- acceptable rate of return.

The area of assumption-making subject to the greatest uncertainty is energy cost escalation. There is, without doubt, uncertainty in all assumptions, but with issues of useful life, salvage value, plant operations (usage rate), taxes and capital costs, we generally have some consistent experience on which to base reasonable assumptions about the future. Not so with energy costs. Almost overnight, energy has become a major production or operating cost for many companies and institutions, and there is no hard evidence that energy costs are stabilizing.

Alternative Systems	Alternative Locations		
	Entire company	Plant A	Plant B
HVAC heat recovery, System #1, 50% level	Alt. #1	Alt. #2	Alt. #3
HVAC heat recovery, System #2 50% level	Alt. #4	Alt. #5	Alt. #6
Burn waste chemicals	Alt. #7	Alt. #8	Alt. #9

CONSTRAINTS: Work is to be done by an outside engineering company for heat recovery and chemicals burning. All process changes to be done by in-plant engineers. Work to be done during routine maintenance shutdown during the summer months.

POSSIBLE DECISION MATRIX FOR COST REDUCTION OF ELECTRIC SPACE HEATING

TABLE E-1

There is no sure way to know future energy costs, but for the purpose of this appendix, the assumption is made that energy costs will escalate by 10 percent to 15 percent annually for the next five years. Wide variations are possible depending on the location of the plant and the energy form. Review this assumption with your local utility representative or fuel supplier.

Wherever possible, assumptions should be tested. This can be done in two ways: (1) direct opinion, and (2) supportive opinion. Direct opinion involves getting a collective agreement from informed and knowledgeable people on the assumption in question. Supportive opinion involves getting the same kind of agreement or opinion on factors that may influence or are related to the assumption in question. Examples of supportive opinion would be a forecast of labor cost increases as support for an assumption of the long-term price increases on a particular type of capital equipment, or the rising cost of fuel and construction costs as an indicator of electric power rates, or the rising price of real estate and construction as an indicator of future rent escalation.

The time to test key assumptions is early in the decision making process, and the list of assumptions made can be used to solicit agreement and involvement from management. For major assumptions, where management does not have experience or expertise, outside experts should be consulted to make the necessary tests.

E.4 ANALYTICAL METHODS FOR SMALL DECISIONS

Even for small decisions, the importance of economics cannot be overstressed. The plant engineer who sees substantial possibilities for energy management must build a case for his management on a dollars and cents basis. He or she must demonstrate, in terms familiar to those who control the finances, that the energy management options are well-reasoned and economically attractive.

Some commonly used criteria for making economic evaluations that are suitable for small decisions are as follows.

A. The Payback Period

The payback period is the period of time it takes the projected savings that are to result from an investment or initial outlay of funds to equal the amount of the investment or outlay. As routinely applied for small decisions, a payback analysis usually ignores interest costs, differences in the expected lives of equipment, and costs or savings beyond the payback period. This is acceptable if the lifetime of the investment, or the payback period is short (one to five years). An illustration of calculations of payback periods for three different types of pumps designed to do the same job, but having different costs and operating efficiencies, is shown in Table E-2. In this payback analysis, pump "C" ranks as the most favorable with a payback period of two years.

B. Life Cycle Costing

Life cycle costing is the comparison of alternatives on the basis of total costs over a comparable time period. In this analysis, total costs are the sum of initial cost, operating and maintenance costs, and replacement costs less salvage value. Table E-3 gives an example of two motors, one with a life of 2 years, and another with an expected life of 3 years. The period of analysis is 6 years, during which the costs of purchasing three "A" motors and two "B" motors, as well as the costs of operation for each, are determined. The advantage of this method of analysis is that it encourages the decision maker to consider annual costs and replacement costs as well as the initial investment.

C. Cost-Benefit Ratio

The cost-benefit ratio is the total cost of a project divided by some measure of the benefit that is realized by that project. This method of comparison is used in instances where the benefits, or results, cannot be conveniently or directly measured in monetary terms. Such an example might arise in comparing two commercial food freezing systems where the output (or benefit) is measured in kilograms of food per hour at a certain level of quality (e.g., flavor retention). System A has

PUMP	COST	ANNUAL COST REDUCTION	PAYBACK PERIOD
A	$800	$130	$\frac{800}{130} = 6$ yrs
B	$1,200	$300	$\frac{1,200}{300} = 4$ yrs
C	$2,000	$1,000	$\frac{2,000}{1,000} = 2$ yrs

RATING THREE PUMPS BY PAYBACK ANALYSIS

TABLE E-2

Item	Motor A	Motor B
Expected life	2 yrs	3 yrs
Initial cost	$400	$500
Annual power cost	$790	$700
Annual maint. cost	$ 22	$ 40
Salvage value (%)	5%	10%
Cost to remove/replace	$ 50	$ 50
Equipment cost, 6 yrs	$1200	$1000
Power cost, 6 yrs	$4740	$4200
Maintenance, 6 yrs	$ 132	$ 240
Remove/replace	$ 150	$ 100
sub total	$6222	$5540
Less salvage	(60)	(100)
Total life cycle cost	$6162	$5440

LIFE CYCLE COST ANALYSIS

TABLE E-3

a different output and quality rating than System B, and each system will have different operating costs and purchase prices. What we need to find out is which freezer gives the most benefit for the least cost.

Defining and measuring the benefit is the creative job of the decision maker. In the example of the food freezers, the benefit for each system might be defined as:

Quality Measured Output = Quality Factor x output

The decision maker will need to make some judgments as to the relative quality rankings of the two systems, such as System A may be given 100 percent quality ranking, and System B only 70 percent.

The cost figures for the cost-benefit analysis can be compiled in any manner that gives a true and total comparison of costs. The life cycle method which was discussed above is a useful method. The resulting ratio of cost divided by benefit (e.g., quality measured output) will give a cost-benefit ratio of dollars per unit of benefit. The lowest number indicates the preferred system.

E.5 ANALYTICAL METHODS FOR LARGE DECISIONS

There are three types of major energy projects that firms deal with:

- Mandatory Projects - projects that must be done to comply with the demands of regulatory agencies;
- Security Projects - projects that are undertaken for the purpose of protection against the possibility of shutdowns or extraordinary loss due to energy shortages; and,
- Discretionary Projects - projects that the firm can select whether or not to pursue without regard to government regulations or the firm's security.

In the first two types of projects, the decisions to proceed are often made without regard to economic benefit, but economic analysis is useful to see what economic effect the project will have on the firm. Discretionary projects, on the other hand, require careful analysis and comparison in advance of decision making.

A. The Time Value of Money

The main principle that underlies analysis for large decisions is that money has a value that is dependent on time. Economists and financial people call this the Time Value of Money, and it is seen most commonly as compounded interest paid on personal savings accounts. Interest compounding has the following effect on the balance in a savings account:

$$\text{Value of Account} = \text{Value today} \times (1 + \text{interest rate})^n \text{ in year "n"}$$

If the value today of the account is \$100 and the interest rate is 7 percent per annum, the value of the account in three years will be:

$$\begin{aligned} V(n=3) &= V(\text{today}) \times (1.07)^3 \\ &= \$100 \times 1.07 \times 1.07 \times 1.07 \\ &= \$100 \times 1.225 \\ &= \$122.50 \end{aligned}$$

In making economic decisions, however, the usual case is that we are given the projections for future costs and profits for different projects, and we know pretty well what initial cost or investment will be required for each project. What we then need is a way to make the future costs and revenues for the different projects comparable to the initial investments. This is where the Time Value of Money and the compound interest formula comes in. We use the compound interest formula in reverse to find the value today (sometimes called present value or present worth) of costs and revenues that are expected to occur in the future. Namely,

$$\begin{aligned} \text{Value (today)} &= \frac{V(\text{future in n years})}{(1+\text{interest rate})^n} \\ &= V(\text{future in n years}) \times \frac{1}{(1+r)^n} \end{aligned}$$

The value (today) is called the present value of V(future) at an interest rate of "r". In the savings example, \$100 is the present value of \$122.50 that is to be received three years hence, using an interest rate of 7 percent per annum.

When calculating present values, the convention is to use the term "discount rate" instead of interest, so that we would say "...\$100 is the present value at a *discount rate* of 7 percent." The expression,

$$\frac{1}{(1 + r)^n}$$

is called the Discount Factor.

B. Present Value Analysis

Present Value Analysis is the comparison of alternative projects on the basis of the net present value of all costs and revenues where the discount rate is given. The procedure is to lay out a calculation sheet with as many columns as there are years (or time periods) in the analysis, compute net revenue or savings for each year (or time period), and determine the present value discount factor applicable for each of the years. The net revenue for each year is multiplied by its corresponding discount factor, and the products for all years are totaled to give a net present value. In this analysis, the initial investment is considered to be a cost that is incurred in the "0" year. An example is shown in Table E-4.

Discount factors for present value analysis are usually taken from tables that are prepared specially for this purpose (Table E-5). The values for the 7 percent example can be found quickly under the 7 percent column.

In present value analysis the project with the highest present value is the most favorable at the assumed discount rate. This ranking, however, may change at different discount rates, so that it is wise to run the analysis for two or three different rates to see if the ranking is sensitive to the rate used. Many firms use a discount rate that approximates their total before-tax cost of capital. Some firms use a blended cost of debt and equity capital, and others use the marginal cost of borrowing.

Present value analysis is usually applied to after-tax cash flow revenues or savings. This means that tax rates, investment tax credits, and the tax effects of the type of depreciation used must be taken into account.

C. Internal Rate of Return

Once we know the amount of investment a project requires, and we have projected its future net after-tax revenue or savings flow, we can calculate what discount rate it will take to make the present value of the future revenue or savings equal to the initial investment amount. This discount rate, which simply makes the present value of future cash flows equal to the initial investment, is called the Internal Rate of Return. It is also referred to as Equalizing Rate of Discount or the Return on Investment (ROI). It is a popular way to compare investments on the basis of the efficiency of capital use. Projects with a high Internal Rate of Return make better use of capital than projects with a low Internal Rate of Return.

The procedure in making a hand calculation of the Internal Rate of Return for a project involves a calculation sheet with the investment amount on the left and the annual net revenue or savings on the right. A trial discount rate is picked from Table E-5 to calculate the present value of the revenue or savings flow. If the present value is larger than the investment, a greater discount rate is needed, and another trial calculation is made with a higher rate. This trial and error process is repeated until the present value of the cash flow stream on the right is made equal to the investment. The discount rate that accomplishes this is the Internal Rate of Return. Table E-6 shows the steps on a project with a 9 percent Internal Rate of Return. Most analyses of this type are done by simple computer programs that are readily available from most computer service bureaus or computer centers. Few large companies today rely on the hand method of making these calculations.

	Year				
	0	1	2	3	4
Revenue/Savings		400	400	400	400
Cost	(1000)	(10)	(10)	(10)	(10)
Net Rev./Saving	(1000)	390	390	390	390
Discount Factor:	$\frac{1}{(1.07)^0}$	$\frac{1}{(1.07)^1}$	$\frac{1}{(1.07)^2}$	$\frac{1}{(1.07)^3}$	$\frac{1}{(1.07)^4}$
	1.000	.935	.873	.816	.763
Present Value (net rev/saving x discount factor)	(1000)+	364 +	340 +	318 +	298
Present Value = 320					

SAMPLE PRESENT VALUE CALCULATION (DISCOUNT RATE OF 7%, INITIAL INVESTMENT = 1000)

TABLE E-4

Year	7%	8%	9%	10%	12%	14%	15%	16%	18%	20%	24%	28%	32%
1	.935	.926	.917	.909	.893	.877	.870	.862	.847	.833	.806	.781	.758
2	.873	.857	.842	.826	.797	.769	.756	.743	.718	.694	.650	.610	.574
3	.816	.794	.772	.751	.712	.675	.658	.641	.609	.579	.524	.477	.435
4	.763	.735	.708	.683	.636	.592	.572	.552	.516	.482	.423	.373	.329
5	.713	.681	.650	.621	.567	.519	.497	.476	.437	.402	.341	.291	.250
6	.666	.630	.596	.564	.507	.456	.432	.410	.370	.335	.275	.227	.189
7	.623	.583	.547	.513	.452	.400	.376	.354	.314	.279	.222	.178	.143
8	.582	.540	.502	.467	.404	.351	.327	.305	.266	.233	.179	.139	.108
9	.544	.500	.460	.424	.361	.308	.284	.263	.226	.194	.144	.108	.082
10	.508	.463	.422	.386	.322	.270	.247	.227	.191	.162.	.116	.085	.062
11	.475	.429	.388	.350	.287	.237	.215	.195	.162	.135	.094	.066	.047
12	.444	.397	.356	.319	.257	.208	.187	.168	.137	.112	.076	.052	.036
13	.415	.368	.326	.290	.229	.182	.163	.145	.116	.093	.061	.040	.027
14	.388	.340	.299	.263	.205	.160	.141	.125	.099	.078	.049	.032	.021
15	.362	.315	.275	.239	.183	.140	.123	.108	.084	.065	.040	.025	.016
16	.339	.292	.252	.218	.163	.123	.107	.093	.071	.054	.032	.019	.012
17	.317	.270	.231	.198	.146	.108	.093	.080	.060	.045	.026	.015	.009
18	.296	.250	.212	.180	.130	.095	.081	.069	.051	.038	.021	.012	.007
19	.276	.232	.194	.164	.116	.083	.070	.060	.043	.031	.017	.009	.005
20	.258	.215	.178	.149	.104	.073	.061	.051	.037	.026	.014	.007	.004
25	.184	.146	.116	.092	.059	.038	.030	.024	.016	.010	.005	.002	.001
30	.131	.099	.075	.057	.033	.020	.015	.012	.007	.004	.002	.001	.000

PRESENT WORTH OF $1

TABLE E-5

Project Investment = $1,450

Investment	Present Value Disc. rate	Amount	Net Savings for years 1	2	3	4
$1,450			100	400	600	800
(Trial #1)	7%	1543	x.935	x.873	x.816	x.763
(Trial #2)	10%	1417	x.909	x.826	x.751	x.683
(Trial #3)	9%	1457	x.917	x.842	x.772	x.708

SAMPLE CALCULATION: INTERNAL RATE OF RETURN

TABLE E-6

Like present value analysis, the Internal Rate of Return is usually based on net cash flow after taxes. Some companies use net cash flow before interest, while others use net cash flow after interest. Whichever method of compiling cash flow projections is used, it is important that the method be consistently applied in all analyses. The net cash flow computation takes into account tax credits, tax deferrals, and the depreciation effect on tax payments.

D. Combined Analysis

Performing both a Present Value and an Internal Rate of Return Analysis for project comparison is recommended. The results of each analysis describe different and important economic characteristics of the project. Internal Rate of Return tells how efficient the project is in its use of capital, while Present Value tells how large the total beneficial impact will be. A project that ranks high on Internal Rate of Return might, in fact, make such a small contribution to the overall results of the firm that it is not worth management attention. We have to remember that both capital and management time are scarce resources, and the best projects will make efficient use of both.

Case Study E-1 is an actual example of an investment analysis made by a large corporation concerning an energy management project. Note that the assumptions are clearly set out, and that the project is measured by Present Value at three different discount rates, by Internal Rate of Return (ROI) for three time horizons, and by Payback Period. As discussed below, Payback Period is a useful measure of project risk.

E.6 EVALUATION OF RISK

Risk is the possibility of loss due to the uncertainty of future events. The evaluation of risk is the process of determining how much detrimental impact unplanned events can have on a particular project. For example, if projects are planned around a future price of electricity at 0.03 $/kWh, what could be the negative impact if electricity costs rise or fall significantly? What is the risk of loss associated with unforeseen electricity price changes?

A. Sensitivity Analysis

Sensitivity Analysis is the analysis of change in the desirability of an investment as the basic assumptions are changed. Normally, this requires determining how sensitive the anticipated Present Value, Internal Rate or Return, or Cost-Benefit Ratio are to changes in those assumptions for which the range of possible values is greatest and for those assumptions to which the economic measures themselves are most sensitive. Future electricity price might be an example of an assumption which has a range of possible values. Initial construction costs might be an example of an assumption to which the project is most sensitive. A project that has a present value that goes from positive to negative with a 10 percent price fluctuation in either energy cost or construction cost would very likely be considered as a risky project.

B. Payback Period

The Payback Period, which was discussed as a tool for small decisions, is a useful measure of risk for both large and small decisions. Projects that require a long time to recover their initial investment are riskier than projects with short recovery periods. Longer payback period means higher risk because there is more opportunity for changes or unforeseen events to adversely affect the project before the project has paid for itself.

C. Break-even Analysis

An important assumption for many projects involves some estimate of the level of operation or the rate of usage. For example, annual operating costs for a motor or an entire plant depend on the assumption of the number of hours per year the motor or plant will be operating. This use rate, or operating level, can change for any one of a great many reasons, and if that change can cause a loss to occur, there is a volume-related risk that needs to be evaluated.

CASE STUDY E-1

ECONOMIC ANALYSIS FOR A MAJOR PROJECT OF HEAT RECOVERY FROM FURNACE EXHAUSTS

The Concept

The objective of this analysis was to determine the relative attractiveness of a $4.3 million investment in heat recovery systems for 15 industrial furnaces.

Background

The rising cost of fuel oil used to fire melting furnaces caused this manufacturer to investigate cost savings that could be realized by fitting 15 of its furnaces with commercially available heat recuperators in the furnace exhaust so that combustion air is preheated and furnace throughput capacity per unit of fuel input is increased.

Findings

The following schedules show pertinent assumptions and cash flow analyses for this project. The calculations for Present Value, Return on Investment (Internal Rate of Return) and Payback Period are done by computer.

Recommendations

Major economic decisions should be made on an after-tax cash flow basis, and more than one measure of economic return should be calculated as the basis of comparison with other projects. Simple computer programs to handle the routine calculations and data print-outs can save much time and assure consistent and accurate handling of the data.

Assumptions

		Present	Proposed
Capital cost (1975) 15 furnaces modified @ $253,000			$4,345,000
Salvage value		none	none
Operating cost savings per year			
R&M @ 2% capital	(86,650)		
Operations, allow 2 @ $20,000	(45,665)		
Supplies, allow	(4,685)		
net	(137,000)	base	$(137,000)
New equipment economic life			20 years
New equipment depreciable life			11 years
Btu fuel requirement per ton remelt product(1)		2.6×10^6	2.2×10^6
Annual Btu fuel requirements		1.884×10^{12}	1.594×10^{12}
Annual Btu fuel savings		base	$290,000 \times 10^6$
Equivalent oil savings per year, bbls.		base	46,400
Annual fuel cost per 10^6 Btu (No. 2 fuel oil)		$2.50	$2.50 + 6% annual escalation
Annual fuel cost		$4,710,000	$4,985,000 + 6% annual escalation
Annual fuel cost savings		base	$ 725,000 + 6% annual escalation
Tax rate		50%	50%
Investment tax credit on capital cost			10%
Inflation rate on operating costs			-0-

(1) 15% fuel savings assumed resulting from application of combustion air preheaters.

RECUPERATOR DATA **SCHEDULE 1**

CASH FLOW
#1 HEAT RECUPERATION
(10^3 DOLLARS)

	1977	1978	1976
VOLUME:			
NO. 2 OIL SAVINGS	0.	290.0	290.0
PRICE:			
NO. 2 OIL SAVINGS (6% ESCALATION)	2.5	2.6	7.6
SALES (SAVINGS)			
NO. 2 OIL SAVINGS	0.	768.5	2193.6
NET SALES (SAVINGS)	0.	768.5	2193.6
GROSS PROFIT	0.	768.5	2193.6
LESS OPERATING EXPENSES:			
OPERATING EXPENSE 1	0.	-137.0	-137.0
TOTAL OPERATING EXPENSES	0.	-137.0	-137.0
LESS TAX DEPRECIATION	0.	724.2	
TAXABLE INCOME BEFORE INT	0.	181.3	2330.6
LESS TAXES	.0	90.7	1165.3
PLUS INVESTMENT TAX CREDIT	0.	434.5	
NET INCOME BEFORE INTEREST	-.0	525.2	1165.3
PLUS TAX DEPRECIATION	0.	724.2	
CASH INCOME (COPAT)	-.0	1249.3	1165.3
INVESTMENTS:			
FIXED ASSETS	-4345.0	0.	.0
NET CASH FLOW BEFORE INTEREST	-4345.0	1249.3	1165.3
CUM CASH FLOW BEFORE INTEREST	-4345.0	-3095.7	12535.8
CUM CASH FLOW BEFORE INTEREST AND WORKING CAPITAL	-4345.0	-3095.7	12535.8
PV OF NET CASH FLOW BEF INT			
10%	2823.1		
12%	1976.4		
15%	1012.2		
5 YEARS - NO ROI			
10 YEAR ROI (%)	13.7		
20 YEAR ROI (%)	19.6		
PAYOUT (YEARS)	5.9		

RECUPERATOR ECONOMIC ANALYSIS

SCHEDULE 2

Break-even Analysis is a method of relating profit (or savings) to the level of use or output. Its purpose is to relate fixed and variable costs to revenue or savings in order to determine the level of output or use where revenue or savings will just cover total costs. The analysis is usually presented graphically as shown in Figure E-1. The figure shows that both revenue (or savings) and total costs are volume dependent and that operating at a volume below the break-even volume incurs losses, while operating above that point means net profit or savings.

Break-even Analysis can be used as a measure of risk by constructing a ratio of planned operating level to the break-even level. This is expressed as "Times Break-even Covered," and is calculated as follows:

$$\text{Times Break-even Covered} = \frac{\text{Planned Volume (usage)}}{\text{Break-even Volume (usage)}}$$

A project whose economic viability was established at a level of operation near the break-even point (Times Break-even Covered is near 1.0) would be risky compared to a project with a Times Break-even Covered equal to 2.0 or 3.0. With a low Times Break-even Covered, a small downward change in expected volume or usage would radically change the desirability of the project.

E.7 COMMUNICATING RESULTS

The best analysis and reasoning can lose its usefulness if the results are not communicated properly. The communication of results should be made with three ideas in mind:

- someone has to use the written communication to make an important decision;
- the decision maker will use the written summary of findings to communicate the results to people less informed than he; and,
- Sometime in the future it may be necessary to refer back to the written summary to understand the methodology and the assumptions.

The communication of results should be as simple and as factual as possible.

Decision making memoranda are written to communicate:

- The Subject - "What is this about?"
- Recommendation - "What do you recommend we do?"
- Reasoning - "Why?"
- Documentation - "On what information is your reasoning based?"

It is not useful for the analyst to simply disgorge large quantities of numbers and statistics; instead, he/she should use his/her skills to help make the best decision. After considerable study the analyst is the one in a position to make a recommendation, and should do so. To substantiate his reasoning he will discuss the options that were considered and the methods of comparison. He will also need to provide documentation on the assumptions that were made, the testing of assumptions (if any) that was done, and any special studies, reports or information which were germane to the findings.

In the end, decisions are usually communicated to people who are affected by them but who are not as informed or knowledgeable as the decision maker or the analyst. How well those decisions are received will depend on the simplicity and the understandability of the communication.

E.8 PRINCIPLES OF PRACTICAL DECISION MAKING FOR ENERGY MATTERS

This appendix has touched on some of the fundamental aspects of economic decision making as it applies to energy management. By way of summary, some useful principles can be extracted to provide a guide for tackling the simplest and the most complex decision situations.

- Alternatives: Decisions start with alternatives. If the alternatives are poor, the decisions will be poor. Be creative in developing alternative courses of action, and think in terms of combinations of the options available.

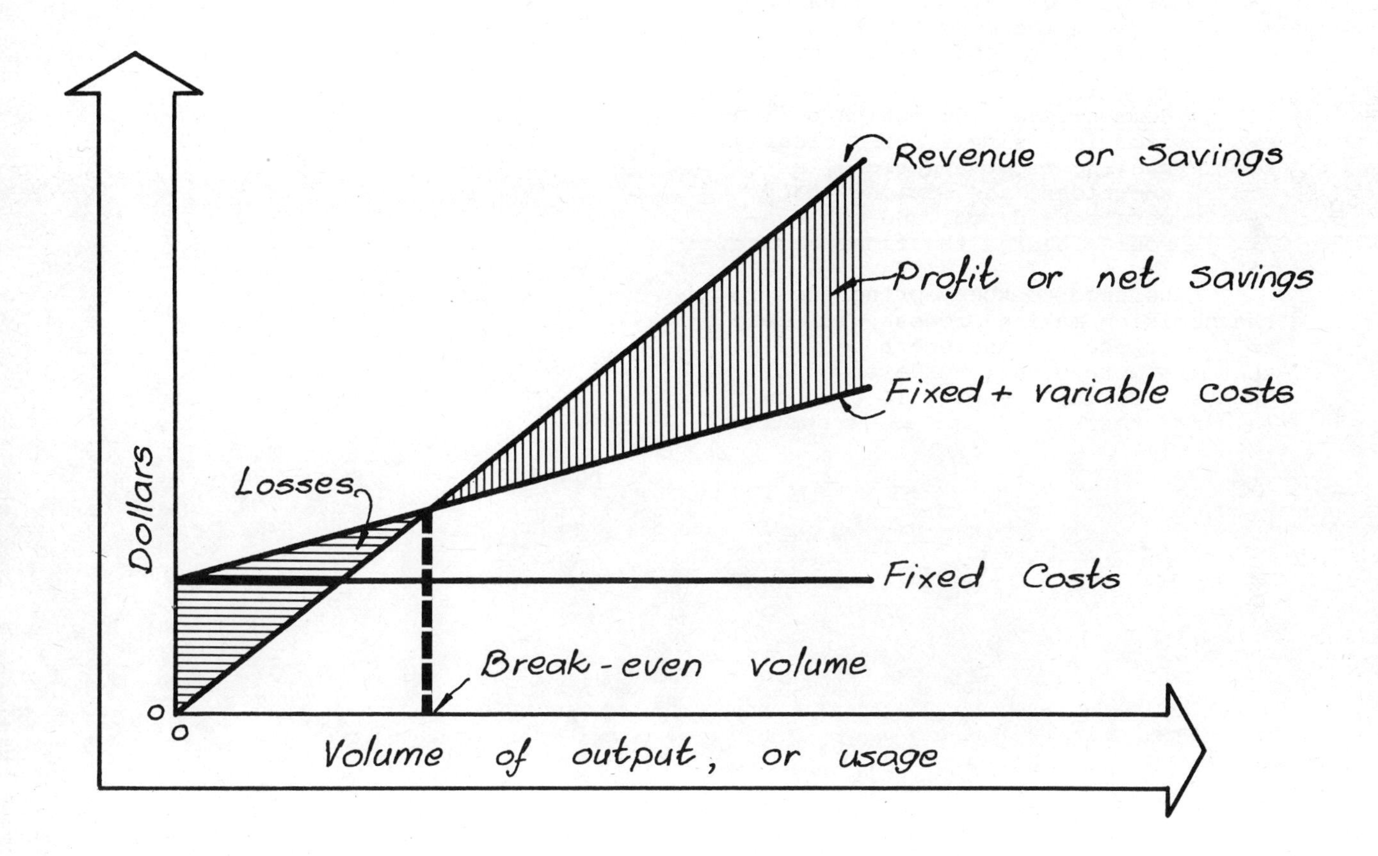

THE CONCEPT OF BREAK-EVEN ANALYSIS

FIGURE E-1

- Assumptions: Bad assumptions are frequently at the heart of colossal mistakes. Take care to list all assumptions and subject them to test.
- Comparison: Compare alternatives using simple, widely used economic criteria. Use more than one method of comparison.
- Risk Evaluation: Measure the risk of different options by testing the sensitivity of findings to changes in key assumptions.
- Communicate: Communicate findings in a simple form, clearly stating recommendations, assumptions, the options that were considered, and the reasoning behind the final choice.

Adherence to these principles in the decision making process will lead to better energy management decisions that are more clearly understood by others. Besides, energy is only one valuable resource; capital is another.

INDEX